The Blue Planet

An Introduction to
Earth System Science

The Blue Planet

An Introduction to Earth System Science

SECOND EDITION

BRIAN J. SKINNER
Yale University

STEPHEN C. PORTER
University of Washington

DANIEL B. BOTKIN
The Center for the Study of the Environment
George Mason University

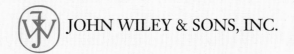

JOHN WILEY & SONS, INC.

Front cover image:	NASA
Back cover images:	lightning — Pete Turner/Image Bank
	coral reef — Fred Bavendam/Minden
	waterfall — Alan Majchrowicz/Peter Arnold
	volcano — Image Makers/Image Bank
	irrigation — William Campbell/Peter Arnold

Acquisitions Editor	Cliff Mills
Developmental Editor	Marian Provenzano
Marketing Manager	Catherine Beckham
Production Editor	Sandra Russell
Cover and Text Designer	Madelyn Lesure
Illustration Editor	Edward Starr
Photo Editor	Jill Hilycord
Illustrations	Precision Graphics & J/B Woolsey Associates

This book was typeset in 10/12 Janson by Ruttle, Shaw & Wetherill, Inc. and printed and bound by Von Hoffmann Press, Inc. The color separations were prepared by Color Associates, Inc. The cover was printed by Phoenix Color Corp. Inc.

The paper in this book was manufactured by a mill whose forest management programs include sustained yield harvesting of its timberlands. Sustained yield harvesting principles ensure that the number of trees cut each year does not exceed the amount of new growth.

This book is printed on acid-free paper. ∞

Library of Congress Cataloging-in-Publication Data
Skinner, Brian J., 1928–
The blue planet: an introduction to earth system science / Brian J. Skinner, Stephen C. Porter, Daniel B. Botkin. -- 2nd ed.

ISBN 978-0-47116114-1 (cloth : alk. paper)

1. Earth.   2. Earth sciences.   I. Porter, Stephen C.   II. Botkin, Daniel B.   III. Title.

QB631.S57   1999
550--dc21

98-31972
CIP

Prited in the United States of America

12 11 10 9

PREFACE

When historians of the future consider the most important achievements of the twentieth century, the chances are that one of their selections will be the development of a holistic view of the Earth. Such a view is a product of the technological and scientific advances that have made it possible to measure the many ways the different parts of the Earth can interact—for example, how events deep inside the Earth can influence events on the Earth's surface, or how small changes in the water temperature of the ocean can bring about changes in the distribution of land plants and how those changes can lead to the evolution of new plant species. Among these scientific advances is the recognition that life affects the environment at a global level. Specifically, our modern industrial society and our huge population are changing the Earth. We can now measure in real time the innumerable ways in which humans through their collective activities are changing the global environment.

This book is an introduction to the science of the holistic view of the Earth. It is about the interactions between the different parts of the Earth—the atmosphere, hydrosphere, biosphere, and geosphere (by which we mean the solid Earth). It is a book that presents a new view of the Earth that has come to be called *Earth system science*.

PURPOSE OF THE BOOK

Earth system science is rapidly changing the way we study and think about the Earth and about life, and as a result, it is changing the way earth science courses are being taught. We have written this book in order to introduce students to the science of the Earth system.

Courses about the Earth and Earth system science are being taught with increasing frequency and in an increasing number of disciplines. Such courses may have titles such as global change, earth science, biospherics, or even global environment, but the approach is increasingly that of studying the Earth as an assemblage of interacting parts.

THE BOOK'S ORGANIZATION

The text begins with a discussion of systems, cycles, and feedbacks as well as examples of several cycles. The second chapter, on the Earth's place in the solar system contrasts the Earth's appearance and structure with those of neighboring planetary bodies. This is followed by a chapter devoted to the Sun, not only because the Sun is the dominant feature of the solar system, but because it supplies the energy that drives most of the surface processes on our planet and that permits life to exist.

Next, in Part Two, we discuss the geosphere and the minerals and rocks that comprise it, the nature of the processes operating deep within the Earth that are inferred indirectly, and the dynamics of the crust, explained in terms of a relatively new, comprehensive theory, the theory of plate tectonics. Having explored the solid Earth beneath our feet, we next examine the layers of water and ice that cover much of its surface—the oceans, streams, groundwater, snow, glaciers, sea ice, and frozen ground—and we explore how some of these different agents erode and shape landscapes on which we live. In the third part of the book we explore the atmosphere, weather, and climate, and we examine the evidence of past changes in climate on various time scales. Having discussed the aspects of the Earth that make it a habitable planet, in Part Four we look at the diversity and dynamics of living organisms—plants, animals, fungi, algae, bacteria—that comprise the biosphere. We focus on the evidence of biological evolution through Earth history that is recorded in fossilferous rocks and on the role that the biosphere plays in the Earth system. In the final section of the book, we look at natural resources that have permitted the development and growth of modern civilization, and we discuss the ways in which human activities contribute to global changes in our environment.

Although we have given careful consideration to the organization of the book, we realize that not all instructors may favor the one we have adopted. Therefore, the parts and chapters have been written so that some reorganization of topics is possible without serious loss of continuity.

THE ARTWORK

Special attention has been devoted to producing artwork and photographs that illuminate discussions in the text. Because no continent or country holds a monopoly on relevant and interesting examples, we have attempted to provide photographs, maps and illustrations from around the world in order to provide a global perspective of Earth system science. The art program has benefited from talented artists who have worked closely with the authors to make their illustrations both attractive and scientifically accurate. More than two hundred photographs and line drawings, all in full color, have been selected.

THE SYSTEMS APPROACH

The key to understanding the Earth as a system of many parts is to appreciate the interactions between those systems. The Earth is, to a very close approximation, a closed system; by this we mean it neither gains nor loses matter. For the sake of study and measurement, the Earth can be divided into a large number of subsystems, all of which are open systems, meaning that matter can move back and forth among them. Earth system science is the study of the Earth as an assemblage of open systems, and the goal of the studies in this book is to eventually understand the interaction among all parts of the assemblage. In this way, the effects throughout the system caused by a perturbation in one part of the assemblage—say a volcanic eruption or a rise in the carbon dioxide content of the atmosphere—can be estimated and forecast.

The traditional way to study the Earth was to consider the various parts in isolation from each other. One group of scientists studied the atmosphere, another group the oceans, another the rocks of the geosphere, and yet another the assemblage of life. Interactions among the groups were rare. Earth system science focuses on the interactions and thereby differs fundamentally from traditional studies of the Earth.

FEATURES

Chapter Opener Essays. Each chapter in the book opens with a topical essay dealing with some aspect of research on the chapter topic. Chapter 8 on the Earth's evolving crust, for example, opens with an essay on research into the slowly changing shape of Greece as a result of tectonic forces.

Closer Look Boxes. Within chapters, specialized and detailed topics are boxed under the heading "A Closer Look." The boxed material can be included or deleted at the discretion of the instructor.

Guest Essays. Many chapters feature guest essays written by a researcher in the field. The essay subjects relate directly to material in the chapter and are intended to provide insights into ongoing research. Essay writers range from scientists working in industry through academic to government scientists.

Summary and Review. Finally, each chapter closes with a summary of in-chapter material, a list of key terms, and questions. The questions are of two kinds: (1) review questions, which relate strictly to the material in the chapter; and (2) discussion questions, which are intended for class or section discussion, sometimes call for a bit of library research, and in most cases raise broader issues than those in the specific chapter to which they are attached.

NEW TO THIS EDITION

This second edition of *The Blue Planet* has been reorganized based on constructive input from users of the first edition. For example, the chapter on materials of the geosphere, rocks and minerals, now follows the chapters on plate tectonics and the internal workings of Earth. The book has been updated throughout, and it includes new line art and many new photographs.

The first chapter, formerly the Introduction, has been restructured and expanded. The chapter now includes discussions of closed and open systems, cycles, feedbacks, and the consequences of the laws of thermodynamics. Case studies of specific cycles that play major roles in the operation of the Earth system, including biogeochemical systems, have been introduced.

A greatly revised chapter on landscape evolution introduces current hypotheses relating erosion to mountain uplift, the role of uplift in climatic change, and the influence of rock weathering on climate. It stresses the important interrelationships of the various Earth systems in creating the Earth's varied landscapes. Accordingly, this chapter is now placed near the end of the book, for it builds on background knowledge of the atmosphere, the oceans, the solid Earth, and life that is gained in earlier chapters.

Also new to this edition, as well as the largest change in the book, is the significant augmentation of the role of the biosphere in the Earth system. This augmentation is the result of the growing recognition among scientists of the effects of life on the atmosphere, hydrosphere, and the solid Earth, the global role of life in the Earth system, and the ways that the entire Earth affects all life. Not only has the number of chapters devoted to the biosphere been increased, but the role of life in affecting air, water, and the solid Earth has been emphasized throughout the text, as illustrated by the opening case study of the first chapter.

SUPPLEMENTS

A full range of supplements to accompany *The Blue Planet*, second edition, is available to assist both the instructor and the student.

Laboratory Manual. Written by Marcia Bjornerud of Lawrence University, Wisconsin, the laboratory manual is divided into six modules with labs available for every part of the text, flexibly arranged to cover all topics. It includes many skill-developing exercises, hands-on activities that students can do locally, and listings of relevant sites on the world wide web.

Instructor's Manual and Test Bank. Written by Barbara Murck of Toronto University, this guide includes a table of contents, chapter summaries, references, and approximately 85 test questions per chapter.

Computerized Test Bank. A computerized test bank is available in both IBM-compatible and Macintosh versions. This easy-to-use test-generating program enables instructors to choose test questions from the printed test bank, print the completed tests for use in the classroom, and save the tests for later use on modification.

Full-Color Overhead Transparencies. The transparencies include over 100 line drawings and tables from the text, edited for maximum classroom effectiveness.

Study Guide. Written by Michael Jordan of Texas A&M University, Kingsville, the study guide stresses processes and the interconnections among the Earth's spheres. It contains questions and exercises to reinforce the systems approach.

Media Resource Manager. A CD-ROM is available to adopters of this text and contains a compilation of photographs and line drawings from the text.

Geosystems Today. This case book and interactive www site (**www.wiley.com/college/blueplanet**) provides students with eight cases from around the world in which to see and explore the interaction of people and their environment. It was authored and developed by Robert Ford of Westminster University and James Hipple of the University of Missouri.

ACKNOWLEDGMENTS

Preparation of a full-color text requires the help and expertise of many people. We are greatly indebted to our publisher and to our editor, Clifford Mills, for supporting us, for understanding when other demands meant we had to put this revision of *The Blue Planet* aside for a time, and for maintaining an encouraging level of confidence that we would finally deliver.

We are indebted to Marian Provenzano for her skillful coordination of the project; she kept things flowing smoothly and did the thousand things we overlooked. Joan Melcher helped edit the manuscript so that the styles of the three authors blended together and avoided redundancies.

Our many colleagues who prepared essays did so with grace and professional acumen. We are very grateful to them. Besides the essay writers, each of whom is identified by name and photo adjacent to their entry, we are extremely grateful for the guidance and judgment provided by colleagues who discussed this project in encounter groups and who reviewed all or part of the manuscript. They are:

David J. Anastasio
Lehigh University

Lisa Barlow
University of Colorado

John M. Bird
Cornell University

James W. Castle
Clemson University

Robert Ford
Westminster College

Karen H. Fryer
Ohio Wesleyan University

H. G. Goodell
University of Virginia

Gregory S. Holden
Colorado School of Mines

Julia Allen Jones
Oregon State University

Dr. Adrienne Larocque
University of Manitoba

Keenan Lee
Colorado School of Mines

Thomas Lee McGehee
Texas A&M University

Dan L. McNally
Bryant College

Chris Migliaccio
Miami-Dade Community College

Barbara Murck
University of Toronto

Robert L. Nusbaum
College of Charleston

Roy E. Plotnick
University of Illinois

Doug Reynolds
Central Washington University

Kathryn A. Schubel
Lafayette College

Lynn Shelby
Murray State University

Leslie Sherman
Providence College

K. Siân Davies-Vollum
Amherst College

Nick Zentner
Central Washington University

ABOUT THE AUTHORS

I was born and raised in small country towns in Australia, and my early rural experiences did a lot to shape the way I think about the Earth and the way I address problems. I learned to look closely at the world around me, for example, and I saw the giant changes in the landscape as unspoiled bushland was cleared for sheep pastures.

By my twelfth birthday I knew that I wished to be a scientist and the science I fixed my mind on was chemistry. Accordingly, when I entered the University of Adelaide in South Australia in 1946, I selected a course of studies leading to a degree in chemistry. Things never go exactly as planned. Because I had a great-grandfather who had what seemed to me a very mysterious profession, mining engineering, and who had worked at the famous copper mines at Moonta in South Australia, I elected to take a course in geology. My motivation was to find out what had interested my ancestor. What I quickly discovered was my own interest in the subject. I completed my degree in chemistry but I also completed the requirements for graduation in geology.

After experience as a member of an exploration team looking for lead and zinc deposits, and employment as a mining geologist at a tin mine in Tasmania, I entered graduate school at Harvard University. When I emerged 3 1/2 years later with a Ph.D. I returned to Australia and started teaching and research on the origin of mineral deposits, a subject that continues to occupy much of my time and energy. An unexpected offer to become a member of the research staff of the U.S. Geological Survey brought me back to the United States in 1960.

After a number of years in the USGS, I moved to the Department of Geology and Geophysics at Yale University, and there I have stayed and worked to the present. It was at Yale that I found opportunities to explore the wider aspects of the earth sciences that increasingly occupied my thoughts. For a number of years I was involved in the space program, first the lunar landing experiments and later the unmanned landing experiments on Mars. With Yale colleagues I worked on problems involving oceanography and climatic change, and on such diverse topics as volcanic gases and economic models of resource depletion. Through exposure to such a diverse range of topics I came to understand that everything on the Earth is interconnected and that the way to appreciate how this wonderful planet of ours works is to understand the system by which the disparate parts interact with each other. It was my search for understanding that led to my participation in the preparation of this book about Earth system science.

Brian J. Skinner

Growing up on the dynamic coast of southern California, I was introduced to geology in action at an early age, an introduction that included being thrown out of bed and across my room by a major earthquake that jolted Santa Barbara in the early 1940s. Although earth science was not part of the curriculum when I attended high school, summers spent trekking in the High Sierra and the Rocky Mountains awakened in me an interest in mountains that gained momentum in college when rock climbing and mountaineering occupied much of my spare time. At Yale, a distinguished faculty introduced me to the science of geology, and after serving aboard a Navy destroyer in the Pacific Fleet, I returned to Yale for graduate study. There I focused on glacial and quaternary geology under the guidance of Professor Richard F. Flint, for many years a prominent author of Wiley geology textbooks. I had the chance to "prove" myself during three long summers of field work in the Brooks Range of Arctic Alaska where I worked at unraveling the puzzling complexities of a thick section of intensely deformed sedimentary rocks, without the benefit of plate tectonics to guide the way, and also studied the glacial history of the range.

A faculty position at the University of Washington provided an ideal setting for someone with my interests in glaciers and the ice ages. Within a day's drive of the campus, our earth science students can be introduced in the field to nearly every topic they will hear or read about in class.

The research on mountain glaciation which I began in western North America expanded to include many of the world's glaciated mountains: the Alps, the Andes, the Himalaya, New Zealand's Southern Alps, and tropical Mauna Kea on the island of Hawaii. In addition to glacial-geologic studies, my work has also involved analyzing rockfall hazards in the Italian Alps, prehistoric cave sites in northern Spain that served as home to ice-age hunters, and prehistoric eruptions of Cascade and Hawaiian volcanoes. My current research includes collaborative studies with Russian colleagues on mountain glaciation in northeastern Siberia and with Chinese colleagues on the monsoon history of central China and Tibet recorded in deposits of wind-blown dust and ancient soils. Such experiences, together with my duties as editor of *Quaternary Research*, an international interdisciplinary journal on the glacial ages, enable me to keep in touch with other earth scientists throughout the world and help me stay abreast of new advances in a rapidly moving scientific field.

Stephen C. Porter

When I was five years old, my sister and I cut out circles of different sizes to represent the Sun and nine planets of our solar system. Then I designed and drew paper spacecraft that had storage places for oxygen, water, and other known requirements for life. We lay the paper Sun and planets on the living room floor and "flew" the paper spacecraft from planet to planet. Thus began my interest in the Earth as a planet and the requirements for long-term space travel (which turns out to be the requirements for an artificial ecosystem).

Throughout junior high school and high school, I read everything I could about all the sciences. My science reading ranged from George Gamow's books about the origin of chemical elements in the universe to the *Radio Amateur's Handbook* to *Animals Without Backbones*, which introduced me to a whole new world of living creatures.

In addition to this love of science, I liked to be outside, swimming in streams and hiking in woods. With high school friends, I hiked and camped in nearby forests on self-organized expeditions. On one of these trips, we awoke one night to discover that we were surrounded by phosphorescent twigs. At the time this seemed magical; now we know that a fungus involved in the decay of woody tissue produces this glow. Experiences like this increased my fascination with life in forests.

At college I majored in physics, but I wanted to do science outside. I asked my professors what science I might pursue, and they told me there were no sciences outside anymore; everything important was done in laboratories. With that (now obviously inaccurate) advice, I followed my other interests.

Eventually, I moved to New Hampshire where I worked as a surveyor's helper in the forests. I became fascinated by questions about why specific species of trees grow where they do and about the relationship between environment and life. About this time, my sister sent me a small paperback book by Eugene Odum called *Ecology*, with a note suggesting that perhaps this was a science that might interest me. Indeed it was; here was the science outside that I had been looking for all my life. At the time, few had heard of ecology or environmental issues, and fewer still had thought about the global environment.

My Ph.D. thesis research concerned photosynthesis in forests and the exchange of carbon dioxide between forests and the atmosphere. My work began to take on a global dimension as I explored the possible effects of a buildup of carbon dioxide in the atmosphere on forests and other forms of life. I was fortunate enough to return to my first scientific fascination in a NASA project that investigated the potential for closed ecological life support systems for long-term space travel. This experience, in turn, led to research on the Earth as a planetary life support system, work that I have continued to pursue. I directed a study of the use of remote sensing of forests to see whether we could determine the amount of organic matter and the stage in development of forests from a satellite. The connection between life and its environment at a global level continues to fascinate me, and this interest made it a natural for me to be a participant in this textbook.

Daniel B. Botkin

BRIEF CONTENTS

CONTENTS

PART TWO
THE EARTH BENEATH OUR FEET

PART THREE
THE EARTH'S BLANKET OF WATER AND ICE

CHAPTER 9

Water on the Land **189**

PART FOUR
THE EARTH'S GASEOUS ENVELOPE

CHAPTER 12

Composition and Structure of the Atmosphere 271

CHAPTER 13

Winds, the Weather, and Deserts 293

PART FIVE
THE DYNAMICS OF LIFE ON EARTH

CHAPTER 15

GUEST ESSAYS

The Blue Planet

An Introduction to
Earth System Science

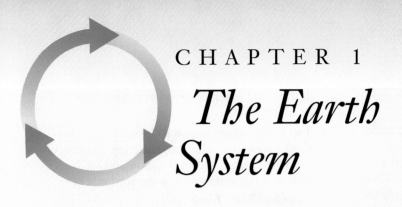

CHAPTER 1

The Earth System

● *Living on the Edge*

Salmon spawn and their young grow in cold northern streams fed by alpine glaciers. During ice ages, streams suitable to salmon are displaced southward by expanding ice sheets. When the ice retreats, southern streams become too warm and the salmon must migrate north again. Salmon have adapted to the long, slow glacial changes in an interesting way; most salmon return to the stream where they were hatched, but about 15 percent stray to new streams. What seems at first glance to be an error of navigation is a necessary strategy for survival of the species.

Salmon deposit their eggs in gravel beds where cool, fresh, water flows rapidly and provides abundant oxygen. Short-term climate effects can strongly influence the success of hatching. In a drought year, stream flow may be too warm and too low to provide sufficient oxygen. On the other hand, in a flood year, torrential waters can stir up the gravel beds and crush the eggs or kill young fish.

Glacial ages, floods, and droughts are just three of many environmental challenges that must be faced by salmon. Now they also have to cope with stress induced by human activities. Salmon once occurred in great abundance worldwide in northern coastal waters—from Siberia and the rivers of western Europe to New England and the provinces of Canada. Many of those stocks are now extinct and others are at risk. Salmon once spawned in the Thames River that flows through London, but overfishing, pollution, and alteration of the river channel ended those runs long ago.

Until the twentieth century, salmon in the Pacific Northwest coped reasonably well with environmental challenges. Now those stocks also are declining, and human activities seem to be the reason. Dams were built in the rivers of the Pacific Northwest beginning in the early decades of the twentieth century. The great Grand Coulee Dam in the Columbia River, largest of the dams, was built without fish ladders, thus cutting off access by salmon to thousands of miles of cool water streams in the northwestern United States and Canada.

We cannot understand the problems facing salmon from an examination of just one part of their environment; they face dangers at every stage of their lives and in many parts of their environment, so we must examine all parts together, as a system. So it is with all aspects of our "Home Planet"; to understand fully how any part of the Earth system works, we must appreciate the dynamic nature of the interrelationships among all the component parts. ●

(opposite) Sockeye salmon spawning in the gravel bed of a cold, glacial meltwater stream in Alaska.

THE EARTH SYSTEM

The global interconnectedness of air, water, rocks, and life has become the focus of much modern scientific investigation and is of great environmental concern. As a result, a new approach to the study of the Earth has taken hold. The traditional way to study the Earth has been to focus on separate units—a population of animals, the atmosphere, an ocean, a single mountain range, soil in some region—in isolation from other units. In the new holistic approach, the Earth is studied as a whole and is viewed as a system of many separate but interacting parts. Examples of the parts are the ocean, the atmosphere, continents, lakes and rivers, soils, plants, and animals; each can be studied separately, but each is more or less dependent on the others. Further consideration reveals that there are numerous interactions between all of the parts. **Earth system science**, then, is the science that studies the whole Earth as a system of many interacting parts and focuses on the changes within and between those parts.

A convenient way to think about the Earth as a system of interdependent parts is to consider it as four vast reservoirs of material with flows of matter and energy between them (Fig. 1.1). The four reservoirs are:

1. The **atmosphere**, which is the mixture of gases—predominantly nitrogen, oxygen, argon, carbon dioxide, and water vapor—that surrounds the Earth

2. The **hydrosphere**, which is the totality of the Earth's water, including oceans, lakes, streams, underground water, and all the snow and ice, but exclusive of the water vapor in the atmosphere

3. The **biosphere**, which is all of the Earth's organisms as well as any organic matter not yet decomposed

4. The **geosphere**, which is the solid Earth, is composed principally of **rock** (by which we mean any naturally formed, nonliving, firm coherent aggregate mass of solid matter that constitutes part of a planet) and **regolith** (the irregular blanket of loose, uncemented rock particles that covers the solid Earth).

Each of the four systems can be further subdivided into smaller, more manageable study units. For example, we can divide the hydrosphere into the ocean, glacier ice, streams, and groundwater.

The Scientific Method

The modern scientific method started in Europe several hundred years ago. Initially, there were no specialties among scientists, and what we call science today was called natural philosophy. By the middle of the nineteenth century, however, so many diverse topics had come under investigation that specialization appeared. Physicists investigated the physical properties of matter and phenomena such as light and magnetism, chemists studied how materials react, biologists studied living things, astronomers the stars, geologists the solid Earth, meteorologists the weather, oceanographers the ocean—on and on it went.

Study of the Earth as a system involves all these specialties. Separate investigations of the oceans, the atmosphere, and the solid Earth are no longer practical. When oceanographers go to sea in research vessels, they are accompanied by geologists who study the seafloor, biologists who study the aquatic life, meteorologists who study the way wind affects the sea surface, and chemists who measure water properties. The way all the different measurements are brought together is through the concept of systems.

As with any science, Earth system science depends on the scientific method. Science is a method of learning and understanding. It advances by application of the **scientific method**, that is, the use of evidence that can be seen and tested by anyone with resources who cares to do so. Although not always a clear-cut process, the scientific method can be viewed as consisting of the following steps.

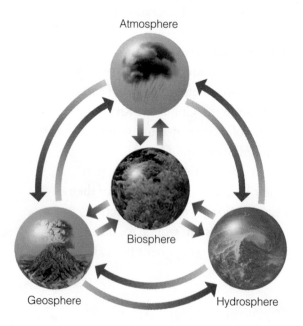

Atmosphere

Biosphere

Geosphere

Hydrosphere

Figure 1.1 Diagrammatic representation of the Earth as a system of interacting parts. Each character represents a reservoir, and each arrow a flow of energy or materials.

1. *Observation*. Scientists acquire evidence that can be measured and observed, for example, the counts of adult salmon returning to a stream.

2. *Formation of a hypothesis*. Scientists try to explain their observations by developing a **hypothesis**—an unproved explanation for the way things happen. For example, the variation in stream water flow is a major determinant of successful salmon spawning.

3. *Testing of hypotheses and formation of a theory*. Hypotheses are used to make predictions about new observations. A comparison of the predictions with the new observation is a *test* of the hypothesis. When a hypothesis has been examined and found to withstand numerous tests, scientists become more certain about it and it becomes a **theory**, which is a generalization about nature. The adjustment of salmon to changing river systems during an ice age is a theory based on repeated observation that some fraction of spawning salmon today do not return to the river of their birth.

4. *Formation of a law*. Eventually, a theory or a group of theories may be formulated into a scientific **law**. A law is a statement that some aspect of nature is always observed to happen in the same way and no deviations have ever been seen. An example of a law is the statement that heat always flows from a hotter body to a cooler one. No exceptions have ever been found. Actually, the flow of heat from a hot body to a cold body is a consequence of an even more fundamental law, the second law of thermodynamics, discussed later in this chapter.

5. *Continual reexamination*. The assumption that underlies all of science is that everything in the material world is governed by scientific laws. Because even theories and laws are open to question when new evidence is found, hypotheses, theories, and laws are continually reexamined. In fact, the key to the scientific method is *disprovability*. An old theory that salmon always return to the river of their birth was disproved by careful observation and by tagging salmon. As a result of this, a new theory was developed. Any theory that cannot, at least possibly, be disproved, is not scientific.

The System Concept

The system concept is a way to break down any large, complex problem into smaller, more easily studied pieces. A **system** can be defined as any portion of the universe that can be isolated from the rest of the universe for the purpose of observing and measuring changes. By saying that *a system is any portion of the universe*, we mean that the system can be whatever the observer defines it to be. That is why a system is only a concept; you choose its limits for the convenience of your study. It can be large or small, simple or complex. You could choose to observe the contents of a beaker in a laboratory experiment. Or you might study a flock of nesting birds, a lake, a small sample of rock, an ocean, a volcano, a mountain range, a continent, or even an entire planet. A leaf is a system, but it is also part of a larger system (a tree), which in turn is part of an even larger system (a forest).

The fact that a system has been *isolated from the rest of the universe* means that it must have a boundary that sets it apart from its surroundings. The nature of the boundary is one of the most important defining features of a system, leading to the three basic kinds of systems—isolated, closed, and open—as shown in Figure 1.2. The simplest kind of system to understand is

Figure 1.2 The three basic types of systems: A. An isolated system. B. A closed system. C. An open system.

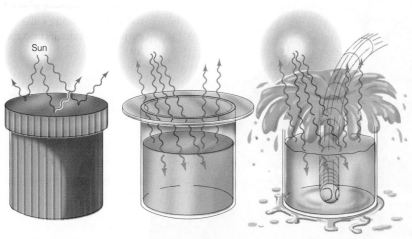

A. Isolated system   B. Closed system   C. Open system

an **isolated system**; in this case the boundary is such that it prevents the system from exchanging either matter or energy with its surroundings. The concept of an isolated system is easy to understand, but such a system is imaginary because although it is possible to have boundaries that prevent the passage of matter, in the real world it is impossible for any boundary to be so perfectly insulating that energy can neither enter nor escape.

The nearest thing to an isolated system in the real world is a **closed system**; such a system has a boundary that permits the exchange of energy, but not matter, with its surroundings. An example of a nearly closed system is the space shuttle, which allows the material inside to be heated and cooled, but is designed to minimize the loss of any material. The third kind of system, an **open system**, is one that can exchange both energy and matter across its boundary. An island on which rain is falling is a simple example of an open system: some of the water runs off via streams or seeps downward to become groundwater, while some is absorbed by plants or evaporates back to the atmosphere (Fig. 1.3).

A New Science and New Tools

A new science requires new tools. Indeed, often a new science arises because new tools allow new kinds of observation and measurement, and these in turn lead to new ways of thinking about some phenomena. Earth system science requires observations of the Earth at large scales and the handling of large amounts of data from many different locations. One of the best known of these tools is satellite **remote sensing**, which has made possible observations at large scales and, in many cases, measurements of factors that could not otherwise have been measured. For example, the ozone hole over Antarctica—the decline in the concentration of ozone high in the atmosphere—is measured by remote sensing as are changes in deserts, forests, and farm lands (see Guest Essay).

When remote-sensing measurements are made from satellites, scientists use the data in many areas of specialization. Satellite observations, above all other ways of gathering evidence, continually remind us that each part of the Earth interacts with, and is dependent on, all other parts. Modern Earth system science was born from the realization of that interdependence and the availability of satellites to make measurements.

New ways to explore previously inaccessible areas of the Earth have also added greatly to our knowledge of the Earth system. For example, small deep-sea submarines allow scientists to travel to the depths of the ocean. These submarines led to the discovery of life near deep-sea vents: that is, entirely new species, food chains, and ecosystems.

Equally important as new methods of measurement are ways to store and analyze the vast amounts of data about the Earth system. **Geographic Information Systems** (commonly called GIS), which are computer-based software programs, allow a large number of data points to be stored along with their locations.

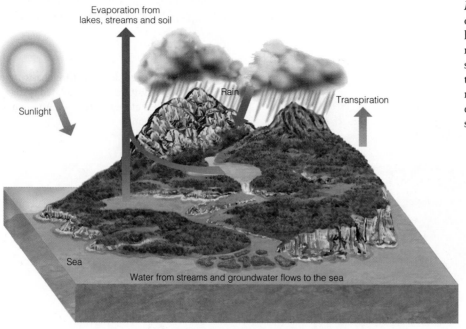

Evaporation from
lakes, streams and soil

Rain

Transpiration

Sunlight

Sea

Water from streams and groundwater flows to the sea

Figure 1.3 Example of an open system. Energy (sunlight) and water (rainfall) reach an island from external sources. The energy leaves the island as long-wavelength radiation; the water either evaporates or drains into the sea.

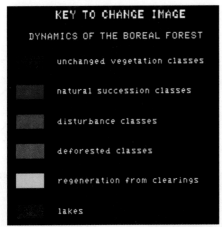

Figure 1.4 Forest fires can be natural or induced by human beings. This figure shows the change in a large area of the Superior National Forest in Minnesota between 1973 and 1983, as observed by the Landsat Satellite. The black boundaries show a central corridor where logging is permitted, surrounded by the Boundary Waters Canoe Area at the top and bottom. This is an area that is protected from all uses except certain kinds of recreation. The bright yellow shows areas that were clear of trees in 1973 but had regenerated to young forest by 1983. Most of this change is due to regrowth following a large fire that burned both inside and outside the wilderness. Red areas were forested in 1973 but cleared in 1983. Most of these are outside the wilderness, and some of these are due to logging (bright red) and some to fire or storms (dark red). Greens show areas that were forested both years.

From these, maps can be produced and sets of information of different kinds can be compared. For example, a satellite remote-sensing image over a forest can be converted to represent stages in the growth of a forest. Two such images from different times can be overlaid and compared, and the changes that have taken place provided as a new, third image (Fig. 1.4).

Living in a Closed System

The Earth is a closed system—or at least very close to such a system (Fig. 1.5). Energy reaches the Earth in abundance in the form of solar radiation. Energy also leaves the system in the form of longer-wavelength infrared radiation. It is not quite correct to say that no matter crosses the boundaries of the Earth system, because we lose a small but steady stream of hydrogen atoms from the upper part of the atmosphere and we gain some extraterrestrial material in the form of meteorites. However, the amount of matter that enters or leaves the Earth system is so minuscule compared with the mass of the system as a whole that for all practical purposes the Earth is a closed system.

The fact that the Earth is a closed system has two important implications.

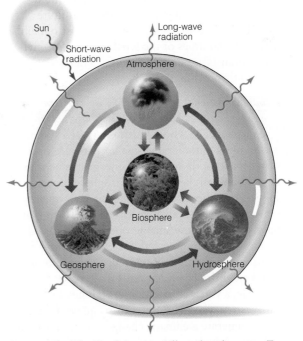

Figure 1.5 The Earth is essentially a closed system. Energy reaches the Earth from an external source and eventually returns to space as long wavelength radiation. Smaller systems within the Earth, such as the atmosphere, biosphere, hydrosphere, and geosphere, are open systems.

Water vapor in atmosphere

Evaporation and transpiration

Rain and snow

Evaporation

Evaporation

Rain

Vegetation, rocks and soil

Lakes and streams

Ocean

Figure 1.6 Depiction of the open system in Figure 1.4 by a box model.

1. *Because the amount of matter in a closed system is fixed and finite,* the mineral resources on this planet are all we have and—for the foreseeable future—all we will ever have. Someday it may be possible to visit an asteroid for the purpose of mining nickel and iron; there may even be a mining space station on the Moon or Mars at some time in the future. For now, however, it is realistic to think of the Earth's resources as being finite and therefore limited.

   A further consequence of a fixed and finite closed system is that waste material must remain within the confines of the Earth system. As environmentalists are fond of saying, "There is no *away* to throw things to."

2. *If changes are made in one part of a closed system, the results of those changes will eventually affect other parts of the system.* The whole Earth is a closed system, but all of its innumerable smaller parts are open systems and both matter and energy can be transferred between them. The atmosphere, hydrosphere, biosphere, and geosphere are all open systems, and every smaller system within them is an open system. These smaller open systems are dynamic and interconnected. When something disturbs one of them, the others also change. Sometimes an entire chain of events may ensue; for example, a volcanic eruption in Indonesia could throw so much dust into the atmosphere that it could generate a climatic change leading to floods in South America and droughts in California, and eventually affect the price of grain in west Africa. One of the main challenges of Earth system science is to understand the dynamic interactions between all of the relevant open systems sufficiently well so that we can accurately predict what the responses will be when some part of a system is disturbed.

Box Models

The storage and movement of materials and energy in a group of interacting systems are commonly depicted in the form of box models. As shown in Figure 1.6, which is a box model representation of Figure 1.3 depiction of the water cycle on an island, water stays on the island for some time before it flows off or is evaporated. The island thus is a **reservoir**, or storage tank, for water in this system. The average length of time water spends in the reservoir is called the **residence time**. The movement represented by the arrows in Figure 1.6 may be fast or slow, and so an essential part of Earth system science is the measurement of rates of movement. Flows between the reservoirs, and even between parts of the same reservoir, never cease, but the rates of flow may change, and when this happens, volumes must change too. One of the keys to understanding the Earth system therefore is an appreciation of why and how reservoir volumes change.

The advantages of box models are simplicity and convenience. A box model can be used to show the following essential features of a system:

1. The rates at which material or energy enter and leave the system.

2. The amount of matter or energy in the system at a certain time.

DYNAMIC INTERACTIONS AMONG SYSTEMS

The causes and effects of disturbances in a complex closed system are very difficult to predict. Consider the anomalously warm ocean conditions called El Niño which occurs every few years off the west coast of South America. El Niños (discussed in greater detail in Chapter 11) are characterized by weakening of tradewinds, suppression of upwelling cold ocean currents, worldwide abnormalities in weather and climatic patterns, and widespread incursions of biological communities into areas where they do not normally occur. These features of El Niños are reasonably well known; what is not known is the triggering event. In other words, the interactions among processes in the atmosphere, hydrosphere, geosphere, and biosphere are so complex, and these subsystems are so closely interrelated, that scientists cannot pinpoint exactly what it is that begins the whole El Niño process. One new hypothesis suggests that El Niño may be a result of ocean–atmosphere interactions due to the difference in viscosity between a liquid and a gas. Another hypothesis suggests the changes originating in the geosphere—in the form of localized heating of ocean water resulting from submarine volcanic activity—may create enough of a thermal imbalance to trigger an El Niño.

From an environmental point of view, the significance of interconnectedness is obvious: when human activities produce changes in one part of the Earth system, their effects—often unanticipated—will eventually be felt elsewhere. When sulfur dioxide is generated by a coal-fired power plant in Ohio or England, it can combine with moisture in the atmosphere and fall as acid rain in northern Ontario or Scandinavia. When pesticides are used in the cotton fields of India, the chemicals can find their way to the waters of the Ganges River and thence to the sea, where some may be ingested and stored in fishes bodies by a process called *bioaccumulation*. The fish, in turn, may be caught and eaten. In this way, pesticides sometimes end up in the breast milk of mothers halfway around the world from the place where they were applied. Such processes can take a long time to happen, and that is why they have been all too easy to overlook in the past.

Feedback

Because energy flows freely in and out of systems, all closed and open systems respond to inputs and have outputs. A special kind of system response, called *feedback*, occurs when the output of the system also serves as an input and leads to changes in the state of the system. A classic example of feedback is a household central heating system (Fig. 1.7). When room temperature cools, a metal strip in the thermostat cools and contacts an electric circuit, turning on the furnace. When the temperature rises, this strip warms and bends away from the electric contact, turning off the furnace. The metal strip senses temperature change and sends a signal to the furnace, hence feedback occurs.

A household central heating system is an example of **negative feedback**: the systems response is in the opposite direction from the output. With **positive feedback**, an increase in output leads to a further increase in the output. A fire starting in a forest provides an example of positive feedback. The wood may be slightly damp at the beginning and not burn well, but once the first starts, wood near the flame dries out and begins to burn, which in turn dries out a great quantity of wood and leads to a larger fire.

Negative feedback is generally desirable because it is stabilizing. It usually leads to a system that remains in a constant condition. Positive feedback, sometimes called the vicious circle, is destabilizing. A serious sit-

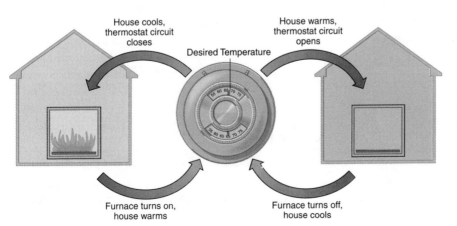

Figure 1.7 A familiar example of negative feedback. A change in temperature in one direction leads the thermostat to send a signal that makes the heating/cooling system change in the opposite direction. Hence, the feedback is negative.

uation can occur when our use of the environment leads to positive feedback. We will discuss positive and negative feedback more in subsequent chapters.

Cycles and Flows

Because material is constantly being transferred from one of the Earth's open systems to another, you may wonder why those systems seem so stable. Why should the composition of the atmosphere be roughly constant for very long periods? Why doesn't the sea become saltier or fresher? Why does rock 2 billion years old have the same composition as rock only 2 million years old? The answers to these questions are the same: many of the Earth's natural processes follow cyclic paths that are stabilized by negative feedback. Materials and energy flow from one system to another, but the systems themselves don't change much because the different parts of the flow paths balance each other. The amounts added equal the amounts removed. This cycling and recycling of materials and the dynamic interactions among subsystems has been going on since the Earth first formed, and it continues today.

A few basic examples can serve to highlight the importance of cycles in the Earth system. They are the *energy cycle*, the *hydrologic cycle*, an example of a *biogeochemical cycle* (of which there are many), and the *rock cycle*. In the discussions that follow, we will briefly consider each of these cycles. It is also possible to extend the concept of cycles to include human-controlled cycles that involve or affect natural processes. Examples of such cycles will be introduced at appropriate places throughout this book.

THE ENERGY CYCLE

The **energy cycle** (Fig. 1.8) encompasses the great "engines"—the external and internal energy sources—that drive the Earth system and all its cycles. We can think of the Earth's energy cycle as a "budget". Energy may be added to or subtracted from the budget and may be transferred from one storage place to another, but overall the additions and subtractions and transfers must balance. When a balance does not exist, the Earth either heats up or cools down until a balance is reached. This has happened in the past, as exemplified by changes in the Earth's average temperature during the ice ages.

Energy in the Earth system differs from matter in one important aspect—matter can be cycled from one reservoir to another, back and forth, endlessly, but energy cannot be endlessly recycled. To understand why this is so we must consider some of the fundamentals about energy.

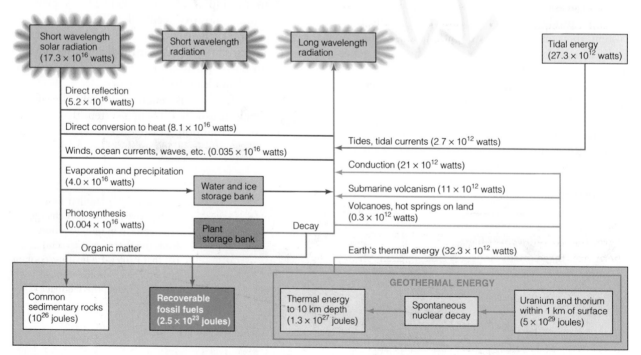

Figure 1.8 The energy cycle. There are three main sources of energy in the cycle: solar radiation, geothermal energy, and tidal energy. Energy is lost from the system though reflection and through degradation and reradiation.

The Laws of Thermodynamics

Energy is a difficult and abstract concept. It is the ability to do work, to move matter. Energy is subject to fundamental natural laws known as the Laws of Thermodynamics. The *first law of thermodynamics* involves the conservation of energy and is stated as follows: *In a system of constant mass, the energy involved in any physical or chemical change is neither created nor destroyed, but merely changed from one form to another.* When the first law was originally discovered, it was not known that matter can be changed to energy according to Einstein's famous equation $e = mc^2$. This is the reason that we now add the words "in a system of constant mass" to the definition of the law, meaning a system where matter is not being transformed to energy.

The *second law of thermodynamics* sometimes confuses people, but it is especially important because it governs all energy flows. According to the second law, *Energy always changes from a more useful, more concentrated form to a less useful, less concentrated form.* Another way of stating the second law is that energy cannot be completely recycled to its original state of usefulness. Complete recycling is impossible because whenever useful work is done, some energy is inevitably converted to heat. The energy needed to collect and recycle all of the energy dispersed as heat requires more energy than can be recovered. In other words, no real process can ever be 100 percent efficient—some heat energy is always lost.

The second law of thermodynamics has many important consequences. One consequence is that the flow of energy involves degradation and increasing disorganization as the energy becomes dispersed as heat and less available. The measure of this disorganization is called **entropy**. All of the energy in a system tends toward a state of increasing entropy. The implications of the laws of thermodynamics for life are discussed in more detail in Chapter 15.

Energy Inputs

The total amount of energy flowing into the Earth's energy budget is more than 174,000 terawatts (or $174,000 \times 10^{12}$ watts).[1] This quantity completely dwarfs the 10 terawatts of energy that humans use. There are three main sources from which energy flows into the Earth system.

Solar Radiation

Incoming short-wavelength solar radiation overwhelmingly dominates the flow of energy in the

Earth's energy budget, accounting for about 99.985 percent of the total. Part of this vast influx powers the winds, rainfall, ocean currents, waves, and other processes in the hydrologic cycle. Another part of solar radiation is used for photosynthesis and is temporarily stored in the biosphere as organic matter. When plants, algae, and bacteria die and are buried, some of the solar energy is stored as coal, oil, and natural gas. When we burn these fossil fuels, we release this stored solar energy.

Geothermal Energy

The second most powerful source of energy, at 23 terawatts or 0.013 percent of the total, is **geothermal energy**, the Earth's internal heat energy. Geothermal energy eventually finds its way to the surface of the Earth, primarily via volcanic pathways. It plays an important part in the rock cycle (discussed below) because it is the source of the energy that uplifts mountains, causes earthquakes and volcanic eruptions, and generally shapes the face of the Earth.

Tidal Energy

The smallest source of energy for the Earth is the energy produced by the interaction of tides and the Earth's rotation. The Moon's gravitational pull lifts a tidal bulge in the ocean; as the Earth spins on its axis, this bulge remains essentially stationary. As the Earth rotates, the tidal bulge runs into the coastlines of continents and islands, causing high tides (Chapter 11). The force of the tidal bulge "piling up" against landmasses acts as a very slow brake, causing the Earth's rate of rotation to decrease slightly. The transfer of tidal energy accounts for approximately 3 terawatts, or 0.002 percent of the total energy budget.

Energy Loss

The Earth loses energy from the energy cycle in two main ways: by reflection, and by degradation and reradiation.

Reflection

About 40 percent of the 174,000 terawatts of incoming solar radiation is simply reflected unchanged, back into space, by the clouds, sea, continents, and ice and snow. For any planetary body, the percentage of incoming radiation that is reflected unchanged is called the *albedo*. A high albedo means a highly reflective surface.

Each material has a characteristic reflectivity. For example, ice is more reflective than rocks or pave-

---

[1]A *watt* is one *joule* per second (1J/s). For a discussion of units, see Appendix A.

ment; water is more highly reflective than vegetation; and forested land reflects light differently than agricultural land. Thus, if large expanses of land are converted from forest to plowed land or from forest to city, the actual reflectivity of the Earth's surface, its albedo, may be altered. Any change in albedo will of course have an effect on the Earth's energy budget.

Degradation and Reradiation

The portion of incoming solar energy that is not reflected back into space, along with tidal and geothermal energy, is absorbed by materials at the surface of Earth, particularly the atmosphere and hydrosphere. This energy undergoes a series of irreversible degradations in which it is transferred from one reservoir to another and converted from one form to another. Because of the second law of thermodynamics, the energy that is absorbed, utilized, transferred, and degraded eventually ends up as heat, in which form it is

reradiated back into space as long-wavelength (infrared) radiation. Weather patterns are a manifestation of energy transfer and degradation. So are ocean currents, the growth of plants, and many other processes of the Earth system.

THE HYDROLOGIC CYCLE

The most familiar cycle is probably the **hydrologic cycle**, which describes the fluxes of water between the various reservoirs of the hydrosphere. We are familiar with these fluxes because we experience them as rain, and flowing streams (Fig. 1.9). Like all the cycles in the Earth system, the hydrologic cycle is composed of pathways, the various processes by which water is cycled around in the outer part of the Earth, and reservoirs, or "storage tanks," where water may be held for varying lengths of time. The total amount of water in

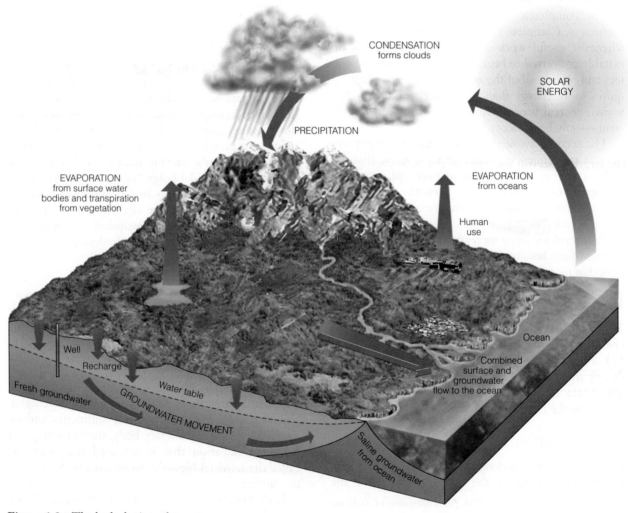

Figure 1.9   The hydrologic cycle.

the hydrologic system is fixed, but there can be quite large fluctuations in the local reservoirs, such as those that cause floods in one area and droughts in another—but on a global scale these fluctuations do not change the total volume of water on the Earth.

Pathways

The movement of water in the hydrologic cycle is powered by heat from the Sun, which causes *evaporation* of water from the ocean and land surfaces. The water vapor thus produced enters the atmosphere and moves with the flowing air. Some of the water vapor condenses and falls as *precipitation* (either rain or snow) on the land or ocean.

Rain falling on land may be evaporated directly or it may be intercepted by vegetation, eventually being returned to the atmosphere through their leaves by a process called *transpiration*, or it may drain off into stream channels, becoming *surface runoff*. Some of it may *infiltrate* the soil, eventually percolating down into the ground to become part of the vast reservoir of *groundwater*. Snow may remain on the ground for one or more seasons until it melts and the meltwater flows away into soils or streams. Snow that nourishes glaciers remains locked up much longer, perhaps for thousands of years, but eventually it too melts or evaporates and returns to the oceans.

Reservoirs

The largest reservoir for water in the hydrologic cycle is the ocean, which contains more than 97.5 percent of all the water in the system. This means that most of the water in the hydrologic cycle is saline, not fresh water—a fact that has important implications for humans because we are so dependent on fresh water as a resource for drinking, agriculture, and industrial uses. Surprisingly, the largest reservoir of fresh water is the permanently frozen polar ice sheets, which contain almost 74 percent of all fresh water. The ice sheets represent a long-term holding facility; water may be stored there for thousands of years before it is recycled. Of the remaining unfrozen fresh water, almost 98.5 percent resides in the next largest reservoir, groundwater. Only a very small fraction of the water passing through the hydrologic cycle resides in the atmosphere or in surface freshwater bodies such as streams and lakes.

In general, there is a correlation between the size of a reservoir and the average time that water stays in that reservoir, known as the *residence time*. Residence time in the large-volume reservoirs, such as the oceans and the ice-caps, is many thousands of years, whereas in the small-volume reservoirs it is short—a few days in the atmosphere, a few weeks in streams and rivers.

THE NITROGEN CYCLE: AN EXAMPLE OF A BIOGEOCHEMICAL CYCLE

A **biogeochemical cycle** describes the movement of any chemical element or chemical compound among interrelated biologic and geologic systems. A biogeochemical cycle involves both, biologic processes such as respiration, photosynthesis, and decomposition as well as nonbiological processes such as weathering, soil formation, and sedimentation in the cycling of chemical elements or compounds. In a biogeochemical cycle, living organisms can be important storage reservoirs for some elements. The cycles of nitrogen, sulfur, oxygen, carbon, and phosphorus are very important because each of these elements is critical for the maintenance of life.

It is difficult to produce a box model, even a highly simplified one, that accurately describes the biogeochemical behavior of an element as it cycles through the Earth system. These cycles potentially involve a wide variety of reservoirs and processes, and elements often change their chemical form as they move through the cycle. This complexity is illustrated by the nitrogen cycle, which we shall discuss briefly here.

Amino acids are essential components of all living organisms. They are given the name *amino* because they contain amine groups (NH_2), in which nitrogen is the key element. Nitrogen therefore is essential for all forms of life. The key to understanding the nitrogen cycle is understanding how nitrogen moves among four of the major reservoirs of the Earth system—the atmosphere, biosphere, hydrosphere, and geosphere. Figure 1.10 shows the reservoirs, the estimated mass of nitrogen in each reservoir, and the paths by which nitrogen moves among the reservoirs.

Nitrogen exists in three forms in nature. In the atmosphere it is present in the elemental form (N_2); but reduced forms such as ammonia (NH_3) and oxidized forms such as nitrate (NO_3) also exist. Only reduced forms of nitrogen can participate in biochemical reactions; N_2 can only be used directly by a few specialized bacteria organisms.

Nitrogen is removed from the atmosphere or made accessible to the biosphere in three ways:

1. Solution of N_2 in the ocean.
2. Oxidation of N_2 by lightning discharges to create

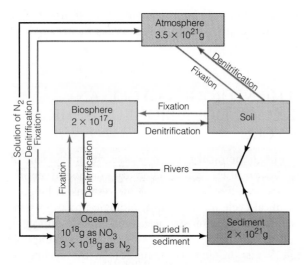

Figure 1.10 The nitrogen cycle.

NO_3, which is rained out of the atmosphere and into the soil and sea. Certain plants, algae and bacteria can reduce NO_3 to NH_3, thereby making nitrogen available to the rest of the biosphere.

3. Reduction of N_2 to NH_3 through the action of nitrogen-fixing bacteria in the soil or sea. The reduced nitrogen is quickly assimilated by the biosphere.

Once nitrogen has been reduced it tends to stay reduced and to remain in the biosphere where it can be reused by other organisms. A small fraction of the reduced nitrogen may be oxidized back to N_2 and returned to the atmosphere, but the main route by which nitrogen returns to the atmosphere, however, is the reduction of nitrate. This pathway is kept open by bacteria that use the oxygen in nitrate during metabolism.

THE ROCK CYCLE AND UNIFORMITARIANISM

Among the many important questions asked by scientists is the question of the relative importance of cumulative small, slow changes like the washing away of soil by an ordinary rainstorm, in contrast to massive, drastic changes like earthquakes and floods. Massive changes are relatively infrequent, but they cause rapid, dramatic changes to the landscape. People remember the floods, hurricanes, landslides, and other great events that change the landscape, but they quickly forget the innumerable small rain showers between the great events. During the seventeenth and eighteenth centuries, before the power of the scientific method became widely appreciated, people hypothesized that all the Earth's features—mountains, valleys, and

oceans—had been produced by a few great catastrophic events. The catastrophes were thought to be so huge that they could not be explained by ordinary processes, and so the supernatural was called upon. This concept came to be known as **catastrophism**. Not only were the catastrophes thought to be gigantic and sudden, but some people also believed they had occurred relatively recently and fit a chronology of catastrophic events recorded in the Bible.

The Rise of a New Theory

During the late eighteenth century, the hypothesis that most features of the geosphere were formed as a result of catastrophies was compared with geological evidence, and found wanting. The person who used the scientific method to assemble the evidence and propose a counter theory was James Hutton (1726–1797), a Scottish physician and "gentleman farmer." Hutton was intrigued by what he saw in the environment around him, especially around Edinburgh, where he lived and studied. He wrote about his observations, offered hypotheses, and then used tests and observational evidence to develop theories. Hutton is widely regarded today as the father of the scientific specialty we now call geology. In 1795 he published a two-volume work titled *Theory of the Earth, with Proofs and Illustrations* in which he introduced his counter theory to catastrophism.

We refer to the complex group of related processes by which rock is broken down as **weathering**, and the processes by which the breakdown products are moved around as **erosion**. Hutton observed the slow but steady effects of weathering and erosion: rock particles are carried great distances by running water and ultimately deposited in the sea. He reasoned that mountains must slowly but surely be eroded away, that new rocks must form from the debris of erosion, and that the new rocks in turn must be slowly thrust up to form new mountains. Hutton couldn't explain what causes mountains to be thrust up, but everything, he argued, moves slowly along in repetitive, continuous cycles. His ideas evolved into what we now call the **Principle of Uniformitarianism**, which states that natural laws do not change and therefore the processes that we see in action today have been operating the same way throughout the Earth's history. We can therefore examine any rock, however old, and compare its characteristics with those of similar rocks forming today in a particular environment. We can then infer that the old rock likely formed in the same sort of environment. In short, *the present is the key to understanding the past.* For example, in many deserts today we can see gigantic sand dunes formed from sand grains transported by the wind. Because of the

A

B

Figure 1.11 The internal structure of sand dunes, ancient and modern, demonstrates the power of uniformitarianism. A. Distinctive pattern of wind-deposited sand grains can be seen in a hole dug in this dune near Yuma, Arizona.

B. The same distinctive pattern in rocks in Zion National Park, Utah, lets us infer that these rocks, too, were once sand dunes.

way they form, the dunes have a distinctive internal structure (Fig. 1.11A). Using the Principle of Uniformitarianism, we infer that any rock composed of cemented grains of sand and having the same distinctive

internal structure as modern dunes (Fig. 1.11B) is the remains of an ancient dune.

Hutton was especially impressed by evidence he saw at Siccar Point in Scotland (Fig. 1.12). Sandstones

Figure 1.12 Siccar Point, Berwickshire, Scotland. The vertical layers of sedimentary rock on the right, originally horizontal, were lifted up into their vertical position. Erosion developed a new land surface that became the surface on which the now gently sloping layers of younger sedi-

ment were laid. The gently sloping layers, which are named the Old Red Sandstone, are 370 million years old. At this locality, in 1788, James Hutton first demonstrated that the cycle of deposition, uplift, and erosion is repeated again and again.

are formed by the cementation of sand grains into solid rocks. Hutton observed sand being deposited in horizontal layers, and he realized that most sandstones must have originally been laid down as horizontal layers. At Siccar Point, however, Hutton could see ancient sandstone layers standing vertical and capped by gently sloping layers of younger sandstone. The boundary between the layers, he pointed out, was an ancient surface of erosion. The now-vertical layers are composed of debris that was eroded, millions of years ago, from an ancient mountain range, transported by streams, deposited on the seafloor, and there formed into new rocks. As a result of mechanisms we shall discuss in Chapter 4, the newly formed rock layers were uplifted, tilted to their present position, and eroded. When erosion had formed a flat surface on the tops of the vertical sandstone layers, a pile of younger erosional debris was deposited on the new surface. Eventually, the younger debris became rock and uplift occurred again, although not much tilting was in evidence during this second stage of uplift. The cycle of uplift, weathering, erosion, transport and deposition, solidification into rock, and renewed uplift that could be deduced from this visible evidence impressed Hutton immensely. He did not use the term **rock cycle** for this sequence of events, but today we do. There is, wrote Hutton, "no vestige of a beginning, no prospect of an end" (1795, *Theory of the Earth*) to the Earth's rock cycle.

Geologists who followed Hutton have been able to explain the Earth's features in a logical manner by using the Principle of Uniformitarianism and the concept of the rock cycle. In so doing, they have also made an outstanding discovery—the Earth is incredibly old. It is clear that most erosional processes are exceedingly slow. An enormously long time is needed to erode a mountain range, for instance, or for huge quantities of sand and mud to be transported by streams, deposited in the ocean, then cemented into new rocks and the new rocks deformed and uplifted to form a new mountain. Slow though it is, this cycle has been repeated many times during the Earth's long history.

Although Hutton never used the terms *Earth system* or *cycle*, he described parts of the Earth system in ways that show he understood both concepts. His concept of a cycle of erosion, transport, deposition, formation of new rock, and uplift is just another way of discussing flows of materials between reservoirs of the geosphere.

The concept of uniformitarianism is important to all branches of science, not just geology. For example, astronomers have developed a powerful theory about the way stars form, pass through a long life cycle, and then die. Because the lifetime of a star is measured in billions of years, it is not possible to make all needed observations by watching a single star. Instead, astronomers study the billions of stars in the sky, observe examples at various stages of development, and find that the cycle of birth, growth, and death follows a predictable pattern. Whenever a new star is examined, uniformitarianism allows the observer to use previous observations to estimate where the new star is in its life cycle.

Rare Events and the Reconsideration of Catastrophism

Uniformitarianism is a powerful principle, but should we abandon catastrophism as a totally incorrect hypothesis? The answer is no, because we now know that events we consider to be catastrophic can be readily explained by well understood, ordinary processes, and therefore by uniformitarianism. These are not the catastrophes perceived by seventeenth-century biblical scholars, who had to call on supernatural forces to explain things. Rather, they are events that can be readily explained but are so large and damaging that they caused catastrophic change.

An example of such a rare event is suggested by recent discoveries of thin but very unusual rock layers at many places around the world. The unusual rock layers are thought to be debris from a huge meteorite striking the Earth (Fig. 1.13). The rock layers are rich in the uncommon metal iridium, which is much more abundant in meteorites than in the Earth's common rocks. The iridium-rich rocks have been discovered in Italy, Denmark, and other places around the world (Fig. 1.14). The hypothesis is that a massive impact occurred about 66 million years ago, likely in the Yucatan area of present-day Mexico, and that, as a result, many forms of life, including many of the dinosaurs, became extinct. The impact is thought to have thrown so much debris into the atmosphere that the air temperature plummeted. Consequently, most animals and many plants could not survive. When the debris settled, it formed a thin, iridium-rich layer wherever sediments were being deposited around the world. This hypothesis is still being tested, and many confusing bits of evidence remain to be explained.

Even more dramatic extinctions than the one 66 million years ago have occurred. The geologic record indicates that almost 90 percent of all plants and animals were driven to extinction about 245 million years ago. No evidence suggests that a meteorite impact caused the great extinction. To the contrary, fragmentary evidence indicates that slow but drastic climate changes resulting from the breakup of a huge super-

Figure 1.13 Meteor Crater, near Flagstaff, Arizona. The crater was created by the impact of a meteorite about 50,000 years ago. It is 1.2 km in diameter and 200 m deep. Note the raised rim and the blanket of broken rock debris thrown out of the crater. Many impacts larger then the Meteor Crater event are believed to have occurred during the Earth's long history.

continent may have been the reason. If so, this is a dramatic example of the interconnecting parts of the Earth's system. When we view the Earth's history as a combination of continual small changes at a wide range of scales, spatial and temporal, as well as a series of repeated but rare events, we have to conclude that uniformitarianism can describe even the rare events and that there is every reason to believe that similar events will occur again in the future. For example, astronomers have already identified a comet that will come close enough to the Earth at some time during the next 1500 years to possibly cause a collision event as big as the one 66 million years ago.

A fascinating but frightening suggestion has been made that a rare event of a different kind may already be happening. Our collective human activities may be changing the Earth so rapidly and so significantly that we may be living through a change similar in magnitude to some of the major ones in the geological record. At present, the suggestion is only a hypothesis; it remains to be tested and thereby proved or disproved. Nevertheless, the very fact that serious scientists are concerned that the hypothesis might prove to

Figure 1.14 This thin, dark layer of rock (marked by the coin) is rich in the rare chemical element iridium and looks out of place in the thick sequence of pale-colored limestones above and below. The iridium-rich layer, here seen in the Contessa Valley, Italy, has been identified at many places around the world and is believed to have formed as a result of a world-circling dust cloud formed by a great meteorite impact about 66 million years ago.

GUEST ESSAY

Satellite Remote Sensing: A Unique Perspective for Studying Earth's Vegetation

Compton Tucker is a physical scientist at NASA's Goddard Space Flight Center in Greenbelt, Maryland. His research involves using satellite remote sensing to study desertification, tropical deforestation, and temperate forest issues. He has a B.S. in biology and M.S. and Ph.D. degrees from the College of Forestry, all from Colorado State University.

As we enter the twenty-first century, satellite technology is providing unique and detailed information about the Earth's vegetation in a variety of ways. Data from Landsat, the French SPOT satellite, the Indian Remote Sensing satellite, and other Earth-viewing high-resolution satellites are providing multispectral 10- to 30-m data about vegetation conditions, land cover, deforestation, pollution sources, crop types, urban sprawl, and numerous other topics of great importance to the rational use of natural resources. More recently, private companies have been providing, upon request, highly detailed 1- to 5-m spatial resolution satellite data.

At the same time, daily satellite data are being acquired at a 1-km spatial resolution from the NOAA series of polar-orbiting satellites, the French Vegetation sensor flown on SPOT-4, and Orbital Sciences Corporation's Sea WiFS satellite. These data are not suited to provide high spatial detail of the Earth; rather, they provide important information on how the Earth's vegetation varies in time. Because each of these satellites images the entire Earth every day, data from these satellites can be combined over several days to yield cloud-free composite images that record information about our planet's vegetation through time.

The combination of higher spatial detail satellite data from Landsat and SPOT, coupled with 1-km time-varying data from NOAA and Sea WiFS, provides a suite of tools for global environmental study. These two information sources enable us to observe and study the global biosphere from several hundred kilometers in space. An analogy to the use of data from multiple satellites is the microscope: the oil-immersion objective for higher detailed study in specific areas coupled with the scanning objective for preliminary analysis.

The remote sensing satellite instruments that collect reflected or emitted radiation from the Earth are all similar: They measure light reflected from the Earth in the visible, near-infrared, and thermal infrared spectral regions. Satellite measurements are frequently used to produce indices that are highly correlated to the photosynthetic capacity of the land vegetation. Furthermore, these indices have the fortuitous feature that all degrading influences (clouds, atmospheric haze, scan angle, etc.) can only decrease these indices. Consequently, by maintaining accurate geographic navigation, satellite data can be combined over several days or weeks by simply selecting the maximum satellite index for each geographic location, producing "seamless" continental and global estimates of photosynthetic capacity.

Examples of Terrestrial Satellite Remote Sensing

Amazon Basin Deforestation Tropical deforestation is a topical scientific issue, with estimates of the annual rate varying from 70,000 km^2 to 165,000 km^2. Recent Landsat thematic mapper research has focused on the Amazon of Brazil, the largest area of tropical forest of our planet (~one-third of the total), where estimates of the annual tropical deforestation rate vary from 20,000 to 80,000 km^2. Using complete Landsat thematic mapper 30 m coverage of the 5,000,000 km^2 area of Brazil's Amazon, coupled with a geographic information system for data management purposes, researchers recently have reported that the 1978 to 1988

be true emphasizes an important fact: human activities are an important part of Earth system science and changes to the Earth and the welfare of the human race are indissolubly linked.

THE BIOLOGICAL DIMENSION

The Earth has three features which together make it unique: the continual rearrangement of continents and oceans by a process called plate tectonics, a sub-

stantial amount of liquid water, and life. These add to the complexity of the Earth system but also seem to contribute to its stability.

One of the new scientific recognitions—new in the past 20 years—is the great extent to which life affects the other major parts of the Earth system. The chemical composition of the Earth's atmosphere is very different from what would be found on a lifeless planet. Suppose you were a traveler from another solar system, from a civilization that had traveled through space to study the planets in our solar system, and suppose the path of your spacecraft took you near

tropical deforestation rate for Brazil's Amazon was ~15,000 km²/yr, substantially lower than previous estimates. The deforestation rate in the Amazon of Brazil varied widely from 1988 to 1997, ranging from ~11,000 km²/yr in 1992 to ~26,000 km²/yr in 1996.

Because these analyses were conducted using a geographic information system, the total area of isolated forest fragments surrounded by deforestation and the total edge or buffer of tropical forest in direct contact with areas of deforestation could be calculated, for these considerations are important in determining the indirect biological diversity impacts of deforestation. Landsat thematic mapper data are the only possible source of satellite data for determining accurate estimates of tropical deforestation. Other satellite data sources were either too expensive (SPOT) or have spatial resolutions of >1 km (NOAA and Sea WiFS), which render them useless for identifying deforestation with dimensions <500 m.

African Grassland and Climate Studies: The Sahel of Africa is a broad transition grassland between the Sahara to the north and the more humid savannas to the south, running from the Atlantic Ocean to the Red Sea, more or less parallel to lines of equal latitude. This area has attracted scientific interest because of periodic drought and concern over possible expansion of the Sahara to the south. Remote sensing research in this area has led to the development of techniques for determining grassland biomass production, which also are used to study large-scale variation in desert extent as well as provide famine early earning. This is an excellent example of the integral linkage between scientific research and day-to-day humanitarian concerns, such as food security.

NOAA polar-orbiting meteorological satellite data have been used since 1980 in the Sahel zone to estimate grassland total biomass production. These data are acquired almost daily and processed into a vegetation index; thermal data are used to identify clouds, and are subsequently combined into a 15-day composite image by selecting the maximum vegetation index value for each grid cell ~1 km square. AVHRR 1-km data have been and are being used to investigate satellite data–grassland biomass relationships, both within Africa and in developed countries, with a high level of accuracy. Once the satellite data–grassland biomass relationship is determined from sufficient specific ground locations, this relationship can be extended through the same ecological region.

NOAA 4-km daily data dating back to 1980 have been processed at an 8-km grid cell size and combined into 10-day vegetation index composite images for the entire continent of Africa. These data have been shown to be highly correlated with precipitation <1000 mm/yr, although different relationships between the satellite data and precipitation exist for different African climatic regions (i.e., East Africa, Sahel Zone, Southern Africa, etc.). These data have been used to investigate expansion and contraction of the major deserts of Africa. Findings to date indicate that all the major deserts of Africa expand some years and contract in others, driven by climatic variation. Continuing this work for 30 to 40 years will provide the baseline data needed to determine if the major arid regions of Africa and elsewhere are expanding as many unsubstantiated reports indicate.

The same 8-km grid cell size time series of satellite-derived data are also used in combination with historical information about crop yields, grazing conditions, severity of droughts, and so on, to provide early warning for food security purposes. Conditions for areas of interest can be monitored through the growing season and compared to the previous 20 years of historical data to identify areas of food or fodder shortfalls, This type of famine early warning is much more objective than types depending on in-country reports, provides rapid and objective information on where to send relief when required, and ensures that the most needy areas are not overlooked. Exactly the same data as those used to investigate desert expansion and contraction are used for food security or famine early warning purposes.

Venus and Mars but not the Earth. Since you understood how solar systems form, you knew that the three inner planets shared a similar origin and history—they had all been formed by the accumulation of material through gravitational attraction. You knew that these three inner planets were within a factor of two of each other in diameter and distance from the Sun. Because the Earth lay between Mars and Venus, you would be comfortable interpolating your observations of those planets to characterize the Earth. Suppose that you made measurements of the atmospheres of Mars and Venus. Their atmospheres are more than 95 percent carbon dioxide and less than 4 percent nitrogen, with traces of oxygen and a small amount of such inert gases as argon. You return to your home base and suggest, because the three planets are so similar, that the atmosphere of the Earth has the same composition as the atmospheres of Mars and Venus. But you would be quite wrong. The Earth's atmosphere is 79 percent nitrogen, 21 percent oxygen, as well as a small amount of carbon dioxide. Only the estimate of the amount of argon would be about right.

The difference between the Earth's atmosphere and that of Mars and Venus is the result of life's processes over billions of years. Photosynthesis by green plants, algae, and photosynthetic bacteria removes carbon dioxide from the atmosphere and adds oxygen. Oxygen is a highly reactive gas that rapidly combines with many other chemical elements and so does not remain in its free form for a long time. To counteract the removal of oxygen by chemical reactions, life acts as an oxygen pump, continuously returning oxygen to the atmosphere. *Free oxygen in the Earth's atmosphere is the result of 3 billion years of photosynthesis and is therefore a product of life.*

One of the peculiar things about our planet is that the geosphere and the atmosphere are in what a chemist calls disequilibrium. By this, a chemist means that if you were able to take the Earth and place it in a completely dark box and leave it for a long time, the composition of the atmosphere would not remain as it is. The oxygen would combine with other elements, including iron, carbon, and nitrogen, and the atmosphere would come to be much like that of Mars and Venus. The atmosphere is one of the profound ways that life has changed the Earth.

If life on the Earth were in a steady state so that the total amount of living organic matter were constant, and the removal of carbon dioxide from the atmosphere by photosynthesis equaled the return of carbon dioxide by decay and respiration, then there would not be a net addition or removal of carbon dioxide to or from the atmosphere. But over geologic ages there has been a slight imbalance, with a slight excess of photosynthesis over decay and respiration. This has not only created the low concentration of carbon dioxide in the atmosphere, the by-product of photosynthesis, and a high concentration of oxygen in the atmosphere, but it has also resulted in the locking up of carbon in the geosphere by deposition in sediments of carbon-bearing minerals such as calcium carbonate (in limestone) and carbon-rich materials such as oil, coal, and organic-rich shales. In the course of this book, we will discover many other ways that life has affected other major components of the Earth system.

The Human Impact

Notice in Figures 1.1 and 1.5 that the biosphere lies at the center of the Earth system diagram. It is placed there for a special reason. Significant changes are now taking place in many of the flows between the biosphere and the other reservoirs, and as a result the reservoirs are changing in sometimes unexpected ways. Some of the changes have become daily news— the ozone hole, the increase of carbon dioxide in the atmosphere, the dispersal of pesticides throughout the

ocean, the rate at which we are consuming nonrenewable resources such as oil, and the extinction of plant and animal species, to name several examples.

We, the human population, are the cause of these and other recent changes. Humans have always changed their local environments, but when the human population was small, these changes happened so slowly that they did not alter the Earth system. Now the population is large and growing ever larger. At the time these words are being written, in 1998, the world's population is nearing 6 billion and increasing by about 90 million each year. There are now so many of us that we are changing the Earth just by being alive and going about our business and in doing so we are taxing the resources of the Earth system.

Many kinds of large animals have, at various times, lived on the Earth. Throughout all of the Earth's long history, however, there has never been such a huge number of large animals as in the human population today. Our collective activities have become so pervasive that there is no place on the Earth we haven't changed. We go almost everywhere to seek the resources we need. In the process, we have made rainfall more and more acidic, we have caused fertile top soil to erode (Fig. 1.15), and we have changed the composition of the soil that remains. We have caused deserts to expand (Fig. 1.16), and we have changed the composition of the atmosphere, the ocean, streams, and lakes. Even the snowflakes that fall on Antarctica bear the imprint of our activities. In short, we are influenc-

Figure 1.15 Tilling cropland in Georgia prior to planting. Tilling disturbs the soil, creating a dust cloud. Wind blows the dust away. Over many years soil loss from plowing and tilling can be significant.

Figure 1.16 Overly intense use of marginal lands causes deserts to expand. Sand dunes advance from right to left across irrigated fields in the Danakil Depression, Egypt.

ing all of the reservoirs and many of the flows, and thereby changing our own environment. And we continue to do so at ever faster rates. We have even coined a special term to describe the changes produced in the Earth system as a result of human activities: **global change**. Measuring, monitoring, and understanding global change is now a topic of intense study by many scientists. Once again, uniformitarianism is their guide: the present is not only the key to understanding the past, but it is also the key to understanding the future.

Global change should not be viewed as necessarily negative. Most human activities have made the world a nicer and friendlier place in which to live. No one could deny that building cities and clearing land for farms causes large changes in the environment, but who would argue that a beautiful city like Paris (Fig. 1.17) is not a proud achievement? Think, too, of the abundant food that flows from modern agriculture. Our ancestors had a much harder time feeding themselves than we do today. To be sure, we have caused some changes to the environment that may one day be dangerous, but most of these changes happened accidentally because we didn't understand the Earth system sufficiently well. When we started burning coal 300 years ago, for instance, carbon dioxide had not

Figure 1.17 The Arc de Triomphe, at the hub of broad, radiating avenues, is a focal point in Paris. The avenue that runs through the center of the arch is the Champs Elysée.

even been discovered so that no one had the slightest idea that someday (i.e., today) the atmosphere would be changed as a result. If, say, the climate becomes warmer because of these changes to the atmosphere, ice in Antarctica might melt, the sea level might rise, and coastal cities might be flooded. Surely those are important consequences that we must consider, but note that we say *might* happen—might, because we do not yet understand the Earth system in enough detail to be sure. A necessary part of Earth system science is therefore an investigation of how the collective actions of the human population are changing the reservoirs and flows, and what the consequences of these changes will be. We address the issues of human changes to the Earth system at many places throughout this book.

SUMMARY

1. Earth system science is the study of the whole Earth viewed as a system of many interacting parts and focuses on the changes within and between the parts.

2. The Earth can be considered as a system of four vast, interdependent reservoirs: the geosphere, the atmosphere, the hydrosphere, and the biosphere.

3. Material moves back and forth from one reservoir to another. Some rates of movement are fast, others slow. If a rate of movement changes, the volumes of the reservoirs adjust in response.

4. Science is a system of learning and understanding that advances by application of the scientific method: observation, formation of a hypothesis, testing of the hypothesis, formation of a theory, more testing, and, in some cases, formation of a law.

5. There are three basic kinds of systems: isolated, closed, and open. The Earth is a close approximation to a closed system, which means energy can enter or leave but that matter can neither enter nor leave. All of the small systems that collectively comprise the Earth system are open systems, which means they are open to the flow of both energy and matter.

6. The energy cycle arises from the flow of energy from two vast energy sources—the Sun's heat energy and the Earth's internal geothermal energy.

7. Energy differs from matter in one important way: energy cannot be endlessly recycled. Energy always changes from a more useful, more concentrated form to a less useful, less concentrated form, and as a result, in all real processes some heat energy is always lost.

8. The hydrologic cycle is powered by energy from the Sun and describes the fluxes of water between the major Earth reservoirs.

9. Biochemical cycles move chemical elements such as nitrogen, which are essential to life, between the major Earth reservoirs.

10. The Principle of Uniformitarianism states that the internal and external processes operating today have been operating throughout Earth's history.

11. Random, massive, but rare events, such as gigantic meteorite impacts, appear to have played an important role in the Earth's history. These events cause catastrophic change in the Earth's appearance but are not attributed to supernatural forces the way the events of the outdated concept called catastrophism were.

12. Three features make the Earth unique among the planets of our solar system—plate tectonics, a large body of liquid water on the surface, and life.

13. The human population has grown so large that our collective activities may be altering many aspects of the Earth system; these collective activities are referred to as global change.

IMPORTANT TERMS TO REMEMBER

atmosphere *2*
biogeochemical cycle *11*
biosphere *2*
catastrophism *12*
closed system *4*
earth system science *2*
energy cycle *8*
entropy *9*
erosion *12*

Geographic Information
   Systems (GIS) *4*
geosphere *2*
geothermal
   energy *9*
global change *19*
hydrologic cycle *10*
hydrosphere *2*
hypothesis *3*

isolated system *4*
law (scientific) *3*
negative feedback 7
open system *4*
positive feedback 7
regolith *2*
remote sensing *4*
reservoir *6*
residence time *6*

rock *2*
rock cycle *14*
scientific method *2*
system *3*
theory *3*
Uniformitarianism,
   Principle of *12*
weathering *12*

QUESTIONS FOR REVIEW

1. What is the scientific method? Illustrate your answer with an example of the scientific method in practice.

2. How does Earth system science differ from physics, biology, or any other specialized area of science?

3. How does the Principle of Uniformitarianism help us to understand the history and workings of the Earth? Explain why this principle can also be used to understand the solar system and the universe.

4. Suggest three human activities that affect the Earth's external activities in a noticeable manner.

5. Identify three human activities in the area where you live that are causing big changes in the environment.

6. What are the principal energy sources that control the Earth system?

7. Why is it not possible to continually recycle energy?

8. What are the differences between a closed and an open system?

9. What consequences arise from the fact that the Earth is a closed system? Would the Earth still be a closed system if we started a colony on Mars and started trading with the colony?

10. Draw a box model for the hydrologic cycle and include the following reservoirs: the ocean, the atmosphere, glacial ice, streams and rivers, underground water.

11. In what form does nitrogen exist in the atmosphere? How is nitrogen removed from the atmosphere and made available to life?

12. What is meant by the term *global change*? How do your own activities contribute to global change?

QUESTIONS FOR DISCUSSION

1. Scientists are currently tracking asteroids (small rocky masses that orbit the Sun) and comets because they are concerned that an asteroid or comet might collide with the Earth sometime over the next few hundred years. What effects might the impact of an asteroid or comet have on the Earth? Which branches of Earth science do you imagine might be most involved in the work on asteroids and comets?

2. Is the suggestion that the extinction of the dinosaurs was due to the impact of a large meteorite a hypothesis or a theory? Research some alternative suggestions about the extinction of the dinosaurs. Which of the suggestions would you call uniformitarianism, which catastrophism?

3. There is currently a vigorous scientific debate about whether human activity is causing global warming. Research some of the hypotheses about global warming and analyze them in terms of the scientific method.

Visit our web site to find several interactive case studies through which you can learn more about the interactions between people and the natural environment.

CHAPTER 2

Fellow Travelers in Space: Earth's Nearest Neighbors

● The Blue Planet

There are nine planets in the solar system and we live on the third planet from the Sun. Five planets are visible to the unaided eye—Mercury, Venus, Mars, Jupiter, and Saturn—but the other three are so distant they can only be seen through telescopes.

Humans have never set foot on any other planet. We may do so some day, but so far the only other place we have reached is the Moon. Unmanned space vehicles have visited all of the planets, landed on two—Venus and Mars—and parachuted into the dense atmosphere of another—Jupiter. Space research has been, and still is, spectacularly successful, and from that research there is a home truth that all of us should remember—the Earth is just a small, isolated planet in orbit around an ordinary, middle-aged star.

The first image of the Earth from space was seen in the 1960s, at the dawn of the "space age." Before that, no one had ever actually seen the Earth as a planet in its entirety. That first image changed our relationship to our planet forever. There it was, the whole strikingly blue planet in one sweeping view, the clouds, the oceans, the polar ice fields, and the continents, all at the same time and in their proper scale. For the first time in human history it was abundantly clear just how small and isolated the Earth actually is, and how unique.

American astronaut Rusty Schweickart went into space with the *Apollo 9* mission in 1969. Here is how he described his reaction to seeing the Earth from space:

For me, having spent ten days in weightlessness, orbiting our beautiful home planet, fascinated by the 17,000 miles of spectacle passing below each hour, the overwhelming experience was that of a new relationship. The experience was not intellectual. The knowledge I had when I returned to Earth's surface was virtually the same knowledge I had taken with me when I went into space. Yes, I conducted scientific experiments that added new knowledge to our understanding of the Earth and the near-space in which it spins. But those specific extensions of technical details I did not come to know about until the data I helped to collect was analyzed and reported. What took no analysis, however, no microscopic examination, no laborious processing, was the overwhelming beauty . . . the stark contrast between bright colorful home and stark black infinity . . . the unavoidable and awesome personal relationship, suddenly realized, with all life on this amazing planet . . . Earth, our home. ●

(opposite) Earthrise seen from the Moon. The reddish-colored land is Africa and the Arabian Peninsula. Antarctica is to the lower right beneath a heavy cloud cover.

ASTRONOMY AND THE SCIENTIFIC REVOLUTION

Why does the Sun rise each day and disappear each evening? For much of human history, people believed that everything in the universe revolved around the Earth and that sunrise and sunset happened because the Sun revolved around the Earth. A universe in which everything revolves around a stationary Earth is called a **geocentric** universe. Today we are all taught from childhood that it is the Moon that revolves around the Earth, and that the Earth revolves[1] around the Sun. Proving the accuracy of what we are taught is a challenging task, and the search for proof that the Earth really does revolve around the Sun was a major factor in the rise of modern science.

Ideas from Antiquity

The Greek civilization of antiquity flowered for 800 years from about 650 B.C. to A.D. 150 and spawned many famous philosophers. The most influential of the philosophers, Aristotle (384–322 B.C.), espoused a geocentric universe. He pictured the Sun, the Moon, the five visible planets, and the stars as being suspended on concentric, hollow spheres that rotate about an imaginary axis extending outward from the two poles of the Earth, with the stars on the outermost sphere. Beyond the star sphere, and invisible to humans, was the realm of the gods (Fig. 2.1).

A few people in Aristotle's time realized that a geocentric universe is not the only way to explain things. The daily rising and setting of the Sun, and the apparent movement of the star sphere across the sky, could as well be explained if the stars were fixed and the Earth rotated on its axis once every day. Similarly, the fact that there are seasons could be explained if the axis of rotation was tilted and the Earth revolved in an orbit annually around the Sun. One Greek philosopher in particular, Aristarchus (312–230 B.C.), favored a Sun-centered, or **heliocentric** system. Aristarchus used two of the branches of mathematics discovered by the Greeks, geometry and trigonometry, to determine the relative sizes of the Sun, the Moon, and the Earth. His measurements indicated a huge Sun and, by comparison, a small Earth and a tiny Moon. It did not make sense to Aristarchus that a huge Sun should

Figure 2.1 The celestial spheres.

rotate around a small Earth, but he was unable to convince people that his heliocentric suggestion might be correct, and so the concept of a geocentric universe continued to be widely accepted until the middle of the sixteenth century, more than fifteen hundred years after the death of Aristotle. By the sixteenth century, belief in a geocentric universe had come to be accepted as a divine fact, and therefore a fact to be protected by the Catholic Church.

The Challenge by Copernicus

The most difficult thing to explain with a geocentric universe concerns the motions of the planets. The five visible planets—Mercury, Venus, Mars, Jupiter, and Saturn—look like stars, but they are stars with a difference because they seem to wander. Indeed, the very name planet comes from *planetai*, Greek for wanderers. The paths of the wanderers, measured against the background of fixed stars, are odd. They move a bit farther east each evening, but periodically they slow down and briefly reverse direction before resuming their eastward motion. The temporary reversal of direction is known as *retrograde motion* (Fig. 2.2).

The geocentric explanation for retrograde motion is as follows: each planet revolves in an orbit around the Earth and *also* follows a small circular orbit (called an *epicycle*) around an imaginary point, as shown in Figure 2.3. The larger the epicycle, the greater the amount of retrograde motion. The person who worked out the geometry of epicycles in greatest detail, and used them to predict planetary positions, was Claudius Ptolemy, a man who lived and worked in

---

[1]In this chapter we use two words for circular motion: revolve and rotate. Revolve means a body moving in an orbit around some central point external to the body; rotate means a body spinning around an axis through the body. The Earth revolves once around the Sun each year and rotates once on its axis in 24 hours.

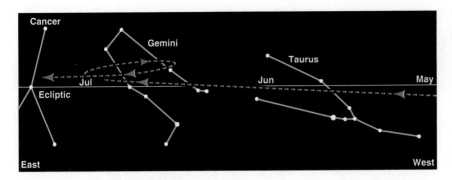

Figure 2.2 The retrograde motion of Mars during June and July 1993. First, the planet moved steadily eastward, passing in front of the constellations Taurus and Gemini. At the eastern edge of Gemini, it suddenly reversed direction and moved westward for a few days before reversing again, passing once more in front of Gemini and continuing in an easterly direction.

Alexandria, Egypt. In about A.D. 150, Ptolemy published the results of his work in the *Almagest*, one of the most important works that has come down to us from the ancient world.

When Nicholaus Copernicus (1473–1543) was a student in the 1490s at the University of Bologna, Italy, he read a Latin translation of the *Almagest* and, like Aristarchus, decided that the heliocentric system made at least as much sense as the geocentric system of Aristotle. Copernicus pointed out that the retrograde motions of planets could be explained in a heliocentric system as a result of differences between the time it takes the Earth to orbit the Sun and the time it takes for any other planet to orbit the Sun, as shown

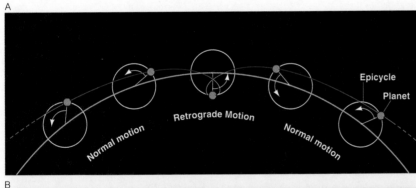

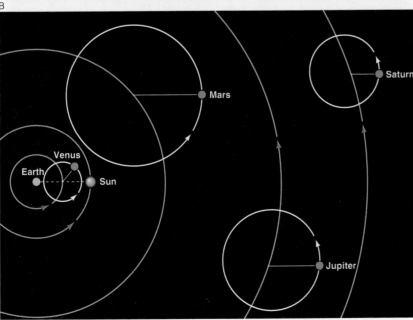

Figure 2.3 The geocentric universe of Aristotle and Ptolemy. A. The planets are imagined to orbit the Earth and to move in smaller orbits, called epicycles, around an imaginary point on the celestial sphere. B. The size of the epicycle determined the amount of retrograde motion.

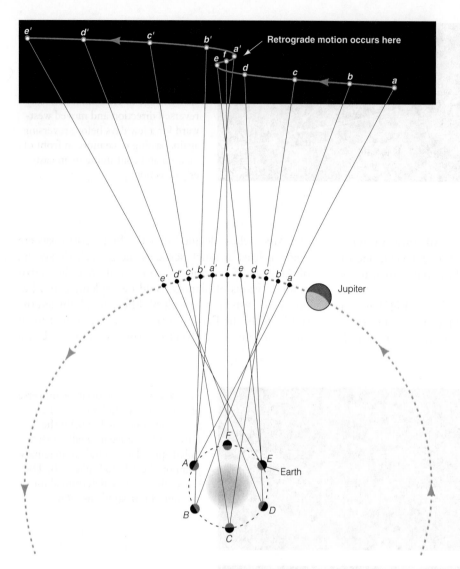

Figure 2.4 The retrograde motion of Jupiter as explained by Copernicus. As Jupiter moves from point *a* to *e'* in its orbit, the Earth moves counterclockwise from *A*, completely around the Sun and back to *A*, then on to point *E*. An observer on the Earth watching the position of Jupiter against the background of fixed stars would see the green curve as the path of Jupiter across the sky.

in Figure 2.4. Furthermore, Copernicus suggested that because Mars has a larger retrograde motion than Jupiter or Saturn, it must be the closest of the three planets to the Earth, while Saturn, with the smallest motion, must be the most distant. This was a hypothesis that could be tested by suitable astronomical measurements.

Copernicus also offered two other major hypotheses. First, he suggested that the positions of the planets at any given time in the future could be accurately predicted by assuming that they move in circular orbits around the Sun. This was another testable hypothesis. As we shall discuss, the hypothesis of a circular orbit turned out to be incorrect, but it led to another hypothesis that was correct and that became an important stepping stone in the scientific revolution. In the second hypothesis, Copernicus revived the old suggestion of Aristarchus that the Earth spins on

its axis and that the daily rising and setting of the Sun is a result of that rotation. Establishing that the Earth really does spin on its axis proved to be quite difficult. A firm demonstration, independent of astronomical observations, was not achieved until 1851 when Jean Foucault, a French physicist, did so, using a special pendulum. By extension, the two hypotheses of Copernicus—an Earth that orbits the Sun and an Earth that spins on its axis—also provided a neat explanation for the seasonality of the Earth's climate; see "A Closer Look: The Seasons."

By espousing a heliocentric system for the planets, a Moon that orbits the Earth, and an Earth that rotates on its axis, Copernicus offered a direct challenge to his church. The Roman Catholic Church had built the idea of a stationary Earth at the center of a geocentric universe into church doctrine. Looking back, we see that, by using the scientific method to question

A CLOSER LOOK

The Seasons

There are seasons because the Earth's spin axis does not stand upright with respect to the plane of the orbit around the Sun. If the Earth stood straight up, meaning that the spin axis was perpendicular to the plane of the orbit, the equator would always face the Sun and there would be no seasons. But the axis is tilted 23.5° away from perpendicular, and the tilt remains in an essentially fixed direction as the Earth moves around the Sun.

Because of the tilt, at one point in the orbit the northern hemisphere faces the Sun, and at another the southern hemisphere faces the Sun (Fig. C2.1). Our calendars are set so that on June 21 (northern hemisphere midsummer day) the Sun is directly overhead at its furthest north position, and on December 21 (southern hemisphere midsummer day) it reaches its furthest south position. This means that the northern hemisphere gets more of the Sun's heat and illumination in June and less in December.

A second factor contributes to seasonality. The Earth's orbit around the Sun is an ellipse, so that the distance between the Earth and the Sun changes day by day. The Earth is closest to the Sun in early January and farthest away in early July. Thus, the Earth overall gets more of the Sun's rays in January, during the southern hemisphere summer, and less overall in July during the northern hemisphere summer.

Although the rotation of the Earth around its axis and the Earth's orbit around the Sun appear to be fixed, there are slight variations that lead to climate changes over time scales of tens of thousands of years. As will be discussed in Chapter 14, axial rotational and orbital changes are the major reasons that continental ice sheets advance and retreat. The Earth's orbital and rotational motions are major features of the Earth system.

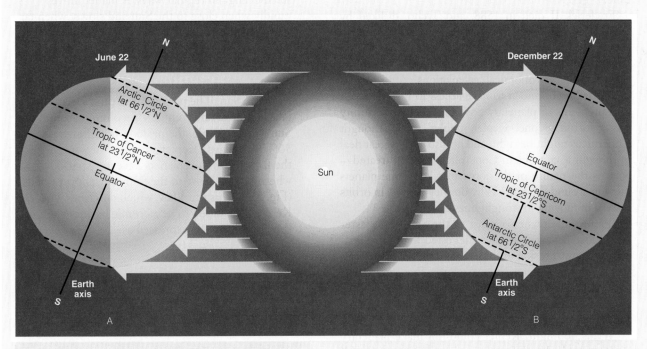

Figure C2.1 The seasons occur because the tilt of the Earth's axis keeps a constant orientation as the Earth revolves around the Sun. A. Summer in northern hemisphere. B. Winter in southern hemisphere.

a topic accepted as doctrine for over a millennium and by offering hypotheses that could be tested, Copernicus sowed the seeds that finally separated science from religion and spawned the continuing scientific revolution that shaped the society in which we live today. It is especially noteworthy that modern science has its roots in astronomical studies and in particular in studies of the motions of the Earth and planets. Earth science is a founding member of modern science.

Kepler and the New Astronomy

When the ideas of Copernicus were published in 1543, they convinced many but not all intellectuals. Among the skeptics was Tycho Brahe (1546–1601). In 1572, with funds from King Frederick II of Denmark, Tycho built the first modern astronomical observatory on the Danish island of Hven, naming it Uraniborg, or Castle of the Heavens. Optical telescopes had not

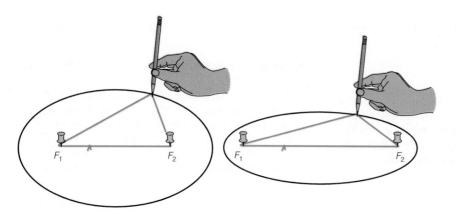

Figure 2.5 How to draw an ellipse. Pin a piece of string at two places and draw a closed figure by pulling the string taut. The two pin points (F_1 and F_2) are the foci of the ellipse. The closer F_1 and F_2 are to each other, the closer an ellipse approaches a circle; the farther apart F_1 and F_2, the greater the eccentricity of the ellipse. Kepler calculated that the orbit of Mars is an ellipse with the Sun at one focus.

been invented in Tycho's day, and so all observations were made with the naked eye. Tycho's measurements of planetary positions, made in part to prove Copernicus wrong, were by far the most accurate up to that time.

Frederick II died in 1588, and Tycho, disliked by Frederick's successor, fell from favor. In 1597 Tycho moved to Prague, and while there he hired a young German mathematician, Johannes Kepler (1571–1630), to carry out astronomical calculations. After Tycho died, Kepler continued to have access to his numerous measurements of planetary positions. The opportunity was a fortunate one. Kepler, unlike Tycho, thought Copernicus might be right, and he also gave a great deal of thought to a problem Copernicus had not treated—what is the nature of the force that keeps the planets moving around the Sun? Why do they revolve in orbits instead of moving in straight lines out into space?

Because the planets closest to the Sun move faster than those far away, Kepler suggested that a mysterious force must reside in the Sun and have a greater effect on closer objects. Today we know that the force is gravity, but in Kepler's day gravity was an unknown concept. Kepler suggested that magnetism might be the force.

Try as he might, Kepler could not make planetary positions calculated from circular orbits agree with Tycho's measurements. Eventually, Kepler tried calculating the position of Mars based on an elliptical orbit (Fig. 2.5), and the calculated positions agreed very closely with the positions observed by Tycho. This is an example of science at its best. Copernicus had offered a hypothesis, and Kepler had used Tycho's measurements to test the hypothesis. Together, Copernicus and Kepler demolished the geocentric idea.

Kepler discovered three laws that describe planetary motions:

1. *The law of ellipses.* The orbit of each planet is an ellipse with the Sun at one focus.

2. *The law of equal areas.* A line drawn from a planet to the Sun sweeps out equal areas in equal times (Fig. 2.6). One consequence of the law of equal areas is that orbital speeds are not uniform but instead change in regular ways. A planet moves rapidly when close to the Sun and slowly when far away from the Sun.

3. *The law of orbital harmony.* For any planet, the square of the orbital period in years is proportional to the cube of the planet's average distance from the Sun. The period is the time a planet takes to make one complete revolution around the Sun. (For example, the period of the Earth is 365.24219 days.) Kepler sought a way to express the fact that distant planets have long periods, whereas those close to the Sun have short periods (Fig. 2.7). This third law, which describes what Kepler considered to be a cosmic harmony among the planets, can be expressed as:

$$p^2 = kd^3$$

where p is the period, k is a constant, and d is the average distance between the planet and the Sun.

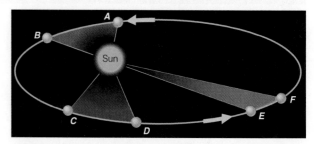

Figure 2.6 Kepler's law of equal areas: Because the orbital speed of any planet varies during the time it takes to complete one orbit, a line connecting the planet and the Sun sweeps out equal areas in equal times. The time it took the planet whose orbit is drawn here to travel from A to B is exactly the same as the time to travel from C to D and from E to F.

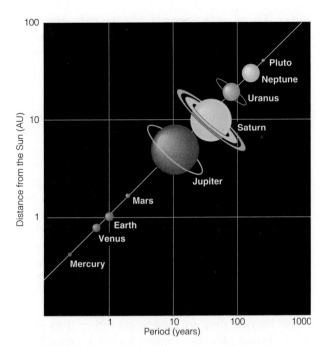

***Figure 2.*7** Kepler's third law relates the distance of a planet from the Sun to the period of the planet's orbit around the Sun. The unit of distance is the *astronomical unit*, the average distance from the Earth to the Sun. Note the gap between Mars and Jupiter. This is the place in the solar system where the asteroids are found. Some scientists believe that the asteroids are simply rocky fragments that did not accrete to form a planet.

Galileo and Newton

Galileo Galilei (1564–1642) was an extraordinary man who made a great many scientific discoveries. In 1609 he constructed a small telescope with a magnification of 30 times. Turning his telescope to the sky, Galileo observed mountains on the Moon, discovered that the Milky Way is a dense mass of stars rather than a band of luminous gases, observed four moons in orbit around Jupiter, observed that Venus, like the Moon, goes through phases from crescent to full, and also changes in apparent size. These observations sealed the fate of the geocentric universe for even the most ardent believers in the concept. For example, because moons revolve around Jupiter, the Earth cannot be the center of the universe, and the fact that Venus has phases and also changes greatly in apparent size can only be explained if Venus and the Earth are in orbit around the Sun. When the Earth and Venus are on the same side of the Sun, Venus is seen as a crescent. When the Earth and Venus are on opposite sides of the Sun, Venus is seen as a full disc, but it is only one-seventh the diameter of the crescent because it is so

far away. In the old Aristotelian-Ptolemaic system, with Venus in orbit around the Earth, the apparent size of Venus should change very little.

Galileo also made major contributions to our understanding of moving bodies. Motion, he pointed out, is due to a force. Once a body is moving, it will stop or change direction only in response to another force. If you drop a ball, it falls because some force (gravity) exerted by the Earth pulls it down. Galileo concluded that gravity pulls all falling bodies with the same acceleration. Therefore, uniform acceleration means that, in the absence of air resistance, all falling bodies, regardless of their mass, reach the same speed and fall the same distance in the same time. Legend has it that Galileo tested this idea by dropping two balls of different weight from the Leaning Tower of Pisa. The balls reached the ground at the same instant. This experiment reversed a very ancient and deep-seated belief that heavy bodies fall faster than light ones. More importantly, it provided one of the key steps by which the law of gravity was discovered and the motions of planets were finally explained. The person who pulled all the pieces together was Isaac Newton (1642–1727).

First turning his attention to mechanics (the branch of physics that deals with mass, velocity, and acceleration), Newton considered the problem of the force that pulls objects toward the Earth. There is a legend that Newton started to think about gravity when he saw an apple fall from a tree. He reasoned that if the force of gravity acts on an apple, that force must also, as he wrote, extend "to the orb of the Moon." The Moon revolves around the Earth instead of moving through space in a straight line because the force exerted by the Earth's gravity continuously exerts a small pull on the Moon (Fig. 2.8). The force that Kepler had misidentified as magnetism was actually gravity. Newton, through his insight, discovered

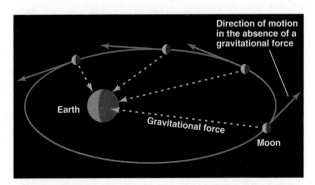

Figure 2.8 The gravitational pull exerted by the Earth on the Moon continuously diverts the Moon's direction of motion from a straight line to a closed ellipse.

one of the universal laws of nature, the **law of gravitation**, which states that *every body in the universe attracts every other body.*

Newton's discovery of the law of gravitation greatly advanced scientific understanding. No longer was it necessary to call on different forces to describe the fall of apples and the motions of planets. More important, Newton managed to put the law into an algebraic form:

$$F = G \frac{M_1 M_2}{R^2}$$

where F is the force of gravitational attraction between two masses separated by a distance R and having masses M_1 and M_2, and G is the gravitational constant.

THE SOLAR SYSTEM

With his crude telescope Galileo could see only the five planets visible to the naked eye. As telescopes improved, however, the other planets of our solar system were discovered. The first of the telescope-discovered planets was Uranus, found by Sir William Herschel (1738–1822) in 1781. Herschel thought he had discovered a new comet, but measurement of the orbit showed it to be a planet. It was soon discovered that Uranus's orbit deviated slightly from the path calculated under the assumption that the only gravitational forces acting on Uranus were those of the Sun and the known planets. Using Newton's law of gravitation to explain the deviations, mathematicians predicted that an undiscovered planet was the cause and suggested where to look for the planet. On September 23, 1846, Johann Galle (1812–1910) discovered the predicted planet, Neptune. The discovery was a great triumph for Newton's law. Early in the twentieth century, evidence of a tiny perturbation in the orbit of Neptune suggested that there might be another undiscovered planet. Several searches were made, but it wasn't until February 18, 1930, that Pluto, the faintest and most distant planet, was finally discovered by Clyde Tombough, a 24-year-old American who had no formal training in astronomy and only a high school diploma. As far as we know, all the planets in the solar system have now been discovered.

The solar system consists of the Sun, nine planets, a vast number of small rocky bodies called asteroids, millions of comets, innumerable small fragments of rock and dust called meteoroids, and 61 known moons. All of the objects in the solar system have smooth, regular orbits, determined by gravitational forces. The planets, asteroids, comets, and meteoroids orbit the Sun, whereas the moons orbit the planets.

The planets can be separated into two groups based on density and closeness to the Sun (Fig. 2.9A). The innermost planets—Mercury, Venus, Earth, and Mars—are small, rocky, and dense (Fig. 2.9B). Because they are similar in composition to our Earth, they are called the **terrestrial planets** (*terra* is Latin for Earth).

The asteroids are also rocky, dense bodies, but they are too small to be called planets. Refer back to Figure 2.7 showing the distances of the planets from the Sun and note the apparent gap between Mars and Jupiter. The asteroids have orbits that fall in this gap, and astronomers hypothesize that they are rocky fragments that failed to join together to make a larger planet.

The planets farther from the Sun than Mars (with the exception of Pluto) are much larger than the terrestrial planets, yet much less dense. They are less dense because they are largely gas rather than solid. These **jovian planets**—Jupiter, Saturn, Uranus, Neptune, and Pluto—take their name from *Jove*, an alternate designation for the Roman god Jupiter.

If you view the solar system from a spaceship, you will notice a striking feature. Almost everything—planets, moons, Sun—revolve and rotate in the same direction. All the planets revolve around the Sun and all the moons revolve around their respective planets in approximately the same plane. If your vantage point is high above the North Pole, the rotations will be counterclockwise. Moons revolve around the planets, and planets revolve around the Sun in the same direction. Furthermore, the Sun and most of the planets (Venus is an exception) rotate around their axes in the same counterclockwise direction. Venus is an exception because, for reasons not fully understood, it rotates slowly in a clockwise direction. The revolutions and rotations of the planets and moons are so regular and consistent that all societies in history have used them as a way of keeping track of the passage of time; see "A Closer Look: Time and the Calendar."

The Origin of the Solar System

Any theory for the origin of the solar system must be able to explain both the remarkably consistent motions of the planets and moons and the grouping of the planets into two classes—terrestrial and jovian. Current hypotheses about the origin of the solar system are as follows.

The Sun is a star approximately 5 billion years old. The universe is at least twice and possibly three times

A.

B.

	Mercury	Venus	Earth	Mars	Jupiter	Saturn	Uranus	Neptune	Pluto
Diameter (km)	4880	12,104	12,756	6787	142,800	120,000	51,800	49,500	6000
Mass (Earth=1)	0.055	0.815	1	0.108	317.8	95.2	14.4	17.2	0.003
Density, g/cm^3 (water=1)	5.44	5.2	5.52	3.93	1.3	0.69	1.28	1.64	2.06
Number of moons	0	0	1	2	16	18	15	8	1
Length of day (in Earth hours)	1416	5832	24	24.6	9.8	10.2	17.2	16.1	154
Period of one revolution around Sun (in Earth years)	0.24	0.62	1.00	1.88	11.86	29.5	84.0	164.9	247.7
Average distance from Sun (millions of kilometers)	58	108	150	228	778	1427	58	4497	5900
Average distance from sun (astronomical units)	0.39	0.72	1.00	1.52	5.20	9.54	0.39	30.06	39.44

Figure 2.9 The planets and their properties. A. The planets, shown in their correct relative sizes and in the correct order outward from the Sun. The Sun is 1.6 million km in diameter, 13 times larger than Jupiter, the largest planet. B. Numerical data concerning the orbits and properties of the planets.

as old as the Sun, which makes the Sun a relatively young star. The origin of the Sun (as well as the planets) was probably similar to the origins of billions of other stars; thus, a lot of the ideas concerning the origin of the solar system come from observations made by astronomers studying star formation. Scientists hypothesize that the solar system formed from a huge, rotating cloud of cosmic gas. Because gases tend to be well mixed, one of the key questions to be answered by such a hypothesis is why the composition of the Sun and the planets is different. Stars, including the Sun, consist largely of the two lightest chemical ele-

ments, hydrogen and helium, with only a tiny amount of heavier elements. Rocky planets like the Earth, Mars, and Venus, on the other hand, consist largely of heavier elements such as carbon, oxygen, silicon, and iron, with very small amounts of hydrogen and helium.

One clue concerning the origin of the solar system is that very ancient stars—those that formed during the earliest moments of the universe—consist predominantly of the lightest chemical element, hydrogen, as well as lesser amounts of helium. From that observation scientists hypothesize that all the heavier chemical elements—carbon, oxygen, silicon, iron, and so on—were somehow made from hydrogen and helium. Stars generate light and heat through nuclear fusion, a process by which hydrogen atoms combine to form helium, and then hydrogen and helium atoms combine through nuclear fusion to form still heavier

A CLOSER LOOK

Time and the Calendar

The calendar is a way of dividing time and is built around the day, month, and year, the lengths of which are determined by three primary astronomical motions. *Rotation* of the Earth, from one sunrise to the next, determines the length of the day; from the *revolution* of the Moon around the Earth we obtain the month; *revolution* of the Earth around the Sun determines the length of the year. Unfortunately, the solar day, solar year, and lunar month are not measured in whole numbers—the solar year is approximately 365.25 days, while the lunar month is approximately 29.5 days. As a result, the calendar is more complicated than it need be and sometimes needs adjusting.

The Length of the Year

The ancient Egyptians, from whom we derive a considerable portion of our calendar, observed that the Sun and the prominent star Sirius appear on the horizon together at daybreak at 365-day intervals. The Egyptian calendar was therefore based on a year of 365 days. Unfortunately, the calendar makers did not take into account the fact that a year measured by the Earth's revolution around the Sun is approximately 365.25 days, so that after four years the calendar was a day off from the year measured by Sirius and the Sun. The Greeks and the Romans adopted the Egyptian solar year of 365 days, and to clear up the problem of the extra quarter day Julius Caesar institutionalized an older Greek idea and decreed that every fourth year would have 366 days. (That is, the calendar year would *leap* ahead by a day and catch up with the solar year.) However, the length of a solar year is not exactly 365.25 days (365 days and 6 hours); it is 365 days, 5 hours, 48 minutes and 46 seconds, making the Julian calendar too long by 11 minutes and 14 seconds, or one day in 128 years. By the time of Pope Gregory XIII, the Julian calendar was 10 days ahead of the seasons recorded by the solar year, and so the Pope ordered 10 days removed from the calendar: the day after Thursday, October 4, 1582 was decreed to be Friday, October 15, 1582. The Gregorian calendar, the one used in much of the world today, corrects further misfits by not having a leap year on centennial years not divisible by 400. Thus, 1900 was not a leap year but 2000 will be. Even the Gregorian correction is not exact, however, and as a result the calendar year is still moving ahead of the solar year, but only by 26 seconds a year, or one day in 3323 years. Sometime about the year 4500 a further correction will be needed.

The Month

The Egyptians and many other early societies organized their first calendars around the phases of the Moon. A new moon rises approximately every 29.5 days, so 12 lunar months determined by new moons is only 354 days, considerably different from a solar year of 365.25 days. When the Egyptian calendar makers finally settled on a solar year for their calendar, they retained an aspect of their ancient lunar calendar and divided the year into 12 months of 30 days each. The remaining 5 days needed to bring the year up to 365 were simply added on at the end of each year. Such a scheme was unacceptable to the Romans, who devised the present scheme of some months having 31 days, some 30, and the second month 28 days (or 29 in leap years).

The names of the months are also Roman in origin. The Romans had a calendar of 10 months: Martius (March), Aprilis (April), Maius (May), Junius (June), Quintilis (later changed to Julius, or July), Sextilis (later changed to Augustus, or August), September (seventh), October (eighth), November (ninth), and December (tenth). When the Romans adopted the Egyptions' 12-month year, they placed the two extra months at the beginning of the year and named them Januarius (January) and Februarius (February), leaving us with an anomaly: the month called the tenth (December) is actually the twelfth.

The Week

The names given to the days of the week lie in seven celestial bodies and the way they are ranked for astrological purposes.

Viewed from the Earth, seven bodies move against the background of fixed stars: Mercury, Venus, Mars, Jupiter, Saturn, the Moon, and the Sun. These bodies were considered to be gods who ruled the heavens and controlled the days. Babylonian and Hindu astronomers ranked the celestial bodies according to the relative ruling powers of the gods as Sun, Moon, Mars, Mercury, Jupiter, Venus, Saturn. The seven, in that order, gave their names to the day of the week of which each was thought to be in charge. In countries where a Romance language, such as Spanish, is spoken, most of the days are still named for the Roman gods. English names are a mixture, with most names being of Saxon origin.

Roman Names	Spanish Names	Saxon Names	English Names
Dies Solis	Domingo	Sun's day	Sunday
Dies Lunae	Lunes	Moon's day	Monday
Dies Martis	Martes	Tiw's day	Tuesday
Dies Mercurii	Miércoles	Woden's day	Wednesday
Dies Jovis	Jueves	Thor's day	Thursday
Dies Veneris	Viernes	Frigg's day	Friday
Dies Saturnii	Sabada	Seterne's day	Saturday

Hours, Minutes, and Seconds

We have inherited the 24-hour day from the ancient Egyptians, who divided the times of daylight and dark into 12 hours each. Because daylight is longer in summer and shorter in winter, the Egyptian hours varied in length throughout the year. An hour of variable length may be satisfactory for a farmer dealing with matters of the field, but it is a great disadvantage for calculations involving time. The Greeks finally cleared up the confusion about 2000 years ago when they divided the time of the Earth's rotation into 24 units of equal duration. To make accurate calculations, the Greeks had to divide up the hour into still smaller units, and to do so they borrowed from the Babylonians. Today we use a counting system based on the number 10, but in ancient Mesopotamia, where the Babylonians lived, a counting system based on the number 60 was in use. The Greeks simply borrowed the Babylonian number system and divided each hour into 60 minutes and each minute into 60 seconds.

chemical elements. We will discuss *nucleosynthesis*—the formation of elements by nuclear fusion—in greater detail in Chapter 3. For the moment, we need only know that the only way elements heavier than helium and lithium can form is by nuclear fusion inside stars. The amounts formed in this way are tiny by comparison with the amounts of hydrogen and helium present in the universe.

In order for rocky planets to form, the heavy elements made inside old stars must somehow get into a cosmic gas cloud. It is hypothesized that this occurs when a massive star explodes in what is called a supernova (Fig. 2.10). Astronomers have discovered and photographed the scattered remains of many exploded stars, and they observe that all of the hydrogen, helium, and heavier elements are scattered into space in a vast, slowly rotating cosmic gas cloud. The next step in the process of forming a solar system is the formation of a new sun and a planetary system from the debris of the cosmic explosion.

We don't know whether all the atoms now in the Sun and in the planets were formed in one ancient star or several, but scientists have estimated that the atoms now in the Sun and the Earth were part of a cosmic cloud about 6 billion years ago. Though thinly spread, the scattered atoms formed a tenuous, slowly rotating cloud of gas. Over a very long period of time, the gas thickened as a result of a slow re-gathering of the thinly spread atoms. The gathering force of the gas was gravity, and as the atoms moved closer together, the gas became hotter and denser as a result of compression. Near the center of the gathering cloud of gas, hydrogen atoms eventually became so tightly compressed, and the temperature so high, that nuclear burning started again and a new sun was born. Surrounding the new sun was a flattened rotating disc of gas and dust, and to this rotating disc the name **solar nebula** is given (Fig. 2.11). Using the largest modern telescopes, astronomers have discovered that many young stars have rotating discs of gas around them.

By the time the Sun started burning, about 5 billion years ago, the cooler outer portions of the solar nebula had become so compressed that the gas started to condense in the same way that ice condenses from water vapor. The condensates were solid particles and these particles became the building blocks of the planets, moons, and all the other solid objects of the solar system. Planets and moons nearest the Sun, where temperatures were highest, consist mostly of substances that can only condense at high temperatures, mainly compounds containing oxygen and elements such as silicon, aluminum, calcium, iron, and magnesium. Farther away from the Sun, where the temperature was lower, sulfur-bearing compounds, water, ice, and methane ice could also condense (Fig. 2.12). Distance from the Sun and condensation temperature explain why the terrestrial planets consist mainly of high-temperature rocky materials that have high densities, whereas the jovian planets and their moons consist primarily of low-temperature, low-density condensates.

Condensation of a cosmic gas cloud is only one piece of the planetary birth story. Condensation formed a cosmic snow of innumerable small rocky fragments. The cosmic snow revolved around the Sun in the same direction as the rotating gas cloud from which the snow particles had formed. One further step was needed in order to form the cosmic snowballs that we call planets. The step involved impacts between fragments of cosmic snow drawn together by gravitational attraction. The growth process—a gathering of more and more bits of solid matter from surrounding space—is called **planetary accretion**. Scientists estimate that condensation of the solar nebula and planetary accretion was complete about 4.56 billion years ago, and when it was complete, the rotational motion of the ancient gas cloud had been preserved in the revolutions and rotations of the Sun, planets, and moons.

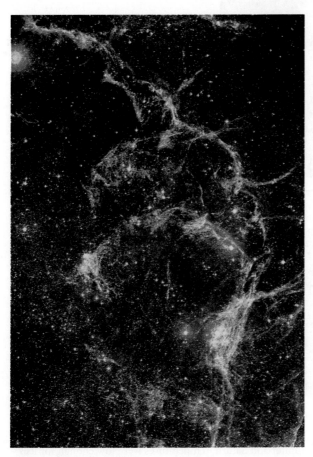

Figure 2.10 A supernova.

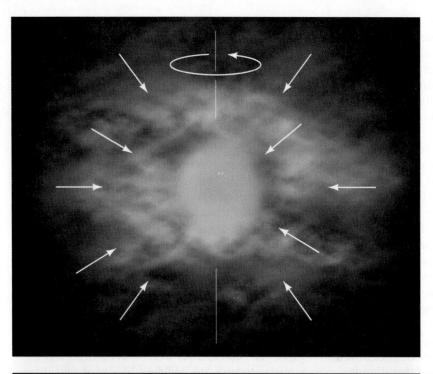

Figure 2.11 Formation of a planetary nebula. The gathering of atoms in space created a rotating cloud of dense gas. The center of the gas cloud eventually became the Sun; the planets formed by condensation of the outer portions of the gas cloud.

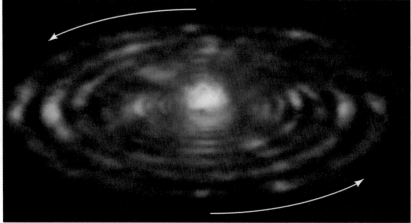

Figure 2.12 Temperature gradient in the planetary nebula. Close to the Sun temperatures reached 2000 K, and only oxides, silicates, and metallic iron and nickel condensed to form planets. Farther away, in the region of Jupiter and Saturn, temperatures were low enough for ices of water, ammonia, and methane to condense.

Evolution of the Planets

Space missions continue to provide evidence proving that all the objects in the solar system formed at the same time and from a single solar nebula. Therefore, the preceding hypothesis of the origin of the solar system agrees reasonably well with the facts that must be explained. Space exploration has also proved that during the final phase of planetary accretion the Moon and the four terrestrial planets became so hot that they began to melt.

Four key factors played the determining roles in the evolution of the terrestrial planets. The first key factor involves melting. When moving bodies collide, the energy of their motions (kinetic energy) is converted to heat energy. As planetary accretion approached a climax about 4.56 billion years ago, bigger and bigger collisions meant that more and more kinetic energy was converted to heat—so much so that the terrestrial planets started to melt. The terrestrial planets probably did not melt completely, but sufficient melting occurred that the heavier, iron-rich liquids sank to the center of the planets, and lighter liquids, rich in light elements essential for life, such as potassium, sodium, phosphorus, aluminum, silicon, and oxygen, floated to the surface.

During and after the phase of partial melting and compositional separation, the Moon and the four terrestrial planets continued to be struck by remaining fragments of the cosmic snow—today we call such fragments meteorites. Although meteorite impacts still occur, the time of nearly continuous, massive impacts ended more than 4 billion years ago. From about 4.0 billion years ago to the present, the terrestrial planets and the Moon have all been independent closed systems, and they have evolved along somewhat different paths. The thick, cloud-encircling atmospheres of the jovian planets obscure details of the evolutionary history of those planets, but at least their earliest history was probably like that of the terrestrial planets.

The second key factor in the evolution of the terrestrial planets involves volcanism. After partial melting, the planets remained hot inside because radioactive elements were and still are present. Every time a naturally radioactive element undergoes a radioactive change, it releases some heat energy. All of the terrestrial planets are cooling down, but the rates of cooling are determined by the sizes of the planets, and the cooling rates vary greatly as a result. The largest planets, Venus and the Earth, are cooling very slowly and therefore are still relatively hot today. One important indication of high internal temperature is volcanism, which continues on the Earth and probably on Venus. There has been volcanic activity on Mars within the past billion years, but Mars is probably not active today. Both the Moon and Mercury, the two smallest bodies, have been volcanically dead for billions of years. As we will discuss later in the book, volcanism has been very important for life. The gases in the atmosphere, for example, escaped from inside the Earth as volcanic emissions.

The third factor that controlled the way the terrestrial planets evolved is their distance from the Sun. The Sun-planet distance determines whether or not H_2O can exist as water and hence whether or not there can be oceans. The two planets closest to the Sun—Mercury and Venus—are too hot for liquid water to exist. Venus does have water vapor in its atmosphere, but the temperature at the surface of Venus is close to 500° C. Mars, which is farther from the Sun than is the Earth, is too cold to have liquid water but does have ice. Liquid water, as will become clear at many places in this book, is an essential condition for life. Scientists sometimes refer to the Earth as a goldilocks planet, in reference to the story of Goldilocks and the three bears—Venus is too hot, Mars is too cold, the Earth is just right!

The fourth factor determining the evolutionary path of a planet is the presence or absence of a biosphere. The hydrosphere and the biosphere play essential roles in biogeochemical cycles that control the composition of the atmosphere. If life had developed on Venus, that planet might have finished with an atmosphere like the Earth's. On the Earth, plants and microorganisms have been the means whereby carbon dioxide and water have been combined, through photosynthesis, to make organic matter and oxygen. The burial of organic matter in sediment removes carbon from the atmosphere and at the same time adds a balancing amount of oxygen to the atmosphere. Because life did not develop on Venus, there is no buried organic matter and all of the CO_2 is still in the atmosphere. Venus suffers from a horrendous greenhouse effect because carbon dioxide is a very effective thermal blanket.

The Earth system and its many parts came into being a long time ago. What that system is today, and how the many parts interact, are very much a product of the Earth's long history and of the two great heat engines that drive it: the solar engine, which has warmed the Earth's surface for the 4.56 billion years of the planet's existence, and the internal heat engine, which drives all the activities of the solid Earth. One vitally important aspect of the history of the Earth system is the chemical differentiation that happened very early as a result of partial melting. The differentiation event led to a concentration of light chemical elements near the surface (Fig. 2.13). The rocks, the soils, the atmosphere, the oceans, are all made from the mixture

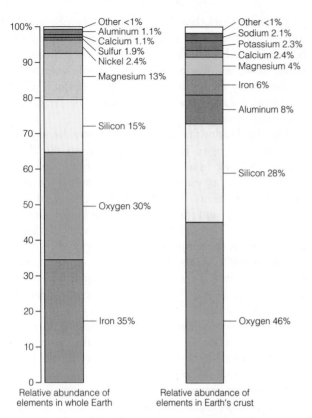

Figure 2.13 The effect of the chemical differentiation event on the distribution of chemical elements in the Earth can be seen by comparing the relative abundances of elements in the whole Earth and in the crust. Differentiation created a light crust depleted in iron and magnesium and enriched in oxygen, silicon, aluminum, calcium, and sodium.

of light chemical elements. The global environment, indeed all of life, is very much determined by the Earth's long-ago chemical differentiation.

THE TERRESTRIAL PLANETS

Each of the terrestrial planets and the Moon have the same gross structure, consisting of three layers distinguished by differences in composition arising from chemical differentiation due to partial melting.

Layers of Different Composition

The structure common to all the planets is most clearly demonstrated in the Earth (Fig. 2.14). At the center is the densest of the three layers, the **core**, a spherical mass composed largely of metallic iron, with smaller amounts of nickel and other elements. The thick shell of dense, rocky matter that surrounds the core is called the **mantle.** The mantle is less dense

than the core but denser than the outermost layer. Above the mantle lies the thinnest and outermost layer, the **crust**, which consists of rocky matter that is less dense than mantle rock.

Each of the terrestrial planets has a core, mantle, and crust, but there are considerable differences in detail, particularly in the crust. For example, it is apparent in Figure 2.14 that the core and the mantle of the Earth have nearly constant thicknesses but the crust is far from uniform and differs in thickness from place to place by a factor of nine. The crust beneath the oceans, the **oceanic crust**, has an average thickness of about 8 km, whereas the **continental crust** averages 45 km in thickness and ranges from 30 to 70 km. The two different kinds of crust are the result of the special internal processes that continually shape and rearrange the Earth's surface, and in particular, plate tectonics (see Chapter 4). The crusts of the other terrestrial planets are thicker than the Earth's crust and approximately uniform in thickness. The uniformity of thickness is an indication that plate tectonics does not, and probably never has, been active on any of the other terrestrial planets.

Because we cannot see and sample either the core or the mantle of a planet, it is valid to ask how we know anything about their composition. The answer is that indirect measurements are used, and again the Earth is used as an example. One way to determine composition is to measure how the density of rock changes with depth below the Earth's surface. We can do this by measuring the speeds with which earthquake waves pass through the Earth, because the speeds are influenced by rock density (see Chapter 5). At some depths, abrupt changes in the speed of earthquake waves indicate sudden changes in density. From the sudden changes, we infer that the solid Earth consists of distinct layers with different densities. Knowing these different densities, we can estimate what the composition of the different layers must be.

Slight compositional variations probably exist within the mantle, but we know little about them. We can see and sample the crust, however, and the sampling shows that, even though the crust is quite varied in composition, its overall composition and density are very different from those of the mantle, and the boundary between them is distinct.

The composition of the core presents the most difficulty. The temperatures and pressures in the core are so great that materials there probably have unusual properties. Some of the best evidence concerning core composition comes from iron meteorites. Such meteorites are believed to be fragments from the core of an asteroid, large enough to be a protoplanet, that was shattered by a gigantic impact early in the history of

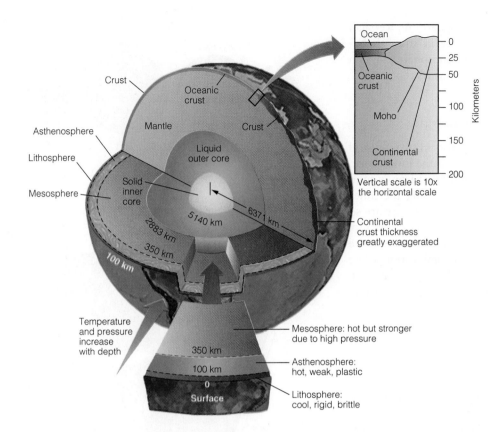

Ocean
Oceanic crust
Moho
Continental crust

0
25
50
100
150
200

Kilometers

Vertical scale is 10x the horizontal scale

Figure 2.14 A sliced view of the Earth reveals layers of different composition and zones of different rock strength. The compositional layers, starting from the inside, are the core, the mantle, and the crust. Note that the crust is thicker under the continents than under the oceans. Note, too, that boundaries between zones of different physical properties—lithosphere (outermost), asthenosphere, mesosphere—do not coincide with compositional boundaries.

the solar system. Scientists hypothesize that this now-shattered protoplanet must have had compositional layers similar to those of the Earth and the other terrestrial planets.

Layers of Different Rock Strength

In addition to compositional layering, the sphere that is our Earth can be divided into three layers based on differences in the strength of the rock making up each layer: the mesosphere, asthenosphere, and lithosphere (Fig. 2.14).

The strength of a solid is controlled by both temperature and pressure. When a solid is heated, it loses strength; when it is compressed, it gains strength. Differences in temperature and pressure divide the mantle and crust into three distinct strength regions. In the lower part of the mantle, the rock is so highly compressed that it has considerable strength even though the temperature is very high. Thus, a solid region of high temperature but also relatively high strength exists within the mantle from the core–mantle boundary (at 2883 km depth) to a depth of about 350 km and is called the **mesosphere** ("intermediate, or middle, sphere") (Fig. 2.14).

Within the upper mantle, from 350 to about 100 km below the Earth's surface, is a region called the **as-thenosphere** ("weak sphere"), where the balance between temperature and pressure is such that rocks have little strength. Instead of being strong, like the rocks in the mesosphere, rocks in the asthenosphere are weak and easily deformed, like butter or warm tar. As far as geologists can tell, the compositions of the mesosphere and the asthenosphere are essentially the same. The difference between them is one of physical properties; in this case, the property that changes is strength.

Above the asthenosphere, and corresponding approximately to the outermost 100 km of the Earth, is a region where rocks are cooler, stronger, and more rigid than those in the plastic asthenosphere. This hard outer region, which includes the uppermost mantle and all of the crust, is called the **lithosphere** ("rock sphere"). It is important to remember that, even though the crust and mantle differ in composition, it is rock strength, not rock composition, that differentiates the lithosphere from the asthenosphere.

The boundary between the lithosphere and the asthenosphere is caused by differences in the balance between temperature and pressure. Rocks in the lithosphere are strong and can be deformed or broken only with difficulty; rocks in the asthenosphere below can be easily deformed. One analogy is a sheet of ice floating on a lake. The ice is like the lithosphere, and the lake water is like the asthenosphere. As we shall see in later chapters, the difference in strength between the lithosphere and the asthenosphere plays an important role in determining the Earth's topography.

Layers of Different Physical State

Metallic iron in the Earth's core exists in two physical states. The solid center of the Earth is the **inner core.** Pressures are so great in this region that iron is solid despite its high temperature. Surrounding the inner core is a zone where temperature and pressure are so balanced that the iron is molten and exists as a liquid. This is the **outer core.** The difference between the inner and outer cores is not one of composition. (The compositions are believed to be the same.) Instead, the difference lies in the physical states of the two: one is a solid, the other is a liquid.

Comparison of the Terrestrial Planets

The terrestrial planets, and possibly the Moon, seem to have had similar early histories. Where ancient surfaces exist, as on the Moon, Mercury, and the southern half of Mars, evidence of a violent period of planetary accretion remains. Each body seems to have experienced a period of partial melting during which a core formed. The striking feature of the various cores, the sizes of which are calculated from the densities of the planets, is how greatly they differ in relative size (Fig. 2.15A). The most remarkable body is Mercury, for on this planet the core is 42 percent of the volume and an estimated 80 percent of the mass. It is not possible, at present, to assert with any certainty whether any of the terrestrial planets besides the Earth have molten or partially molten cores. The molten outer core and the relatively rapid rotation of the Earth give rise to the Earth's strong magnetic field. Magnetic fields do exist on the other terrestrial planets, but they are much weaker than the Earth's field.

Spacecraft have landed on the Moon, Mars, and Venus, and on those bodies we have been able to make direct measurements of the crust. Fly by missions to Mercury reveal that a crust is present there too. The existence of a core and a crust suggests a mantle, and the necessary measurements have been made on the Moon, Mars, and Venus to establish that indeed each has a mantle. We can be reasonably sure, therefore, that the structures of all the terrestrial planets are similar.

Whether or not each terrestrial planet has a lithosphere, an asthenosphere, and a mesosphere is a more

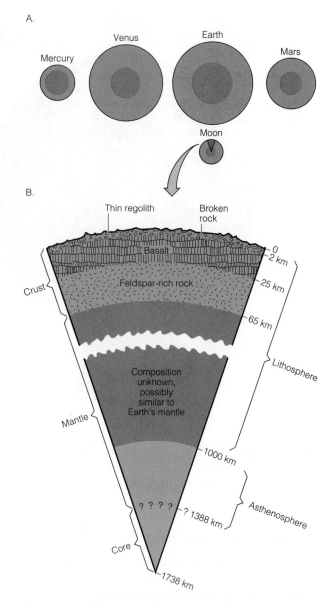

Figure 2.15 The internal structures of the Moon and the terrestrial planets. A. Comparative sizes of the cores. Mercury, nearest the Sun, where only the highest temperature materials could condense, has a huge core. Mars, farthest away from the Sun, has a small core. B. Structure of the Moon. Crust composition is known with certainty only in the vicinity of the astronauts' landing sites.

difficult question to answer. Simple observation reveals that rocks on the surface of each planet fracture and deform as they do on the Earth, and this evidence indicates that a lithosphere is present. Astronauts left instruments on the Moon to measure the properties of moonquakes and from those measurements the presence of an asthenosphere can be inferred (Fig. 2.15B), but the presence of a mesosphere seems unlikely.

Measurements made on Mars have determined that an asthenosphere exists there, too, but for Venus and Mercury it is possible only to infer the existence of an asthenosphere. What little evidence we have suggests that asthenospheres and lithospheres probably are present in each terrestrial planet but that the asthenosphere of the Earth is unusually near the surface and hence the lithosphere is unusually thin. Very likely, the Earth is such a dynamic planet *because* its lithosphere is thin. The other terrestrial planets seem to have much thicker lithospheres and to be much less dynamic than the Earth.

Venus, the Earth, and Mars are large enough that their gravitational fields have been able to retain the atmospheres formed as a result of melting and outgassing. Mercury and the Moon are too small to have held on to the gases given off, and so they lack atmospheres.

Given that the terrestrial planets are similar in composition and overall structure, an intriguing question still to be answered is, "Why is there abundant life on Earth but no life, so far as we know, on Mars, or Venus?" The answer probably does not lie in the chemical differentiation of the planets. Rather, it probably lies in two factors: first, the size of the Earth and the fact that it has a still hot interior, which leads to volcanism and a continual addition of new nutrients in the form of lava to the surface; and second, the distance of the Earth from the Sun is such that H_2O can exist as water, ice, and water vapor.

Figure 2.16 Europa, smallest of the four large moons of Jupiter. Europa has a low density, indicating it contains a substantial amount of ice. The surface is mantled by ice to a depth of 100 km. The fractures indicate that some internal process must be disturbing and renewing the surface of Europa. The dark material (here appearing red) in the fractures apparently rises up from below. The cause of the fracturing is not known. The image was taken by *Voyager 2* in July 1979.

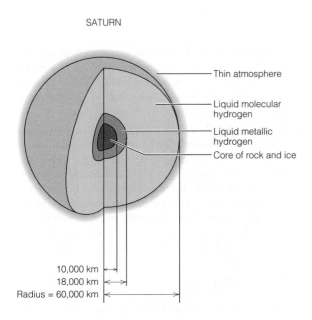

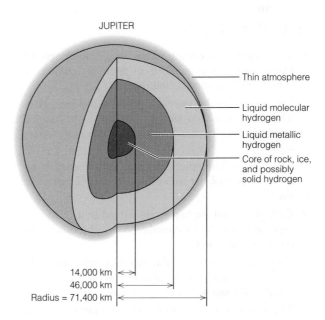

Figure 2.17 Comparison of the probable interior structures of Jupiter and Saturn.

THE JOVIAN PLANETS

We cannot see anything that lies below the thick blankets of atmosphere that cover the jovian planets. Therefore, we can only hypothesize about their internal structure, based on remote-sensing measurements. For example, we can calculate that the masses of Jupiter and Saturn are so great that none of their atmospheric gases has been able to escape their gravitational pulls. This is true even for the two lightest gases, hydrogen and helium, which made up the bulk of the planetary nebula. This means that the two largest jovian planets must have bulk compositions that are about the same as the composition of the solar nebula from which they formed. For example, the composition of Jupiter is estimated to be 74 percent hydrogen, 24 percent helium, and 2 percent heavy elements.

Because the moons of the jovian planets are rocky with thick sheaths of ice (Fig. 2.16), it is presumed that a rocky mass resides at the center of each planet. The rocky cores of Jupiter and Saturn may be as large as 20 or more earth masses. Surrounding the rocky cores is possibly a layer of ice, analogous to the ice sheaths seen on the moons (Fig. 2.17).

Pressures inside the jovian planets must be enormous, and this leads to the hypothesis that deep in the interiors hydrogen may be so tightly squeezed that it is condensed to a liquid. Proceeding inward from the outer atmosphere, which consists mostly of hydrogen gas, we hypothesize that a point is soon reached where a thick layer of liquid hydrogen is present. Still deeper inside Jupiter and Saturn, pressures equivalent to 3 million times the pressure at the surface of the Earth are reached, and under such conditions the electrons and protons of hydrogen become less closely linked and hydrogen becomes metallic; a layer of molten metallic hydrogen is the result. In Jupiter it is possible that pressures may even reach values high enough for solid metallic hydrogen to form a sheath around the ice core.

Neptune and Uranus are thought to be similar to Jupiter and Saturn, although neither is large enough for pressures to be high enough to form metallic hydrogen. Pluto, the planet farthest from the Sun and the smallest of the jovian planets, is little larger than the Moon. Pluto lacks the massive atmosphere of the larger jovian planets and has a density of 2.06 g/cm^3, intermediate between the high densities of the terrestrial planets and the low densities of the other jovian planets. The most probable explanation for the density of Pluto is that the planet has a structure like those of the moons of Jupiter and Saturn — a rocky center but a thick outer layer of ice. Pluto differs sufficiently from the other planets that astronomers hypothesize that it formed in some other solar system and was captured by the Sun's gravitational field sometime after the formation of our solar system.

SUMMARY

1. The solar system consists of the Sun, nine planets, 61 known moons, vast numbers of asteroids, millions of comets, and innumerable meteoroids.

2. The planets revolve around the Sun in elliptical orbits. The moons revolve around the planets, also in elliptical orbits. All revolve counterclockwise when viewed from above.

3. The orbits of the other planets and all moons are approximately coplanar with the orbit of the Earth about the Sun.

4. Each planet and the Sun rotate around an axis, and except for Venus, the rotation direction is the same as the revolution direction.

5. The solar system formed through the condensation of a solar nebula followed by planetary accretion and was completed about 4.56 billion years ago.

6. The evolutionary history of a planet is controlled by its size, distance from the Sun, and the presence or absence of life.

7. The planets can be divided into two groups: the terrestrial planets, the four nearest the Sun, each a small, rocky mass with a high density; and the jovian planets, the five outermost planets, each, with the exception of Pluto, large and gassy.

8. The terrestrial planets are compositionally zoned into a metallic core, a mantle, and a crust. Terrestrial planets also have layers that differ in strength. In the case of the Earth, the outermost layer, about 100 km thick, is hard and rigid and is called the lithosphere. Beneath the lithosphere is a region about 250 km thick called the asthenosphere where rocks are soft, weak, and easily deformed. Beneath the asthenosphere is the mesosphere, a region where high pressures keep rocks strong despite the high temperatures.

9. At the center of each terrestrial planet is a metallic core composed mostly of iron but with admixtures of nickel and other elements. The Earth's core has an inner component that is solid and an outer component that is molten.

10. The jovian planets, with the exception of Pluto, the smallest, are shrouded by thick atmospheres rich in hydrogen and helium. The cores of the jovian planets are inferred to be rocky, like a terrestrial planet, to be surrounded by a thick layer of ice, and above the ice liquid hydrogen, which grades outward to the hydrogen-rich atmosphere. Pluto has only a rocky core and a sheath of ice.

IMPORTANT TERMS TO REMEMBER

asthenosphere *37*	gravitation, law of *30*	mantle *36*	planetary accretion *33*
continental crust *36*	heliocentric *24*	mesosphere *37*	solar nebula *33*
core *36*	inner core *38*	oceanic crust *36*	terrestrial planet *30*
crust *36*	jovian planet *38*	outer core *38*	
geocentric *24*	lithosphere *37*		

QUESTIONS FOR REVIEW

1. What are the differences between a geocentric and a heliocentric solar system?

2. What is retrograde motion of the planets? How did the explanations of Ptolemy and Copernicus for retrograde motion differ?

3. What great conceptual advance did Kepler make concerning the orbits of planets?

4. Why do planets move in orbits around the Sun rather than through space along straight paths?

5. The existence of the two outermost planets, Neptune and Pluto, were predicted prior to their discoveries. What was the basis of the predictions?

6. Describe the two primary motions of the Earth with respect to the Sun.

7. What is a solar nebula, and what role may a nebula have played in the formation of the solar system?

8. The planets can be divided into two distinctly different groups; what is the basis of such a division? What are the names of the planets in each group?

9. Terrestrial planets are compositionally layered. Describe the layering and explain how it is thought to have formed.

10. What is a lithosphere? an asthenosphere? Sketch a section through the Earth showing the positions of the lithosphere, asthenosphere, and mesosphere.

11. Three major factors are thought to control the evolutionary history of a planet; what are the factors and what controls do they exert?

Questions for A Closer Look

1. Why does the Earth have seasons?

2. What astronomical motions are used to measure the passage of time?

3. Why did Julius Caesar and Pope Gregory XIII adjust the calendar?

QUESTIONS FOR DISCUSSION

1. If it were possible to send an unmanned spaceship to a nearby star in order to inspect the planets in orbit around the star, what kind of measurements would you advise the space scientists to try and obtain as the spaceship flies past each planet?

2. Astronomers have observed a number of supernovas in recent years. Research their findings and discuss the influence of the discoveries on hypotheses about the origins of planets.

To learn more about the use of satellite-based remote sensing in environmental monitoring, visit our web site.

CHAPTER 3

The Sun, Giver of Life

● *How Many Suns in the Universe?*

With the exception of the planets and their moons, each point of light in the night sky—each "star"—is a sun or a collection of suns, called a galaxy. On a clear, dark night, it's easy to count as many as 5000 stars. With an ordinary pair of binoculars, almost a million stars are visible, and with the aid of the most powerful telescopes, the number of stars that can be seen rises to the billions. The number of stars turns out to be so large that no one has ever tried to make an exact count. The best we can do is estimate.

Even a cursory look at the night sky reveals that the visible stars are not evenly distributed. When telescopes were invented several hundred years ago, one of the earliest discoveries scientists made was that stars occur in clusters. We now call each large cluster of stars a **galaxy.** A further discovery made with telescopes is that our Sun (and all the planets that circle it) are in a galaxy called the Milky Way.

The number of stars in the Milky Way is mind boggling enough, but modern telescopes reveal an estimated 100 billion other galaxies in the universe! Most of the other galaxies are so far distant from us that they appear as single, tiny, fuzzy-looking stars through all but the largest telescopes. With those largest telescopes, however, astronomers can confirm that each fuzzy "star" is indeed a galaxy. Now, multiply the estimated number of galaxies (10^{11}) by the average number of stars in a galaxy (10^9). The result, a hundred billion billion (10^{20}), is an estimate of the minimum number of stars in the sky and therefore of the minimum number of suns in the universe. The reason no one has managed to make an exact count of all the suns is obvious—trying to do so would be like trying to count all the sand grains on all the beaches, all the river banks, and all the deserts in the world! ●

(opposite) The object in the constellation Andromeda known as M31 is a huge spiral galaxy consisting of millions of stars. The galaxy is 2.2 million light-years from the Earth. Two smaller galaxies, N6C205 and M32, are also visible, one on each side of M31.

THE LIFE-GIVING PROPERTIES OF THE SUN

The Sun is a vast ball of gas, and at its center is a huge nuclear reactor. The light and heat generated by this nuclear reactor control the Earth's climate and make the Earth a habitable planet. It is because three parts of the Earth system—the biosphere, hydrosphere, and atmosphere—derive their energy from the Sun that the Earth system works the way it does. If the amount of light and heat were either more or less, our planet would be a very different place.

Because it is situated 150 million kilometers from the Sun, the Earth receives just the right amount of light and heat to support life. Venus and Mercury, the two planets closest to the Sun, are too hot and too dry for life to exist. Mars is so far from the Sun that it is too cold for life and also too dry because H_2O does not exist there as water, only as ice and water vapor. Mars may once have been a warmer planet and may once have had liquid water on the surface. If the evidence in favor of an earlier, more hospitable Mars turns out to be correct, forms of life may once have existed there, but it is most unlikely that life exists there today.

With only a few exceptions, the energy of sunlight is the force that sustains life on the Earth. The Sun can be viewed as both a giver and supporter of life. Plants, algae, and certain bacteria utilize the Sun's energy to produce organic matter in a process called photosynthesis. Other organisms obtain the Sun's energy indirectly when they feed on plants, algae, or photosynthetic bacteria. The Sun's energy is transferred through a food web as one set of organisms feeds on others.

Life is possible on the Earth because of the Sun's energy; could life therefore exist on suitable planets in other solar systems? We don't know, but a lot of research has been carried out seeking an answer. As a result of space research, it seems unlikely that life now exists anywhere else in our solar system. There is a possibility that life may have existed on Mars at some time in the distant past but space research has not found any evidence of life on Mars today. The situation with respect to planets in other solar systems is much more complicated. Astronomers have recently discovered planets associated with a few of the closer stars, but they are large planets that are very close to those suns and probably much too hot for life to exist. We can barely see the smaller, outer planets in our solar system, so it is not surprising that astronomers are still trying to prove that potentially habitable Earth-like planets actually exist around other stars. Although scientists hypothesize that most suns have planets in orbit around them and speculate that among the billions of stars some must have Earth-like planets, proving the point will be very difficult. Nevertheless, such speculations raise one of the great scientific challenges: Does Earth-like life exist elsewhere in the Universe? Or, to pose the question another way: Is the Earth system really unique?

We do not know much about planets around other Suns, but we do know a great deal about our Sun. We can even estimate when and how the Sun will die—and therefore when the Earth system will die. Our knowledge comes from analyzing the Sun's energy output. We can also analyze the energy output from other stars and then, as discussed later in the chapter, we can use the Principle of Uniformitarianism to estimate how long the nuclear fusion reactor has been operating in the Sun, how steadily it sends out light and heat, and how long it will continue to do so in the future.

THE SUN'S VITAL STATISTICS

Our Sun is an ordinary, medium-sized, middle-aged, run-of-the-mill star with properties and characteristics identical to those of billions of other ordinary, medium-sized, middle-aged stars.

Size

The Sun is vastly larger than any other body in our solar system. At approximately 1.4 million km, the diameter of the Sun is 109 times the diameter of the Earth. We say approximately because it is hard to measure the edge of a ball of gas exactly. In the Sun's case, the edge is generally considered to be the limit of the glowing, visible sphere, even though there is a transparent blanket of gas several thousand kilometers deep outside of the glowing sphere.

The Moon is about 382,000 km from the Earth. If we were to draw a sphere the size of the Sun centered on a point at the center of the Earth, the edge of the sphere would be about 315,000 km beyond the Moon. Because of its vast size, the Sun's mass, 2×10^{30} kg, is 300,000 times greater than the Earth's mass. However, because the Sun is entirely gas, its density is only one-fourth that of the Earth.

Apparent Motion

The Earth revolves around the Sun, but to an observer on the Earth it is the Sun that *appears* to revolve around the Earth. Because of the Earth's daily

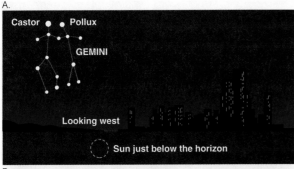

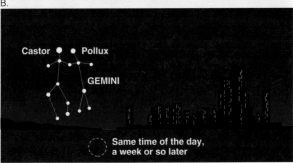

Figure 3.1 An easy way to prove that the Sun moves relative to the more distant stars. A. Just after sunset, noting the exact time and your exact location, pick a bright constellation close to the western horizon near the point where the Sun sets. The constellation shown here is Gemini, close to the horizon in July. In other months choose other constellations. B. Observe your constellation a week later, at the same time of the evening as the initial observation. Your stars will be closer to the horizon and therefore closer to the Sun. After a few weeks, your constellation will set before the Sun, so that it is absent when you look at the sky just after sunset.

rotation about its axis, the Sun appears to arc across the sky from east to west every day. Thus, this first apparent motion of the Sun is a result of the Earth's rotation. If the stars were visible during the day, you would notice another apparent motion. Relative to the background of stars, all of which are so distant they appear fixed in space, the daily arc traveled by the Sun is in a different place in the sky from one day to the next. This second apparent motion of the Sun relative to the fixed stars is due to the Earth's revolution around the Sun. The way to see this relative movement for yourself is to pick out a group of bright stars—perhaps a recognizable constellation—located close to the western horizon soon after sunset. Note the exact time, your exact location, and the distance from the horizon to your star group. Wait a week or two and then look at the same stars, at exactly the same hour of the evening, and with you standing in exactly the same spot. You will notice that the stars

have moved closer to the horizon and therefore closer to the position of the Sun (Fig. 3.1).

Each distinctive star pattern in the sky is called a **constellation**. The constellations are mostly named for animals and mythical characters, and most of the names we use today we inherited from the ancient Greeks and Babylonians. The sky is divided into 88 constellations. Even though not all stars are part of a constellation, the constellations are a convenient way to divide the sky for purposes of location. For example, we can describe Castor and Pollux as the two bright stars in the constellation Gemini. (This is the constellation shown in Fig. 3.1.)

Relative to the stars, the Sun appears to move to the east and, in one year, to move completely across the sky and return to its initial position. This means that the Sun appears to make a complete circuit of 360° in a year, or about 1° a day. Actually, it is the Earth that moves in its orbit around the Sun once every year—we observe the Sun's apparent motion relative to the Earth when in fact it is the other way around. One observation made by the ancient Babylonians is that the Sun's apparent eastward motion through the sky takes it through the same 12 constellations each year. You can make the same observations yourself if, at regular intervals throughout the year, you observe where the Sun sets relative to the stars. The 12 constellations through which the Sun passes are called collectively the **zodiac** (Fig. 3.2), a Greek term meaning circle of animals.

Remember that the Sun's motion against the background of fixed stars is an *apparent* motion. The body that actually moves is the Earth in its orbit around the Sun. The plane of the Earth's orbit around the sun is called the **ecliptic**, and so the path of the Sun through the constellations of the zodiac is the trace of the ecliptic on the background of fixed stars. If the Earth's axis of rotation were exactly perpendicular to the plane of its orbit, the ecliptic would be a straight line. But the axis of rotation is tilted 23.5° away from the perpendicular, and so the trace of the ecliptic is a smoothly curving line (Fig. 3.3).

Energy Output

The total amount of energy radiated outward each second by the Sun or any other star is called its **luminosity**. Because the Sun radiates energy equally in all directions, only a tiny fraction of the total energy emitted reaches the Earth. An Earthbound scientist wanting to determine the Sun's luminosity therefore must do some calculating.

Artificial satellites that orbit the Earth get their energy, via solar panels, from the Sun. From such satel-

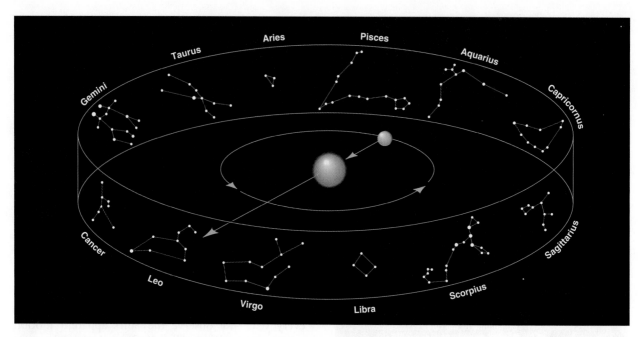

Figure 3.2 The apparent motion of the Sun as seen by a viewer on the Earth. The Sun appears to move when viewed against the background of fixed stars. The Sun's apparent motion takes it through the constellations of the zodiac.

lites we know that solar energy reaches the Earth at a rate of 1370 watts per square meter (W/m²) of surface. When energy continuously passes through, or continuously falls on, a unit area, we say there is an energy **flux** through or on that area. The energy flux reaching the Earth from the Sun is therefore 1370 W/m². The energy flux can be used to calculate the Sun's luminosity in the following manner. Picture an imaginary sphere with a radius equal to the Earth–Sun distance (1.5×10^{11} m) and centered on the Sun (Fig. 3.4). Such a sphere has a surface area of 2.8×10^{23} m². The energy flux through every square meter on the

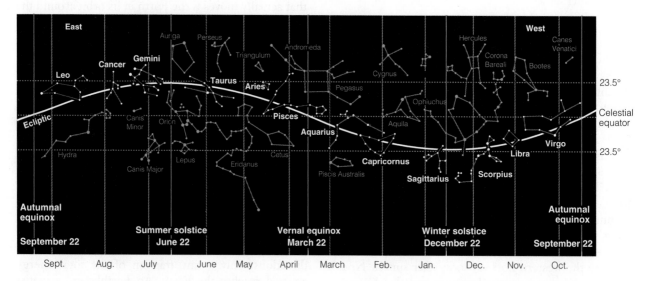

Figure 3.3 The ecliptic and the constellations of the zodiac. The ecliptic is curved because the Earth's axis of rotation is tilted at 23.5° to the plane of the Earth's orbit. A viewer on the Earth is therefore sometimes above the plane of the ecliptic, sometimes below. The maximum point above the ecliptic for a northern hemisphere viewer is the summer solstice on June 21, when the midday Sun is at its most northerly point. The winter solstice is December 21, when the Sun is at its most southerly point. The times when the Sun is directly over the equator at midday are the vernal (spring) and autumnal equinoxes.

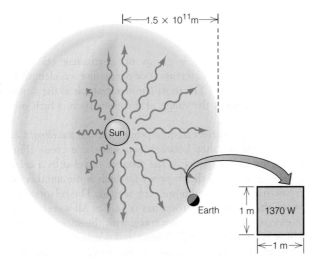

Figure 3.4 To measure the Sun's luminosity, create an imaginary sphere that is centered on the Sun and has a radius equal to the average Earth-Sun distance, 1.5×10^{11} m. The inside surface of this imaginary sphere would capture all of the energy radiated by the Sun. Energy from the Sun reaches the Earth at a rate of 1370 W/m², so every square meter of the inside surface of the imaginary sphere must receive energy at the same rate. Calculate the number of square meters in the sphere's surface from the formula: surface area = $4\pi r^2$, multiply by 1370 W/m², and you have the Sun's luminosity, 3.8×10^{26} W.

surface of the sphere is 1370 watts. The total energy output of the Sun must therefore be the number of square meters multiplied by the flux:

$$2.8 \times 10^{23} m^2 \times 1370 \ W/m^2$$

or

$$3.8 \times 10^{26} \text{ watts}$$

which is the Sun's luminosity.

The Earth receives only a tiny fraction of this luminosity. Viewed from the Sun, the Earth is a disc with a radius of 6.4×10^6 m (Fig. 3.5). The surface area of this disc is 13×10^{14} m², and so the luminosity that hits the whole Earth is $1.3 \times 10^{14} m^2 \times 1370 \ W/m^2 = 1.8 \times 10^{17}$ watts. Thus, the Earth receives only one 2 billionth of the total solar output of energy! Although this fraction is tiny, it is sufficient to supply all the energy needed to drive the Earth's external processes and to keep the biosphere growing healthily.

SOURCE OF THE SUN'S ENERGY

Nuclear reactions inside stars involve the fusion of lightweight chemical elements, particularly hydrogen, to form heavier elements such as helium and carbon. The fusion process, which happens only at exceedingly high temperatures, converts matter to energy. The first person to have an inkling about the energy stored in atoms was Albert Einstein, who, in 1905, showed that matter and energy are connected through the now famous equation $E = mc^2$, where E is energy, m is mass, and c is the speed of light in a vacuum. Nuclear fusion converts some of the mass of an atom into energy. Nuclear fusion has been achieved on the Earth, but only in an uncontrolled manner in hydrogen bombs. If it could be done so that energy is released in a controlled manner, society would have a nearly limitless energy supply because so much hydrogen is available on Earth. Extensive research is therefore being done on nuclear fusion.

There are many possible fusion reactions, but the Sun and most other stars produce their energy by two of them: the proton-proton (PP) chain and the carbon-nitrogen-oxygen (CNO) chain. In both processes

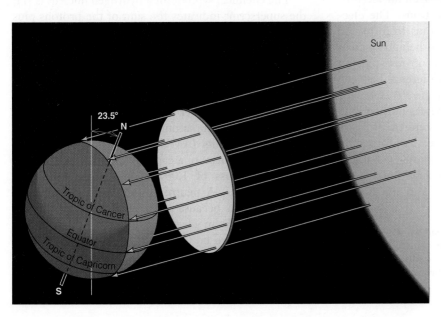

Figure 3.5 Energy from the Sun passes through an imaginary disc that has a diameter equal to the Earth's diameter. The flux of energy through the disc is 1370 watts per square meter. The amount of energy that hits a square meter on the Earth's surface is maximum at the point where the incoming radiation is perpendicular to the Earth's surface (that is, where the Sun is directly overhead at midday). This point changes daily because the Earth's axis is tilted at 23.5° to the ecliptic. The most northerly point is reached on June 21 (the summer solstice); the most southerly point is reached on December 21 (the winter solstice).

A CLOSER LOOK

Electromagnetic Radiation

Whenever an electrically charged particle is accelerated, it radiates energy in the form of **electromagnetic radiation.** Light is the most familiar form of electromagnetic radiation, but X-rays, γ-rays, infrared rays, and radio waves are also electromagnetic radiation—all these many forms differ

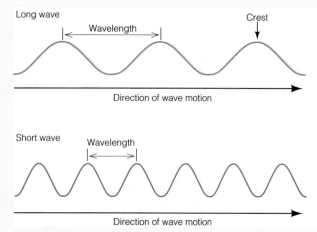

Figure C3.1   The properties of waves. The wavelength is the distance from one crest to the next. The wave frequency is the number of crests that pass a given point each second. If both a short wave and a long wave move at the same speed, the short wave has the higher frequency.

only in wavelength. A group of electromagnetic rays arranged in order of increasing or decreasing wavelength is called a **spectrum**. The most familiar example is the *visible spectrum*, which is the range of wavelengths to which our eyes are sensitive.

Waves have three essential properties: the *wavelength λ*, which is the distance between two successive crests (Fig. C3.1); the *speed v*, which is the distance traveled by a crest in one second; and the *frequency f*, which is the number of crests that pass a given point each second. The relation between the three wave properties is $f\lambda = v$. All wavelengths of electromagnetic radiation travel with the speed of light, which in vacuum is 299,793 km/s and is usually designated *c*. For electromagnetic radiation, therefore, $f\lambda = c$. Because all electromagnetic radiation has exactly the same speed, *c*, it is possible to refer to electromagnetic radiation in terms of either wavelength or frequency.

Wavelengths of electromagnetic waves are usually measured in meters (Fig. C3.2). The unit of frequency is the *hertz* (Hz). A frequency of 1 Hz is one wave crest passing a given point each second. Scientists prefer to use the term *cycle* rather than wave crest when referring to frequency. One cycle per second is just another way of saying one wave crest per second.

Electromagnetic radiation can be described equally well in terms of waves or in terms of packets of radiant energy called *quanta* or *photons*. Sometimes it is more convenient to deal with the wave properties of the radiation; at other times it is better to deal with the packets of energy properties. The relationship between these two "forms" of electromagnetic radiation is $E = hf$, where *E* is the energy of a

the net result is the same—four hydrogen nuclei (protons) fuse while absorbing two electrons. The electrons combine with two of the protons to form two neutrons, and the result is a single helium nucleus containing two protons and two neutrons[1] plus an enormous amount of energy. The difference between the PP and CNO chains is that in the PP chain the protons fuse directly to helium, whereas in the CNO chain the process has intermediate steps that involve carbon, nitrogen, and oxygen, as well as protons. In the Sun the PP chain accounts for about 88 percent of the energy produced and the CNO chain the remaining 12 percent.

---

[1]As you probably recall from high school chemistry, atoms are made up of protons and neutrons, bunched together in the nucleus, and electrons which move in orbits around the nucleus. The structure and properties of atoms are discussed more fully in Chapter 5.

The chemical symbol for a hydrogen nucleus is ^1_1H; the superscript indicates the sum of the protons plus neutrons in the nucleus (in this case a sole proton), whereas the subscript indicates the number of protons, in this case also one. Helium is written ^4_2He, which indicates a total of four particles in the nucleus: two of them are protons, as we can read from the subscript, and so the other two must be neutrons.

The mass of an atom is expressed in terms of atomic mass units (AMU), where 1 AMU is one-twelfth of the mass of a carbon-12 atom ($^{12}_6C$). A proton has a mass of 1.00758 AMU and a neutron 1.00893 AMU. A neutron is slightly more massive than a proton because it is formed when a proton combines with an electron. The mass of one ^4_2He should therefore be $(2 \times 1.00758) + (2 \times 1.00893) = 4.03302$ AMU. When the mass of a helium nucleus is determined, however, it is found to be only 4.00260

photon, f is the frequency of the corresponding electromagnetic wave, and h is a constant known as Planck's constant. When f is measured in hertz, h is equal to 6.63×10^{-34} J.s, and E is in joules.

The energy of a photon corresponding to a 1000-kHz wave, in the middle of the AM radio band, is

$$E = (6.63 \times 10^{-34} \text{ J.s}) \ (1.000 \times 10^3/\text{s})$$
$$= 6.63 \times 10^{-28} \text{ J}$$

By contrast, the energy of a photon that has a frequency of 6×10^{14} Hz, which has a wavelength in the middle of the visible light range, is 3.98×10^{-19} J; a photon in the γ-ray range, which has a frequency of 10^{23} Hz, is 6.63×10^{-11} J. Obviously, the higher the frequency (and therefore the shorter the wavelength of the corresponding electromagnetic wave), the greater the amount of energy carried by a photon.

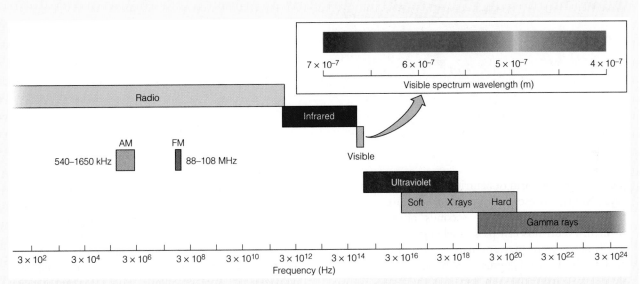

Figure C3.2 The electromagnetic spectrum. Because all electromagnetic waves travel with the same speed (the speed of light, 3.0×10^8 m/s), they can be discussed in terms of either frequency or wavelength.

AMU. Some of the mass has been lost, and it is this lost mass that is converted to energy according to Einstein's $E = mc^2$. The amount of energy released by the fusion of four ${}^1_1$H to produce one ${}^4_2$He is small, about 4.2×10^{-12} J, but 4.5×10^6 metric tons of hydrogen are converted to helium every second in the Sun, which explains why the Sun's luminosity is 3.8×10^{26} W.

Because a huge amount of its hydrogen is continuously being converted to helium, the Sun will eventually run out of hydrogen fuel. Fortunately, the Sun's supplies of hydrogen are enormous, and scientists calculate that there is enough in the Sun's interior to keep the nuclear fusion reactor operating for another 4 to 5 billion years.

Proton-proton fusion requires a temperature of at least 8×10^6 K, and the CNO fusion chain requires a temperature of 15×10^6 K. Although we cannot see into the interior of the Sun where fusion is occurring,

scientists are reasonably certain that, because the CNO chain is operating, the temperature at the center of the Sun is at least 15×10^6 K.

Although most of the Sun's energy comes from the fusion of hydrogen (the lightest element) to helium (the second lightest element), the process of fusing lighter atoms to form heavier ones does not stop with helium. All atoms heavier than helium have been formed by fusion inside stellar nuclear reactors. As a result, all of the heavy atoms in the Earth, including those in our bodies, were made by nuclear fusion in some ancient star.

The energy released by fusion reactions in the Sun is in the form of *gamma rays* (γ-rays), which are extremely short electromagnetic waves (see "A Closer Look: Electromagnetic Radiation"), and *neutrinos*, which are electrically neutral, nearly massless particles that move at the speed of light. Of the total energy, 2

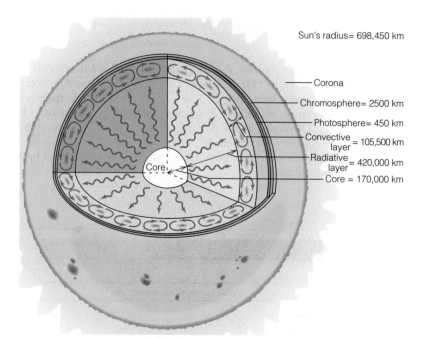

Sun's radius= 698,450 km

Corona

Chromosphere= 2500 km

Photosphere= 450 km

Convective layer = 105,500 km

Radiative layer = 420,000 km

Core = 170,000 km

Core

Figure 3.6 A model of the Sun's interior. Energy is created in the core when hydrogen is fused to helium. This energy flows out from the core by radiation through the radiative layer, by convection through the convective layer, and by radiation from the surface of the photosphere, which is the portion of the Sun we see.

percent is in the form of neutrinos and 98 percent is γ-rays. Neutrinos escape from the Sun's core so easily that they escape about 2 seconds after they are formed. Neutrinos can pass, unchanged, through the Earth and do not play any part in bringing solar energy to the Earth. Gamma rays, however, cannot easily get free from the Sun, but it is the γ-rays, as we shall see later in the chapter, that are responsible for the energy that reaches the Earth.

STRUCTURE OF THE SUN

Figure 3.6 shows that the Sun consists of six concentric layers—four inner regions that make up the sphere we see and two gaseous outer layers that we cannot see.

The Sun's *core*, the site of all the nuclear fusion reactions, is about 170,000 km in radius. The temperature of the core ranges from 8×10^6 K at the margin to 15×10^6 K at the center. The composition of the core is estimated to be about 62 percent helium and 38 percent hydrogen by mass.

Surrounding the core is a region that is very hot but not hot enough for fusion to occur. Stretching from 170,000 to 590,000 km, when measured out from the center, this region is called the *radiative layer*. The energy released in the core moves across the radiative layer by radiation, and it is this layer that makes the escape of energy from the Sun such a slow process. It is electrons in the radiative layer that absorb the γ-radiation and make the layer opaque.

Above the radiative layer is the *convective layer*, from 590,000 to 695,500 km, across which energy moves by convection. Both the radiative and the convective layer have compositions that are little changed from the composition of the original solar nebula and about the same as that of Jupiter, which is 72 percent hydrogen, 26 percent helium, and 2 percent heavier elements, by mass. In a sense the radiative and convective layers are kept gassy and are prevented from collapsing into the core by the intense pressure created by γ-radiation attempting to move outward.

Above the convective layer is the surface layer, the portion of the Sun we see. Called the *photosphere*, it is an intensely turbulent zone that emits the light that reaches the Earth (Fig. 3.7). The photosphere is about 450 km thick and has an average temperature of about 5800 K, ranging from 8000 K at its boundary with the convective layer to 4000 K at its outer edge.

The photosphere passes into the *chromosphere*, a low-density layer of very hot gas about 2500 km thick. The chromosphere has such a low density that it is transparent to light passing through and therefore is very difficult to see.

The chromosphere merges into the outermost layer of the Sun, the *corona*, a zone of even lower density gas than the chromosphere. The corona, which grades off into space, is quite hot but, like the chromosphere, is not visible because it has such a low density. Because both the chromosphere and the corona are transparent, we are able to observe and measure them only during a solar eclipse, when light from the body of the Sun is obscured by the Moon (Fig. 3.8).

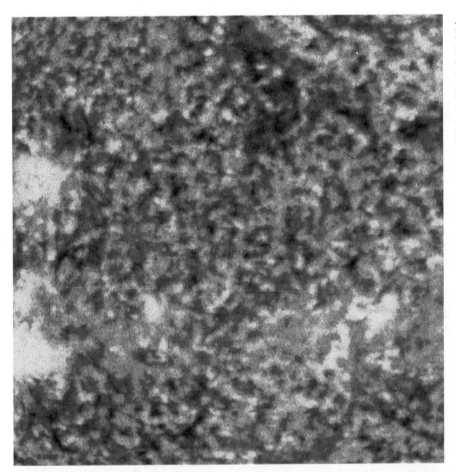

Figure 3.7 Granules up to 1500 km across on the surface of the Sun's photosphere. Granules are the tops of huge, upward-rising bubbles and columns of intensely hot gas. The darker regions between the bright granules are places where cooler gas flows back into the photosphere.

Figure 3.8 The Sun's chromosphere and corona can be clearly seen only during a solar eclipse, when the Moon blocks the light coming from the photosphere. This photo, taken with a special camera during an eclipse in 1988, shows faintly glowing gas streaming hundreds of thousands of kilometers out from the corona.

THE SOLAR SPECTRUM

The radiation energy released in the Sun's core by the PP and CNO fusion chains has a frequency of about 10^{23} Hz, in the γ-ray range. Such radiation has a very short wavelength and is extremely energetic. As the γ-rays move out through the radiative layer, they are repeatedly absorbed and re-emitted by electrons and in the process converted to longer-wavelength, lower-energy radiation. No energy is lost in the process; it is just parceled out into a greater number of less energetic rays. By the time the radiation reaches the photosphere, it has a frequency in the range from 10^{14} to 1.5×10^{15} Hz or, as more commonly designated, wavelengths in the range 3×10^{-6} m to 2×10^{-7} m.

Note in Figure 3.9 that the energy flux from the Sun varies with the wavelength, the peak being close to the wavelength of yellow light. The shape of the Sun's spectral curve (which is a graph of wavelength versus flux) is interesting because it matches almost exactly the spectrum of radiation given off by a perfectly black body heated up to 5800 K, the average temperature of the photosphere.

When a piece of metal is heated in an intensely hot flame, the metal first starts to glow a dull red. Then, as it gets hotter and hotter, it becomes more brightly red, then orange, yellow, white, and finally bluish white. The color of the metal at any given time while it is being heated is a measure of the temperature, and this relationship has many practical uses. Blacksmiths, for example, use color to estimate when the temperature of a piece of iron is high enough for the task in hand.

If we measure the electromagnetic radiation emitted by the hot metal, we will find that the spectral curve has the same shape as the curve in Figure 3.9.

Any body, no matter what its composition, that has a spectral curve similar to Figure 3.9 is called a **blackbody radiator.** The term *blackbody* seems confusing when applied to something that is glowing brightly and emitting electromagnetic radiation. In fact, the term refers to the *radiation-absorbing* properties of a body, and a perfect blackbody is one that absorbs all light that strikes it and reflects none. If you directed a very powerful beam of light at the Sun, almost none would be reflected back, making the Sun a nearly perfect black body. (Note that the two terms *black body* and *blackbody radiator* mean the same thing and are interchangeable.)

There are several important points to remember about blackbody radiators:

1. The hotter the radiating body, the shorter the wavelength of the radiation peak (Fig. 3.10). The peak of radiation for the Sun is at a wavelength of about 5×10^{-7} m, corresponding to a temperature of 5800 K.

2. All objects, no matter what their temperature, emit electromagnetic radiation and have a spectral curve approximating, at least to some degree, the curve for a blackbody radiator. The Earth, for instance, has an average surface temperature of about 290 K, and the radiation it emits has a peak at a wavelength of 1×10^{-5} m, much longer than the Sun's peak wavelength of 5×10^{-7} m. The Earth's peak wavelength is in the infrared region of the electromagnetic spectrum, and so the Earth's radiation cannot be seen by the human eye.

3. The hotter an object is, the more energy it radiates. We know this is true from such simple observations as the amount of energy given off by a hot

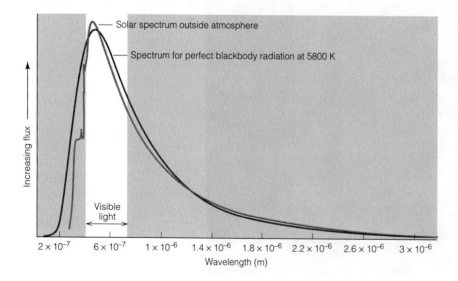

Visible light

2 × 10⁻⁷   6 × 10⁻⁷   1 × 10⁻⁶   1.4 × 10⁻⁶   1.8 × 10⁻⁶   2.2 × 10⁻⁶   2.6 × 10⁻⁶   3 × 10⁻⁶

Wavelength (m)

Figure 3.9 The Sun's spectrum is nearly identical to that of a perfect blackbody radiator. The minor differences occur because gases in the chromosphere and corona selectively absorb some wavelengths of the electromagnetic radiation emitted by the Sun.

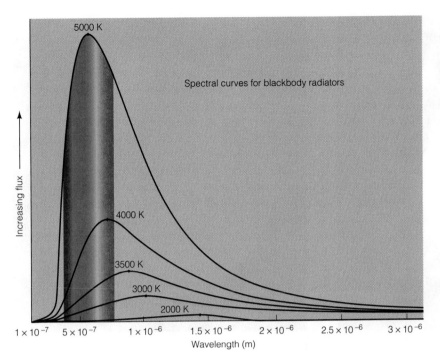

Figure 3.10 The energy flux from blackbody radiators at different temperatures. Note how the radiation peak moves to shorter wavelengths as the temperature increases. The area under any one curve is the total flux of energy emitted by a radiator at a given temperature. Note that the higher the temperature, the greater the flux.

stove versus a cold stove. The same relationship is apparent in Figure 3.10, where we see that the higher the temperature of the black body, the greater the energy flux at any given wavelength.

Solar Radiation at the Earth's Surface

The solar spectrum in Figure 3.9 is the spectrum measured in space, far above the Earth's atmosphere. The spectrum measured at sea level is a lot different because the electromagnetic radiation has had to pass through the atmosphere, and gases in the atmos-

phere—principally oxygen, water vapor, and carbon dioxide—selectively absorb certain wavelengths. The curve in Figure 3.11 shows how absorption by the atmosphere modifies the radiation from the Sun that reaches the Earth's surface.

THE ACTIVE SUN

So far, we have been discussing the Sun as if it were a reliable, smoothly operating body. Most of the time the Sun does work smoothly, and astronomers tend to

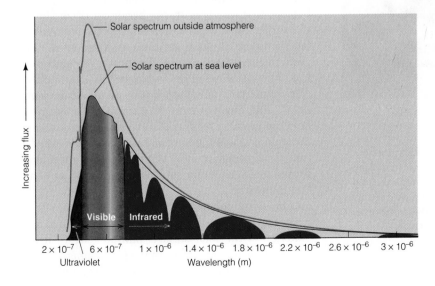

Figure 3.11 The outer-space and sea-level spectra of solar radiation. The two curves are different because gases in the atmosphere selectively absorb some of the wavelengths of emitted radiation.

A.

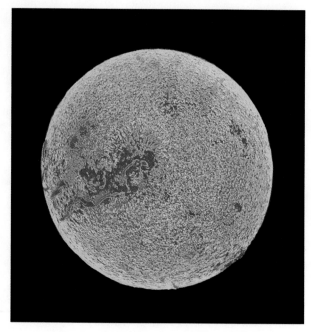

B.

C.

Figure 3.12 A period of the active Sun. A. A vast fiery prominence of hot gas bursting out from the photosphere through the chromosphere and corona. B. A huge sunspot with an unusual spiral structure breaking through the photosphere in 1982. C. Both prominences and sunspots cause streams of protons to flow out into space. When the protons hit the Earth's outer atmosphere, they create the beautiful electrical effects called auroras. This aurora was seen in a far northern latitude, and at the moment it was photographed a meteor flashed across the sky.

refer to it under such conditions as a quiet Sun. At other times, however, and generally for short periods, the Sun becomes intensely turbulent and erupts with huge fiery prominences and vast sunspots (Fig. 3.12). Such times are referred to as periods of the active Sun, and the places where the events occur are called *active regions*.

The development of active regions has two principal causes. Differential rotation, the first cause, arises because the Sun is a rotating ball of gas and rotates faster at the equator (once every 25 days) than at its poles (once every 31 days). Differential rotation is ever-present in the Sun, but much of the time the turbulence that results is unseen because it occurs below the surface of the photosphere. The second, and by far the more important, cause is magnetism. The gases in the Sun consist of electrically charged particles. The vast flows of gas inside the Sun are, in effect, electric currents, and electric currents create magnetic fields. The Sun is therefore a magnet. Its magnetic field can be distorted by the differential rotation or by turbulence inside the Sun, and this distortion can disrupt the flowing gases. When the internal turbulence breaks through to the surface, a period of an active Sun follows.

The most important active Sun phenomenon, as far as the Earth is concerned, is sunspots—huge dark blotches on the solar surface. Sunspots are relatively cool regions on the surface of the photosphere, and so they appear dark by comparison with the rest of the

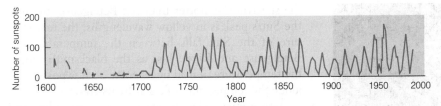

Figure 3.13 The sunspot cycle over the past 400 years. Note the period before 1700, when, for reasons that are not understood, very few sunspots were observed. Sunspots have reached a maximum about every 11 years since 1700, and there is also a suggestion of some sort of cycle on a 55- to 57-year time scale. Because the pre-1700 period of low sunspot activity coincides with a prolonged cool period that is sometimes called the Little Ice Age, some scientists have speculated that sunspot activity and climate are connected somehow.

photosphere. (Even so, they are intensely hot.) When a sunspot starts to form, the granular surface of the photosphere separates, and a tiny dark spot that is intensely magnetic appears and starts to grow. Exactly how and why sunspots form is not clearly understood, but the fact that they seem to occur in cycles of about 11 years (Fig. 3.13), which is the period calculated by astronomers for turbulent interactions between the solar magnetic field and differential rotation, suggests they may be a normal part of the Sun's activities. Even more important, because the Sun's magnetic field influences the Earth's outer atmosphere, many experts believe that sunspots influence the climate on the Earth.

Changes in Luminosity

Astronomers monitor the Sun's luminosity very carefully. A 1 percent decrease in the flux of electromagnetic radiation, from 1370 W/m² to 1356 W/m², is estimated to reduce the Earth's average temperature by 1 K. Similarly, a 1 percent increase in luminosity to 1384 W/m² would probably increase the average temperature by 1 K. A change of 1 K may seem small, but even a mighty volcanic eruption such as the eruption of Tamboro, a volcano in Indonesia, which put so much debris into the atmosphere that scientists call 1816 the "year without a summer," probably did not cause the Earth's average temperature to drop by any more than 1 K. Clearly, changes in the Sun's luminosity have the potential to cause significant changes in the Earth system.

Exact measurements of luminosity are difficult, but indications are that since about 1980 the Sun's luminosity has decreased by about 0.3 percent, or 4 W/m². Where climate changes are concerned, therefore changes in luminosity, as well as natural and human-engendered changes to the atmosphere, must be considered (see Chapter 20).

Changes in the Sun's luminosity have been monitored only during the twentieth century, and thus questions of long-term changes remain unanswered. As we shall see later in this chapter, a star the size of the Sun commences life with a luminosity about 10 percent lower than the present solar luminosity. This means one of two things. Perhaps the early Earth was much colder than today's Earth. Geological evidence does not support a cold Earth hypothesis so it seems likely that the Earth system of the time adjusted in some way to keep the surface warm. If the carbon dioxide level of the atmosphere was higher, for example, the atmosphere would have been a better thermal blanket and the surface would have been kept warm. In the carbon dioxide hypothesis scientists suggest that, as the sun's luminosity slowly increased, the biosphere removed carbon dioxide from the atmosphere, thus keeping a balance between incoming radiation and a comfortable climate.

OTHER SUNS

Because the stars are so far away from the Earth,[2] it is not possible to measure how big they are or to see all the detail we can see on the Sun. Almost everything we know about stars comes by way of the electromagnetic radiation they emit. The way we decode and interpret the messages carried by starlight depends to a large degree on our understanding of how the Sun works. Fortunately, a lot of information can be gathered from some straightforward measurements.

[2] Astronomical distances are measured by the time it takes for light to travel between the two points being measured; 1 light-minute is 18×10^9m, and the Sun–Earth distance is only 8.3 light-minutes. The star nearest to the Earth, Alpha Centauri, is 4 light-years away (1 ly is 9.5×10^{15}m). When we look through a telescope at a star a billion light-years away, we are seeing light that left the star a billion years ago; we are looking back in time.

Figure 3.14 The constellation Orion (the hunter). The reddish star at the upper left is Betelgeuse; the bright, bluish-white star at the lower right is Rigel. Because this photo is a time exposure, many faint stars not visible to the eye are shown.

Star Color and Luminosity

When you look at the stars on a clear night, two things are quickly apparent. The first observation is that the colors are not all the same; they range from red through yellow to bluish-white. For example, in Orion, one of the most familiar and easily recognized constellations, there is a bright but distinctly red star called Betelgeuse and an equally bright star called Rigel that is a striking bluish-white (Fig. 3.14).

The second observation is that stars vary greatly in their brightness. Some, like Sirius, in the contellation Canis Major, and Rigel, blaze out quickly and draw attention because they are so bright. If you look closely, however, you can find stars so faint you can hardly be sure whether or not they are there.

Like the Sun, all other stars are blackbody radiators. As shown in Figure 3.10, the peak wavelength of a blackbody radiator is determined by temperature. The higher the temperature, the shorter the wavelength and the bluer the star. Conversely, the lower the temperature, the longer the wavelength and the redder the star. Rigel, with its peak at the blue end of

the spectrum, is hotter than red Betelgeuse. Because the Sun's peak is in yellow wavelengths, the temperature of the Sun falls between the temperatures of Betelgeuse and Rigel. Just as the blacksmith judges the temperature of iron by its color, so it is possible to judge the temperature of a star by its color.

Astronomers classify stars based on color and hence temperature (Table 3.1 and Fig. 3.15). Each color—or as it is more commonly called, spectral class—is further split into ten subdivisions ranging from 9 (hottest) to 0 (coolest). The Sun is a G5 star.

In order to measure a star's luminosity, we need to know the Earth-star distance. This distance is difficult to measure, but it can be determined for stars out to a distance of 300 light-years from the Earth using a system called *parallax*. You can demonstrate parallax very easily by holding a pencil perpendicular to the floor and at arm's length, and alternately opening and closing each eye. The pencil, viewed against a fixed background, appears to move from side to side (Fig. 3.16A). A parallax measurement of the distance to a nearby star uses the Earth at two opposite points on its orbit for the two "eyes," and very distant stars as the fixed background. The method is illustrated in Figure 3.16B. The diameter of the Earth's orbit is known; the angle subtended by the star to the two points of observation is measured, and from the data it

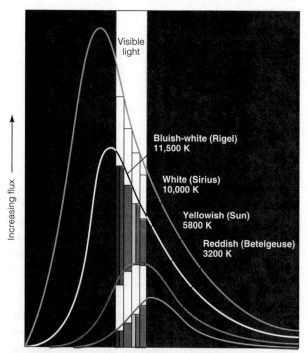

Figure 3.15 Stars as blackbody radiators. Note that only for yellowish stars is the radiation peak in the visible range. For reddish, white, and bluish-white stars, the radiation peaks lie outside the visible range. We can see such stars because they do emit some radiation in the visible range.

Table 3.1 The Spectral Classes of Stars

Spectral class	Color	Surface Temperature	Example
O	Bluish-white	Greater than 30,000 K	Naos
B	Bluish-white	11,000–30,000 K	Rigel
A	Bluish-white	7500–11,000 K	Sirius
F	White to bluish-white	6000–7500 K	Canopus
G	White to yellowish-white	5000–6000 K	Sun
K	Yellowish-orange	3500–5000 K	Aldebaran
M	Reddish	Less than 3500 K	Betelgeuse

is possible to calculate the distance to the star. For very distant stars, where the angles are tiny, measurements are imprecise. That is why, at present, only stars out to 300 light-years can be measured with any degree of accuracy.

Once the distance to a star is known, its luminosity can be calculated. Remember that luminosity is the total amount of energy emitted by a star each second, and to get an accurate measurement we must know how far away a star is. The process is the same as mea-suring the luminosity of the Sun as discussed earlier. First, the energy flux of the star is measured through a telescope (see "A Closer Look: Telescopes"). Then the surface area of a sphere with a radius equal to the Earth–star distance is calculated just as we did for the Sun in Figure 3.4. Since flux is energy/second/unit area, the total energy emitted by a star each second is easily calculated.

Once the temperature and luminosity of a star are known, they can be compared with the values for

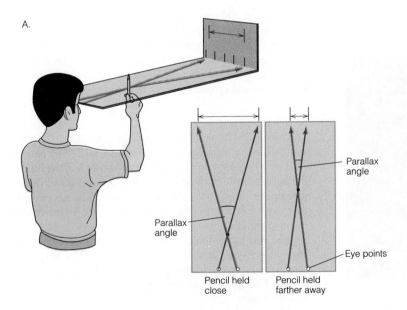

A.

Parallax angle

Parallax angle

Pencil held close

Pencil held farther away

Parallax angle

Eye points

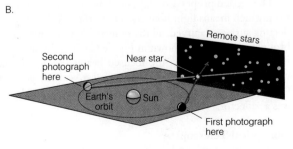

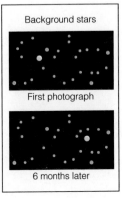

B.

Second photograph here

Near star

Remote stars

Earth's orbit

Sun

First photograph here

Background stars

First photograph

6 months later

Figure 3.16 Parallax used to measure star distances. A. An example of parallax. Alternately shut your left and right eyes, and the pencil will appear to move relative to a fixed background. B. As the Earth goes around the Sun, a near star appears to move against the background of more distant stars. By observing the near star at six-month intervals, knowing that the base of the green triangle is the diameter of the Earth's orbit, and measuring the angle of the shift, the distance between the near star and the Earth can be calculated.

A CLOSER LOOK

Telescopes

Astronomy is an observational science. The only way hypotheses concerning stars can be tested is through observation, and because stars are so faint and so distant, the observations have to be made with telescopes.

All telescopes have one function: they gather and concentrate electromagnetic radiation. Those that gather visible light are called *optical telescopes*, those that gather radio waves are *radio telescopes*, and *infrared* and *ultraviolet telescopes* gather infrared and ultraviolet waves, respectively.

Optical Telescopes
Optical telescopes use either of two properties to gather and concentrate light. The first property is **refraction**, which means the path of a beam of light is bent when the beam crosses from one transparent material to another (Fig. C.3.3A). Refraction occurs because the speed of light is different in different media. Remember that *all* electromagnetic radiation travels with the same speed in a vacuum. In water, glass, or any other transparent medium, the speed of light is less than in a vacuum. Consider what happens when

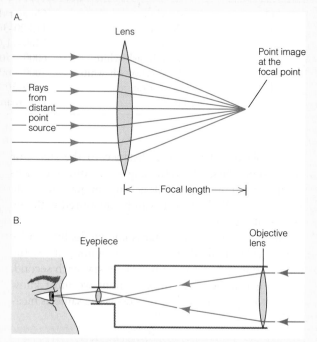

Figure C3.4   The principle of a refracting telescope. A. A simple lens refracts rays to a sharp focal point. B. A telescope gathers light through the *objective* lens and focuses the image at the focal point. An additional lens, called an *eyepiece*, is used to view the image. The combination of objective lens and eyepiece is the telescope.

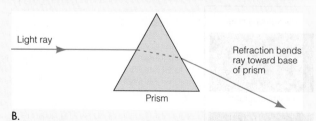

Figure C3.3   Examples of refraction. A. Refraction of light causes a drinking straw to appear to be bent at the air–water boundary. B. Refraction of light through a prism.

a beam of light is traveling through air at some initial speed and then hits a glass prism, as in Figure C3.3B. The part of the beam that hits the glass first slows down, but the rest of the beam is still traveling at the initial fast speed and therefore catches up with the slow-moving part. The net effect is to rotate the direction in which the entire beam moves through the glass.

The way refraction is used to construct a *refracting telescope*, or *refractor*, is shown in Figure C3.4. The first telescopes made were refractors, and it was with a simple refractor that Galileo made his epochal observations of the moons around Jupiter.

The second property of light used in telescopes is **reflection**, which means light bounces off a surface. When the reflecting surface is curved, all the reflected light beams can be brought to a focus, as shown in Figure C3.5(A). By using either of two geometries, the focused beams can be viewed with an eyepiece, thus creating a *reflecting telescope* (Fig. C3.5B). The telescope Newton made to observe the planets was a reflector.

Nonoptical Telescopes
Galileo, Newton, and many generations of later astronomers had only their eyes to look through telescopes

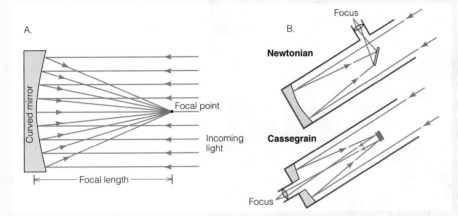

A.

Curved mirror

Focal point

Incoming
light

Focal length

B.

Focus

Newtonian

Cassegrain

Focus

Figure C3.5 The principle of a reflecting telescope.
A. A curved mirror reflects incoming light rays to a focal point. B. A reflecting telescope requires an eyepiece to examine the image at the focal point. Two geometries, Newtonian and Cassegrain, are used to prevent the head of the observer from blocking incoming light rays.

and see the stars and planets. Today's astronomers rarely look through telescopes. Instead they substitute other light-sensitive devices for their eyes. Such devices, which can be photographic films, solid-state photoelectric chips like those in solar-powered calculators, or photomultiplier tubes, are more sensitive than eyes and have an additional advantage: they can measure the *amount* of incoming electromagnetic radiation. Such measurement is essential in determining luminosity.

Unique information about stars can be obtained at almost every wavelength in the electromagnetic spectrum. Astronomers therefore seek to "look at" the sky at many different wavelengths, not only those in the visible region. Electromagnetic radiation that penetrates the atmosphere can be studied with *ground-based*, nonoptical telescopes. In the case of *radiotelescopes*, which operate in a wavelength range of about 10^{-6} to 10^2 m, the "mirror" is a large metal dish, and the detector is a radio receiver placed at the focal point (Fig. C3.6). Near-infrared wavelengths, from about 10^{-6} to 10^{-3} m, are gathered and focused in an *infrared* telescope using a mirror, just as in an optical telescope, but the detector is a chip of metallic germanium, which is sensitive to infrared rays.

As is clear from Figure 3.11, some wavelengths are absorbed by the atmosphere. To "see" the sky at those wavelengths, it is necessary to put *space* telescopes outside the atmosphere. Wavelengths in the ultraviolet, from 10^{-10} to 10^{-7} m, and in the far infrared, beyond about 4×10^{-5} m, are partly absorbed by the atmosphere. Telescopes operating in these wavelength ranges are either lifted aloft by huge balloons or placed in orbit by rockets.

The best possible place to view the sky is completely outside the atmosphere. Recently, a large optical telescope, called the Hubble Telescope, with a mirror 2.4 m in diameter, was placed in orbit 400 km above the Earth. Despite some initial mechanical problems that astronauts were able to correct, Hubble has been an extraordinary success. Indeed, it has been such a success that plans are now being drawn up to place a telescope on the far side of the Moon, where it will be free not only from the atmosphere but also from all the electromagnetic radiation from the Earth. Space astronomy will probably become a standard way of operating in the twenty-first century.

Figure C3.6 A radiotelescope in Owens Valley, California, operated by astronomers of the California Institute of Technology. Radio waves from space hit the curved dish and are focused on the receiver mounted on the top of the four legs. The receiver, which works like a radio, can be tuned to receive different wavelengths.

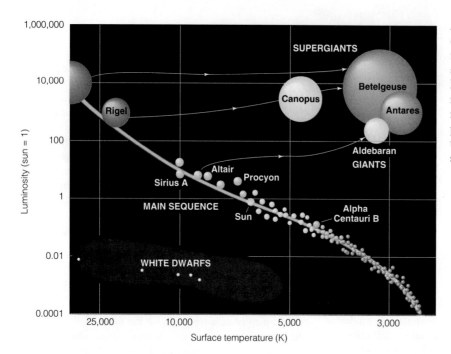

Figure 3.17 A Hertzsprung-Russell diagram of star luminosity versus surface temperatures. The vertical axis is a comparative one based on the Sun having a luminosity of 1. The horizontal axis is reversed from the normal order, with values of surface temperature increasing to the left. Note that the Sun is a middle-range, main-sequence star.

other stars. One convenient way to make a comparison is through the **Hertzsprung-Russell diagram (H-R)**, a plot of luminosity versus temperature.

Hertzsprung-Russell Diagrams

The H-R diagram was devised early in the twentieth century by two astronomers, Ejnar Hertzsprung (1873–1967), a Dane, and Henry N. Russell (1877–1957), an American, with the two men working independently of each other. Although the spectral class, and hence the temperature, of a star can be measured no matter how far away the star is, it is possible to measure luminosity only at known distances. The H-R plot was therefore developed from stars in our galaxy that are near enough for the distance to be measured by parallax.

About 85 percent of the stars plotted on an H-R diagram fall on or close to a smooth curve (Fig. 3.17) that astronomers refer to as the **main sequence.** The rest of the stars fall into two groups—one above, the other below the main sequence.

Luminosity is a function of both star temperature and star size. At a given temperature, the larger the star, the greater the luminosity. All the stars that plot above the main sequence are very luminous. However, based on color, astronomers know that these stars are cooler than main-sequence stars of equal luminosity. For example, the red star Antares in Figure 3.17 has the same luminosity as the blue star Rigel but is cooler. There is only one conclusion to draw: in order to have such high luminosities at lower temperatures, the stars above the main sequence must be very large;

they are giants. About 2 to 3 percent of all stars are giants. Depending on their size, they may be referred to as **red giants** or, in the case of the very largest stars, such as Betelgeuse, super giants. Betelgeuse is so large that if it were the Sun, its edge would be beyond the edge of Mars and the Earth would be inside a huge star!

Below the main sequence is a small group of stars that are much less luminous and therefore much smaller than the main sequence stars. Although the stars in this group do not all fall into the white color class (class A), they have come to be called **white dwarfs.**

Astronomers discovered the importance of the H-R diagram a long time ago: it can be used to explain the history of a star. All stars have lives; they are born, they age, and they die. Their life cycle has much to do with their size at any given time.

Stellar Evolution

Stars have long lives, and the smaller a star, the longer it can live. Very massive stars live for hundreds of millions of years, intermediate-mass stars like the Sun live up to 10 billion years, and the smallest stars can live for 20 billion years or longer. The reasons for the differences lie in the balance of forces inside a star.

As we have previously pointed out, all stars are huge, hot balls of gas. The hot gas does not drift off into space because the force of gravity continuously pulls it *inward*. The greater the mass of gas in a star, the stronger the inward pull. In the absence of a counterbalancing *outward* force, gravity would cause a star to collapse into a small, dense mass. The principal force that counterbalances gravity is electromagnetic radiation. The outward flux of electromagnetic radiation generated by nuclear fusion creates an outward push that prevents gravitational collapse; the balance between the inward gravitational force and the outward radiation force determines the size of the star. A fusion reactor requires fuel, and just as an automobile no longer runs when its gasoline supply is used up, so a fusion reactor can no longer operate when its nuclear fuel is depleted. When its fuel supply runs low, the balance of forces in a star changes, and as a result the size, temperature, and luminosity all change too. When and how the changes occur are a function of how much fuel was present to start with and how fast it was used up.

Because star lifetimes are vastly longer than human lifetimes, it is reasonable to ask how we humans can ever decipher star histories. The answer is provided by the Principle of Uniformitarianism. Among the billions of stars in the sky, astronomers can find examples of every mass, every luminosity, and every stage of star life. Understanding how a star ages is like deciphering the stages of a human life by studying the population of a large town for a few weeks. The result is a composite, an *average* life, rather than the life of a single individual. In the case of stars, the problem is made a little more complex because star masses differ by a factor of a thousand and the life of a small star differs from the life of a large star.

The mass of the Sun (one solar mass, 1 S) is used as a measure of star masses. Masses smaller than about 0.1 S are too small for the temperature in the core to get hot enough for nuclear fusion to start, and so below 0.1 S there are what we call "almost stars". The planet Jupiter is an example of an almost star. At the other end of the size range, stars more massive than 100 S generate such intense radiation that gravity is overcome by the outward radiation forces and the star simply blows itself apart.

The Life Cycle of a 1-S Star

The life of a star the size of the Sun begins with the gravitational compression of a mass of gas. Just as the air in a bicycle pump becomes heated as a result of compression, so does star gas heat up as a result of gravitational compression. When the temperature at the center of a compressed young star reaches 8×10^6 K, PP fusion commences.

Initially, gravitational compression exceeds the radiation counterforce, and a newly burning star continues to shrink. Within about 50 million years, a balance is reached between the gravitational and radiation forces, and the star attains a stable luminosity and temperature that place it on the main sequence of the

H-R plot. The evidence that stars spend most of their lives on the main sequence is straightforward—most of the stars in the sky plot on the main sequence. Stars appear to be born and to die at about the same rate, yet 85 percent of all stars plot on the main sequence. That can happen only if stars have, through most of their lives, a temperature and luminosity that plot on the main sequence.

The nuclear fusion reactor of a 1-S star converts hydrogen to helium. When the hydrogen in the core is used up, nuclear fusion in the core ceases, gravity asserts control, and the now helium-rich core contracts. However, there is still abundant hydrogen surrounding the core in the radiative layer. As the core collapses and becomes even hotter, therefore, a shell of hydrogen in the inner part of the radiative layer starts the nuclear fusion process in what is called **shell fusion**. Core collapse heats the star interior by compression; increasing temperature speeds up the rate of nuclear fusion in the radiative-layer shell. As a consequence, such an immense amount of heat is generated that the star expands. Such a shell-fusion star moves off the main sequence and becomes a red giant.

Despite the fact that a red giant is very large, its helium-rich core continues to contract even after shell fusion commences. Eventually, the core temperature is hot enough for helium fusion to start by a process called the triple-alpha reaction, in which three helium nuclei fuse to form carbon. When all the helium fuel in the core is used up, shell fusion in the radiative layer starts again, but this time it is the fusion of helium. Inside the shell, the now carbon-rich core continues to contract gravitationally. Shell fusion of helium is a violent process accompanied by great explosions, and each explosion blasts star gas out into space. Finally, all that remains of a 1-S star is the carbon-rich core surrounded by a slowly diminishing shell of helium; when the shell fusion of helium ceases, the star starts to die.

The mass of the carbon-rich core of a 1-S star is too small for contraction to raise the temperature to the point at which carbon can fuse to heavier elements. The carbon-rich core becomes a white dwarf, a small, very dense star that is slowly cooling. On the H-R plot a white dwarf plots below the main sequence. As cooling proceeds, a white dwarf loses its luminosity, moves toward the lower right-hand corner of the H-R plot, and eventually becomes a dead star called a *black dwarf*. The star's life cycle is now complete.

The scenario just described is approximately that which our Sun will follow. There is nothing for humans to fear, however, because the hydrogen fuel in the Sun's core is sufficient to keep the Sun on the main sequence for at least 4 billion more years.

The Life Cycle of a 0.25-S Star

It is not possible to know the full life cycle of a small-mass star. Such stars are so long-lived, 20 billion years or more, that there hasn't been enough time for them to evolve off the main sequence. The universe itself is only 15 billion years old!

A small-mass star is born in the same way as a 1-S star—by gravitational contraction and the onset of the PP cycle. However, gravity in a small-mass star is so weak that the core where nuclear fusion commences is small; as a consequence, the small amount of electromagnetic radiation released just counteracts the gravitational contraction force. Such stars have low luminosities and low temperatures and plot on the lower-right-hand end of the main sequence. The hydrogen fuel in a small-mass star is used up so slowly that it lasts an incredibly long time. The fuel will eventually be depleted, however, and shell fusion, core contraction, and helium fusion will follow. Beyond that stage, however, observation tells us nothing about what will happen in the future.

The Life Cycle of a 5-S Star

Details in the life of a very massive star differ from those of a 1-S star. The first difference is that the initial gravitational contraction of a massive star is so intense that the temperature in the star's core is soon high enough for the main hydrogen fusion reaction to be the CNO cycle rather than the PP cycle. Both cycles work, of course, but with two cycles active, massive stars burn fuel very rapidly, have very high luminosities, plot on the upper-left-hand end of the main sequence, and quickly deplete their fuel. As a result, massive stars have short lives on the main sequence.

Once off the main sequence, massive stars go though the same steps of burnout, core contraction, shell fusion of hydrogen, core fusion of helium, burnout, and contraction of a carbon-rich core. The next event, however, is very different. As the carbon core contracts and shell fusion of helium proceeds, a temperature is reached where carbon starts to fuse to heavier elements. This happens because the core is much larger than the core in a 1-S star, and this is the environment in which all the heavy elements now in the Earth are believed to have formed. Such heavy-element formation is thought to happen in a flash and to release so much energy that the star blows up in a *supernova*. After a supernova, what remains of the core is crushed into an immensely dense mass, a *black hole*. Scientists speculate that the matter now in the Sun and the planets was blasted into space in one or more supernovas about 10 billion years ago. In a very real sense, all the atoms in our bodies and in everything around us are stardust from an ancient supernova.

SUMMARY

1. The Sun, which is so hot it is gaseous throughout, has a diameter that is 109 times the diameter of the Earth but a density that is only a quarter that of the Earth.

2. Viewed from the Earth, the Sun *appears* to make a complete circuit of 360 against the background of fixed stars. The apparent motion of the Sun is due to the Earth's orbit around the Sun. The constellations through which the Sun's apparent motion carries it are the constellations of the zodiac.

3. The rate at which energy leaves the Sun in the form of electromagnetic radiation (that is, the luminosity) is 3.8×10^{26} watts. The rate at which energy reaches the Earth is only 1.8×10^{17} watts, or $1370 \ W/m^2$.

4. The source of the Sun's energy is nuclear fusion, which occurs in the core and involves the fusion of four hydrogen nuclei to produce one helium nucleus plus energy. Eighty-eight percent of the Sun's energy arises from the PP chain and 12 percent from the CNO chain.

5. The Sun has an internal structure. At the center is the core, the site of nuclear fusion. Surrounding the core is a radiative layer, which is a gas containing atomic particles through which electromagnetic radiation moves very slowly. Beyond the radiative layer is the convective layer, then successively outward, the photosphere, chromosphere, and the corona.

6. The Sun is a blackbody radiator with a temperature of 5800 K.

7. During periods of an active Sun, sunspots—areas of great turbulence— appear on the surface of the photosphere, and gigantic prominences burst out into the chromosphere. The causes of the sunspots and prominences are the differential speed of rotation of the Sun's gas—faster at the equator, slow at the poles—and the Sun's magnetic field.

8. A plot of luminosity versus blackbody temperature of stars is called a Hertzsprung-Russell (H-R) plot. An H-R plot is used to follow the life cycle of a star.

9. Stars spend most of their lifetimes on the main sequence of an H-R plot. As fuel burns out, they move off the main sequence and become red giants. In Sun-sized stars, the red giant phase is followed by a white dwarf phase. In a star much larger than our Sun, the red giant phase is followed by a supernova, which is a tremendous explosion that destroys the star.

IMPORTANT TERMS TO REMEMBER

blackbody radiator *52*
constellation *45*
ecliptic *45*
electromagnetic radiation
   48

flux *46*
galaxy *43*
Hertzsprung-Russell
   diagram (H-R) *60*
luminosity *45*

main sequence *60*
red giants *60*
reflection *58*
refraction *58*
shell fusion *61*

spectrum *48*
white dwarfs *60*
zodiac *45*

QUESTIONS FOR REVIEW

1. Describe how you can demonstrate that the Sun follows a regular path against the background of fixed stars. Why do we refer to the Sun's motion as an apparent motion?

2. What are constellations, and what is special about the constellations of the zodiac?

3. Explain the source of the Sun's energy. What is luminosity and how do astronomers measure it?

4. The temperature in the core of the Sun reaches 15×10^6 K, yet the temperature of the surface of the Sun is only 5800 K. Explain.

5. Sketch a cross section through the Sun and label the various layers.

6. What is meant by a blackbody radiator? What is the difference in the radiation spectrum of a blackbody radiator at 6000 K and one at 2000 K?

7. Why and how does the Sun's spectrum of electromagnetic radiation in space differ from that measured at the surface of the Earth?

8. What is a Hertzsprung-Russell diagram? Where does the Sun plot on such a diagram?

9. Briefly describe how a Hertzsprung-Russell diagram is used to follow the life cycle of a star.

Questions for A Closer Look

1. What is the spectrum of electromagnetic radiation?

2. What are the three essential properties of waves?

3. What is the relationship between the frequency and the wavelength of electromagnetic radiation?

4. Describe how the two kinds of optical telescopes work.

5. Name some types of nonoptical telescopes. Do they work on the principle of refraction or reflection?

6. Explain why astronomers believe it is important to have telescopes in orbit around the Earth or even mounted on the Moon.

QUESTIONS FOR DISCUSSION

1. Do you consider the launching of more sophisticated space telescopes worthwhile? Why or why not? If yes, what kind of information would you hope to gain from such exploration?

2. When was the most recent burst of sunspot activity? (You will need to do some research to find out.) Was any change in climate associated with the activity?

3. Do you think it may someday be possible to generate energy on the Earth in the same way the Sun generates energy? What advantages or drawbacks do you foresee if fusion energy is attained?

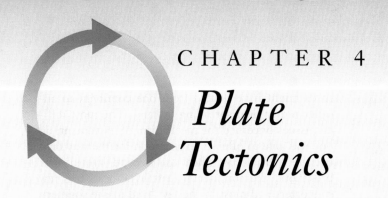

CHAPTER 4
Plate Tectonics

● *Darwin and the Rock Cycle*

When Leonardo da Vinci, in 1508, discovered fossil seashells high in the mountains of Italy, he realized that he must be looking at an ancient seafloor. But how and why were the shells so high up? Had the ocean once covered the mountains, or had the seafloor been locally uplifted? Because seashells are not found everywhere across the land surface, he reasoned that local uplift must be the answer. As he demonstrated many times through his common sense and clear reasoning, Leonardo was an exceptional person.

When rocks are uplifted, they are exposed to weathering and erosion. Three centuries after Leonardo, James Hutton incorporated the idea of uplift and erosion into the concept of the rock cycle, and then Charles Darwin, whose name is forever entwined with evolution, made an observation that was a key piece of the rock cycle concept.

In the 1830s Darwin made a globe-encircling voyage on the HMS *Beagle*. In southern Argentina Darwin saw vast sediment-covered plains that stretch eastward from the Andes to the Atlantic Ocean. The sediments had apparently been formed by erosion of the Andes, but how, he wondered, could "any mountain chain have supplied such masses, and not have been utterly obliterated"?

On February 20, 1835, on the Chilean coast of South America, Darwin experienced a great earthquake. When he investigated the effects of the quake, he discovered "putrid mussel shells still adhering to rocks ten feet above high water mark." The ground had been elevated by the earthquake, and this provided an explanation for an earlier observation Darwin had made near Valparaiso, Chile. There he had found loose mussel shells 1300 feet above sea level. The Andes, he realized, were being pushed up bit by bit, earthquake by earthquake. Although Darwin could not explain the nature of the "force which has upheaved the mountains," he immediately understood that continual slow elevation answered the question he had posed in Patagonia— tectonic forces continue to raise mountains as the erosive forces of the rock cycle wear them away.

One final piece of the rock cycle puzzle remained to be answered. It was the question Darwin had posed, What is the "force which has upheaved the mountains"? A century would pass before the answer was supplied. The answer is plate tectonics. ●

(opposite) Steep sided spires of the Towers of Paine, in the southern Andes Mountains of Chile. It was in the Andes that Charles Darwin realized that a balance existed between tectonic forces that elevate the Andes, and weathering forces which slowly break them down.

MOVING CONTINENTS: A BRIEF HISTORY

By the middle of the nineteenth century, the idea that the Earth's crust is subject to large vertical movements (uplift) was widely accepted. A little bit of evidence–mainly the parallelism of the coastlines on either side of the Atlantic Ocean–suggested the possibility of lateral (sideways) movements too, but a hypothesis of moving continents (continental drift, as it came to be called) was so poorly supported by evidence available at the time that most scientists found the idea unacceptable.

Frank B. Taylor, an American geologist, offered a hypothesis in 1910 concerning lateral movements; they do occur, he suggested, and furthermore, two ancient continental masses, one over the South Pole and the other over the North Pole, had broken up and the pieces slid slowly to their present sites. This hypothesis, he argued, provided a neat explanation for the formation of mountain ranges and midocean ridges. Taylor's hypothesis did not gain many followers.

A more persuasive, and in the long run more convincing, proponent of continental drift was Alfred Wegener, a German scientist. Wegener, who was unaware of Taylor's work, published a book in 1914 in which he tried to explain such phenomena as the parallelism of the Atlantic coastlines and the fact that similar plant and animal fossils can be found on different continents. At some time in the distant past, Wegener suggested, all of the world's landmasses were formed together in a single huge continent, so that plants and animals could spread freely. To this ancient, huge continent Wegener gave the name Pangaea (pronounced Pan-jeé-ah, meaning "all lands"). According to Wegener's hypothesis, Pangaea was somehow disrupted and its fragments (the continents of today) slowly drifted to their present positions. Proponents of the theory likened the process to the breaking up of a sheet of ice that floats in a pond. The broken pieces, they argued, should all fit back together again, like pieces of a jigsaw puzzle.

Although Wegener presented impressive evidence that continental drift may have happened, the hypothesis was not widely accepted in his lifetime because no one could explain how a solid, rocky continent could possibly overcome friction and slide across the oceanic crust. The process, said his critics, is like trying to slide two sheets of coarse sandpaper past each other.

Wegener died in 1930. Debate about continental drift slowed down because some of the supporting evidence gathered by Wegener was found, on close examination, to be open to doubt—for example, plant seeds might have floated from one continent to another rather than spreading by land. The most difficult issue concerned the mechanism of movement; no one was able to offer an explanation for the way drift occurred. By 1939, when the Second World War broke out, the continental drift hypothesis had few supporters and most of the few lived in the southern hemisphere where some of the best supporting evidence was to be found. What revived the debate on a worldwide basis in the 1950s and what led, eventually, to the hypothesis of plate tectonics and an explanation of how continents can move, were discoveries about the Earth's magnetism made possible by technologies developed as a result of the war. Wegener was right, the continents do move, but not for the reasons he presented—they move as a consequence of a process called *plate tectonics*.

THE GEOSPHERE RESERVOIR

The solid Earth, the geosphere, largest of the four major reservoirs of the Earth system, may seem to be constant and unchanging, but nothing could be further from the truth. Constancy is an illusion. Measurements show that the solid Earth is ever changing, that mountains are slowly rising, continents are drifting, and ocean basins are continually changing their shapes and sizes. Plate tectonics and the rock cycle, it is now recognized, are the principal mechanisms by which such changes happen to the face of the Earth. Through plate tectonics the Earth's surface is slowly but continually being renewed. Through weathering and erosion the Earth's surface is continually being worn down. Understanding plate tectonics and the rock cycle is the key to understanding the geosphere's role in the Earth system. To begin our study of both plate tectonics and the rock cycle, we must first examine the way heat energy moves inside the Earth.

Transfer of Heat

Everything that happens in and on the Earth requires energy. The flows of material between the reservoirs of the Earth system all involve energy, particularly heat energy. Heat can be transferred in three ways:

1. **Conduction** is the process by which heat can move through solid rock, or any other solid body, without changing the shape of the solid. Conduction is the way heat moves along the metal handle of a hot saucepan, but it does not cause the movement of hot material from one place to another.

2. **Convection** is the process by which hot, less dense materials rise upward and are replaced by

Figure 4.1 Convection in a saucepan full of water. Heated water expands and rises. As the hot water rises, it starts to cool, flow sideways and sinks, eventually to be reheated and pass again through the convection cell.

cold, downward-flowing and sideways-flowing materials to create a **convection current** (Fig. 4.1).

3. **Radiation** is the process in which heat passes through a gas, a liquid, or even a vacuum. Radiation is the way heat reaches the Earth from the Sun.

Heat energy must come from somewhere. As we discussed in Chapter 1, two main heat sources power the systems of the Earth: the Sun and the Earth's internal heat, known as **geothermal energy**. It is geothermal energy that drives plate tectonics.

Energy from the Earth's Interior

Volcanic eruptions, unlike winds, are unrelated to the Sun's energy output. No matter how hot it gets on a summer's day, the Sun's heat is insufficient to melt rocks, and even frigid Antarctica has active volcanoes. We therefore reason that the heat energy needed to form the molten lava that spews from a volcano must come from somewhere inside the Earth.

This reasoning is not difficult to test. If you went down into a mine and measured rock temperatures, you would find that the deeper you went, the higher the temperature would become. The increase in temperature as you go deeper is called the **geothermal gradient.** We use this gradient to make a deduction based on the second law of thermodynamics, that heat always flows from a warmer place to a cooler one. We deduce that heat energy must be flowing outward from the hot interior of the Earth toward the cool surface. Careful measurements made in mines and drill holes around the world show that the geothermal gradient varies from place to place, ranging from 15° to 75°C/km, (95°F/mi to 269°F/mi) but becomes less pronounced with depth. Far inside the Earth we calculate that the gradient is only about 0.5°C/km (1.4°F/mi). By extrapolation, we calculate further that the Earth's core must be as hot as the surface of the Sun—5000°C (9032°F). Measurements also establish that the heat flow is greatest in those places where there is volcanic activity. We can conclude, therefore, that volcanism is indeed caused by the Earth's internal heat energy.

The heat energy that flows out through solid rocks does so by conduction. But because volcanoes obviously involve the movement of hot material, we have to conclude that at least some heat energy moves inside the Earth by convection.

The hypothesis that convection occurs in a seemingly rigid solid body like the Earth may seem odd, but flow in a solid can be observed in any glacier. Glaciers flow slowly downvalley partly because solid ice at the bottom is deformed by the weight of ice above. Tests show that rocks, like glaciers, don't have to melt before they can flow. Rocks, if sufficiently hot, can flow like sticky liquids, although the rates of flow are exceedingly slow. The higher the temperature, the weaker a rock is and the more readily it will flow. Slow convection currents of rock are possible deep inside the Earth because the interior is very hot. Convection currents bring masses of hot rock upward from the Earth's interior. The hot rock flows slowly up, spreads sideways, and eventually sinks downward as the moving rock cools and becomes more dense.

If you watch a pot of boiling pudding or sauce, you will see that a thin, hard film or skin forms on top of the fluid, where it is coolest. This film tends to ride around on the convecting fluid underneath. The same is true of the Earth: the lithosphere, or outer 100 km (approximately) of the Earth, is a cold outer layer lying on top of hot, convecting material (Fig. 4.2). The thickness of the lithosphere relative to that of the Earth as a whole is about the same as that of the skin of an apple relative to the whole apple, or the glass sphere of a lightbulb relative to the whole bulb. Heat reaches the bottom of the lithosphere by convection; it passes through the lithosphere by conduction. Figure 4.3 shows the geothermal gradient through the continental and oceanic lithosphere; they are different because oceanic crust is thinner than continental crust. At a depth of about 100 km, where the temperature is about 1300°C, rock strength has declined so greatly that convection is possible. The change in the geothermal gradient at about 100 km depth marks the place where heat moves by convection (below 100 km) and starts moving by conduction (above 100 km).

Think about an exceedingly thin layer of cold, brit-

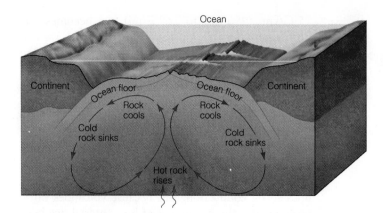

Figure 4.2 Convection as it is hypothesized to occur in the Earth. Though much slower than convection in a saucepan full of water (see Fig. 4.1), the principle is the same. Hot rock rises slowly from deep inside the earth, cools, flows sideways and sinks. The rising hot rocks and sideways flow are thought to be the factors that control the positions of ocean basins and continents, which means that convection determines the shape of the Earth's surface.

tle material riding and jostling around on top of a hot, mobile, convecting fluid. What can happen to the cold boundary layer? It can break, of course, and that is exactly what has happened to the rocky outer layer of the Earth. The lithosphere has broken into a number of jagged, rocky pieces called *plates*, which range from several hundred to several thousand kilometers in width (Fig. 4.4).

The lithospheric plates are riding around on an underlying layer of hot, ductile, easily deformed material called the *asthenosphere*, or "weak layer." Some of the

lithospheric plates are composed primarily of oceanic crustal material, whereas others are composed primarily of continental material. If it were possible to remove all the water from the ocean and view the dry Earth from a spaceship, we would see that the continents stand, on average, about 4.5 km above the floor of the ocean basins (Fig. 4.5). Continental crust is relatively light (density 2.7 g/cm^3), whereas oceanic crust is relatively heavy (density close to 3.2 g/cm^3). Because the lithosphere is floating on the weak asthenosphere, the plates capped by light continental crust

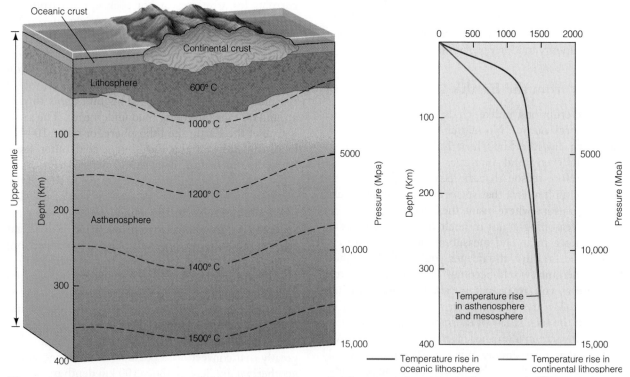

Figure 4.3 Temperature increases with depth in the Earth. A. Dashed lines are isotherms, lines of equal temperatures. Note that temperature increases more slowly with depth under continental crust then under oceanic crust.

B. The same information as shown in A but in graph form. The earth's surface is at the top, so depth (and corresponding pressure) increases downward. Temperature increases from left to right.

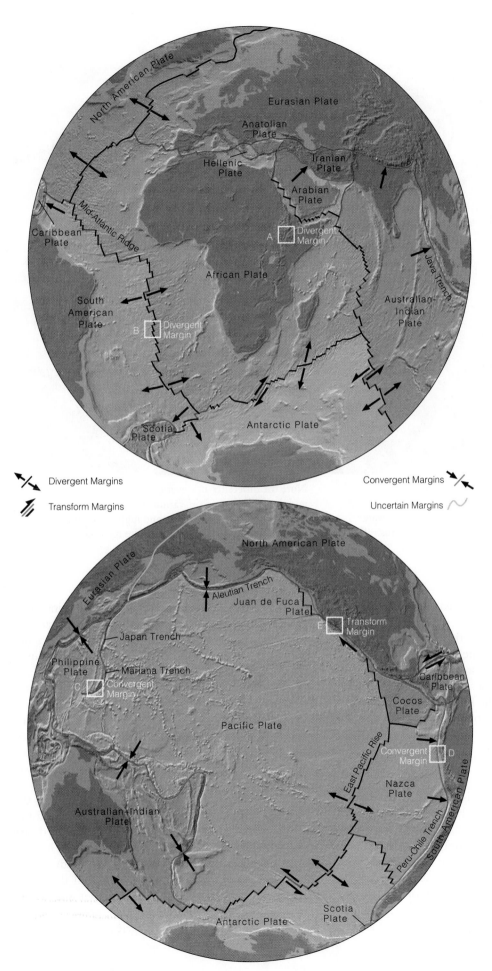

Figure 4.4 Six large plates and a number of smaller ones comprise the Earth's surface. They are moving very slowly in the directions shown by the arrows. The labels A, B, C, D and E correspond to the different types of plate margins discussed in the text and illustrated in Figure 4.17.

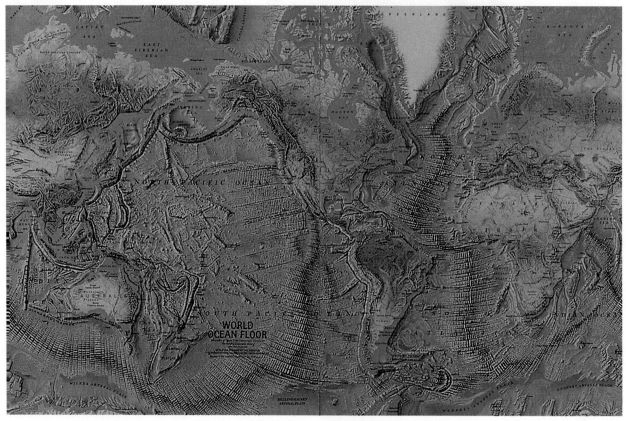

Figure 4.5 Topography of continents and the seafloor. If we could drain all the water from the ocean, we would see the seafloor as vast, flat areas, long chains of underwater mountains, and deep trenches like those east of Australia. Note the difference in elevation between the continents and the seafloor, and the various features such as midocean ridges, deep oceanic trenches and high mountain ranges, all of which can be explained by plate tectonics.

stand high, whereas those capped by heavy oceanic crust sit lower. The Earth's bimodal topography, by which we mean the high continental topography as opposed to the lower oceanic topography, is apparently unique in the Solar System. Because the Earth's bimodal topography is a consequence of plate tectonics, we hypothesize that the Earth is the only body in our Solar System on which plate tectonics is active.

Plate Tectonics and the Face of the Earth

Convection within the Earth is constantly moving the plates of lithosphere and slowly changing the Earth's surface. Mountains such as the Alps or Appalachians that seem changeless to us are only transient wrinkles when viewed from the perspective of geologic time. Mountain ranges grow when fragments of moving lithosphere collide and heave masses of twisted and deformed rock upward. Then the ranges are slowly worn away, leaving only the eroded roots of an old

mountain range to record the ancient collision (Fig. 4.6). The continents are still slowly moving at rates up to 10 cm a year, sometimes bumping into each other and creating a new mountain range and sometimes splitting apart so that a new ocean basin forms. The Himalaya is a range of geologically young mountains that began to form when the Indian subcontinent collided with Asia about 45 million years ago. The Red Sea is a young ocean that started forming about 30 million years ago when a split developed between the Arabian Peninsula and Africa as the two landmasses began to move apart.

But it is not just the continents that move, it is the entire lithosphere. The continents, the ocean basins, and everything else on the surface of the Earth are moving along like passengers on large rafts. The rafts are huge plates of lithosphere that float on the underlying convecting material. As a result, all the major features on the Earth's surface, whether submerged beneath the sea or exposed on land, arise as a direct result of the motion of lithospheric plates.

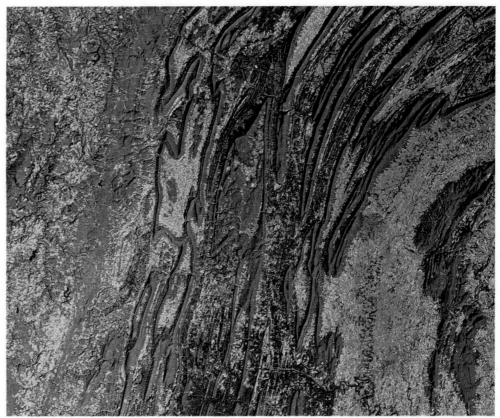

Figure 4.6   Space image of the deeply eroded Appalachian Mountains in central Pennsylvania. The mountains were formed as a result of a collision between two masses of continental crust several hundred million years ago. The collision folded, twisted and fractured the originally flat-lying layers of sedimenta-ry rocks. Red color indicates dense forest vegetation covering ridges of rocks that are relatively resistant to erosion. Blue areas are agricultural lands in valleys formed by erosion of rocks that have relatively little resistance to erosion.

Such motions involve complicated events, both seen and unseen, all of which are embraced by the term *tectonics*, derived from the Greek work, *tekton*, which means carpenter or builder. Tectonics is the study of the movement and deformation of the lithosphere. The branch of tectonics that deals with the processes by which the lithospheric plates move and interact with one another is called **plate tectonics**.

Plate tectonics provides a unifying theory that can be used to explain hundreds of years of independent observations of the processes of rock formation, mountain building, and terrain modification. It also provides an effective framework for our discussion of geologic processes that affect people and form the environment in which we live. After this brief introduction to plate tectonics, let's return to the story of how Wegener's hypothesis of continental drift was revived and how it led, finally, to the theory of plate tectonics. The story is a nice example of the scientific method.

MAGNETISM AND THE REVIVAL OF THE CONTINENTAL DRIFT HYPOTHESIS

Wegener's hypothesis of continental drift was largely abandoned by the 1940s because the movement of an entire continent seemed to be physically impossible. However, as happens so often in science, the hypothesis was revived because of accidental discoveries in another field—studies of the Earth's magnetism.

The source of the Earth's magnetism lies in the molten outer core. As a result of the Earth's rotation, the molten iron of the outer core flows continually around the solid inner core. The flowing stream of molten iron causes an electrical current to flow in the outer core, and the electrical current in turn creates the magnetic field. Because the magnetism is a result of the Earth's rotation, the north magnetic pole is

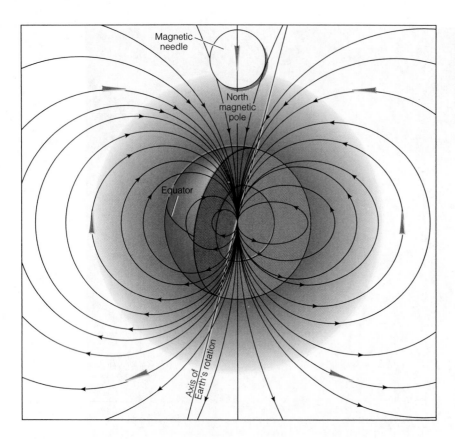

Figure 4.7 The Earth is surrounded by a magnetic field generated by the magnetism of the core. A magnetic needle, if allowed to swing freely, will always line up parallel to the magnetic field and point to the magnetic north pole.

close to the North Pole and the south magnetic pole is close to the South Pole.

The Earth, being a gigantic magnet, is surrounded by an invisible magnetic field that permeates everything placed in the field. If a small magnet is allowed to swing freely in the Earth's magnetic field, the magnet will become oriented so that its axis points to the Earth's magnetic north pole (Fig. 4.7). This is true for all places on the Earth. All free-swinging magnets will point to the north magnetic pole.

Certain rocks became permanent magnets as a result of the way they formed, and like free-swinging magnets they point to the Earth's north magnetic pole. Investigation of the properties of natural magnetism in rocks led to the revival of the continental drift hypothesis in the following manner.

Magnetism in Rocks

Magnetite and certain other iron-bearing minerals can become permanently magnetized. This property arises because the electrons spinning around an atomic nucleus create a tiny atomic magnet. In minerals that can become permanent magnets, the atomic magnets line up in parallel arrays and reinforce each

other. In nonmagnetic minerals, the atomic magnets are oriented in random directions.

Above a temperature called the **Curie point**, the thermal agitation of atoms is such that permanent magnetism of the kind found in magnetite is impossible. The Curie point for magnetite is 580°C. Below 580°C adjacent atomic magnets reinforce each other (Fig. 4.8). When the Earth's magnetic field permeates the magnetite, all magnetic domains (regions in which the atomic magnets point in the same direction) parallel to the Earth's magnetic field become larger and expand at the expense of adjacent, nonparallel domains. Quickly, the parallel domains become predominant, and a permanent magnet is the result.

Consider what happens when lava cools. All the minerals crystallize at temperatures above 700°C—well above the Curie point of magnetite. As the crystallized lava continues to cool and the temperature drops below 580°C, all the magnetite grains in the rock become tiny permanent magnets having the same polarity as the Earth's field. Grains of magnetite locked in a lava cannot move and reorient themselves the way a freely swinging magnet can. As long as that lava lasts (until it is destroyed by weathering or metamorphism), it will carry a record of the Earth's magnetic field at the moment it cooled below 580°C.

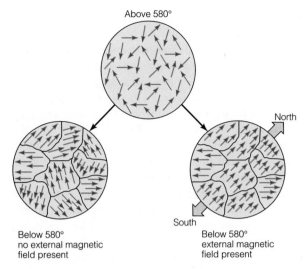

Figure 4.8 Magnetization of magnetite. Above 580°C (the Curie point), the vibration of atoms is so great that the magnetic poles of individual atoms, shown as arrows, point in random directions. Below 580°C, atoms in small domains reinforce one another and form tiny magnets. In the absence of an external field, the domains are randomly oriented. In the presence of a magnetic field, most domains tend to be parallel to the external field and the material becomes permanently magnetized.

Sedimentary rocks can also acquire weak but permanent magnetism through the orientation of magnetic grains during sedimentation. As sedimentary grains settle through ocean or lake water, or even as dust particles settle through the air, any magnetite particles act as freely swinging magnets and orient themselves parallel to the Earth's magnetic field. Once locked into a sediment, the grains make the rock a weak permanent magnet.

Apparent Polar Wandering

During the 1950s, scientists started using ancient lavas to measure past directions of the Earth's magnetism. **Paleomagnetism** is the magnetism in rocks that records the directions of ancient magnetic fields at the time of rock formation, just as a free-swinging magnet indicates the direction of today's magnetic field. Two pieces of information can be obtained from paleomagnetism. The first is the direction of the magnetic field at the time the rock became magnetized. The second provides the data needed to determine how far the magnetic poles lay from the point of rock formation. This is the *magnetic inclination*, which is the angle from the horizontal assumed by a freely swinging bar magnet. Note in Figure 4.9 that inclination varies regularly

with latitude, from zero at the magnetic equator to 90° at the magnetic pole. The paleomagnetic inclination is therefore a record of the place between the pole and the equator (that is, the *magnetic latitude*) where the rock was formed. Once we know the magnetic latitude of a rock and the direction of the Earth's magnetic field at the time the rock was formed, we can determine the position of the magnetic poles at that time.

Geophysicists studying paleomagnetic pole positions during the 1950s found evidence suggesting that the poles wandered all over the globe. They referred to the strange plots of paleopole positions as *apparent polar wandering*. The geophysicists were puzzled by this evidence because the Earth's magnetic poles and the poles of the Earth's rotation axis should always be close together. Determination of the magnetic latitude of any rock should therefore be a good indication of the geographic latitude at which the rock was formed. When it was discovered that the path of apparent polar wandering measured in North America differed from that in Europe (Fig. 4.10), geophysicists were even more puzzled. Somewhat reluctantly, they concluded that, because it is unlikely that the magnetic poles moved, the continents—and with them the magnetized rocks—must have moved. In this way, the hypothesis of continental drift was revived, but a mechanism to explain how the movement occurred was still lacking.

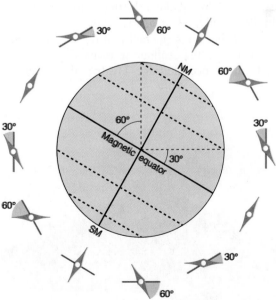

Figure 4.9 Change of magnetic inclination with latitude. The solid red diamonds show the magnetic inclinations of a free-swinging magnet. The solid blue line indicates a horizontal surface at each point.

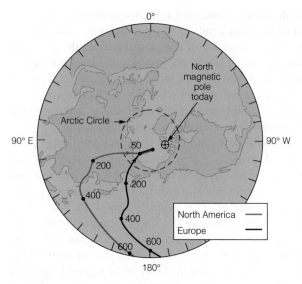

Figure 4.10 Apparent path of the north magnetic pole through the past 600 million years. Numbers are millions of years before the present. The curve determined from paleomagnetic measurements in North America (red) differs from that determined from measurements in Europe (black). Wide-ranging movement of the pole is unlikely; therefore, scientists conclude that it was the continents, not the pole, that moved.

Seafloor Spreading

Help came from an unexpected quarter. All the early debate about continental drift, and even the data on apparent polar wandering, had centered on evidence drawn from the continental crust. But if continental crust moves, why shouldn't oceanic crust move too?

In 1962 Harry Hess of Princeton University hypothesized that the topography of the seafloor could be explained if the seafloor were moving sideways, away from the oceanic ridge. His hypothesis came to

be called **seafloor spreading**, and strong evidence in favor of it was soon found.

From studies of paleomagnetism, geophysicists had discovered an extraordinary and still poorly understood phenomenon—some rocks contain a record of reversed magnetic polarity. That is, some lavas indicate a south magnetic pole where the north magnetic pole is today, and vice versa (Fig. 4.11). Just why the Earth's poles reverse polarity is not yet understood, but the fact that they do provided some very interesting information. The ages of lavas can be accurately determined using radioactive dating. (See Chapter 8). By a combination of radioactive dating and magnetic polarity measurements in thick piles of lava extruded over several million years, it has been possible to determine when magnetic polarity reversals have occurred. Figure 4.12 shows the detailed record of magnetic reversals for the past 20 million years, and ongoing work is detailing evidence of changes for nearly 200 million years.

The Hess hypothesis of seafloor spreading postulated that oceanic crust moved sideways, away from the oceanic ridge, and that basaltic magma rose from the mantle and formed new oceanic crust along the ridge. Hess could not explain what made the oceanic crust move, but he nevertheless proposed that it did and that as a consequence the oceanic crust far from any ridge must be older than crust nearer the ridge. A powerful test of the Hess hypothesis was proposed in the 1960s by three geophysicists: Frederick Vine (who was a student at the time), Drummond Matthews (Vine's mentor), and Lawrence Morley (a Canadian scientist). The Vine-Matthews-Morley suggestion

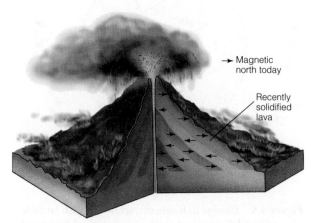

Figure 4.11 Lavas retain a record of the polarity of the Earth's magnetic field at the instant they cool through the Curie point. A pile of lava flows, like those in the volcanoes of the Hawaiian islands, may record several field reversals.

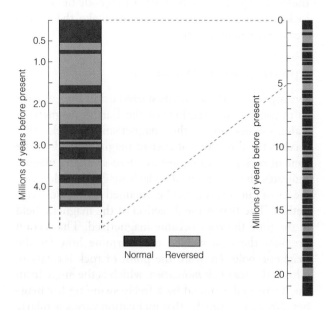

Figure 4.12 Polarity reversals during the past 20 million years.

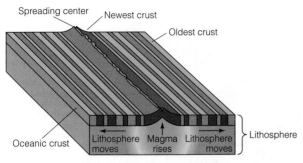

Figure 4.13 Schematic diagram of oceanic crust. Lava extruded along a spreading center at the midocean ridge forms new oceanic crust. As the lava cools, it becomes magnetized with the polarity of the Earth's magnetic field. Successive strips of oceanic crust have opposite polarities.

concerned paleomagnetism, magnetic polarity reversals, and the oceanic crust.

When lava is extruded at the oceanic ridge, the rock it forms becomes magnetized and acquires the magnetic polarity that exists at the time. Thus, oceanic crust should contain a record of when the Earth's magnetic polarity was reversed. The oceanic crust should be, in effect, a very slowly moving magnetic tape recorder. In fact, two oceanic tape recorders commence at the midocean ridge, one on each side of the ridge. In these "recorders," successive strips of oceanic crust are magnetized with normal and reversed polarity (Fig. 4.13). It was a straightforward matter to match this magnetic pattern with a record of magnetic polarity reversals, such as that shown in Figure 4.12. The magnetic striping allowed the age of any place on the seafloor to be determined.

Because the ages of magnetic polarity reversals had been so carefully determined, magnetic striping provided a means of estimating the speed with which the seafloor had moved. In some places, such movement is remarkably fast: as high as 9 cm/yr.

PLATE TECTONICS

Proof that the seafloor moves was the impetus needed for the emergence of the theory of plate tectonics which, as mentioned previously, states that segments (plates) of the Earth's hard, outermost shell (the lithosphere) move slowly sideways. Two essential points in formulating the theory were, first, the concept of the asthenosphere, hypothesized many years earlier in order to explain the bimodal distribution of the Earth's topography. About a century ago it was suggested that the outermost layer of the Earth (the lithosphere) floats on the weak asthenosphere. Low density continental lithosphere floats higher than high density ocean lithosphere. The flotational property of the lithosphere on the asthenosphere is called *isostasy*.

The second point was that the rigid lithosphere is strong enough to form coherent slabs (plates) that can slide sideways over the weak, underlying asthenosphere. These two points answered the main objection to Wegener's hypothesis—movement must occur with minimal resistance from friction. The crust—both oceanic and continental—is part of the lithosphere, so one consequence of plate tectonics is that as the plates of lithosphere move, the crust on the plate is rafted along as a passenger. Continents move, to be sure, but they do so only as portions of larger plates.

The plate tectonics theory provided a solution for one of the most puzzling aspects of seafloor spreading. If, as the theory of seafloor spreading required, new oceanic crust was created along a midocean ridge, either the Earth's surface must be expanding and the ocean basins getting larger, or else an equal amount of old crust must be getting destroyed. The answer to the puzzle was provided by previously unexplained regions inside the Earth, called Benioff zones, where very-deep-seated earthquakes occur. We will discuss deep earthquakes more fully in Chapter 5. For the moment let us accept the fact that Benioff zones mark the places where lithosphere capped by old, cold oceanic crust is sinking into the asthenosphere (Fig. 4.14). Destruction of old oceanic crust and creation of new oceanic crust are in balance—another example of a cyclic process.

Earthquake studies quickly provided evidence to support the hypothesis that the lithosphere is broken

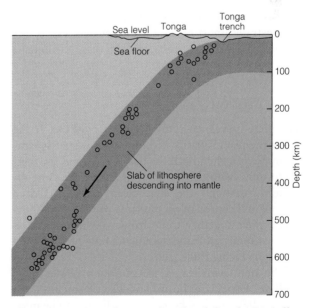

Figure 4.14 Deep-focus earthquakes define the Benioff zone near the island of Tonga in the Pacific Ocean. Each circle represents a single earthquake in 1965. The earthquakes are generated by downward movement of a comparatively cold slab of lithosphere.

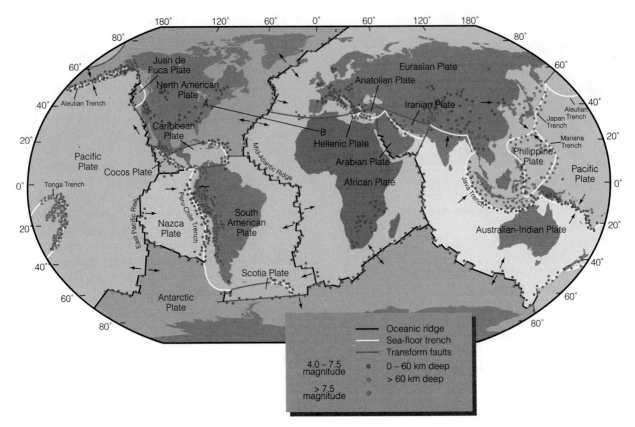

Figure 4.15 Earthquakes outline plate margins. Six large plates of lithosphere and several smaller ones are present. Each plate moves slowly but steadily in the direction shown by the arrows. The profile shown in Figure 4.19 lies along the line A-B.

into six large and many small plates. As Figure 4.15 demonstrates, most of the Earth's seismic activity occurs in sharply defined belts, and it is these earthquake belts that outline the plates.

Plate Motions

As a plate moves, everything on it moves too. If the top of the plate is partly oceanic crust and partly continental crust, then both the ocean floor and the continent move with the same speed and in the same direction. Although the first clear evidence that a seafloor and continent on the same plate of the lithosphere move together came from paleomagnetism, a series of remarkable measurements now provide even more convincing evidence.

The new evidence of plate motion comes from satellites. Using laser beams bounced off satellites, the distance between two points on the Earth can be measured with an accuracy of about 1 cm. Thus we can monitor any change in distance between, for example, Los Angeles on the Pacific plate and San Francisco on the North American plate. Therefore, by making dis-

tance measurements several times a year, we can measure present-day plate velocities directly. As seen in Figure 4.16, plate speed based on satellite measurements closely agrees with speeds calculated from paleomagnetic measurements. The agreement implies that the plates move steadily rather than by starts and stops. For additional discussion of plate speeds, see "A Closer Look: How Fast Do Plates Move?"

Plate Margins

Plates move as individual units, and interactions between plates occur along their edges (Fig. 4.17). Plate interactions are distinctively expressed by volcanism and earthquakes, and it is by studying both phenomena that scientists have deciphered most of what we know about plate margins:

1. **Divergent margins**, also called **spreading centers** because they are fractures in the lithosphere where two plates move apart. They are characterized by weak and shallow earthquakes.

2. **Convergent margins** occur where two plates are moving toward each other. Along convergent margins, either one plate must sink beneath the other creating a Benioff zone, in which case we refer to the margin as a **subduction zone,** or else continental crust on the two plates collides, in which

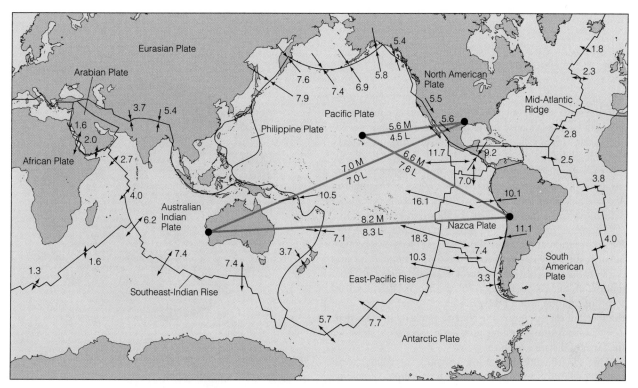

Figure 4.16 Present-day plate speeds in centimeters per year, determined in two ways. Numbers along the mido-cean ridges are average speeds indicated from paleomagnetic measurements. A speed of 16.1, as shown for the East Pacific Rise, means that the distance between a point on the Nazca plate and a point on the Pacific plate increases, on the average, by 16.1 cm each year in the direction of the arrows. The long red lines connect stations used to determine plate motions by means of satellite laser ranging (L) techniques. The measured speeds between stations are very close to the average speeds estimated from magnetic measurements (M).

case we refer to the margin as a **collision margin.** Convergent margins have earthquakes that are sometimes very deep and frequently very strong.

3. **Transform fault margins** are fractures in the lithosphere where two plates slide past each other, grinding and abrading their edges as they do so. Earthquakes can be either shallow or deep, they tend to be very frequent, and many are very strong.

PLATE TECTONICS AND THE EXTERNAL STRUCTURE OF THE EARTH

The beauty of plate tectonics is that it provides explanations for all the major features we see at the Earth's surface. These features are most easily visualized by considering the different kinds of plate margins: divergent, convergent, and transform fault.

Divergent Margins

Curious as it may seem, divergent plate margins start life on a continent and become an ocean. The sequence of events is illustrated in Figure 4.18.

The reasons a continent splits and a new ocean forms have to do with heat escaping from the Earth. Huge continental masses are thermal blankets that slow down the escape of heat from the interior. A plate capped by a large continent, such as the African plate, slowly heats up from below, expands, and eventually splits to start a cycle of spreading.

The structure of ocean basins can now be explained as the result of the formation of divergent margins. Modern shorelines don't coincide exactly with the splits that make the boundaries between continental crust and oceanic crust. This is because some ocean water spills out of the ocean basin onto the continent (Fig. 4.19). The boundaries between continental and oceanic crust are therefore covered by water, and today's shorelines are actually on the continents. As a result, each continent is surrounded by a flooded margin of continental crust that is of variable width and known as the **continental shelf**. The geological edge of the ocean basin is not the shoreline; rather, it is the place where oceanic crust joins the continental crust.

A CLOSER LOOK

How Fast Do Plates Move?

You might think, intuitively, that all points on a plate move with the same speed, but your intuition would be incorrect. It would be correct only if the plates were flat and moved over a flat asthenosphere (like plywood floating on water). Tectonic plates are pieces of a shell on a spherical Earth; in other words, they are curved, not flat. Any movement on the surface of a sphere is a rotation about an axis of the sphere. A consequence of such rotation is that different parts of a plate move with different speeds, as shown in Figure C4.1.

The plate in Figure C4.1 moves independently of the Earth's rotation and rotates about an axis of its own, colloquially called a *spreading axis*. In the figure, point *P*, where the spreading axis reaches the surface, is a *spreading pole*. The motion of each of the Earth's plates can be described in terms of rotation around the plate's own spreading axis, and the speed of each point on the plate depends on how far that point is from the spreading pole—the speed is greatest at the farthest distance from the pole.

One consequence of rotation around a spreading axis is that the width of new oceanic crust bordering a spreading center increases with distance from the spreading pole (Fig. C4.2).

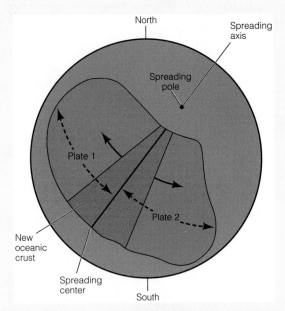

Figure C4.2 The width of new oceanic crust increases away from the spreading pole.

Some plates rotate faster about their spreading axes than others, and the reason has to do with the load of the lithosphere. Continental crust is thicker than oceanic crust; consequently, lithosphere capped by continents seems to protrude deeper into the asthenosphere than does lithosphere capped by oceanic crust. The extra protrusion seems to slow things down. Plates that carry lots of continental crust, such as the African, North American, and Eurasian plates, move relatively slowly. Plates without any continental crust, such as the Pacific and Nazca plates, move relatively rapidly.

The fastest moving plates are the Pacific and Nazca plates. The point on the Pacific plate that is farthest from the spreading axis (the fastest moving point on a plate) moves at a speed of 9 cm/y, about twice as fast as a fingernail grows. The Nazca plate is adjacent to the Pacific plate, and the fastest moving point on the Nazca plate also has a speed of 9 cm/y. This means that the distance between two points, one on the Pacific plate and one on the Nazca plate, grows larger by 18 cm/yr.

From the symmetrical spacing of magnetic time lines on the two sides of a spreading center (Fig. C4.3), it appears that a spreading center is fixed and that both plates move away from it at equal rates, but this is a case where appearances deceive. The same pattern of magnetic time lines would be observed if the African plate were stationary and both the Mid-Atlantic Ridge and the North American plate were moving westward. In fact, all that can be deduced from magnetic time lines is the *relative speed* of two plates. In

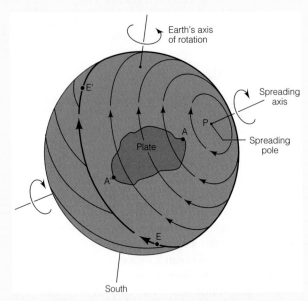

Figure C4.1 Movement of a curved plate on a sphere. The movement of each plate of lithosphere on the Earth's surface can be described as a rotation about the plate's own spreading axis. Point *P* has zero speed because it is the fixed point around which rotation occurs. Point *A'*, at the edge of the plate closest to the equator *EE'*, has a high speed. Point *A*, closest to the pole, has a low speed.

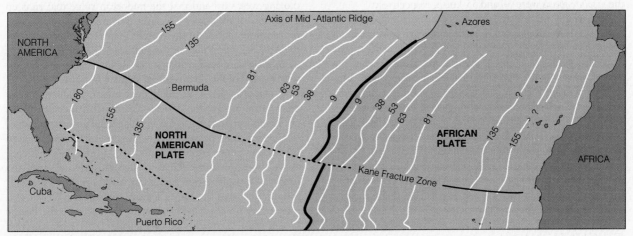

Figure C4.3 Age of the ocean floor in the central North Atlantic, deduced from magnetic striping. Numbers give ages in millions of years before the present.

order to measure the *absolute speed* of a plate, an external frame of reference is needed.

A familiar example of absolute versus relative motion occurs when one automobile overtakes another. If observers in the two automobiles could see only each other and not any fixed objects outside their cars, they could judge only the *difference* in speed between the two cars. For example, one car could be traveling at 50 km/h and the overtaking car at 55 km/h, but all the observers could determine was that the *relative speed* difference was 5 km/h. On the other hand, if

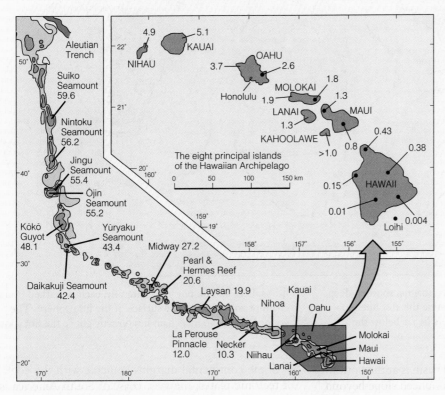

Figure C4.4 Hawaiian chain of volcanoes, showing the increase in age from southeast to northwest.

the observers measured speed with respect to a stationary reference, such as the ground, they would determine that the *absolute speeds* were 50 and 55 km/h, respectively.

We would be constrained to determine only relative plate speeds if a fixed framework did not exist. Fortunately, a framework does exist. During the nineteenth century, the American geologist James Dwight Dana observed that the age of volcanoes in the Hawaiian chain, some now submerged beneath the sea, increases from southeast to northwest (Fig. C4.4). Apparently, a long-lived magma source (a *hot spot*) lies somewhere deep in the mantle. As the Pacific plate moves, a volcano can remain in contact with the magma source for only about a million years. A chain of volcanoes should therefore be a consequence of lithosphere moving over a fixed hot spot. If long-lived hot spots exist in the mantle, they can provide a series of fixed points against which to measure absolute plate speeds.

More than a hundred hot spots have now been identified (Fig. C4.5). Using them for reference, scientists have determined that the African plate is nearly stationary (evidenced by the fact that volcanoes there seem to be very long lived). Because the African plate is almost completely surrounded by spreading centers, and because the relative speeds of the plates on either side of Africa are the same, we must conclude that the Mid-Atlantic Ridge in the southern Atlantic Ocean is moving westward and that the midocean ridge that runs up the center of the Indian Ocean is moving eastward. Because the absolute motion of the African plate is zero or nearly so, the Mid-Atlantic Ridge in the southern Atlantic Ocean must be moving westward at the rate of about 2 cm/y, and the absolute speed of the South American plate must be 4 cm/y.

A plate that has a spreading edge but no subduction zone must grow in size; the African and North American plates are examples. To keep things in balance, plates with subduction zones must be slowly shrinking. Most of the modern subduction zones are to be found around the Pacific Ocean along the edge of the Pacific plate, and thus much of the oceanic lithosphere now being destroyed is in the Pacific. It follows then that the Indian Ocean, the Atlantic Ocean, and most other oceans must be growing larger, while the Pacific Ocean must be steadily getting smaller. It is estimated that about 200 million years in the future the Pacific Ocean will have disappeared and Asia and North America will have collided as a result. The assembly of a new Pangaea will then be underway.

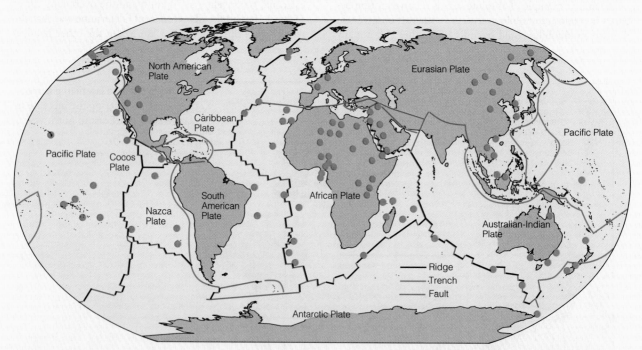

Figure C4.5 Long-lived hot spots (magma sources) deep in the mantle can be used to determine the absolute motions of plates. Because the hot spots lie far below the lithosphere and do not move laterally, each is marked by a chain of volcanoes on the surface of the lithosphere. The youngest volcano in a chain lies directly above the hot spot.

The geological edge of an ocean basin is at the bottom of the **continental slope**, a pronounced slope beyond the seaward margin of the continental shelf.

The **continental rise** lies at the base of the continental slope. It is a region of gently changing slope where the oceanic crust meets the continental crust.

Some continental margins coincide with the edges of tectonic plates; the west coast of South America is an example. Other continents sit in the middle of plates, and their margins are far from plate edges; North America and Africa are examples. Regardless of today's configurations, the margins of all continents

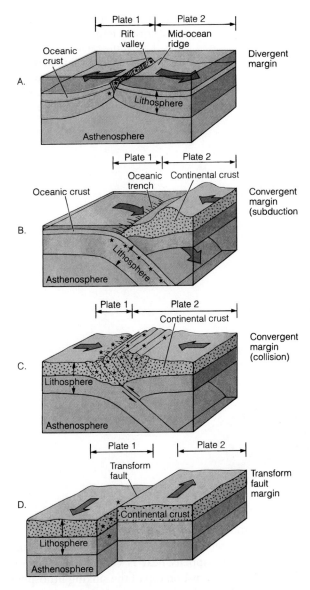

Figure 4.17 The various kinds of plate margins. Stars indicate earthquake foci. A. Divergent margin (also called spreading center). The topographic expression is a midocean ridge formed as a result of volcanism. Earthquakes are shallow and weak. B. Convergent margin-subduction. The topographic expression is a deep trench. Earthquakes are deep—down to 700 km—and very strong. C. Convergent margin-collision. The topographic expression is a mountain range. Earthquakes down to 300 km and sometimes strong. D. Transform fault margin. No characteristic topographic expression, but margin is often marked by a long, thin valley. Earthquakes down to 100 km and often strong.

have, at some time in the geological past, coincided with plate margins.

Beyond the continental slope and rise lies the rarely seen world of the deep ocean floor. Teams of oceanographers and seagoing geologists used deep diving submarines and other new devices to sound and sample the ocean bottom. As a result of this work, we now know almost as much about the seafloor as we do about the land surface.

Large, flat areas known as **abyssal plains** are a major topographic feature of the seafloor and lie adjacent to the continental rise (Fig. 4.19). These plains generally are found at depths of 3 to 6 km below sea level and range in width from about 200 to 2000 km. They are most common in the Atlantic and Indian oceans, which have large, mud-laden rivers entering them. Abyssal plains form as a result of the mud settling through the ocean water and burying the original seafloor topography beneath a blanket of fine debris.

Convergent Margins

Convergent margins are the places where lithosphere capped by oceanic crust sinks into the asthenosphere; the process is called subduction. The geographic features of a subduction zone are shown in Figure 4.20. Particularly striking is the deep-sea trench that marks the place where oceanic-capped lithosphere bends and sinks into the asthenosphere, and an arc-shaped chain of volcanic islands (called an *island arc* or a *magmatic arc*) formed above the sinking lithosphere. Trenches are the deepest parts of the ocean. Besides the prominent trench and the island arc, three less prominent features are present above subduction zones: the fore-arc ridge and the fore-arc basin lie between the trench and the magmatic arc, and the back-arc basin which lies on the opposite side of its magmatic arc from the trench. A *fore-arc ridge* is commonly underlain by a zone of smashed and shattered rock called a *mélange* that causes a local thickening of the crust. A *fore-arc basin* is a low-lying region between the fore-arc ridge and the island arc. Behind the island arc is another shallow basin, the *back-arc basin*. The islands of Sumatra and Java in Indonesia are an example of a present-day magmatic arc that is flanked by a fore-arc ridge and a back-arc basin (Fig. 4.21).

In order to understand how and why a subduction zone develops and thus how a convergent margin begins, it is necessary to consider what happens to a plate of lithosphere as it moves away from a spreading center.

Near a spreading center, the lithosphere is thin, and its boundary with the asthenosphere is close to the surface (Fig. 4.22). This thinning happens because magma rising toward the spreading center heats the lithosphere so that only a thin layer near the top retains the strength properties of the lithosphere.

As the lithosphere moves away from the spreading center, it cools and becomes denser. In addition, the boundary between the lithosphere and the asthenosphere becomes deeper, and as a result the lithosphere becomes thicker and the asthenosphere thinner. Fi-

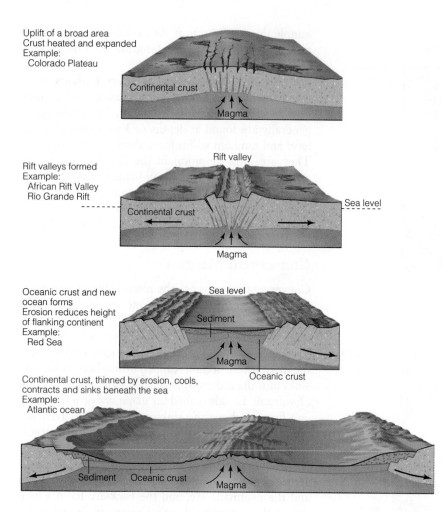

Uplift of a broad area
Crust heated and expanded
Example:
  Colorado Plateau

Continental crust

Magma

Rift valleys formed
Example:
  African Rift Valley
  Rio Grande Rift

Rift valley

Continental crust

Sea level

Magma

Oceanic crust and new
ocean forms
Erosion reduces height
of flanking continent
Example:
  Red Sea

Sea level

Sediment

Magma

Oceanic crust

Continental crust, thinned by erosion, cools,
contracts and sinks beneath the sea
Example:
  Atlantic ocean

Sediment   Oceanic crust   Magma

Figure 4.18 The rifting of continental crust to form a new ocean basin. The rifting can cease at any stage. It is not necessarily correct to conclude that the Rio Grande or the African Rift valley, for example, will someday open to form new oceans.

nally, about 1000 km from the spreading center, the lithosphere reaches a constant thickness and is much cooler and denser than the hot, weak asthenosphere below. Eventually, the cool lithosphere breaks and starts to sink downward. Like a conveyor belt, old lithosphere with its capping of oceanic crust sinks into the asthenosphere and eventually into the mesosphere.

As the moving strip of lithosphere sinks slowly through the asthenosphere, it passes beyond the region where geologists can study it directly. Consequently, what happens next is conjecture. The thin layer of oceanic crust on top of the sinking lithosphere starts to melt at a depth of about 100 km and becomes magma; some of which reaches the surface to form the volcanoes of the magmatic arc. Figure 4.23 is an example of a present-day magmatic arc of volcanoes parallel to, but about 150 km from, the trench that marks

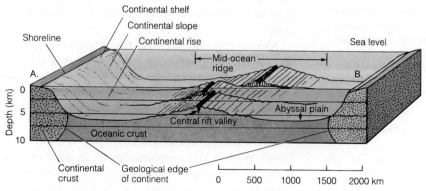

Shoreline

Continental shelf
Continental slope
Continental rise

Mid-ocean
ridge

Sea level

A.

B.

Depth (km)

0

5

10

Central rift valley

Abyssal plain

Oceanic crust

Continental
crust

Geological edge
of continent

0   500   1000   1500   2000 km

Figure 4.19 Portion of the Atlantic Ocean, showing the major topographic features.

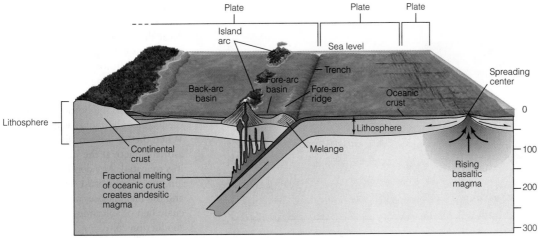

Figure 4.20 Structure of tectonic plates at a convergent margin. Along the line of subduction, an oceanic trench is formed, and sediment deposited in the trench, as well as sediment from the sinking plate, is compressed and deformed to create a mélange of shattered and crushed rock shaped as a fore-arc ridge. The sinking oceanic crust eventually reaches the temperature where melting commences and forms andesitic magma, which then rises to form an arc of volcanoes on the overriding plate. On the side of the island arc away from the trench, tensional forces lead to the development of a back-arc basin.

the place where the Nazca plate sinks below the South American plate. The Sierra Nevada of California is an example of an old magmatic arc, presenting evidence that the western margin of North America was once a subduction zone like the western margin of South America today.

When the sinking rate of a subducting plate is faster than the forward motion of the overriding plate, part of the overriding plate can be subjected to tensional (pulling) stress. The leading edge of the over-

riding plate must remain in contact with the subduction edge or else a huge void will open. What happens is that the overriding plate grows slowly larger at a rate equal to the difference in velocities between the two plates. Most commonly, this process is manifested by a thinning of the crust and the formation of a back-arc basin behind and parallel to the island arc (Figs. 4.20 and 4.21).

Collision Zones

Because it is less dense than the mantle, continental crust is too buoyant to be dragged down into a subduction zone. Therefore, when the two members of a converging pair of plates are capped by continental crust, the eventual result is a collision, as Figure 4.24 shows.

A collision zone marks the disappearance of an ocean and the formation of spectacular mountain ranges in its place. The Alps, the Urals, the Himalaya, and the Appalachians are the results of continental collisions, and therefore each is the graveyard of an ancient ocean basin. Because continental crust cannot sink into the mantle, much of the evidence concerning ancient plates and their motions is recorded in the bumps and scars of past continental collisions.

Transform Fault Margins

Transform faults are great vertical fractures that cut right down through the entire lithosphere. One transform fault margin much in the public eye because of

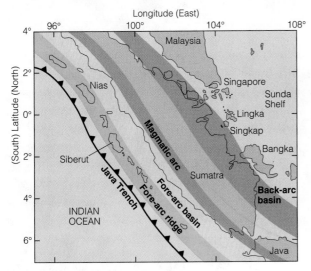

Figure 4.21 Map of portion of Indonesia showing the positions of the major topographic features in a present-day convergent margin.

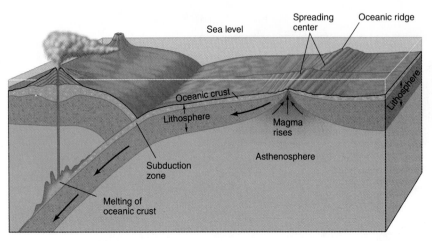

Figure 4.22 Schematic diagram showing the major features of a plate. Near the spreading center, where the temperature is high because of rising magma, the lithosphere is thin. Away from the spreading center, the lithosphere cools, becomes denser and also thicker, and so the lithosphere-asthenosphere boundary is deeper. When the lithosphere sinks into the asthenosphere at the subduction zone, it is reheated. At a depth of about 100 km, the oceanic crust starts to melt, and the magma rises and forms an arcuate belt of volcanoes parallel to the subduction zone.

the threat of earthquakes along it is the San Andreas Fault in California (Fig. 4.25). This fault, which runs approximately north-south, separates the North American plate on the east, on which San Francisco sits, from the Pacific plate on the west, on which Los Angeles sits. The Pacific plate is moving in a northerly direction, and the North American plate is moving in a southerly direction. As the two plates grind and scrape past each other, Los Angeles is slowly moving north and San Francisco is moving south. At times the plate edges grab and lock, and as they do the rocks on both sides flex and bend. When the locked section breaks free, the flexed rock snaps suddenly and an earthquake occurs.

Eventually, many millions of years in the future, Los Angeles and San Francisco will be adjacent. Then, as the two plates continue to move, the fragment of continental crust on which Los Angeles sits will become a long thin island. The trip will end when the future "Los Angeles Island" reaches the subduction zone along the northern edge of the Pacific plate and the island collides with Alaska and the Aleutian islands.

Figure 4.23 An example of a present-day convergent margin. This chain of steep-sided volcanoes in Ecuador sits above the subduction zone where the Nazca plate sinks below the western edge of the South American plate. Several snow-capped volcanoes are visible in this aerial photograph.

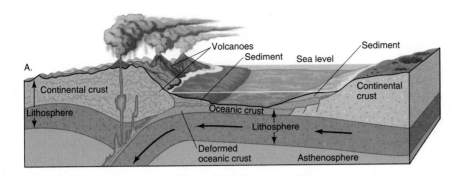

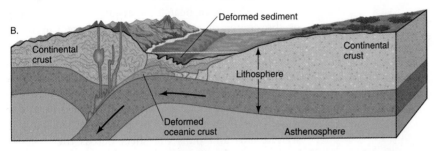

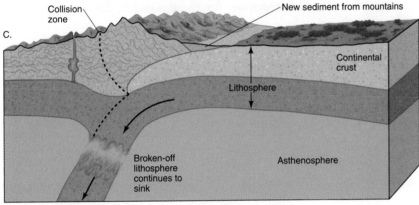

Figure 4.24 Mountains form when two masses of continental crust collide. A. Subducting oceanic lithosphere compresses and deforms sediments at the edge of continental crust on overriding plate (left). Sediments at the edge of continental crust on subducting plate (right) are undeformed. B. Collision. Sediment at the edge of continental crust on subducting plate is deformed and welded onto already deformed continental crust on overriding plate. C. After collision. The leading edge of the subducting plate breaks off and continues to sink. The two continental masses are welded together, and a mountain range stands where once there was ocean.

CAUSE OF PLATE TECTONICS

Just as Wegener felt sure that continents had drifted but could not explain how, today we are sure that plates move and that convection currents play a role. But we are still unable to say exactly what role. The situation is analogous to recognizing the shape, color, size, and speed capability of an automobile and knowing that gasoline supplies the energy needed for movement but not knowing how the gasoline makes the engine work. Until the driving mechanism is explained, plate tectonics must remain only an approximate description. Meanwhile, we can hypothesize about the causes of the motion and test the hypotheses by making detailed calculations based on the laws of nature.

The lithosphere and asthenosphere are closely bound together. If the asthenosphere moves, the lithosphere must move too, just as the movement of sticky molasses moves a piece of wood floating on the surface of the molasses. Conversely, movement of the lithosphere causes movement in the asthenosphere. Such is our state of uncertainty, however, that we cannot yet separate the relative importance of the two effects. However, on two points we are quite certain: (1) the lithosphere has energy of motion and (2) the source of this energy is the Earth's internal heat. We know, too, as discussed at the beginning of the chapter, that the heat energy reaches the base of the lithosphere by convection in the mesosphere and asthenosphere. What has not yet been figured out is the precise way the heat energy brought up by convection causes plates to move. Nevertheless, all scientists who have studied the problem are in agreement on one point: convection keeps the asthenosphere hot and weak by bringing up heat from deep in the mantle and

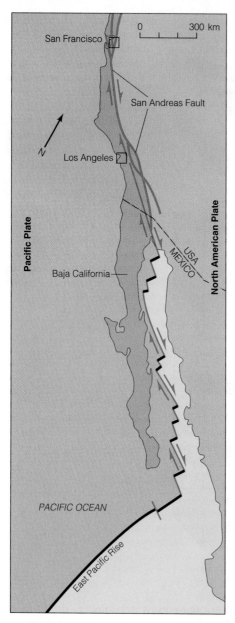

Figure 4.25 The San Andreas Fault is a transform fault margin that separates the Pacific plate from the North American plate. Directions of motion are shown by the arrows. Los Angeles, on the Pacific plate, is moving north, while San Francisco is moving south, bringing the two cities closer together at a speed of 5.5 cm/y; they will be adjacent about 10 million years in the future.

the core. In this sense at least, convection is essential for plate tectonics.

Movement of the Lithosphere

Three forces might play a role in moving the lithosphere. The first is a push away from a spreading center. Rising magma at a spreading center creates new lithosphere and in the process pushes the plates away from each other (Fig. 4.26A). Once the process is started, it tends to keep itself going. The problem with this hypothesis is that pushing involves compression, but the existence of rifts along a midocean ridge indicates a state of *tension* (the opposite of compression).

A second way lithosphere could be made to move is by dragging. Proponents of the dragging idea point out that lithosphere breaks and starts to sink through the asthenosphere because cold lithosphere is more dense than the hot asthenosphere. Because rock is a poor conductor of heat, they argue, the temperature at the center of a tongue of lithosphere can be as much as 1000°C cooler than the mantle at depths of 400 to 500 km. This means that a sinking tongue of lithosphere will continue to be more dense than rock it is sinking through and it must exert a pull on the entire plate. This is like a heavy weight that hangs over the side of a bed and is tied to the edge of a sheet. The weight falls and pulls the sheet across the bed. To compensate for the descending lithosphere, rock in

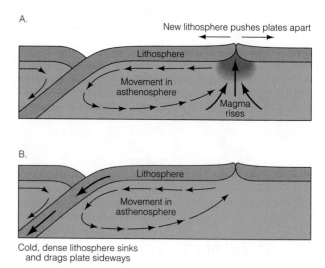

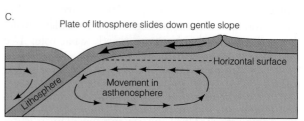

Figure 4.26 Three suggested mechanisms by which lithosphere might move over the asthenosphere. A. Magma rising at a spreading center exerts enough pressure to push the plates of lithosphere apart. B. A tongue of cold, dense lithosphere sinks into the mantle and drags the rest of the plate behind it. C. A plate of lithosphere slides down a gently inclined surface of asthenosphere.

the asthenosphere must flow slowly back toward the spreading center (Fig. 4.26B).

Both the pushing and the dragging mechanism have problems, however. Plates of lithosphere are brittle, and they are much too weak to transmit large-scale pushing and pulling forces without major deformation occurring in their middle. Deformation is not present, however, and mid-plate earthquakes, which would be expected for a plate undergoing deformation, are infrequent.

The third possible mechanism for lithosphere movement is for the whole plate to slide downhill away from the spreading center. The lithosphere grows cooler and thicker away from a spreading center. As a consequence, the boundary between lithosphere and asthenosphere slopes away from the spreading center. If the slope is as little as 1 part in 3000, the lithosphere's own weight could cause the lithosphere to slide at a rate of several centimeters per year (Fig. 4.26C).

At present, there is no way to choose among the three proposed mechanisms. Calculations suggest that each operates to some extent, so that the entire process is possibly more complicated than we now imagine. The prevailing idea at present is that subduction starts when old, cold lithosphere breaks and begins to sink, pulls on the plate, and starts the movement. Once movement starts, downhill slide and ridge push combine to keep the movement going. Only future research will resolve the question.

THE ROCK CYCLE AND THE PLATE TECTONIC CYCLE

The **plate tectonic cycle** involves the cycling of material through the mantle via the oceanic crust. Hot magma rises from the mantle and forms new oceanic crust at a midocean ridge; the oceanic crust moves laterally away from the middle and finally sinks ino the mantle at a subduction zone. The rate at which new oceanic crust is created is equal to the rate at which it is consumed. The energy source that drives the plate tectonic cycle is the Earth's internal heat.

As discussed in Chapter 1, the **rock cycle** involves the cycling of the material through the continental crust as a result of repeated weathering, erosion, transport, deposition, formation of new rock, uplift, weathering, erosion, and so on. The energy source that drives the rock cycle is the Sun's heat. James Hutton and the other early geologists who discovered the rock cycle were completely unaware of the remarkable processes of plate tectonics. But now we realize that plate tectonics answers many of the questions those early geologistis could not answer. We now also realize that two cycles, the rock cycle and the plate tectonic cycle, continually cycle material through the geosphere.

Figure 4.27 is a box diagram for the two cycles of the geosphere and shows how the rock cycle interacts with the plate tectonic cycle. It is clear from Figure

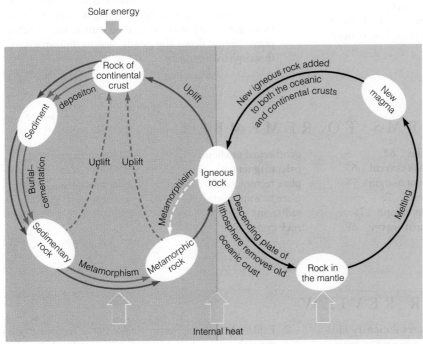

Figure 4.27 The rock cycle, an interplay of processes driven both by the Sun's energy and by the Earth's internal heat, interacts with the plate tectonic cycle, which slowly but continually cycles material through the mantle.

4.27 that even though the rock and plate tectonic cycles move very slowly, they must be involved with everything that goes on at the Earth's surface. In short the rock and plate tectonic cycles are the most fundamental of the cycles of the Earth system. We will return to various aspects of the two cycles in the next four chapters.

SUMMARY

1. Alfred Wegener proposed a hypothesis of continental drift in the early years of the twentieth century but because he could not explain how continents could move his hypothesis was not widely accepted.

2. Studies of the Earth's magnetism led to a revival of the continental drift hypothesis in the 1950s, and this eventually led to the plate tectonics hypothesis in the 1960s.

3. The source of the Earth's magnetism is the molten outer core.

4. As minerals such as magnetite, which can become permanent magnets, cool through the Curie temperature, they acquire magnetism with the polarity, inclination, and direction of the Earth's field.

5. Paleomagnetism of lavas and sedimentary rocks reveals that either the magnetic poles have wandered in the past (apparent polar wandering) or the continents have drifted. The latter is the correct conclusion.

6. Seafloor spreading is a hypothesis that new oceanic crust is created at midocean ridges by magma rising from deep inside the Earth.

7. The seafloor spreading hypothesis was proved correct by the record of magnetic polarity reversals recorded in the oceanic crust.

8. The lithosphere is broken into six large and many smaller plates, each about 100 km thick and each slowly moving over the top of the weak asthenosphere beneath it.

9. Three kinds of margins are possible between plates. Divergent margins (spreading centers) are those where new lithosphere forms; plates move away from them. Convergent margins (subduction zones) are lines along which lithosphere capped by oceanic crust is subducted back into the mantle and where continental crust on adjacent plates collides to form a collision margin. Transform fault margins are lines where two plates slide past each other.

10. Divergent margins form when a continent splits because of thermal stresses. As the split grows wider, an ocean forms.

11. Collision zones are the places where oceans disappear.

12. The continental shelf is the flooded margin of a continent. The geological edge of a continent is the bottom of the continental slope, where oceanic crust meets continental crust.

13. A deep-sea trench forms where lithosphere sinks into the asthenosphere as a result of subduction. Parallel to the trench and offset by about 150 km, a line of volcanoes forms an island arc. The volcanoes form because subducted oceanic crust starts melting at a depth of about 100 km.

14. Plates move as a combination of pull exerted by the sinking, cold tongue of lithosphere, by a push exerted by magma rising at the spreading center, and by lithosphere sliding down the gently sloping lithosphere–asthenosphere boundary.

IMPORTANT TERMS TO REMEMBER

abyssal plain *81*	convection *66*	geothermal gradient *67*	seafloor spreading *74*
collision margin *77*	convection current *67*	paleomagnetism *73*	spreading center *76*
conduction *66*	convergent margin *76*	plate tectonics *71*	subduction zone *76*
continental rise *80*	Curie point *72*	plate tectonic cycle *87*	transform fault margin *77*
continental shelf *77*	divergent margin *76*	radiation *67*	
continental slope *80*	geothermal energy *67*	rock cycle *87*	

QUESTIONS FOR REVIEW

1. Who was Alfred Wegener, and what revolutionary idea did he suggest? Why were scientists reluctant to accept Wegener's idea when it was first proposed?

2. Explain how the apparent wandering of magnetic poles throughout geologic history can be used to help prove continental drift.

3. How does lava carry a record of the Earth's magnetic field?

4. What are the main features of seafloor spreading? What critical test proved that the seafloor does move?

5. What is a "plate" in plate tectonics?

6. Briefly describe the three kinds of plate margins.

7. Describe what happens when two plates topped by oceanic crust converge. Compare your description with what happens when the converging is between two plates capped by continental crust.

8. How do satellite measurements help confirm the theory of plate tectonics?

9. Identify the major topographic features of the ocean floor, and state how they are related to tectonic plates.

10. How can earthquakes be used to outline the shape of plates and locate plate margins?

11. Draw a cross section through the lithosphere at a convergent plate margin and mark the positions of the fore-arc basin, the back-arc basin, and the magmatic arc.

12. The Himalaya and the Alps are said to be "graveyards" of ancient oceans. Why?

13. What causes earthquakes along a transform fault?

14. Where does the energy come from to cause the movement of tectonic plates?

15. What are the current theories of the causes of plate tectonics?

16. Why do so many volcanoes occur around the Pacific rim?

Questions for A Closer Look

1. What is a spreading pole?

2. Why does the speed differ from place to place on a plate?

3. Where on a plate is the speed a maximum? Where a minimum?

4. How do we know that the Pacific Ocean must be getting smaller?

5. What is the difference between the relative speed of a plate and its absolute speed?

6. How are the absolute speeds of plates determined?

7. The place where New York City now stands was once attached to North Africa at approximately the position of Marrakech in Morocco. New York City and Marrakech are now 5700 km apart. The North Atlantic plate is moving away from the African plate at a speed of 2 cm/y. How long has it taken for the Atlantic Ocean to reach its present width?

QUESTIONS FOR DISCUSSION

1. In the vicinity of Los Angeles, the Pacific plate is moving northerly relative to the North American plate at a speed of 5.5 cm/y. Determine how long it will be before Los Angeles and San Francisco are side by side. Draw a map of the way the west coast of North America might look when Los Angeles and San Francisco are side by side. Now draw a map 10 million years after the two towns are adjacent.

2. Although plate tectonics is manifested by such disasters as massive earthquakes and volcanic eruptions, it is nevertheless said that plate tectonics is a good thing for our planet. Discuss why this is so.

3. What key observations would you plan to make if you were sending a spaceship to another planet and wished to find out whether plate tectonics operated on the planet? Assume for the purpose of discussion that the spaceship cannot land and so all observations have to be made remotely.

At our web site, you will find interactive case studies with information about earthquakes, volcanism, and the relationship between plate tectonics and the landscape.

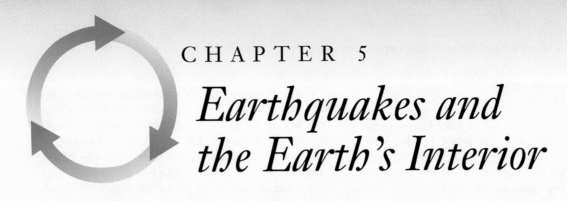

CHAPTER 5

Earthquakes and the Earth's Interior

● *Bad and Good News about Earthquakes*

There is a saying among engineers that earthquakes don't kill people, buildings do. Shaking ground may make people fall down and falls may break legs and arms, but they don't kill. However, shaking ground can make buildings collapse, and collapsing buildings can definitely kill.

The worst earthquake of the twentieth century (to date) happened on July 28, 1976. At 3:45 A.M., while 1 million inhabitants of T'ang Shan, China, slept, a 7.8 magnitude quake leveled the city. Hardly a building was left standing, and the few that did withstand the first quake were destroyed by a second, magnitude 7.1, which struck at 6:45 P.M. the same day. When the wreckage of T'ang Shan was cleared, 240,000 people were dead. Losses were large because most of the buildings had not been constructed to withstand an earthquake. They had unreinforced brick walls. When the ground started to shake, the walls collapsed, the roofs caved in, and the sleeping inhabitants were crushed.

Earthquakes are caused by sudden releases of stored elastic energy in the Earth, and the source of most of that energy is plate tectonics. One of the great challenges scientists face today is to understand the dynamics of the Earth system sufficiently well so that the timing and probable size of an earthquake can be accurately predicted and human safety assured. Chinese scientists have been more successful than most in predicting quakes, but the T'ang Shan quake gave no recognizable warning signs and was completely unexpected. We still have a long way to go.

Not everything about earthquakes is bad—in fact, some things about them are so important for learning about how the Earth works that they might even be called good. For example, earthquakes can be used to study the Earth's interior. The way the Earth vibrates after a large quake is controlled in large degree by the properties of the rocks inside. Used in this way, earthquake vibrations are like the *X* rays a doctor uses to study the inside of a human body—they are the probes we use to sense and measure the world beneath our feet. As we saw in Chapter 4, earthquakes also play an essential role in outlining tectonic plates and in deciphering the mechanisms by which plates are formed at spreading edges and consumed by subduction at subduction zones.

In this chapter we discuss how earthquakes occur and how the vibrations they cause can be used to assemble a picture of the Earth's interior. We also discuss how earthquakes are measured and how they cause damage. ●

(opposite) Collapse of a bank building in Kobe, Japan, as a result of an earthquake in 1995. Built on a filled-in estuary, the collapse illustrates the danger of siting high rise buildings on soft, unstable sediment in an area subject to earthquakes.

EARTHQUAKES

Place a thin piece of wood across your knees, press both ends downward, and the wood bends. Stop pressing and the wood springs back to its original shape. Any change of shape or size that disappears when the deforming forces are removed is called **elastic deformation.** The muscle energy you used to bend the wood doesn't disappear—it is stored as elastic energy in the wood. When the bending force is removed, it is this elastic energy that restores the wood to its original shape. Consider what happens, however, when the pressure is so great that the elastic limit is exceeded and the wood breaks with a sudden snap. The elastic energy is now converted in part to heat at the breakage point, in part to the sound waves that make the snapping noise, and in part to vibrations in the wood.

Earthquake vibrations are the same kind of vibrations that you feel when the wood breaks. When the Earth quakes, it is as if a huge log has broken somewhere inside the Earth. The more energy released, the stronger the quake. Just how elastic energy is stored and built up in the Earth, however, and how and why it is suddenly released continue to be the subjects of intensive research.

Of course, breaking logs are not the cause of earthquakes. However, the sudden breaking of a mass of rock or the sudden slippage of two rock masses past each other, has the same effect as breaking a log. The most widely accepted theory concerning the origin of earthquakes involves slipping rock masses and the elastic rebound theory.

Origin of Earthquakes

When slippage of rock occurs along a fracture in a rock, the fracture is called a **fault**. The cause of most earthquakes is thought to be sudden movement along faults, but it cannot be that simple. Some earthquakes are millions of times stronger than others. The same amount of energy that in one case is released by thousands of tiny slips and tiny earthquakes is, in another case, stored and released in a single immense earthquake. The **elastic rebound theory** suggests that, if fault surfaces are rough, so that they lock rather than slip easily past one another, the rocks on either side of the fault will bend and in bending will store elastic energy, just as the bending of a piece of wood stores elastic energy. When a locked fault finally does slip, the bent rocks will rebound to their original shapes, and the stored energy is released so rapidly that it causes an earthquake.

The first measurements supporting the elastic rebound theory came from studies of the San Andreas Fault in California, a vertical fracture passing through the entire lithosphere. The San Andreas Fault separates the Pacific plate to the west from the North American plate on the east. The two plates are sliding in opposite directions. In places the sliding is smooth and continuous, but at other places the fault is locked. During long-term field observations in central California, beginning in 1874, scientists from the U.S. Coast and Geodetic Survey determined the precise position of many points both adjacent to and distant from a part of the fault that was locked (Fig. 5.1). As time passed, changes in the relative positions of these points revealed that the Earth's crust on each side of the fault was slowly being bent. The fault was locked and did not slip. On April 18, 1906, near San Francisco, the two sides of the locked fault shifted abruptly. The elastically stored energy was released as the rock masses moved and the bent crust rebounded, thereby creating a violent earthquake. A subsequent examination of the new positions of the points revealed that the rocks on each side of the fault were no longer bent. The part of the fault where the 1906 earthquake occurred is now locked again and eventually another earthquake will occur when the lock is broken. The evidence is clear—faults move in recurring steps, and as a consequence earthquakes are also recurrent.

The 1906 San Francisco earthquake caused an enormous amount of damage, but it also started a great deal of research. From that research, our present-day understanding of how earthquake damage can be minimized and how earthquakes can be used to study the internal structure of the Earth, has developed. Before proceeding with a discussion of these points, let's expand our discussion of the way earthquake vibrations travel through rock.

Seismic Waves

The point where energy is first released is called the **earthquake focus**. Because most earthquakes are caused by movement along a fault, the focus is not a point but rather a region that may extend for several kilometers (Fig. 5.2). Because the focus generally lies at some depth below the Earth's surface, it is more convenient to identify an earthquake site from the **epicenter**, which is the point on the Earth's surface that lies vertically above the focus (Fig. 5.2). The usual way to describe the location of an earthquake focus is to state its depth and the location of its epicenter.

When an earthquake occurs, the elastically stored energy is carried outward from the focus to other

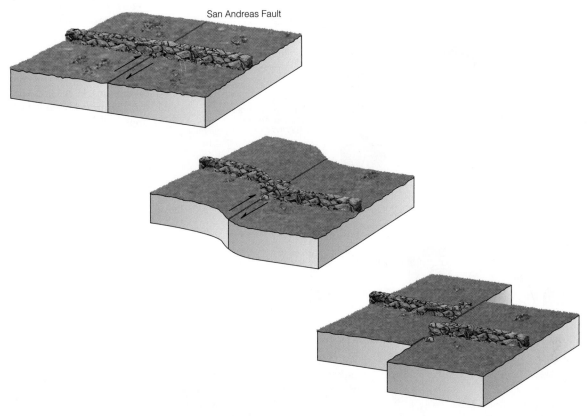

Figure 5.1 An earthquake is caused by sudden release of elastic energy stored in rocks. This sketch is based on surveys near the San Andreas Fault, California, before and after the earthquake of 1906. A stone wall crosses the fault and is slowly distorted as the rock beneath it is bent and elastically strained. After the earthquake, two segments of the wall are offset 7 m.

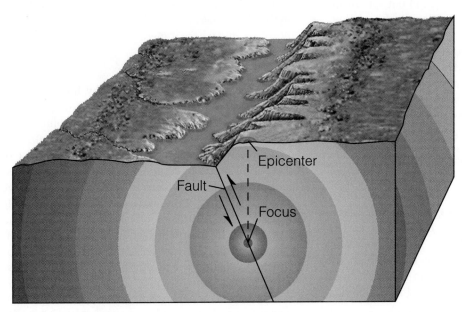

Figure 5.2 The focus of an earthquake is the site of first movement on a fault and the center of energy release. The epicenter of an earthquake is the point on the Earth's surface that lies vertically above the focus.

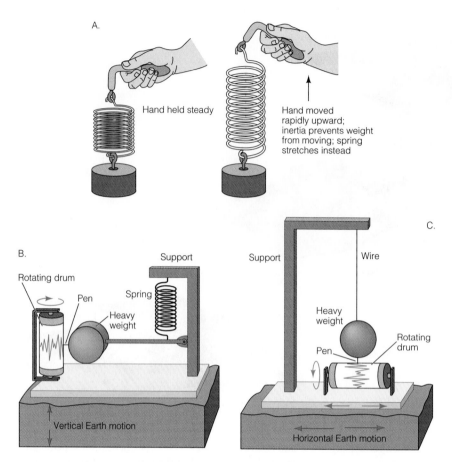

Figure 5.3 Seismographs make use of inertia, which is the resistance of a stationary weight to sudden movement. A. The principle of the seismograph. B. A seismograph for measuring vertical motions. C. A seismograph for measuring horizontal motions.

parts of the Earth by vibrations. The vibrations, which are also called **seismic**[1] **waves**, spread out spherically in all directions, just as sound waves spread out spherically in all directions from a sound source.

Seismic waves are elastic disturbances, and so unless their elastic limit is exceeded, there is no deformation in the rocks through which the waves pass, and as a consequence there is no permanent record of the vibrations. Seismic waves must therefore be recorded while the rock is still vibrating. For this reason, many continuous recording devices that can detect seismic waves, called *seismographs* (Fig. 5.3), have been installed around the world.

There are several kinds of seismic waves, and they belong to two families. **Body waves** travel outward in all directions from the focus and have the capacity to travel through the Earth's interior. **Surface waves**, on the other hand, travel around but not through the Earth; they are guided by the Earth's surface. Body waves are analogous to light and sound waves, both of

which travel outward in all directions from a source. Surface waves, analogous to ocean waves, travel only on the Earth's *solid* surface, both where it meets the atmosphere and where it meets the ocean.

Body Waves

Rocks can be elastically deformed by body waves in two ways: by a change in shape (like bending or twisting a piece of wood) or by a change in volume (like squeezing a tennis ball or blowing up a balloon).

Body waves that cause volume changes consist of alternating pulses of compression (squeezing) and expansion (stretching) acting in the direction of wave travel (Fig. 5.4A). Sound waves are a familiar example of compression/expansion waves. A sound wave passes through air by alternating compressions and expansions of the air. Our ears sense the pulses of compression and expansion, and our brains transform these pulse signals into sound. Compression/expansion waves can pass through gases, liquids, and solids. That is why we can hear sounds not only in the air but also through the walls of houses and under water. Compression/expansion waves can pass easily through

[1]"Seismic" means caused by an earthquake and comes form the Greek verb *seiein*, to shake.

rocks. They have the greatest velocity of all seismic waves—6 km/s (3.7 mi/s) is a typical value near the Earth's surface—and they are the first to be recorded by a seismograph after an earthquake. They are therefore called **P** (for **primary) waves.**

Body waves that deform materials by change of shape are called shear waves. Liquids and gases don't have shapes; they simply flow freely to fill any container we put them in. Therefore, liquids and gases cannot transmit waves that depend on a change in shape, and so shear waves can be transmitted only by solids. As a shear wave travels through a material, each particle in the material is displaced perpendicular to the direction of wave travel (Fig. 5.4B). A typical velocity for shear waves in rocks near the Earth's surface is 3.5 km/s (2.2 mi/s). Because shear waves are slower than P waves and reach a seismograph later, they are called **S** (for **secondary) waves.**

Seismic body waves behave like light waves, which is to say that, in addition to being able to pass through a medium, they can also be reflected and refracted. *Reflection* is the familiar phenomenon of light bouncing off a mirror or other shiny surface, and body waves are reflected by numerous boundaries in the Earth. The less familiar process, *refraction*, as discussed in the "Closer Look: On Telescopes," in Chapter 3, occurs when the speed of a wave changes as it passes from one medium to another and this change causes the wave path to bend (see Fig.C3.3).

Surface Waves

Surface waves are a little more difficult to understand than body waves. They are due to the same phenomenon that makes a bell ring. When a bell is rung, the surface of the bell rises and falls in complex ways and makes the air around the bell vibrate too. The sounds we hear from a ringing bell come from the vibrating air. So too with the Earth. An earthquake makes the whole Earth ring like a bell—though, of course, the vibrations don't create sounds we can hear. When the Earth "rings," the surface rises and falls, owing to bell-like vibrations. To an observer on the surface of the Earth, the bell-like vibrations appear to be waves, just like P and S waves, and are recorded as such on a seismograph in the same way that body waves are recorded. Surface waves travel more slowly than P and S waves, and in addition they pass around the Earth

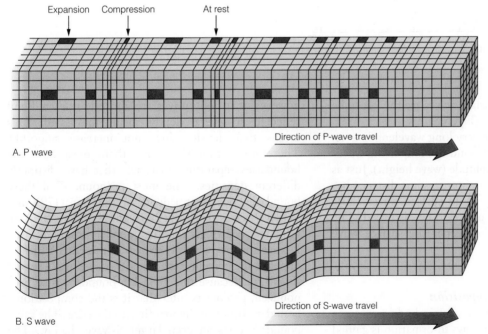

Figure 5.4 Body waves of the P (compression/expansion) and S (shear) types. A. The body waves known as P waves cause volume changes in the rock the wave is passing through by alternate compressions and expansions. An individual point in a rock moves back and forth parallel to the direction of P-wave propagation. As wave after wave passes through, a square repeatedly expands to a rectangle, returns to a square, contracts to a smaller rectangle, returns to a square, and so on. B. The body waves called S waves cause a change in rock shape because they result in a shearing motion. Any individual point in the rock moves up and down, perpendicular to the direction of S-wave propagation. A square repeatedly changes to a parallelogram, then back to a square.

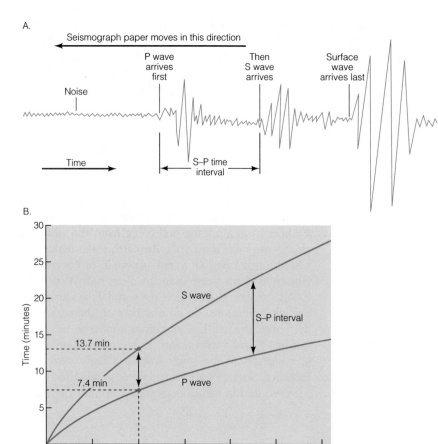

A.

B.

Figure 5.5 Travel times of P body waves, S body waves, and surface waves. A. Typical record made by a seismograph. All three types of waves leave the earthquake focus at the same instant. The fast-moving P waves reach the seismograph first, and some time later the slower-moving S waves arrive; the delay in arrival times is proportional to the distance traveled by these two waves. The surface waves travel more slowly than either P or S waves. B. Seismologists use a travel-time graph for P and S waves to locate an epicenter. For example, when a station measures the S-P time interval to be 13.7 min − 7.4 min = 6.3 min, they know the epicenter is 4000 km away.

rather than through it. Surface waves are the last to be detected by a seismograph (Fig. 5.5).

Surface waves can have very long wavelengths—up to hundreds of kilometers—and the longer the wavelength the greater the amplitude (wave height). Just as an ocean wave disturbs the water to some distance beneath the ocean surface, so do surface waves cause rocks below the Earth's surface to be disturbed. The greater the amplitude, the deeper the surface wave motion reaches.

Layers of Different Composition

Reflection and refraction of seismic body waves are one of the most important ways information is gained about the different compositional layers in the Earth.

The speed of body waves is determined, in part, by the density of the rocks they are passing through and in part by other properties such as the strength, or rigidity. The higher the density, the greater the speed. If the Earth had a homogeneous composition and if rock density just increased smoothly with depth as a result of increasing pressure, body wave velocities would also increase smoothly. Measurements reveal,

however, that body waves are abruptly refracted and reflected at several depths inside the Earth. This means that density does not increase smoothly. Therefore, within the Earth there must be some boundaries separating materials that have distinctly different densities. The most pronounced of these boundaries occurs at a depth of 2883 km (1791 mi). When P waves reach the 2883-km boundary, they are refracted so strongly that the boundary is said to cast a P-wave shadow, which is an area of the Earth's surface opposite the epicenter where no P-waves are observed (Fig. 5.6). Because this 2883-km boundary is so pronounced, geologists infer that it is the compositional boundary between the mantle and the core. The same boundary casts an even larger S-wave shadow. The reason here is not refraction, but the fact that S waves cannot travel through liquids. Therefore, the huge S-wave shadow lets us conclude that at least the outer portion of the core must be liquid.

Seismic waves cannot directly tell us the composition of the core, but they help us to deduce what the composition might be. Seismic-wave speeds indicate that rock density increases slowly from about 2.7 g/cm^3 at the top of the crust to about 5.5 g/cm^3 at the

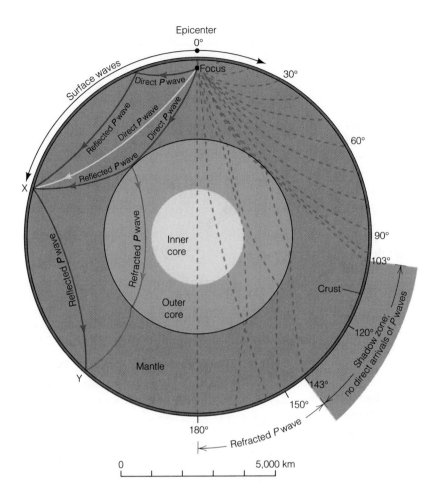

Figure 5.6 Refraction and reflection of body waves. On the left-hand side of the figure are shown various paths of P waves in the Earth, which is made up of concentric spheres of different compositions. Seismographs at some places (locations *X* and *Y*, for example) receive both direct P waves and reflected and refracted P waves. The right-hand side shows paths of P waves moving out from an epicenter at 0°. Reflection and refraction of P waves at the core–mantle boundary create a P-wave shadow from 103° to 143°.

base of the mantle. The average density of the whole Earth is 5.5 g/cm³. Therefore, to balance the relatively low density of the crust and upper mantle, the core must be composed of material having a density of at least 10 g/cm³. The only common substance that comes close to fitting this requirement is iron with a little admixture of nickel.

Early in the twentieth century, the existence of a compositional boundary between the Earth's crust and mantle was demonstrated by Andrija Mohorovičić, a Croatian scientist. Mohorovičić noticed that, for earthquakes whose focus lay within 40 km (25 mi) of the surface, seismographs about 800 km (500 mi) from the epicenter recorded two sets of body waves that arrived at the seismograph at different times. He concluded that the set that arrived second must have traveled from the focus to the station by a direct path through the crust, whereas the set that arrived first must have been refracted into rock that was denser than crustal rock. These refracted waves, moving through the denser zone, traveled faster within that zone and so reached the surface first (Fig. 5.7). Mohorovičić hypothesized that a distinct compositional boundary must separate the crust from this underlying

zone of denser composition. Crust thickness ranges from 30 to 70 km (19 to 43 mi) in continental regions but is only 8 km (5 mi) thick beneath the oceans (Fig. 2.14). Scientists now refer to this boundary as the **Mohorovičić discontinuity** and recognize it as the boundary that marks the base of the crust (or, put another way, the top of the mantle). The feature is commonly called the **M-discontinuity** and in conversation is shortened still further to **moho**.

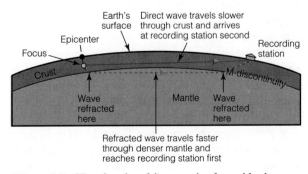

Figure 5.7 Travel paths of direct and refracted body waves from shallow-focus earthquake to nearby seismograph station.

Layers of Different Strength

As mentioned previously, density is not the only rock property that affects the speed of seismic waves. Rock strength also plays a role. Rock strength is an expression of the elasticity, and this in turn can be equated to the tendency of a rock to fracture (called brittleness; higher brittleness means higher elasticity), as opposed to a tendency to deform and flow like putty (called ductility; higher ductility = lower brittleness = lower elasticity). Rock strength, which is strongly affected by temperature and pressure, has a marked effect on the speed of both body and surface waves. The more ductile a rock, the lower the speed.

Studies of seismic wave speeds at various depths have given us the following picture of our Earth's interior. In addition to the boundary between the crust and the mantle (the moho), and between the mantle and core, there are three strength boundaries. The first boundary is about 100 km (62 mi) below the Earth's surface and separates brittle rocks above from ductile rocks below. This is the lithosphere–asthenosphere boundary (refer back to Fig. 2.14). At about 350 km (220 mi) there is a diffuse region separating very ductile rock above and somewhat less ductile rock below. This is the region of the asthenosphere–mesosphere boundary. Finally, there is a boundary between molten iron above and solid iron below. This is the outer core–inner core boundary.

Locating the Epicenter

The location of an earthquake's epicenter can be determined from the arrival times of the P and S waves at a seismograph. The farther a seismograph is from an epicenter, the greater the time difference between the arrival of the P and S waves (Fig. 5.5B). After using a graph like that shown in Figure 5.5B to determine how far an epicenter lies from a seismograph, the seismologist draws a circle on a map around the station with a radius equal to the calculated distance to the epicenter. The exact position of the epicenter can be determined when data from the three or more seismographs are available—the center lies where the circles intersect (Fig. 5.8).

If a local earthquake is recorded by several nearby seismographs, the focal depth can be determined in the same way that the epicenter is determined, by using the P-S time intervals. For distant earthquakes, the following method is employed. Note in Figure 5.6 that a seismograph at position X would record both a direct P wave and a P wave reflected from the Earth's surface (labeled PP). The direct P wave, having a shorter path length, would arrive before the PP wave. The travel-time difference between them is a measure of the focal depth of the earthquake.

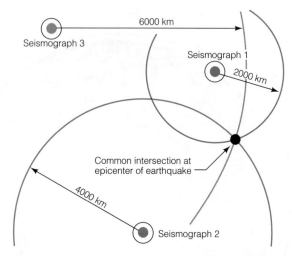

Figure 5.8 Locating an epicenter. The effects of an earthquake are felt at three seismograph stations. The time interval between the arrival of the first P and the first S waves depends on the station-epicenter distance. The following distances are calculated by using the curves in Figure 5.5B.

	Time Calculated	Interval	Distance
Seismograph 1	8.8 min − 4.7 min =	4.1 min	2000 km
Seismograph 2	13.7 min − 7.4 min =	6.3 min	4000 km
Seismograph 3	17.5 min − 9.8 min =	7.7 min	6000 km

On a map, a circle of appropriate radius is drawn around each station. The epicenter is where the three circles intersect.

Location of an epicenter and focal depth are only part of the information that can be read from the seismograph records. Of equal importance is the calculation of the amount of energy released during an earthquake or, as it is commonly stated, the magnitude of the earthquake. To see how magnitudes are calculated, see "A Closer Look: Earthquake Magnitudes."

EARTHQUAKE RISK

Most people in the United States think immediately of California when earthquakes are mentioned. However, the most intense earthquakes to jolt North America in the past 200 years were centered near New Madrid, Missouri. Three earthquakes of great size occurred on December 16, 1811, and January 23 and February 7, 1812. The exact sizes of these earthquakes are unknown because instruments to record them did not exist at the time. However, judging from the local damage and from the fact that tremors were felt and minor damage occurred as far away as New York and South Carolina, it is estimated that the largest of these quakes was larger than the one that leveled San Francisco in 1906.

A CLOSER LOOK

Earthquake Magnitudes

Very large earthquakes (of the kind that destroyed San Francisco in 1906; T'ang Shan, China, in 1976; parts of Mexico City in 1985; and parts of San Francisco again in 1989) are, fortunately, relatively infrequent. In earthquake-prone regions, very large earthquakes occur about once a century. They occur more frequently in some areas and less frequently in others, but a century is an approximate average.

This frequency rate means that the time needed to build up elastic energy to a point where the strength of a locked fault is exceeded is about 100 years. Small earthquakes may occur along a fault during this time as a result of local slippage, but even so, elastic energy is accumulating because most of the fault remains locked. When the lock is broken and an earthquake occurs, the elastic energy is released during a few terrible minutes. By careful measurement of elastically strained rocks along the San Andreas Fault, seismologists have found that about 100 joules (J) of elastic energy can be accumulated in 1 m^3 of deformed rock. This is not very much—100 J is equivalent to only about 25 calories of heat energy—but when billions or trillions of cubic meters of rock are strained, the total amount of stored energy can be enormous. The amount of elastically stored energy released during the Loma Prieta earthquake of 1989 was about 10^{15} J, and the 1906 San Francisco earthquake released at least 10^{17} J!

The Richter Magnitude Scale
Measurements of the bending of elastically deformed rocks before an earthquake, and of those same rocks after an earthquake has released the deforming force, can provide an accurate measure of the amount of the energy released. The task is very time-consuming, however, and all too frequently the pre-earthquake measurements are not available. Therefore, seismologists estimate the energy released by measuring the amplitudes of seismic waves. The **Richter magnitude scale**, named after Charles F. Richter, the seismologist from the California Institute of Technology who developed it, is defined by the maximum amplitudes of seismic waves (that is, the heights of the waves on a seismogram) 100 km from an epicenter. Because wave signals vary in strength by factors of a hundred million or more, the Richter scale is logarithmic, which means it is divided into steps called magnitudes, starting with magnitude 1 and increasing upward. Each unit increase in magnitude corresponds to a tenfold increase in the amplitude of the wave signal. Thus, a magnitude 2 signal has an amplitude that is ten times larger than a magnitude 1 signal, and a magnitude 3 is a hundred times larger than a magnitude 1 signal.

We can see from Figure C5.1 how a Richter magnitude is calculated. The energy of a seismic wave is a function of both its amplitude and the duration of a single wave oscillation, T. Divide the maximum amplitude, X, measured in steps of 10^{-4} cm on a suitably adjusted seismograph, by T, measured in seconds. Then add a correction factor, Y, for the distance between the epicenter and the seismograph, determined from the S-P wave interval. The ratio X/T is a measure of the maximum energy reaching the seismograph. The formula is

$$M = \log X/T + Y$$

where M is the Richter magnitude.

One Richter magnitude scale unit corresponds to a tenfold increase in X. But the energy increase is proportional to X^2, which is to say, a hundredfold. However, the duration of a single oscillation differs greatly from one earthquake to another. In particular, the most energetic earthquakes have a higher proportion of long-duration waves. As a result, the energy increase corresponding to one Richter scale unit increase, when summed over the whole range of waves in a wave record, is only a thirtyfold increase. Thus, the difference in energy released between an earthquake of magnitude 4 and one of magnitude 7 is $30 \times 30 \times 30 = 27,000$ times!

How big can earthquakes get? The largest recorded to date have Richter magnitudes of about 8.6, which means they release about as much energy as 10,000 atom bombs of the kind that destroyed Hiroshima at the end of World War II. It is possible that earthquakes do not get any larger than this because rocks cannot store more elastic energy. Before they are deformed further, they fracture and so release the energy.

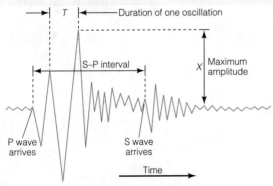

Y is a correction factor that depends on the distance of the seismograph from the epicenter. It is calculated from the S–P interval.

Figure C5.1 Measurements used for determining the Richter magnitude (*M*) from a seismograph record.

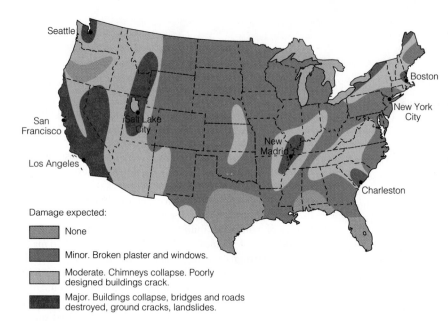

Damage expected:

- None
- Minor. Broken plaster and windows.
- Moderate. Chimneys collapse. Poorly designed buildings crack.
- Major. Buildings collapse, bridges and roads destroyed, ground cracks, landslides.

Figure 5.9 Seismic-risk map of the United States based on quake intensity. The map does not indicate earthquake frequency. For example, frequency is high in southern California but low in eastern Massachusetts. Nevertheless, when earthquakes do occur in eastern Massachusetts, they can be as severe as the more frequent quakes in southern California.

Based on known geological structures (mainly faults) and on the location and intensity of past earthquakes, the National Oceanographic and Atmospheric Administration prepared the seismic-risk map shown in Figure 5.9.

Earthquake Disasters

Every year the Earth experiences hundreds of thousands of earthquakes. Fortunately, only one or two either are large enough or close enough to major population centers to cause loss of life. Certain areas are known to be earthquake-prone, and special building codes in such places require structures to be as resistant as possible to earthquake damage. All too often, however, an unexpected earthquake will devastate an area where buildings are not adequately constructed, as in the T'ang Shan earthquake discussed at the beginning of this chapter. Other examples are the earthquake that destroyed parts of the center of Mexico City and killed 9500 people in 1985 (Fig. 5.10), and the one that struck in Armenia in 1988, killing an estimated 25,000 people (Fig. 5.11).

Historically, seventeen earthquakes are known to have caused 50,000 or more deaths apiece (Table 5.1). The most disastrous one on record occurred in 1556 in Shaanxi Province, China, where an estimated 830,000 people died. Many of those people lived in cave dwellings excavated in a soft, wind-deposited sediment called loess, which collapsed as a result of the quake. Since 1900, there have been 42 earthquakes worldwide in which 500 or more people have died.

Figure 5.10 A building that was not constructed to withstand expected earthquakes, the Hotel DeCarlo, was one of the buildings that collapsed during the earthquake that struck Mexico City in 1985. Proper building design can minimize damage. Nearby buildings of sturdier construction withstood the shaking.

Table 5.1 Earthquakes During the Past 800 Years
That Have Caused 50,000 or More Deaths

Place	Year	Estimated Number of Deaths
Silicia, Turkey	1268	60,000
Chihli, China	1290	100,000
Shaanxi, China	1556	830,000
Shemaka, Azerbaijan	1667	80,000
Naples, Italy	1693	93,000
Catania, Italy	1693	60,000
Beijing, China	1731	100,000
Calcutta, India	1737	300,000
Lisbon, Portugal	1755	60,000
Calabria, Italy	1783	50,000
Messina, Italy	1908	160,000
Gansu, China	1920	180,000
Tokyo and Yokohama, Japan	1923	143,000
Gansu, China	1932	70,000
Quetta, Pakistan	1935	60,000
T'ang Shan, China	1976	240,000
Iran	1990	52,000

Earthquake Damage

The dangers of earthquakes are profound, and the havoc they can cause is often catastrophic. Their effects are of six principal kinds. The first two, ground motion and faulting, are primary effects, and they cause damage directly. The other four effects are secondary and cause damage indirectly as a result of processes set in motion by the earthquake.

1. Ground motion results from the movement of seismic waves, especially surface waves, through surface-rock layers and soil. The motion can damage and sometimes completely destroy buildings. Proper design (including such features as a steel framework and a foundation tied to bedrock) can do much to prevent such damage, but in a very strong earthquake even the best buildings may suffer some damage.

2. Where a fault breaks the ground surface, buildings can be split, roads disrupted, and any feature that crosses or sits on the fault broken apart.

3. A secondary effect, but one that is sometimes a greater hazard than moving ground, is fire. Ground movement displaces stoves, breaks gas lines, and loosens electrical wires, thereby starting fires. Ground motion also breaks water mains, which usually means that there is no water available to put out fires. In the earthquakes that struck San Francisco in 1906 and Tokyo and Yokohama in 1923, more than 90 percent of the building damage was caused by fire.

4. In regions of steep slopes, earthquake vibrations may cause soil to slip and cliffs to collapse. This is particularly true in Alaska, parts of southern California, China, and hilly places such as Iran and Turkey. Houses, roads, and other structures are destroyed by rapidly moving soil flows.

5. The sudden disturbance of water-saturated sediment and soil can turn seemingly solid ground to a liquidlike mass of quicksand. This process is called

Figure 5.11 When a magnitude 6.8 earthquake struck Armenia on December 7, 1988, poorly constructed buildings with inadequate foundations collapsed like houses of cards. The principal cause of collapse was ground motion.

Figure 5.12 Gaping fissures in a residential area of Anchorage, Alaska, formed during the 1964 earthquake. The fissures result from liquefaction and failure of weak subsurface rocks.

liquefaction, and it was one of the major causes of damage during the earthquake that destroyed much of Anchorage, Alaska, on March 27, 1964 (Fig. 5.12), and the earthquake that caused apartment houses to sink and collapse in Niigata, Japan, that same year.

6. Finally, there are **seismic sea waves**, called **tsunami**. A tsunami is often erroneously called a "tidal" wave, but it has nothing to do with the tides. Tsunami are started by the sudden movement of the seafloor due to an earthquake and have been particularly destructive in the Pacific Ocean. Seismic sea waves travel at speeds up to 950 km/h (590 mi/h) and have wavelengths up to 200 km (124 mi). Amplitudes of the waves are so low they can rarely be seen in the ocean, but as a wave approaches the shore the water is piled up rapidly to heights of 30 m (100 ft) or more. About 5 h after a severe submarine earthquake near Unimak Island, Alaska, in 1946, for instance, a tsunami struck Hawaii. The wave had traveled at a speed of 800 km/h (500 mi/h). Although the amplitude of the wave in the open ocean was less than 1 m (1.1 yd), the amplitude increased dramatically as the wave approached land, so that when it hit Hawaii, the wave had a crest 18 m (60 ft) higher than normal high tide. This destructive wave demolished nearly 500 houses, damaged a thousand more, and killed 159 people.

Modified Mercalli Scale

Because damage to the land surface and to property is so important, the scale of earthquake-damage intensity (called the **modified Mercalli scale**) is based on the amount of vibration people feel during low-magnitude quakes and the extent of building damage during high-magnitude quakes. The correspondence between Mercalli intensity and Richter magnitude is listed in Table 5.2.

Table 5.2   Earthquake Magnitudes, Frequencies for the Entire Earth, and Damaging Effects

Richter Magnitude	Number per Year	Modified Mercalli Intensity Scale[a]	Characteristic Effects of Shocks in Populated Areas
<3.4	800,000	I	Recorded only by seismographs
3.5–4.2	30,000	II and III	Felt by some people who are indoors
4.3–4.8	4,800	IV	Felt by many people; windows rattle
4.9–5.4	1,400	V	Felt by everyone; dishes break, doors swing
5.5–6.1	500	VI and VII	Slight building damage; plaster cracks, bricks fall
6.2–6.9	100	VII and IX	Much building damage; chimneys fall; houses move on foundations
7.0–7.3	15	X	Serious damage, bridges twisted, walls fractured; many masonry buildings collapse
7.4–7.9	4	XI	Great damage; most buildings collapse
>8.0	One every 5–10 yr	XII	Total damage; waves seen on ground surface, objects thrown in the air

[a]Mercalli numbers are determined by the amount of damage to structures and the degree to which ground motions are felt. These depend on the magnitude of the earthquake, the distance of the observer from the epicenter, and whether an observer is in or out of doors.

EARTHQUAKE PREDICTION

Some of the most dreadful natural disasters have been caused by earthquakes. It is hardly surprising, therefore, that a great deal of research around the world focuses on earthquakes. The hope is that through research we will be able to improve our forecasting ability.

Because China has suffered so many terrible earthquakes, Chinese scientists have tried everything they can think of to predict quakes. They have even observed animal behavior, and on one occasion animals did successfully foretell a quake. In July 18, 1969, zookeepers at the People's Park in Tianjin observed highly unusual animal behavior. Normally quiet pandas screamed, swans refused to go near water, yaks did not eat, and snakes would not go into their holes. The keepers reported their observations to the earthquake prediction office, and at about noon on the same day a magnitude 7.4 earthquake struck.

There have been many informal reports of strange animal behavior before earthquakes, but the Tianjin quake is the only well-documented case. Unfortunately, most quakes do not seem to be preceded by anything odd. Although scientists haven't given up on animals, they measure lots of other things besides animal behavior.

Most research on earthquake prediction today is based on the properties of elastically strained rocks such as rock magnetism, electrical conductivity, and porosity. Even simple observations, such as the level of water in a well, might indicate a porosity change. Tilting of the ground or slow rises and falls in elevation may also indicate that strain is building up. Most significant are the small cracks and fractures that can develop in severely bent rock. These openings can cause swarms of tiny earthquakes called foreshocks that may be a clue that a big quake is coming. One of the most successful cases of earthquake prediction, made by Chinese scientists in 1975, was based on slow tilting of the land surface, on fluctuations in magnetism, and on the swarms of small foreshocks that preceded the 7.3 Richter magnitude quake that struck the town of Haicheng. Half the city was destroyed, but authorities had evacuated more than a million people before the quake. As a result, only a few hundred were killed.

In places where earthquakes are known to occur repeatedly, such as around the margins of the Pacific Ocean, geologists can sometimes discern recurrence patterns. If such a pattern suggests a recurrence interval of, say, a century, it may be possible to predict where and when a large quake may happen. Certainly it is possible to monitor such areas closely when a big quake is thought to be due. Furthermore, studies of recurrence patterns have identified a number of *seismic gaps* around the Pacific rim (Fig. 5.13). These are

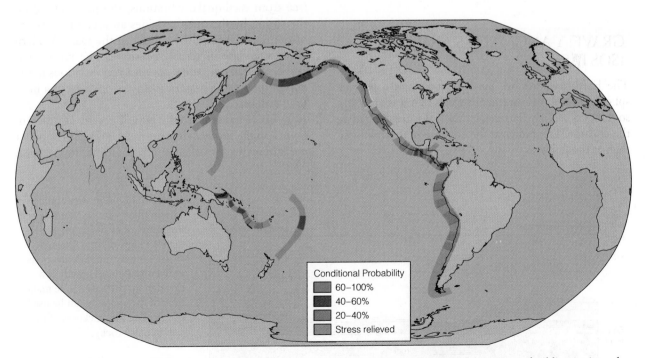

Conditional Probability
60–100%
40–60%
20–40%
Stress relieved

Figure 5.13 Seismic gaps in the circum-Pacific belt. In the areas indicated, earthquakes of magnitude 7.0 or greater are known to have occurred a long time ago but have not occurred in recent times. Strain is now building up in each seismic gap, raising the probability that a large quake will occur before the year 2000.

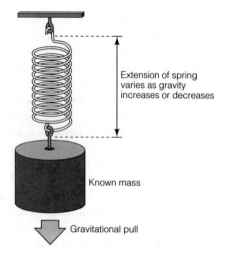

Figure 5.14 A gravimeter is a heavy piece of metal suspended on a sensitive spring. The weight exerts a greater or lesser pull on the spring as gravity changes from place to place, extending the spring more or less. The weight and spring are contained in a vacuum together with exceedingly sensitive measuring devices.

places where, for one reason or another, earthquakes have not occurred for a very long time and where elastic strain has been steadily increasing. Seismic gaps receive a lot of research attention because they are considered the places most likely to experience large earthquakes.

GRAVITY ANOMALIES AND ISOSTASY

The Earth *looks* round, but in fact it's not a perfect sphere; careful measurement reveals that it is an ellipsoid that is slightly flattened at the poles and bulged at the equator. The radius at the equator is 21 km longer than at the poles.

Because the gravitational pull between two objects is inversely proportional to the square of the distance between their centers of mass, the pull exerted by the Earth's gravity on a body at the Earth's surface is slightly greater at the poles (because there the body is closer to the center of the Earth) than at the equator. Recall from high school science that your weight is a measure of how strongly the Earth's gravitational force is pulling your body toward the center of the Earth. Thus, a man who weighs 90.5 kg at the North Pole would observe his weight decreasing to 90 kg simply by traveling to the equator. If the weight-conscious traveler made very exact measurements as he traveled, however, he would observe that his weight changed irregularly rather than smoothly. From this irregular change, he could conclude that the pull of gravity must change irregularly. If the traveler went one step further and carried a sensitive device called a *gravimeter* (or *gravity meter*) for measuring the pull of gravity at any locality, he would indeed find an irregular variation. From those irregular variations, a great deal of important information about the interior of the Earth can be deduced.

Gravity Anomalies

A gravimeter, which is similar to an inertial seismograph, consists of a heavy mass suspended by a sensitive spring (Fig. 5.14). When the ground is stable and free from earthquake vibrations, the pull exerted on the spring by the mass provides an accurate measure of the Earth's gravitational pull on the mass. Modern gravimeters are incredibly sensitive. The most accurate devices in operation can measure variations in the pull of the Earth's gravity as tiny as one part in a hundred million.

In order to compare the pull of gravity at different points on the Earth, scientists must correct gravimeter measurements for changes in latitude and topography.

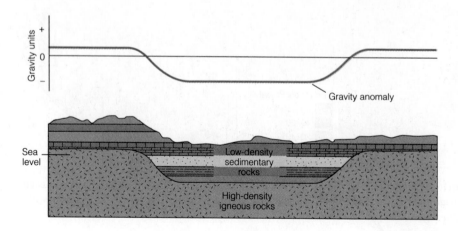

Figure 5.15 A gravity anomaly: a basin filled with low-density sedimentary rocks sitting on a basement of high-density igneous rocks. Gravity measurements reveal a pronounced gravity low throughout the basin. The magnitude of the anomaly can be used to calculate the thickness of the sedimentary rocks.

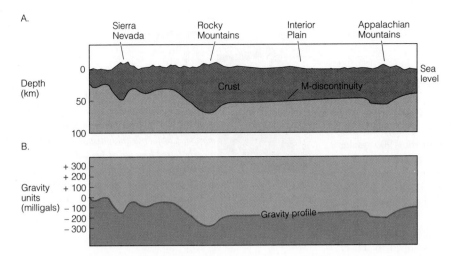

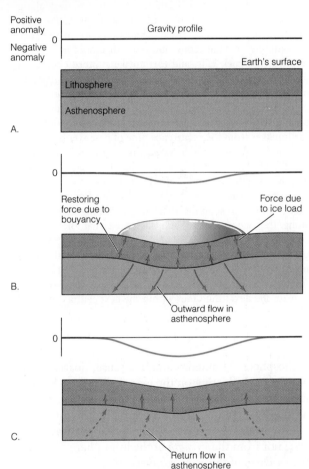

Figure 5.16 Profile of the crust beneath the United States. A. Thickness of the crust is determined from measurements of seismic waves. The crust is thicker beneath major mountain masses, such as the Sierra Nevada, the Rocky Mountains, and the Appalachians. B. A gravitational profile. There are distinct negative gravity anomalies over the Sierra, the Rockies, and the Appalachians due to the masses of low-density rocks that lie beneath these topographic highs.

The idea behind the corrections is to know the pull of gravity at a constant distance from the center of the Earth. Then, if the rock mass between the gravimeter and the center of the Earth were everywhere the same, the adjusted figures for the force of gravity might be expected to be the same at every place on the Earth. In fact, the adjusted figures reveal large and significant variations called **gravity anomalies.** Anomalies are due to bodies of rock having different densities, and a great deal of important information can be derived from them. A simple example of an anomaly is shown in Figure 5.15.

The thickness of the crust beneath the United States, as determined from seismic measurements of the moho, is shown in profile in Figure 5.16A. Beneath the three major mountain systems (the Appalachians, the Rockies, and the Sierra Nevada), the crust is thicker than in the nonmountainous regions of the country. The crust beneath the mountains resembles icebergs that have high peaks above the waterline but also massive roots below. The accuracy of this analogy is demonstrated by the gravity profile across the United States, shown in Figure 5.16B. Gravity measurements almost everywhere across the continent are less then expected considering the distance from the center of the Earth, and they are least, where the crust is thickest. We refer to such regions of low gravity pull as regions of negative gravity anomalies. The anomalies are caused by the masses of low-density rock that are the roots of the mountains, just as the basin of low-density sediment produces the gravity anomaly shown in Figure 5.15.

The reason why mountains stand so high on the landscape provides some interesting insights into the Earth's physical properties. Wherever a mountain range occurs, the lithosphere is locally thickened. Mountains stand high because they are made up of low-density rocks and the thickened lithosphere is supported by the buoyancy of the easily deformed as-

Figure 5.17 Depression of the lithosphere and asthenosphere by a continental ice sheet. A. Prior to formation of the ice sheet, there is no gravity anomaly. B. When the ice sheet forms, it depresses the lithosphere and the asthenosphere. At some depth in the asthenosphere, material must slowly flow outward to accommodate the sagging lithosphere. C. When the ice melts, buoyancy slowly restores the lithosphere and asthenosphere to their original levels. A negative gravity anomaly continues until the depression is removed. The viscosity of the asthenosphere controls the rate of flow and therefore the rate of recovery.

GUEST ESSAY

Rethinking Earthquake Prediction

Jeffrey Park is an associate professor of geology and geophysics at Yale University. His interest in earthquakes dates from the February 9, 1971 earthquake in San Fernando, California (magnitude 6.5), which nearly tossed him out of his bed in the predawn twilight. Professor Park studies the properties of the Earth's deep interior with earthquake waves, and the relation between mantle flow and plate tectonic motions.

Perhaps seismologists were unlucky that the great 1906 San Francisco earthquake was one of the first large seismic events to be studied thoroughly. Although the 1906 quake was probably a typical example of the largest earthquakes that occur along the San Andreas Fault, residents in California may face greater risks from the more numerous smaller earthquakes, whose behavior is only now becoming understood.

For most of this century, earthquake prediction has depended on the elastic rebound theory developed by H. F. Reid. Based largely on field studies of ground motion along the San Andreas Fault after the great 1906 earthquake, Reid's theory holds that strain gradually builds in the rock surrounding a fault zone until a "threshold strain" is reached, the rock fails, and earthquake rupture occurs. In the 1960s plate tectonics theory allowed geophysicists to predict the rate at which relative motion occurs along plate boundaries. If the threshold-strain idea of the elastic rebound theory is correct, one can predict the timing of future earthquakes from the recurrence time of past events.

The San Andreas Fault system divides the North American Plate from the Pacific Plate, extends nearly the entire length of California, and is positioned to pose a serious hazard to the urban areas of San Francisco and Los Angeles. It came as no surprise in the 1970s that trenching experiments along the San Andreas Fault suggested that there was a characteristic interval (125 to 225 years) between Magnitude 8+ earthquakes along the southern part of the fault near Los Angeles. Because the last such earthquake near Los Angeles occurred in 1857 near Fort Tejon, seismologists anticipate the next "Big One" within the next century. Around

the world, seismologists forecasted earthquakes in "seismic gaps," locations here plate tectonic theory predicts movement, but where no large earthquake has occurred in decades or centuries.

Still, this does not offer us "short-term" earthquake prediction, which would warn residents and public officials that a damaging earthquake is likely to occur in years or months, not decades. In an ambitious short-term earthquake prediction experiment, a large concentration of geophysical instruments was installed near Parkfield, California, to measure ground motion, tilt, strain, water level, and other suspected earthquake precursors, or phenomena that might help predict an earthquake in the short term. Parkfield was selected for this experiment because it lies beside a special segment of the San Andreas Fault, where the first break of the 1857 Fort Tejon earthquake was thought to have occurred. Since that event, earthquakes of magnitude 6 or so have occurred in the Parkfield segment in 1881, 1901, 1922, 1934, and 1966, suggesting a regular recurrence time of 20 to 25 years. Seismologists extrapolated the sequence to forecast another magnitude 6 earthquake in 1988, with an esti-

---

thenosphere. Mountains are, in a sense, floating. It is not the crust floating on the mantle, however. Rather, it is the lithosphere (all of the crust plus the uppermost part of the mantle) that floats on the asthenosphere. Strange as it may seem, topographic variations arise not from the *strength* of the lithosphere but from the *weakness* of the asthenosphere.

Isostasy

The flotational balance among segments of the lithosphere is referred to as **isostasy**. The great ice sheets of the last ice age provide an impressive demonstration of isostasy. The weight of a large continental ice sheet, which may be 3 to 4 km thick, will depress the lithosphere. When the ice melts, the land surface slowly rises again. The effect is very much like pushing a block of wood into a bucket of thick, viscous oil.

When you stop pushing, the wood slowly rises to an equilibrium position determined by its density. The speed of its rising is controlled by the viscosity of the oil. Glacial depression and rebound of the lithosphere mean that rock in the asthenosphere must flow laterally when the ice depresses the lithosphere, and then must flow back again when the ice melts away (Fig. 5.17). From the fact that the land surface in parts of northeastern Canada and Scandinavia is still rising, even though most of the ice that covered these areas during the last ice age had melted away 7000 years ago, we infer that the flow must be slow and therefore that the asthenosphere must be extremely viscous.

Continents and the mountains on them are composed of low-density rock, and they stand high because they are thick and light; ocean basins are topographically low because the oceanic crust is composed of denser rock. Isostasy and the fact that the continen-

mated five-year margin of error. Nature's repeatability failed in this case, however, as no earthquakes as large as magnitude 5 occurred near Parkfield in the ten-year period that ended December 30, 1992, when the Parkfield "prediction window" closed. Elastic rebound theory, at least in its simplest form, had failed.

This was not the only challenge to the assumptions of would-be earthquake predictors. In the ten-year Parkfield prediction window, several earthquakes in central and Southern California forced seismologists to rethink their hazard assessments. Plate boundaries within continents are rarely sharp, and the associated deformation can be spread in a zone more than a hundred kilometers wide. As part of this complicated deformation, geologists estimate that the Los Angeles basin, south of the San Andreas Fault, is being compressed by 7 mm per year. This roughly north–south shortening leads to motion along thrust faults within the basin, which gives rise to the east–west-trending ranges of hills that partition the Los Angeles metropolitan area. In 1987 a magnitude 5.9 earthquake struck Whittier, California, within the Los Angeles metropolitan area and far south of the San Andreas Fault, along a thrust fault more than 10 km beneath the Montebello Hills. Serious property damage was caused in a localized region near the fault. In 1983 a similar, but larger (magnitude 6.5), earthquake devastated the town of Coalinga in central California. On January 17, 1994, yet another such earthquake (magnitude 6.6) ruptured under Northridge, a Los Angeles suburb, causing property damage estimated at over $15 billion.

The type of faulting in these events raised concern. Rupture occurred along "blind-thrust" faults, where a dipping fault surface lies within deeply buried sedimentary rock. Such faults have no expression on the surface, aside from an anticlinal upwarp of the rock layers above the fault. Blind thrusts have estimated recurrence intervals of thousands rather than hundreds of years, so they give little indication of their destructive potential in the historical record. In fact, the concept of recurrence interval may be inappropriate for blind-thrust earthquakes, for faulting on these geological structures may depend on the interaction of many tectonic movements within the Los Angeles basin. Scientists are now mapping blind-thrust faults in the basin with equipment and techniques similar to those used to discover petroleum and natural gas deposits.

On June 28, 1992, the San Andreas Fault was upstaged by a magnitude 7.3 earthquake near Landers, California, nominally within the North American Plate in the Mojave Desert. The largest earthquake in California since the 1906 San Francisco quake, the Landers quake ruptured a string of surface faults that previously were not thought to be linked. One dramatic result from the Landers event is the precise measurement of its total surface motion using Global Positioning System (GPS) satellite receivers. Under favorable conditions, a GPS receiver can determine its absolute position with an accuracy better than a centimeter by referencing itself to Earth-orbiting satellites. GPS-receiver networks in principle can measure the strain buildup and release associated with major earthquakes. If large-scale flexing of the crust along a fault zone occurs prior to large earthquakes, these types of measurements may provide seismologists with early warning signs. Seismologists hope to understand the earthquake process better by combining these new types of observations with traditional earthquake studies, thereby making short-term earthquake prediction more realistic.

---

tal crust is less dense than the oceanic crust are the reason the Earth has continents and ocean basins.

The important point to be drawn from this discussion of isostasy is that the lithosphere acts as if it were "floating" on the asthenosphere. (*Floating* is not exactly the correct word because the Earth is solid, but the lithosphere is buoyant and acts as though it were floating.) Sometimes gravity measurements suggest that a mountain has been pushed up so rapidly it is top-heavy and has too little root of low-density rock to counterbalance its upper mass. Sometimes it is observed that low-density crust has been dragged down so rapidly that it forms a root without a mountain mass above it. These and many other situations lead to local gravity anomalies. That the anomalies do not become very large suggests that the Earth is always moving toward an isostatic balance. Indeed, isostasy is the principal explanation for vertical motions of the Earth's surface, just as plate tectonics is the principal explanation for lateral motions.

The importance of the asthenosphere in determining the shape of the Earth's surface cannot be overemphasized. The asthenosphere has the properties it does because the balance between temperature and pressure is such that rock in the asthenosphere is very close to melting. That, apparently, is why the asthenosphere is so weak and why seismic wave speeds in the asthenosphere are low. The asthenosphere is also vitally important for another reason—wherever temperatures exceed rock melting temperatures, magma (molten rock) is formed in the asthenosphere. The magma rises to the surface and spews forth from volcanoes. The formation and eruption of magma are a vital aspect of the Earth's dynamic system because new rock is slowly, but continuously, added to the Earth's surface from deep in the interior.

SUMMARY

1. Abrupt movement of faults that release elastically stored energy is thought to cause earthquakes: this is known as the elastic rebound theory.

2. Earthquake vibrations are called seismic waves and are measured with seismographs.

3. Energy released at an earthquake's focus radiates outward as two kinds of body waves: P waves (*Primary* waves, which are compression expansion waves) and S waves (*Secondary* waves, which are shear waves). Earthquake energy also causes the surface of the Earth to vibrate because of surface waves.

4. From the study of seismic-wave refraction and reflection, scientists infer the internal structure of the Earth by locating discontinuities in its composition and physical properties. A pronounced compositional boundary occurs between the mantle and the outer core.

5. At the mantle/crust interface is a pronounced seismic discontinuity called the Mohorovičić discontinuity. Crust thickness ranges from 30 to 70 km in continental regions but is only 8 km beneath oceans.

6. The core has a high density and is inferred to consist of iron plus small amounts of nickel and other elements.

The outer core must be molten because it does not transmit S waves. The inner core is solid.

7. From a depth of 100 km to 350 km is a zone of low seismic-wave speed. This low-speed zone is the asthenosphere. The lithosphere, which is rigid and on average 100 km thick, overlies the asthenosphere and consists of the upper part of the mantle and all of the crust. Below the asthenosphere is the mesosphere, a zone of higher seismic wave speed than in the asthenosphere.

8. The amount of energy released during an earthquake is calculated on the Richter magnitude scale.

9. Earthquakes cause damage in six different ways: ground motion; faulting, fire, land movement and slope collapse, liquefaction, and by tsunami.

10. The outer portions of the Earth are in approximate isostatic balance; in other words, like huge icebergs floating in water, the lithosphere "floats" on the asthenosphere.

11. When parts of the lithosphere are not in flotational equilibrium, gravity anomalies occur.

IMPORTANT TERMS TO REMEMBER

body wave *94*
earthquake focus *92*
elastic deformation *92*
elastic rebound theory *92*
epicenter *92*
fault *92*

gravity anomaly *105*
isostasy *106*
M-discontinuity *97*
Modified Mercalli scale *102*
moho *97*

Mohorovičić discontinuity *97*
P (primary) wave *95*
Richter magnitude scale *99*
S (secondary) wave *95*

seismic sea wave *102*
seismic wave *94*
surface wave *94*
tsunami *102*

QUESTIONS FOR REVIEW

1. Explain how most earthquakes are thought to occur and why there seems to be a limit on earthquake magnitude.

2. What is the relation between an earthquake focus and the corresponding epicenter?

3. How are seismic waves recorded and measured? How would you locate an epicenter from seismic records? Explain how a focus depth is determined.

4. What are the differences between seismic body waves and surface waves? Identify two kinds of body waves and explain the differences.

5. Earthquakes can cause damage in many ways; name four. Where on the Earth was the most disastrous

earthquake on record, and how did the people die? Where was the biggest known earthquake in the United States?

6. What are reflection and refraction, and how do they affect the passage of seismic waves? How can refraction and reflection be used to define the mantle–crust boundary? The core–mantle boundary?

7. Briefly describe how seismic waves can be used to infer that the outer core is molten while the inner core is solid. What evidence indicates that the core is made largely of metallic iron?

8. Describe the Earth's three compositional layers.

9. The Earth is layered with respect to rock strength into five zones. Name and describe these zones.

10. What is the relationship between the crust, the mantle, and the lithosphere?

11. Why are ocean basin places of low elevation on the Earth's surface and continents are places of high elevation?

12. How do gravity anomalies arise and how can they be measured?

13. Describe some evidence that proves that isostasy is operating in the Earth. How is the Earth's surface topography related to isostasy?

14. Draw an east-west profile of the crust under the United States and indicate how isostasy plays a role in what you have drawn.

Questions for A Closer Look

1. How much energy can be stored in a cubic meter of elastically strained rock? How much energy can be released during a single big earthquake event?

2. Explain how seismologists use the Richter magnitude scale to estimate the energy released during an earthquake.

3. Why is the Richter magnitude scale logarithmic, and how big are the steps between magnitudes?

QUESTIONS FOR DISCUSSION

1. The lithosphere of the Moon is about 1000 km thick, and the asthenosphere is only about 380 km thick. Would you expect isostasy to operate on the Moon? Could your hypothesis be tested by measuring gravity anomalies with an orbiting space ship? Might gravity anomalies be something to measure if a space ship were sent to investigate planets around some other sun?

2. Research the current work being done on earthquake prediction. How closely in time would a prediction of an earthquake have to be in order to be useful—a few hours; a few days; a few weeks? Be sure to explain your reasoning.

 Our web site will provide you with an interactive case study about earthquake hazards.

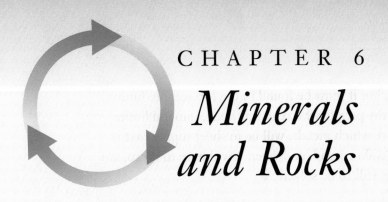

CHAPTER 6

Minerals and Rocks

● *Minerals: A Linchpin of Society*

Most minerals that are abundant in the Earth's crust have neither commercial value nor any particular use. Those few minerals that are the raw materials of industry, which we call *ore minerals*, tend to be rare and hard to find—gold, for instance, or sphalerite, the main zinc mineral. From the ore minerals we get the metals to make our machines and the ingredients for chemicals and fertilizers. Our modern society is totally dependent on an adequate supply of ore minerals. Without them we could not build planes, cars, televisions, or computers. Industry would falter and living standards would decline.

Can the ore minerals in the Earth's crust sustain both a growing population and a high standard of living for everyone? This difficult question has many experts puzzled. The minerals they think most about are those used as sources of such important metals as lead, zinc, and copper. Metals, the experts point out, begin the chain of resource use. Without metals, we cannot make machines. Without machines we cannot convert the chemical energy of coal and oil to useful mechanical energy. Without mechanical energy, the tractors that pull plows must grind to a halt; trains and trucks must stop running; and indeed our whole industrial complex must become still and silent.

Experts cannot tell how long the Earth's supplies of ore minerals will last because there is no way to "see" inside the Earth and know exactly what is there. Optimists point to the great success our technological society has enjoyed over the past two centuries as ever more remarkable discoveries have been made. Improved prospecting will keep up the success story, they insist. However, if mineral supplies do become limited, the experts suggest we will find ways to get around the limits by recycling, by substitution, and by the discovery of new technologies.

Many scientists have a more pessimistic opinion. Technologically advanced societies have faced mineral resource limits in the past, they point out, but the solution has always been to import new supplies rather than develop substitutes or effective recycling measures. England, for instance, was once a great supplier of metals. Today, the minerals are mined out, most of the mines are closed, and English industry runs on raw materials imported from abroad. The United States, too, was once self-sufficient in most minerals and an exporter of many. Slowly the United States has become a net importer and now relies on supplies from such countries as Australia, Chile, South Africa, and Canada. Today no large industrial country can supply its own mineral needs. The only country that might be able to do so is Russia. But eventually the Russian mines will be depleted, and so too will the mines of Australia and other countries. Where then does society turn?

(opposite) Carving up a mountain to recover a valuable metal. An aerial view of an open pit gold mine, in Montana.

The answer to this question is not obvious, but it must be found in the foreseeable future. It is highly likely that, within the lifetimes of the people who read this book, mineral limitations will occur. Which minerals, and therefore which metals, will be in short supply first is still an open question. How society will cope and respond, and when it will have to do so, are just two of the great social and scientific issues still to be solved. ●

MINERALS AND THEIR CHEMISTRY

The word "mineral" means different things to different people, but for scientists who study the Earth it has a very specific connotation. A **mineral** is any naturally formed, solid chemical element or chemical compound having a definite composition and a characteristic crystal structure. To take just one example, diamond is a mineral. It is naturally formed, it is a solid, it is made of the chemical element carbon, and all the carbon atoms are packed together in a regular and characteristic geometric array called the crystal structure of diamond.

Coal resembles diamond in that it is largely carbon, but coal is not a mineral; it is a rock. In addition to its carbon, coal contains many chemical compounds, and its composition varies from sample to sample so that it does not have a specific composition. Nor does coal have a characteristic crystal structure.

Rocks are aggregates of minerals. They are nature's books and in them is recorded the story of the way the Earth works: how continents move, how mountains form and then erode. Minerals are the words used in nature's books. Minerals and rocks go together naturally, but before we can study rocks, we need to know something about minerals, and the most direct way to introduce minerals is through an examination of their two most important characteristics:

1. Composition, which is the kinds of chemical elements present and their proportions; and
2. Crystal structure, which is the way in which the atoms of the elements are packed together in a mineral.

Because most minerals contain several chemical elements, we commence our discussion by reviewing the way in which elements combine to form compounds.

Elements and Atoms

Chemical elements are the most fundamental substances into which matter can be separated by ordinary chemical means. For example, table salt is not an element because it can be separated into sodium and chlorine. Neither sodium nor chlorine can be further broken down chemically, however, and so each is an element.

Each element is identified by a symbol, such as H for hydrogen and Si for silicon. Some symbols, such as that for hydrogen, come from the element's English name. Other symbols come from other languages. For example, iron is Fe from the Latin *ferrum*, and sodium is Na from the Latin *natrium*. All of the naturally occurring elements and their symbols, together with all of the man-made elements, are listed in Appendix B.

Even the tiniest piece of a pure element consists of a vast number of identical particles of that element called atoms. An **atom** is the smallest individual particle that retains the distinctive properties of a given chemical element. Atoms are so tiny they can be seen only by using the most powerful microscopes ever invented, and even then the image is imperfect because individual atoms are only about 10^{-10} m in diameter.

Atoms are built up from *protons* (which have positive electrical charges), *neutrons* (which, as their name suggests, are electrically neutral), and *electrons* (which have negative electrical charges that balance exactly the positive charges of protons). Protons and neutrons join together to form the core, or *nucleus*, of an atom. Electrons are much smaller than protons or neutrons; in an atom they move in a distant and diffuse cloud around the nucleus (Fig. 6.1).

Protons give a nucleus a positive charge, and the number of protons in the nucleus of an atom is called the *atomic number* of the atom. The number of protons in the nucleus (in other words, the atomic number) is what gives the atom its special characteristics and what makes it a specific element. Thus, any and all atoms containing one proton in the nucleus are atoms of hydrogen; atoms containing two protons in the nucleus are helium; and so on. All atoms having the same atomic number are atoms of the same element. The atomic numbers of all elements are listed in Appendix B.

Because neutrons are electrically neutral, they cannot change the atomic number of an element. Neutrons can change the mass of an atom, however, and the sum of the neutrons plus protons in the nucleus of

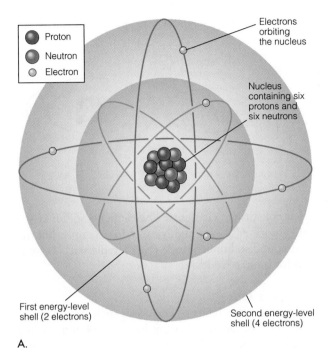

A.

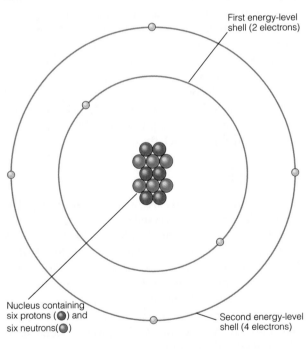

B.

Figure 6.1 Schematic diagram of an atom of the element carbon. The nucleus contains six protons and six neutrons. Electrons orbiting the nucleus are confined to specific orbits called energy-level shells. A. Three-dimensional representation showing the first energy-level shells. The first shell can contain two electrons, the second eight. B. Two-dimensional representation of the carbon atom to show the number of protons and neutrons in the nucleus and the number of electrons in the energy-level shells. The first energy-level shell is full because it contains two electrons. The second shell contains four electrons and so is half full.

an atom is the *mass number*. As we learned in Chapter 3, where the processes of element manufacture in stars was discussed, the number of protons in the nucleus (the atomic number) is designated by a subscript before the chemical symbol, whereas the sum of the protons plus neutrons (the mass number) is indicated by a superscript: helium is written $^4_2$He. Most elements have several **isotopes**; these are atoms with the same atomic number and hence the same chemical properties, but different mass numbers. Carbon, for example, has three naturally occurring isotopes, $^{12}_6$C, $^{13}_6$C, and $^{14}_6$C. Some isotopes are radioactive, which means they transform spontaneously to another isotope of the same element or to an isotope of a different element. Among the carbon isotopes only $^{14}_6$C is radioactive, and it transforms to an isotope of nitrogen, $^{14}_7$N by the spontaneous transformation of a neutron to a proton. There are 25 naturally occurring radioactive isotopes.

Energy-Level Shells

Electrons have specific energy levels, and we refer to the levels as **energy-level shells**. The maximum number of electrons that can have the energy characteristic of a given energy-level shell is fixed. As shown in Figure 6.1, energy-level 1 can only accommodate 2 electrons; level 2, however, can accommodate up to 8 electrons; level 3, 18; and level 4, 32.

Ions

An energy-level shell filled with electrons is very stable and is like an evenly loaded boat. To fill their outermost energy-level shell and so reach a stable configuration, atoms either share or transfer electrons among themselves. In its natural state, an atom is electrically neutral because the positive electrical charge of its protons is exactly balanced by the negative electrical charge of its orbiting electrons. When an electron is transferred as part of the stabilizing of energy-level shells, this balance of electrical forces is upset. An atom that loses an electron has lost a negative electrical charge and is left with a net positive charge. An atom that gains an electron has a net negative charge. An atom that has excess positive or negative charges caused by electron transfer is called an **ion**. When the excess charge is positive (meaning that the atom gives up electrons), the ion is called a **cation**; when negative (meaning an atom adds electrons), the ion is an **anion.**

The most convenient way to indicate ionic charges is with superscripts. For example, Ca^{2+} is a cation (calcium) that has given up two electrons, and F^- is an anion (fluorine) that has accepted an electron.[1] Because the formation of ions involves only the energy-

---

[1]Note that when the ionic charge is 1, we omit the number. The symbol F^- means F^{1-}, and Li^+ means Li^{1+}. Most atoms are present in the Earth as ions rather than electrically neutral atoms.

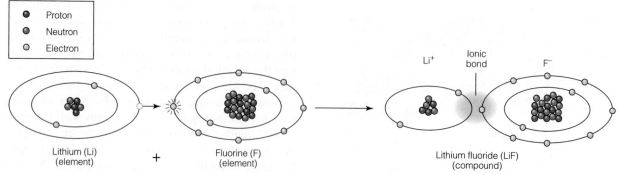

Figure 6.2 To form the compound lithium fluoride, an atom of the element lithium combines with an atom of the element fluorine. The lithium atom transfers its lone outer-shell electron to fill the fluorine atom's outer shell, creating an Li^+ cation and a F^- anion in the process. The electrostatic force that keeps the lithium and fluorine ions together in the compound lithium fluoride is an ionic bond.

level shell electrons and not the protons and neutrons in the nucleus, it is common practice to omit the atomic and mass number symbols for reactions involving ions.

Compounds

Chemical compounds form when one or more kinds of anion combine with one or more kinds of cation in a specific ratio. For example, hydrogen cations (H^+) combine with oxygen anion (O^{2-}) in the ratio 2:1, making the compound H_2O (water). In a compound, the sum of the positive and negative charges must be zero.

The formula of a compound is written by putting the cations first and the anions second. The numbers of cations or anions are indicated by subscripts, and for convenience the charges of the ions are usually omitted. Thus, we write H_2O rather than $H_2^+O^{2-}$.

Figure 6.2 presents an example of the way electron transfer leads to formation of a compound for the elements lithium and fluorine. A lithium atom has energy-level 1 occupied by two electrons but has only one electron in level 2, even though level 2 can accommodate eight electrons. The lone electron in level 2 can easily be transferred to an element such as fluorine, which already has seven electrons in level 2 and needs only one more to be completely filled. In this fashion, both a lithium cation and a fluorine anion end up with filled shells, and the resulting positive charge on the lithium and negative charge on the fluorine draw, or bond, the two ions together.

Lithium and fluorine form the compound lithium fluoride, which is written LiF to indicate that for every Li ion there is one F ion. The combination of one Li ion and one F ion is called a molecule of lithium fluoride. A *molecule* is the smallest unit that retains all the properties of a compound. Properties of molecules are quite different from the properties of

their constituent elements. The elements sodium (Na) and chlorine (Cl) are highly toxic, for example, but the compound sodium chloride (NaCl, table salt) is essential for human health.

Complex Ions

Sometimes two kinds of ions form such strong bonds with each other that they act like a single ion. Such a strongly bonded pair is called a *complex ion*. Complex ions act in the same way as single ions, forming compounds by bonding with other ions of opposite charge. For example, carbon and oxygen combine to form the complex carbonate anion (CO_3)$^{2-}$. The carbonate anion then bonds with cations such as Na^+ and Ca^{2+} to form compounds such as Na_2CO_3 and $CaCO_3$. Some other important complex ions in nature are the sulfate (SO_4)$^{2-}$, phosphate (PO_4)$^{3-}$, nitrate (NO_3)$^-$ and silicate (SiO_4)$^{4-}$ anions.

Crystal Structure

The ions in most solids are organized in the regular, geometric patterns of a crystal structure, like eggs in a carton, as shown in Figure 6.3. Solids that have such a crystal structure are said to be *crystalline*. Solids that lack a crystal structure are *amorphous* (Greek for "without form"). Glass and amber are examples of amorphous solids. All minerals are crystalline, and the crystal structure of a mineral is a unique property of that mineral. All specimens of a given mineral have identical crystal structure.

Before proceeding, let's review what is meant by the term *mineral*. To be called a mineral, a substance must meet four requirements:

1. It must be *naturally formed.* This excludes the vast numbers of substances that can be produced in the laboratory but are not found in nature.

2. It must be a *solid.* This excludes all liquids and gases.

3. It must have a *specific chemical composition.* This excludes solids, like glass, that have a continuous composition range that cannot be expressed by an exact chemical formula. This requirement for a

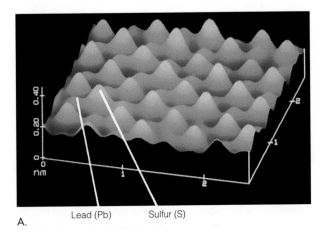

Lead (Pb)   Sulfur (S)

A.

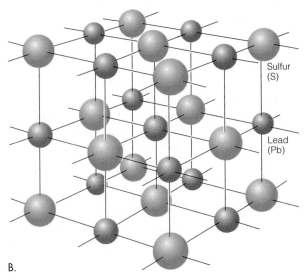

Sulfur
(S)

Lead
(Pb)

B.

Figure 6.3 The arrangement of ions in the most common lead mineral, galena (PbS). Lead, the Pb part, is a cation with a charge of $2+$, and sulfur, S, is an anion with a charge of $2-$. To maintain a charge balance between the ions, there must be an equal number of Pb and S ions in the structure. Ions are so small that a cube of galena 1 cm on its edge contains 10^{22} ions each of lead and sulfur. A. Ions at the surface of a galena crystal revealed with a scanning-tunneling microscope. Sulfur ions are the large lumps, lead the smaller ones. B. The packing arrangement of ions is repeated continuously through a crystal. The ions are shown pulled apart along the black lines to demonstrate how they fit together.

specific compound means that minerals are either chemical compounds or chemical elements.

4. It must have a *characteristic crystal structure*. This excludes amorphous materials.

Common Minerals

Scientists have identified approximately 3500 minerals, and the number is rising slowly because a few new ones are found every year. Most occur in the rocks of the continental crust, but a few have been identified only in meteorites, and two new ones were discovered in the Moon rocks brought back by the astronauts. The total number of minerals may seem large, but it is tiny compared with the astronomically large number of ways a chemist can combine naturally occurring elements to form compounds. The reason for the disparity between nature and chemical experiment becomes apparent when we consider the relative abundances of the chemical elements. As Table 6.1 shows, only 12 elements occur in the continental crust in amounts equal to or greater than 0.1 percent. Together these 12—usually referred to as the abundant elements, with all others referred to as the scarce elements—make up 99.23 percent of the continental crust mass. The continental crust is constructed, therefore, of about 40 minerals, most of which contain one or more of the 12 abundant elements. Notice in Table 6.1 that with two exceptions, carbon and nitrogen, the elements essential for life are abundant elements—these are oxygen, phosphorus, hydrogen, and potassium. Both carbon and nitrogen are almost abundant elements, which is fortunate, because it means that life on the Earth is not limited by a supply of the key elements.

Minerals containing scarce elements do occur, but only in small amounts, and those small amounts form only under special and restricted circumstances. Some scarce elements, such as hafnium, are so rare that they are not known to form minerals under any circumstances; they occur only as trace impurities in common minerals.

As Table 6.1 shows, two elements, oxygen and silicon, make up more than 70 percent of the continental

Table 6.1 The Most Abundant Chemical Elements in the Continental Crust

Element	Ion	Percent by Weight
Oxygen (O)	O^{2-}	45.20
Silicon (Si)	Si^{4+}	27.20
Aluminum (Al)	Al^{3+}	8.00
Iron (Fe)	Fe^{2+} and Fe^{3+}	5.80
Calcium (Ca)	Ca^{2+}	5.06
Magnesium (Mg)	Mg^{2+}	2.77
Sodium (Na)	Na^+	2.32
Potassium (K)	K^+	1.68
Titanium (Ti)	Ti^{4+}	0.86
Hydrogen (H)	H^+	0.14
Manganese (Mn)	Mn^{2+} and Mn^{4+}	0.10
Phosphorus (P)	P^{3+}	0.10
All other elements		0.77
Total		100.00

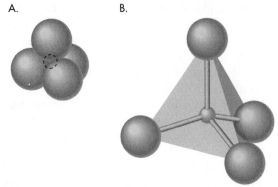

A.   B.

Figure 6.4 The tetrahedron-shaped silicate anion SiO_4^{4-}. A. Anion with the four oxygens touching each other in natural position. Silicon (dashed circle) occupies central space. B. Exploded view showing the relatively large oxygen anions at the four corners of the tetrahedron, equidistant from the relatively small silicon cation.

crust. Oxygen forms a simple anion, O^{2-}, and compounds that contain the O^{2-} anion are called oxides. Silicon forms a simple cation, Si^{4+}, and oxygen and silicon together form a strong complex ion, the **silicate anion** $(SiO_4)^{4-}$. Minerals that contain the silicate anion are complex oxides, and to distinguish them from simple oxides they are called **silicates**. The compound MgO is an oxide, but Mg_2SiO_4 is a silicate.

Figure 6.5 Polymerization of complex silicate anions. A. A polymer chain in which each silicate anion shares two of its oxygens with adjacent anions. A geometric representation of the chain is on the right. The formula of each basic unit in the chain is $(SiO_3)^{2-}$. B. Double polymer chain for which the formula of the basic unit is $(Si_4O_{11})^{6-}$.

Silicates are the most abundant of all minerals, and simple oxides are the second most abundant group. Other mineral groups, all of them important but all less common than silicates and oxides, are sulfides, which contain the simple anion S^{2-}, carbonates $(CO_3)^{-2}$, sulfates $(SO_4)^{-2}$, and phosphates $(PO_4)^{-3}$.

The Silicates

The silicate anion $(SiO_4)^{4-}$ has the shape of a tetrahedron. The four relatively large oxygen anions surround and bond to the much smaller silicon cation as shown in Figure 6.4. All silicates contain the silicate anion as an integral part of the crystal structure. In many silicates, however, the anions actually join together by sharing their oxygens and so form chains, sheets, and three-dimensional networks of tetrahedra (the process is called polymerization). How this is done is shown in Figures 6.5 and 6.6. Polymerization plays a major role in determining the properties of silicates.

The silicates are by far the most abundant minerals in the continental crust, and among them the feldspars are the predominant variety—approximately 60 percent of all minerals in the Earth's crust are feldspars (Fig. 6.7). Indeed, the very name reflects how common feldspars are. The name is derived from two Swedish words, *feld* (field) and *spar* (mineral). Early Swedish miners were familiar with feldspar in their mines. The miners were also farmers, and they found the same minerals in the rocks they had to clear from their fields before they could plant crops. Struck by the abundance of feldspar, the miners chose a name

A.

B.

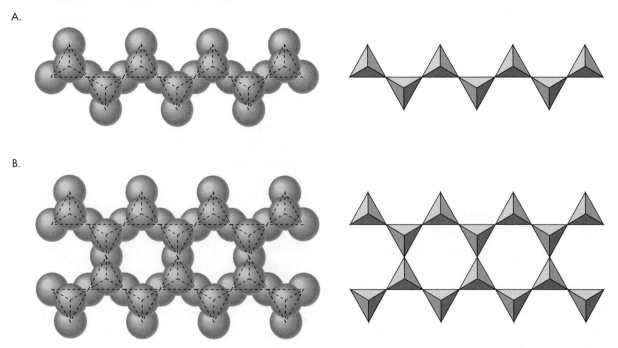

Mineral	Formula	Cleavage	Structure
Olivine	$(Mg,Fe)_2SiO_4$	None	Isolated tetrahedra 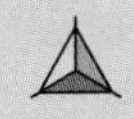
Garnet	$Mg_3Al_2(SiO_4)_3$	None	
Pyroxene	$CaMg(SiO_3)_2$	Two planes at 90°	Chain of tetrahedra 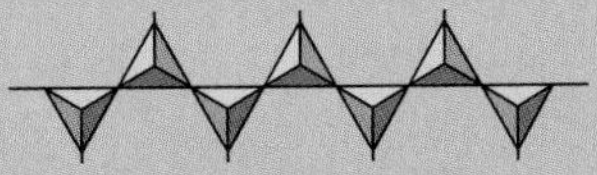
Amphibole	$Ca_2Mg_5(Si_4O_{11})_2(OH)_2$	Two planes at 120°	Double chain of tetrahedra 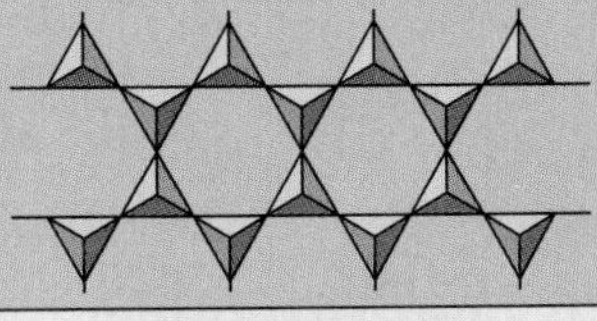
Mica	$KAl_2(Si_3Al)O_{10}(OH)_2$	One plane	Sheet of tetrahedra 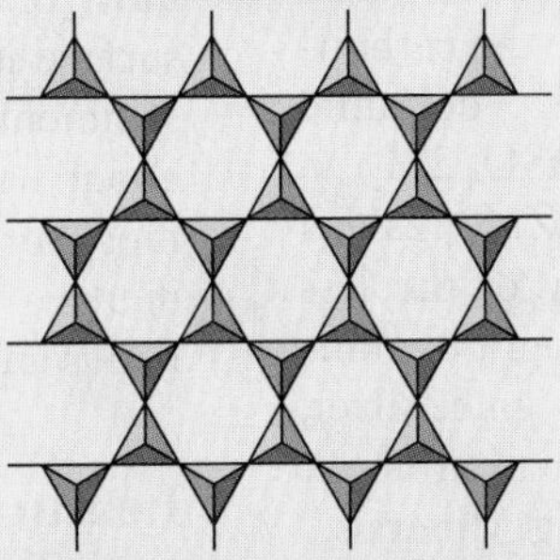
Clay	$Al_4Si_4O_{10}(OH)_8$		
Feldspar	$KAlSi_3O_8$	Two planes at 90°	Three-dimensional network too complex to be shown by a two-dimensional drawing
Quartz	SiO_2	None	

Figure 6.6 Summary of the way silicate anions polymerize to form the common silicate minerals. The most important polymerizations are those that produce chains, sheets, and three-dimensional networks. Note the relationship between crystal structure and cleavage. (Cleavage is discussed in "A Closer Look: Identifying Minerals" on page 120).

Figure 6.7 The two most common minerals in the Earth's crust. Crystals of feldspar (green) and quartz (gray) from Pikes Peak, Colorado. This specimen is 20 cm across.

to indicate that their fields seemed to be growing an endless crop of the minerals.

The second most abundant mineral in the crust is the silicate called quartz. Feldspar and quartz together account for 75 percent of the continental crust. All the silicates added together make up 95 percent or more of both the continental crust and the oceanic crust, and an even larger percentage of the mantle.

The Nonsilicates

The nonsilicate minerals are widespread and may at first sight be thought to be more abundant than they actually are. Three oxides of iron—hematite (Fe_2O_3), magnetite (Fe_3O_4), and goethite ($FeO \cdot OH$)—are estimated to be the most abundant nonsilicates. Other important non silicate mineral groups are the carbonate minerals, calcite ($CaCO_3$) and dolomite ($CaMg(CO_3)_2$), the sulfate gypsum ($CaSO_4 \cdot 2H_2O$), and the sulfides pyrite (FeS_2), sphalerite (ZnS), galena (PbS), and chalcopyrite ($CuFeS_2$). Many of the less common nonsilicates are the minerals sought by miners for the production of metals such as gold, silver, iron, copper, and zinc. For more information about minerals, see "A Closer Look: Identifying Minerals." Here you will find a discussion of mineral properties, how minerals are identified by using those properties, and tables of the properties of common minerals.

ROCKS

The definition of a rock, given in Chapter 1, is any naturally formed, nonliving, firm, and coherent aggregate mass of solid matter that constitutes part of a planet. The word "mineral" does not appear in the definition because rocks can be made of materials that are not minerals, such as natural glass (in the rock called obsidian), or bits of organic matter (in the rock called coal). Nevertheless, most rocks are made either entirely or almost entirely of minerals, and the relationship between kinds of rocks and the minerals they contain requires closer attention.

There are three large families of rock, each defined by the process that forms the rocks:

1. **Igneous rock** (named from the Latin, *igneus* meaning fire) is formed by the cooling and consolidation of magma.

2. **Sedimentary rock** (from the Latin *sedimentum*, meaning a settling) is formed either by chemical precipitation of material carried in solution in sea, lake, or river water, or by deposition of mineral particles transported in suspension by water, wind, or ice.

3. **Metamorphic rock** (from the Greek *meta*, meaning change, and *morphe*, meaning form; hence, change of form) is either igneous or sedimentary rock that has been changed as a result of high temperatures, high pressures, or both. Metamorphism, the process that forms metamorphic rock, is analogous to the process that occurs when a potter fires a clay pot in an oven. The tiny mineral grains in the clay undergo a series of chemical reactions as a result of the increased temperature. New compounds form, and the formerly soft clay molded by the potter becomes hard and rigid.

The crust is mainly igneous rock or metamorphic derived from igneous. However, as Figure 6.8 shows, most of the rock that we actually see at the Earth's surface is sedimentary. The difference arises because sediments are products of all the changes brought about by reactions between rainfall, ice, and the atmosphere, changes we refer to as *weathering*. Sediments are draped as a thin veneer over the largely igneous and metamorphic crust below.

Features of Rocks

At first glance, rocks seem confusingly varied. Some are distinctly layered and have pronounced, flat surfaces covered with a silicate mineral called mica. Others are coarse and evenly grained and lack layering; yet, they may contain the same kinds of minerals present in the layered, micaceous (the adjective form of "mica") rock. Studying a large number of rock specimens soon makes it clear that no matter what kind of rock is being examined—sedimentary, metamorphic,

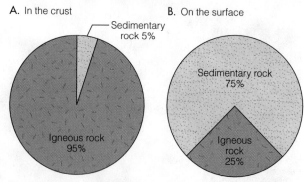

Figure 6.8 Relative amounts of sedimentary and igneous rock. A. The great bulk of the crust consists of igneous rock (95%), with sedimentary rock (5%) forming a thin veneer at the surface. B. Because the sedimentary rock veneer covers so much of the Earth's surface, it is mainly what we see. Thus, 75 percent of the surface is sedimentary rock. Igneous formations pushing through the sedimentary veneer account for the other 25 percent.

A CLOSER LOOK

Identifying Minerals

Mineral properties are determined by composition and crystal structure. It is not necessary, however, to analyze a mineral for its chemical composition or determine its crystal structure in order to discover its identity. Once we know which properties are characteristic of which minerals, we can use those properties to identify the minerals. The properties most often used to identify minerals are crystal form, growth habit, cleavage, luster, color, hardness, and specific gravity. Appendix C lists the properties of common minerals.

Crystal Form and Growth Habit

Ice fascinated the ancient Greeks. When they saw glistening needles of ice covering the ground on a frosty morning, they were intrigued by the fact that the needles were six-sided and had smooth, planar surfaces. Greek philosophers made many discoveries about the branch of mathematics called geometry, but they could not explain how three-dimensional, geometric solids could apparently grow spontaneously. The ancient Greeks called ice *krystallos*, and the Romans latinized the name to *crystallum*. Eventually, the word **crystal** came to be applied to any solid body that grows with planar surfaces. The planar surfaces that bound a crystal are called **crystal faces**, and the geometric arrangement of crystal faces, called the **crystal form,**[*] became the subject of intense study during the seventeenth century.

Seventeenth-century scientists discovered that crystal form could be used to identify minerals, but some aspects of crystal form were difficult for them to explain. Why, for example, did the size of crystal faces differ from sample to sample. Under some circumstances, a mineral may grow as a thin crystal; in other cases, the same mineral may grow as a fat one, as Figure C6.1 shows. Superficially, the two crystals of quartz in Figure C6.1 look very different, and this photograph illustrates that neither crystal size nor crystal face size is a unique property of a mineral.

The person who solved the mystery was a Danish physician, Nicolaus Steno. In 1669 Steno demonstrated that the unique property of crystals of a given mineral is not the relative face sizes, but rather the angles between the faces. It is this angle that gives each mineral a distinctive crystal form. The angle between any designated pair of crystal faces is constant, he wrote, and is the same for all specimens of a mineral, regardless of overall shape or size. Steno's discovery that interfacial angles are constant is made clear by the numbering in Figure C6.1. The same faces occur on both crystals. All the sets of faces are parallel: face 1 on the left is parallel to face 1 on the right, face 2 is parallel to face 2, and

[*]Crystal form refers to the arrangement of the crystal faces; crystal structure refers to the geometric packing of atoms in a crystal. We can macroscopically and microscopically observe crystal form, but we can only "see" crystal structure with X rays.

Figure C6.1 Because these two crystals are both quartz, they have the same crystal form. Although the sizes of the individual faces differ markedly between the two crystals, each numbered face on one crystal is parallel to an equivalent face on the other crystal. It is a fundamental property of crystals that, as a result of the internal crystal structure, the angles between adjacent faces are identical for all crystals of the same mineral.

so forth. Therefore, the angle between any two equivalent faces must be the same on both crystals.

Steno speculated that constant interfacial angles must be a result of internal order, but the ordered particles—ions—were too small for him to see. Proof of internal order was only achieved in 1912 when Max von Laue, a German scientist, demonstrated, by use of X-rays, that crystals are made up of ions packed in fixed geometric arrays, as shown in Figure 6.3.

Crystals form only when a mineral can grow freely in an open space. Crystals are uncommon in nature because most minerals do not form in open, unobstructed spaces. Compare Figures C6.1 and C6.2. The crystals in Figure C6.1 grew freely into an open space, and so, well-developed crystal faces were able to form. The quartz in Figure C6.2, however, grew irregularly, without developing crystal faces, because it grew in an environment restricted by the presence of other minerals. We call such irregularly shaped mineral particles *grains*. Using X-ray techniques, it is easy to show that in both a crystal of quartz and an irregularly shaped grain of quartz, all the atoms present are packed in the same strict crystal structures. That is, both the quartz crystals and the irregular quartz grains are crystalline.

Every mineral has a characteristic crystal form. Some have such distinctive forms that we can use the property as an identification tool without having to measure angles between faces. For example, the mineral pyrite (FeS_2) is commonly (but not always) found as intergrown cubes (Fig. C6.3) with markedly striated faces. Cube-shaped crystals with striated faces are a reliable way to identify pyrite.

Figure C6.2 Quartz grains (colorless) that grew in an environment where other grains prevented development of well-formed crystal faces. The amber-colored grains are iron carbonate ($FeCO_3$). Compare with Figure C6.1, which shows quartz crystals that grew in open spaces, unhindered by adjacent grains.

A few minerals develop distinctive growth habits when they grow in restricted environments, and these growth habits can be used for identification. For example, Figure C6.4 shows asbestos, a variety of the mineral serpentine that characteristically grows as fine, elongate threads.

Figure C6.3 Distinctive external shape of pyrite, FeS_2. The characteristic shape of pyrite is crystals with faces at right angles and with pronounced striations on the faces. The largest crystals in the photograph are 3 cm on an edge. The specimen is from Bingham Canyon, Utah.

Cleavage

The tendency of a mineral to break in preferred directions along bright, reflective planar surfaces is called **cleavage.**

If you break a mineral with a hammer or drop a specimen on the floor so that it shatters, you will probably see that the

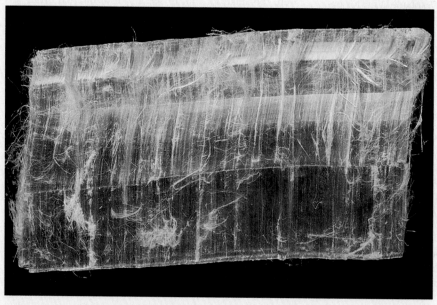

Figure C6.4 Some minerals have distinctive growth habits, even though they do not develop well-formed crystal faces. The mineral chrysotile sometimes grows as fine, cottonlike threads that can be separated and woven into fireproof fabric. When chrysotile is used for this purpose, it is referred to as asbestos.

Figure C6.5 Relation between crystal structure and cleavage. Halite, NaCl, has well-defined cleavage planes; it always breaks into fragments bounded by perpendicular faces.

broken fragments are bounded by cleavage surfaces that are smooth and planar, just like crystal faces. In exceptional cases, such as sodium chloride which is the mineral halite (NaCl), as shown in Figure C6.5, all of the breakage surfaces are smooth planar surfaces. (Don't confuse crystal faces and cleavage surfaces, however, even though the two often look alike. A cleavage surface is a breakage surface, whereas a crystal face is a growth surface.)

Many common minerals have distinctive cleavage planes. One of the most distinctive is found in mica (Fig. C6.6). Clay also has a distinctive cleavage; that is why it feels smooth and slippery when rubbed between the fingers.

Luster

The quality and intensity of the light reflected from a mineral produce an effect known as **luster**. Two minerals with almost identical color can have quite different lusters. The most important lusters are described as *metallic*, like that on a polished metal surface, and *nonmetallic*. Nonmetallic lusters are divided into *vitreous*, like that on glass; *resinous*, like that of resin; *pearly*, like that of pearl; and *greasy*, as if the surface were covered by a film of oil.

Color and streak

The color of a mineral, though often striking, is not a reliable means of identification. Color is determined by several factors, one of which is chemical composition, and even trace amounts of chemical impurities can produce distinctive colors.

Color in opaque minerals having a metallic luster can be very confusing because the color is partly a property of grain size. One way to reduce errors of judgment where color is concerned is to prepare a **streak**, which is a thin layer of powdered mineral made by rubbing a specimen on a nonglazed porcelain plate, called a 'streak plate'. The powder gives a reliable color effect because all the grains in a powder streak are very small and so the grain-size effect is reduced. Red streak characterizes hematite (Fe_2O_3), even though the specimen looks black and metallic (Fig. C6.7).

Hardness

The term **hardness** refers to the relative resistance of a mineral to being scratched. It is a distinctive property of minerals. Hardness, like crystal form and cleavage, is governed by crystal structure and by the strength of the bond-

Figure C6.6 Perfect cleavage of mica (variety muscovite) is illustrated by the planar flakes into which this specimen is being split. The cleavage flakes suggest leaves of a book, a resemblance embodied in the term *books of mica*.

Table C6.1 Mohs' Scale of Relative Hardness[a]

	Relative Number in the Scale	Mineral	Hardness of Some Common Objects
Hardest	10	Diamond	
	9	Corundum	
	8	Topaz	
	7	Quartz	
	6	Potassium feldspar	
			Pocketknife; glass
	5	Apatite	
	4	Fluorite	
			Copper penny
	3	Calcite	
			Fingernail
	2	Gypsum	
Softest	1	Talc	

[a]Named for Friedrich Mohs, an Austrian mineralogist, who chose the 10 minerals of the scale.

ing forces that hold the atoms of the crystal together. The stronger the forces, the harder the mineral.

Relative hardness values can be assigned by determining the ease or difficulty with which one mineral will scratch another. Talc, the basic ingredient of most baby ("talcum") powder, is the softest mineral known, and diamond is the hardest. A scale called the *Mohs' relative hardness scale* is divided into 10 steps, each marked by a common mineral (Table C6.1). These steps do not represent equal intervals of hardness, rather, any mineral on the scale will scratch all other minerals on the scale that have a lower number. Minerals on the same step of the scale can only scratch each other.

Figure C6.7 Color contrast between hematite and a hematite streak. Massive hematite is opaque, has a metallic luster, and appears black. On a porcelain plate, however, this mineral gives a red streak.

Density and Specific Gravity

We know that two identical baskets have different weights when one is filled with feathers and the other with rocks. The property that causes this difference is **density**, or the average mass per unit volume. The units of density are grams per cubic centimeter (g/cm^3).

Because density is difficult to measure accurately, we usually measure a property called specific gravity instead. **Specific gravity** is the ratio of the weight of a substance to the weight of an equal volume of pure water. Specific gravity is a ratio of two weights, and so it does not have any units. Because the density of pure water is 1 g/cm^3, the specific gravity of a mineral is numerically equal to its density.

Steps to Follow in Identifying Minerals

The following steps, used in conjunction with Table C6.2 and Appendix C, will help you identify common minerals.

1. Decide whether the mineral has a metallic or nonmetallic luster. If the mineral has a metallic luster, use the streak, hardness, and cleavage to decide which mineral it is.

2. If it has a nonmetallic luster, determine whether it is harder or softer than the blade of a pocket knife. (If harder, the mineral will scratch the blade; if softer, the blade will scratch the mineral.)

3. Once you determine hardness relative to the knife, decide whether the sample is dark or light in color. Go to the appropriate section of the table and use the cleavage data to determine which mineral you have.

Table C6.2   Reference Chart for the Identification of Common Minerals and a Guide to the Rock Types in which the minerals might be found

Metallic Luster[a]

Mineral	Streak	Rock Type[b]
Chalcopyrite	Greenish yellow	O, I, M
Galena	Lead gray	O
Hematite	Reddish brown	O, M, S
Limonite	Yellow to brown	S, W
Magnetite	Black	I, M, S
Pyrite	Brass yellow	O, M, I, S
Sphalerite	Yellow to brown	O

Non-Metallic Luster[c]

Mineral	Cleavage	Rock Type
A. Harder Than a Knife Blade		
Dark Colored		
Amphibole	Perfect, two planes at 120°	I, M
Garnet	None	M, I
Olivine	None	I
Pyroxene	Perfect, two planes at 90°	I, M
Quartz	None	I, M, O, S
Light Colored		
Feldspar	Perfect, two planes at 90°	I, M
Quartz	None	I, M, S, O
B. Softer Than a Knife Blade		
Dark Colored		
Chlorite	Perfect, one plane	M, S
Hematite (earthly variety)	None	O, S, M
Limonite (earthly variety)	None	W, S
Mica (var. biotite)	Perfect, one plane	I, M, S
Light Colored		
Apatite	Poor, one plane	I, M, S
Calcite	Perfect, three planes	S, M, O, I
Clay (var. kaolinite)	Perfect, one plane	W, S
Dolomite	Perfect, three planes	S, M, O
Fluorite	Perfect, four planes	O, S
Gypsum	Perfect, one plane	S, W
Halite	Perfect, three planes at 90°	S
Mica (var. muscovite)	Perfect, one plane	I, M, S, O
Talc	Perfect, one plane	M, S

[a]See Table C.1 in Appendix C for additional properties.

[b]I = igneous, M = metamorphic, O = ore, S = sedimentary, W = weathering product.

[c]See Table C.2 for additional properties.

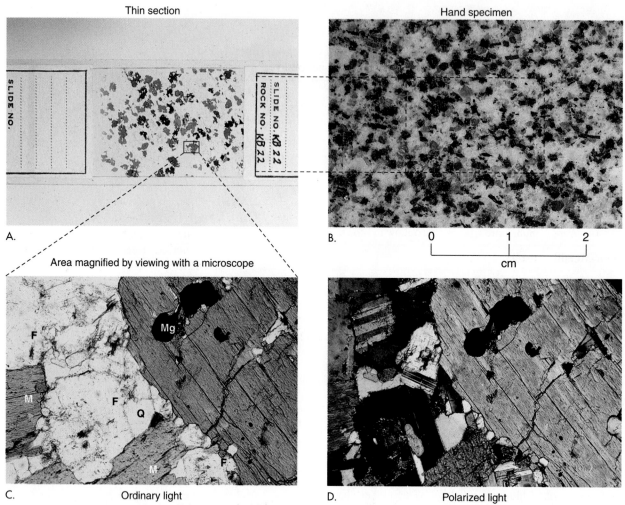

Figure 6.9   Polished surfaces and thin slices reveal textures and mineral assemblages to great advantage. The specimen here is an igneous rock containing quartz (Q), feldspar (F), amphibole (A), mica (M), and magnetite (Mg). A. A thin slice mounted on glass. The slice is 0.03 mm thick, and light can pass through the minerals. B. A polished surface. The dashed rectangle indicates the area used to make the thin slice shown in part A. C. An area of the thin slice as viewed under a microscope. The magnification is 25x. D. The same view as in part C seen through polarizers in order to emphasize the shapes and orientations of individual grains.

or igneous—the differences between samples can be described in terms of two features.

The first feature is **texture**, by which is meant the overall appearance a rock has because of the size, shape, and arrangement of its constituent mineral grains. For example, the grains may be flat and parallel to each other, giving the rock a pronounced platy texture—like a pack of playing cards. In addition, the various minerals may be unevenly distributed and concentrated into specific layers. The rock texture is then both distinctly layered and platy. Specific textural

terms are used for each rock family and will be introduced at the appropriate place in subsequent chapters.

Commonly, examination of a microscopic texture requires the preparation of a *thin section* of rock that must be viewed through a microscope. A thin section is prepared by first grinding and polishing a smooth, flat surface on a small fragment of rock. The polished surface is glued to a glass slide, and then the rock is ground away until the glued fragment is so thin that light passes through it easily. A polished surface and a thin section are shown in Figure 6.9.

The second feature used in identifying rocks is the kinds of minerals present. A few kinds of rock contain only one mineral, but most rocks contain two or more minerals. The varieties and abundances of minerals present in a rock, commonly called the **mineral assemblage** of the rock, are important pieces of information for interpreting how the rock was formed.

The rock in Figure 6.9 is an igneous rock called granite. The mineral assemblage is quartz, feldspar, amphibole, mica (variety, biotite), and magnetite. The texture is typical of granites and would be described as granitic, meaning the grains are uniform in size, intricately interlocked, irregular in shape, and randomly distributed. The minerals found most commonly in the three rock families are listed in Table 6–2.

What Holds Rocks Together?

The mineral grains in some kinds of rock are held together with great tenacity, whereas in other kinds of rock the grains are easily broken apart. The most tightly bound rocks are igneous and metamorphic because both types contain intricately interlocked mineral grains. During the formation of igneous and metamorphic rocks, the growing mineral grains crowd against each other, filling all spaces and forming an intricate, three-dimensional jigsaw puzzle. A similar interlocking of grains holds together steel, ceramics, and bricks.

The forces that hold the grains of sedimentary rocks together are less obvious. Sediment is a loose aggregate of particles, and it must be transformed into sedimentary rock. The two ways sediment becomes sedimentary rock are:

1. *Deposition of a cement.* Water circulating slowly through the open spaces in a sediment deposits new materials such as calcite, quartz, and goethite, which cement the sediment grains together.

2. *Recrystallization.* As a consequence of the geothermal heat, temperature increases with depth. Thus, as layer after layer of sediment is deposited, the deeper layers of sediment are subjected to rising temperatures. In response to increased temperatures, mineral grains in deeply buried sediment begin to recrystallize, and the growing grains interlock and form strong aggregates. The process is the same as when ice crystals in a snow pile recrystallize to form a compact mass of ice.

Both mineral assemblage and texture reflect the conditions under which a rock formed. In the next chapter we will see how these properties can be used to recover information from the most abundant and most important family of rocks, the igneous rocks.

FROM ROCK TO REGOLITH

Whether rapid or slow, rocks of all kinds are physically broken up and chemically altered throughout the zone where the geosphere, hydrosphere, biosphere, and atmosphere mix. This zone extends from the ground surface downward to whatever depth air and water penetrate. Within it, the rock constitutes a porous framework full of fractures, cracks, and other

Table 6.2 Minerals Most Commonly Found in the Three Rock Families

Rock Family	Common Minerals
Igneous	Feldspar, quartz, olivine, amphibole, pyroxene, mica, magnetite
Sedimentary	Clay, chlorite, quartz, calcite, dolomite, gypsum, goethite, hematite
Metamorphic	Feldspar, quartz, mica, chlorite, garnet, amphibole, pyroxene, magnetite

openings, some of which are very small but all of which make the rock vulnerable to attack by aqueous solutions. Given sufficient time, the result is conspicuous decomposition and disintegration of the rock, processes known collectively as **weathering**.

We have all seen weathering in action. We may have visited a cemetery and strained to read the inscription on an old marble tombstone so modified by weathering that the characters were barely legible (Fig. 6.10). Or we may have been seated around a roaring campfire and suddenly been struck by flying rock fragments as a rock next to the fire exploded, because it was composed of minerals that expand at different rates when heated. Such examples show that weathering can involve both chemical and physical processes.

Chemical Weathering

The minerals of igneous and metamorphic rocks that have crystallized within the Earth's crust at high pressure and temperature are chemically unstable at the lower temperatures and pressures at the surface. When such rocks are uplifted and eventually exposed, therefore, their mineral components are chemically changed into new, more stable minerals. **Chemical weathering**, then, is the decomposition of rocks and minerals as chemical reactions transform them into new chemical combinations that are stable at or near the Earth's surface.

The principal agent of chemical weathering is a weak solution of carbonic acid (H_2CO_3), formed as falling rainwater dissolves small quantities of atmospheric carbon dioxide (Table 6.3, Eq. 1). As the water moves downward and laterally beneath the ground surface, additional carbon dioxide is dissolved from decaying vegetation. Thus, chemical weathering is the result of interactions of the atmosphere, the hydrosphere, and the biosphere to produce the weakly acidic solution that attacks the upper part of the geosphere.

The chemical reaction that decomposes the common rock-forming mineral potassium feldspar pro-

Figure 6.10 Because marble is composed of soluble calcite, this marble tombstone standing in a New England cemetery since the early nineteenth century shows the corrosive effects of the carbonic acid present in rainwater. Over the years the rock surface has been slowly dissolved, making the once sharply chiseled inscription illegible.

vides a good example of chemical weathering (Table 6.3, Eq. 2). A molecule of carbonic acid dissociates in water to form a hydrogen ion (H^+) and a bicarbonate ion [$(HCO_3)^-$]. H^+ ions enter the potassium feldspar and replace potassium ions (K^+), which then leave the crystal and pass into solution. Water combines with the remaining aluminum silicate molecule to create *kaolinite*, a new clay mineral not present in the original rock.

Weathering of a rock, which contains feldspars and Fe-Mg bearing minerals, produces clay minerals and *goethite*, a weathering product of the common iron mineral magnetite (Table 6.3, Eq. 3; Fig. 6.11). When a granite weathers, clay minerals and goethite are also produced from the feldspar and mica it contains. However, the quartz, because it is resistant to chemical weathering, survives unaltered as a weathering product.

Physical Weathering

Sometimes regolith consists of fragments identical to the adjacent bedrock. However, the mineral grains are unweathered or only slightly weathered, indicating little or no evidence of chemical alteration. Instead, the fragments must have experienced **physical weathering,** which is the disintegration (physical breakup) of rocks.

A variety of natural physical processes are effective in physical weathering: (1) a rock mass buried deep beneath the land surface is subjected to immense confining pressure. However, as erosion gradually removes the overlying rock, the pressure is reduced and the buried rock mass adjusts by expanding upward. In

---

Table 6.3 Common Chemical Weathering Reactions

1. Production of carbonic acid by solution of carbon dioxide:

H_2O	+	CO_2	$\rightarrow$	H_2CO_3	$\rightarrow$	H^+	+	$(HCO_3)^-$
Water		Carbon dioxide		Carbonic acid		Hydrogen ion		Bicarbonate ion

2. Hydrolysis of potassium feldspar:

$4KAlSi_3O_8$	+	$4H^+$	+	$2H_2O$	$\rightarrow$	$4K^+$	+	$Al_4Si_4O_{10}(OH)_8$	+	$8SiO_2$
Potassium feldspar		Hydrogen ions		Water		Potassium ions		Kaolinite		Silica

3. Oxidation of iron (Fe^{2+}) oxide to form goethite:

$4Fe^{2+}O$	+	$2H_2O$	+	O_2	$\rightarrow$	$4Fe^{3+}O{\cdot}OH$
Iron oxide		Water		Oxygen		Goethite

4. Dissolution of carbonate rock by carbonic acid:

$CaCO_3$	+	H_2CO_3	$\rightarrow$	Ca^{2+}	+	$2(HCO_3)^-$
Calcium carbonate		Carbonic acid		Calcium ion		Bicarbonate ions

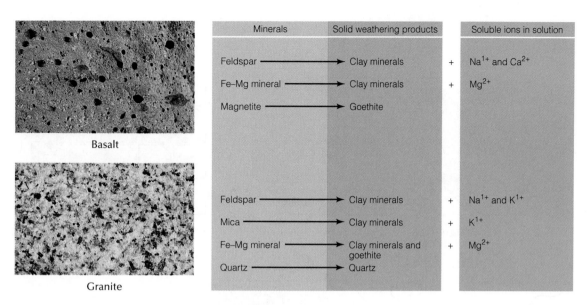

Minerals		Solid weathering products		Soluble ions in solution
Feldspar	⟶	Clay minerals	+	Na^{1+} and Ca^{2+}
Fe–Mg mineral	⟶	Clay minerals	+	Mg^{2+}
Magnetite	⟶	Goethite		
Feldspar	⟶	Clay minerals	+	Na^{1+} and K^{1+}
Mica	⟶	Clay minerals	+	K^{1+}
Fe–Mg mineral	⟶	Clay minerals and goethite	+	Mg^{2+}
Quartz	⟶	Quartz		

Basalt

Granite

Figure 6.11 When basalt is chemically weathered, its minerals are converted to new clay minerals and to goethite. Soluble ions are carried away in groundwater. When granite weathers, clay minerals, goethite, and soluble ions are also produced, as well as grains of quartz, a mineral that is resistant to chemical decay.

the process, sheetlike fractures develop parallel to the surface (Fig. 6.12A). (2) Ions in the groundwater moving through fractured rock can precipitate out to form salts. The enormous forces exerted by salt crystals growing in cavities or along mineral grain boundaries of a rock can easily lead to rupture or disaggregation (Fig. 6.12B). (3) When water freezes, its volume increases about 9 percent. If water freezes in a confined crack, the resulting stresses can be so great that the rock is wedged apart. (4) Fire, too, can be a very effective agent of weathering. An intense forest or brush fire can overheat the outer part of a rock, causing it to expand, fracture, and break away. Repeated fires can

Figure 6.12A This granite outcrop in Yosemite National Park, California, displays sheetlike joints, giving a stepped appearance to the mountain slope. The jointing is thought to result from progressive removal of overlying rock, leading to reduced pressure. This causes expansion of the uppermost rock, which fractures along planes parallel to the land surface.

Figure 6.12B Granite on the side of Gondola Ridge in Antarctica is so intensely weathered that it resembles Swiss cheese. Such cavernous weathering is produced by crystallization of salt in small cavities and along grain boundaries.

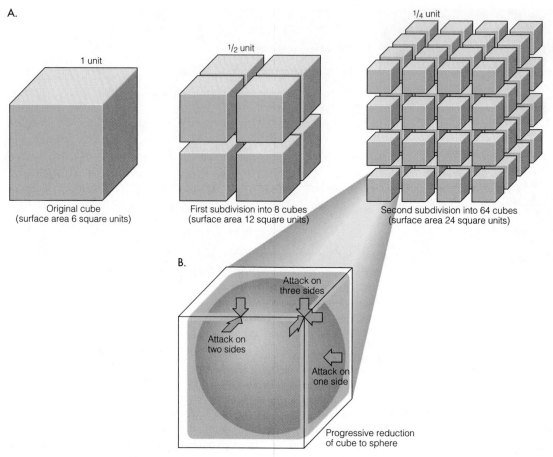

Figure 6.13 Weathering causes progressive subdivision of rocks. A. Each time a cube is subdivided by slicing it through the center of each edge, the aggregate surface area doubles, thereby increasing the effectiveness of chemical attack. B. Solutions moving along joints that separate cube-shaped blocks of rock attack corners, edges, and sides at rates that decline in that order, because the number of corresponding surfaces under attack are 3, 2, and 1, respectively. Corners therefore become rounded, and eventually the blocks are reduced to spheres. Once a spherical form has been reached, chemical attack is distributed over the entire surface, and no further change in form occurs.

thereby significantly reduce the size of rocks. (5). Finally, plant roots extending along cracks can slowly wedge the rock apart.

Although physical weathering is distinct from chemical weathering, the two processes generally work hand in hand, and their effects are inseparably blended. The effectiveness of chemical weathering increases as the exposed surface area increases, and surface area increases greatly whenever a large unit is divided into successively smaller units (Fig. 6.13). Repeated subdivision leads to a remarkable result. Whereas one cubic centimeter of rock has a surface area of 6 cm^2 (0.9 in^2), when subdivided into particles the size of the smallest clay minerals, the total surface area now exposed to weathering increases to nearly 40 million cm^2 (6.2 million in^2).

SOILS

The physical and chemical weathering of solid rock is the initial step in soil formation. However, soil also contains organic matter mixed with the mineral component. This organic part is an essential part of the definition of **soil**: the part of the regolith that can support rooted plants.

The organic matter in soil is derived from the decay of dead plants and animals. Living plants are nourished by the nutrients released from decaying organisms, as well as by the nutrients released during weathering. Plants draw these nutrients upward, in water solution, through their roots. Therefore, throughout their life cycle, plants are directly involved in the manufacture of the fertilizer that will nourish future generations of plants. These activities are an integral part of a continuous cycling of nutrients through the regolith and biosphere. With its partly mineral, partly organic composition, soil forms an important bridge between the geosphere and the teeming biosphere.

The Soil Profile

As bedrock and regolith weather, soil gradually develops from the surface downward, producing an identifiable succession of nearly horizontal weathered zones called **soil horizons**. Each horizon has distinctive physical, chemical, and biological characteristics. Although soil horizons may resemble a sequence of deposits, or layers, they are not strata. Instead they represent physical, chemical, and biological changes to the regolith. Collectively, the soil horizons constitute a **soil profile** (Fig. 6.14).

Soil profiles generally display two or more horizons. The uppermost horizon of some profiles consists of decomposing organic matter (*O horizon*). If an O horizon is absent, then an *A horizon* generally is the

uppermost horizon. Typically, it is dark gray or black because decomposed plant and animal tissues are mixed with the mineral matter. The A horizon has lost some of its original substance through the downward transport of clay particles and, most important, through the chemical removal of soluble minerals. A light-colored *E horizon*, sometimes present beneath the A horizon in acidic soils, is often developed beneath evergreen forests. The *B horizon* underlies the surface horizon(s) and commonly has a brownish or reddish color. This horizon is enriched in clay or iron and aluminum hydroxides produced by the weathering of minerals within the horizon and is also transported downward from the A horizon. The B horizon often has a distinct structure that causes it to break into blocks or prisms. A *K horizon*, present in some arid-region soils beneath the B horizon, is densely impregnated with calcium carbonate that coats all mineral grains and constitutes up to 50 percent of the volume of the horizon. The *C horizon* is the deepest horizon and constitutes the parent regolith in various stages of weathering, but it lacks the distinctive properties of the A and B horizons. Oxidation in the C horizon generally imparts to it a light yellowish-brown color.

Figure 6.14 Soils vary across the landscape, as shown by this example of three soil profiles from forest, grassland, and desert regions. Differences are explainable in terms of regolith composition, slope steepness, vegetation cover, soil biota, climate, and the time required to develop the profile.

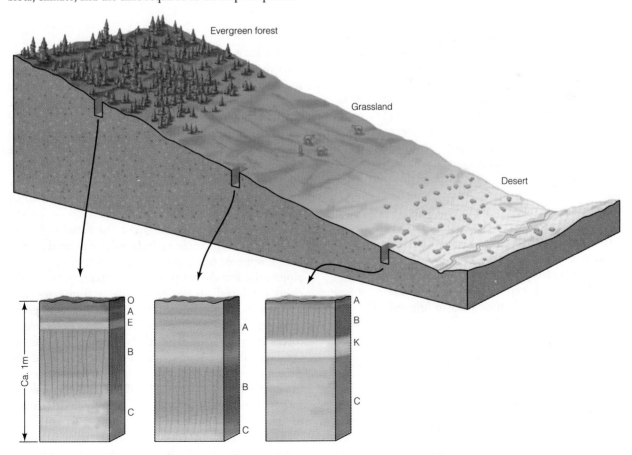

GUEST ESSAY

Asbestos Science and Politics

"Asbestos" is a commercial rather than a mineralogical term applied to a variety of fibrous silicate minerals of different chemical compositions. An unusual combination of useful properties such as the ability to divide into fine fibers, high strength and flexibility, high chemical and mechanical durability, low thermal and electric conductivity, and relative incombustibility, allows its use in industrial products and processes. Roofing and flooring materials, automobile brake linings, cement and mortar, and building insulation commonly contain asbestos.

Asbestiform describes the tendency of any mineral to break into fine fibers. Many minerals have asbestiform habits under some physical conditions; however, only six asbestiform minerals attract commercial interest. These six minerals—chrysotile, amosite, crocidolite, fibrous anthophyllite, fibrous tremolite, and fibrous actinolite—are all referred to as asbestos. The majority of asbestos (95%) is chrysotile, $Mg_3Si_2O_5(OH)_4$, also known by the generic name serpentine, which is formed by the linking of silicate tetrahedra (Fig. 6.6) to form sheets. In chrysotile, the sheets of tetrahedra curl up to form hollow cylinders, thus producing the asbestiform habit. Most of the chrysotile for commercial asbestos comes from deposits in Canada. The five

Jill S. Schneiderman is Associate Professor of Geology at Vassar College. She served as the Geological Society of America's Congressional Science Fellow during the 104th Congress in the Office of the Senate Minority Leader, Tom Daschle.

other minerals that form asbestos—amosite, crocidolite, fibrous anthophyllite, fibrous tremolite, and fibrous actinolite—are all amphiboles; amphiboles are silicate minerals of variable composition that have a crystal structure formed of two chains of silicon tetrahedra linked together (Figs. 6.5 and 6.6). These five minerals are, respectively:

$$Fe_7^{2+}Si_8O_{22}(OH)_2,$$
$$Na_2Fe_3^{2+}Fe_2^{3+}Si_8O_{22}(OH)_2,$$
$$Mg_7Si_8O_{22}(OH)_2, Ca_2Mg_5Si_8O_{22}(OH)_2, \text{ and}$$
$$Ca_2(Fe^{2+},Mg)_5Si_8O_{22}(OH)_2.$$

The double-chain structure also gives rise to the asbestiform habit.

Soil Types

An astute observer traveling across the landscape will note that soils are not everywhere the same (Fig. 6.14). Different soils result from the influence of six factors: climate, vegetation cover, soil organisms, regolith composition, topography, and time. A soil forming under prairie grassland, for example, differs from soil in an evergreen forest or that of a tropical rainforest. The character of a soil may change abruptly as we move from one rock type to another or from a gentle to a steep slope, and it also will change with the passage of time. Such differences make it possible for soil scientists to classify and map soils across the landscape in much the same way that geologists classify and map rocks. For more details on soil types and their distribution refer to Appendix D.

REVISITING THE CYCLES OF THE GEOSPHERE

James Hutton recognized that the cycle of uplift of the continental crust, weathering, erosion, formation of new rock, uplift, weathering, and so on—the rock cycle—has been operating for all of most of the

Earth's long history. Because weathering and erosion involve the hydrosphere, atmosphere, and biosphere, the rock cycle of the geosphere is directly involved with the processes of the other spheres. Because the mass of the continental crust is large, the average length of time any given mass takes to complete the rock cycle is long. Time estimates vary, and they are difficult to make, but the average age of all rock in the continental crust—and therefore the average time for material to pass through the rock cycle—seems to be about 650 million years.

As discussed in Chapter 4 and illustrated in Figure 4.27, the plate tectonic cycle plays an essential role in the geosphere and therefore in the rock cycle. For example, the forces that produce the basins where sediment produced by the rock cycle accumulates, and that cause mountains to be uplifted and thereby eroded as part of the rock cycle, are the same internal forces that drive the plate tectonic cycle. What is not yet clear from our discussions of the cycles of the geosphere is how the plate tectonic cycle influences the composition of the hydrosphere. It does so in the following way. The magma that rises from the mantle to form new oceanic crust at a midocean ridge becomes hot igneous rock that reacts with seawater. Some constituents of the hot rock, such as calcium,

During the 1960s researchers conducted studies on rates of mesothelioma, or cancer of the lining of the chest or abdomen, which is nearly always fatal, among insulation workers. The results convincingly showed that exposure to asbestos caused this disease. Consequently, strict regulations to control the amount of airborne asbestos fibers were enacted to protect those individuals occupationally exposed to asbestos. However, epidemiological studies during the 1970s suggested that the type of fiber comprising asbestos affects the degree of health risk. In particular, some researchers concluded that chrysotile asbestos is far less carcinogenic than amphibole asbestos.

Currently, researchers studying the health hazards of asbestos remain severely divided; one group maintains that chrysotile asbestos poses no health risk in nonoccupational environments such as schools and houses. In fact, they state that since 95 percent of the asbestos used commercially in the United States is chrysotile asbestos, the health risk posed by asbestos in buildings is much less than other environmental health hazards like radon or tobacco smoke. Hence, they say that chrysotile asbestos should be regulated differently than amphibole asbestos. Furthermore, since current regulations also protect the occupants of buildings that contain asbestos by requiring asbestos removal, they advocate removal of only amphibole asbestos. The other group maintains that fiber-type studies are uncertain and that epidemiological data show that exposure to asbestos, whether it is chrysotile asbestos or amphibole asbestos, leads to asbestosis, a disease that causes scarring of the lungs, and lung cancer. They advocate the removal of both chrysotile asbestos and amphibole asbestos from buildings.

An important question further complicates this issue: what acts as a greater airborne health risk, leaving asbestos in place or disturbing it through removal? Because of its fibrous structure, asbestos easily becomes an airborne particle, readily dispersed to be inhaled and ingested. Therefore, removal of asbestos may produce a health hazard that otherwise would not exist if asbestos remained entombed in ceiling tiles, cement, and mortar. Answers offered to the question posed are frequently motivated by politics and economics. For example, school districts that have paid enormous sums of money to remove asbestos from school buildings will be able to recoup their costs from manufacturers if indeed it turns out that all asbestos is deemed a substantial health hazard. Asbestos manufacturers stand to gain by being absolved of responsibility for removal costs if chrysotile asbestos is as innocuous as some claim. Certainly, the debate about asbestos has created a huge industry in asbestos removal, and thousands of legal cases regarding asbestos have kept lawyers busy and prosperous. Whose interests are being served? Whatever the outcome, a knowledge of mineralogy makes the debate comprehensible.

---

are dissolved in the seawater, and some constituents already in the seawater, such as magnesium, are deposited in the igneous rock. Thus, via the reactions between hot, newly formed crust and seawater, the plate tectonic cycle plays a role in determining the composition of seawater.

The plate tectonic cycle also plays a role in determining the composition of the atmosphere and thereby the viability of the biosphere. All magma contains some dissolved gas. During an eruption the dissolved gas bubbles out of the magma and mixes with the atmosphere. Because a lot of magma originates in the mantle and reaches the Earth's surface as a result of plate tectonics, eruptions are the means by which gas is transferred from the mantle to the atmosphere.

There have been times in the Earth's long history when so many volcanoes were erupting, and so much gas was being added to the atmosphere, that very long-lasting climatic effects resulted. An unusually high rate of volcanic activity occurred between 135 and 115 million years ago when a vast submarine lava plateau was formed by eruption in the southwest Pacific. So much carbon dioxide was released during eruption that it is estimated that the atmospheric concentration of carbon dioxide was about 20 times higher than today's level. As a result, the global temperature rose about 10°C and several million years of scorching climates followed, causing drastic changes in the distribution of plants and animals.

UNIFORMITARIANISM AND THE RATE OF THE ROCK CYCLE

As mentioned in Chapter 1, it was James Hutton who recognized that the same external and internal processes occurring today have been operating throughout the Earth's long history and therefore that the present is the key to the past. This is the Principle of Uniformitarianism.

During the nineteenth century, with the Principle of Uniformitarianism generally accepted, geologists tried to estimate how long the rock cycle has been going on by estimating the thickness of all sediments laid down through geological time. They assumed that the principle applied to process rates as well as the processes themselves, and hence that deposition rates have always been constant and equal to today's rates. Thus, these early geologists thought it would be a simple matter to estimate the time needed to produce all the sediments. The results, we now know, were greatly in error; one reason for the error was the assumption of rate constancy. The more we learn of the Earth's history and the more accurately we determine the timing of past events through radiometric dating (Chapter 8), the clearer it becomes that cycle rates have not always been the same.

The rate of the rock cycle has changed through time in part because the Earth is very slowly cooling as its internal heat leaks away. The Earth's internal temperature is maintained by natural radioactivity. Because radioactive isotopes transform spontaneously to nonradioactive isotopes, the Earth's natural radioactivity is slowly declining. Early in its history, the Earth must therefore have contained more radioactive atoms than there are today, and so more heat must have been produced than is now produced. Therefore processes, being driven by the Earth's internal heat, must have been more rapid than they are today. It is possible that 3 billion years ago oceanic crust was created at a faster rate than now, that tectonic plates moved faster, that volcanoes were more active, and that continental crust was uplifted and eroded at a faster rate. Any or all of these actions would cause both the rock cycle and the plate tectonic cycle to speed up.

At the same time, the rates of external processes such as weathering and erosion have also varied. Long-term changes in rates seem to have occurred because of slow increases in the luminosity of the Sun and because of the gradual slowing of the Earth's rotation. (Scientists estimate that 600 million years ago there were 400 days in the year and 2 billion years ago there were 450.) In other words, even though the cycles of the geosphere have been continuous, they have not maintained constant rates through time. Therefore, we turn next to a consideration of magma and what it tells us about the rates of the Earth's internal activities today.

SUMMARY

1. Minerals are naturally formed, solid chemical elements or compounds having a definite composition and a characteristic crystal structure. Crystal structure is the geometric array of atoms in a crystalline solid.

2. Minerals are formed through the bonding together of cations and anions of different chemical elements.

3. Silicates are the most common minerals (95% of the crust), followed by oxides, carbonates, sulfides, sulfates, and phosphates.

4. The basic building block of silicate minerals is the silicate tetrahedron, a complex anion in which an Si^{4+} ion is bonded to four O^{2-} ions. The four O^{2-} ions sit at the apexes of a tetrahedron, with the Si^{4+} at its center. Adjacent silicate tetrahedra can bond together to form polymers by sharing oxygens.

5. The feldspars are the most abundant group of minerals in the Earth's crust (60%). Quartz is the second most common mineral in the crust (15%).

6. The principal properties used to characterize and identify minerals are crystal form, growth habit, cleavage, luster, color and streak, hardness, and specific gravity.

7. There are three families of rocks: sedimentary, igneous, and metamorphic. Igneous are the most common kinds of rock in the continental crust.

8. The differences between rocks can be described in terms of mineral assemblage and texture.

9. Sediment is transformed to sedimentary rock by cementation or recrystallization of the sediment particles.

10. Weathering involves the physical breakup and chemical alteration of rock, and it occurs in the zone at the Earth's surface where the lithosphere, hydrosphere, biosphere, and atmosphere interact.

11. Physical and chemical weathering processes generally operate together, and their effects are inseparably mixed.

12. Soil is weathered regolith capable of supporting plants. A soil profile consists of successive horizons that develop from the surface downward.

13. Differences in soils result from variations in climate, vegetation cover, soil organisms, composition of the regolith, topography, and the length of time during which a profile has developed.

14. The rock cycle arises from the interactions of the Earth's internal and external processes. Igneous rock is eroded, creating sediment, which is deposited in layers that become sedimentary rock. When sedimentary rock is buried, changes in temperature and pressure cause it to convert to metamorphic rock. Eventually, temperatures and pressures may become so high that metamorphic rock melts and forms new magma. The magma rises, forms new igneous rock, and the cycle is repeated.

15. The rock cycle in the oceanic crust interacts with that in the continental crust through the agency of plate tectonics.

16. The Principle of Uniformitarianism is an accurate guide to understanding geological processes throughout the Earth's long history, but the Principle is not a guide to the rates of geological processes—many rates have varied considerably.

IMPORTANT TERMS TO REMEMBER

anion *113*
atom *112*
cation *113*
chemical weathering *125*
cleavage *120*
crystal *119*
crystal face *119*
crystal form *119*
crystal structure *119*

density *122*
element (chemical) *112*
energy-level shell *113*
hardness (of a mineral)
  121
igneous rock *118*
ion *113*
isotope *113*
luster *121*

metamorphic rock *118*
mineral *112*
mineral assemblage *124*
physical weathering *126*
sediment *118*
sedimentary rock *118*
silicate anion $(SiO_4)^{4-}$ *116*
silicates *116*
soil *128*

soil horizon *129*
soil profile *129*
specific gravity *122*
streak *121*
texture (of a rock) *124*
weathering *125*

QUESTIONS FOR REVIEW

1. What is a mineral? Give three reasons why the study of minerals is important.

2. Approximately how many common minerals are there? Which is the most common one?

3. Name five minerals that are found in the area in which you live. Are any minerals mined in the area in which you live? What are they and what are they mined for?

4. Describe the structure of the silicate anion.

5. Describe how silicate anions join together to form silicate minerals.

6. Describe the polymer structure of pyroxenes, micas, and feldspars.

7. Name five minerals that are not silicates and name the anion each contains.

8. Can a rock be uniquely defined on the basis of its mineral assemblage? If not, what additional information is needed?

9. Describe two ways by which loose aggregates of sediment are transformed into sedimentary rock.

10. What holds together the mineral grains in metamorphic and igneous rocks?

11. Why are sedimentary rocks so abundant at the Earth's surface when igneous rocks make up most of the crust?

12. Why does the physical breakup of rock increase the effectiveness of chemical weathering?

13. Explain why a soil profile formed in a semiarid grassland differs from one developed in a wet tropical forest.

14. What is the rock cycle? How does oceanic crust interact with continental crust through the rock cycle?

QUESTIONS FOR DISCUSSION

1. Within the room in which you are sitting, identify all the objects that are derived in some way from minerals. What would happen to society if all mining were stopped?

2. Would you expect other planets in the solar system to have the same kinds of minerals as on the Earth? How about planets around other suns? Be sure to say why you think there may be similarities or differences.

3. Find a cemetery at least 100 years old and carefully examine the surfaces of gravestones for evidence of weathering. Classify the gravestones according to rock type and date of emplacement. Which rock types weather most rapidly and which least rapidly? What factors likely control the degree of weathering that you can see?

*To learn more about
soils and soil erosion,
visit our web site.*

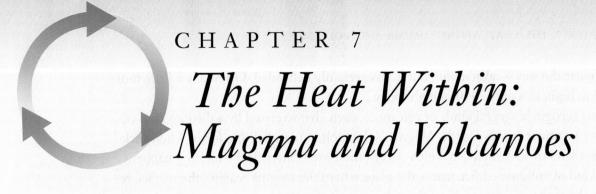

CHAPTER 7

The Heat Within: Magma and Volcanoes

● *Violent Eruptions*

During the summer of 1883 an apparently dormant Indonesian volcano called Krakatau started to emit steam and ash. Krakatau was an island off the western end of Java. On Sunday, August 26, activity increased, and on the next day Krakatau blew up: the island disappeared. As a telegram of the time tersely reported, "Where once Mount Krakatau stood, the sea now plays." Noise from the paroxysmal explosion was heard on an island in the Indian Ocean, 4600 km away. As the island blew apart it created *tsunami* that moved out from the site of the explosion and crashed into the shores of Java and Sumatra, the two closest Indonesian islands. Thirty-six thousand people lost their lives.

The effect of Krakatau was felt around the world. About 20 km³ of volcanic debris was ejected during the eruption, some blasted as high as 50 km into the stratosphere. Within 13 days the stratospheric dust had encircled the globe, and for months there were strangely colored sunsets—sometimes green or blue and other times scarlet, or flaming orange. One November sunset over New York City looked so like the glow from a massive fire that fire engines were called out. The suspended dust made the atmosphere so opaque to the Sun's rays that the temperature around the Earth dropped an estimated 0.5° during 1884. It was five years before the atmosphere cleared and the climate returned to normal.

In March 1980, the sudden, violent eruption of Mount St. Helens, a long quiescent volcano in the state of Washington, reminded us once again of the enormous magnitude of volcanic forces. Mount St. Helens was known to have been active in historic times, but it had not erupted for more than 200 years. Then, early in 1980 people living near Mount St. Helens began reporting frequent small earthquakes. On March 27 steam and volcanic ash puffed from the summit.

Monitoring by geologists working for the U.S. Geological Survey quickly revealed that Mount St. Helens was swelling like a balloon: the north face was watched especially closely because by early May it was bulging outward at a rate of 1.5 m/day. From an observation post several kilometers north of the volcano, geologists, stationed at Vancouver, Washington, mounted a round-the-clock watch. On Sunday, May 18, 1980, David A. Johnson was on duty, and at 8.32 A.M. he shouted into his microphone, "Vancouver, Vancouver, this is it." They were Johnson's last words. A devastating eruption was under way. A gigantic mass of volcanic particles and very hot gases blasted out sideways, directly toward David Johnson. No trace of Johnson or the observation post has ever been found. At least 62 other people were killed by the eruption, but the total would have been much higher had not authorities heeded early warnings by geologists and kept people far away. Mount St. Helens didn't ex-

(opposite) A column of hot volcanic ash rises from Mount St. Helens, Washington, during the violent eruption of May 18, 1980. Sixty-three people died as a result of the eruption.

actly disappear the way Krakatau did, but it was certainly beheaded. Originally a little more than 2900 m high, it is only 2490 m high today.

Scientists recognize several kinds of volcanoes, each characterized by a distinct kind of magma and a distinct eruption style. Magma is the molten material that forms when rock melts. What kind of rock melts, in which part of the Earth's interior the melting happens, and what kind of volcanic edifice marks the place where the magma reaches the surface reveal much about the processes taking place deep inside the Earth.

It is important to study volcanism and the deep-seated processes that give rise to it because volcanism plays an important role in the Earth system. Volcanic eruptions, particularly large ones such as the Krakatau eruption of 1883, can change climates around the world. Through volcanism, events that happen deep inside the Earth influence what happens on the Earth's surface. And through volcanism, new fresh rock is brought to the Earth's surface, where it becomes weathered and forms new, rich soil. ●

PROPERTIES OF MAGMA

One of the best ways to learn about volcanoes and the Earth's interior is to study magma, the material that volcanoes erupt. Magma is sometimes described as molten rock, but in fact it is a much more complex material. A complete definition for **magma** is the mixture of molten rock, suspended mineral grains, and dissolved gases that forms in the crust or mantle when temperatures are sufficiently high. Magma reaches the Earth's surface through a **volcano**, which is a vent from which magma, solid rock debris, and gases are erupted. The term *volcano* comes from the name of the Roman god of fire, Vulcan, and it conjures up visions of streams of **lava**—magma that reaches the Earth's surface—pouring out over the landscape. Some lava does flow as hot streams, but magma can also be erupted as clouds of tiny, red-hot fragments, as was the case at Mount St. Helens.

Volcanoes are the only places we can see and study magma, and so we start this chapter by gaining some insight into volcanoes and the properties of magma. By observing lava, it is possible to draw three important conclusions concerning magma:

1. Magma is characterized by a *range of compositions* in which silica (SiO_2) is always predominant.

2. Magma has the properties of a liquid, including the *ability to flow*. This is true even though most magma is a mixture of suspended crystals, dissolved gases, and molten rock (often referred to as *melt*), and in some instances almost as stiff as window glass.

3. Magma is characterized by *high temperatures*.

Composition

Magma composition is determined by the common chemical elements in the Earth—silicon (Si), aluminum (Al), iron (Fe), calcium (Ca), magnesium (Mg), sodium (Na), potassium (K), hydrogen (H), and oxygen (O). Because O^{2-} is by far the most abundant anion and is therefore the anion that balances the charges on all the cations, scientists usually express magma composition in terms of charge-balanced oxides, such as SiO_2 and Al_2O_3. The most abundant component of magma is silica, SiO_2.

Small amounts of gas (0.2 to 5% by weight) are dissolved in all magma and play an important role in eruptive processes. The principal gas is water vapor, which, together with carbon dioxide, accounts for more than 98 percent of all gases emitted from volcanoes. The remaining 2 percent is nitrogen, chlorine, sulfur, and argon.

Magma gases are also called volcanic gases. They are important in Earth system science because:

1. Volcanic gases, and especially carbon dioxide and sulfur dioxide, influence the composition of the atmosphere and thereby the climate.

2. The rate at which gas bubbles out of a magma controls the violence of an eruption—rapid bubbling means a violent and therefore hazardous eruption.

3. Violent, gas-driven eruptions, such as the eruptions of Krakatau and Mount St. Helens mentioned in the chapter opening essay, can blast such massive amounts of volcanic dust into the atmosphere that the global temperature can drop: the

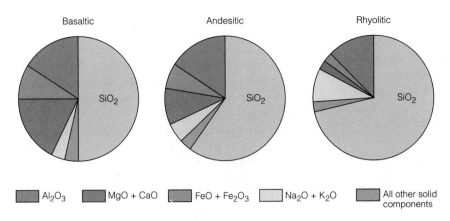

Figure 7.1 The average compositions of the solid part of the three principal kinds of magma. In addition to the solid materials, the magmas also contain dissolved gases. Basaltic magma has a low content of dissolved gas; andesitic and rhyolitic magmas tend to be very gassy.

Basaltic · Andesitic · Rhyolitic

SiO_2

Al_2O_3 | $MgO + CaO$ | $FeO + Fe_2O_3$ | $Na_2O + K_2O$ | All other solid components

drop was 0.5°C in the case of Krakatau but could be 1°C or more for a huge eruption.

Three distinct types of magma are more common than all others: basaltic, andesitic, and rhyolitic (Fig. 7.1).

1. Basaltic magma contains about 50 percent SiO_2 and very little dissolved gas. The two common igneous rocks derived from basaltic magma are **basalt** and **gabbro**.[1]

2. Andesitic magma contains about 60 percent SiO_2 and a lot of dissolved gas. **Andesite** and **diorite** are the common igneous rocks derived from andesitic magma.

3. Rhyolitic magma contains about 70 percent SiO_2 and the highest gas content. **Rhyolite** and **granite** are the common igneous rocks derived from rhyolitic magma.

The three magmas are not formed in equal abundance. Approximately 80 percent of all magma erupted by volcanoes is basaltic, with andesitic and rhyolitic each about 10 percent. Hawaiian volcanoes, such as Kilauea and Mauna Loa, are basaltic. Mount St. Helens and Krakatau are both andesitic volcanoes, and the now dormant volcanoes at Yellowstone National Park are rhyolitic. As we saw in Chapter 4, the locations of many volcanoes are closely related to plate margins.

Viscosity

Dramatic pictures of lava flowing rapidly down the side of a volcano prove that some magma is very fluid.

Basaltic lava moving down a steep slope on Mauna Loa in Hawaii has been clocked at 16 km/h. Such fluidity is rare, however, and flow rates are more commonly measured in meters per hour or even meters per day. As suggested by the scene in Figure 7.2, which shows basaltic lava destroying a house in Hawaii, flow rates are usually slow enough so that people can easily get out of the way.

The property that causes a substance to resist flowing is **viscosity**. The more viscous a magma, the less fluid it is. Magma viscosity depends on temperature and composition, especially the SiO_2 content. The higher the content, the more viscous the magma. For this reason, rhyolitic magma is always more viscous than basaltic, and andesitic has a viscosity intermediate between the two. As we shall see, magma viscosity has a lot to do with the violence of an eruption.

Figure 7.2 An advancing tongue of basaltic lava setting fire to a house in Kalapana, Hawaii, during an eruption of Kilauea volcano in June 1989. Flames at the edge of the flow are due to burning lawn grass.

[1]Basalt and gabbro contain the same minerals; the two rocks differ only in grain size. Basalt contains small mineral grains, whereas gabbro contains coarse grains. Andesite and rhyolite are fine-grained rocks; diorite and granite are their respective coarse-grained equivalents. See "A Closer Look: Naming Igneous Rocks" for more information on rock names.

Temperature

Magma temperature can sometimes be measured during a volcanic eruption. Because volcanoes are dangerous places and because scientists who study them are not eager to be roasted alive, measurements must be made from a distance using optical devices. Magma temperatures determined in this manner during eruptions range from 1000° to 1200°C.

The higher the temperature, the lower the viscosity of a magma and the more readily it flows. In Figure 7.3, the smooth, ropy-surfaced lava on which the geologist is standing, called *pahoehoe* (a Hawaiian word pronounced pa-hó-e-hó-e), formed from a hot, very fluid basaltic magma. The rubbly, rough-looking lava piled up at the center and left formed from a cooler basaltic magma that had a higher viscosity. Hawaiians call this rough lava *aa* (pronounced ah'-ah).

No matter how hot a magma is when it exits a volcano, the lava soon cools, becomes more viscous, and eventually it is so viscous it slows to a complete halt.

ERUPTION OF MAGMA

Magma, like most other liquids, is less dense than the solid matter from which it forms. Therefore, once

Figure 7.3 Lava flow rate is controlled by viscosity, which in turn is controlled by temperature. They have the same basaltic composition. The formation on which the geologist is standing is pahoehoe lava formed from a very hot, low-viscosity, and therefore fast-moving lava erupted in 1959. The upper flow (the one being sampled), which is relatively cool and therefore very viscous and slow moving, is an aa lava erupted from Kilauea volcano in 1989.

formed, lower-density magma exerts an upward push on any enclosing higher-density rock and slowly forces its way up. There is, of course, a reverse pressure on a rising mass of magma owing to the weight of all the overlying more dense rock. Because this pressure is proportional to depth, it decreases as a magma rises upward.

Pressure controls the amount of gas a magma can dissolve—more gas dissolved at high pressure, less at low. Gas dissolved in a rising magma acts the same way as gas dissolved in soda water. When a bottle of soda is opened, the pressure inside the bottle drops, gas comes out of solution, and bubbles form. Gas dissolved in an upward-moving magma also comes out of solution and forms bubbles as the pressure decreases. What happens to these bubbles determines whether an eruption will be explosive or nonexplosive.

Nonexplosive Eruptions

People tend to regard any volcanic eruption as a hazardous event that should be avoided. However, geologists have discovered that basaltic volcanoes, such as those in Hawaii, are comparatively safe because they generally erupt nonexplosively.

The differences between nonexplosive and explosive eruptions are largely a function of magma viscosity and dissolved gas content. Nonexplosive eruptions are characteristic of low-viscosity magmas and low-dissolved gas levels. Basaltic magma has a lower SiO_2 content, a higher temperature, and therefore a lower viscosity than andesitic or rhyolitic magmas; it also has a lower content of dissolved gas than either of the other magma types. Eruptions of basaltic magma are rarely explosive.

Even though they are nonexplosive, basaltic magma eruptions can be spectacular. Gas bubbles in a low-viscosity basaltic magma will rise rapidly upward, like the gas bubbles in a glass of soda water. If basaltic magma moves rapidly upward to the Earth's surface, so that the pressure exerted by overlying rock drops quickly, gas can come out of solution so rapidly that when the magma starts erupting the froth of bubbles can cause spectacular fountaining (Fig. 7.4). When fountaining dies down because most of the dissolved gas has come out of solution and escaped, the hot, fluid lava emerging from the vent flows rapidly downslope (Fig. 7.5). As the lava cools and continues to lose dissolved gases, its viscosity increases and the character of flow changes. The very fluid initial lava forms thin pahoehoe flows, but with increasing viscosity the rate of movement slows and the cooler, stickier lava is transformed into a slow-moving aa flow. Thus, during a single, nonexplosive, Hawaiian-type eruption, pahoehoe and aa may be formed from the same batch of magma.

***Figure* 7.4** Fountaining starts an eruption of a basaltic volcano in Iceland. Use of a telephoto lens foreshortens the field of view. The geologists who are measuring the height of the fountain (200 m) are many hundreds of meters away from the erupting lava.

***Figure* 7.5** This stream of low-viscosity (and therefore very hot), basaltic lava moving smoothly away from an eruptive vent demonstrates how fluid and free flowing lava can be. The temperature of the lava is about 1100°C. The eruption occurred in Hawaii in 1983.

As a hot, low-viscosity basaltic lava cools and the viscosity increases, gas bubbles find it increasingly difficult to escape. When the lava finally solidifies to rock, the last-formed bubbles become trapped and their form is preserved. These bubble holes are called *vesicles*, and the texture they produce in an igneous rock is said to be *vesicular*.

Explosive Eruptions

Viscous magmas—both andesitic and rhyolitic—have higher silica content and erupt at lower temperatures than basaltic magma; they also tend to have high dissolved-gas contents; the combination of composition, temperature, and gas content is a recipe for an explosive eruption. As the gas-charged, viscous magma rises, the gas comes out of solution and bubbles form, but they cannot escape from the sticky, viscous magma. If the rate at which the magma rises (and hence the rate at which bubbles form) is rapid, the bubbles can shatter a viscous magma into a cloud of tiny, red-hot fragments.

A fragment of hot, shattered magma, or any other fragment of rock ejected during an explosive volcanic eruption, is called a **pyroclast** (named from the Greek words *pyro*, meaning fire, and *klastos*, meaning broken). Geologists refer to a deposit of unconsolidated (loose) pyroclasts as **tephra**, a Greek term for ash;

when the pyroclasts in a deposit of tephra are consolidated (cemented together), the result is a **pyroclastic rock**. The terms used to describe tephra of different size are listed in Table 7.1 and illustrated in Figure 7.6.

Before proceeding, let's review the different materials erupted from volcanoes.

1. Lava is magma that oozes out of a volcano and flows over the landscape.

2. Pyroclasts are broken bits of rock and hot fragments of viscous magma shattered as a result of gas escape. An unconsolidated deposit of pyroclasts is called tephra.

***Table* 7.1** Names for Tephra and Pyroclastic Rock

Average Particle	Tephra (unconsolidated material)	Pyroclastic Rock (consolidated material)
> 64 mm	Bombs	Agglomerate
2-64 mm	Lapilli	Lapilli tuff
<2 mm	Ash[a]	Ash tuff

[a]The word *ash* is misleading because it means, strictly, the solid matter left after something flammable, such as wood, has burned. However, the fine tephra thrown out by volcanoes during violent eruptions looks so much like true ash that it has become customary to use the word for this material as well.

A.

B.

C.

***Figure* 7.6**   Tephra. A. Large spindle-shaped pyroclasts up to 50 cm in length cover the surface of a tephra cone on Haleakala volcano, Maui. B. Intermediate-sized tephra called lapilli cover the Kau Desert, Hawaii. The coin is about 1 cm in diameter. C. Volcanic ash, the smallest-sized tephra, blankets a farm in Oregon following the eruption of Mount St. Helens in 1980.

3. Volcanic gases, which are mainly water vapor, carbon dioxide and sulfur, are emitted before, during, and after the eruption of lava or pyroclasts.

Eruption Columns and Tephra Falls

As rising magma approaches the Earth's surface, the rapid drop in pressure causes dissolved gas to bubble furiously, like a violently shaken bottle of soda. As a result of the bubbling, a viscous magma can break into a mass of hot, glassy pyroclasts; the resulting mixture of hot gas and pyroclasts produces a violent upward thrust that culminates in an explosive eruption. The hot, turbulent mixture of gas and pyroclasts rises rapidly in the cooler air above the volcano to form an *eruption column* that may reach as high as 45 km (28 mi) in the atmosphere. The opening figure for this chapter shows a huge eruption column rising from Mount St. Helens. The rising, buoyant column is driven by heat energy released from hot, newly formed pyroclasts. At a height where the density of the col-

umn equals that of the surrounding atmosphere, the column begins to spread laterally to form a mushroom-shaped eruption cloud.

As an eruption cloud begins to drift with the upper atmospheric winds, the pyroclasts fall out and eventually accumulate on the ground as tephra deposits. During exceptionally explosive eruptions, tephra can be carried as far as 1500 km or more. Some eruption columns reach such great heights that winds are able to transport the pyroclasts and gases completely around the world. This was the case in the eruption of Krakatau. As mentioned in this chapter's opening essay, such atmospheric pollution, by blocking incoming solar energy, can lower average temperatures at the land surface for a year or more and cause spectacular sunsets as the Sun's rays are refracted by the airborne particles.

Pyroclastic Flows

A hot, highly mobile flow of tephra that rushes down the flank of a volcano during a major eruption is called a *pyroclastic flow*. Such a flow, which is often referred to by the French term *nuée ardente* (glowing cloud), occurs when the mixture of red-hot tephra and searing gases is too dense to rise upward. Pyroclastic flows are among the most devastating and lethal forms of volcanic eruptions. Observations of historic pyroclastic flows show that they can travel 100 km or more from source vents and reach velocities of more than 700 km/h. One of the most destructive, on the Caribbean island of Martinique in 1902, rushed down the flanks of Mount Peleé volcano and overwhelmed the city of Saint Pierre, instantly killing 29,000 people. The term *nuée ardente* was coined by the French geologists who investigated the disaster at Saint Pierre.

Lateral Blasts

The 1980 eruption of Mount St. Helens displayed many features of a typical large, explosive eruption. Nevertheless, the magnitude of the event caught geologists by surprise. The events leading to this eruption are shown diagrammatically in Figure 7.7. As magma moved upward under the volcano, the northern flank of the mountain began to bulge upward and outward. Finally, the bulging flank became unstable, broke loose, and quickly slid toward the valley as a gigantic landslide of rock and glacier ice. The landslide exposed the mass of hot magma in the core of the volcano. With the lid of rock removed, dissolved gases bubbled so furiously that a mighty blast resulted, blowing a mixture of pulverized rock, pyroclasts, and hot gases sideways as well as upward. The sideways blast, initially traveling at the speed of sound, roared across the landscape, killing David Johnson and others in the blast zone. Within the devastated area which extends as much as 30 km from the crater and covers some 600 km², trees were blasted to the ground and covered with hot debris.

Although Mount St. Helens provides the best documented recent example of a lateral blast, a closely similar eruption of Bezmianny volcano in Kamchatka (Eastern Siberia) in 1956 also produced a devastating lateral blast, a high eruption column, and associated pyroclastic flows.

Types of Volcanoes

The name *volcano* is applied to any vent hole or opening from which lava, pyroclasts, or gases are erupted. The volcanic shape has a lot to do with the kind of magma erupted and with the relative proportions of lava and pyroclasts. There are three shape-related terms more specific than the general term *volcano*: shield volcano, tephra cone volcano, and stratovolcano.

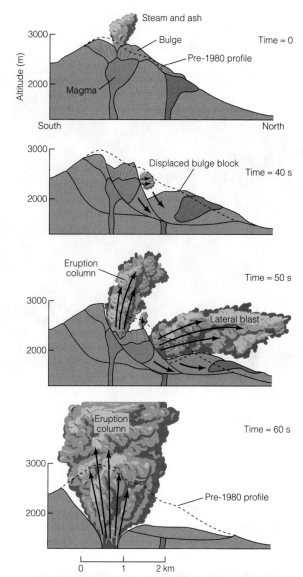

Figure 7.7   Sequence of events leading to the eruption of Mount St. Helens on May 18, 1980. Time ≈0: Earthquakes and then puffs of steam and ash indicate that magma is rising: the north face of the mountain bulges alarmingly. Time ≈40s: an earthquake shakes the mountain and the bulge breaks loose and slides downward. This reduced the pressure on the magma and initiated the lateral blast that killed David Johnson. Time ≈50s: the violence of the eruption causes a second block to slide downward, exposing more of the magma and initiating an eruption column. Time ≈60s: the eruption increases in intensity. The eruption column carries volcanic ash as high as 19 km into the atmosphere.

The kind of volcano that is easiest to visualize is one built up of successive flows of very fluid lava. Such lavas are capable of flowing great distances down gentle slopes and of forming thin sheets of nearly uniform thickness. Eventually, the pile of lava builds up a **shield volcano**, which is a broad, roughly dome-shaped mound with an average surface slope of only 5°

Figure 7.8   Mauna Kea, a 4200-m-high shield volcano on Hawaii, as seen from Mauna Loa. Note the gentle slopes formed by highly fluid basaltic lava. The view is almost directly north. A pahoehoe flow is in the foreground on the northeast flank of Mauna Loa.

Figure 7.9   Tephra cones. A. Two small tephra cones forming as a result of an eruption of andesitic lava in Kivu, Zaire. Arcs of lights are caused by the eruption of red-hot lapilli and bombs. B. Tephra cone in Arizona built from lapilli-sized basaltic tephra. Note the small basaltic lava flow coming from the base of the cone.

A.

B.

Figure 7.10 Mount Fuji, Japan, a snow-clad giant that towers over the surrounding countryside, displays the classic profile of a stratovolcano.

near the summit and about 10° on the flanks (Fig. 7.8). Shield volcanoes are characteristically formed by the eruption of basaltic lava; the proportions of ash and other pyroclasts are small. Hawaii, Tahiti, Samoa, the Galapagos, and many other oceanic islands are the upper portions of large shield volcanoes.

Rhyolitic and andesitic volcanoes tend to eject a large proportion of pyroclasts and therefore to be surrounded by layers of tephra. As the debris showers down, a steep-sided volcano, composed entirely of tephra and therefore called a **tephra cone,** builds up around the vent (Fig. 7.9). The slope of the cone is determined by the size of the pyroclasts.

Large, long-lived volcanoes, particularly those of andesitic composition, emit a combination of lava flows and pyroclasts. **Stratovolcanoes** are defined as steep conical mounds consisting of layers of lava and tephra. The volume of tephra may equal or exceed the volume of the lava. The slopes of stratovolcanoes, which may be thousands of meters high, are steep like those of tephra cones. The slope is about 30°, near the summit of a stratovolcano, and 6° to 10° at the base.

The beautiful, steep-sided cones of stratovolcanoes are among Earth's most picturesque sights (Fig. 7.10). The snow-capped peak of Mount Fuji in Japan has inspired poets and writers for centuries. Mount Rainier and Mount Baker in Washington and Mount Hood in Oregon are majestic examples in North America.

Many large volcanoes, especially shield and stratovolcanoes, are marked near their summit by a large depression. This is a **caldera,** a roughly circular, steep-walled basin several kilometers or more in diameter. Calderas form after the partial emptying of a magma chamber in a volcanic eruption (Figs. 7.11 and 7.12). Rapid ejection of magma during an eruption can leave the magma chamber empty or partly empty. The now-unsupported roof of the chamber slowly sinks under its own weight, like a snow-laden roof on a shaky barn, dropping downward on a ring of steep vertical fractures. Subsequent eruptions commonly occur along these fractures. Crater Lake, Oregon, occupies a circular caldera 8 km in diameter, formed after an explosive pyroclastic eruption about 6600 years ago (Fig. 7.12). The volcano that erupted has been posthumously named Mount Mazama.

Sometimes lava reaches the Earth's surface through a vent that is an elongate fracture in the crust rather than through a small circular opening. Extrusion of lava from an extended fracture is called a *fissure eruption.* (In this case the fissure is the volcano.) Such eruptions, which are often very dramatic, are characteristically associated with basaltic magma, and the lavas resulting from a fissure eruption on land tend to spread widely and to create flat lava plains.

An eruption in 1783, in Iceland, known as the Laki eruption, was a fissure eruption, occurring along a

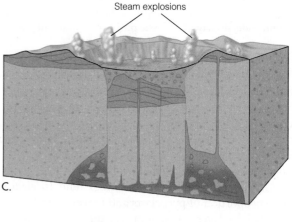

Figure 7.11 Crater Lake, Oregon, occupies a caldera 8 km in diameter that crowns the summit of a once lofty stratovolcano, posthumously called Mount Mazama. Wizard Island is a small tephra cone that formed after the collapse that created the caldera.

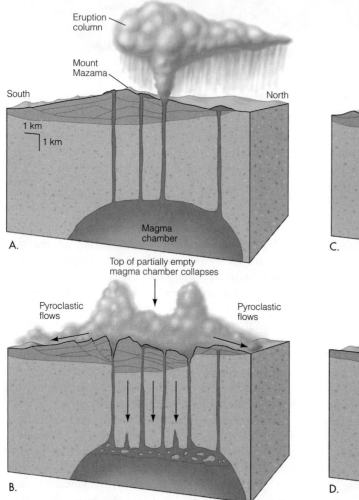

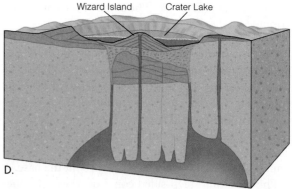

Figure 7.12 Sequence of events that formed Crater Lake following the eruption of Mount Mazama 6600 years ago. A. Eruptive column of tephra rises from the flank of Mount Mazama. B. The eruption reaches a climax. Dense clouds of ash fill the air, and the hot pyroclastic flows sweep down the mountain side. C. The top of Mount Mazama collapses into the partly empty magma chamber, forming a caldera 10 km in diameter. D. During a final phase of eruption, Wizard Island formed. The water-filled caldera is Crater Lake, shown in Figure 7.11.

fracture 32 km long. Lava flowed 64 km outward from one side of the fracture and nearly 48 km outward from the other side. Altogether the Laki eruption covered an area of 588 km². The volume of the lava extruded was 12 km³, making this the largest lava flow in historic times. It was also one of the most deadly, destroying homes and food supplies and covering vast areas of farmlands. Famine followed and 9336 people died.

There is good evidence that larger fissure eruptions occurred in prehistoric times. The Roza flow, a great prehistoric sheet of basaltic lava in eastern Washington State, can be traced over 22,000 km² and shown to have a volume of 650 km³.

VOLCANIC HAZARDS

Volcanic eruptions are not rare events. Every year about 50 volcanoes erupt somewhere on Earth. Eruptions of basaltic volcanoes are rarely dangerous, and basaltic eruptions are far more common than andesitic and rhyolitic eruptions. Though less common, pyroclastic eruptions from andesitic or rhyolitic stratovolcanoes, such as Mount St. Helens and Krakatau, do occur and can be disastrous, since millions of people live on or close to stratovolcanoes.

Eruptions of stratovolcanoes present five kinds of hazards:

1. Hot, rapidly moving pyroclastic flows (*nuée ardente*) and lateral blasts may overwhelm people before they can run away. The tragedies of Mount Peleé in 1902 and Mount St. Helens in 1980 are examples.

2. Tephra and hot, poisonous gases may bury people or suffocate them. Such a tragedy occurred in A.D.79 when Mount Vesuvius, the supposedly dormant volcano in southern Italy, burst to life. First, hot, poisonous gases killed people in the nearby Roman city of Pompeii, and then tephra buried them (Fig. 7.13).

3. Tephra can be dangerous long after an eruption has ceased. Rain or meltwater from snow can loosen tephra piled on a steep volcanic slope and start a deadly mudflow. In 1985, following a small and otherwise nondangerous eruption of the Colombian volcano Nevado del Ruíz, massive mudflows were formed by melting glaciers. The mudflows moved swiftly down the mountain and killed 20,000 people.

4. Violent undersea or coastal eruptions can cause tsunami. Set off by the eruption of Krakatau, tsunami killed more than 36,000 coast dwellers on Java and Sumatra.

Figure 7.13 Evidence of an ancient disaster. Casts of bodies of five citizens of Pompeii, Italy, killed during the eruption of Mount Vesuvius in A.D. 79. Death was caused by poisonous gases, and then the bodies were buried by lapilli. Over the centuries, the bodies decayed, but the body shapes were imprinted in the tephra blanket. Excavators who discovered the imprints carefully recorded them with plaster casts.

5. A tephra eruption may wreak such havoc on agricultural land and livestock that people die from famine. Tephra can also overwhelm cities and other sites of human activities. A recent eruption of Mount Pinatubo in the Philippines, which caused the destruction of a nearby U.S. airforce base, is an example.

Since A.D.1800 there have been 18 volcanic eruptions in which a thousand or more people died (Table 7.2). It is certain that other violent and dangerous eruptions will occur in the future. Likely candidates for dangerous eruptions in the United States are volcanoes in the states of Oregon, Washington, and Alaska. Potentially dangerous volcanoes are also to be found in Japan, the Philippines, New Guinea, New Zealand, Indonesia, the countries of Central and South America, and the Caribbean islands. To some extent, volcanic hazards can be anticipated provided experts can gather data before, during, and after eruptions. The experts can then advise civil authorities when to implement hazard warnings and when to move endangered populations to areas of lower risk.

Table 7.2 Volcanic Disasters Since A.D. 1800 in Which a Thousand or More People Lost Their Lives

Volcano	Country	Year	Primary Cause of Fatalities			
			Pyroclastic Eruption	Mudflow	Tsunami	Famine
Mayon	Philippines	1814	1,200			
Tambora	Indonesia	1815	12,000			80,000
Galunggung	Indonesia	1822	1,500	4,000		
Mayon	Philippines	1825		1,500		
Awu	Indonesia	1826		3,000		
Cotopaxi	Ecuador	1877		1,000		
Krakatau	Indonesia	1883			36,417	
Awu	Indonesia	1892		1,532		
Soufriere	Saint Vincent	1902	1,565			
Mount Peleé	Martinique	1902	29,000			
Santa Maria	Guatemala	1902	6,000			
Taal	Philippines	1911	1,332			
Kelud	Indonesia	1919		5,110		
Merapi	Indonesia	1930	1,300			
Lamington	Papua-New Guinea	1951	2,942			
Agung	Indonesia	1963	1,900			
El Chichón	Mexico	1982	1,700			
Nevado del Ruíz	Colombia	1985		25,000		

Source: From a Report by the Task Group for the International Decade of Natural Disaster Reduction. Published in *Bull. Volcan. Soc. Japan,* Series 2, Vol. 35, #a, 1990, pp. 80–95, 1990.

AFTER AN ERUPTION

When a volcanic eruption spreads lava or tephra across the land, a blanket of fresh new rock is found. In this manner volcanism renews the land surface. Because volcanism occurs on each of the terrestrial planets, surface renewal by volcanism is an important planetary process.

Surface renewal is an especially important process for life on Earth. New rock means new supplies of the fertilizer elements needed by plants. Lava and volcanic ash, when subjected to weathering, produce very fertile soils. Some of the richest agricultural land in Italy, near Naples, has volcanic soil developed on tephra. Japan, the North Island of New Zealand, the Hawaiian islands, the Philippines, and Indonesia are other places where rich volcanic soils produce high agricultural yields.

It is remarkable how quickly the land recovers after an eruption. Within a year of the Mount St. Helens eruption, trees and other plants had started to sprout in the area devastated by the lateral blast of 1980; animals started to return to the area to graze as soon as the plants were big enough.

Agricultural land can also be worked very quickly after an eruption. Although the eruption of Mount Pinatubo caused great distress to the local population, local farmers planted crops in the volcanic ash as soon as the eruption stopped. Plants can even grow on recently erupted lava. In Hawaii, papaya trees have been reported to grow on basaltic lava within two years after the lava was erupted.

From the human viewpoint, volcanism has bad features and good. The bad are the hazards and dangers associated with eruptions and the effects of volcanic dust on the climate. The good is the production of **volcanic rock** from which rich volcanic soils develop.

INTRUSION OF MAGMA

Now that we have considered what happens to magma when it erupts on the Earth's surface, we turn to the question of how magma works its way upward and how igneous rocks are formed from magma.

Beneath every volcano lie a complex of chambers and channelways through which magma reaches the surface. Naturally, we cannot study the magmatic channels of an active volcano, but we can look at ancient cones that have been laid bare by erosion, as seen in Figure 7.14. What we find is that these ancient channelways are filled by igneous rock because they are the underground sites where magma solidified.

All bodies of what we call intrusive igneous rock, regardless of shape or size, are called **plutons**, after

A.

B.

Figure 7.14 Shiprock, New Mexico. A. The conical tephra cone that once surrounded this volcanic neck has been removed by erosion. B. Diagram of the way the original volcano may have appeared prior to erosion.

Pluto, the Greek god of the underworld. The magma that formed a pluton did not originate where we now find the pluton. Rather, the magma was squeezed upward from the place where it formed, thereby intruding the overlying rock (thus, the term *intrusive igneous rock*).

Plutons are given special names depending on shape and size (Fig. 7.15):

1. Dikes, sills, and laccoliths: tabular, parallel-sided sheets of igneous rock. **Dikes** cut across the layering of the intruded rock, **sills** are parallel to the layering, and **laccoliths** are sills that cause the intruded rocks to bend upward.

2. Volcanic pipes and necks: a **volcanic pipe** is a cylindrical conduit of igneous rock below a volcanic vent. A **volcanic neck** is a pipe laid bare by erosion (Fig. 7.14).

3. Batholiths and stocks: intrusive igneous bodies of irregular shape that cut across the layering of the intruded rock. **Stocks** are small (less than 10 km in maximum dimension), and **batholiths** are huge (up to 1000 km in length and 250 km wide).

THE ORIGIN OF MAGMAS AND IGNEOUS ROCKS

We come now to the most difficult but also one of the most interesting questions concerning magma, volcanoes, and igneous rocks: where do magmas form, and why are there three major kinds of magma (basaltic, andesitic, and rhyolitic)? A great many hypotheses have been proposed over the years, but only two have stood the repeated test of investigation and experimentation.

The first hypothesis arose from the pioneering studies carried out by a Canadian scientist, Norman L. Bowen, early in the twentieth century. Bowen discovered by laboratory experiment that, as a magma cools, minerals crystallize in a specific sequence. Furthermore, he discovered that the first-formed minerals later react with the cooler, remaining magma to form different minerals. These reactions, he discovered, also follow a definite sequence. The sequence of reactions is now called **Bowen's reaction series**.

Bowen reasoned that his reaction series could account for the different magmas in the following way.

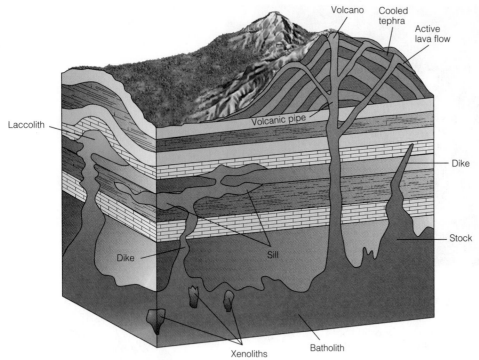

Figure 7.15 Diagrammatic section of part of the crust showing the various forms assumed by plutons.

Suppose that the only primary magma is basaltic magma, formed deep in the mantle, and suppose further that mineral grains, once formed in this primary magma, are somehow separated from the remaining magma. What kind of rock, Bowen asked himself, might form from the separated minerals, and what kind of magma might remain? (See "A Closer Look: Naming Igneous Rocks.") The first mineral to crystallize from basaltic magma is olivine, and olivine contains a lower amount of SiO_2 than the magma. This means that by separating olivine, the remaining magma contains more SiO_2. Bowen called the separation process *fractional crystallization* (Fig. 7.16), and he used his laboratory research to demonstrate that by separating minerals and magma at different stages in the crystallization and reaction process, a single magma could be changed from basaltic to andesitic and rhyolitic, as shown in Figure 7.17.

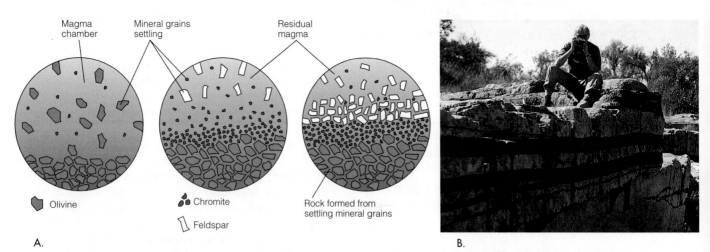

Figure 7.16 Fractional crystallization. A. Grains of three minerals—olivine, chromite (chromium-iron oxide), and feldspar—settle one after the other to the bottom of a magma chamber, producing three types of rocks whose compositions differ considerably from that of the parent magma. B. Layers of plagioclase (light gray) and chromite (black) formed by fractional crystallization in the Bushveld Igneous Complex, South Africa.

A CLOSER LOOK

Naming Igneous Rocks

Igneous rock forms by the cooling and solidification of magma. **Extrusive igneous rocks** are those formed by solidification of lava; **intrusive igneous rocks** are those formed when magma solidifies within the crust or mantle. Both extrusive and intrusive igneous rocks are classified and named on the basis of rock texture and mineral assemblage.

Naming by Texture
The most obvious textural feature of an igneous rock is the size of its mineral grains. Lava cools so rapidly that mineral grains do not have sufficient time to become large. As a result, extrusive igneous rocks are fine-grained (individual grains being less than 2 mm in diameter). Some lavas cool so rapidly that the rocks they form are glassy. Figures C7.1A and B are examples of a glassy and a fine-grained igneous rock, respectively.

Intrusive igneous rock tends to be coarse-grained because magma that solidifies in the crust or mantle cools slowly and has sufficient time to form large mineral grains. Figure C7.1C is an example of a coarse-grained igneous rock.

One special texture involves a distinctive mixture of large and small grains. Rock of such a texture is called a **porphyry**, meaning an intrusive igneous rock consisting of coarse mineral grains scattered through a mixture of fine mineral grains, as shown in Figure C7.1D. The large grains in a porphyry are formed in the same way those of any other coarse-grained igneous rocks are formed—by slow cooling of magma in the crust or mantle; the fine-grained mass that encloses the coarse grains provides evidence that partly solidified magma moved quickly upward. In the new setting, the magma cooled rapidly, and as a result, the later mineral grains are all tiny.

Naming by Mineral Assemblage
When magma or lava of a given composition solidifies, the mineral assemblage that forms is the same for intrusive and extrusive rocks; the only differences are textural. Once the

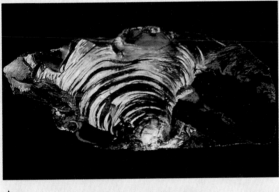

A.

B.

C.

D.

Figure C7.1 Different textures in igneous rock. A. Obsidian, a wholly glassy igneous rock (extrusive). B. Basalt, a fine-grained igneous rock (extrusive). C. Gabbro, a coarse-grained igneous rock (intrusive). D. Basalt porphyry (extrusive). Sample A has the composition of a rhyolite (Table C7.1), but B, C, and D have the same mineral assemblage—feldspar (white), pyroxene (dark green to black), and olivine (pale brown).

texture of an igneous rock has been determined, therefore, specimens are named on the basis of mineral assemblage, as shown in Table C7.1 and Figure C7.2.

All common igneous rocks are composed of one or more of these six minerals or mineral groups: quartz, feldspar, mica (both muscovite and biotite), amphibole, pyroxene, and olivine. Although mineral assemblages are gradational, the common igneous rocks can be divided into four families based on one key feature: the presence or absence of quartz and olivine (Table C7.1).

Varieties of Pyroclastic Rocks

There is an old saying that pyroclasts are igneous on the way up and sedimentary on the way down. As a result, pyroclastic rocks are transitional between igneous and sedimentary. They are called **agglomerates** when tephra is bomb sized,

or **tuffs** when the pieces are either lapilli or ash. The igneous origin of a pyroclastic rock is indicated by the name for the mineral assemblage. For example, we refer to a rock of appropriate mineral assemblage as an andesite tuff.

Tephra is converted to pyroclastic rock in two ways. The first, and most common, way is through the addition of a cementing agent, such as quartz or calcite introduced by groundwater. Figure C7.3A is an example of a rhyolitic tuff formed by cementation. The second way tephra is transformed to pyroclastic rock is through the welding of hot, glassy, ash particles. When ash is very hot and plastic, the individual particles can fuse together to form a glassy pyroclastic rock. Such a rock is called **welded tuff** (Fig. C7.3B).

Uses of Igneous Rock

Igneous rocks have many uses. Granites, diorites, and gabbros are very attractive when polished and are widely used

Figure C7.2 Three coarse-grained igneous rocks. Compare their mineral assemblages by using Table C7.1. Note the change in color from granite (left), which is light colored because it is rich in feldspar and quartz, through diorite (center), to gabbro (right), which is quartz-free and rich in pyroxene and olivine and therefore darker in color. Each specimen is 7 cm across.

The second hypothesis for the origins of the three major magma types, which arose through the work of many people, is the reverse of the Bowen hypothesis. Instead of seeking the answer in cooling processes (the Bowen approach), one can look at melting processes—*fractional melting* rather than fractional crystallization. When a rock starts to melt, the first liquid to form contains more SiO_2, Al_2O_3, K_2O, and

Na_2O than the parent rock. As so often happens, both hypotheses turn out to be correct in part. As discussed later, fractional melting does indeed seem to be the reason there are three major kinds of magma. However, many of the subtle variations in igneous rocks are best explained by the reaction series identified by Bowen.

Table C7.1   Mineral Assemblages of Common Igneous Rocks

		Name[a]	
Key Feature	**Mineral Assemblage**	**Coarse (intrusive)**	**Fine (extrusive)**
Quartz yes, olivine no	Quartz, feldspar, muscovite, biotite, amphibole	Granite	Rhyolite
Quartz no, olivine no	Feldspar, amphibole, pyroxene, biotite	Diorite	Andesite
Quartz no, olivine yes	Feldspar, pyroxene, olivine, biotite	Gabbro	Basalt
Quartz no, olivine yes	Olivine (abundant), pyroxene, feldspar	Peridotite	(none)

[a]When a rock has a porphyritic texture, we use the name determined by the mineral assemblage as an adjective and the term for the texture of the groundmass as the noun. For example, if the groundmass is fine, we call it a rhyolite porphyry; if the groundmass is course, we call it a granite porphyry.

for facing buildings and for ornamental purposes. Basalt is a very tough rock, and fragments of crushed basalt are widely used as the aggregate in concrete and for roadways. Crushed basalt of commerce is sometimes called trap rock. Rhyolite and andesite are sometimes used for roadways, but when rhyolite is glassy (obsidian) it is also used for jewelry and other ornamental purposes.

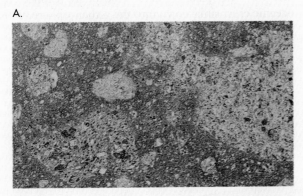

A.

B.

Figure C7.3   Two ways of forming a pyroclastic rock. A. Rhyolite tuff, formed by cementation of lapilli and ash, from Clark County, Nevada. B. Welded tuff from the Jemez Mountains, New Mexico. The dark patches are glassy fragments flattened during welding. Note the fragments of other rocks in the specimen. Both samples are 4 cm across.

Fractional Melting and Magma Types

Evidence supporting the hypothesis that the three types of magma originate through fractional melting comes both from laboratory experiment and from the distribution of the different kinds of volcanoes on our planet. A close relationship exists between plate tectonics and the volcano locations. A summary of present thinking about the distribution is presented in Figure 7.18 and in the following discussion, which should be read with frequent references to this figure.

It has long been known that volcanoes that erupt rhyolitic magma are abundant on the continental crust but are not known on the oceanic crust. This observation suggests that the processes that form rhy-

MAGMA COMPOSITION | MINERALS FORMING | ROCK TYPE

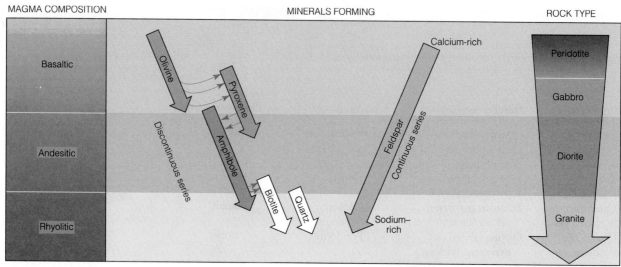

Figure 7.17 Bowen's reaction series demonstrates how the cooling and crystallization of a primary magma of basaltic composition, through reactions between mineral grains and magma followed by separation of mineral grains and magma, can change from basaltic to andesitic to rhy- olitic. Bowen identified two series of reactions: a *continuous series* in which one mineral, feldspar, changes from an initial calcium-rich form to a sodium-rich one, and a *discontinuous series* in which minerals change abruptly, for example from olivine to pyroxene.

olitic magma do not occur in the mantle and must be confined within the continental crust. (If the processes that form rhyolitic magma did occur in the mantle, the rhyolitic magma would rise to the surface regard- less of the kind of crust above and therefore be found in volcanoes on oceanic crust.)

Volcanoes that erupt andesitic magma are found on both the oceanic crust and the continental crust. This suggests that andesitic magma must form in the man- tle and rise up regardless of the nature of the overly- ing crust. However, andesitic volcanoes have a re- stricted geographic distribution. For example, as shown in Figure 7.19, a ring of andesitic volcanoes surrounds the Pacific, forming the so-called Ring of Fire. The Ring of Fire, which geologists also call the *Andesite Line*, is exactly parallel to the plate subduction margins, the places where lithosphere capped by oceanic crust sinks down into the asthenosphere. This suggests that

1. andesitic magma forms as a consequence of plate tectonics, and

2. andesitic magma must be formed by the fractional melting of subducted oceanic crust.

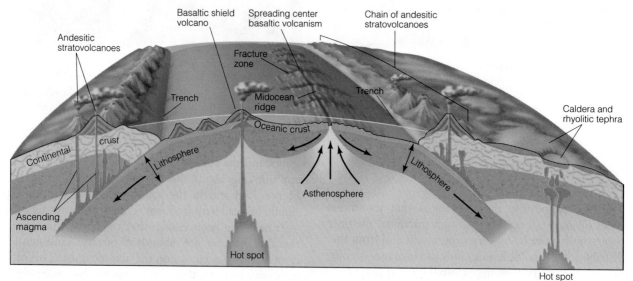

Figure 7.18 Diagram illustrating the locations of the major kinds of volcanoes in a plate-tectonic setting.

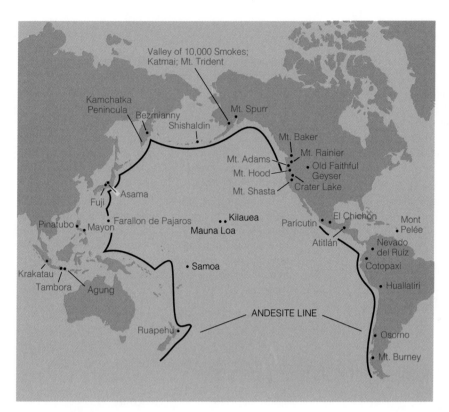

Figure 7.19 The Ring of Fire around the Pacific Ocean basin is formed by andesitic volcanoes. The Ring of Fire is coincident with subduction zones where lithosphere capped with oceanic crust is being subducted into the asthenosphere. Volcanoes in the Pacific Ocean, such as Mauna Loa, erupt basaltic magma but not andesitic magma.

Volcanoes that erupt basaltic magma also occur on both the oceanic and the continental crust. The source of basaltic magma like the source of andesitic magma, must, therefore, be in the mantle. The geographic distribution of basaltic volcanoes does not, however, seem to be related to specific features of the crust or plate tectonics, such as subduction zones. This suggests that basaltic magma must be formed by the melting of the mantle itself and that the magma must rise up regardless of what lies above. The most likely hypothesis for the generation of basaltic magma, therefore, is that it is formed by melting of the mantle as a result of deep-seated convection currents.

The three magma types are now thought to arise in the following ways:

1. Basaltic magma forms by fractional melting of rock in the mantle. The mantle is not thought to be gas-rich, and that is the reason basaltic magma is usually gas-poor.

2. Andesitic magma forms by the fractional melting of oceanic crust (which is basaltic in composition) once that crust has been subducted into the mantle. Oceanic crust is in contact with seawater, and some of the water is carried down with the subducted crust. When fractional melting of this subducted oceanic crust starts at a depth of 80 to 100 km, gas-rich (mainly water vapor) andesitic magma is the result.

3. The continental crust is andesitic in composition. Rhyolitic magma forms by fractional melting of continental crust of andesitic composition. Rhyolitic magma tends to be gas-rich because rocks of the continental crust contain both water vapor and carbon dioxide, and during fractional melting these gases become concentrated in the magma. The vast outpouring of the rhyolitic lava and tephra that created Yellowstone National Park, for example, apparently occurred because the base of the continental crust was heated by basaltic magma in the mantle. The heating caused fractional melting and the production of rhyolitic magma from the andesitic continental crust. Scientists are now testing the hypothesis that the entire continental crust may have formed by a sequence of fractional melting processes—first to form a basaltic oceanic crust, then to form andesitic continental crust by subduction.

It should be apparent from the topics discussed in this chapter that the formation of magma and the subsequent intrusion or eruption of the magma play an essential role in the rock cycle. In the next chapter we examine the long-term history of the crust and the evidence proving that the rock cycle has been an essential part of the Earth system for a very long time.

SUMMARY

1. There are three predominant kinds of magma: basaltic, andesitic, and rhyolitic.

2. The variables that influence the physical properties of magma most are temperature and SiO_2 content. High formation temperature and low SiO_2 content result in fluid, nonviscous magma (basaltic). Lower formation temperature and high SiO_2 contents result in viscous magma (andesitic and rhyolitic).

3. Volcano sizes and shapes depend on the kind of material erupted, viscosity of the lava, and explosiveness of the eruptions.

4. Low-viscosity magma, low in SiO_2, erupts as fluid lavas that build gently sloping shield volcanoes.

5. Viscous magma, rich in SiO_2, erupts mainly as pyroclasts and builds steep-sided tephra cones or stratovolcanoes.

6. Igneous rock may be intrusive (meaning it formed within the crust) or extrusive (meaning it formed on the surface). The texture and grain size of igneous rock indicate how rapidly, and where, the rock cooled.

7. Basaltic magma forms by fractional melting of rock in the mantle. Andesitic magma forms during subduction by fractional melting of basalt in oceanic crust. Rhyolitic magma forms by fractional melting of rock in the continental crust.

8. Two opposing hypotheses have been proposed to explain the origin of the three predominant magmas: Bowen's hypothesis called on fractional crystallization of a single parent magma, whereas the opposing hypothesis calls on fractional melting. Of the two, the fractional melting hypothesis is the most widely accepted.

IMPORTANT TERMS TO REMEMBER

agglomerate *150*
andesite *137*
basalt *137*
batholith *147*
Bowen's reaction series *147*
caldera *143*
dike *147*
diorite *137*
extrusive igneous rock *149*

gabbro *137*
granite *137*
intrusive igneous rock *149*
laccolith *147*
lava *136*
magma *136*
pluton *146*
porphyry *149*
pyroclast *139*

pyroclastic rock *139*
rhyolite *137*
shield volcano *141*
sill *147*
stock *147*
stratovolcano *143*
tephra *139*
tephra cone *143*
tuff *150*

viscosity *137*
volcanic neck *147*
volcanic pipe *147*
volcanic rock *146*
volcano *136*
welded tuff *150*

QUESTIONS FOR REVIEW

1. What's the difference between magma and lava? between lava and pyroclasts? between pyroclasts and tephra?

2. Is the major oxide component of magma SiO_2, MgO, or Al_2O_3? Describe the effect of SiO_2 content on magma fluidity. What effect does temperature have on viscosity?

3. Where inside the Earth does basaltic magma form? What does the term *fractional melting* mean, and what role does it play in formation of basaltic magma?

4. What is the origin of andesitic magma? With what kind of volcanoes are andesitic eruptions associated?

5. What is the origin of rhyolitic magma? Where do you expect to find rhyolitic volcanoes?

6. What are the differences between a shield volcano and a stratovolcano? between a tephra cone and a stratovolcano?

7. The island of Tahiti in the mid-Pacific Ocean is a volcanic island. What kind of volcano formed Tahiti?

8. How might it be possible for fractional crystallization to produce more than one kind of igneous rock from a single magma?

9. Why does a shield volcano like Mauna Loa in Hawaii have a gentle surface slope, whereas a stratovolcano such as Mount Fuji in Japan has steep sides?

10. How do calderas form?

11. Name some ways in which volcanoes affect life on Earth.

12. Why are some volcanic eruptions violent and others not?

13. What is the difference between a dike and a sill? a batholith and a stock?

QUESTIONS FOR REVIEW

1. What are the distinguishing features of pyroclastic rocks? How might you tell the difference between a rhyolite that flowed as a lava and a rhyolitic tuff formed as a result of a pyroclastic eruption?

2. How does a welded tuff differ from a cemented tuff? Could both rocks form as a result of eruptions from the same volcano?

3. If you were vacationing near Mount St. Helens and picked up an igneous rock, how would you identify a sample that had the following qualities? Texture—fine-grained; mineral assemblage—feldspar + amphibole + pyroxene + biotite.

4. On another vacation, this time in Hawaii, you find an igneous rock with the following characteristics: texture—a porphyry, ground-mass coarse-grained; mineral assemblage—feldspar + pyroxene + olivine + biotite. What name would you give to the rock?

QUESTIONS FOR DISCUSSION

1. Spaceships have landed on Venus, Mars, and the Moon. In each case basaltic igneous rocks have been found, but rhyolitic or andesitic rocks, common on the Earth, have not been found. What hypothesis can you suggest about the evolution of Venus, Mars, and the Moon to explain this observation? Can you suggest why the Earth seems to be so different?

2. There are two different hypotheses to explain the sudden disappearance of the dinosaurs. Both suggest that some massive event (or sequence of events) so changed the atmosphere that large animals like dinosaurs could not survive. One hypothesis is that the impact of a giant meteorite was the cause; the other hypothesis is that a series of great volcanic eruptions was the culprit. Suggest ways by which the two hypotheses could be tested.

3. Do some research to see if Mount Vesuvius is still an active volcano. When did it last erupt, and what danger does it pose for people who live near the volcano? How many people live in the area?

At our web site, you can learn more about life in the East African Rift, an active volcanic region.

CHAPTER 8

The Earth's Evolving Crust

● *Tectonics, Erosion, and the Rock Record*

Recently, a team of Greek and British scientists discovered that tectonic forces are stretching Greece and slowly making it grow larger!

A century ago the distances between a series of Greek survey monuments were measured very accurately. In 1988 a scientific team remeasured the distances and found that Greece is now a meter longer. They also discovered that Greece is being twisted so that the southern end, the Peloponnesus, is moving to the southwest relative to the rest of Greece. The reason for the stretching and twisting is that a slice of Mediterranean seafloor is being slowly forced under Greece.

Because Greece is being stretched, we conclude that rock in the Greek crust is being deformed. There is nothing unique or unusual about this conclusion. Evidence that rocks can be deformed is easy to find. If you look at a photograph of the Alps, the Rockies, the Appalachians, or any other mountain range, you will see that once horizontal layers of sedimentary rocks are now tilted and bent.

As we saw in Chapter 4, the source of energy for tectonic forces is the Earth's heat energy. The huge, slow, convective flows of hot rock in the mesosphere and asthenosphere continuously buckle and warp the lithosphere. It is those convective forces that are ultimately the cause of the rock deformation we observe in mountain ranges and that are stretching Greece. Convection is slowly but steadily changing the external face of the Earth, the geosphere reservoir, and in response to those changes the other reservoirs, the hydrosphere, atmosphere, and biosphere, have to adjust and change. Where slow, long-term changes to the Earth systems are concerned, it is the geosphere reservoir that calls the shots.

The bumping, grinding, twisting, and stretching of the crust caused by tectonic forces have been continuous throughout the Earth's long history. However, just as tectonic forces have caused mountains to be raised, so has weathering broken down and worn away the uplifted rocks, and then water, wind, and ice have transported the debris and formed sediment. Sediment becomes sedimentary rock, and, when continents collide, sedimentary rock becomes metamorphic rock. The history of all the tectonic bumps and grinds, indeed the long history of the Earth itself, is read both from deformed rocks and from the debris of erosion, which is all of the sedimentary rocks and the metamorphic rocks formed from them. We therefore begin this chapter on the evolution of the Earth's crust by discussing how the ages of rocks are determined. Then we turn to sediments and sedimentary rocks, followed by metamorphism and metamorphic rocks. The chapter ends with a discussion of the structure of continents, including how they have grown and changed. ●

(opposite) Mount Rundle, near Banff, Alberta, in the Rocky Mountains of Canada. The mountain consists of tilted and twisted strata that are being slowly raised by tectonic forces. Mass wasting, glacial scouring, and stream transport are slowly eroding the mountain at about the same rate it is rising.

SEDIMENTARY STRATA

Like a perpetually restless housekeeper, nature is ceaselessly sweeping regolith off the solid rock beneath it, carrying the sweepings away and depositing them as sediment in river valleys, lakes, and innumerable other places. We can see sediment being transported by trickles of water after a rainfall and by every wind that carries dust. The mud on a lake bottom, the sand on a beach, even the dust on a windowsill is sediment. Because erosion and deposition of rock particles take place almost continuously, we find sediment nearly everywhere. By looking closely, it is possible to see that sediment, regardless of where it is deposited, is piled layer on layer. Studies of the layers reveal a great deal about the way the Earth works.

Sedimentary **stratification** results from the arrangement of sedimentary particles in layers (Fig. 8.1). Each sedimentary **stratum** (plural = **strata**) is a distinct layer of sediment that accumulated at the Earth's surface. The layered arrangement of strata in a body either of sediment or of sedimentary rock is referred to as **bedding**. Each **bed** in a succession of strata can be distinguished from adjacent beds by differences in thickness or some character such as size or shape of particles, or color.

Stratigraphy

The historical information that geologists work with is largely in the form of layered sedimentary rocks that crop out at the Earth's surface or that can be penetrated by drilling. Examine the rocks shown in Figure 8.1. You will see distinct differences in the thicknesses and colors of the many layers. The differences arise from changes in the environment as the sediments accumulated. Because sedimentary rocks carry important clues about past environments at the Earth's surface, the sequence and age of strata provide the basis for reconstructing much of the Earth's environmental history.

The study of strata is called **stratigraphy**. Two straightforward and simple, but nevertheless very powerful, laws underlie stratigraphy. The first is the **law of original horizontality**, which states that sediments are deposited in strata that are horizontal or nearly so and parallel to the Earth's surface. From this generalization we can infer that rock layers now inclined, or even buckled and bent, must have been disturbed since the time they were deposited.

Figure 8.1 Multicolored sedimentary rocks in Capital Reef National Park, Utah. Each layer is a separate stratum.

The second law is the **principle of stratigraphic superposition**, which states that in any sequence of sedimentary strata the order in which the strata were deposited is from the bottom to the top. The red strata at the bottom of Capital Reef (Fig. 8.1) are older than the light-brown stratum at the top of the reef.

Breaks in the Stratigraphic Record

A pile of strata deposited layer after layer without any interruption is said to be *conformable*. Commonly, however, there are substantial breaks or gaps in a sedimentary record. These represent times of nondeposition to which the term **unconformity** is applied. An unconformity records a change in either environmental conditions that caused deposition to cease for a considerable time, or erosion that resulted in loss of part of an earlier-formed depositional record, or a combination of both.

There are three important kinds of unconformities. The first, labeled (1) in Figure 8.2, is a *nonconformity*, where strata overlie igneous or metamorphic rocks. The second and most obvious is the *angular unconformity*, which is marked by angular discontinuity between older and younger strata. It is labeled (2) in Figure 8.2. An angular unconformity implies that the older strata were deformed and then truncated by erosion before the younger layers were deposited across them. The outcrop at Siccar Point, discussed in Chapter 1 (Fig 1.11) and used by James Hutton in his hypothesis of the rock cycle, is obviously an angular unconformity. The second kind of unconformity is called a *disconformity*; it is an irregular surface of erosion between parallel strata. The surface numbered (3) in Figure 8.2 is a disconformity. A disconformity implies a cessation of sedimentation, as well as erosion, but no tilting.

A study of unconformities brings out the close relationship among tectonics, erosion, and sedimentation. (See Chapter 9 for an additional discussion of this topic.) All of the Earth's land surface is a potential surface of unconformity. Some of today's surface will be destroyed by erosion, but some will be covered by sediment and preserved as a record of the present landscape. For example, the Swiss Alps, which were elevated by plate-tectonic movements, are now being eroded away. Meanwhile, the eroded material is being carried away by streams and deposited in the Mediterranean Sea. The Mediterranean seafloor was once dry land, but tectonic forces depressed it, just as tectonic forces elevated the Alps. A surface of unconformity separates the young, river-transported sediments and the older rocks of the seafloor on which the sediments are being piled. In a sense, accumulation in one place compensates for destruction in another. As James Hutton recognized, unconformities provide powerful evidence that interactions between the geosphere on the one hand and the atmosphere, hydrosphere, or biosphere on the other hand have been going on throughout the Earth's long history.

Stratigraphic Correlation

Any distinctive stratum or group of strata that differ from the strata above and below are given a name. The strata in Figure 8.1, for example, are called the Navajo Sandstone. Early in the nineteenth century an English land surveyor, William Smith, while surveying for the construction of new canals, realized that distinctive sedimentary strata throughout western England lay, as he put it, "like slices of bread and butter" in a definite, unvarying sequence. He became familiar with the characteristics of each layer, especially the fossils each contained, and with the sequence of the layers. By looking at a specimen of sedimentary rock collected from anywhere in southern England,

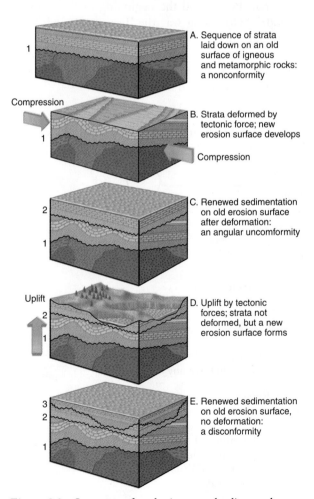

Compression

Compression

Uplift

A. Sequence of strata laid down on an old surface of igneous and metamorphic rocks: a nonconformity

B. Strata deformed by tectonic force; new erosion surface develops

C. Renewed sedimentation on old erosion surface after deformation: an angular uncomformity

D. Uplift by tectonic forces; strata not deformed, but a new erosion surface forms

E. Renewed sedimentation on old erosion surface, no deformation: a disconformity

Figure 8.2 Sequence of geologic events leading to the three kinds of unconformity: (1) nonconformity; (2) angular unconformity; and (3) disconformity.

he could name the stratum from which it had come and, of course, the position of the stratum in the sequence.

Smith did not believe that his discovery reflected any particular scientific principle; he considered it purely practical. Nevertheless, it opened the door to the correlation of sedimentary strata over increasingly wide areas. *Correlation* means the determination of equivalence in age of the succession of strata found in two or more different areas. Smith initially correlated strata over distances of several kilometers and later over tens of kilometers. By means of fossils in the sedimentary rocks, it ultimately became possible to correlate through hundreds and then thousands of kilometers.

THE GEOLOGIC COLUMN

Geologists deal with two kinds of time, relative and absolute. Relative time is the order in which a sequence of past events occurred. Absolute time is the time in years ago, when a specific event happened.

One of the great successes of the nineteenth-century geologists was the demonstration, through stratigraphic correlation, that the relative ages of stratigraphic sequences are the same on all continents. Through worldwide correlation, those nineteenth-century geologists assembled a **geologic column**, which is a composite columnar section containing in chronological order the succession of known strata, fitted together on the basis of their fossils or other evidence of relative age.

Standard names have evolved for the subdivisions of the geologic time units corresponding to the rock units of the geologic column. The units of the geologic time scale, which, like the geologic column, can be used worldwide, are eons, eras, periods, and epochs as shown in Figure 8.3 and Table 8.1.

The scientists who worked out the geologic column and time scale were challenged by the question of absolute time. They knew the relative time order in which strata of the geologic column had formed, but they also wished to know whether the sediments in the strata had accumulated during the same length of time. They sought answers to questions such as these: "How much time elapsed between the end of the Cambrian Period and the beginning of the Permian Period?" "How long was the Tertiary Period?" Absolute ages must be determined in order to answer such questions as the age of the Earth, the age of the ocean, how fast mountain ranges rise, and how long humans have inhabited the Earth.

Table 8.1 Origin of Names for Periods of the Paleozoic, Mesozoic, and Cenozoic Eras, and the Epochs of the Quaternary and Tertiary Periods

Era	Period	Epoch	Origin of Name
Cenozoic	Quaternary[a]	Holocene	Greek for wholly recent
		Pleistocene	Greek for most recent
		Pliocene	Greek for more recent
	Tertiary[a]	Miocene	Greek for less recent
		Oligocene	Greek for slightly recent
		Eocene	Greek for dawn of the recent
		Paleocene	Greek for early dawn of the recent
Mesozoic	Cretaceous		Latin for chalk, after chalk cliffs of southern England and France
		↑	
	Jurassic	Epoch	Jura Mountains, Switzerland and France
	Triassic	Names	Threefold division of rocks in Germany
Paleozoic	Permian	Used	Province of Perm, Russia
	Pennsylvanian	Only	State of Pennsylvania
	Mississippian	by	Mississippi River
	Devonian	Specialists	Devonshire, county of southwest England
	Silurian	↓	Silures, ancient Celtic tribe of Wales
	Ordovician		Ordovices, ancient Celtic tribe of Wales
	Cambrian		Cambria, Roman name for Wales

[a]Derived from the eighteenth- and nineteenth-century geological time scale that separated crustal rocks into a fourfold division of Primary, Secondary, Tertiary, and Quaternary, based largely on relative degree of lithification and deformation.

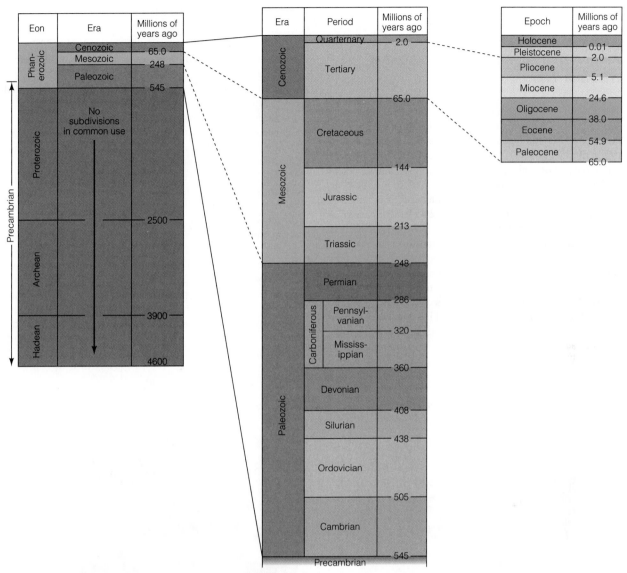

Eon	Era	Millions of years ago
Phanerozoic	Cenozoic	65.0
	Mesozoic	248
	Paleozoic	545
Precambrian Proterozoic	No subdivisions in common use	2500
Archean		3900
Hadean		4600

Era	Period	Millions of years ago
Cenozoic	Quarternary	2.0
	Tertiary	
Mesozoic		65.0
	Cretaceous	
	Jurassic	144
	Triassic	213
Paleozoic		248
	Permian	286
	Carboniferous Pennsyl-vanian	320
	Mississ-ippian	360
	Devonian	
		408
	Silurian	438
	Ordovician	
		505
	Cambrian	
	Precambrian	545

Epoch	Millions of years ago
Holocene	0.01
Pleistocene	2.0
Pliocene	
	5.1
Miocene	
	24.6
Oligocene	38.0
Eocene	
	54.9
Paleocene	
	65.0

Figure 8.3 The geologic time scale. Absolute ages obtained from radiometric dates. Note that the Pennsylvanian and Mississippian Periods are equivalent to the Carboniferous Period of Europe. The time boundary between the Archean and Hadean is uncertain, for no rocks of the Hadean Eon are known on the Earth. Hadean rocks are known to exist on other planets in the solar system.

The discovery of radioactivity in 1896 provided a reliable way to measure absolute geologic time. Radioactivity is a process that runs continuously, that is not reversible, that operates the same way and at the same speed everywhere, and that leaves a continuous record without any gaps. For a discussion of how radioactivity is used to measure absolute ages, see "A Closer Look: Radioactivity and the Measurement of Absolute Time."

SEDIMENT AND SEDIMENTARY ROCK

There are two families of sediment, clastic and chemical. The principal difference between them is the way the sediment is transported.

Clastic sediment (from the Greek word *klastos*, meaning broken) is simply bits of broken rock and minerals that are moved as solid particles. Any individual particle of clastic sediment is a *clast*, and clasts tend to be the rock-forming minerals, such as quartz and feldspar, that are most durable during weathering.

Chemical sediment is transported in solution and deposited when the dissolved minerals are precipitated.

A CLOSER LOOK

Radioactivity and the Measurement of Absolute Time

We learned in Chapter 6 that most chemical elements have several naturally occurring isotopes (atoms with the same atomic number and hence the same chemical properties, but different mass numbers). Most isotopes found in the Earth are stable and not subject to change. However, a few, such as carbon-14 (^{14}C), are radioactive because of an instability in the nucleus and will transform spontaneously to either a more stable isotope of the same chemical element or an isotope of a different chemical element—^{14}C, for example, expels an electron from its nucleus and transforms to ^{14}N.

The rate of transformation—*radioactive decay* rate as it is now more commonly called—is different for each isotope. Careful study of radioactive isotopes in the laboratory has shown that decay rates are unaffected by changes in the chemical and physical environment. Thus, the decay rate of a given isotope is the same in the mantle or a magma as it is in a sedimentary rock. This is a particularly important point because it leads to the conclusion that rates of radioactive decay are not changed by geologic processes and therefore can be used to measure absolute time.

All decay rates follow the same basic law that is depicted in Figure C8.1. The law of radioactive decay, stated in words, is that the *proportion* of parent atoms that decay during each unit of time is always the same. The number of decaying parent atoms continuously decreases, while the number of daughter atoms continuously increases.

The rate of radioactive decay is measured by the *half-life*, which is the time needed for the number of parent atoms to be reduced by one-half. For example, if the half-life of a radioactive isotope is 1 hour and we started an experiment with a mineral containing 1000 radioactive atoms, only 500 parent atoms would remain at the end of an hour and 500 daughter atoms would have formed. At the end of a second hour there would be 250 parent and 750 daughter atoms, and after hour 3, 125 parents and 875 daughters. The half-lives of radioactive isotopes used to measure absolute geologic times are thousands to millions of years long, but the decay law is the same for all isotopes regardless of the length of the half-life.

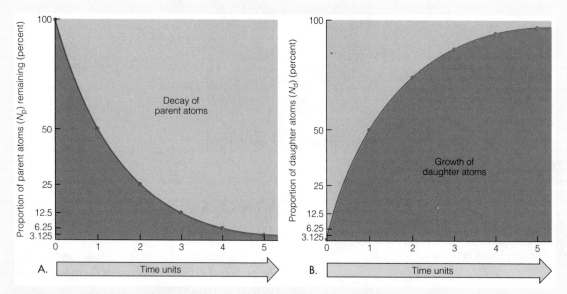

Figure C8.1 Curves illustrating the basic law of radioactivity. A. At time zero, a sample consists of 100 percent radioactive parent atoms. During each time unit, half the atoms remaining decay to daughter atoms. B. At time zero, no daughter atoms are present. After one time unit corresponding to a half-life of the parent atoms, 50 percent of the sample has been converted to daughter atoms. After two time units, 75 percent of the sample is daughter atoms, and 25 percent is parent atoms. After three time units, the percentages are 87.5 and 12.5, respectively. Note that at any given instant N_p, the number of parent atoms remaining, plus N_d, the number of daughter atoms, equals N_o, the number of parent atoms at time zero.

In the graphic illustration of radioactive decay in Figure C8.1, the time units marked are half-lives. Of course, the time units are of equal length, but at the end of each unit the number of parent atoms, and therefore the radioactivity of the sample, has decreased by exactly one-half of the value at the beginning of the unit. Figure C8.1 also shows that the growth of daughter atoms just matches the decline of parent atoms. When the number of remaining parent atoms (N_p) is added to the number of daughter atoms (N_d), the result is N_o, the number of parent atoms that a mineral sample started with. That fact is the key to the use of radioactivity as a means of measuring geologic time and determining ages.

Potassium-Argon ($^{40}K/^{40}Ar$) Dating

We have selected one of the naturally radioactive isotopes, potassium-40 (^{40}K), to illustrate how the absolute time of formation of certain minerals can be determined. Potassium has three natural isotopes: ^{39}K, ^{40}K, and ^{41}K. Only one, ^{40}K, is radioactive, and its half-life is 1.3 billion years. The decay of ^{40}K is interesting because two different decay schemes occur. Twelve percent of the ^{40}K atoms decay to ^{40}Ar, an isotope of the gas argon. The remaining 88 percent of the ^{40}K atoms decay to ^{40}Ca. It is important to know that the fraction of ^{40}K atoms decaying to ^{40}Ar is always 12 percent; the percentage is not affected by changes in physical or chemical conditions.

When a potassium-bearing mineral crystallizes from a magma, or grows within a metamorphic rock, it includes some ^{40}K in its crystal structure. As soon as the mineral is formed, ^{40}Ar and ^{40}Ca daughter atoms start accumulating in the mineral, because they are trapped, like the parent ^{40}K atoms, in the crystal structure. Because the ratio of ^{40}Ar to ^{40}Ca daughter atoms is always the same, it is only necessary to measure either ^{40}Ar or ^{40}Ca daughter atoms in order to know how many ^{40}K atoms have decayed. It is more convenient to measure ^{40}Ar because argon is an element that can be measured very accurately.

All that has to be done to determine the absolute time of eruption of an extrusive igneous rock is to select a potassium-bearing mineral such as an amphibole, a mica, or a feldspar, and measure the amount of parent ^{40}K that remains, as well as the amount of trapped ^{40}Ar. Because the half-life of ^{40}K is known, it is a straightforward matter to calculate the radiometric age—the length of time a mineral has contained its built-in radioactivity clock. What is actually measured, of course, is the time since the mineral formed in a cooling magma. Because the time of mineral formation is effectively the same as the time at which the extrusive igneous rock was formed, the mineral age and the rock age are the same.

Absolute Time and the Geologic Time Scale

Many naturally radioactive isotopes can be used for radiometric dating, but six predominate in geologic studies. These are the two radioactive isotopes of uranium and the single radioactive isotopes of thorium, potassium, rubidium, and carbon. These isotopes occur widely in different minerals and rock types, and they have a wide range of half-lives, so that many geologic materials can be dated radiometrically.

Through the various methods of radiometric dating, geologists have determined the dates of solidification of many bodies of igneous rock. Many such bodies have identifiable positions in the geologic column, and as a result, it becomes possible to date, approximately, a number of the sedimentary layers in the column.

The standard units of the geologic column consist of sedimentary strata containing characteristic fossils, but the typical rocks from which radiometric dates (other than ^{14}C dates) are determined are igneous rocks. It is necessary, therefore, to be sure of the relative time relations between an igneous body that is datable and a sedimentary layer whose fossils closely indicate its position in the column.

Figure C8.2A and B shows how the ages of sedimentary strata are approximated from the ages of igneous bodies. In Figure C8.2A, a sequence of sedimentary strata containing fossils of known relative ages are separated by an unconformity and two disconformities. Intrusive igneous rock A cuts strata 1 and 2 but is truncated by the disconformity at the top of stratum 2. Thus, A must be younger than strata 1 and 2 but older than stratum 3, which was laid down on the erosion surface at the top of stratum 2 and contains weathered fragments of A among the sedimentary particle. Similarly, the combination of dikes and sills that make up the intrusive igneous complex B are truncated by the disconformity at the top of stratum 3, and they must be younger than stratum 3 but older than stratum 4. Lava flow C above the disconformity at the top of stratum 3 must also be younger than stratum 3 and younger than the dike–sill complex B. Lava flow C must be older than stratum 4, however, because it is covered by stratum 4, and lava flow D must be even younger because it overlies stratum 4.

From the radiometric dates of the igneous bodies and the relative ages of the geologic relations shown in Figure C8.2A, inferences can be drawn about the ages of the sedimentary strata as shown in Figure C8.2B.

Through a combination of geologic relations and radiometric dating methods, twentieth-century scientists have been able to fit a scale of absolute time to the geologic column worked out in the nineteenth century. The scale is being continuously refined, and so the numbers given in Figure 8.3 should be considered the best available now. Further work will make them more accurate.

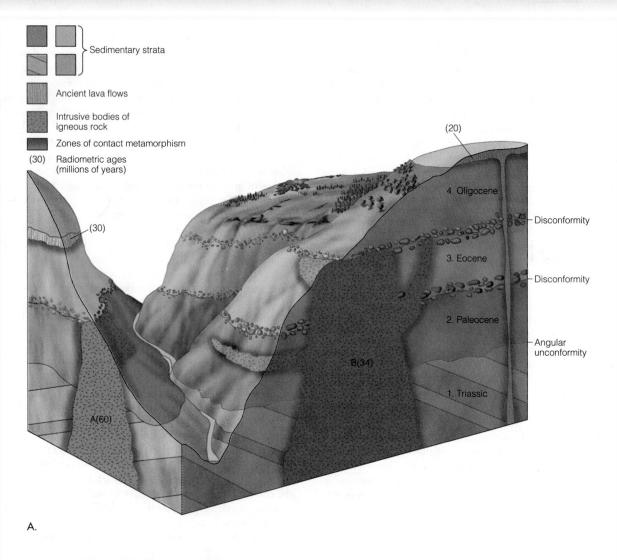

A.

Stratum	Age (Millions of years)	Interpretation
4	<34 (younger than B) <30 (younger than C) >20 (older than D)	Age lies between 20 and 30 million years
3	<60 (younger than A) >34 (older than B) >30 (older than C)	Age lies between 34 and 60 million years
2	>60 (older than A)	Age of both is greater than 60 million years
1	>60 (older than A)	

B.

Figure C8.2 The application of radiometric dating to the geologic column. For method, see the text discussion. A. Idealized section showing sedimentary strata, angular unconformity, disconformities, lava flows, and igneous intrusions. Numbers in brackets are the radiometric ages of the igneous rocks determined by the potassium-argon method. B. Interpretation of the absolute ages of the four groups of sedimentary strata labeled 1 to 4.

A. Sediment

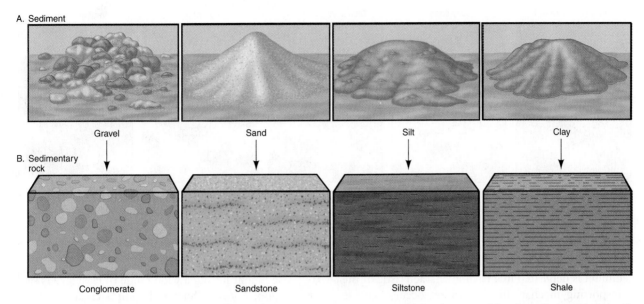

B. Sedimentary rock

Gravel   Sand   Silt   Clay

Conglomerate   Sandstone   Siltstone   Shale

Figure 8.4 Principal kinds of clastic sediment and the sedimentary rocks formed from them. A. Sediment is classified and named for the sizes of the clasts. B. Rock is formed from clastic sediment.

The transformation of sediment to sedimentary rock is called *lithification* (from the Greek *lithos*, meaning stone; hence stone-making). As discussed in Chapter 6, lithification happens either by the addition of a cement or by recrystallization of the sediment particles to a firm, coherent mass.

Clastic Sediment and Clastic Sedimentary Rock

Clast size is the primary basis for classifying clastic sediment and clastic sedimentary rock. Clastic sediment can be divided into four main classes, which from coarsest to finest are gravel, sand, silt, and clay (Fig. 8.4). Gravel is further classified on the basis of dominant clast size into boulder gravel, cobble gravel, and pebble gravel (Table 8.2). The names of the clastic sedimentary rocks corresponding to the various clastic sediments are **conglomerate, sandstone, siltstone, and shale** as listed in Figure 8.4 and Table 8.2.

Clastic sediment is transported in many ways. It may slide or roll down a hillside under the pull of gravity or be carried by a glacier, by the wind, or by flowing water. In each case, when transport ceases, the

Table 8.2. Definition of Clastic Particles, Together with the Sediments and Sedimentary Rocks Formed from Them

Name of Particle	Size Limits of Diameter (mm)[a]	Names of Loose Sediment	Name of Consolidate Rock
Boulder	More than 256	Boulder gravel	Boulder conglomerate[c]
Cobble	64 to 256	Cobble gravel	Cobble conglomerate[c]
Pebble	2 to 64	Pebble gravel	Pebble conglomerate
Sand	1/16 to 2	Sand	Sandstone
Silt	1/256 to 1/16	Silt	Siltstone
Clay[b]	Less than 1/256	Clay	Shale

[a]Note that size limits of sediment classes are powers of 2, just as are memory limits in microcomputers (for example, 2K, 64K, 256K, 512K).

[b]Clay, used in the context of this table, refers to particle size. The term should not be confused with clay minerals, which are definite mineral species.

[c]If the clasts are angular, the rock is called a *breccia* rather than a conglomerate.

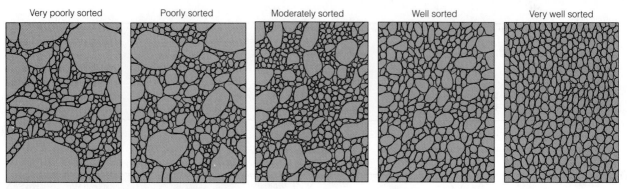

Very poorly sorted   Poorly sorted   Moderately sorted   Well sorted   Very well sorted

Figure 8.5 Clastic sediment ranges from very poorly sorted to very well sorted depending on the extent to which the constituent grains are of equal size.

sediment is deposited in a fashion characteristic of the transporting mechanism. Deposition occurs because of a drop in energy. Sediment transported by wind or water is deposited when the moving air or flowing water slows to a speed at which clasts can no longer be carried. In a general way, grain size in sediment moved by wind or water is related to the speed of the transporting agent: the faster the speed, the larger the clasts that can be moved. When the speed fluctuates, clast sorting occurs. For example, a rapidly flowing river will remove all the fine particles, leaving behind only the largest clasts. As the speed of the flowing water slows, smaller and smaller clasts are dropped, and well-sorted sediment is the result (Fig. 8.5).

Figure 8.6 Varves deposited in a glacial-age lake in southern Connecticut. Each pair of layers in a sequence of varves represents an annual deposit. Light-colored silty layers were deposited in summer, and the dark-colored clayey layers accumulated in winter.

A sediment having a wide range of clast size is said to be poorly sorted. Such sediment is created, for example, by rockfalls, by the sliding of debris down hillslopes, by slumping of loose deposits on the seafloor, by mudflows, and by deposition of debris from glaciers or from floating ice. Some poorly sorted sediments are given specific names. (For example, *till* is a poorly sorted sediment of glacial origin, while the corresponding rock is a *tillite*.)

Some clastic sediment displays a distinctive *alternation* of parallel layers having different clast sizes (Fig. 8.6). Such alternation suggests that some naturally occurring rhythm has influenced sedimentation. A pair of such sedimentary layers deposited over the cycle of a single year is termed a **varve** (Swedish for cycle). Varves are most common in deposits of high-latitude or high-altitude lakes, where there is a strong contrast in seasonal conditions. In spring, as the ice of nearby glaciers starts to melt, the inflow of sediment-laden water in a lake increases and coarse sediment is deposited throughout the spring and summer. With the onset of colder conditions in the autumn, streamflow decreases and ice forms over the lake surface. During autumn and winter, very fine sediment that has remained suspended in the water column slowly settles to form a thinner, darker layer above the coarse, lighter-colored summer layer. Varved lake sediments are common in Scandinavia and New England where they formed beyond the retreating margins of Ice Age glaciers.

Cross bedding refers to sedimentary beds that are inclined with respect to a thicker stratum within which they occur (Fig. 8.7). Cross beds consist of clasts coarser than silt and are the work of turbulent flow in streams, wind, or ocean waves. As they are moved along, the clasts tend to collect in ridges, mounds, or heaps in the form of ripples, waves, or dunes that migrate slowly forward in the direction of the current. Clasts accumulate on the downstream slope of the pile to produce beds having inclinations as great as 35°. The direction in which cross bedding is inclined tells the direction in which the related current of water or air was flowing at the time of deposition.

Many bodies of sediment contain **fossils**, the remains of plants and animals that died and were incorporated and preserved as the sediment accumulated. Sometimes the form of an original plant or animal is preserved (Fig. 8.8), but more commonly the remains are broken and scattered. As we will see in later chapters, especially Chapters 15 to 17, fossils in sedimentary rocks provide important evidence about the history of the biosphere.

Chemical Sediment and Chemical Sedimentary Rock

Chemical sediment forms when dissolved substances precipitate. One common example of precipitation involves the evaporation of seawater or lake water; as the water evaporates, dissolved matter is concentrated, and salts begin to precipitate out as chemical sediment. Chemical sediment formed as the result of evaporation is called an *evaporite*. The most important are halite (NaCl) and gypsum (CaSO$_4$·2H$_2$O); the corresponding rocks are salt and gyprock, respectively. Most of the salt we use in our cooking is mined from beds of salt formed by evaporation.

When chemical sediment forms as a result of biochemical reactions in water, the resulting sediment is said to be *biogenic.* One common example of a biogenic reaction involves the formation of calcium carbonate shells by clams and other aquatic animals. The clams extract both calcium and carbonic acid from the water and use them to lay down layers of solid calcium carbonate. In effect, they make a home for themselves by a biogenic reaction. When the animals die, their shells become biogenic sediment.

Limestone and **dolostone**, containing the minerals calcite and dolomite, respectively, are the most important of the biogenic rocks. Limestone composed entirely of shelly debris is called *coquina*. Other limestones consist of cemented reef organisms *(reef limestone)*, the compacted carbonate shells of minute floating organisms *(chalk)*, and accumulations of tiny round, calcareous accretionary bodies (ooliths) that are 0.5 to 1 mm in diameter *(oolitic limestone)*.

Biogenic sediment can form both in the sea and on land. Trees, bushes, and grasses contribute most of the biogenic material on land. In water-saturated environments, such as bogs or swamps, plant remains accu-

Figure 8.7 Ancient cross-bedded sand dunes that have been converted to sedimentary rock that crops out near Kanab, Utah. The inclination of the cross beds shows that the ancient prevailing winds were blowing from left to right.

mulate to form **peat**, a sediment with a carbon content of about 60 percent. Peat is the initial stage in the development of the biogenic sedimentary rock we call **coal.**

As peat is buried beneath more plant matter and accumulating sand, silt, or clay, both temperature and pressure rise. As millions of years pass, the increased temperature and pressure bring about a series of

Figure 8.8 The fossil remains of a kauri pine lie next to the skeleton of a fossil fish on a bedding plane of a 175-million-year-old shale in Australia.

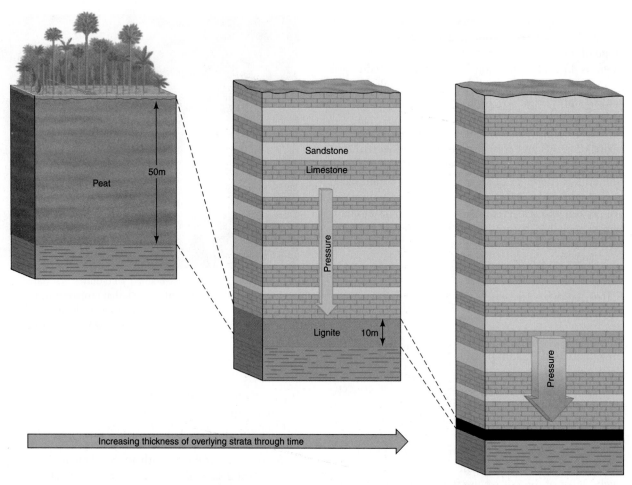

Figure 8.9 Plant matter in peat (a biogenic sediment) is converted to coal (a biogenic sedimentary rock) by decomposition and increasing pressure and temperature as overlying sediments build up. By the time a layer of peat 50 m thick is converted to bituminous coal, its thickness has been reduced by 90 percent. In the process, the proportion of carbon has increased from 60 to 80 percent.

changes. The peat is compressed, water is squeezed out, and the gaseous organic compounds such as methane (CH_4) escape, leaving an increased proportion of carbon. The peat is thereby converted into *lignite* and eventually into *bituminous coal* (Fig. 8.9).

METAMORPHISM: NEW ROCKS FROM OLD

Metamorphic rocks are of particular interest because the changes they undergo happen in the solid state. When tectonic plates move and crustal fragments collide, rocks are squeezed, stretched, bent, heated, and changed in complex ways. Even when a rock has been altered two or more times, however, vestiges of its earlier forms are usually preserved because the changes occurred in the solid state. Solids, unlike liquids and gases, tend to retain a "memory" of the events that changed them. In many ways, therefore, metamorphic rocks are the most complex but also the most interesting of the rock families. In them is preserved the story of all the collisions that have happened to the crust. Deciphering the record is an exceptional challenge for geologists. For example, when continental masses collide because of moving plates, distinctive kinds of metamorphic rocks form along the plate edges. Therefore, it is possible to determine where the boundaries of ancient continents once were by studying the distribution of metamorphic rocks. Geologists also use evidence derived from metamorphic rocks to determine how long plate tectonics has been active on the Earth. So far the evidence suggests that plate tectonics has been operating for at least 2 billion years and probably 3 billion years.

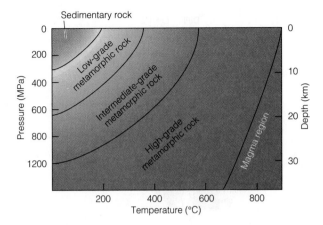

Figure 8.10 Ranges of temperature and pressure (equivalent to depth) under which metamorphism occurs in the crust. At lowest temperatures and pressure, sediment is converted to sedimentary rock. At the highest temperature and pressure, melting commences and the result is magma rather than rock.

The Limits of Metamorphism

Metamorphism describes changes in mineral assemblage and texture in sedimentary and igneous rocks subjected to temperatures above 200°C and pressure in excess of about 300 MPa (the pressure caused by a few thousand meters of overlying rock). There is, of course, an upper limit to metamorphism because at sufficiently high temperatures rock will melt. Remember then: metamorphism refers only to changes in solid rock, not to changes caused by melting. Changes due to melting involve igneous phenomena, as discussed in Chapter 7.

Low-grade metamorphism refers to metamorphic processes that occur at temperatures from about 200°C to 320°C and at relatively low pressures (Fig. 8.10). High-grade metamorphism refers to metamorphic processes at high temperature (above about 550°C) and high pressure. Intermediate-grade metamorphism lies between low and high grade.

Controlling Factors in Metamorphism

In a simplistic way, you can think of metamorphism as cooking. When you cook, what you get to eat depends on what you start with and on the cooking conditions. So too with rocks; the end product is controlled by the initial composition of the rock and by the metamorphic (or cooking) conditions. The chemical composition of a rock undergoing metamorphism

plays a controlling role in the new mineral assemblage; so do changes in temperature and pressure. The ways in which temperature and pressure control metamorphism are not entirely straightforward, however, because they are strongly influenced by such factors as the presence or absence of fluids, how long a rock is subjected to high pressure or high temperature, and whether the rock is simply compressed or is twisted and broken as well as compressed during metamorphism.

Chemical Reactivity Induced by Fluids

The innumerable open spaces between grains in a sedimentary rock and the tiny fractures in many igneous rocks are called pores, and all pores are filled by a watery fluid. The fluid is never pure water, for it always has dissolved in it small amounts of gases such as CO_2 and salts such as $NaCl$ and $CaCl_2$, as well as traces of all the mineral constituents present in the enclosing rock. At high temperature the fluid is more likely to be a vapor than a liquid. Regardless of its composition or state, the *pore fluid* (for that is its best designation) plays a vital role in metamorphism.

When the temperature of, and pressure on, a rock undergoing metamorphism change, so does the composition of the pore fluid. Some of the dissolved constituents move from the pore fluid to the new minerals growing in the metamorphic rock. Other constituents move in the other direction, from the minerals to the pore fluid. In this way the pore fluid serves as a transporting medium that speeds up chemical reactions in much the same way that water in a stew pot speeds up the cooking of a tough piece of meat.

Pressure and Temperature

When a mixture of flour, salt, sugar, yeast, and water is baked, the high temperature causes a series of chemical reactions—new compounds are formed, and the result is a loaf of bread. When rocks are heated, new minerals grow and the result is a metamorphic rock. In the case of the rocks, the cooking is brought about by the Earth's internal heat. Rocks can be heated by burial, by a nearby igneous intrusion, or by a thickening of the crust owing to collision. But burial, collision, and intrusion can also cause a change in pressure. Therefore, whatever the cause of the heating, metamorphism can rarely be considered to be entirely due to the rise in temperature. The effects attributable to changing temperature and pressure must be considered together.

When discussing metamorphic rocks, scientists often use the term *stress* in place of pressure. They do so because stress has the connotation of direction.

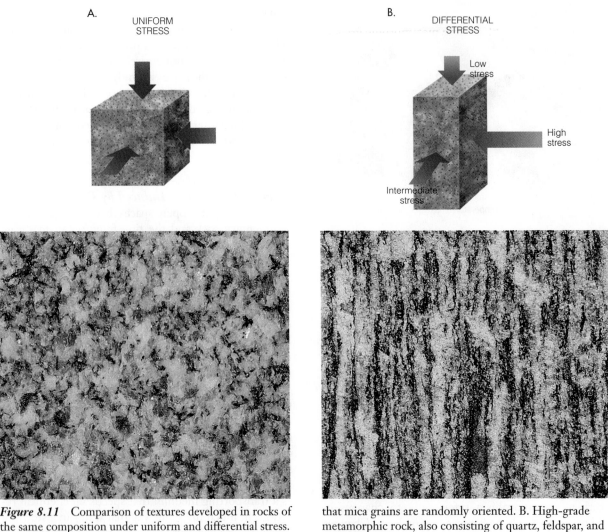

Figure 8.11 Comparison of textures developed in rocks of the same composition under uniform and differential stress. A. Granite, consisting of quartz, feldspar, and mica (the dark mineral) that crystallized under a uniform stress. Note that mica grains are randomly oriented. B. High-grade metamorphic rock, also consisting of quartz, feldspar, and mica, that crystallized under a differential stress. Mica grains are parallel, giving the rock a distinct foliation.

Rocks are solids, and solids can be squeezed more strongly in one direction than another; that is, stress in a solid, unlike stress in a liquid, can be different in different directions. The textures in many metamorphic rocks record **differential stress** (meaning not equal in all directions) during metamorphism. In contrast, igneous rocks have textures formed under **uniform stress** (meaning equal in all directions) because igneous rocks crystallize from liquids.

The most visible effect of metamorphism in a differential stress field involves the texture of silicate minerals, such as micas and chlorites, that contain polymerized $(Si_4O_{10})^{4-}$ sheets. Compare Figures 8.11A and B. Figure 8.11A is a granite that has a typical texture of randomly oriented mineral grains that grew in a uniform stress field. Figure 8.11B is a metamorphic rock containing the same minerals as in A, but this rock formed in a differential stress field. Note that in Figure 8.11B all the biotite (mica) grains (black) are parallel, giving the rock a distinctive texture.

In a metamorphic rock containing sheet-structure minerals, the sheets are oriented perpendicular to the direction of maximum stress, as shown in Figure 8.11B. The parallel sheets produce a planar texture called **foliation**, named from the Latin word *folium*, meaning leaf. Foliated rock tends to split into thin flakes.

It is important to understand that stress can be high or low. It is the magnitude of the stress that determines a mineral assemblage (refer back to Chapter 6 for a discussion of mineral assemblages). Texture, on the other hand, is controlled by differential versus uniform stress. To avoid confusion, geologists often use the term *stress* to discuss texture and *pressure* to discuss mineral assemblages and metamorphic grades.

Metamorphic Mineral Assemblages

Metamorphism produces new mineral assemblages as well as new textures. As temperature and pressure rise, one new mineral assemblage follows another. For any given rock composition, each assemblage is characteristic of a given range of temperature and pressure. A few of these minerals are found rarely (or not at all) in igneous and sedimentary rocks. Therefore their presence in a rock is usually evidence enough that the rock has been metamorphosed. Examples of these metamorphic minerals are chlorite, serpentine, talc, and the three Al_2SiO_5 minerals andalusite, kyanite, and sillimanite. Figure 8.12 illustrates the way mineral assemblages change with the grade of metamorphism as a shale is metamorphosed.

Kinds of Metamorphism

The processes that result from changing temperature and pressure, and that cause the metamorphic changes observed in rocks, can be grouped under the terms *mechanical deformation* and *chemical recrystallization*. Mechanical deformation includes grinding, crushing,

and the development of foliation (Fig. 8.13). Chemical recrystallization includes all the changes in mineral composition, in the growth of new minerals, and the losses of H_2O and CO_2 that occur as rock is heated and squeezed. Different kinds of metamorphism reflect the different levels of importance of the two processes. The two most important kinds of metamorphism are burial and regional metamorphism.

Burial Metamorphism

Sediments, together with interlayered pyroclastics, may attain temperatures in excess of 200°C or more when buried deeply in a sedimentary basin and thus subjected to metamorphism. Abundant pore water is present in buried sediment, and this water speeds up chemical recrystallization and helps new minerals to grow. Because water-filled sediment is weak and acts more like a liquid than a solid, however, the stress during **burial metamorphism** tends to be uniform. As a result, burial metamorphism involves little mechanical deformation, and the metamorphic rock that results lacks foliation. The texture looks like that of an essentially unaltered sedimentary rock, even though the mineral assemblages in the two are completely different from one another.

Burial metamorphism is the first stage of metamorphism in deep sedimentary basins, such as deep-sea trenches on the margins of tectonic plates, in the great

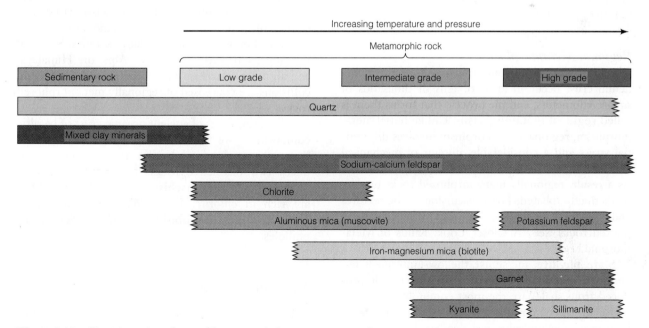

Figure 8.12 Changing mineral assemblages as a shale is metamorphosed from low to high grade. Kyanite and sillimanite have the same composition (Al_2SiO_5) but different crystal structures. They are found only in metamorphic rocks.

A.

B.

Figure 8.13 Development of foliation in a granite by mechanical deformation—the mineral assemblage is unchanged, but the texture changes. From Groothoek, South Africa. A. Undeformed granite consisting of quartz, feldspar, and mica. The dark patch in the center of the field of view is a fragment of amphibolite (a metamorphic rock consisting largely of amphibole) that fell into the magma during intrusion. Foliation is not present. B. The original granitic texture has been completely changed, and the granite has been metamorphosed to a gneiss with a distinct foliation. The dark streak above the hammer handle was originally an inclusion of amphibolite like that in A. The amphibolite fragment has been crushed and stretched by the differential stress during the metamorphic processes.

piles of sediment that accumulate at the foot of the continental shelf along passive continental margins, such as the east coast of North America, and off the mouths of great rivers. Burial metamorphism is known to be happening today in the great pile of sediments accumulated in the Gulf of Mexico, off the mouth of the Mississippi River.

Regional Metamorphism

The most common metamorphic rocks of the continental crust occur in areas of tens of thousands of square kilometers, and the process that forms them is called **regional metamorphism.** Unlike burial metamorphism, regional metamorphism involves differential stress and a considerable amount of mechanical deformation in addition to chemical recrystallization. As a result, regionally metamorphosed rocks tend to be distinctly foliated. For a discussion of the textures and mineral assemblages of regionally metamorphosed rocks, see "A Closer Look: Kinds of Metamorphic Rock."

Slate, **phyllite**, and **schist**, the low-, intermediate-, and high-grade metamorphic rocks respectively produced from shale, are the most common varieties of regionally metamorphosed rocks. **Gneiss** is also a high-grade metamorphic rock produced from shale; it has more quartz and feldspar and less mica than is present in schist. Regionally metamorphosed rocks are usually found in mountain ranges formed as a result of either subduction or collision between fragments of continental crust. During both subduction and collision between continents, sedimentary rock along the margin of a continent is subjected to intense differential stresses. The foliation that is so characteristic of slates, phyllites, schists, and gneisses is a consequence of those intense stresses. Regional metamorphism is therefore a result of plate tectonics. Rocks of the Appalachian Mountains, the Alps, the Himalaya, and all other mountain ranges formed by continental collisions are composed of regionally metamorphosed rocks.

When segments of ancient oceanic crust of basaltic composition are incorporated into the continental crust as a result of subduction, metamorphism produces two distinctive rocks. Low-grade metamorphism produces **greenschists**, so named because they are rich in chlorite. Intermediate-grade metamorphism produces **amphibolites**, which are rich in amphiboles.

Metamorphic Facies

A famous Finnish geologist, Pennti Eskola, pointed out in 1915 that the same metamorphic mineral assemblages are observed again and again. This led him to propose a concept known as **metamorphic facies.**

A CLOSER LOOK

Kinds of Metamorphic Rock

Metamorphic rocks are named partly on the basis of texture, partly on mineral assemblage. The most widely used names are those applied to metamorphic derivatives of shales, sandstones, limestones, and basalts. This is because shales, sandstones, and limestones are the most abundant sedimentary rock types, whereas basalt is by far the most abundant igneous rock.

Metamorphism of Shale

Slate   The low-grade metamorphic product of shale is *slate* (Fig. C8.3). The minerals usually present in shale include quartz, clays, calcite, and feldspar. Slate contains quartz, feldspar, and mica or chlorite. Although a slate may still look like a shale, the tiny mica and chlorite grains give slate a distinctive foliation called slaty cleavage. The presence of slaty cleavage is clear proof that a

rock has changed from a sedimentary rock (shale) to a metamorphic rock (slate).

Phyllite   Continued metamorphism of a slate to an intermediate grade of metamorphism produces both larger grains of mica and a changing mineral assemblage; the rock develops a pronounced foliation (Fig. C8.3) and is called *phyllite* (from the Greek *phyllon*, a leaf). In a slate it is not possible to see the new grains of mica with the unaided eye, but in a phyllite they are just large enough to be visible.

Schist and Gneiss   Still further metamorphism beyond that which produces a phyllite leads to a coarse-grained rock with pronounced schistosity, called *schist* (Fig. C8.3). The most obvious differences between slate, phyllite, and schist are in grain size (Fig. C8.4). At the high grades of metamorphism characteristic of schists, minerals may start to segregate into separate bands. A high-grade rock with coarse grains and pronounced foliation,

A.

B.

C.

D.

Figure C8.3   Progressive metamorphism of shale and the development of foliation. A. Slate from Bangor, Pennsylvania. Individual mineral grains are too small to be visible. Slaty cleavage records the beginning of metamorphism. B. Phyllite from Woodbridge, Connecticut. Mineral grains are just visible. Foliation is more pronounced. C. Schist,

from Manhattan, New York. Mineral grains are now easily visible and foliation is pronounced. D. Gneiss, from Uxbridge, Massachusetts. Quartz and feldspar layers (light) are segregated from mica-rich layers (dark). Foliation is pronounced.

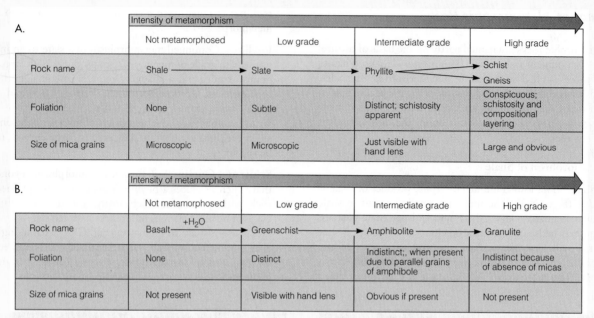

Figure C8.4 Progressive metamorphism of shale and basalt. Both foliation and mica grain size change as a result of increasing temperature and differential stress.

but with layers of micaceous minerals segregated from layers of minerals such as quartz and feldspar, is called a *gneiss* (pronounced nice, from the German *gneisto*, meaning to sparkle) (Fig. C8.3).

Metamorphism of Basalt

Greenschist   The main minerals in basalt are olivine, pyroxene, and feldspar, each of which is anhydrous. When a basalt is subjected to metamorphism under conditions where H_2O can enter the rock and form hydrous

minerals, distinctive mineral assemblages develop. At low grades of metamorphism, mineral assemblages such as chlorite + feldspar + epidote + calcite form. The resulting rock is equivalent in metamorphic grade to a slate but has a very different appearance. It has pronounced foliation as a phyllite does, but it also has a very distinctive green color because of its chlorite content: it is termed *greenschist*.

Amphibolite and Granulite   When a greenschist is subjected to an intermediate grade of metamorphism, chlorite is replaced by amphibole; the resulting rock is generally coarse grained and is called an *amphibolite*. Foliation is present in amphibolites but is not pronounced

(The term *facies* comes from the Latin for *face*, or appearance.) According to the metamorphic facies concept, for a given range of temperature and pressure, and for a given rock composition, the assemblage of minerals formed during metamorphism is always the same. Based on mineral assemblages, Eskola defined a series of pressure and temperature ranges that he called metamorphic facies.

To help you understand Eskola's idea, another analogy with cooking is appropriate; think of a large roast of beef. When it is carved, one sees that the center is rare, the outside is well done, and in between is a region of medium rare meat. The differences occur because the temperature was not uniform throughout. The center, rare-meat facies is a low-temperature facies; the outside, well-done facies is a high-tempera-

ture one. The composition of beef varies little, if at all, from roast to roast, and so the facies must depend not on composition but on the temperature. So too with rocks, although in any rock of given composition, pressure as well as temperature determines mineral assemblage.

Figure 8.14 shows the principal metamorphic facies, together with geothermal gradients to be expected under three different geological conditions and therefore the conditions typical of each facies.

Plate Tectonics and Metamorphism

One of the triumphs of plate tectonics is that, for the first time, it explains the distribution of metamorphic

A.

Figure C8.5 Texture of nonfoliated metamorphic rocks seen in thin section and viewed in polarized light. Notice the interlocking grain structure produced by recrystallization during metamorphism. Each specimen is 2 cm across.

B.

A. Marble, composed entirely of calcite. All vestiges of sedimentary structure have disappeared. B. Quartzite. Arrows point to faint traces of the original rounded quartz grains in some of the grains.

because micas and chlorite are usually absent. At highest grade metamorphism, amphibole is replaced by pyroxene and an indistinctly foliated rock called a *granulite* develops.

Metamorphism of Limestone and Sandstone

Marble and **quartzite** are the metamorphic derivatives of limestone and sandstone, respectively. Neither limestone nor quartz sandstone (when pure) contains the necessary ingredients to form sheet- or chain-structure minerals. As a result, marble and quartzite commonly lack foliation.

Marble   Marble consists of a coarsely crystalline, interlocking network of calcite grains. During recrystallization of a limestone, the bedding planes, fossils, and other features of sedimentary rocks are largely obliterated. The end result, as shown in Figure C8.5A, is an even-grained

rock with a distinctive, somewhat sugary texture. Pure marble is snow white and consists entirely of pure grains of calcite. Such marbles are favored for marble gravestones and statues in cemeteries, perhaps because white is considered to be a symbol of purity. Many marbles contain impurities such as organic matter, pyrite, goethite, and small quantities of silicate minerals, which impart various colors.

Quartzite   Quartzite is derived from sandstone by the filling in of the spaces between the original grains with silica, and by recrystallization of the entire mass (Fig. C8.5B). Sometimes, the ghostlike outlines of the original sedimentary grains can still be seen, even though recrystallization may have rearranged the original grain structure completely.

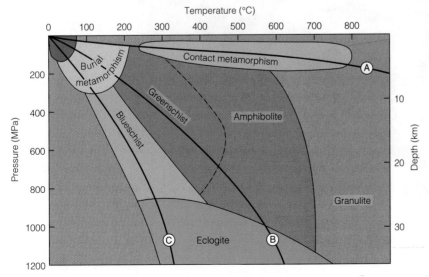

Figure 8.14 Metamorphic facies plotted with respect to temperature and pressure. Curve A is a typical thermal gradient around an intrusive igneous rock that causes low-pressure metamorphism. Curve B is a normal continental geothermal gradient. Curve C is the geothermal gradient developed in a subduction zone.

facies in regionally metamorphosed rocks. To repeat what we said above, regional metamorphism occurs at a convergent plate boundary, as shown in Figure 8.15.

The temperatures and pressures characteristic of *blueschist* facies in regional metamorphism are reached when crustal rocks are dragged down by a rapidly subducting plate. Under such conditions, pressure increases more rapidly than temperature, and as a result the rock is subjected to high pressure and relatively low temperature. Rocks subjected to blueschist facies metamorphism are widespread in the Coast Ranges of California. Blueschist metamorphism is probably happening today along the subducting margin where the Pacific plate plunges under the coast of Alaska and the Aleutian islands.

The metamorphic conditions characteristic of *greenschist* and *amphibolite facies* metamorphism occur where crust is either thickened by continental collision or heated by rising magma. Continental collision is the most common setting for regional metamorphism, and rocks formed in this way are observed throughout the Appalachians and the Alps. Such metamorphism is known to be occurring today beneath the Himalaya, where the continental crust is thickened by collision, and beneath the Andes, where it is both thickened by subduction and heated by rising magma. If the crust is sufficiently thick, rocks subjected to amphibolite facies or higher grade metamor-

phism can reach temperatures at which partial melting commences and metamorphism passes into magma formation.

PLATE TECTONICS, CONTINENTAL CRUST, AND MOUNTAIN BUILDING

Continental crust is simply a passenger being rafted on large plates of lithosphere. It is, however, a passenger that is buffeted, stretched, fractured, and altered by the ride.

Each bump between two crust fragments forms a mountain belt of metamorphic rocks, each grind a transform fault, and each stretch a rift valley. Scars left in the continental crust by bump-and-grind tectonics are evidence of former plate motions. That this continental evidence exists is fortunate because the most ancient known crust in the ocean is only about 180 million years old. Thus, the only direct evidence concerning geological events more ancient than 180 million years comes from the continental crust.

Before beginning our discussion on how the continental crust has grown and been shaped by plate tectonics, it is helpful to look first at the large-scale structure of continents.

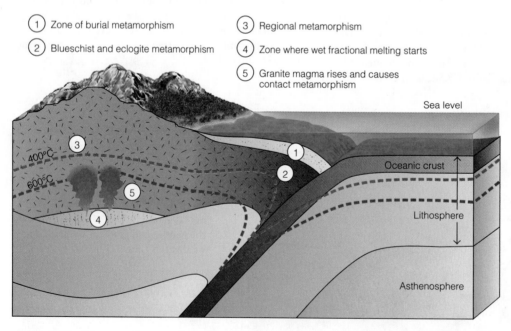

Figure 8.15 Diagram of a convergent plate boundary, showing the different regions of metamorphism. Dashed lines indicate temperature contours.

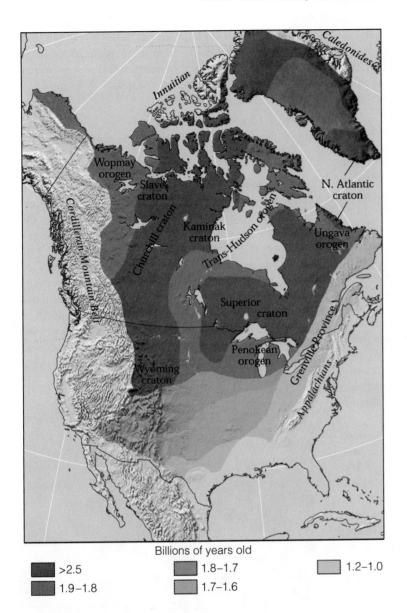

Billions of years old

>2.5	1.8–1.7	1.2–1.0
1.9–1.8	1.7–1.6	

Figure 8.16   The North American cratons and associated orogens. The Grenville orogen is about 1 billion years old, whereas the Caledonide, Appalachian, Cordilleran, and Innuitian orogens are each younger than 600 million years. The assemblage of cratons and orogens, all older than 1.8 billion years, which are surrounded by the five young orogens, is the Canadian Shield.

Regional Structures of Continents

Cratons and Orogens

On the scale of a continent, two kinds of structural units can be distinguished in the continental crust: cratons and orogens. A **craton** is a core of very ancient rock (Fig. 8.16). The term is applied to those ancient parts of the Earth's crust that have attained isostatic stability. Rocks within cratons have been deformed, and the deformation is invariably ancient, but how ancient cratons formed is still a matter of intense research.

Draped around cratons are the second kind of crustal building unit, **orogens**, which are elongate regions of crust that have been intensely bent and fractured during continental collisions. Crust in an orogen is commonly thicker than crust in a craton, and many orogens—even some very old ones—have not yet attained isostatic equilibrium. Orogens are the eroded roots of ancient mountain ranges that formed as a result of collisions between cratons. Orogens differ from each other in age, history, size, and details of origin; however, all were once mountainous terrains, and all are younger than the cratons they surround. Only the youngest orogens are mountainous today. Ancient orogens, now deeply eroded, reveal their history through the kinds of metamorphic rock they contain and the way the rocks are twisted and deformed.

An assemblage of cratons and ancient orogens is called a **continental shield.** North America has a huge continental shield at its core, and around the shield are five young orogens: the Grenville, Appalachian, Caledonide, Innuitian, and Cordilleran orogens (Fig. 8.16). Because the North American shield crops out principally in Canada (especially Ontario and Quebec), but is covered mostly by younger, flat-lying sedimentary rocks in the United States, geologists often refer to it as the Canadian Shield. That portion of a continental shield that is covered by a thin layer of younger sedimentary rocks is called a *stable platform*.

Through careful mapping, geologists have identified several ancient cratons and orogens in the Canadian Shield. All the craton rocks are older than 2.5 billion years. Such rocks can be observed in many places in eastern Canada, but in the United States they crop out only in small regions around Lake Superior and in Wyoming. By drilling through the stable platform that covers most of the U.S. portion of the shield, geologists have discovered that cratons and ancient orogens similar to those that surface in eastern Canada lie below much of the central United States and part of western Canada.

The three small cratons, Slave, Wyoming, and Kaminak, and the three larger cratons, Churchill, Superior, and North Atlantic, shown in Figure 8.16, were once minicontinents. By about 1.6 billion years ago, these minicontinents had become welded together to form the assemblage of cratons and ancient orogens (in other words, the Canadian Shield) we see in North America today. Collisions between the three small cratons and the large Churchill craton did not form large orogens, but each time two of the larger cratons collided, an orogen was formed between them. For example, when the Superior and Churchill cratons collided, the Trans-Hudson orogen formed. The existence of ancient collision belts—orogens—between the cratons of the Canadian Shield is the best evidence available to support the idea that plate tectonics operated at least as far back as 2 billion years ago. This means that the geosphere portion of the Earth system must have been working much as it works today, for *at least* 2 billion years.

Continental Margins

The fragmentation, drifting, and welding together of pieces of continental crust are direct consequences of plate tectonics. Various combinations of these processes are responsible for the five types of continental margins we know of today: passive, convergent,

collision, transform fault, and accreted terrane. Before we discuss additional evidence for plate tectonics, it will be helpful to review briefly the features associated with each of the continental margins.

Passive Continental Margins

A passive continental margin is one that occurs in the stable interior of a plate. The Atlantic Ocean margins of the Americas, Africa, and Europe are passive. The eastern coast of North America, for example, is in the stable interior of the North American plate, far from the plate margins. Passive continental margins develop when a new ocean basin forms by the rifting of continental crust, as illustrated in Figure 4.12. This process is happening today in the Red Sea, which is a young ocean with an active spreading center running down its axis (Fig. 8.17). New, passive continental margins have formed along both edges of the Red Sea.

Passive continental margins are places where great thicknesses of sediment accumulate. The kinds of sediment deposited are distinctive, and the Red Sea pro-

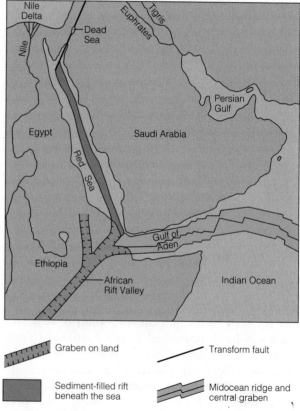

Figure 8.17 Three spreading centers meet at a triple junction. Two, the Gulf of Aden and the Red Sea, are actively spreading, and there are passive continental margins along the adjacent coastlines. The African Rift appears to be a failing rift that will not develop into an open ocean.

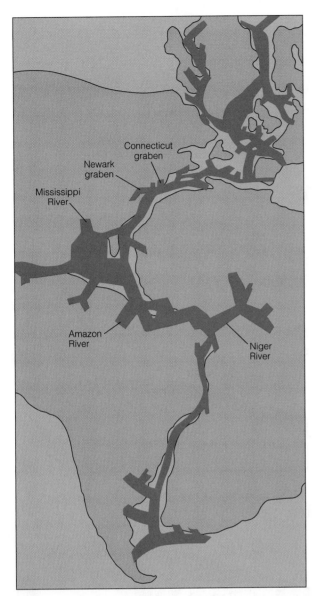

Figure 8.18 Map of a closed Atlantic Ocean showing the rifts that formed when Pangaea was split by a spreading center. The rifts on today's continents are now filled with sediment. Some of them serve as the channelways for large rivers.

vides an example. Deposition commenced with clastic, nonmarine sediments followed by chemical sediments (rock salt) and then clastic marine shales. The sequence apparently arises in the following manner. Basaltic magma, associated with formation of the new spreading center that splits the continent, heats and expands the lithosphere so that a plateau forms with an elevation of as much as 2.5 km above sea level (Fig. 4.12). Tensional stress breaks the crust along the plateau and forms a rift, with the result that there is a

pronounced topographic relief between the plateau and floor of the rift. The earliest rifting of what is now the Red Sea must have looked much the way the African Rift Valley looks today. Before the rift floor sank low enough for seawater to enter, clastic nonmarine sediments, which formed rocks such as conglomerates and sandstones, were shed from the steep valley walls and accumulated in the rift. Associated with these sediments are basaltic lavas, dikes, and sills, all formed by magma rising up the fractures. As the rift widened, a time was reached where seawater entered. The early flow was apparently restricted, and the water was shallow, resembling a shallow lake more than an ocean. The rate of evaporation would have been high, and as a result strata of rock salt were laid down on top of the clastic nonmarine sediments. Finally, as rifting continued and the depth of the seawater increased, normal clastic marine sediments were deposited. This is the stage the Red Sea is in today. Eventually, as further rifting exposes new oceanic crust, the Red Sea will evolve into a younger version of the Atlantic Ocean.

Notice in Figure 8.17 that the Gulf of Aden, the Red Sea, and the northern end of the African Rift Valley meet at angles of 120°. Such a meeting point formed by three spreading centers is called a *plate triple junction*. Two of the centers, the Gulf of Aden and the Red Sea, are active and still spreading. The third, the African Rift Valley, is apparently no longer spreading and possibly will not evolve into an ocean. What will remain on the African continent is a long, narrow rift filled primarily with nonmarine sediment. The formation of triple junctions with one arm not developing into an ocean is apparently a characteristic feature of passive continental margins. This can be seen Figure 8.18, which shows the reassembled positions of the continents that today flank the Atlantic Ocean. Note that some of the world's largest rivers, such as the Niger, the Amazon, and the Mississippi, flow down valleys formed by undeveloped rifts associated with the opening of the Atlantic Ocean. This is a compelling example of one way in which the geosphere portion of the Earth system influences the hydrosphere.

Continental Convergent Margins

A continental convergent margin is one where the edge of a continent coincides with a convergent plate margin along which oceanic lithosphere is being subducted beneath continental lithosphere. The Andean coast of South America is an example. On this coast, the Nazca plate (capped by oceanic crust) is being subducted beneath the South American plate (capped

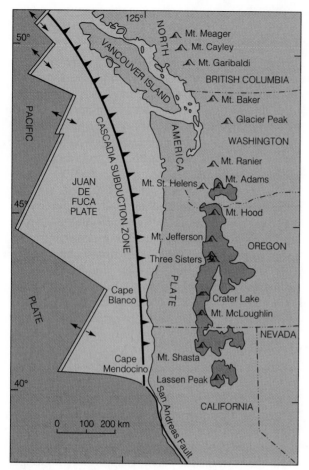

Figure 8.19 Volcanoes of the Cascade Range, a continental volcanic arc above a subduction zone. Each volcano has been active during the last 2 million years. Magma to form the volcanoes comes from partial melting of oceanic crust on the Juan de Fuca plate as it is subducted beneath the North American plate.

by continental crust). Fractional melting of the subducted Nazca plate produced the andesitic magma that formed the Andes (a continental volcanic arc).

Subduction produces intense deformation of a continental margin (together with characteristic magmatic activity and a distinctive style of metamorphism and deformation in sediments deposited in the trench). The tectonic setting in which sediments are subjected to high-pressure, low-temperature metamorphism is a subduction zone (curve C in Figure 8.14).

The most distinctive feature of a continental convergent margin is the continental volcanic arc. Modern examples are the chains of volcanoes in the Andes and the Cascade Range (Fig. 8.19). Where the volcanoes have been eroded and the deeper parts of the un-

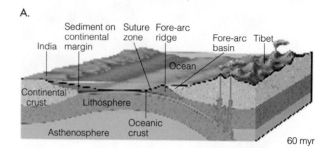

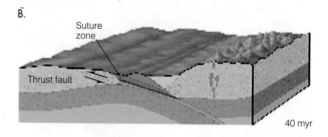

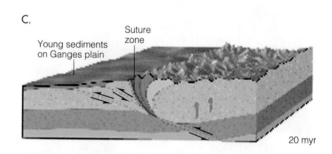

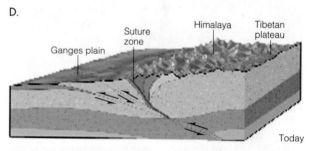

Figure 8.20 Collision between two fragments of continental crust shown schematically for the collision between India and Tibet. A. India, on the left, moves north. Sixty million years ago an ocean still separated India and Tibet. B. India and Tibet start to collide about 40 million years ago. Sediment is buckled and fractured, and the lithosphere is thickened. C. The collision starts to elevate the Himalaya about 20 million years ago. The downward-moving plate of lithosphere capped by oceanic crust breaks off and continues to sink. D. The edge of the remaining segment of the plate on which India sits, and which is capped by buoyant continental crust, is partly thrust under the edge of the overriding plate on which Tibet sits, causing further elevation of the collision zone. The process is continuing and the Himalaya are still rising.

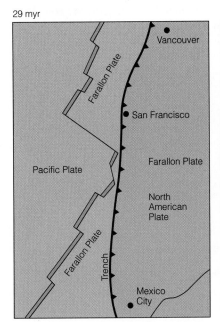
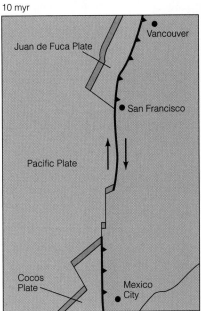
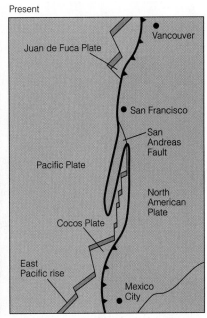

Figure 8.21 Origin of the San Andreas Fault. Twenty-nine million years ago, the edge of North America overrode a portion of the spreading center separating the Pacific plate from the Farallon plate, creating two smaller plates in the process, the Cocos plate and the Juan de Fuca plate. The San Andreas Fault is the transform fault that connects the remaining pieces of the severed spreading ridge.

derlying magmatic arc exposed, granitic batholiths can be observed. They are remnants of the magma chambers that once fed stratovolcanoes far above. The Sierra Nevada of California is an example of an eroded continental volcanic arc.

Continental Collision Margins

A continental collision margin is one where the edges of two continents, each on a different plate, come into collision (Fig. 8.20). A modern example is the line of collision between the Australian-Indian plate and the Eurasian plate. India, on the Australian-Indian plate, and Asia, on the Eurasian plate, have collided and the Alpine-Himalayan mountain chain is the result.

When continental crust is carried on a plate that is being subducted beneath a continental convergent margin, the two continental fragments must eventually collide. The collision sweeps up and deforms any sediment that accumulated along the margins of both continents and forms a mountain system characterized by metamorphism of the sediment together with intense fracturing and folding of strata.

All modern continental collision margins are young orogens characterized by soaring mountain systems.

Occurring in great arc-shaped systems a few hundred kilometers wide, these mountain systems commonly reach several thousand kilometers in length. Within a system, strata are compressed, fractured, folded, and crumpled, commonly in an exceedingly complex manner. Metamorphism and igneous activity are always present. Examples are widespread: the Alps, the Himalaya, and the Carpathians are all young mountain systems still being actively uplifted, whereas the Appalachians and the Urals are older systems that are slowly being eroded down.

Transform Fault Margins

A transform fault continental margin occurs when the margin of a continent coincides with a transform fault boundary of a plate. The most striking example of a modern transform fault continental boundary is the western margin of North America, from the Gulf of California to San Francisco, where it is bounded by the San Andreas Fault.

The San Andreas Fault apparently arose when the westward-moving North American continent overrode part of the midocean ridge (the East Pacific Rise), as shown in Figure 8.21. The San Andreas is the transform fault that connects the two remaining segments of the old spreading center.

Accreted Terrane Margins

An **accreted terrane** continental margin is a former continental convergent margin or continental transform fault margin that has been further modified by

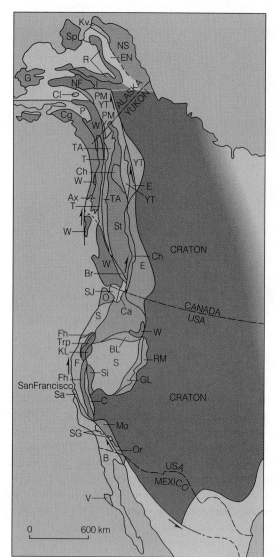

Figure 8.22 The western margin of North America is a complex jumble of terranes accreted during the last 200 million years. Some terranes, such as Wrangellia (W), were broken up during accretion and now occur in several different fragments.

Symbol	Name
Ax	Alexander
B	Baja
BL	Blue Mountains
BR	Bridge River
C	Calaveras
Cg	Chugach
Ch	Cache Creek
Cl	Chulitna
E	Eastern assemblages
En	Endicott
F	Franciscan and Great Valley
Fh	Foothills Belt
Gl	Golconda
I	Innoko
Kl	Klamath Mountains
Kv	Kagvik
Mo	Mohave
NF	Nixon Fork
NS	North Slope
O	Olympic
P	Peninsular
PM	Pingston and McKinley
R	Ruby
RM	Roberts Mountains
S	Siletzia
Sa	Salinian
SG	San Gabriel
Si	Northern Sierra
SJ	San Juan
Sp	Seward Peninsula
St	Stikine
T	Taku
TA	Tracy Arm
Trp	Western Triassic and Paleozoic of Klamath Mountains
V	Vizcaino
W	Wrangelia
YT	Yukon-Tanana

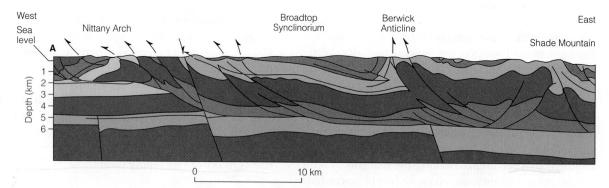

Figure 8.23 A slice through the Valley and Ridge Province of the Appalachians in Pennsylvania. Colors represent different rock types. The prominent purple unit, originally flat-lying but now contorted and fractured, is a stratum of limestone approximately 500 million years old. Rocks on the extreme right hand of the lower half of the diagram (brown and yellow) are igneous and metamorphic. Heavy black lines, including those that are

the addition of rafted-in, exotic fragments of crust. They are the most complex of the five kinds of continental margin. The western margin of North America from central California to Alaska is an example (Fig. 8.22).

Plate motion can raft fragments of crust tremendous distances. Eventually, any fragment not consumed by subduction is added (accreted) to a larger continental mass. Some of the fragments form when they are sliced off the margin of a large continent by a transform fault, much as the San Andreas Fault is slicing a fragment off North America today. Other combinations of volcanism, rifting, fracturing, and subduction can also form fragments of crust that are too buoyant to be subducted. In the western Pacific Ocean, there are many such small fragments of continental crust; examples include the island of Taiwan, the Philippine islands, and the many islands of Indonesia. Each fragment, called a *terrane*, is a geological entity characterized by distinctive rocks.

Mountain Building

Today's great mountain ranges are the orogens that have formed during the last few hundred million years. They are such distinctive and impressive features that we close this chapter by briefly describing one of the most beautiful and carefully studied mountain systems, the Appalachians.

The Appalachians are a mountain system 2500 km long that borders the eastern and southeastern coasts of North America and continues offshore, as eroded remnants, beneath the sediment of the modern continental shelf. The sedimentary strata in the system contain mud cracks, ripple marks, fossils of shallow-water organisms, and, in places, freshwater materials such as coal. Evidence is strong that sediment was deposited on the continental shelf of an old passive continental margin. The sedimentary strata, which thicken from west to east, are underlain by a basement of metamorphic and igneous rocks (Fig. 8.23).

Most but not all of the old strata of the Appalachians have now been deformed. Today, if we approach the central Appalachians from the west, we see first the former sediment occurring as essentially flat-lying, undisturbed strata. These strata were too far from the line of collision to suffer any deformation. Continuing eastward, we notice that the same strata thicken and we reach the point where the effects of the collision become apparent—the strata become gently bent into wavelike folds. Many of Pennsylvania's oil pools are found in these gently folded strata.

Proceeding east, we see the folds becoming less gentle and the development of gently inclined fractures until, finally, we reach the core of the Appalachians. Here the ancient basement rocks and the deep-water sediments deposited long ago on the old continental rise have been pushed upward and can be examined. The strata are increasingly metamorphosed, and deformation becomes more intense as the line of collision is reached.

The Appalachians have a complex history that started more than 600 million years ago, as demonstrated in Figure 8.24. Notice that three collision events are postulated and that an ancient ocean disappeared as a result of the collision about 350 million years ago. Today the old mountain system is being slowly eroded away, and the sediment is being deposited along the passive continental margin of eastern North America.

In Figure 8.24A you can see that 600 million years ago a passive continental margin bounded proto-

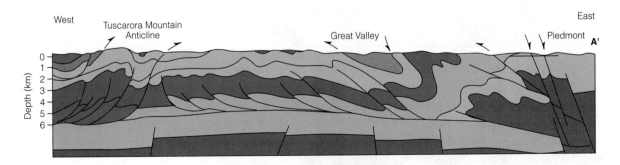

curved, are fractures. Arrows indicate direction of movement along the fractures. Note that the slice runs from A on the west to A' on the east, so the left-hand edge of the section on page 183 joins the right-hand edge of the section on page 182.

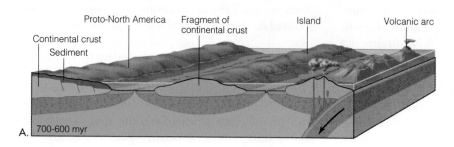

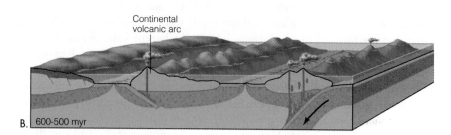

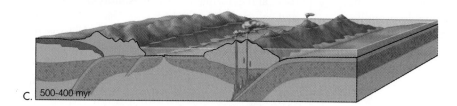

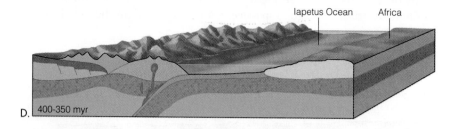

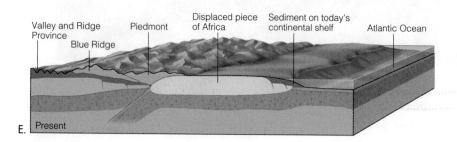

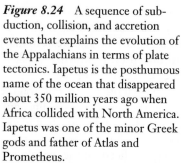

Figure 8.24 A sequence of subduction, collision, and accretion events that explains the evolution of the Appalachians in terms of plate tectonics. Iapetus is the posthumous name of the ocean that disappeared about 350 million years ago when Africa collided with North America. Iapetus was one of the minor Greek gods and father of Atlas and Prometheus.

A. Six hundred million years ago a small fragment of continental crust (an island) lay offshore of North America. Beyond the island were a subduction zone and an arc of volcanoes. B. A new subduction zone starts beneath the island about 500 million years ago. C. Between 400 and 500 million years ago the island collides with North America (these rocks can be seen today in North Carolina and Virginia), and the Appalachians start to form. D. Between 350 and 400 million years ago the arc of volcanoes collides with North America. The Iapetus Ocean slowly closes as North Africa approaches and eventually collides. E. When Pangaea broke apart about 200 million years ago, a fragment of Africa remained attached to North America. The passive continental margin so formed (today's margin) bounds the eastern edge of North America.

North America. Eventually, 400 to 500 million years ago, the passive margin became a convergent margin as a result of subduction starting. It is now thought that all passive margins become continental convergent margins when old oceanic lithosphere fractures close to the join between oceanic crust and continental crust and subduction commences. If this hypothe-sis is correct, at some unknown time in the future a new subduction zone will form along the Atlantic margin of North America, the Atlantic will slowly start to close, and eventually a new mountain system will form when Africa and Europe collide with North America

SUMMARY

1. Stratification results from the arrangement of sedimentary particles in layers. Each bed in a succession of strata is distinguished by its distinctive thickness, and the shape, size, and color of its sedimentary particles.

2. Two basic laws underlie stratigraphy: the law of original horizontality and the principle of stratigraphic superposition.

3. Substantial breaks or gaps in the sedimentary record are called unconformities. They record changes in environmental conditions of deposition or erosion and loss of earlier-formed sediment.

4. Unconformities record the close relationship among tectonics, erosion, and sedimentation.

5. The geologic column, a composite section of strata fitted together on the basis of relative age, has been carried around the world by correlation of strata based on the fossils in sedimentary rocks.

6. The absolute ages of strata in the geologic column have been determined by radiometric dating.

7. There are two families of sediment, clastic and chemical. Clastic sediment is material transported as solid bits of rocks and minerals. Chemical sediment forms when material is transported in solution and then deposited. If the cause of deposition is biochemical, a chemical sediment is called a biogenic sediment.

8. Sediment is lithified to sedimentary rock by cementation or recrystallization of the sediment particles.

9. Various arrangements of the particles in strata are seen in parallel strata, cross strata, paired strata, and poorly sorted layers.

10. Clastic sedimentary rocks, like sediments, are classified on the basis of predominant particle size. Conglomerate, sandstone, siltstone, and shale are the rock equivalents of gravel, sand, silt, and clay, respectively.

11. Metamorphism involves changes in mineral assemblage and rock texture and occurs in the solid state as a result of changes in temperature and pressure.

12. Mechanical deformation and chemical recrystallization are the processes that affect rock during metamorphism.

13. The presence of pore fluid greatly speeds up metamorphic reactions.

14. Foliation, as expressed by a direction of easy breakage in a metamorphic rock, arises from parallel growth of minerals formed during metamorphism.

15. Metamorphism can be explained by plate tectonics. Burial metamorphism occurs within the thick piles of sediment at the foot of continental slopes and in the submarine fans off the mouths of the world's great river systems; regional metamorphism is found in regions of subduction and continental collision.

16. Regional metamorphism, which involves both mechanical deformation and chemical recrystallization, is the result of plate tectonics. Regionally metamorphosed rocks are produced along subduction and collision edges of plates.

17. Two major structural units can be discerned in the continental crust. Cratons are ancient portions of the crust that are tectonically and isostatically stable. Separating and surrounding the cratons are orogens of highly deformed rock, marking the sites of former mountain ranges.

18. An assemblage of cratons and deeply eroded orogens that forms the core of a continent is a continental shield.

19. There are five kinds of continental margins: passive, convergent, collision, transform fault, and accreted terrane.

20. Passive margins develop by rifting of the continental crust. The Red Sea is an example of a young rift, and the Atlantic Ocean is a mature rift.

21. Continental convergent margins are the locale of belts of metamorphic rock, chains of stratovolcanoes (magmatic arc), and belts of granitic batholiths.

22. Collision margins are the locations of mountain systems. Transform fault margins occur where the edge of a continent coincides with the transform fault boundary of a plate.

23. Accreted terrane margins arise from the addition of blocks of crust brought in by subduction and transform fault motions.

IMPORTANT TERMS TO REMEMBER

accreted terrane   *181*
bed   *158*
bedding   *158*
burial metamorphism   *171*
chemical sediment   *161*
clastic sediment   *161*
continental shield   *178*
craton   *177*
cross bedding   *166*
differential stress   *170*
foliation   *170*
fossil   *167*
geologic column   *160*

metamorphic facies   *172*
metamorphism   *169*
original horizontality
   (law of)   *158*
orogen   *177*
regional metamorphism
   172
strata (singular = stratum)
   158
stratification   *158*
stratigraphic superposition
   (principle of)   *159*
stratigraphy   *158*

unconformity   *159*
uniform stress   *170*
varve   *166*

Sedimentary Rocks

coal   *167*
conglomerate   *165*
dolostone   *167*
limestone   *167*
peat   *167*
sandstone   *165*
shale   *165*
siltstone   *165*

Metamorphic Rocks

amphibolite   *172*
gneiss   *172*
greenschist   *172*
marble   *175*
phyllite   *172*
quartzite   *175*
schist   *172*
slate   *172*

QUESTIONS FOR REVIEW

1. What two laws underlie the study of stratigraphy? What conclusion would you draw if you observed a pile of sedimentary strata in which the strata are vertical?

2. How and why do breaks occur in stratigraphic sequences, and what significance do they have for determining the history of the Earth?

3. How are strata correlated from place to place?

4. What is the difference between the relative and absolute age of a stratum?

5. What is the geologic column? Name the four eons of the column.

6. Starting from the oldest, name in order the periods of the Paleozoic Era.

7. Name the two families of sediments and describe the basic difference between them.

8. On what basis are clastic sediments and sedimentary rocks classified?

9. Describe two chemical reactions that can lead to precipitation of chemical sediments.

10. If you picked up a sedimentary rock containing rounded clasts about 5 cm in diameter, what name would you give to the rock?

11. What role does the biosphere play in the formation of sediment? Name a biogenic sedimentary rock and suggest how it may have formed.

12. What would you call a sedimentary rock composed entirely of broken bits of plant matter?

13. What features in a sediment or sedimentary rock are responsible for stratification?

14. Briefly describe the factors that control metamorphism.

15. What is foliation? The presence of foliation is a sure clue that a rock has been metamorphosed. Why?

16. A distinctly foliated rock contains quartz, potassium feldspar, garnet, and sillimanite; under what grade of metamorphism would it have formed?

17. What is regional metamorphism?

18. What is the geological setting of regional metamorphism? Name two places in the world where regional metamorphism is probably happening today.

19. What is burial metamorphism? Suggest some place on the Earth where it is probably happening today.

20. What is the metamorphic facies concept and how does it help in the study of metamorphic rocks?

21. Name three minerals that are found only in metamorphic rocks.

22. In what way does the presence of a pore fluid influence the speed with which metamorphism proceeds?

23. What are cratons and how do they differ from orogens? Name three orogens in North America that are less than a billion years old.

24. What is the origin of the Cascade Range? the Appalachians?

25. What evidence indicates that plate tectonics has been operating for at least the last 1.8 billion years of Earth history?

26. Name the five kinds of continental margins and describe how they form.

27. Describe the sequence of events that leads to the opening of a new ocean basin flanked by two passive continental margins.

28. With what kinds of continental margin is mountain building associated?

29. How does an accreted terrane margin form? Name a continental margin that was modified by terrane accretion.

1 — 1000   3 — 250
2 — 500   4 — 125

Closer Look

ALL

QUESTIONS FOR REVIEW

1. What is the law of radioactive decay?

2. If you started with 2000 radioactive atoms and the half-life of the atoms is one week, how many radioactive atoms would remain after four weeks?

3. How can naturally occurring radioactive atoms be used to determine the absolute ages of minerals?

4. A grain of mica from a granite contains 5000 atoms of ^{40}K and 600 atoms of ^{40}Ar. What is the absolute age of the mica? *Note*: the half-life of ^{40}K is 1.3 billion years.

5. Describe how radiometric ages are used to determine the absolute age of the geologic column.

6. Describe the changes in a shale as it metamorphosed successively from a slate, to a phyllite, to a schist.

7. What is a greenschist? Why does a greenschist differ from a slate even though both rocks form at the same grade of metamorphism?

8. Why do most marbles and quartzites lack foliation?

9. How does a quartzite differ from a sandstone?

ALL

QUESTIONS FOR DISCUSSION

1. What kind of sediment can you identify in the area in which you live? Identify the source of the sediment, how the sediment was transported, and where and why it was deposited.

2. Discuss the importance of the atmosphere, hydrosphere, and biosphere in the formation of sediment. What kind of sediment would you expect on a planet such as Mars that lacks a hydrosphere and a biosphere?

3. Discuss how the geosphere reservoir interacts with the other three reservoirs of the Earth system. Why is it that even though events in the geosphere, such as the raising of a mountain range, happen very slowly, while events in the atmosphere, hydrosphere and biosphere happen rapidly, the geosphere plays a dominant role in long-term changes in the other reservoirs?

An interactive case study available at our web site will help you learn more about sedimentary rock strata and the petroleum resources they may contain.

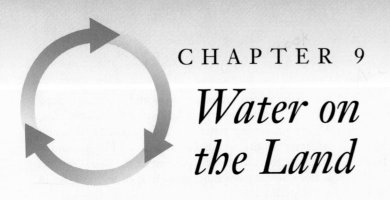

CHAPTER 9

Water on the Land

● *Restoring the Colorado River's Ecosystem*

When John Wesley Powell and his companions made the first transit of the Grand Canyon by boat in 1879, the Colorado River was a mighty through-flowing stream with a series of dangerous rapids. The far-sighted Powell envisaged dams across the river that someday would provide abundant water to the arid lands of the American Southwest. Today, like most western rivers, the Colorado is no longer entirely free-flowing. In 1963, the federal government built the Glen Canyon Dam, designed to provide both water and electrical energy to nearly 20 million people. In the years before the dam was raised, the river annually carried millions of tons of sediment into the Grand Canyon where it created bars and beaches that supported the vegetation on which myriad animals depended for food.

Like other dams built across rivers throughout the continent, the Glen Canyon Dam disrupted the natural aquatic habitat. It halted the supply of fresh sediment from upstream, reducing the sediment load that led early settlers to claim that the river was "too thick to drink, and almost thick enough to plow." Furthermore, until recently, release of the impounded waters was timed to accommodate the fluctuating needs of consumers for power to run air conditioners on hot afternoons, or to turn on electric stoves and ovens for the evening meal. Aquatic populations find it difficult to adjust to frequent highly variable changes in water level, and so it is not surprising that the river became biologically poorer.

Alarmed by the serious changes in the ecosystem resulting from the dam, and alert to studies that emphasized the need of riverine ecosystems to have a sustained natural flow of water, scientists proposed that an experimental flood be released to assess its effect in restoring natural conditions along the stream system. In the spring of 1996, a week-long release of water from the reservoir created an artificial flood. By autumn, the government reported that several major waterways and former beaches had been restored and that nutrient-rich sediment had been made available to plants and animals. Based on these studies, the government has released new guidelines for government-managed dams designed to allow water to flow more naturally. ●

(opposite) Experimental release of Colorado River water from Glen Canyon Dam in 1996 to simulate a natural flood.

WATER AND THE HYDROLOGIC CYCLE

Liquid water makes the Earth unique in the solar system. Seen from space, the Earth appears mostly blue and white because of its cover of water, snow, ice, and clouds. Although water has been detected on other bodies of the solar system, it does not appear to be present as a liquid anywhere except on our planet. The surface of Venus is so hot that water exists only as vapor, while very low temperature and pressure at the surface of Mars mean that water can exist there only as vapor or as ice. The surface of Ganymede, the largest of Jupiter's moons, is so frigid that it is covered by a thick "lithosphere" of ice.

In Chapter 1, we named the four reservoirs that make up the Earth system—the geosphere, atmosphere, hydrosphere, and biosphere—and defined the hydrosphere as the part containing the totality of the Earth's water, exclusive of atmospheric water vapor. Although this definition is straightforward, it is not strictly accurate, for water is present in all four parts of the Earth system. In other words, the hydrosphere overlaps with the geosphere, atmosphere, and biosphere. It is mainly for convenience of discussion that we distinguish the hydrosphere as the "water sphere," for in it resides the bulk of the Earth's water.

Most of the Earth's water, more than 97 percent, resides in the oceans. Next in importance are the myriad bodies of frozen water (snow and ice) that occupy the high mountains and polar latitudes of our planet. All the remaining water—including that in lakes and streams, in the atmosphere, and in the ground—amounts to only about 1 percent of the total. Yet this is the water we are most conscious of and rely on in our daily lives.

The physical state of water is controlled by temperature and pressure. At high temperatures or low pressures, water vapor is the stable state for H_2O, whereas ice forms at low temperatures or high pressures. The air pressure at sea level and that at the top of Mount Everest represent the extremes of air pressures at the Earth's surface. Surface temperatures range from about −100°C to +50°C (−148°F to 122°F). Within these limits of temperature and pressure, water can exist naturally in all three states of matter—solid (ice), liquid (water), or gas (water vapor). In a view across Lemaire Channel in Antarctica (Fig. 9.1), we can witness water in three states: as seawater, as ice in glaciers and floating sea ice, and as clouds that have condensed from water vapor in the air.

The movement of water among the four Earth system reservoirs constitutes the hydrologic cycle, which was introduced in Chapter 1 (Fig. 1.9). As we noted there, these movements are powered by the Sun's heat and involve evaporation, condensation, precipitation, transpiration, and surface runoff, as well as infiltration into the ground. Although water is always in a state of

Figure 9.1 Glaciers meet the sea along Lemaire Channel on the Antarctic Peninsula, producing an array of icebergs scattered across the ocean surface.

movement and is continuously being cycled from one reservoir to another, the total volume of water in each reservoir is approximately constant over short time intervals. Over lengthy intervals, however, the volume of water in the different reservoirs can change dramatically. During glacial ages, for example, vast quantities of water are evaporated from the oceans and precipitated on land as snow. The snow slowly accumulates to build ice sheets that are thousands of meters thick and cover vast areas where none exists today. At the culmination of a glacial age, the amount of water removed from the oceans is so large that world sea level falls about 120 m, and the expanded glaciers increase the ice-covered area of the Earth by more than 300 percent.

An important consequence of the hydrologic cycle is the varied landscapes of the Earth (Chapter 19). The erosional and depositional effects of streams, waves, and glaciers, coupled with the tectonic movements of crustal rocks, have produced a diversity of landscapes that make the Earth's surface unlike that of any other planet in the solar system. In its effect on erosion and sedimentation, the hydrologic cycle is intimately related to the rock cycle. Furthermore, it is a key component of an array of biogeochemical cycles that control the composition of the atmosphere and influence all living creatures on the Earth.

STREAMS AND THEIR CHANNELS

Almost anywhere we travel over the land surface, we can see evidence of the work of running water. Even in places where no rivers flow today, we are likely to find sediments and landforms that tell us water has been instrumental in shaping the landscape. Most of these features can be related to the activity of streams that are part of complex drainage systems.

Drainage systems evolve because a significant portion of the water falling on the land as precipitation collects and moves downslope in the general direction of the nearest ocean. The Earth's drainage systems thus form a fundamental part of the hydrologic cycle: water is evaporated from the oceans into the atmosphere, a portion is precipitated on the land surface, and part of this travels across the land on its way back to the sea. Enroute, streams erode the land, transport and deposit sediment, and support complex ecosystems that rely on a dependable supply of water.

The average annual rainfall on the area of the United States is equivalent to a layer of water 76 cm (30 in) thick equally covering this same land surface. Of this layer, 45 cm (17.7 in) returns to the atmosphere by evaporation and transpiration (the release of moisture from the pores of a plant), and 1 cm (0.4 in)

infiltrates the ground; the remaining 30 cm (11.9 in) forms **runoff**, the portion of precipitation that flows over the land surface. By standing outside during a heavy rain, you can see that water initially tends to move downslope in broad, thin sheets called *overland flow*. You will also notice, however, that after traveling a short distance overland flow begins to concentrate into well-defined channels, thereby becoming *streamflow*. Runoff is a combination of overland flow and streamflow.

A **stream** consists of water that flows downslope along a clearly defined natural passageway. As it moves, the water transports particles of sediment and dissolved substances (Fig. 9.2). The passageway is the stream's **channel**, and the sediment constitutes the bulk of its **load**, which is the total of all the sediment and dissolved matter that the stream transports. Geologists refer to the sediment load as **alluvium**. Every stream or segment of a stream is surrounded by its **drainage basin**, the total area that contributes water to the stream. The line that separates adjacent drainage basins is a **divide**. Drainage basins range in size from less than a square kilometer to vast areas of subcontinental dimension. In North America, the huge drainage basin of the Mississippi River encompasses an area that exceeds 40 percent of the area of the contiguous United States (Fig. 9.3). It should come as no surprise that the area of any drainage basin is proportional to both the length of the stream that drains the basin and the average annual volume of water that moves through the drainage system.

The Stream as a Natural System

A stream channel is an efficient conduit for carrying water. The size and shape of any particular channel cross section reflects several important controlling factors. They include the erodibility of the rock or sediment across which the stream flows, as well as the average volume of water passing through the channel cross section. Some very small streams are about as deep as they are wide, whereas very large streams often have widths many times greater than their depths.

If we measure the vertical distance that a stream channel descends between two points along its course, we will have obtained a measure of the stream's **gradient** between those points. The average gradient of a steep mountain stream may reach or exceed 60 m/km (330 ft/mi), whereas near the mouth of a large river the gradient may be less than 0.1 m/km (0.5 ft/mi).

Overall, the gradient of a river decreases downstream but not always smoothly, as any white-water rafter or kayaker can attest. A local change in gradient may occur, for example, where a channel passes from

Figure 9.2 Maroon Creek, in Colorado's White River National Forest, produces a succession of small rapids where it flows over and between boulders scattered along its gravelly channel.

resistant rock into more erodible rock, or where a landslide or lava flow forms a natural dam across the channel. At such places, water may tumble rapidly through a stretch of rapids or form a waterfall where it plunges over a steep drop.

A stream is a complex natural system, the behavior of which is controlled by five basic factors:

1. The *average width* and *depth* of the channel
2. The channel *gradient*
3. The *average velocity* of the water
4. The **discharge**, which is the quantity of water passing a point on a stream bank during a given interval of time

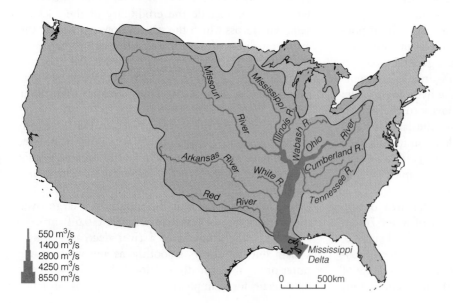

Figure 9.3 The drainage basin of the Mississippi River encompasses a major portion of the central United States. In this diagram, the width of the river and its major tributaries reflect discharge rates.

550 m³/s
1400 m³/s
2800 m³/s
4250 m³/s
8550 m³/s

0   500km

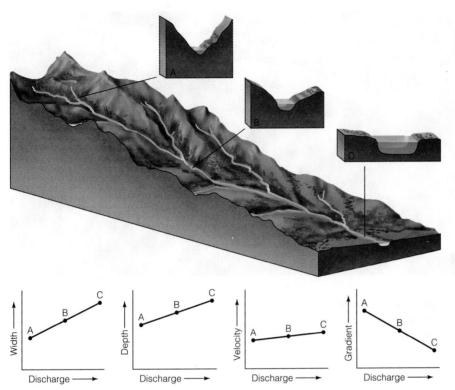

Figure 9.4 Changes in stream properties along a river system. Discharge increases as new tributaries join the main stream. Channel width and depth are shown by cross sections A, B, and C. Graphs show the relationship of discharge to channel width and depth, to velocity, and to channel gradient at the same three cross sections.

5. The *sediment load* (The dissolved component of the load generally has little effect on stream behavior.)

A stream experiences a continuous interplay among these factors. Measurements of natural streams show that as discharge changes, velocity or channel shape, or both, also change. This relationship can be expressed by the formula

$$
\begin{array}{ccc}
\text{Discharge} & = & \text{Cross-sectional} \times \text{Average} \\
\text{(m}^3\text{/s)} & & \text{area of channel} \quad \text{velocity} \\
& & \text{(width} \times \text{average depth)} \quad \text{(m/s)} \\
& & \text{(m}^2\text{)}
\end{array}
$$

The variable factors in this equation are interdependent, which means that when one changes, one or more of the others will also change. Changes in these stream variables commonly occur during a major storm that leads to an increase in discharge. With increased discharge, the velocity also typically increases. This can cause the stream to erode and enlarge its channel, rapidly if it flows on alluvium and much more slowly if it flows on bedrock. This erosion continues until the increased discharge can be accommodated by an enlarged channel and by faster flow. When discharge decreases, the channel dimensions decrease as some of the load is dropped. Typically, the velocity also decreases. In these ways channel width,

channel depth, and velocity continuously adjust to changing discharge, and an approximate balance among the various factors is maintained.

Traveling along a stream from its head to its mouth, we can see orderly adjustments occurring along the channel. For example, (1) the width and depth of the channel increase, (2) the gradient decreases, (3) velocity increases, and (4) discharge increases (Fig. 9.4). The fact that velocity increases downstream seems to contradict the common observation that water rushes down steep mountain slopes and flows smoothly over nearly flat lowlands. However, the physical appearance of a stream may not be a true indication of its velocity. Discharge is low in the headward reaches of a stream, and average velocity is also low because of the frictional resistance caused by the water passing over a very rough stream bed. Here, where the flow is *turbulent* (agitated or disorderly), the water moves in many directions rather than uniformly downstream. Discharge increases downstream as each **tributary** (a stream joining a larger stream) introduces more water, and the cross-sectional area of the channel increases to accommodate the increased volume. Despite a progressive decrease in slope, velocity also increases downstream because of a progressive increase in discharge, a decrease in frictional resistance, due to the decreasing roughness of the stream bed, and a more uniformly directed flow along the channel.

Figure 9.5 A meandering stream near Phnom Penh, Cambodia. Light-colored point bars, composed of gravelly alluvium, lie opposite steep banks on the outside of meander bends. Two oxbow lakes, the product of past meander cutoffs, lie adjacent to the present channel.

Meandering Channels

From an airplane, it is easy to see that streams vary considerably in size and shape. Straight channel segments are rare and generally occur for only brief stretches before the channel assumes a sinuous shape. In many streams, the channel forms a series of smooth, looplike bends with similar dimensions (Fig. 9.5). Such a looplike bend of a stream channel is called a **meander**, after the Menderes River (in Latin, *Meander*) in southwestern Turkey which is noted for its winding course. Meanders occur most commonly in channels cut in fine-grained alluvium that have gentle gradients. The meandering pattern reflects the way in which a river minimizes resistance to flow and dissipates energy as uniformly as possible along its course.

Try to wade or swim across a meandering stream, and it quickly becomes apparent that the velocity of the flowing water is not uniform. Velocity is lowest along the perimeter of the channel because this is where the water encounters the greatest frictional resistance to flow. The maximum velocity along a straight channel segment is found in midchannel. However, wherever the water rounds a bend, the zone of highest velocity swings toward the outside of the channel (Fig. 9.6).

The nearly continuous shift, or migration, of a meander is accomplished by erosion on the outer banks of the meander loops. Along the inner side of each meander loop, where the water is shallow and velocity is lowest, sediment accumulates to form a distinctive *point bar* (Fig. 9.5). Collapse of the stream banks oc-

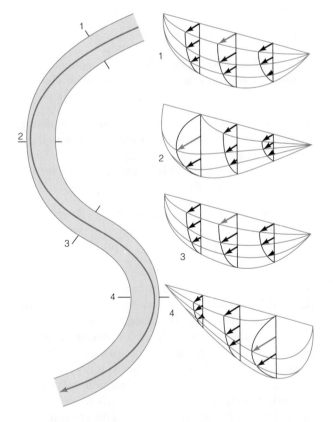

Figure 9.6 Velocity distribution in cross sections through a sinuous channel. (Lengths of arrows indicate relative flow velocities.) Where the channel is relatively straight (sections 1 and 3), the zone of highest velocity (red arrow) lies near the surface and toward the middle of the stream. At bends (sections 2 and 4), the maximum velocity swings toward the outer bank and lies below the surface.

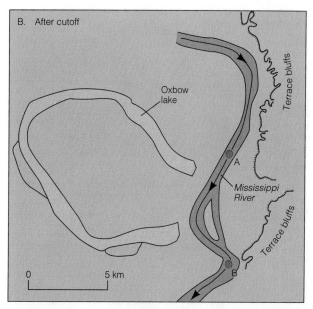

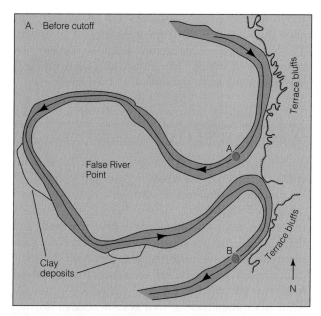

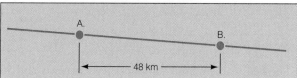

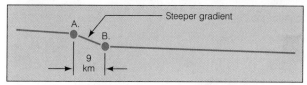

Figure 9.7 Cutoff of a meander loop of the Mississippi River in Louisiana. A. The downvalley migration of the loop encircling False River Point was halted when the channel segment on the south side of the loop encountered a body of clay in the alluvium. The next meander loop upstream continued to advance and finally cut off the entire loop surrounding False River Point. B. Over its new, shorter path, the stream had a steeper gradient than the abandoned course, and it developed a new pattern in which the single channel was replaced by two channels with an island between them.

curs most frequently along the outer side of a meander bend. There the high current velocity impinges on the channel margin, causing erosion and undercutting of the stream bank. In this way, meanders tend to migrate slowly down a valley, progressively removing and adding land along the banks.

Wherever a segment of a meander that is cutting into sandy alluvium encounters less erodible sediment, such as clay, downvalley migration of the meander will be slowed. Meanwhile, the upstream segment of the meander, migrating more rapidly through erodible sandy sediment, may intersect and cut into the slower moving downstream segment (Fig. 9.7A). When this happens, the stream bypasses the channel loop between the upstream and downstream segments, cutting it off and converting it into an arcuate *oxbow lake*. Because the new course is shorter than the older course, the channel gradient is steeper there and the overall stream length is shortened (Fig. 9.7B).

Nearly 600 km (373 mi) of the Mississippi River channel has been abandoned through such cutoffs since 1776. However, the river has not been shortened appreciably because the shortening due to cutoffs was balanced by channel lengthening as other meanders were enlarged. Nevertheless, because along much of the river's course the midpoint of the stream defines the boundary between adjacent states, cutoffs can shift land from one state to another.

Braided Channels

The intricate geometry of a **braided stream** resembles the pattern of braided hair, for the water repeatedly divides and reunites as it flows through two or more adjacent but interconnected channels (Fig. 9.8). For example, a stream unable to transport all the available load tends to deposit the coarsest and densest sediment to form a bar, which locally divides the flow and concentrates it in the deeper segments of channel to either side of the bar. As a bar builds up, it may emerge above the stream surface as an island and become stabilized by vegetation that anchors the sediment and inhibits erosion.

Large braided rivers typically have numerous shallow channels that change size and shift position as the stream erodes laterally and deposits sediment. Although at any moment the active channels may cover no more than 10 percent of the width of the entire

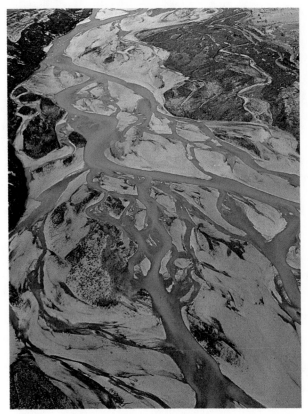

Figure 9.8 The shifting channels of the Rakaia River, flowing from glaciers in New Zealand's Southern Alps, form a braided pattern that is constantly changing form.

channel system, within a single season all or most of the surface sediment may be reworked by the laterally shifting channels.

A braided pattern tends to form in streams having highly variable discharge and easily erodible banks that can supply abundant sediment load to the channel system. Streams of meltwater issuing from glaciers typically have a braided pattern because the discharge

varies both daily and seasonally, and the glacier supplies the stream with large quantities of sediment. The braided pattern, therefore, seems to represent an adjustment by which a stream becomes more efficient in transporting an overabundance of sediment.

A STREAM'S LOAD

A stream has potential energy that can be used to erode the rocks across which it flows and to transport the resulting sediments downstream. A stream's load consists of three parts. Two of these, coarse particles that move along the stream bed (the **bed load**) and fine particles that are suspended in the water (the **suspended load**), comprise the sediment load. In addition, streams carry dissolved substances (the **dissolved load**) that are chiefly a product of chemical decomposition of exposed rock, as well as suspended or floating organic debris.

Bed Load

The bed load generally amounts to between 5 and 50 percent of the total sediment load of most streams. The average rate at which bed-load particles move is less than that of the water, for the particles are not in constant motion. Instead, they move discontinuously by rolling or sliding. Where forces are sufficient to lift a particle of the bed load off the stream bed, it may move short distances by saltation, a motion that is intermediate between suspension and rolling or sliding. **Saltation** involves the progressive forward movement of a particle that travels in short, intermittent jumps along arcuate paths (Fig. 9.9). Saltation continues as long as currents are sufficiently rapid or turbulent to lift particles and carry them downstream.

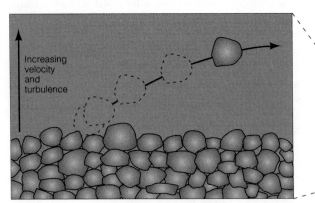

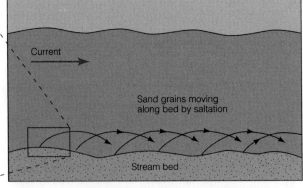

Figure 9.9 A sandy bed load moves by saltation when sand grains are carried up into a stream at places where turbulence locally reaches the bottom or where suspended grains impact other grains on the bed. Once raised into the flowing water, the grains are transported along arc-shaped trajectories as gravity pulls them toward the stream bed where they impact other particles which, in turn, are set in motion.

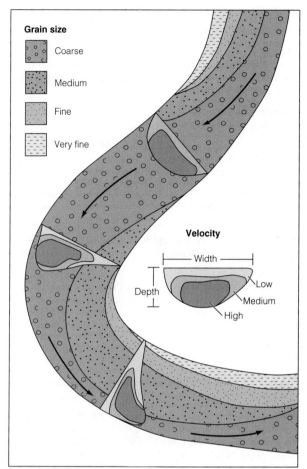

Grain size

Coarse

Medium

Fine

Very fine

Velocity

Width

Depth

Low

Medium

High

Figure 9.10 Relationship between bed-load grain size and velocity in a section of meandering channel. The coarsest sediment is associated with the zone of highest velocity; on the outside of a bend, both coarse grains and fast-moving water lie adjacent to the stream bank, but between bends both lie in the center of the channel. The finest sediment is associated with the zone of lowest velocity which lies on the inside of a bend. At such places, sediment accumulates to form point bars.

The distribution of bed-load sediment in a stream is generally related to the distribution of water velocity within the channel (Fig. 9.10). Coarse-grained sediment is concentrated where the velocity is high, whereas finer grained sediment is relegated to zones of progressively lower velocity.

Suspended Load

The muddy character of many streams results from particles of silt and clay moving in suspension. Most of the suspended load is derived from fine-grained regolith washed from areas unprotected by vegetation and from sediment eroded and reworked by the stream from its own banks. China's Yellow River derives its color from the great load of yellowish silt it erodes from the thick unconsolidated deposits of wind-blown dust that cover much of its basin (Fig. 9.11).

Because upward-moving currents within a turbulent stream exceed the velocity at which particles of silt and clay can settle toward the bed under the pull of gravity, such particles tend to remain in suspension longer than they would in nonturbulent waters. They settle and are deposited only where velocity decreases and turbulence ceases, as in a lake or in the sea.

Dissolved Load

Even the clearest stream water contains dissolved substances that constitute part of its load. Seven ions comprise the bulk of the dissolved content of most rivers: bicarbonate [$(HCO_3)^-$], calcium (Ca^{2+}), sulfate [$(SO_4)^{2-}$], chloride (Cl^-), sodium (Na^+), magnesium (Mg^{2+}), and potassium (K^+). In some streams the dissolved load may represent only a small percentage of the total load. However, streams that receive large contributions of underground water generally have higher dissolved loads than those whose water comes mainly from surface runoff.

Downstream Changes in Grain Size and Composition

The size of the particles a stream can transport is related mainly to velocity. Therefore, we might expect the average size of sediment to increase in the downstream direction as velocity increases. In fact, the opposite is true. Sediments normally decrease in coarseness downstream. In mountainous headwaters of large

Figure 9.11 A large suspended load, eroded from extensive deposits of unconsolidated wind-blown silt, gives the Yellow River a very muddy appearance.

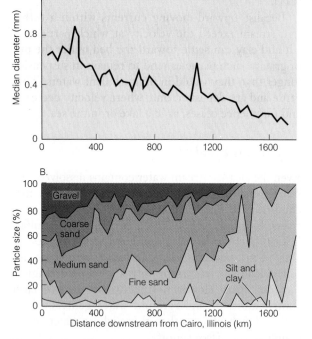

Figure 9.12 Change in sediment size along the Mississippi River downstream from Cairo, Illinois. A. Over the lower 1600 km of the channel, median diameter decreases from 0.8 to 0.2 mm. B. At Cairo, about 30 percent of the stream's load is gravel, 62 percent is sand, and 8 percent is silt and clay. At a point 1600 miles downstream, the sediment is almost entirely fine sand, silt, and clay.

rivers, tributary streams mostly flow through channels floored with coarse sediment that may include boulders a meter or more in diameter. Fine sediment is easily moved, even by streams having low discharge, and so it is readily carried downstream, leaving the coarser sediment behind. When a large stream eventually reaches the sea, its bed load may consist mainly of sediment no coarser than sand. We can see such a progressive downstream change in sediment size along the channel of the Mississippi River below Cairo, Illinois (Fig. 9.12).

Large streams generally cross a variety of exposed rocks, and the amount and composition of the load they carry changes as tributaries introduce new sediment. The Nile River provides a good example. Flowing through lower Egypt toward its delta, the Main Nile includes water contributed by three major tributaries: the White Nile, the Blue Nile, and the Atbara (Fig. 9.13). The White Nile contributes nearly a third

Figure 9.13 Change in sediment composition along the Nile River. A. Map of the Nile River and its principal tributaries. B. Discharge, load, and amphibole/pyroxene ratio of the Main Nile and its principal tributaries. The White Nile, which flows from Lake Victoria, contributes less than a third of the Nile's discharge and only 3 percent of its load. The greatest percentage of discharge and load is supplied by the Blue Nile, which originates in the highlands of Ethiopia. Different percentages of minerals are contributed by each tributary, causing a change in sediment composition of the Main Nile as each tributary joins it.

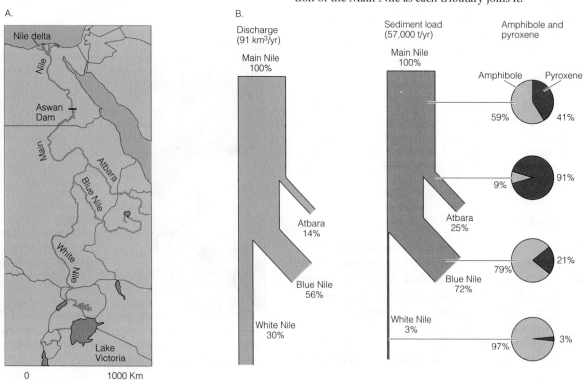

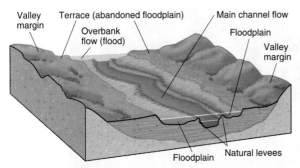

Figure 9.14 The major landforms of an alluvial valley.

of the total discharge but is responsible for only 3 percent of the sediment load in the Main Nile. In the mineral component of this load, the ratio of amphibole (eroded from metamorphic bedrock of the central African Plateau) to pyroxene is 97:3. The Blue Nile, which drains the highlands of Ethiopia, contributes more than half the discharge and nearly three quarters of the bed load. Its amphibole-to-pyroxene ratio is 79:21, reflecting the volcanic character of the source region. The more northerly Atbara contributes 14 percent of the discharge and a quarter of the load. In this stream, pyroxene is abundant, and the amphibole to pyroxene ratio is 9:91. These differing mineral components mix together as they enter the Main Nile, resulting in an amphibole-to-pyroxene ratio of 59:41. This ratio largely reflects the major contribution of amphibole-rich sediment from the two southerly tributaries, which together account for 82 percent of the load of the Main Nile.

Landforms Resulting from Stream Deposition

Distinctive stream deposits form along channel margins, valley floors, mountain fronts, and the shores of lakes and oceans, for these all are places where stream energy changes. The lower Mississippi River and other large, smoothly flowing streams like it typically deposit well-sorted layers of coarse and fine particles as they swing back and forth across a wide valley. During floods, as sediment-laden water flows out of the completely submerged channel, the depth, velocity, and turbulence of the water decrease abruptly at the channel margins, where the coarsest part of the suspended load is deposited to form a *natural levee* (Fig. 9.14). Farther away, finer silt and clay settle out across the stream's floodplain, a relatively flat region of valley floor that is periodically inundated by floodwater.

Most stream valleys contain terraces, which are floodplains abandoned when the stream cuts downward to a lower level (Figs. 9.14 and 9.15). Typically, such downcutting occurs in response to tectonism or to a change in discharge, load, or gradient. In many stream valleys, terraces lying at various levels record a complex history of alternating deposition and erosion.

A large, swift stream flowing down a steep mountain valley can transport an abundant load of coarse sediment, but on leaving the valley the stream loses

Figure 9.15 Alluvial terraces adjacent to Cave Stream, South Island, New Zealand, record former floodplains that were abandoned when the stream incised its channel and reached a new level.

Figure 9.16 A symmetrical alluvial fan has formed at the margin of Death Valley, California, where a stream channel emerges from a steep mountain canyon.

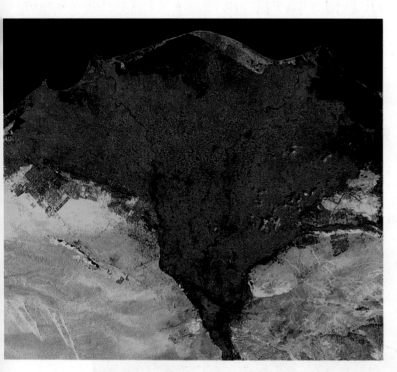

Figure 9.17 Delta of the Nile River, along the Mediterranean coast of Egypt. The reddish color in this vertical satellite image denotes vegetation growing on the fertile delta sediments. The delta and the Nile are bounded by a desert landscape of bare rock and shifting sands

energy, usually because of a change in gradient, velocity, or discharge. Its transporting power therefore decreases, and it deposits part of its sediment load. No longer constrained by valley walls, the stream can shift laterally back and forth across more gentle terrain. The resulting deposit, an **alluvial fan**, is a fan-shaped body of alluvium at the base of an upland (Fig. 9.16). Alluvial fans are common landforms along the base of most arid and semiarid mountain ranges. Some fans are so large and closely spaced that they merge to form a broad piedmont surface that slopes away from the base of the mountains.

When stream water enters the standing water of the sea or a lake, its speed diminishes rapidly, decreasing its ability to transport sediment. The water deposits its load in the form of a **delta**, so named because the deposit may develop a crudely triangular shape that resembles the Greek letter delta (Δ) (Fig. 9.17). Many of the world's largest streams, among them the Nile, the Ganges-Brahmaputra, the Huang He (Yellow River), the Amazon, and the Mississippi, have built massive deltas at their mouths. Each delta has its own peculiarities, determined by such factors as the stream's discharge, the character and volume of its sediment load, the shape of the adjacent bedrock coastline, the offshore topography, and the strength and direction of currents and waves.

FLOODS

The uneven distribution of rainfall through the year causes many streams to rise seasonally in flood. A *flood* occurs when a stream's discharge becomes so great that it exceeds the capacity of the channel, and water overflows the stream banks (Fig. 9.18). People affected by floods are often surprised and even outraged at what the rampaging stream has done to them. However, geologic studies of flood deposits show clearly that floods are normal and expected events.

As discharge increases during a flood, so does velocity. This enables a stream to carry a greater load, as well as larger particles. The collapse of the large Saint Francis Dam in southern California in 1928 provides an extreme example of the exceptional force of floodwaters. When the dam gave way, the impounded water was released and rushed down the valley as a gigantic flood, moving blocks of concrete weighing as much as 9000 metric tons (20 million pounds) through distances of more than 750 m (0.5 mi). Because natural floods are also capable of moving very large objects as well as great volumes of sediment, they are able to accomplish considerable geologic work.

Major floods—well outside a stream's normal flood range—occur infrequently, perhaps only once in several centuries. Even greater floods, evidence for which we can find in the geologic record, can be viewed as catastrophic events that occur rarely even on geologic time scales. In the 1920s, geologist J Harlen Bretz began a study of a curious landscape associated with a broad basalt plateau in eastern Washington State, locally called the Channeled Scabland. This landscape consists of dry coulees (canyons) with steep cliffs marking sites of former huge waterfalls, plunge pools, potholes, deep rock basins carved in the basalt, massive gravel bars containing enormous boulders (Fig. 9.19A), deposits of gravel in the form of gigantic ripples (Fig. 9.19B), and scoured land that extends hundreds of meters above valley floors. After studying this array of features, Bretz was led inescapably to conclude that they could be accounted for only by a truly gigantic flood, far larger than any historic flood. Later, the source of the necessary enormous volume of floodwater was resolved with the discovery that the continental ice sheet covering western Canada during the last glaciation dammed the Clark Fork River, thereby creating a huge lake in the vicinity of Missoula, Montana. The ice-impounded lake contained between 2000 and 2500 km³ (480 and 600 mi³) of water. When the glacier dam failed, water was released rapidly from the basin, as though a plug had

A.

B.

Figure 9.18 A pair of satellite images shows the region where the Missouri River joins the Mississippi River near St. Louis, Missouri. A. In a typical summer (July 1988) and B. during the disastrous flood of July 1993 when weeks of torrential rains caused the streams to overflow protective

levees and inundate numerous towns and vast areas of farmland. Losses, amounting to billions of dollars, included destroyed crops, closure of water treatment plants, severely damaged roads and bridges, and the destruction of entire communities.

A.

B.

Figure 9.19 Features attributable to catastrophic flooding in the Columbia Plateau region. A. Large boulder transported and deposited by floodwaters beyond the mouth of Grand Coulee, a major channel excavated by successive floods of glacial meltwater. B. Huge ripple marks formed by raging floodwaters as they swept around a bend of the Columbia River. Composed of coarse gravel, the ripples are up to several meters high, and their crests are as much as 100 m apart.

been pulled from a gigantic bathtub. The only possible exit route lay across the Channeled Scabland region and down the Columbia River to the sea. Subsequent studies have shown that the array of features scattered throughout the Scabland region provide dramatic evidence that such floods occurred repeatedly and that the geologic work they accomplished was immense.

WATER IN THE GROUND

Less than 1 percent of the water in the hydrosphere lies beneath the land surface as **groundwater**, defined as all the water occupying openings in bedrock and regolith in the water-saturated portion of the upper lithosphere. Although the percentage of groundwater in the hydrologic system is small, it is 35 times larger than the volume of all the water in freshwater lakes or flowing in streams and nearly a third as large as the water contained in all the world's glaciers and polar sea ice. More than half of all groundwater, including most of the water that is usable, occurs within about 750 m (2460 ft) of the Earth's surface. The volume of water in this zone is estimated to be equivalent to a layer of water approximately 55 m (180 ft) thick spread over the world's land areas.

Groundwater operates continuously as a small but integral part of the Earth's water cycle (Fig. 1.9). Part of the water evaporated from the oceans falls on the land as rain, seeps into the ground, and enters the groundwater system. Some of this slowly moving underground water reaches stream channels and contributes to the water they carry to the ocean, while some of it flows directly into coastal marine waters.

The Water Table

Much of what we know about the occurrence of groundwater has been learned from the accumulated experience of generations of people who have dug or drilled millions of wells. This experience tells us that a bore hole penetrating the ground ordinarily passes first into a zone in which open spaces in regolith or bedrock are filled mainly with air (Fig. 9.20). This is the **zone of aeration** (also called the *unsaturated zone*, for although water may be present, it does not saturate the ground). The hole then enters the **saturated zone**, the underlying zone in which all openings are filled with water. We call the upper surface of the saturated zone the **water table**. Whatever its depth, the water table is a significant surface, for it represents the upper limit of all readily usable groundwater.

In humid regions, the water table is a subdued imitation of the ground surface above it, being high beneath hills and low at valleys. If all rainfall were to cease, the water table would slowly flatten and gradually approach the levels of the valleys; seepage of water into the ground would diminish, then cease, and the streams in the valleys would dry up. In times of drought, when rain may not fall for several weeks or even months, wells can dry up as the water table falls below their bottoms. It is repeated rainfall, dousing the ground with fresh supplies of water, that maintains the water table at a normal level.

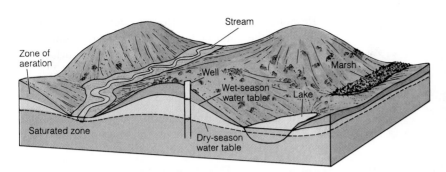

Figure 9.20 In a typical groundwater system, the water table separates the zone of aeration from the saturated zone and fluctuates in level with seasonal changes in precipitation. Corresponding fluctuations are seen in the water level in wells that penetrate the water table. Lakes, marshes, and streams occur where the water table intersects the land surface. In shape, the water table is a subdued imitation of the overlying land surface.

Movement of Groundwater

Most groundwater within a few hundred meters of the surface is in motion. Unlike the swift flow of rivers, which is measurable in kilometers per hour, groundwater moves so slowly that velocities are expressed in centimeters per day or meters per year. The reason for this contrast is easily explained. Whereas the water of a stream flows unimpeded through an open channel, groundwater must move through small, constricted passages, often along a tortuous route. Therefore, the flow of groundwater to a large degree depends on the nature of the rock or sediment through which it moves.

Porosity and Permeability

The amount of water that can be contained within a given volume of rock or sediment depends on the **porosity**, the percentage of the total volume of a body of bedrock or regolith that consists of open spaces, or *pores* (Fig. 9.21). In some well-sorted sands and gravels, the porosity may exceed 20 percent, while some very porous clays have a porosity of more than 50 percent. In sediments and sedimentary rocks, porosity is affected by the sizes, shapes, and arrangement of the rock particles, as well as by the extent to which the pores are filled with cementing substances (Fig. 9.21*C*). The porosity of dense igneous and metamorphic rocks is generally low, except when joints and fractures are common.

Permeability is a measure of how easily a solid allows fluids to pass through it. A rock or sediment of very low porosity is also likely to have low permeability. A well-sorted gravel, with large openings, is more permeable than sand and can yield large volumes of water. However, high porosity does not necessarily

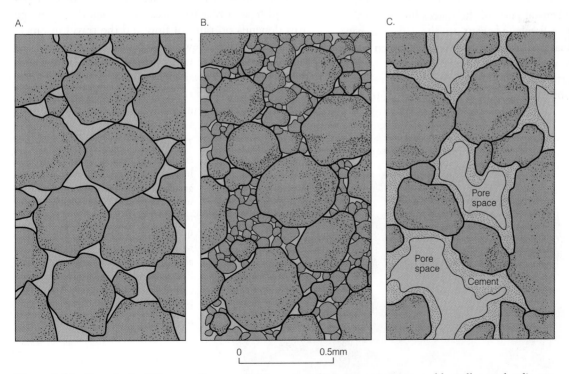

Figure 9.21 Porosity in different sediments. A. A porosity of 30 percent in a reasonably well-sorted sediment.
B. A porosity of 15 percent in a poorly sorted sediment in which fine grains fill spaces between larger grains.
C. Reduction in porosity in an otherwise very porous sediment due to cement that binds grains together.

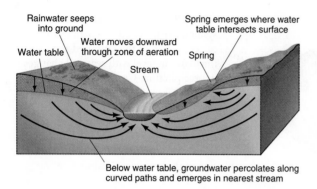

Figure 9.22 Paths of groundwater flow in a humid region in uniformly permeable rock or sediment. Long, curved arrows represent only a few of many possible paths. Springs are located where the water table locally intersects the land surface.

mean a corresponding high permeability, because the size and continuity of the openings influence permeability in an important way. For example, cement deposited between grains can restrict flow of water between pores, thereby reducing permeability.

Movement of Groundwater

The movement of groundwater is similar to the flow of water when a saturated sponge is squeezed gently. Water moves slowly through small open spaces along parallel, threadlike paths. Movement is easiest through the central parts of the spaces but diminishes to zero immediately adjacent to the sides of each space. Normally, flow velocities range between half a meter a day and several meters a year. The highest rate yet measured in the United States, in exceptionally permeable material, was only about 250 m/yr (820 ft/yr).

Responding to gravity, groundwater flows from areas where the water table is high toward areas where it is lowest. In other words, it flows toward surface streams or lakes, or toward the ocean (Fig. 9.22). Much of it flows along innumerable long, curving paths that go deeper through the ground. However, most of the groundwater entering a stream travels along shallow paths not far beneath the water table.

Recharge and Discharge Areas

Replenishment of groundwater, referred to as **recharge,** occurs as rainfall and snowmelt enter the ground in *recharge areas*. These are areas of the landscape where precipitation seeping into the ground reaches the saturated zone (Fig. 9.23). The water then moves through the groundwater system to *discharge areas*, areas where subsurface water emerges as springs or is discharged to streams, lakes, ponds, or swamps. The extent of a recharge area is invariably larger than that of the discharge area. In humid regions, recharge areas encompass nearly all the landscape except streams and their adjacent floodplains. In more arid regions, recharge occurs mainly in mountains and in alluvial deposits that border them. In such regions, recharge also occurs along channels of major streams that are underlain by permeable alluvium through which water leaks downward and recharges the groundwater.

The time it takes for water to move through the ground from a recharge area to the nearest discharge area depends on rates of flow and the travel distance. It may take only a few days, or possibly thousands of years in cases where water moves through the deeper parts of the groundwater system (Fig. 9.23).

Along the world's coastlines, groundwater can flow directly into the ocean through porous rocks and sedi-

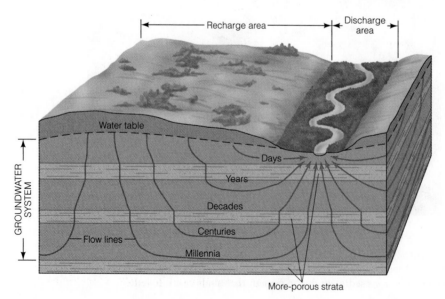

Figure 9.23 Distribution of recharge and discharge areas in a humid landscape. The purple lines are possible pathways that groundwater may take from recharge area to discharge area. The times required along various pathways are labeled and depend on the permeability of the rock or regolith along the path and on the distance traveled. Downward and upward movement of groundwater is faster and more direct in the most porous strata.

ments by means of submarine groundwater discharge. Along portions of the coastline in southeastern United States, such discharge is comparable to the measured discharge from rivers; similar examples of discharge through porous limestone are known in the Mediterranean and Red seas. In the Persian Gulf, up-welling plumes of discharged groundwater were a source of fresh water for ancient mariners.

Springs

A **spring** is a flow of groundwater emerging naturally at the ground surface. The simplest kind of spring is one occurring where the land surface intersects the water table. Small springs are found in all kinds of rocks, but almost all large springs flow from lava, limestone, or gravel. A vertical or horizontal change in permeability is a common reason for the localization of springs. Often this change involves the presence of an **aquiclude**, a body of impermeable or distinctly less permeable rock lying adjacent to a

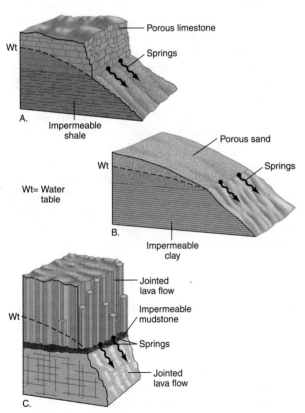

Figure 9.24 Examples of springs formed in different geologic conditions. A. A spring discharges water at the contact between a porous limestone and an underlying impermeable shale. B. Springs lie at the contact between a porous sandy unit and an underlying impermeable clay. C. Springs issue along the contact between a highly jointed lava flow and an underlying impermeable mudstone.

permeable one (Fig. 9.24). If a porous sand or lava overlies a relatively impermeable clay aquiclude, water percolating downward will flow laterally when it reaches the underlying clay and will emerge as a spring where the stratigraphic boundary between the two units intersects the land surface.

Aquifers

If we wish to find a reliable supply of groundwater, we search for an **aquifer** (Latin for water carrier), a body of rock or regolith sufficiently permeable to conduct economically significant quantities of groundwater to springs or wells. Bodies of gravel and sand generally are good aquifers, for they tend to be highly permeable and often are very extensive. Many sandstones are also good aquifers. However, in some sandstone bodies, a cementing agent between the grains reduces the diameter of the openings, thereby reducing permeability and decreasing their potential as aquifers.

An aquifer having a water table is called an *unconfined aquifer*. When water is pumped from a new well in an unconfined aquifer, the rate of withdrawal initially exceeds the rate of local groundwater flow. This imbalance in flow rates creates a **cone of depression** in the water table immediately surrounding the well (Fig. 9.25). The locally steepened slope of the water table will eventually increase the flow of water to the well. Once the rate of inflow balances the rate of withdrawal, the slope of the water table stabilizes, but it will change if either the rate of pumping or the rate of recharge changes. In most small domestic wells, the cone of depression is hardly discernible. Wells pumped for irrigation and industrial uses, however, withdraw so much water that such a cone can become very wide and deep and can lower the water table in all wells of a district.

About 30 percent of the groundwater used for irrigation in the United States is obtained from the High Plains aquifer, an unconfined groundwater system that lies at shallow depths beneath the High Plains, to the east of the Rocky Mountains (Fig. 9.26). The aquifer is tapped by about 170,000 wells and is the principal source of water for a major agricultural region that encompasses 20 percent of the irrigated land of the country. The aquifer consists of a number of sandy and gravelly rock units in which groundwater can readily flow. Its saturated thickness averages about 65 m (213 ft). The water table slopes gently from west to east, and water flows through the aquifer at an average rate of about 30 cm/day (12 in/day). Recharge comes from precipitation and through seepage from streams.

Development of groundwater for irrigation in the High Plains was spurred by severe regional drought in

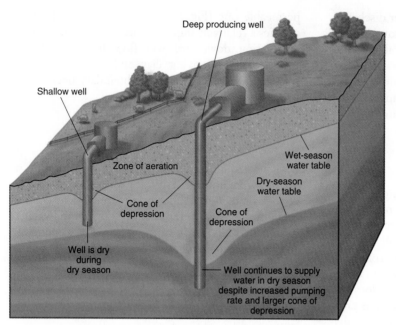

Figure 9.25 Effect of seasonal changes in precipitation on the position of the water table. During the wet season, recharge is high and the slope of the water table is relatively steep, so that water is present both in a shallow well and in a deeper well upslope. During the dry season, the water table falls, the slope of the water table decreases, and the shallow well is dry. The deeper well continues to supply water, but increased pumping during the dry season enlarges the cone of depression.

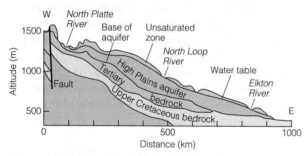

Figure 9.26 The High Plains aquifer is an example of an unconfined aquifer. A. This section across southeastern Wyoming and central Nebraska shows the eastward slope of the water tale and the relation of the aquifer to underlying rock units.

the 1930s and again in the 1950s. Annual recharge of the High Plains aquifer from precipitation is much less than the amount of water being withdrawn, and so the inevitable result has been a long-term fall of the water table. Furthermore, a dramatic increase in pumping rates has led to serious declines in water level. In parts of Kansas, New Mexico, and Texas, the thickness of the saturated zone has declined by more than 50 percent. The resulting decreased water yield and increased pumping costs have led to major concern about the future of irrigated farming on the High Plains.

Artesian Systems

In contrast to an unconfined aquifer, a *confined aquifer* is bounded above and below by bodies of impermeable rock or sediment, or by distinctly less permeable

rock or sediment than that of the aquifer. Water that enters a confined aquifer in an upland recharge area flows downward under the pull of gravity. As it reaches greater depths, the water comes under increasing *hydrostatic pressure* (pressure due to the weight of water at higher levels in the zone of saturation). If a well is drilled to the aquifer, the difference in pressure between the water table in the recharge area and the level of the well intake will cause water to rise in the well. Such an aquifer is called an **artesian aquifer**, and the well is called an *artesian well* (Fig. 9.27). Similarly, a freely flowing spring supplied by an artesian aquifer is an *artesian spring*. The term *artesian* comes from a French town, Artois (called Artesium by the Romans), where artesian flow was first studied. Under unusually favorable conditions, water pressure can be great enough to create fountains that rise as much as 60 m (200 ft) above ground level.

Chemistry of Groundwater

Analyses of many wells and springs show that the elements and compounds dissolved in groundwater consist mainly of chlorides, sulfates, and bicarbonates of calcium, magnesium, sodium, and potassium. We can trace these substances to the common minerals in the rocks from which they were weathered. As might be expected, the composition of groundwater varies from place to place according to the kind of rock present. In much of the central United States, the water is rich in calcium and magnesium bicarbonates that have been dissolved from the local carbonate bedrock. Taking a bath in such water, termed *hard water*, can be frustrating because soap does not lather easily and a crustlike ring forms in the tub. Hard water also leads to deposi-

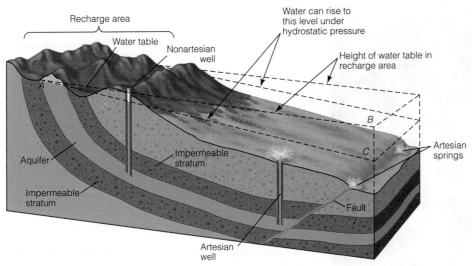

Figure 9.27 Two conditions are necessary for an artesian system: a confined aquifer and water pressure sufficient to make the water in a well rise above the aquifer. The water in a nonartesian well rises to the same height as the water table in the recharge area (line *AB*), minus an amount determined by the loss of energy in friction of percolation. Thus, the water can rise only to the line *AC*, which slopes downward and away from the recharge area. In the artesian well downslope, water flows out at the surface without pumping, for the well top lies below line *AC*.

tion of scaly crusts in water pipes, eventually restricting water flow. By contrast, water that contains little dissolved matter and no appreciable calcium is called *soft water*. With it, we can easily get a nice soapy lather in the shower.

Groundwater flowing through rocks that have a high arsenic or lead content may dissolve these toxic elements, making it dangerous to drink. Water circulating through sulfur-rich rocks may contain dissolved hydrogen sulfide (H_2S) which, though harmless to drink, has the disagreeable odor of rotten eggs. In some arid regions, the concentration of dissolved sulfates and chlorides is so great that the groundwater is unusually noxious.

THE GEOLOGIC WORK OF GROUNDWATER

Slowly moving groundwater has the capacity to perform an enormous amount of geologic work. In regions underlain by rocks that are highly susceptible to chemical weathering, groundwater has created distinctive landscapes that are among the most unusual on our planet.

Dissolution of Carbonate Rocks

As soon as rainwater infiltrates the ground, it begins to react with minerals in regolith and bedrock and weathers them chemically. An important part of chemical weathering involves mineral and rock constituents passing directly into solution, a process known as **dissolution**. Limestone, dolostone, and marble—the common carbonate rocks—are readily attacked by dissolution. Although carbonate minerals are nearly insoluble in pure water, they are readily dissolved by rainwater charged with CO_2, which becomes a dilute solution of carbonic acid (Table 6.3). The result is impressive. When carbonate rocks weather, nearly all their volume can be dissolved away in slowly moving groundwater.

Caves and Sinkholes

Carbonate caves come in many sizes and shapes, and they often contain spectacular formations on their walls, ceilings, and floors (Fig. 9.28). Although most caves are small, some are of exceptional size. The Carlsbad Caverns in southeastern New Mexico include one chamber 1200 m long, 190 m wide, and 100 m high (3940, 625, and 330 ft, respectively). Mammoth Cave in Kentucky consists of interconnected chambers with an aggregate length of at least 48 km (30 mi).

Limestone caves form as circulating groundwater slowly dissolves the carbonate rock. The usual sequence of development is thought to involve (1) initial dissolution along a system of interconnected open joints and bedding planes by percolating groundwater, (2) enlargement of a cave passage along the most fa-

Figure 9.28 An explorer in Lechuguilla Cave, a limestone cave in the Carlsbad Caverns region of New Mexico, examines the bizarre formations produced as carbonate precipitated from dripping and flowing water during past millennia.

vorable flow route by water that fully occupies the opening, (3) deposition of carbonate formations on the cave walls while a stream occupies the cave floor, and (4) continued deposition of carbonate on the walls and floor of the cave after the stream has stopped flowing. Although geologists have argued for years whether caves form in the zone of aeration or in the saturated zone, available evidence favors the idea that most caves are excavated in the shallowest part of the saturated zone, along a seasonally fluctuating water table.

In contrast to a cave, a **sinkhole** is a large dissolution cavity that is open to the sky. Some sinkholes are caves whose roofs have collapsed, whereas others are formed at the surface. Those produced by cave collapse can form abruptly. As a result, they pose a potential hazard for people whose houses or property may suddenly disappear into a widening conical depression tens of meters across (Fig. 9.29).

Karst Landscapes

In some regions of exceptionally soluble rocks, sinkholes and caves are so numerous that they combine to form a distinctive topography characterized by many small, closed basins and intervening ridges or pinnacles. In this kind of landscape, streams disappear into the ground and eventually reappear elsewhere as large springs. Such terrain is called **karst** (German for bare, stony ground) after the classic karst region of Slovenia (former northwestern Yugoslavia). This remarkable landscape of closely spaced sinkholes and subsurface drainage has resulted from the dissolution of pure limestone.

Several factors control the development of karst landscapes: the topography must permit the flow of groundwater through soluble rock under the pull of gravity, precipitation must be adequate to maintain the groundwater system, and soil and plant cover must supply an adequate amount of carbon dioxide. A warm and moist climate promotes carbon dioxide production, and therefore dissolution. Although karst terrain is found throughout a wide range of latitudes and at various altitudes, it often is best developed in moist

Figure 9.29 Most of a city block in Winter Park, Florida disappeared into a widening crater as this sinkhole formed in underlying carbonate bedrock.

Figure 9.30 Steep limestone pinnacles up to 200 m high, surrounded by flat expanses of alluvium, form a spectacular karst landscape around the Li River near Guilin, China.

temperate to tropical regions underlain by thick and widespread soluble rocks.

One of the most famous and distinctive of the world's karst regions lies in southeastern China, where vertical-sided limestone peaks stand up to 200 m (660 ft) high (Fig. 9.30). This dramatic landscape has inspired both classical Chinese painters and present-day photographers.

LAKES

A **lake** is any body of inland water of appreciable size that occupies a depression in the Earth's surface. A majority of the world's lakes are found in high latitudes and in mountains. Canada contains nearly half of the world's lakes (Fig. 9.31), because former continental ice sheets carved depressions in the exposed

Figure 9.31 A multitude of lakes occupies glacially scoured basins in north-central Canada. Some long, narrow lakes are aligned along faults or other structural features of the Precambrian bedrock.

GUEST ESSAY

River Aesthetics—A Janus Perspective

Janus, the Roman god of doors and gates, had two faces, allowing him to look in opposite directions. His namesake month, January, affords views of both the past and the future. Nondeities among us, and society in general, tend not to concentrate their attention on Janus's multidirectional manner. Historians look to the past, whereas developers and economists perhaps look to the future. However, land regulators, aquatic ecologists, and many others involved in the conflict between use and aesthetics of drainage basins and their river systems might profit from eyes of Janus, to see both back and ahead—to remember and revere the unspoiled, and to recognize and minimize the inevitable encroachments into riparian habitats by an expanding human population.

Aesthetics is a concept of personal preference, a judgment, sense, or love of beauty. The qualities that render a landscape or a river aesthetically rewarding to someone vary with experiences. My sense of aesthetics, as applied to natural systems, tempers the idealism of childhood perceptions—perhaps an innate recognition of beauty—with the experiences and awarenesses of an earth scientist. As it did for Thoreau, it includes a spiritual emotion akin to holism, an awareness of complete systems of interdependent parts as opposed to a concentrated attention on an individual or isolated part of the system. I look back and cherish the natural wonders I perceived as a child; I look forward and question whether those perceptions, if faithful, can be experienced again. Despite apprehensions, I remain optimistic that the sights of a youthful Janus can be compatible with his seasoned view, encompassing the realities of modern society.

Why the optimism? Because a stream is a highly dynamic core of a much larger, more complex organism: the drainage basin. In *Le Phenomene Humain*, the French geologist/cleric, Pierre Teilhard de Chardin, likened human society to an evolving organism, one whose history necessarily helps direct its future. The interplay of physical and biological processes in a drainage basin in turn permits analogy to a discrete, active organism, one that is most dynamic where water and energy are concentrated. When one part of the organism is injured, regardless of cause, natural or inflicted, the system adjusts and heals, and stream channels, the arteries of the organism, generally show the greatest and fastest recuperative powers. To me, the beauty of a stream, its aesthetics, is displayed by its vibrance, its energy, a life inseparable from and nourished by all other parts of the drainage basin or watershed.

This organism is exceedingly complex but is conceptualized by earth scientists as elements or interacting systems of soil and rock, water and air, plants and animals. If one element suffers disturbance, all elements suffer disturbance.

W. R. Osterkamp is a geomorphologist with the National Research Program, U.S. Geological Survey. He has undergraduate degrees in geology and chemistry from the University of Colorado, and an M.S. and Ph.D. in geology from the University of Arizona. His principal research interests include stream-channel dynamics, geomorphic-vegetative relations, and methods for tracing the movement and storage of sediment.

Recovery by one system is shared by recovery of all systems. But as with other organisms, healing requires circulation. And it is the arteries of circulation that most enrich and become enriched, conveying nutrients and toxins.

Terms typically employed to describe the aesthetic quality of a stream and its adjacent riches seem inappropriate when applied to a scientific view of a watershed-organism. An aesthetic view of a stream may express ideal qualities based on beauty, but an aesthetically pristine, timeless, or unrestrained stream constantly faces real threats to its health. Objective measures of the character of a stream channel are easily collected, but the beauty, not utility, of the channel must remain ephemeral if other parts of the organism become diseased. A river distributes nourishment, and it discharges wastes, both of which are derived from other parts of the being. Its health rests in the balance. Floods, fires, infections, and disturbances are components of the history of any drainage basin, and none is timeless. No organism is free of wastes or decay, and thus none is pristine.

A drainage basin and its streams, like all organisms, is dynamic and is constantly changing according to the energy available and stresses imposed on it. If the basin that feeds the rivers is healthy, its component systems unstressed and adjusted to each other, the stream, too, is healthy—it has beauty and aesthetic value. In mimicking Janus, we must not lose sight of our highly individualized perceptions and the qualities that have always been associated with the beauty of natural flowing water. Similarly, as scientifically literate members of society, we must look ahead and realize that change induced anywhere in a drainage basin induces change everywhere. Conservation practices in all parts of a drainage basin are necessary if we want to minimize harmful human impacts and aesthetic impairment to the select circulatory parts of the drainage organisms. Perhaps the effects of conservation are not apparent immediately, or within a human life span, but ultimately change and balance will be assured. Janus, that is society, must continue to see clearly in all directions if both the health of stream channels and the aesthetic value such health provides are truly to be protected.

rocks and left piles of glacially transported sediment that created innumerable hollows and natural dams. In addition to glaciation, lakes also are formed by volcanism (for example, crater and caldera lakes [Fig. 7.11], lava-dammed lakes), tectonism (downfaulted or downdropped terrain), stream cutoffs (oxbow lakes), dissolution and collapse of carbonate rocks, landslide dams, tributary alluvial fans that dam stream valleys, thawing of permafrost to form thaw lakes (Chapter 10), and coastal processes (freshwater lagoons). Although most lakes contain fresh water, many lakes in arid and semiarid regions have a large dissolved salt content (saline lakes). The salty Caspian Sea in central Asia is the lake with the largest area (144,000 km²), whereas Siberia's freshwater Lake Baikal is the deepest (1742 m, or 5715 ft).

Lakes are predominantly fed by runoff and direct precipitation, and their level commonly reflects a balance between freshwater input from streams and direct precipitation, outflow, and evaporation. In a small basin, the level of a lake may be controlled by the water table, with water table fluctuations controlling lake level fluctuations.

Lakes tend to be transitory features on the landscape. Few are more than a million years old, and most are no older than the end of the last glacial age (about 12,000 to 14,000 years ago; Chapter 14). A lake may drain away if its outlet becomes deeply eroded, or it may disappear if the climate leads to a negative water balance (more water lost by evaporation and outflow than added by inflow and direct precipitation). Small lakes may disappear with a fall in the water table. A lake may also gradually become shallow and disappear through the filling of its basin with organic and inorganic sediment, forming a swamp or bog and eventually a meadow.

The history of existing lakes is deciphered by studying sediment cores. These contain a record of changing water chemistry, sediment input, and biologic productivity, typically related to changes in local or regional climate and to human activity (Chapters 14 and 20).

SUMMARY

1. Streams are the chief means by which water returns from the land to the sea. They help shape the Earth's surface and transport sediment to the oceans.

2. A drainage basin encompasses the area supplying water to the stream system that drains it. Its area is closely related to the stream's length and annual discharge.

3. Discharge, velocity, and channel cross-sectional area are interrelated, so that when discharge changes, one or both of the other factors also changes to restore equilibrium. As discharge increases downstream, velocity and the width and depth of the channel increase.

4. Straight channels are rare. Meandering channels commonly form on gentle slopes and where sediment load is small to moderate. Braided patterns develop in streams that have variable discharge and a large bed load.

5. Stream load is the sum of bed load, suspended load, and dissolved load. Bed load comprises as much as 50 percent of the total load and moves by rolling, sliding, and saltation. Most suspended load is derived from erosion of fine-grained regolith or from stream banks. Streams that receive large contributions of underground water commonly have higher dissolved loads than those deriving their discharge principally from surface runoff.

6. The average grain size of a stream's load tends to decrease downstream as particles are reduced in size by abrasion and as the sediment is sorted by weight. Changes in composition of a stream's load reflect changes in the lithology of rocks across which the stream flows.

7. Natural levees, alluvial fans, and deltas form where stream energy changes and a stream's load is deposited. Stream terraces record changing depositional and erosional regimes, commonly related to tectonism or to changes in load, discharge, or gradient.

8. Floods result when discharge exceeds the capacity of a stream's channel. Streams experiencing large floods are capable of transporting great loads of sediment as well as very large boulders. Exceptional floods, well outside a stream's normal range, occur very infrequently.

9. Although water may be present in the zone of aeration, it does not saturate the ground. In the underlying saturated zone all openings are filled with water. The water table marks the boundary between the zone of aeration and the saturated zone and in humid regions is a subdued imitation of the ground surface above it.

10. Groundwater flows far more slowly than the water in surface streams, normally at rates between half a meter a day and several meters a year.

11. In moist regions, groundwater seeps downward under the pull of gravity. It flows from areas where the water table is high toward areas where it is low.

12. Groundwater is replenished in recharge areas and moves downward and laterally to emerge in discharge areas where it forms springs, streams, lakes, ponds, or swamps.

13. Water pumped from a new well in an unconfined aquifer initially creates a cone of depression in the water table immediately surrounding the well.

14. Major supplies of groundwater are found in aquifers, among the most productive of which are porous sand, gravel, and sandstone. An unconfined aquifer has a water table, whereas a confined aquifer is bounded above and below by impermeable, or distinctly less permeable, rocks or sediments.

15. Because of high hydrostatic pressure, water of an artesian aquifer may flow freely out the top of a well that is lower than the recharge area.

16. Dissolution of carbonate rocks can lead to the formation of caves and distinctive karst landscapes

17. Lakes tend to be ephemeral features on the landscape. Their level reflects a balance between freshwater input from streams and direct precipitation, outflow, and evaporation.

IMPORTANT TERMS TO REMEMBER

alluvial fan *200*	delta *200*	lake *209*	sinkhole *208*
alluvium *191*	discharge *192*	load *191*	spring *205*
aquiclude *205*	dissolution *207*	meander *194*	stream *191*
aquifer *205*	dissolved load *196*	permeability *203*	suspended load *196*
artesian aquifer *206*	divide *191*	porosity *203*	tributary *193*
bed load *196*	drainage basin *191*	recharge *204*	water table *202*
braided stream *195*	gradient *191*	runoff *191*	zone of aeration *202*
channel *191*	groundwater *202*	saltation *196*	
cone of depression *205*	karst *208*	saturated zone *202*	

QUESTIONS FOR REVIEW

1. How does streamflow differ from overland flow, and how are these two kinds of surface flow related?

2. How do a stream's channel dimensions (depth, width) and velocity adjust to changes in discharge?

3. Why does stream velocity generally increase downstream, despite a decrease in stream gradient?

4. Studies of meandering streams show that a relationship exists between the width of a channel and the diameter of meander loops: the larger the channel width, the larger the loop diameter. Why should this be true?

5. Why, if velocity increases downstream, does sediment in transport typically decrease in size in that direction?

6. If you were to collect samples of sandy, gravelly alluvium along the length of a large stream and determine the composition of the gravel and mineralogy of the sand, you would likely see obvious changes in these properties in the downstream direction. What factor or factors might explain such changes?

7. Explain the occurrence of alluvial fans and deltas on the basis of the standard stream discharge equation.

8. How is a flood related to changes in a stream's discharge and channel dimensions?

9. In what ways does the flow of groundwater differ from the flow of water in a stream?

10. What determines how long it takes water to pass from a recharge area to a discharge area?

11. What causes a cone of depression to form around a producing well?

12. Why are sandstones generally better aquifers than siltstones or shales?

13. For what reason does water rise to or above the ground surface in an artesian well?

14. Why do you suppose that many waste treatment facilities pass contaminated water through sand in their purification process?

15. What is the origin of "hard" water in regions of carbonate bedrock?

16. Describe how a large carbonate cave is likely to evolve.

17. What factors control the surface level of a lake?

QUESTIONS FOR DISCUSSION

1. Let's assume you are considering the purchase of a 30-year-old house adjacent to a large river that experiences seasonal flooding. What information might you seek to evaluate the possibility that an unusually large flood could inundate the house sometime during the next half century?

2. Find out where your community or city derives its supply of fresh water. Will the existing source of supply prove adequate if the population doubles during the next several decades? If not, what other potential sources exist?

3. A large area on a hillside has been suggested as a landfill site for garbage generated by a small nearby city. You are asked for a geologic appraisal of the site to determine if local subsurface water supplies might be affected. What geologic factors would you investigate and why?

Interactive case studies at our web site will provide you with further information about flooding, saline lakes and environmental hazards associated with surface water contamination.

CHAPTER 10

The World
of Snow and Ice

● *Glacial Water for Arid Lands*

As world population increases and our insatiable demand for fresh water continues to rise, a critical problem we face is how to meet this demand. Some imaginative people think the answer may lie in a vast, unutilized resource: Antarctic icebergs.

Large icebergs are plentiful around Antarctica, where an estimated 1000 km³ (240 mi³) of glacier ice breaks off as icebergs each year. The English explorer James Cook (1728–1779) was among the first to recognize the potential for using the fresh water locked up in these icebergs. In 1773, while his ship lay off the coast of Antarctica, his sailors hoisted 15 tons of berg ice on board, and Cook noted in his log that "this is the most expeditious way of watering I ever met with."

Icebergs also occur in Arctic waters, but they are far less numerous and on average much smaller than Antarctic ones. Today, scientists visualize towing large Antarctic icebergs north to supply water to the arid coasts of the Middle East, South America, Africa, Australia, and the southwestern United States. A French engineering firm, commissioned by Saudi Arabia, has already studied the feasibility of towing huge icebergs north to the Red Sea, where they could supply water for drinking and agricultural use.

As impressive as such a proposal sounds, the technological problems are formidable. How can blocks of ice hundreds of meters thick and several square kilometers in area be moved thousands of kilometers through warm and often stormy seas and still have enough mass to make the venture economically feasible? At an average speed of 0.5 m/s (1 knot), a huge berg 1 to 2 km (0.6 to 1.2 mi) long would require 70 days to be towed 3000 km (1860 mi) through relatively cool waters to Australia, but the outermost 60 m (200 ft) of ice would likely melt away during the trip. An unprotected iceberg of this size is unlikely to survive the long trip (up to a year) through much warmer waters to southern California or the Middle East. To avoid excessive loss, the ice would have to be insulated against melting and evaporation and protected against collapse and disintegration. In addition, the towing vessels or propulsion systems large enough to move huge icebergs over great distances have yet to be developed. Although these technological challenges appear immense, the day may come when ice that has been stored in Antarctic glaciers for tens of thousands of years will bring life-sustaining waters to the Earth's low-latitude, arid lands. ●

(opposite) Because of the density of glacier ice, 90 percent of a typical Antarctic iceberg lies below sea level. The technology required to tow such a huge ice mass to low-latitude arid lands, where it might be used as a source of fresh water, does not yet exist.

THE EARTH'S COVER OF SNOW AND ICE

Among the planets of our solar system, only the Earth has a bluish color, imparted by the vast expanse of world ocean. Nevertheless, the polar regions of the northern and southern hemispheres appear largely white because of an extensive, perennial cover of snow and ice. In the northern hemisphere, much of the ice floats as a thin sheet on the Arctic Ocean, whereas in the southern hemisphere it consists of a vast glacier system overlying the continent of Antarctica and adjacent islands and seas as well as floating sea ice that extends far beyond the coastline. The vast mantle of polar ice is linked in important ways to the oceans and the atmosphere. It is involved in the generation of cold, dense water that drives deep ocean circulation (Chapter 11), and it is an important factor in world climate (Chapter 13). It is also an extremely dynamic part of our planet, for its expanse fluctuates seasonally as well as on geologic time scales.

The part of the Earth's surface that remains perennially frozen constitutes the **cryosphere** *(cold or frozen sphere)*. It includes not only glaciers and sea ice, but also vast areas of frozen ground that lie beyond the limits of glaciers. At present, glaciers cover about 10 percent of the Earth's land surface, while perennially frozen ground covers an additional 20 percent. Thus, nearly a third of the Earth's land area belongs to the cryosphere.

SNOW AND THE SNOWLINE

The winter snow that must be shoveled from sidewalks and delays the morning commute also plays an important role in the Earth's climate system. Its highly

Figure 10.1 A map of average snow cover in the northern hemisphere (expressed as percentage of land area covered by snow) during December 1992 is based on data received from a microwave sensor aboard an orbiting satellite. The greatest snow cover lies in regions of continental climate in middle to high latitudes (northern North American and northeastern Asia) and high-altitude regions such as the Tibetan Plateau of central Asia.

Figure 10.2 The lower limit of snow in late spring forms an irregular line across the flank of Mount Cook, the highest peak in New Zealand's Southern Alps. As the weather warms and the snow melts, the snow limit rises to its highest level at the end of the summer. This late-summer limit marks the annual snowline. Above the snowline, most of the ground remains snow-covered all year.

reflective surface bounces sunlight back into space, thereby reducing surface air temperature. When the snow melts, it becomes a major source of water for rivers and moisture for agricultural soils. However, the timing or amount of snowfall can affect people adversely. For example, a heavy late-winter snowfall can generate widespread flooding, while a snowfall deficit can lead to water rationing during the summer.

Annual Snow Cycle

Prior to the mid–1960s, estimates of variations in continental snow cover were obtained from limited ground-based measurements. Today, variations in snow depth and area are monitored by satellites.

During a typical year, northern hemisphere snow cover first appears in northern Alaska and northeastern Siberia in mid-September to mid-October (Fig. 10.1). Through November, the snow cover expands southward and begins to thicken. By December, the expanding snow cover reaches southern Russia, central Europe, and the northern United States, and snow blankets nearly all of the high Tibetan Plateau. From December through March, the snowpack thickens in continental interiors, but its southern limit begins to retreat as air temperature rises. The snowpack then recedes rapidly northward during late spring, and by mid-June the remaining snow is confined mainly to high mountains and to lands bordering the cold Arctic Ocean.

The Snowline

If we view a high mountain at the end of the summer, just before the earliest autumn snowfall, we commonly will see a snowy zone on its upper slopes. The lower boundary of this zone is the **snowline**, which is defined as the lower limit of perennial snow (Fig. 10.2). Above the snowline, part of the past winter's snow survives the warm temperatures of summer, along with any snow from earlier winters that persisted through previous summers. In detail, the snowline is an irregular surface, its shape controlled both by variations in the thickness of the winter snowpack and by local topography. When viewed from a distance, however, the snowline appears as a line delimiting snow-covered land from snow-free land.

The altitude of the snowline and its horizontal position on the landscape typically change from year to year depending on the weather. Although a number of climatic factors are involved, the two principal ones are winter snowfall, which affects total snow accumulation, and summer temperature, which influences melting.

In the polar regions, the amount of annual snowfall

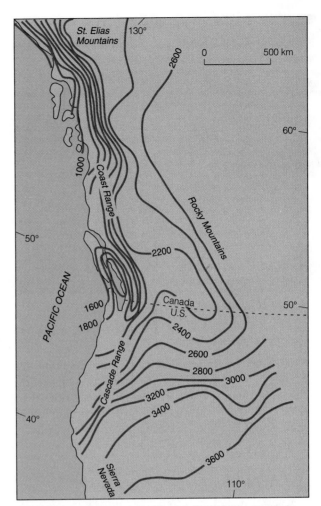

Figure 10.3 Contours (in meters) show the regional altitude of the snowline throughout northwestern United States, British Columbia, and southern Alaska for a representative balance year. The surface defined by the contours rises steeply inland from the Pacific coast in response to increasingly drier climate, and more gradually from north to south in response to progressively higher mean annual temperatures.

is generally very low because the air is too cold to hold much moisture. Nevertheless, because summer temperatures are also low, little melting occurs and the snowline generally lies at low altitudes. Mean summer temperatures increase toward the equator, and so the altitude of the snowline also rises toward the equator, but not uniformly (Fig. 10.3). Because its level is also controlled by precipitation, the snowline rises inland as precipitation decreases away from ocean moisture sources. This effect is especially noticeable if we trace the snowline inland from a midlatitude continental coast. For example, winter snowfall is very high in the Coastal Ranges of southern Alaska and British Columbia adjacent to the Pacific Ocean source of moisture, and the snowline lies as low as 600 m (ca. 1200 ft). However, it rises steeply inland to 2600 m (ca.

Figure 10.4 A small cirque glacier below a mountain summit in Alaska's Denali National Park.

Figure 10.5 Dark bands of rock debris delineate the boundaries between adjacent tributary ice streams that merged to form Kaskawulsh Glacier, a large valley glacier in Yukon Territory, Canada.

8500 ft) in the Rocky Mountains as precipitation decreases by more than half. The snowline is highest on the lofty summits of central Asia [in places more than 6000 m (19,700 ft)], which lie farther than any other high mountains from an oceanic source of moisture.

GLACIERS

Wherever the amount of snow falling each winter is greater than the amount that melts during the following summer, the snow grows progressively thicker. Gradually, the deeper snow recrystallizes into denser ice under the increasing weight of overlying snow. When the snow and ice become so thick that the pull of gravity causes the frozen mass to move, a glacier is born. Accordingly, we define a **glacier** as a permanent body of ice, consisting largely of recrystallized snow, that shows evidence of slow downslope or outward movement owing to the stress of its own weight.

Glaciers vary considerably in shape and size. We recognize several fundamental types:

1. The smallest, a *cirque glacier* (Fig. 10.4), occupies a protected, bowl-shaped depression (a **cirque**) on a mountainside that is produced by glacial erosion.

2. A cirque glacier that expands outward and downward into a valley becomes a *valley glacier* (Fig. 10.5). Many of the Earth's high mountain ranges contain glacier systems that include valley glaciers tens of kilometers long, each heading in one or several cirques (Fig. 10.6).

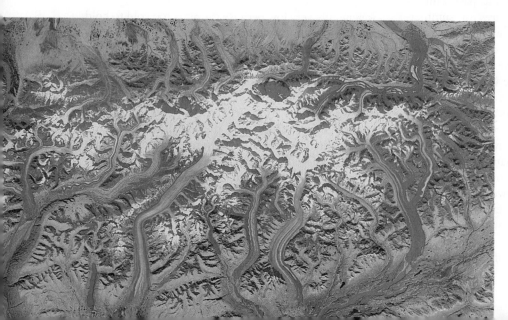

Figure 10.6 A vertical satellite view of the valley-glacier complex that covers much of Denali National Park in south-central Alaska. Mount McKinley, the highest peak in North America, lies near the center of the glacier-covered region.

***Figure 10.*7** Several ice caps cover areas of high land on Iceland. Vatnajökull, in the southeastern part of the island, is the largest ice cap (8300 km²) and overlies an active volcano.

3. Along some high-latitude sea coasts, nearly every large valley glacier occupies a deep **fjord**, the drowned seaward end of a glacier-carved bedrock trough. Such glaciers are called *fjord glaciers*.

4. *Ice caps* cover mountain highlands or low-lying land at high latitudes and generally flow radially outward from their center (Fig. 10.7).

5. Huge continent-sized *ice sheets* overwhelm nearly all the land surface within their margins (Fig. 10.8). Modern ice sheets, which are confined to Greenland and Antarctica, include about 95 percent of the ice in existing glaciers and reach thicknesses of 3000 m (> 9800 ft) or more.

6. Floating *ice shelves* hundreds of meters thick occupy large embayments along the coasts of Antarctica (Fig. 10.8); smaller ones are found among the Canadian Arctic islands.

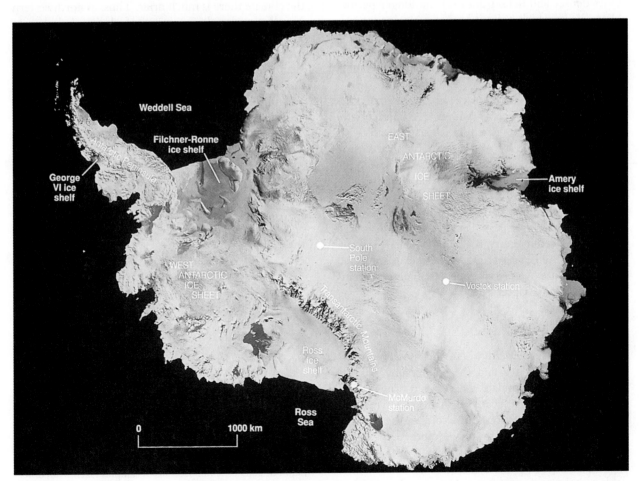

Figure 10.8 Satellite view of Antarctica. The East Antarctic Ice Sheet overlies the continent, while the much smaller West Antarctic Ice Sheet covers a volcanic island arc and surrounding seafloor. Major ice shelves occupy large coastal embayments. The ice-covered regions of Antarctica nearly equal the combined areas of Canada and the conterminous United States.

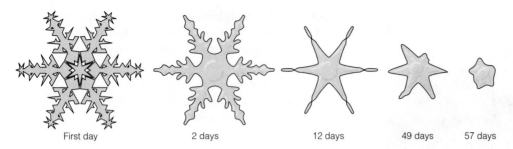

First day   2 days   12 days   49 days   57 days

Figure 10.9 As a snowflake is slowly converted to a granule of old snow, melting and evaporation cause its delicate points to disappear. The resulting meltwater refreezes, and vapor condenses near the center of the crystal, making it denser.

How Glaciers Form

Newly fallen snow is porous and has a density less than a tenth that of water. Air easily penetrates the pore spaces, and the delicate points of each snowflake gradually evaporate. The resulting water vapor condenses, mainly in constricted places near a snowflake's center. In this way, the fragile ice crystals slowly become smaller, rounder, and denser, and the pore spaces between them disappear (Fig. 10.9).

Snow that survives a year or more gradually becomes denser and denser until it is no longer permeable to air, at which point it becomes **glacier ice**. Although now a rock, glacier ice has a far lower melting temperature than any other naturally occurring rock, and its density—about 0.9 g/cm³—means it will float in water.

Further changes take place as the ice becomes buried deeper and deeper within a glacier. Figure 10.10 shows a core obtained by Russian glaciologists drilling deep in the East Antarctic Ice Sheet at Vostok Station. As snowfall adds to the glacier's thickness, the increasing pressure causes initially small grains of glacier ice to grow until, near the base of the ice sheet, they reach a diameter of 1 cm (0.4 in) or more. This increase in grain size is similar to what happens in a fine-grained sedimentary rock that is carried down deep within the Earth's crust. As the rock is subjected to high pressure over a long time, large mineral grains slowly develop (Chapter 8).

Distribution of Glaciers

A glacier can develop anywhere the snowline intersects the land surface and the topography permits snow and ice to accumulate. This explains why glaciers are found not only at sea level in the polar regions, where the snowline is low, but also near the equator, where some lofty peaks in New Guinea, East Africa, and the Andes rise above the snowline. As we might expect, most glaciers are found in high latitudes, the coldest regions of our planet. However, because low temperatures also occur at high altitudes, many glaciers also can exist in lower latitudes on high mountains.

Low temperature is not the only limiting factor determining where glaciers can form, for a glacier must receive a continuing input of snow. Proximity to a moisture source is therefore another requirement. The abundance of glaciers in the coastal mountains of northwestern North America, for example, is related mainly to the abundant precipitation received from air masses moving landward from the Gulf of Alaska. Farther inland in the same latitude zone, the Rocky Mountains contain fewer and smaller glaciers because the climate there is much drier. Thus, in northwestern North America the existence of glaciers is linked to the interaction of several Earth systems: tectonic forces that have produced high mountains; the adjacent ocean, which provides an abundant source of moisture; and the atmosphere, which delivers the moisture to the land as snow.

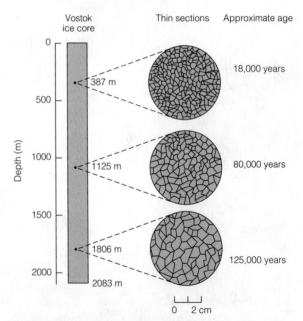

Vostok ice core   Thin sections   Approximate age

Depth (m)

0

387 m   18,000 years

500

1000   1125 m   80,000 years

1500

1806 m   125,000 years

2000

2083 m

0   2 cm

Figure 10.10 A deep ice core drilled at Russia's Vostok Station penetrates through the East Antarctic Ice Sheet to a depth of 2083 m. Thin-sections of samples taken from different depths in the core show a progressive increase in the size of ice crystals, the result of slow recrystallization as the thickness and weight of overlying ice slowly increase with time.

Warm and Cold Glaciers

Glaciers obviously are cold, for they consist of ice and snow. However, when we drill holes through glaciers in a variety of geographic environments, we find a large range in ice temperatures. This temperature range allows us to divide glaciers into warm and cold types. The difference between them is important, for ice temperature is a major factor controlling how glaciers behave.

Ice throughout a warm glacier, more commonly called a **temperate glacier**, can coexist in equilibrium with water, for the ice is at the **pressure melting point**, which is the temperature at which ice can melt at a particular pressure (Fig. 10.11A). Such glaciers are restricted mainly to low and middle latitudes. At high latitudes and high altitudes, where the mean annual air temperature lies below freezing, the temperature in a glacier remains below the pressure melting point, and little or no seasonal melting occurs (Fig. 10.11B). Such a cold glacier is commonly called a **polar glacier.**

If the temperature of snow crystals falling to the surface of a temperate glacier is below freezing, how does ice throughout the glacier reach the pressure melting point? The answer lies in the seasonal fluctu-

A.

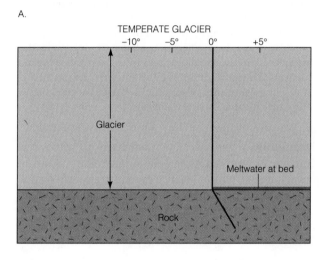

B.

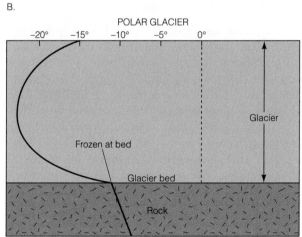

Figure 10.11 Temperature profiles through temperate and polar glaciers. A. Ice in a temperate glacier is at the pressure melting point from surface to bed. The terminus is rounded, as illustrated by Pré de Bar Glacier in the Italian Alps, because melting occurs at the surface. B. Ice in a polar glacier remains below freezing, and the ice is frozen to its bed. Subfreezing temperatures inhibit melting at the terminus, which forms a steep cliff of ice, as illustrated by Commonwealth Glacier in Antarctica.

ation of air temperature and in what happens when water freezes to form ice. In summer, when air temperature rises above freezing, solar radiation melts snow and ice at the glacier's surface. The meltwater percolates downward, where it encounters freezing temperatures and it therefore freezes. When changing state from liquid to solid, each gram of water releases 335 J of heat. This released heat warms the surrounding ice and, together with heat flowing upward from the solid earth beneath the glacier, keeps the temperature of the ice at the pressure melting point.

Why Glaciers Change Size

Nearly all high-mountain glaciers have shrunk substantially in recent decades, exposing extensive areas of valley floor that only a century ago were buried beneath thick ice. Other glaciers have remained relatively unchanged, however, and a few have even expanded. To understand why glaciers advance and retreat, and why glaciers in the same region can show dissimilar behavior, we need to examine how a glacier responds to a gain or loss of mass.

Annual Balance of a Glacier

The mass of a glacier is constantly changing as the weather varies from season to season and, on longer time scales, as local and global climates change. We can think of a glacier as being analogous to a checking account in a bank. The balance in the account at the end of the year is the difference between the amount of money added during the year and the amount removed. The balance of a glacier's account is measured in terms of the amount of snow added, mainly in the winter, and the amount of snow (and ice) lost, mainly during the summer. The additions are called **accumulation**, and the losses are **ablation.** The total in the account at the end of a year—in other words, the difference between accumulation and ablation—is a measure of the glacier's **mass balance** (Fig. 10.12). The account may have a surplus (a positive balance) or a deficit (a negative balance), or it may have exactly the same amount at the end as at the beginning of a year.

If a glacier is viewed at the end of the summer ablation season, two zones are generally visible on its surface (Fig. 10.13). An upper zone, the **accumulation area**, is the part of the glacier covered by remnants of the previous winter's snowfall and is an area of net gain in mass. Below it lies the **ablation area**, a region of net loss where bare ice and old snow are exposed because the previous winter's snow cover has melted away.

When, over a period of years, a glacier gains more mass than it loses, its volume increases. The front, or

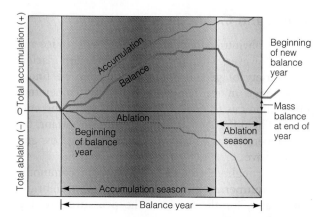

Figure 10.12 Accumulation and ablation determine glacier mass balance (heavy line) over the course of a balance year. The balance curve, obtained by summing values of accumulation (positive values) and ablation (negative values), rises during the accumulation season as mass is added to the glacier and then falls during the ablation season as mass is lost. The mass balance value at the end of the balance year reflects the difference between annual mass gain and mass loss.

terminus, of the glacier is then likely to advance as the glacier grows. Conversely, a succession of years in which negative mass balance predominates will lead to retreat of the terminus. If no net change in mass occurs, the terminus is likely to remain relatively stationary.

The Equilibrium Line

The **equilibrium line** marks the boundary between the accumulation area and the ablation area (Fig. 10.13) It lies at the level on the glacier where net mass loss equals net mass gain. The equilibrium line on temperate glaciers coincides with the local snowline. Being very sensitive to climate, the equilibrium line fluctuates in altitude from year to year and is higher in warm, dry years than in cold, wet years (Fig. 10.14). Because of this sensitivity, we can use the altitude of the equilibrium line to estimate a glacier's mass balance without having to make detailed field measurements of accumulation and ablation. Thus, fluctuations in the altitude of the equilibrium line over time can provide us with a measure of changing climate.

How Glaciers Move

One way to prove that glaciers move is to walk out onto a glacier near the end of the summer and carefully measure the position of a surface boulder with respect to some fixed point beyond the glacier margin. Remeasure the boulder's position a year later and you will find that the boulder may have moved up to several meters in the downglacier direction. Actually, it is

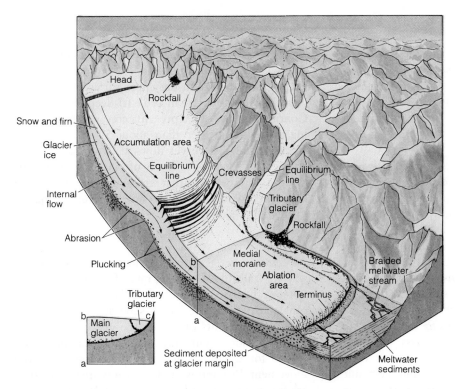

Figure 10.13 Main features of a valley glacier. The glacier has been cut away along its center line so that only half is shown. Crevasses form where the glacier flows over an abruptly steepened slope. Arrows show the local directions of ice flow. A band of rock debris forms a medial moraine that marks the boundary between the main glacier and a tributary glacier joining it from a lateral valley.

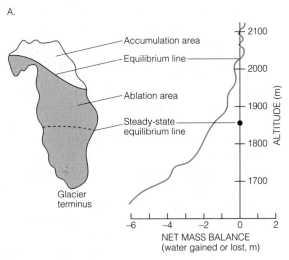

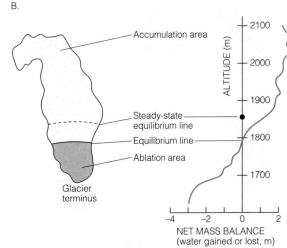

Figure 10.14 Maps of South Cascade Glacier in Washington's Cascade Range at the end of two successive balance years showing the position of the equilibrium line relative to the position it would have under a balanced (steady-state) condition. The curves plot mass balance as a function of altitude. During the first year, A, a negative balance year, the glacier lost mass and the equilibrium line was high (2025 m). In the following year, B, a positive balance year, the glacier gained mass and the equilibrium line was low (1795 m).

the ice that has moved, carrying the boulder along for the ride.

What causes a glacier to move may not be immediately obvious, but we can find clues by examining the ice and the terrain on which it lies. These clues tell us that ice moves in two primary ways: by internal flow and by sliding of the basal ice across rock or sediment.

Internal Flow

When an accumulating mass of snow and ice on a mountainside reaches a critical thickness, the mass begins to deform and flow downslope under the pull of gravity. The flow takes place mainly through movement within individual ice crystals, which are subjected to higher and higher stress as the weight of the overlying snow and ice increases. Under this stress, ice crystals are deformed by slow displacement (termed *creep*) along internal crystal planes in much the same way that cards in a deck of playing cards slide past one another if the deck is pushed from one end (Fig. 10.15). As the compacted, frozen mass begins to move, stresses between adjacent ice crystals cause some to grow at the expense of others, and the resulting larger crystals end up with their internal planes oriented in the same direction. This alignment of crystals leads to increased efficiency of flow, because the internal creep planes of all crystals now are parallel.

In contrast to deeper parts, where the ice flows as a result of internal creep, the surface portion of a glacier

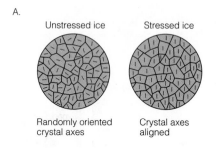

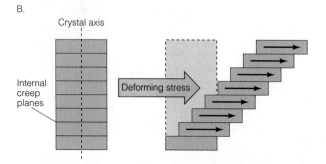

Figure 10.15 Internal creep in the ice crystals of a glacier. A. Randomly oriented ice crystals in the upper layers of a glacier are reorganized under stress so that their axes are aligned. B. When stress is applied to an ice crystal, creep along internal planes causes slow deformation.

has relatively little weight on it and is brittle. Where a glacier passes over an abrupt change in slope, such as a bedrock cliff, the surface ice cracks as tension pulls it apart. When a crack opens up, it forms a **crevasse**, a deep, gaping fissure in the upper surface of a glacier, generally less than 50 m (165 ft) deep (Fig. 10.13). Continuous flow of ice prevents crevasses from forming at depths greater than about 50 m. Because it cracks at the surface yet flows at depth, a glacier is analogous to the upper layers of the Earth, which include a surface zone that cracks and fractures (the lithosphere) and a deeper zone in the upper mantle that can flow slowly.

Basal Sliding

Ice temperature is very important in controlling the way a glacier moves and its rate of movement. Meltwater at the base of a temperate glacier acts as a lubricant and permits the ice to slide across its *bed* (the rocks or sediments on which the glacier rests) (Fig. 10.16). In some temperate glaciers, such sliding accounts for up to 90 percent of the total observed movement. By contrast, polar glaciers are so cold they are frozen to their bed. Their motion largely involves internal deformation rather than basal sliding, and so their rate of movement is greatly reduced.

Ice Velocity

Measurements of the surface velocity across a valley glacier show that the uppermost ice in the central part of the glacier moves faster than ice at the sides, similar to the velocity distribution in a river (Figs. 10.16 and 9.5). The reduced rates of flow toward the glacier margins are due to frictional drag of the ice against the valley walls. A similar reduction in flow rate toward the bed is observed in a vertical profile of velocity (Fig. 10.16).

Although snow piles up in the accumulation area each year, and melting removes snow and ice from the ablation area, a glacier's surface profile does not change much because ice is transferred from the accumulation area to the ablation area. In the accumulation area, the mass of accumulating snow and ice is pulled downward by gravity, and so the dominant direction of movement is toward the glacier bed. However, the ice does not build up to ever greater thickness because a downglacier component of flow is also present. Ice flowing downglacier replaces ice being lost from the glacier's surface in the ablation area, and so in this area the flow is upward toward the surface (Fig. 10.13). Ice crystals falling as snowflakes on the glacier near its head therefore have a long path to follow before they emerge near the terminus. Those falling close to the equilibrium line, on the other hand, travel only a short distance through the glacier before reaching the surface again.

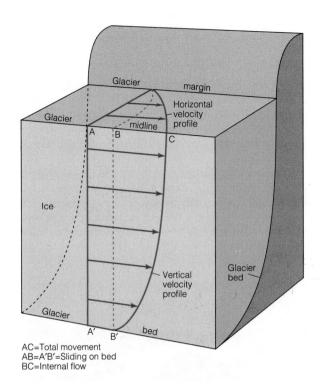

Figure 10.16   Three-dimensional view through half of a temperate glacier showing horizontal and vertical velocity profiles. Glacier movement is due partly to internal flow and partly to sliding of the glacier across its bed.

Even if the mass balance of a glacier is negative and the terminus is retreating, the downglacier flow of ice is maintained. Retreat does not mean that the ice-flow direction reverses; instead, it means that the rate of flow downglacier is insufficient to offset the loss of ice at the terminus.

In most glaciers, flow velocities range from only a few centimeters to a few meters a day, or about as fast as the rate at which groundwater percolates through crustal rocks. Hundreds of years have elapsed since ice now exposed at the terminus of a very long glacier fell as snow near the top of its accumulation area.

Response Lags

Advance or retreat of a glacier terminus does not necessarily give us an accurate picture of changing climate because a time lag occurs between a climatic change and the response of the glacier terminus to that change. The lag reflects the time it takes for the effects of an increase or a decrease in accumulation above the equilibrium line to be transferred through the slowly moving ice to the glacier terminus. The length of the lag depends both on the size of a glacier and on the way the ice moves; the lag will be longer for large glaciers than for small ones and for polar glaciers than for temperate ones. Temperate glaciers of modest size (like those in the European Alps) have response lags that probably range from several years to a decade or more. This lag time can explain why, in any one region that has glaciers of different sizes, some glaciers may be advancing while others are either stationary or retreating. A glacier may also be out of phase with nearby glaciers for reasons unrelated to climate (see "A Closer Look: Earthquakes, Rockfalls, and Glacier Mass Balance").

Calving Glaciers

During the last century and a half, many coastal Alaskan glaciers have receded at rates far in excess of typical glacier retreat rates on land. Their dramatic recession is due to frontal **calving**, which is defined as the progressive breaking off of icebergs from the front of a glacier that terminates in deep water. Although the base of a fjord glacier may lie far below sea level along much of its length, its terminus can remain stable as long as it is resting (or "grounded") against a shoal (a shallow submarine ridge) (Fig. 10.17). How-

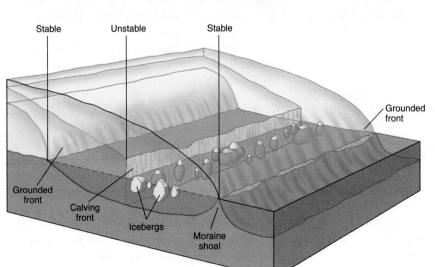

Figure 10.17   The terminus of a fjord glacier remains stable if it is grounded against a shoal (submarine ridge), but if the glacier retreats into deeper water, calving will begin. The unstable terminus then retreats at a rate that depends mainly on water depth. Once it becomes grounded farther up the fjord, the terminus is stable again.

A CLOSER LOOK

Earthquakes, Rockfalls, and Glacier Mass Balance

The great Prince William Sound earthquake of March 27, 1964, which resulted in widespread death and destruction throughout southern Alaska, triggered the collapse of a massive mountain buttress above Sherman Glacier in the Chugach Mountains. The resulting landslide spread debris across 8.5 km² (3.3 mi²) of the glacier surface, covering about a third of the ablation area (Fig. C10.1). Before the earthquake, the mass balance of the glacier was slightly negative, the annual loss of ice in the ablation area was about

4 m (4.4 yd), and the terminus was retreating about 25 m/year (27 yd/yr). Within a few years following the earthquake, the insulating debris cover (averaging 1.3 m, or 1.4 yd thick) reduced the annual melting to only a few cm. The resulting change in mass balance ended the glacier's rapid recession.

The abrupt addition of a debris cover in the ablation area will cause a glacier to respond differently than other nearby glaciers that respond only to climate. Thus, a sudden rockfall event could explain why a glacier would behave nonsynchronously compared to other glaciers in the same climatic environment.

Figure C10.1 A vast sheet of rocky debris covers the lower ablation zone of Sherman Glacier following the collapse of a large mountain buttress during the 1964 Alaska earthquake. The debris cover impeded melting of the underlying ice, leading to a positive mass balance and subsequent advance of the glacier terminus.

ever, if the terminus retreats off the shoal, water will replace the space that had been occupied by ice. With the glacier now terminating in water, conditions are right for calving. Because the base of a glacier within a fjord descends in altitude in the upfjord direction, the water becomes progressively deeper as the calving terminus retreats. The deepening water leads to faster retreat because the greater the water depth, the faster the calving rate. Once started, calving will continue rapidly and irreversibly until the glacier front recedes

into water too shallow for much calving to occur, generally near the head of the fjord.

Icebergs produced by calving glaciers constitute an ever-present hazard to ships in subpolar seas. In 1912, when the S.S. *Titanic* sank after striking an iceberg in the North Atlantic ocean, the detection of approaching bergs relied on the sharpness of sailors' vision. Today, with sophisticated electronic equipment, large bergs can generally be identified well before an encounter. Nevertheless, ice has a density of 0.9, so that

90 percent of an iceberg lies under water, making it difficult to detect (Chapter Opener, p. 214). In coastal Alaska, where calving glaciers are commonplace, icebergs pose a potential threat to huge oil tankers. For this reason, Columbia Glacier, which lies adjacent to the main shipping lanes from Valdez at the southern end of the Alaska Pipeline, is being closely monitored as its terminus pulls steadily back and multitudes of bergs are released.

Glacier Surges

Although most glaciers slowly grow or shrink as the climate fluctuates, some experience episodes of unusual behavior marked by rapid movement and dramatic changes in size and form. Such an event, called a **surge**, is unrelated, or only secondarily related, to a change in climate. When a surge occurs, a glacier seems to go berserk. Ice in the accumulation area begins to move rapidly downglacier, producing a chaos of crevasses and broken pinnacles of ice in the ablation area. *Medial moraines*, which are bands of rocky debris marking the boundaries between adjacent tributary glaciers (Fig. 10.5), are deformed into intricate patterns (Fig. 10.18). In some cases, a glacier terminus has advanced up to several kilometers during a surge. Rates of movement as great as 100 times those of non-surging glaciers and averaging as much as 6 km (3.7 mi) a year have been measured.

Figure 10.18 Contorted medial moraines of Susitna Glacier in the Alaskan Range provide striking evidence of periodic surges during which tributary ice streams advance at rates far greater than those of adjacent nonsurging glaciers.

The cause of surges is still not fully understood, but available evidence supports a reasonable hypothesis. We know that the weight of the ice can produce high pressure in water trapped at the base of a glacier. Over a period of years, this steadily increasing pressure may cause the glacier to separate from its bed. The resulting effect is similar to the way a car hydroplanes on a wet road surface. According to this hypothesis, as the ice is floated off its bed, its forward mobility is greatly increased and it moves rapidly forward before the water escapes and the surge stops.

Glaciated Landscapes

Skiers racing down the steep slopes at Alta, Mammoth, or Whistler and rock climbers inching their way up the cliffs of Yosemite Valley, the granite spires of Mont Blanc, or the icy monoliths of the southern Andes owe a debt to the ancient glaciers that carved these mountain playgrounds. The scenic splendor of these and most of the world's other high mountains is the direct result of glacial sculpturing. Over other vast areas of central North America and northern Europe, farmers gain their livelihood from productive soils developed on widespread glacial sediments left by former continental ice sheets. In all, fully 30 percent of the Earth's land area has been shaped by glaciers in the recent past.

Glacial Erosion and Sediment Transport
In shaping the land surface over which it moves, a glacier acts like a plow, a file, and a sled. As a plow, it

scrapes up weathered rock and soil and plucks out blocks of bedrock. As a file, it rasps away firm rock. As a sled, it carries away the sediment acquired by plowing and filing, together with rock debris that falls from adjacent slopes.

Unlike a stream, part of a glacier's coarse load can be carried at its sides and even on its surface. A glacier can carry very large rocks and can transport large and small pieces side by side without segregating them according to size and density. Thus, sediments deposited directly by a glacier are neither sorted nor stratified.

The load of a glacier typically is concentrated at its base and sides because these are the areas where glacier and bedrock are in contact. The coarse fraction of the load is derived partly from fragments of rock plucked from the lee (downglacier) side of outcrops over which the ice flows, a process called glacial plucking. Generally, such fragments are bounded by joints along which the rock has fractured. A significant component of the basal load of a glacier consists of very fine sand and silt grains informally called *rock flour.* If we examine such particles under a microscope, we find that they have sharp, angular surfaces that are produced by crushing and grinding.

Small rock fragments embedded in the basal ice scrape away at the underlying bedrock and produce long, nearly parallel scratches called **striations** (Fig. 10.19). Larger rock fragments that the ice drags across a bedrock surface cut *glacial grooves,* aligned in the direction of ice flow. Rock flour in the basal ice acts like fine sandpaper and can polish the rock until it has a smooth, reflective surface.

The bulk of the rock debris visible on the surface of valley glaciers arrived there by rockfalls from adjacent cliffs. If a rockfall reaches the accumulation area, the flow paths of the ice (Fig. 10.13) will carry the debris downward into the glacier and then upward to the surface in the ablation area. If rocks fall onto the ablation area, the debris will remain at the surface and be carried along by the moving ice. Where two glaciers join, rocky debris at their margins merges to form a medial moraine (Figs. 10.5B and 10.13).

Glacial Sculpture

In mountainous regions, cirques are among the most common and distinctive landforms produced by glacial erosion (Fig. 10.20A). The characteristic bowl-like shape of a cirque is the result of frost-wedging, combined with plucking and abrasion at the glacier bed. As cirques on opposite sides of a mountain grow larger, they intersect to produce sharp-crested ridges. Where three or more cirques have sculptured a mountain mass, the result can be a high, sharp-pointed peak, a classic example of which is the Matter-horn in the Swiss/Italian Alps (Fig. 10.19).

A valley that has been shaped by glaciers differs from ordinary stream valleys in having a distinctive U-shaped cross profile and a floor that often lies below the floors of tributary valleys (Fig. 10.20B). Streams commonly descend as waterfalls, or cascades, as they flow from the tributary valleys into the main valley. The long profile of a glaciated valley floor may possess steplike irregularities and shallow basins related to the spacing of joints in the rock, which influences the ease of glacial plucking, or to changes in rock type along the valley. Finally, the valley typically heads in a cirque or group of cirques.

Fjords deeply indent the mountainous, west-facing coasts of Norway, Alaska, British Columbia, Chile, and New Zealand (Fig. 10.21). Typically shallow at their seaward end, fjords become deeper inland, implying deep glacial erosion. Sognefjord in Norway, for example, reaches a depth of 1300 m (4260 ft), yet near its seaward end the water depth is only about 150 m (495 ft).

Glacial erosion is also responsible for countless lakes that lie inside the limit of the last glaciation. Among the largest are the huge lakes that form an arc

Figure 10.19 A deglaciated bedrock surface beyond Findelen Glacier in the Swiss Alps displays grooves and striations etched by rocky debris in the base of the moving glacier when it overlay this site. In the background rises the Matterhorn, a glacial horn sculpted by glaciers that surround its flanks.

A.

B.

Figure 10.20 Typical landforms of glaciated mountains. A. A steep-walled cirque, carved in granitic rock of Monte Bianco (Mont Blanc) in northern Italy, was occupied by a glacier during the last glacial age. B. A deep U-shaped valley in the southern Coast Range of British Columbia, Canada, was carved during repeated invasions of ice-age glaciers that left the valley walls smoothed and abraded to a height of nearly 2 km above the valley floor.

Figure 10.21 Trekkers atop the Pulpit, a spectacular vantage point far above Lysefjord, can look far inland toward the source region of the glacier that carved this fjord, typical of numerous others that indent the rocky western coast of Norway.

across southern and western Canada and include the Great Lakes, Lake Winnipeg, Lake Athabaska, Great Slave Lake, and Great Bear Lake.

Eroding ice sheets sometimes mold smooth, nearly parallel ridges of sediment or bedrock, called *drumlins*, which are elongated parallel to the direction of ice flow (Fig. 10.22). Drumlins, like the streamlined bodies of supersonic airplanes that are designed to reduce air resistance, offer minimum resistance to glacier ice flowing over and around them.

Glacial Deposits   A moving glacier carries with it rock debris eroded from the land over which it is passing or dropped on the glacier surface from adjacent cliffs. As the debris is transported past the equilibrium line and ablation reduces ice thickness, the debris begins to be deposited. Some of the basal debris is plastered directly onto the ground as till (Chapter 8). Some also reaches the glacier margin, where it is released by the melting ice and either accumulates there or is reworked by meltwater that transports it beyond the terminus where it is deposited as *outwash*.

A ridgelike accumulation of sediment built up along the margin of a glacier is an **end moraine** (Fig. 10.23). An end moraine built at the terminus of a glacier is a *terminal moraine*, and one constructed along the side of a mountain glacier is a *lateral moraine*. End moraines form as sediment is bulldozed by a glacier advancing across the land, as loose surface debris on a glacier slides off and piles up along the glacier margin, or as debris melts out of ice and accumulates along the edge of a glacier. They range in height from a few meters to hundreds of meters. The great thickness of some end moraines results from the repeated accretion of sediment from debris-covered glaciers during successive ice advances.

When rapid melting greatly reduces a glacier's thickness in its ablation area, ice flow may virtually cease. Sediment deposited by meltwater streams flowing over or beside such immobile ice will slump and collapse as the supporting ice slowly melts away, leaving a hilly, often chaotic surface topography. Among the landforms associated with such terrain are *kames*, small hills of stratified sediment, and *kettles*, closed basins created by the melting away of a mass of underlying glacier ice. Landscapes marked by numerous kettles and kames are clear evidence of stagnant-ice conditions (Fig. 10.24).

Glaciers and People

Modern humans have evolved during the glacial ages under conditions that fluctuated between those like the present and those of the last ice age when vast

Figure 10.22   Drumlins in Dodge County, Wisconsin, each shaped like the inverted hull of a ship, are aligned parallel to the flow direction of the continental ice sheet that shaped them during the last glaciation.

Figure 10.23   Tumbling Glacier on Mount Robson in British Columbia, Canada, has retreated up-slope from a sharp-crested terminal moraine that it constructed during the nineteenth and twentieth centuries.

areas of the northern hemisphere continents were covered with thick ice sheets, climates were cooler, and the skies were dustier. When the latest ice sheets retreated, people invaded the deglaciated landscapes and colonized them. Today, a substantial fraction of the world's population lies within or near the limit of former glaciers, and this population is directly af-

fected by the legacy of ice-age glaciers. The rich agricultural lands of the American Midwest and central Europe are a gift of the glaciers that deposited sediments rich in mineral nutrients and of strong winds that eroded and deposited fine dust across the landscape. Important subsurface groundwater supplies are mined from glacial gravels deposited along former

Figure 10.24   Lake-filled kettles are scattered over the surface of an end-moraine complex in the lake district of

central Chile that formed at the end of the last glaciation when debris-covered stagnant ice slowly melted away.

meltwater stream systems. In the northern United States and parts of Europe, glacial sands and gravels are among the most valuable economic mineral resources, and in Alaska and Russia, such sediments are mined for placer gold, silver, and platinum. Vacationers and alpinists are attracted to glaciated mountains the world over for their spectacular scenery and recreational opportunities. Where modern glaciers contribute to local or regional water supplies, the glaciers help moderate the flow of water throughout the year, releasing it to streams during warm, dry summer months. Nearby streams flowing from basins that lack glaciers may experience strong seasonal fluctuations in streamflow, including serious summer droughts.

Glaciers as Environmental Archives

Trapped in the snow that piles up each year in the accumulation area of a glacier is evidence of both local and global environmental conditions. The evidence includes physical, chemical, and biological components that can be extracted in a laboratory and studied as a record of the changing natural environment. The oldest ice in most cirque and valley glaciers is several hundred to several thousand years old, but large ice caps and ice sheets contain ice that dates far back into the ice ages. The record they contain is often unique, and it is of critical importance for understanding how the atmosphere, oceans, and biosphere have changed over hundreds of thousands of years.

If we dig a pit many meters deep in the accumulation area of a glacier and look closely at the snow, a cyclic layering can be seen. In each layer, the relatively clean snow that records a succession of winter snowfalls passes upward to a darker layer that contains dust and refrozen meltwater, a record of relatively dry summer weather. Depending on the local accumulation rate, one or several such annual layers may be exposed in our pit. Because digging a pit into a glacier is an inefficient way to examine the stratigraphy, glaciologists drill cores of ice that can be extracted and returned to a laboratory for analysis. Some drilling operations have penetrated to the base of the thick Greenland and Antarctic ice sheets, while others have focused on high-latitude and high-altitude ice.

Ice cores have proved a boon for atmospheric scientists who would like to know whether the concentrations of important atmospheric trace gases like carbon dioxide and methane, which are trapped in air bubbles in the ice, have fluctuated as the climate changes (Chapter 14). Measurements of oxygen isotopes in the ice can tell us the air temperature when the snow accumulated on the glacier surface. Ice cores also provide a record of major volcanic eruptions that generate sulfur dioxide gas which, combined with water, accumulates as a layer of acid snowfall on gla-

ciers. High concentrations of fine dust in ice layers dating back to the last ice age show that the windy climate of glacial times was also extremely dusty. Tiny fragments of organic matter and fossil pollen grains trapped in the ice layers can tell us about local vegetation composition near a glacier and can be radiocarbon-dated to provide ages for the enclosing ice. Because these natural historical archives are trapped in annual ice layers that can be read like the pages of a book, they offer an unparalleled, detailed look at past surface conditions on our planet.

SEA ICE

Approximately two-thirds of the Earth's permanent ice cover floats as a thin veneer of **sea ice** at the ocean surface in polar latitudes (Fig. 10.25). Despite its vast extent, sea ice comprises only about 1/1000 of the Earth's total volume of ice.

How Sea Ice Forms

Once the ocean surface cools to the freezing point of seawater, slight additional cooling leads to ice formation. The first ice to form consists of small crystalline platelets and needles up to 3 or 4 mm (0.1 or 0.2 in) in diameter that collectively are termed *frazil ice*. As more ice crystals form, they produce a soupy mixture at the ocean surface. In the absence of waves or turbulence, the crystals freeze together to form a continuous cover of ice 1 to 10 cm (0.4 to 4 in) thick. If waves are present, the crystals form rounded, pancake-like masses up to 3 m (10 ft) in diameter that eventually weld together into a continuous sheet of sea ice. Once a continuous ice cover forms, the cold atmosphere is no longer in contact with the seawater, and sea-ice growth then proceeds by the addition of ice to the sea-ice base. In the Arctic, over the course of a yearly cycle, about 45 cm (18 in) of ice is lost from the ice surface, but an equal amount is added to the base. As a result, an ice crystal added to the sea ice at its base will move upward through the ice column with an average velocity of about 45 cm/yr (18 in/yr) until it reaches the surface and melts away.

Sea-Ice Distribution and Zonation

The contrasting geography of the Earth's polar regions leads to important differences in the distribution of sea ice in the two hemispheres. The South Pole lies near the middle of the Antarctic continent, which is covered by a vast, thick ice sheet, whereas the North Pole lies in the middle of the deep Arctic Ocean basin. The open Southern Ocean adjacent to Antarctica contrasts with the largely land-locked Arc-

Figure 10.25 A glaciological field camp on a slab of sea ice in the southern Beaufort Sea (ca. 76°N Lat., 150°W Long.) As ice broke up at the end of summer, a fissure (left foreground) split the camp in two, causing one segment of the camp to drift 400 m away from the one shown here.

tic Ocean, which is connected to the world ocean only by relatively narrow straits. In the Antarctic region, sea ice forms a broad ring around the continent and adjacent ice-covered archipelago, a ring that varies in width with the seasons. At its greatest northward extent in winter it covers 20 million km² (7.7 million mi²), but it shrinks to only 4 million km² (1.5 million mi²) in summer (Figs. 10.26A and 10.27). By contrast, the Arctic Ocean is ice-covered most of the year, and several marginal seas (i.e., the Sea of Japan, the Sea of Ohkotsk, the Bering Sea, Davis Strait, Hudson Bay)

are partially or wholly ice-covered during the winter. At its minimum extent in August, Arctic sea ice covers about 7 million km² (2.7 million mi²), whereas during its winter maximum it expands to 14 million km² (5.4 million mi²) (Fig. 10.26B).

Scientists commonly categorize sea-ice zones as being either perennial or seasonal. The *perennial ice zone* contains sea ice that persists for at least several years (multiyear ice). In the Arctic, this zone lies north of 75° latitude and contains about two-thirds of all perennial sea ice. Near the center of the basin, the ice

Figure 10.26 Seasonal extent of sea ice in A. southern hemisphere and B. northern hemisphere.

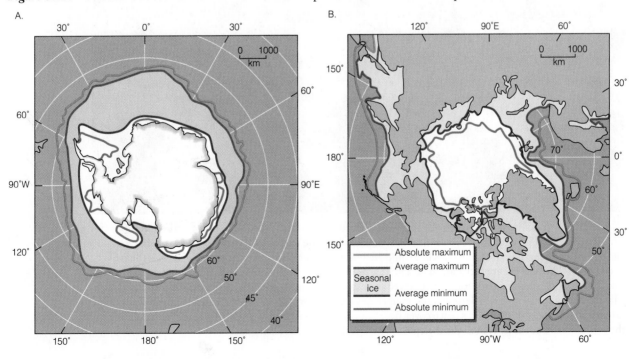

Figure 10.27 Seasonal variations in the sea-ice cover around Antarctica in a typical year. The ice is least extensive during the summer months (January–March) but steadily increases, reaching a maximum in winter (July–September). At the time of maximum sea ice (September), a large polynya has developed northeast of the Weddell Sea.

has an average thickness of 3 to 4 m (10 to 13 ft) and an age of up to at least several decades. In the Antarctic, multiyear ice is restricted to semi-enclosed seas (the Ross, Weddell, and Bellinghausen seas), where it reaches a thickness of up to 5 m (16.5 ft) but an age of less than five years. In the *seasonal ice zone*, the ice cover varies annually. In the Arctic, ice of this zone is less than 2 m (6.5 ft) thick where undeformed, but deformation within the pack ice often increases thickness substantially. In the Southern Ocean, the limit of seasonal ice shifts, on average, through 10° of latitude. Here, the ice front retreats poleward in summer largely in response to heat derived from the ocean

water, whereas in the Arctic, surface melting in summer is a major factor in the retreat of the ice margin.

Sea-Ice Motion

Sea ice is in constant motion, driven by winds and ocean currents. Annual changes in sea-ice extent, character, and motion can now be studied using radar imaging systems carried on orbiting satellites (Fig. 10.28). Average drift rates in the Arctic Ocean are about 7 km/day (4.3 mi/day), whereas in the Greenland and Bering seas velocities reach 15 km/day (9.3 mi/day). Each year, about 10 percent of the Arctic Sea ice moves south into the Greenland Sea, where it eventually breaks up and melts away. Sea ice generally moves clockwise around Antarctica, but a large gyre in the Weddell Sea, east of the Antarctic Peninsula, causes the drifting ice to pile up to form a large region of multiyear ice.

Stresses resulting from diverging movement of the thin ice cover cause it to break, exposing the underly-

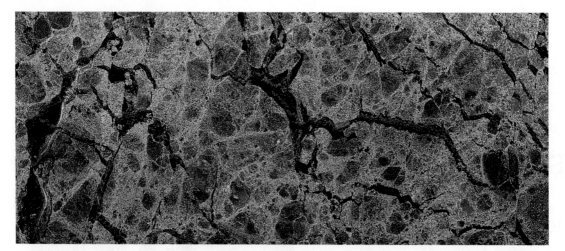

Figure 10.28 Satellite radar image of the seasonal sea-ice cover in the Weddell Sea, Antarctica (58.2°S, 21.6°E). Such satellite data provide scientists with details about the ice pack they cannot see any other way and show that the large expanse of sea ice is comprised of many smaller rounded ice floes. Ice floes (bluish-gray), about 70 cm thick, are surrounded by a jumble of deformed ice pieces (reddish tinge) that are up to 2 m thick. The more extensive dark zones are covered by a layer of smooth, level ice less than 70 cm thick. The winter cycle of ice growth and deformation often causes this ice cover to split apart, exposing open leads. The extent and thickness of the polar sea-ice cover have important implications for global climate by regulating the loss of heat from the ocean to the cold polar atmosphere.

ing water. Such a linear opening, called a *lead*, tends to be long and narrow and may extend for many kilometers. An exceptionally large lead may grow to become a huge area of open water called a *polynya* (Fig. 10.27). Because of the large temperature gradient between the air and seawater in a lead, the water loses heat rapidly, causing a new, thin cover of ice to form quickly. As a result, the fractured ice pack becomes a changing complex mosaic of new ice and older ice. Although the exposure of surface water to the atmosphere permits substantial amounts of solar energy to reach the upper ocean, such open water commonly comprises less than 1 percent of the total area of the winter sea-ice cover.

Early explorers who tried to reach the North Pole by crossing the Arctic ice pack quickly found it rough going. The ice is not a vast smooth surface; rather, it is broken by numerous *pressure ridges*, formed when the shifting, fractured ice converges, shears, and piles up, in much the same way that converging lithospheric plates produce mountain chains on the continents. Beneath each pressure ridge is a submerged *keel* of deformed ice, much like the keel of a sailboat, up to five times as thick as the overlying ridge. Estimates suggest that as much as 40 percent of the mass of Arctic sea ice is contained in such deformation features. In the Antarctic, pressure ridges are far less common because prevailing winds and currents tend to disperse the pack ice, shifting it away from the continent at rates as high as 65 km/day (40 mi/day).

Sea Ice in the Earth System

Interactions among sea ice, ocean, and atmosphere in the seasonal ice zone are believed to influence ocean structure and circulation. Because it is only a few meters thick, sea ice is very sensitive to temperature changes in the overlying atmosphere and in the ocean water below. In turn, the ice cover affects both the atmosphere and ocean in important ways. The growth of sea ice increases salinity at the top of the mixed layer of the ocean (Chapter 11) when salt is excluded as seawater freezes. Conversely, when the ice melts, the salinity is decreased as fresh water is added to the ocean surface. Thus, continual variations in the extent of sea ice cause corresponding variations in the salinity near the top of the water column.

Exclusion of salt as seawater freezes leads to the production of cold, saline (and therefore dense) water on the continental shelves. This dense water spills downward off the shelves into the ocean basins to produce deep water and bottom water. The process is enhanced in the marginal Antarctic seas, where offshore winds generate polynyas: Here, rapid ice growth under extremely cold conditions produces large quantities of dense water that likely is the source of much

of the Antarctic Bottom Water (Chapter 11). Similar processes operating in the Greenland and Norwegian seas are responsible for producing North Atlantic Deep Water, which is crucial to maintaining the global thermohaline circulation system (Chapter 11).

Both its rapid response to changing conditions and its direct influence on the atmosphere and oceans make sea ice an important component of the Earth's climate system. The floating cover of ice effectively isolates the ocean surface from the atmosphere, thereby cutting off the exchange of heat between these two reservoirs; the more extensive the ice cover, the stronger the effect. At the same time, the ice surface is highly reflective (i.e., it has a high **albedo**) and bounces incoming solar radiation back into space. The high albedo makes the ice-covered polar regions far colder than if the same areas were covered with water, which has a lower albedo than ice. As a result, the climate of the polar oceans more closely resembles that of large continental ice sheets than it does a typical oceanic region in lower latitudes.

The steep temperature gradient between low latitudes and the polar regions is of major importance to atmospheric circulation. If the climate were to become colder, causing the sea ice to expand in area, the result would be a positive feedback. The increased area of sea ice would increase the total planetary albedo, leading to further cooling. Conversely, if the ice cover shrinks or disappears, significant disruption in the pattern of atmospheric circulation might occur.

That sea ice influences global climate leads to some important questions: How stable is the ice pack in the land-locked Arctic basin? What would it take to remove the thin ice cover completely, and how long would it take? In a time of climatic warming (Chapter 14), might we expect the ice cover to disappear suddenly, thereby causing abrupt changes in climate in some of the Earth's most densely populated regions? Climate models suggest that in a warming climate, the warming at high latitudes will be several times that at low and middle latitudes. A possible scenario indicates that, as the Arctic warms up, we can expect a gradual shrinking of the ice pack followed by a discontinuous transition to ice-free conditions. We can easily calculate how much of a change would lead to the disappearance of perennial Arctic ice. It could occur, for example, with a 3° to 5°C increase in annual temperature, a 25 to 30 percent increase in solar radiation reaching the ice surface, a 15 to 20 percent decrease in summer albedo (brought about by increased surface melting), or a significant change in cloudiness. Although exactly how a change in albedo would take place is not known, modeling suggests that once the albedo reaches a sufficiently low value, the shift to an ice-free Arctic Ocean would occur rapidly, measurable in years rather than in decades.

PERIGLACIAL LANDSCAPES AND PERMAFROST

Land areas beyond the limit of glaciers where low temperature and frost action are important factors in determining landscape characteristics are called **periglacial** zones. Periglacial conditions are found over more than 25 percent of the Earth's land areas, primarily in the circumpolar zones of each hemisphere and at high altitudes.

Permafrost

A common feature of periglacial regions is perennially frozen ground, also known as **permafrost**—sediment, soil, or even bedrock that remains continuously below freezing for an extended time (from two years to tens of thousands of years). The largest areas of permafrost occur in northern North America, northern Asia, and the high, cold Tibetan Plateau (Fig. 10.29). It has also been found on many high mountain ranges, even including some lofty summits in tropical and subtropical latitudes. The southern limit of continuous permafrost in the northern hemisphere generally lies where the annual air temperature is between −5 and −10°C (23 and 14°F).

Most permafrost is believed to have originated during either the last glacial age or earlier glacial ages. Remains of woolly mammoth and other extinct ice-

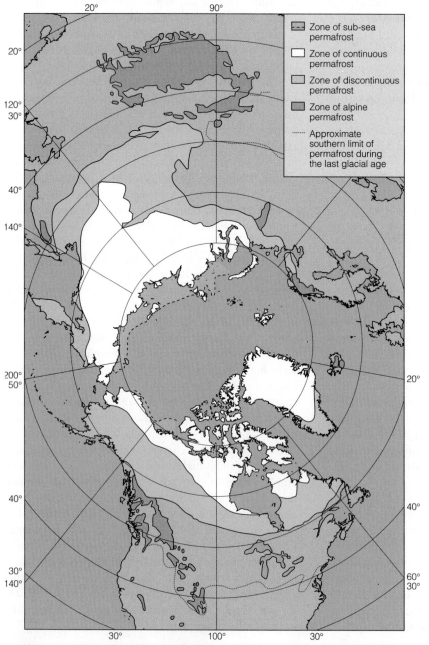

Figure 10.29 Distribution of permafrost in the northern hemisphere. Continuous permafrost lies mainly north of the 60th parallel and is most widespread in Siberia and Arctic Canada. Extensive alpine permafrost underlies the high, cold plateau region of central Asia. Smaller isolated bodies occur in the high mountains of the western United States and Canada.

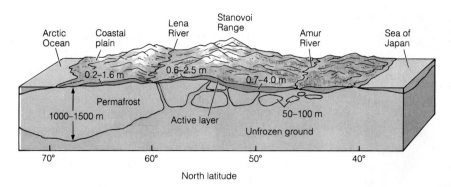

Figure 10.30 Diagrammatic transect across northeastern Siberia (vertical scale is greatly exaggerated) showing distribution and thickness of permafrost and thickness of the active layer. Thick, continuous permafrost under the Arctic coastal plain thins southward where it becomes discontinuous in response to warmer mean annual air temperature. The active layer increases in thickness southward due to warmer summer temperatures.

age animals found well preserved in frozen ground indicate that permafrost existed at the time of their death.

The depth to which permafrost extends depends not only on the average air temperature but also on the rate at which heat flows upward from the Earth's interior and on how long the ground has remained continuously frozen. The maximum reported depth of permafrost is about 1500 m (4900 ft) in Siberia (Fig. 10.30). Thicknesses of about 1000 m (3300 ft) in the Canadian Arctic and at least 600 m (2000 ft) in northern Alaska have been measured. These areas of very thick permafrost all occur in high latitudes outside the limits of former ice sheets. The ice sheets would have insulated the ground surface and, where thick enough, caused ground temperatures beneath them to rise to the pressure melting point. On the other hand, open ground unprotected from subfreezing air temperatures by an overlying ice sheet could have become frozen to great depths during prolonged cold periods.

Living with Permafrost

In permafrost terrain, a thin surface layer of ground that thaws in summer and refreezes in winter is known as the *active layer*. In summer this thawed layer tends to become very unstable. The permafrost beneath, however, is capable of supporting large loads without deforming. Many of the landscape features we associate with periglacial regions reflect movement of regolith within the active layer during annual freeze/thaw cycles.

Permafrost presents unique problems for people living on it. If a building is constructed directly on the surface, the warm temperature developed when the building is heated is likely to thaw the underlying permafrost, making the ground unstable (Fig. 10.31). Arctic inhabitants learned long ago that they must place the floors of their buildings above the land surface on pilings or open foundations so that cold air can circulate freely beneath, thereby keeping the ground frozen.

Wherever a continuous cover of low vegetation on a permafrost landscape is ruptured, melting can begin. As the permafrost melts, the ground collapses to form impermeable basins containing ponds and lakes. Thawing can also be caused by human activity, and

Figure 10.31 This cabin in central Alaska settled more than a meter in eight years as permafrost beneath its foundation thawed.

Figure 10.32 The Alaska Pipeline carries petroleum from the North Slope oil fields near Prudoe Bay southward across two mountain ranges en route to the port of Valdez. To increase the ease of flow through the pipe, the oil is heated. Because much of the pipeline route lies across permafrost terrain, in many sectors the huge pipe is suspended above ground on large piers to keep it from melting the frozen ground beneath.

the results can be environmentally disastrous. Large wheeled or tracked vehicles crossing Arctic tundra can quickly rupture it. The water-filled linear depressions that result from thawing can remain as features of the landscape for many decades.

The discovery of a commercial oil field on the North Slope of Alaska in the 1960s generated the need to transport the oil southward by pipeline to an ice-free port. The company formed to construct the pipeline was faced with some unique problems. In order for the sticky oil to flow through a pipeline in the frigid Arctic environment, the oil had to be heated. However, an uninsulated, heated pipe in the frozen ground could melt the surrounding permafrost. Even if the pipe were insulated before placing it underground, the surface vegetation cover would be disrupted, likely leading to melting and instability. For these reasons, along much of its course the pipeline was constructed on piers above ground, thereby greatly reducing the possibility of ground collapse (Fig. 10.32)

SUMMARY

1. The seasonal snow cover in the northern hemisphere appears in the Arctic during early autumn, grows in thickness as it expands southward to reach a late-winter maximum, and then retreats rapidly during the spring.

2. The snowline marks the lower limit of perennial snow; its altitude is controlled mainly by precipitation and summer temperature.

3. Glaciers, which constitute the bulk of the ice in the cryosphere, are permanent bodies of moving ice that consist largely of recrystallized snow. They can form only at or above the snowline, which is close to sea level in the polar regions and rises to high altitudes in the tropics.

4. Ice in a temperate glacier is at the pressure melting point, and liquid water exists at the base of the glacier; ice in a polar glacier is below the pressure melting point and is frozen to the rock on which it rests.

5. The mass balance of a glacier is measured in terms of accumulation (gain) and ablation (loss). The equilibrium line separates the accumulation area from the ablation area and marks the level on the glacier where net gain is balanced by net loss.

6. Temperate glaciers move as a result of internal flow and basal sliding. In polar glaciers, motion is much slower and involves only internal flow. Surges involve extremely rapid flow, probably related to excessive amounts of water at the base of a glacier.

7. A fjord glacier with its base below sea level will begin an irreversible calving retreat if its terminus becomes ungrounded and recedes into deepening water upfjord.

Retreat ends only when the glacier again terminates in water too shallow for appreciable calving to occur.

8. Ice cores extracted from polar glaciers contain natural archives of changing environmental conditions in the form of changing oxygen isotope ratios, samples of ancient atmospheric gases, dust concentrations, acid (volcanic) fallout, and organic particles.

9. Sea ice thinly covers vast areas of polar ocean and is highly sensitive to changes in climate and ocean conditions. Perennial sea ice persists for at least several years and reaches thicknesses of 3 to 4 m, whereas seasonal ice that forms and disappears annually is less than 2 m thick. Convergent deformation produces linear pressure ridges where ice is unusually thick.

10. When seawater freezes, dense, cold, saline water is produced that sinks into the deep ocean where it forms deep water and bottom water. The high albedo of sea ice influences global climate, helping to create a steep pole-to-equator temperature gradient.

11. Permafrost, a common feature of periglacial zones, is confined mainly to areas where annual air temperature is at least $-5°C$. It reaches maximum thicknesses of at least 1500 m and is believed to have formed during glacial ages in subfreezing landscapes not covered by continental ice sheets.

12. Permafrost can present unique engineering problems, for thawing commences when the vegetation cover is broken, leading to collapse and extreme instability of the ground surface.

IMPORTANT TERMS TO REMEMBER

ablation *222*

ablation area *222*

accumulation *222*

accumulation area *222*

albedo *235*

calving *225*

cirque *218*

crevasse *224*

cryosphere *216*

end moraine *230*

equilibrium line *222*

fjord *219*

glacier *218*

glacier ice *220*

mass balance *222*

periglacial *236*

permafrost *236*

polar glacier *221*

pressure melting point *221*

sea ice *232*

striations *228*

surge *226*

temperate glacier *221*

terminus *222*

QUESTIONS FOR REVIEW

1. How are glaciers related to the snowline?

2. Describe the steps in the conversion of snow to glacier ice.

3. What characteristics distinguish temperate glaciers from polar glaciers?

4. Why does the position of the equilibrium line provide a rough estimate of a glacier's mass balance?

5. Why is there commonly a time lag between a change of climate and the response of a glacier's terminus to the change?

6. In what ways does ice temperature influence the way a glacier moves?

7. Describe the unique motions of surging and calving glaciers. How are their fluctuations related to climate?

8. Make a list of the distinctive erosional and depositional features that would enable you to differentiate a landscape primarily shaped by streams from one shaped by glaciers.

9. What factors control the distribution and thickness of sea ice?

10. How does sea ice influence climate? ocean circulation?

11. What is the active layer in permafrost terrain, and how does it form?

12. Where, and why, would you expect to find permafrost at latitudes of less than 40°?

13. Describe what potential foundation problems a home builder might encounter in northern Alaska if the contractor were to clear the building site of vegetation and begin construction on the exposed ground surface.

Question for A Closer Look

1. Why might a large mass of rock debris falling onto the accumulation area of a glacier cause the glacier terminus to advance?

QUESTIONS FOR DISCUSSION

1. Suggest ways in which changes in the solid Earth, such as uplift, earthquakes, and volcanism, might affect the distribution and fluctuations of glaciers.

2. One way to estimate the altitude of a valley glacier's equilibrium line is to calculate the median altitude of the glacier (midway in altitude between the glacier's head and terminus). Obtain topographic maps from your library (for example, the glacier-clad mountains of southern Alaska, British Columbia, Alberta, or Washington State) and calculate the equilibrium-line alti-

tudes of several valley glaciers using this method. What differences do you obtain for nearby glaciers within a single mountain range? How might the differences be explained?

3. Huge pools of petroleum likely underlie the Arctic continental shelves of Alaska, Canada, and Russia. In exploiting such petroleum resources, what problems might be encountered that are related to the presence of sea ice and permafrost?

CHAPTER 11

The World Ocean

● *Polynesian Navigators*

More than 1300 years ago, Polynesian explorers set out from Havai'i (now Raiatea, in the Society islands) in great double-hulled canoes across the vast unknown expanse of the North Pacific Ocean. By chance, they discovered, and subsequently colonized, the Hawaiian Islands. A long canoe voyage across the uncharted ocean must have required an exceptional degree of navigational skill. How did the ancient Polynesians repeatedly traverse great ocean distances without the aid of a compass or other modern navigational aids?

Like competent modern sailors, the Polynesian navigators became familiar with the winds and current systems that affected their vessels. In sailing from Havai'i to Hawaii, a canoe crossed three wind systems and the surface ocean currents related to them. Havai'i lay in the southeast tradewinds belt, where prevailing winds blow from the southeast, and also in the region of a great westward-flowing ocean current that lies just south of the equator (Fig. 11.1). The initial part of the voyage therefore involved sailing somewhat east of north across this belt until, just north of the equator, a belt of light variable winds and an eastward-flowing current must be passed. Still farther north, a canoe encountered the northeast tradewinds and another westward-flowing current system that helped carry it northwestward toward Hawaii. By observing the winds and currents, ancient navigators could get an idea of their approximate geographic position.

By keeping his canoe consistently oriented with respect to major ocean swells, a Polynesian navigator could maintain his course across the otherwise featureless ocean. At night, he could steer with respect to a succession of stars rising above the horizon; as one star rose too high to steer by, a newly rising star would then be tracked. Polynesian navigators knew the rising and setting points of more than 150 stars, and these points served them well in lieu of a compass.

The ancient Polynesians also had ways of anticipating a landfall. For example, clouds tend to accumulate around high islands, and the base of the clouds often is illuminated by light reflected off a bright, sunlit lagoon. The bright glow of the clouds can be visible long before land is seen. Plant debris drifting on the ocean surface may be evidence of land to windward. Finally, birds returning home at dusk after a day of fishing can point the way toward land over the horizon.

The Polynesians lacked a written language and committed to memory all the information needed to retrace a lengthy ocean voyage. They relied on a lifetime of personal experience, an intimate knowledge of the visible universe, and a clear understanding of the interrelationship of land, sea, atmosphere, and living organisms. ●

(opposite) Early Polynesian explorers, navigating by stars, winds, and currents, cross the vast expanse of the central Pacific Ocean and discover the island of Hawaii.

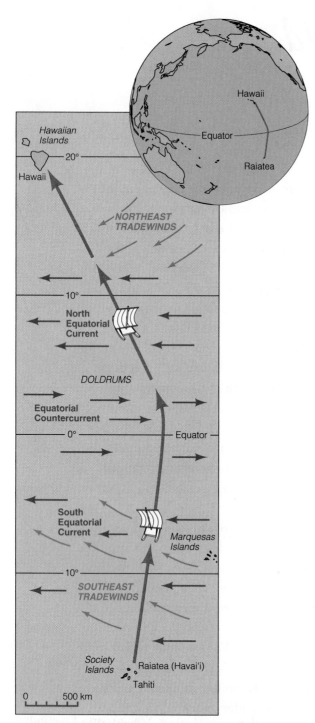

Figure 11.1 Likely route of discovery of the Hawaiian Islands. Double-hulled canoes, pushed by the southeast tradewinds, sailed slightly east of north from Havai'i (Raiatea) in the Society islands across the South Equatorial Current. Moving northward across the Equatorial Countercurrent in the Doldrums, the canoes then picked up the North Equatorial Current and the northeast tradewinds, which carried them toward landfall at the island of Hawaii.

THE OCEANS

Imagine, if you can, a dry Earth, devoid of water. The Earth's surface would appear far different from the one familiar to us. Viewed from an orbiting spacecraft, a dry Earth would no longer have a bluish color (requiring another title for this book), the land would lack a vegetation cover, and no clouds would obscure the surface. The Earth would appear as desolate as the rocky surfaces of the Moon or Mars. We would see the high-standing continents ending where their bordering continental slopes meet a great expanse of empty seafloor. As we learned in Chapter 4, this primary topography reflects the contrasting densities and thicknesses of continental (less dense, thicker) and oceanic (denser, thinner) lithosphere. Circling the Earth, we would see several vast interconnected basins, each floored with oceanic crust and rimmed with continental crust.

If these huge basins were now slowly filled with water, the scene would be transformed. The rising water would initially fill the deepest parts of the basins, creating a number of shallow seas, but as the water level continued to rise, these seas would merge to form a larger and larger ocean that eventually would creep up the slopes of the continental margins. With the ocean basins filled to capacity, more than two-thirds of the Earth's surface would now be covered by water and the Earth would be a unique planet in the solar system, for it would have become the Blue Planet.

Under the ocean, beyond the continental slopes, lies the remote world of the deep ocean floor. With devices for sounding the sea bottom and for sampling its sediment, teams of physical oceanographers and marine geologists have explored the ocean floor and greatly expanded our knowledge of the submarine regions. Scuba-diving geologists have visited, photographed, and mapped areas of seafloor at depths as great as 70 m (230 ft), and observers in specially designed submersible crafts have descended more than 6 km (3.7 mi) to visit the greatest depths of the ocean floor. In recent years, available depth soundings have been combined with high-resolution marine gravity measurements from Earth-orbiting satellites to produce new and accurate topographic maps of the ocean floors (Fig. 11.2).

Because of this intensive research involving many nations, we are gradually coming to understand the oceans. The romanticist in each of us may regret that beliefs and legends built up through more than 3000 years of human history—monsters, mermaids, strange and threatening sea gods, fabled cities and castles that sank into watery deeps—have vanished. These and other poetic visions have faded away as scientific

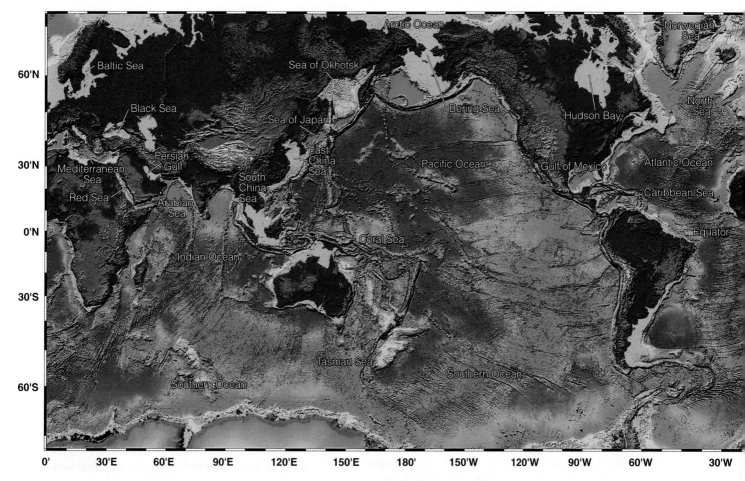

Figure 11.2 Shaded relief map of the oceans based on high-resolution marine gravity data obtained by satellite altimetry and on ship depth soundings, showing major and minor basins of the world ocean.

knowledge has steadily increased. In return, however, that knowledge has helped us appreciate the fragile environment of the oceans, which is responsible for a large part of the Earth's biological heritage.

Ocean Geography

Seawater covers 70.8 percent of the Earth's surface. The land comprising the remaining 29.2 percent is unevenly distributed. This uneven distribution is especially striking when we compare two views of the globe: one from a point directly above Great Britain and the other from a point directly above New Zealand (Fig. 11.3). In the first view, more than 46 percent of the viewed hemisphere is land, whereas in the second view it is more than 88 percent water. The uneven distribution of land and water plays an important role in determining the paths along which water circulates in the open ocean and its marginal seas.

Most of the water on our planet is contained in three huge interconnected basins—the Pacific, Atlantic, and Indian oceans (Fig. 11.2). All three are connected with the Southern Ocean, a body of water south of 50° S latitude that completely encircles Antarctica. Collectively, these four vast interconnected bodies of water, together with a number of smaller ones, are often referred to as the *world ocean*. The smaller water bodies connected with the Atlantic Ocean include the Mediterranean, Black, North, Baltic, Norwegian, and Caribbean seas, the Gulf of Mexico, and Baffin and Hudson bays. The Persian Gulf, Red Sea, and Arabian Sea are part of the Indian Ocean, while among the numerous marginal seas of the Pacific Ocean are the Gulf of California, Bering Sea, Sea of Okhotsk, Sea of Japan, and the East China, South China, Coral, and Tasman seas. All these seas and gulfs vary considerably in shape and size; some are almost completely surrounded by land, whereas others are only partly enclosed. Each owes its distinctive geography to plate tectonics, for this ongoing process has led to the creation of numerous small basins both in and adjacent to the major ocean basins.

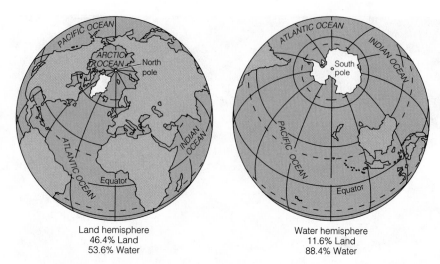

Land hemisphere
46.4% Land
53.6% Water

Water hemisphere
11.6% Land
88.4% Water

Figure 11.3 The unequal distribution of land and ocean can be seen if we view the Earth from above Britain and above New Zealand. In the first view, land covers nearly half the hemisphere, whereas in the other nearly 90 percent of the hemisphere is water.

Depth and Volume of the Oceans

Before the present century, little was known about the depth of the oceans. Water depths were determined from soundings made with either a weighted hemp line or a strong wire lowered from a ship. Although this technique proved satisfactory and relatively rapid in shallow water, it could take 8 to 10 hours to recover a weighted wire in water thousands of meters deep. By the close of the nineteenth century, about 7000 measurements had been made in water more than 2000 m (6500 ft) deep, and fewer than 600 in water deeper than 9000 m (29,500 ft). In the 1920s, ship-borne acoustical instruments called *echo sounders* were developed to measure ocean depths. An echo sounder generates a pulse of sound and accurately measures the time it takes for the echo bouncing off the seafloor to return to the instrument. Because the speed of sound traveling through water is known, the water depth beneath a ship can be calculated.

Over the past 70 years, the oceans have been crossed many thousands of times by ships carrying echo sounders. As a result, the topography of the seafloor and the depth of the overlying water column are known in considerable detail for all but the most remote parts of the ocean basins. The greatest ocean depth yet measured (11,035 m; 36,205 ft) lies in the Mariana Trench near the island of Guam in the western Pacific. This is more than 2 km (6500 ft) farther below sea level than Mount Everest rises above sea level. Based on recent satellite measurements, the average depth of the sea is about 4500 m (14,760 ft) compared to an average height of the land of only 750 m (2460 ft). The present volume of seawater is about 1.35 billion cubic kilometers (324 million mi^3); more than half this volume resides in the Pacific Ocean. We say *present* volume because the amount of water in the ocean fluctuates over thousands of years, mainly be-cause of the growth and melting of continental glaciers (Chapters 10 and 14).

Age and Origin of the Oceans

The Earth's oldest rocks include sedimentary strata that were deposited by water and are similar to strata we see being deposited today. From such evidence we are sure that, as far back in history as we can see, which is 3.95 billion years, the Earth has had liquid water on its surface. We can be reasonably certain, therefore, that ocean waters began to accumulate on the Earth's surface sometime between 4.6 billion years ago, when the Earth formed, and 3.95 billion years ago, when the oldest known sedimentary rock was made.

Where the water to create the oceans came from is still an open question, however. The most likely answer is that it condensed from steam produced during primordial volcanic eruptions. Because volcanic activity has persisted throughout the Earth's history, the volume of the world ocean likely has increased through time.

THE SALTY SEA

About 3.5 percent of average seawater, by weight, consists of dissolved salts. This is enough to make the water undrinkable. If these salts were precipitated, they would form a layer about 56 m (183 ft) thick over the entire seafloor.

Ocean Salinity

Salinity is the measure of the sea's saltiness, expressed in per mil (‰ = parts per thousand). The salinity of

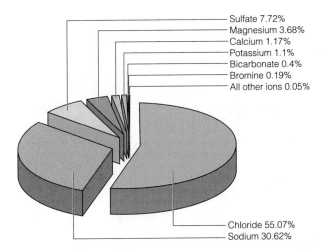

Sulfate 7.72%
Magnesium 3.68%
Calcium 1.17%
Potassium 1.1%
Bicarbonate 0.4%
Bromine 0.19%
All other ions 0.05%

Chloride 55.07%
Sodium 30.62%

Figure 11.4 Principal ions in seawater. More than 99.9 percent of the salinity of seawater is due to eight ions, the two most important of which (Na+ and Cl-) are the constituents of common salt.

seawater normally ranges between 33 and 37‰. The principal elements that contribute to this salinity are sodium and chlorine. Not surprisingly, when seawater is evaporated, more than three quarters of the dissolved matter is precipitated as common salt (NaCl). However, seawater contains most of the other natural elements as well, many of them in such low concentrations that they can be detected only by extremely sensitive analytical instruments. As can be seen in Figure 11.4, more than 99.9 percent of the salinity is caused by only eight ions.

Where do these ions come from? Each year streams carry 2.5 billion tons of dissolved substances to the sea. As exposed crustal rocks interact with the atmosphere and the hydrosphere (rainwater), cations are leached out and become part of the dissolved load of streams flowing to the sea. The principal anions in seawater, on the other hand, are believed to have come from the mantle. Chemical analyses of gases released during volcanic eruptions show that the most important volatiles are water vapor (steam), carbon dioxide (CO_2), and chloride (Cl^{1-}) and sulfate (SO_4^{2-}) anions. If released to the atmosphere, these two anions dissolve in atmospheric water and return to the Earth in precipitation, much of which falls directly into the ocean. Part of the remainder is carried to the sea dissolved in river waters. However, volcanic gases are also released directly into the ocean from submarine eruptions, and these are likely a major source of the anions in ocean water. Other sources of ions include dust eroded from desert regions and blown out to sea, and gaseous, liquid, and solid pollutants released through human activity either directly into the oceans or carried there by streams or polluted air.

The quantity of dissolved ions added by rivers over the billions of years of Earth history far exceeds the amount now dissolved in the sea. Why, then, doesn't the sea have a higher salinity? The reason is that chemical substances are being removed at the same time they are being added. Some elements, such as silicon, calcium, and phosphorus, are withdrawn from seawater by aquatic plants and animals to build their shells or skeletons. Potassium and sodium are absorbed and removed by clay particles and other minerals as they settle slowly to the seafloor. Still others, such as copper and lead, are precipitated to form sulfide minerals in claystones and mudstones rich in organic matter. Because these and other processes of extraction are essentially equal to the combined inputs, the composition of seawater remains virtually unchanged.

Has the ocean always been salty? The best evidence of the sea's past saltiness is the presence, in marine strata, of salts precipitated by the evaporation of seawater. Marine strata containing salts that were concentrated by the evaporation of seawater are common in young sedimentary basins, but they are not known from rocks older than about a billion years. Possibly this is because ancient marine deposits consisting of soluble salts have been completely removed from the geologic record through the slow dissolving action of percolating groundwater.

Salinity of Surface Waters

The salinity of surface waters is closely related to latitude (Fig. 11.5). The most important factors affecting salinity are (1) evaporation (which removes water and leaves the remaining water saltier), (2) precipitation (which adds fresh water, thereby diluting the seawater and making it less salty), (3) inflow of fresh (river) water (which makes the seawater less salty), and (4) the freezing and melting of sea ice. (When seawater freezes, salts are excluded from the ice, leaving the unfrozen seawater saltier.) As one might expect, salinity is high in the latitudes where the Earth's great deserts lie. In these zones evaporation exceeds precipitation, both on land and at sea. In a restricted sea, like the Mediterranean, where there is little inflow of fresh water, surface salinity exceeds the normal range; in the Red Sea, which is surrounded by desert, salinity reaches 41‰. Salinity is lower near the equator because precipitation is high and cool water, which rises from the deep sea and sweeps westward in the tropical eastern Pacific and eastern Atlantic oceans, reduces evaporation. It is also low at high latitudes that are rainy and cool. Up to 100 km (62 mi) offshore from the mouths of large rivers, the surface ocean water can be fresh enough to drink.

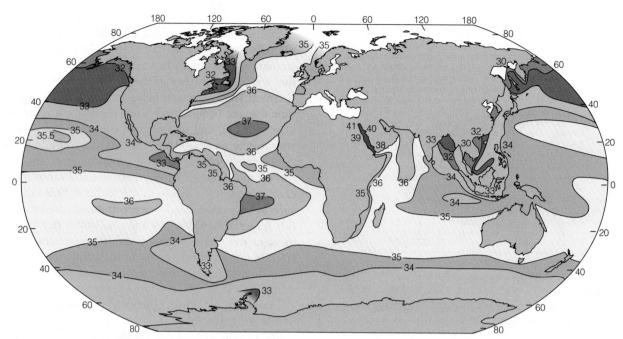

Figure 11.5 Average surface salinity of the oceans. High salinity values are found in tropical and subtropical waters where evaporation exceeds precipitation. The highest salinity has been measured in enclosed seas like the Persian Gulf, the Red Sea, and the Mediterranean Sea. Salinity values generally decrease poleward, both north and south of the equator, but low values also are found off the mouths of large rivers.

TEMPERATURE AND HEAT CAPACITY OF THE OCEANS

An unsuspecting tourist from Florida who decides to take a swim on the northern coast of Britain quickly learns how varied the surface temperature of the ocean can be. A map of global summer sea-surface temperature displays pronounced east–west temperature belts, with *isotherms* (lines connecting points of equal temperature) approximately paralleling the equator (Fig. 11.6). The warmest waters during August exceed 28°C (82°F) and occur in a discontinuous belt between about 30° N and 10° S latitude in the zone where received solar radiation is at a maximum. In winter, when the zone of maximum received solar radiation shifts southward, the belt of warm water also moves south until it is largely below the equator. Waters become progressively cooler both north and south of this belt, and reach temperatures of less than 10°C (50°F) poleward of 50° N and S latitude. The average surface temperature of the oceans is about 17°C (63°F), while the highest temperatures (>30°C or 86°F) have been recorded in restricted tropical seas, such as the Red Sea and the Persian Gulf.

The ocean differs from the land in the amount of heat it can store. For a given amount of heat absorbed, water has a lower rise in temperature than nearly all other substances, i.e., it has a high *heat capacity*. Because of water's ability to absorb and release large amounts of heat with very little change in temperature, both the total range and the seasonal changes in ocean temperatures are much less than what we find on land. For example, the highest recorded land temperature is 58°C (136°F), measured in the Libyan Desert, and the lowest, measured at Vostok Station in central Antarctica, is −88°C (−126°F). The range, therefore, is 146°C. By contrast, the highest recorded ocean temperature is 36°C (97°F), measured in the Persian Gulf, and the coldest, measured in the polar seas, is −2°C (28°F), a range of only 38°C.

The annual change in sea-surface temperatures is 0–2°C in the tropics, 5–8°C in middle latitudes, and 2–4°C in the polar regions. Corresponding seasonal temperature ranges on the continents can exceed 50°C. Coastal inhabitants benefit from the mild climates resulting from this natural ocean thermostat. Along the Pacific coast of Washington and British Columbia, for example, winter air temperatures seldom drop to freezing. East of the coastal mountain ranges they can plunge to −30°C (−22°F) or lower. In the interior of a continent, summer temperatures may exceed 40°C (104°F), whereas along the ocean margin they typically remain below 25°C (77°F). Here, then, is a good example of the interaction of the hydrosphere, atmosphere, land surface, and biosphere: ocean temperatures affect the climate, both over the ocean and over the land, and climate ultimately is a major factor in controlling the distribution of plants and animals.

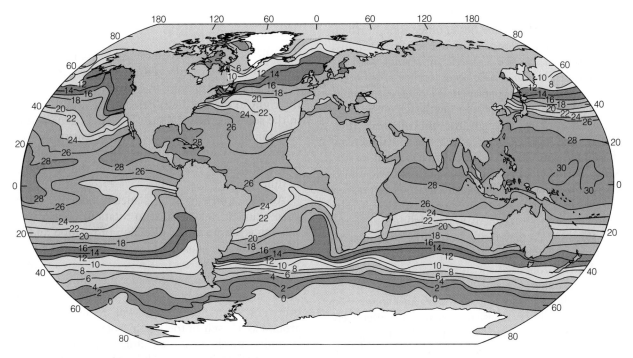

Figure 11.6 Sea-surface temperatures in the world ocean during August. The warmest temperatures (≥28°C) are found in the tropical Indian and Pacific oceans. Tempera- tures decrease poleward from this zone, reaching values close to freezing in the north and south polar seas.

VERTICAL STRATIFICATION OF THE OCEANS

The physical properties of seawater vary with depth. To help understand why this is so, think about shaking up a bottle of oil-and-vinegar salad dressing and then set it on a table. After a minute or two, the oil will rise to the top of the bottle and the vinegar will settle to the bottom. The two ingredients become stratified because they have different densities. The less dense oil floats on the denser vinegar. The oceans also are vertically stratified as a result of variations in the density of seawater. Seawater becomes denser as its tem- perature decreases and as its salinity increases. Gravity pulls dense water downward until it reaches a level where the surrounding water has the same density. These density-driven movements lead both to stratifica- tion of the oceans and to circulation in the deep ocean.

Oceanographers recognize three major depth zones in the ocean (Fig. 11.7). A *surface zone*, typically extending to a depth of 100 to 500 m (330 to 1640 ft), consists of relatively warm water (except in polar lati- tudes, where the surface zone is absent). This zone is also referred to as the *mixed layer* because winds, waves, and temperature changes cause extensive mix- ing in it.

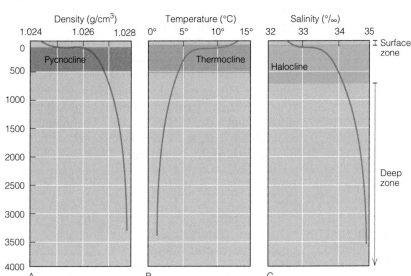

Figure 11.7 Depth zones in the ocean. Below the surface zone lies another zone in which the ocean-water properties ex- perience a significant change with in- creasing depth. This zone is variously known as A. the *pycnocline*, a zone of in- creasing density. B. the *thermocline*, a zone of decreasing temperature; and C. the *halocline*, a zone of increasing salinity. Still lower lies the *deep zone*, where wa- ters are dense as a result of their low temperature and high salinity.

Below the surface zone lies another zone in which the ocean-water properties of temperature, salinity, and density experience a significant change with increasing depth. This zone goes by three different names, one for each property. In the open ocean, temperature commonly decreases markedly, then more slowly, downward through the **thermocline** (Fig. 11.7B). The *halocline*, marked by a substantial increase of salinity with depth, is found over much of the North Pacific Ocean and in other high-latitude waters where solar heating of the ocean surface is diminished and precipitation is relatively high (Fig. 11.7C). The rapid increase in water density that defines the *pycnocline* (Fig. 11.7A) may result from a decrease in temperature, an increase in salinity, or both.

Below the zone that encompasses the thermocline, halocline, and pycnocline lies the *deep zone*, which contains about 80 percent of the ocean's volume. In low and middle latitudes, the pycnocline effectively isolates water of the deep zone from the atmosphere, but in high latitudes, where the surface zone is absent, water of the deep zone lies in direct contact with the atmosphere.

OCEAN CIRCULATION

When Christopher Columbus set sail from Spain in 1492 to cross the Atlantic Ocean in search of China, he took an indirect route. Instead of sailing due west, which would have made his voyage shorter, he took a longer route southwest toward the Canary Islands, and then west on a course that carried him to the Caribbean islands where he first sighted land. In choosing this course, he was following the path not only of the prevailing winds but also of surface ocean currents. Instead of fighting the westerly winds and currents at 40° N latitude, he drifted with the Canary Current and North Equatorial Current, as the northeast tradewinds filled the sails of his three small ships.

Surface Currents of the Open Ocean

Surface ocean currents, like those Columbus followed, are broad, slow drifts of surface water set in motion by the prevailing surface winds. Air that flows across the sea drags the water slowly forward, creating a current of water as broad as the current of air, but rarely more than 50 to 100 m (165 to 330 ft) deep. The ultimate source of this motion is the Sun, which heats the Earth unequally, thereby setting in motion the planetary wind system (Chapter 13). Thus, ocean circulation results from the interplay of several key elements of the Earth system: (1) radiation from the Sun provides heat energy to the atmosphere; (2) nonuniform heating generates winds; and (3) the winds, in turn, drive the motion of the ocean's surface water.

The Coriolis Effect

The direction taken by ocean currents is also influenced by the **Coriolis effect**, a phenomenon by which all bodies moving in a given direction veer to the right in the northern hemisphere and to the left in the southern hemisphere. The effect is named for the nineteenth-century French scientist, Gustav Gaspard de Coriolis, who in 1835 first explained how the Earth's rotation influenced the movement of fluids.

An object on a rotating planetary body has an angular velocity (velocity due to rotation). The angular velocity is always in the direction of rotation and is a minimum at the equator and a maximum at the poles. It may seem odd to say that angular velocity is a minimum at the equator, but consider the following: Imagine a stone tower built exactly at the North Pole; every 24 hours the tower will rotate completely around as a result of the Earth's rotation (Fig. 11.8A). A similar tower on the equator, however, would not rotate at all; rather, it would describe an end-over-end motion. At any latitude between the equator and the pole, some rotation and some end-over-end motion occur. For this reason the Coriolis effect, which is due to rotation, is latitude-dependent and reaches a maximum at the poles.

A body that moves on the Earth has two velocity components—the velocity of forward motion and the angular velocity of rotation. In order for a moving body to maintain the same velocity as its location on the Earth, its angular velocity would have to change continually. A change in velocity is an acceleration. The *Coriolis acceleration* is the angular acceleration that would be needed for a moving object to stay on track with respect to the rotating frame of reference, in this case the Earth.

Because the angular acceleration is usually absent or insufficient, the *Coriolis effect* occurs and this, as mentioned above, is a deflection in the path of a moving object toward the right in the northern hemisphere and to the left in the southern hemisphere.

Every moving body is subject to the Coriolis effect. One of the most dramatic demonstrations of the effect happened during World War I when the German army bombarded Paris from a distance of 120 km (75 mi) with a gun called "Big Bertha." The gunners discovered that their shots were falling 1 to 2 km (0.6 to 1.2 mi) to the right of the target. The reason for their poor shooting was their failure to account and correct for a deflection in the trajectory due to the Coriolis effect.

The magnitude of the Coriolis effect varies with latitude and with the speed of the moving body. The latitude effect arises from the variation in angular velocity which, as was discussed previously, reaches a maximum at the poles and a minimum at the equator.

A.

B.

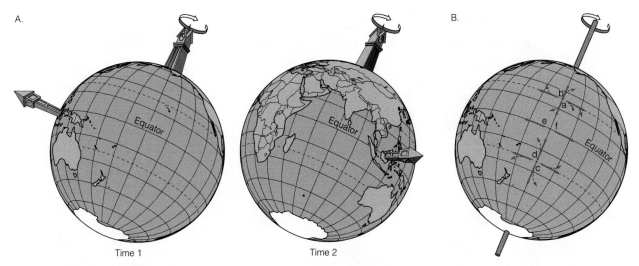

Time 1   Time 2

Figure 11.8 Coriolis Effect A. A body at the pole rotates completely around every 24 hours, while a body on the equator goes end-over-end but does not rotate. The face on the tower at the pole rotates with respect to an external observer, whereas a tower on the equator always faces the same direction. B. On the rotating Earth, an object freely floating on the ocean in the northern hemisphere (*a, b*) is deflected by the Coriolis effect to the right, whereas in the southern hemisphere (*c, d*) it is deflected to the left. A moving object at the equator (*e, f*) is not deflected.

The Coriolis effect therefore reaches a maximum at the poles and is zero at the equator.

If a freely floating object on the ocean moves away from the pole (Fig. 11.8B, *a*), its angular velocity about the pole will be slower than that of the water. This causes the object to lag behind the rotation, and so it will be deflected in a clockwise direction (to the right). If such an object is moving toward the pole (*b*), its angular velocity about the pole will be faster than the water, resulting in a counterclockwise deflection (again toward the right). Regardless of the direction of movement, an object in the northern hemisphere will be deflected to the right. In the southern hemisphere, the deflection is to the left (*c, d*), while at the equator the effect disappears (*e, f*). Although the Coriolis effect does not cause ocean currents, it deflects them once they are in motion.

Current Systems

Low-latitude ocean regions in the tradewind belts are dominated by the warm, westward-flowing North and South Equatorial currents (Fig. 11.9). In their midst, and lying in the doldrums belt of light, variable winds, is the eastward-flowing Equatorial Countercurrent.

Each major ocean current is part of a large subcircular current system called a **gyre**. Within each gyre different names are used for different segments of the current system. Figure 11.9 shows the Earth's five major ocean gyres, two each in the Pacific and Atlantic oceans and one in the Indian Ocean. Currents in the northern hemisphere gyres circulate in a clockwise direction, whereas those in the southern hemisphere circulate counterclockwise.

In each major ocean basin, westward-flowing equatorial currents are deflected poleward as they encounter land. Each current is thereby transformed into a *western boundary current* that flows generally poleward, parallel to a continental coastline. In the North Atlantic Ocean this current is called the Gulf Stream, whereas in the North Pacific it is the Kuroshio Current. In the South Atlantic the Brazil Current parallels the South American coast. In the Pacific and Indian oceans the corresponding currents are the East Australian Current and the Mozambique Current.

On reaching the belt of westerly winds, the Kuroshio Current changes direction to form the North Pacific Current on the poleward side of the North Pacific gyre. In the Atlantic, the Gulf Stream passes eastward into the northeast-flowing North Atlantic Current. In the southern hemisphere, the poleward moving waters of the Brazil, East Australian, and Mozambique currents enter the Antarctic Circumpolar Current that circles the Earth near latitude 60° S.

At the southeastern ends of the southern hemisphere gyres, cool southern waters move northward along western continental coasts forming the West Australian Current in the eastern Indian Ocean, the Humboldt Current along the southwestern coast of South America, and the Benguela Current off the southwestern coast of Africa.

The northern Indian Ocean exhibits a unique circulation pattern in which the direction of flow changes seasonally with the monsoons (Chapter 13). During the summer, strong and persistent monsoon winds blow the surface water eastward, whereas in winter, winds from Asia blow the water westward.

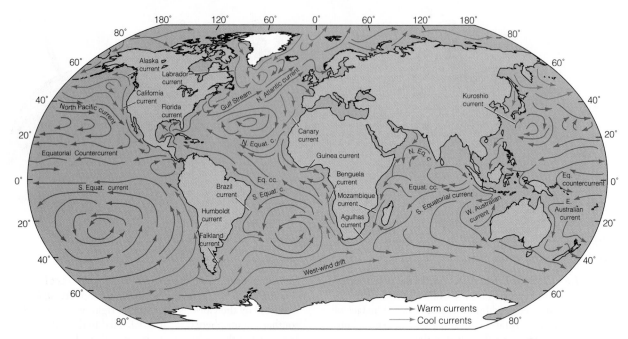

Figure 11.9 Surface ocean currents form a distinctive pattern, curving to the right (clockwise) in the northern hemisphere and to the left (counterclockwise) in the southern hemisphere. The westward flow of tropical Atlantic and Pacific waters is interrupted by continents, which deflect the water poleward. The flow then turns away from the poles and define the middle-latitude margins of the five great midocean gyres.

Ekman Transport

In 1893, Norwegian explorer Fridtjof Nansen (1861–1930) began an epic voyage across the frozen Arctic Ocean in his now-famous vessel, the *Fram*. Frozen in the shifting pack ice, the ship slowly drifted poleward, and then southward, eventually emerging into navigable water nearly three years later. Nansen's observations disclosed something totally unexpected: the floating pack ice moved in a direction 20 to 40° to the right of the prevailing wind. This phenomenon was subsequently explained mathematically by V. W. Ekman, who postulated that wind blowing across the ocean affects the uppermost layers of the water column, producing a net water flow that is at an angle to the wind. We can see an example of this effect in the relationship between the westward-flowing North Equatorial Current and the northeast tradewinds. The surface water layer dragging on the water immediately beneath sets the lower layer in motion, and the process continues downward. Internal friction, however, causes a decrease in current velocity with increasing depth. In addition, the Coriolis effect shifts each successive, slower moving layer farther to the right, producing a spiraling current pattern (called the *Ekman spiral*) when seen from above (Fig. 11.10). At a depth of about 100 m (330 ft), the current has reversed 180° from the direction of the surface wind, and its speed has dropped to only a small percentage of the surface current velocity. The average flow over the full depth of the spiral, called the **Ekman transport**, moves at 90° to the wind direction.

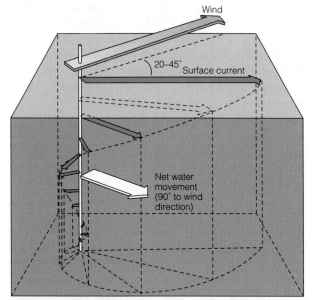

Figure 11.10 Wind blowing across the ocean in the northern hemisphere affects the surface water, producing a net flow 20–45° to the right of the wind direction. The surface water drags on the water immediately beneath, setting it in motion, and so on down the water column. Internal friction steadily reduces the current velocity with depth, and the Coriolis effect shifts each successively slower moving layer farther to the right, thereby producing an *Ekman spiral*. The average flow over the full depth of the spiral is termed the *Ekman transport* and is directed at 90° to the wind direction.

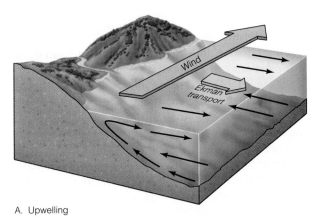

A. Upwelling

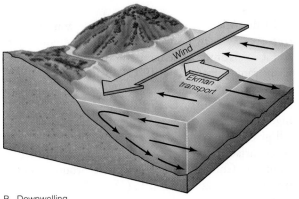

B. Downwelling

Figure 11.11 Winds blowing parallel to a coast exert a drag on the surface water, forcing it away from or toward the land, depending on wind direction. If the net Ekman transport is away from the land, rising subsurface water replaces water moving offshore, producing upwelling. If the net transport is toward the shore, the surface water thickens and sinks, producing downwelling.

Upwelling and Downwelling

Near coasts, Ekman transport can lead to vertical movement of ocean water. Winds blowing parallel to the coast can drag a layer of surface water tens of meters thick toward or away from land, depending on wind direction (Fig. 11.11). If the net transport is away from land, subsurface waters flow upward and replace the water moving away, a process called **upwelling**. If the net Ekman transport is toward the

coast, the surface water thickens and sinks in a process known as **downwelling**. Important areas of upwelling occur along west-facing low-latitude continental coasts (e.g., Oregon/California and Ecuador/Peru), where cold waters, rich in nutrients and originating at depths of 100 to 200 m (330 to 655 ft), support productive fisheries.

Major Water Masses

The water of the oceans is organized vertically into major water masses, stratified according to density. The identity and sources of these masses have been determined by studying the salinity and temperature structure of the water column at many places. The Atlantic Ocean provides a good example (Fig. 11.12).

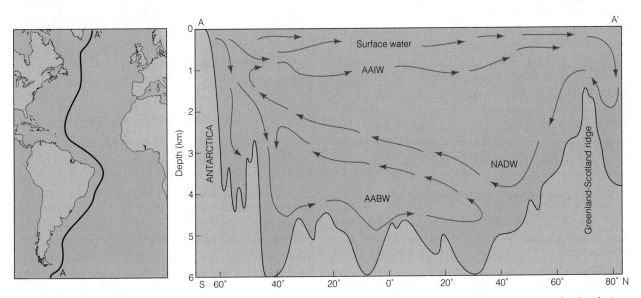

Figure 11.12 Transect along the western Atlantic Ocean showing water masses and general circulation pattern. North Atlantic Deep Water (NADW) originates near the surface in the North Atlantic as northward-flowing surface water cools, becomes increasingly saline, and plunges to

depths of several km. As NADW moves into the South Atlantic, it rises over denser Antarctic Bottom Water (AABW), which forms adjacent to the Antarctic continent and flows into the North Atlantic as Antarctic Intermediate Water (AAIW) at a mean depth of about 1 km.

In the Atlantic, water in the surface zone forms a *central water mass* north and south of the equator to about 35° latitude. The temperature of this water typically ranges from 6 to 19°C (43 to 66°F), and the salinity ranges from 34 to 36.5‰. Cooler subarctic and subantarctic surface water masses are found at high latitudes where cool temperatures and high rainfall give rise to colder, less saline waters. The largest polar surface water mass, flowing as the Antarctic Circumpolar Current, moves clockwise around Antarctica. Its temperature is 0–2°C (32–36°F), and its salinity is 34.6–34.7‰.

The central water mass of the Atlantic overlies an *intermediate water mass* that extends to a depth of about 1500 m (4920 ft). The Antarctic Intermediate Water mass (AAIW), the most extensive such body of water, originates as cold subantarctic surface water that sinks and spreads northward across the equator to about 20° N latitude. Its temperature ranges from 3 to 7°C (37–45°F), and its salinity lies within the range of 33.8–34.7‰. Water entering the Atlantic from the Mediterranean Sea is so saline (37–38‰) that it flows over a shallow sill at Gibraltar (2400 m) and downward beneath intermediate water to spread laterally over much of the ocean basin.

In the North Atlantic, the deep ocean consists of a *deep-water mass* that extends from the intermediate water to the ocean floor. This dense, cold (2–4°C, or 36–39°F), saline (34.8–35.1‰) **North Atlantic Deep Water (NADW)** originates at several sites near the surface of the North Atlantic, flows downward, and spreads southward into the South Atlantic.

The deepest, densest, and coldest water in the Atlantic is the *bottom water mass* that forms off Antarctica and spreads far northward. In the Pacific it reaches as far as 30° N latitude. Because of its greater density, Antarctic Bottom Water (AABW) flows beneath North Atlantic Deep Water. It forms when dense brine, produced during the formation of winter sea ice in the Weddell Sea adjacent to Antarctica, mixes with cold circumpolar surface water and sinks into the deep ocean. This dense water has an average temperature of −0.4°C (31°F) and a salinity of 34.7‰.

The Global Ocean Conveyor System

The sinking of dense cold and (or) saline surface waters provides a link between the atmosphere and the deep ocean. It also propels a global **thermohaline circulation** system, so called because it involves both the temperature and salinity characteristics of the ocean waters. This circulation can be traced from the North Atlantic southward toward Antarctica and into the other ocean basins (Fig 11.13). The largest mass of NADW begins to form in the Greenland and Norwegian seas where relatively warm and salty surface

water entering from the western North Atlantic cools, becomes denser, and sinks into a confined basin north of a submarine ridge connecting Scotland and Greenland (Fig. 11.14). The dense water then spills over low places along the ridge and plunges down into the deep ocean to form NADW. Warm, salty surface and intermediate water is drawn toward the North Atlantic to compensate for the south-flowing deep water. It is the heat lost to the atmosphere by the warm surface water of the Gulf Stream that maintains a relatively mild climate in northwestern Europe.

In the South Atlantic, south-flowing NADW enters the Antarctic Circumpolar Current of the Southern Ocean from which it spreads into the Indian and Pacific oceans. Surface and intermediate water flowing into the South Atlantic via the southern tip of Africa replaces the deep water moving out of the Atlantic basin. Meanwhile, Antarctic Bottom Water flowing into the southernmost Atlantic moves northward, slowly wells up, mixes with overlying NADW, and flows back toward Antarctica as Circumpolar Deepwater (Fig. 11.12A). These movements complete an important segment of the global system of ocean circulation, the Atlantic thermohaline circulation cell, which acts like a great conveyor belt.

Other circulation cells, linked to the Atlantic cells via the Antarctic Circumpolar Current and also driven by density contrasts related to temperature and salinity, exist in the Pacific and Indian oceans (Fig. 11.13). In concert, they help move water along the global ocean conveyor system, slowly replenishing the waters of the deep ocean. NADW is estimated to form at a rate of 15–20 million m³/s (equal to about 100 times the rate of outflow of the Amazon River), whereas Antarctic Bottom Water forms at a rate of about 20–30 million m³/s. Together these water masses could replace all the deep water of the world ocean in about 1000 years.

El Niño/Southern Oscillation

The fishing grounds off the coast of Peru, among the richest in the world, are sustained by upwelling cold waters filled with nutrients. Periodically, a mass of unusually warm water appears off the coast, an event that Peruvians refer to as El Niño because it commonly appears at Christmas time, the season of the Christ Child (El Niño). During El Niño years, the tradewinds slacken, upwelling is markedly reduced, and the fish population declines, accompanied by a great die-off of the coastal bird population, which depends on the fish for food. The Peruvian fishery is among the most important in the world, and so the occurrence of an El Niño event constitutes a local economic catastrophe. Coincident with the Peruvian El Niño conditions, very heavy rains fall in normally

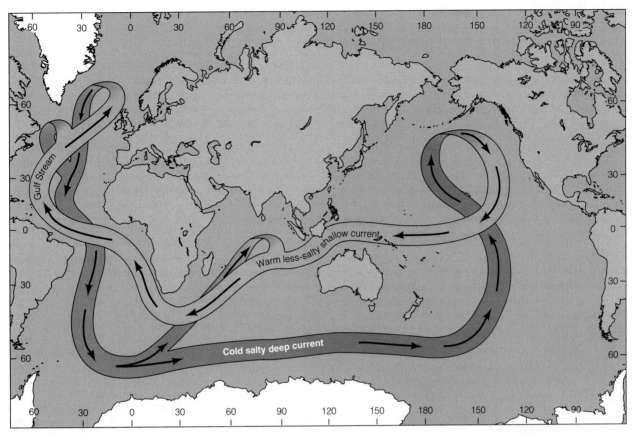

Figure 11.13 The major thermohaline circulation cells that make up the global ocean conveyor system are driven by exchange of heat and moisture between the atmosphere and ocean. Dense water forming at a number of sites in the North Atlantic spreads slowly along the ocean floor, eventually to enter both the Indian and Pacific oceans before slowly upwelling and entering shallower parts of the thermohaline circulation cells. Antarctic Bottom Water (AABW) forms adjacent to Antarctica and flows northward in fresher, colder circulation cells beneath warmer, more saline waters in the South Atlantic and South Pacific. It also flows along the Southern Ocean beneath the Antarctic Circumpolar Current to enter the southern Indian Ocean. Warm surface waters flowing into the western Atlantic and Pacific basins close the great global thermohaline cells.

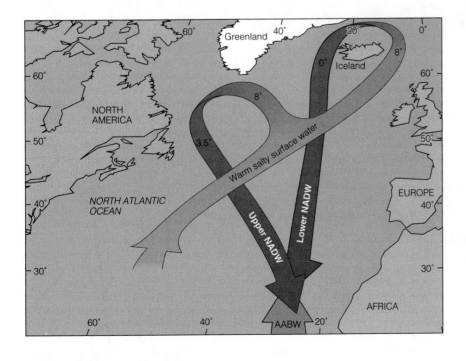

Figure 11.14 North Atlantic Deep Water (NADW) forms when the warm, salty water of the Gulf Stream/North Atlantic Current cools, becomes increasingly saline due to evaporation, and plunges downward to the ocean floor. The densest water then spills over the Greenland–Scotland ridge and flows southward as Lower NADW. Less dense water forming between Greenland and North America moves southeastward as Upper NADW and overrides denser Lower NADW. Because both water masses are less dense than northward-flowing Antarctic Bottom Water (AABW), they pass over it on their southward journey (see also Fig. 11.13).

arid parts of Peru and Ecuador, Australia experiences drought conditions, anomalous cyclones appear in Hawaii and French Polynesia, the seasonal rains of northeastern Brazil are disrupted, and the Indian monsoon may fail to appear. During exceptional El Niño years, climates over much of Africa, eastern Asia, and North America are affected. In North America, unusually cold winters can result in the northern United States, while the Southeast becomes wetter; in California abnormally high rainfall can produce major flooding and widespread landsliding.

El Niño has been experienced by generations of Peruvians, but its broader significance was recognized only recently. In the late 1960s, a link was made between cyclic El Niño events and changing atmospheric pressure anomalies over the equator, anomalies that had earlier been referred to as the *Southern Oscillation*. Today, **El Niño/Southern Oscillation (ENSO)** is regarded as an extremely important element in the Earth's year-to-year variations in climate. We now recognize that El Niño recurs erratically, but on average about every four years, and that its effect on climates is felt over at least half the Earth. It presents us with an especially instructive example of the close interaction between the Earth's atmosphere, hydrosphere, and biosphere. When an El Niño event occurs, it not only involves the tropical oceans and atmosphere, but it also directly affects precipitation and temperature on major land areas, thereby also impacting plants and animals.

Although many details of the El Niño phenomenon remain under study, the general mechanism is reasonably well understood. During normal years, the tradewinds blowing across the Pacific pile up a large pool of warm water in the west, which contrasts with cooler water that wells up in the eastern tropical Pacific (Fig. 11.15A). The warm water promotes a large center of heavy rainfall around Indonesia. An El Niño event begins with a slackening of the tradewinds. This leads to expansive warming of surface waters in the central and eastern Pacific as the warm water that had accumulated in the western Pacific moves back in the direction of South America. The eastward movement of warm water causes the zone of high rainfall to shift

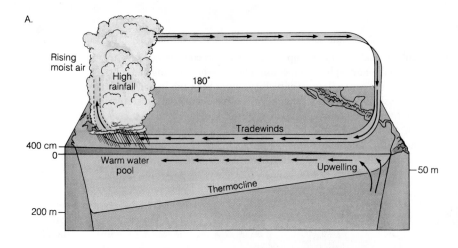

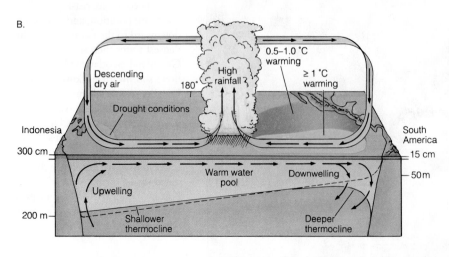

Figure 11.15 El Niño cycle. A. During normal years, persistent tradewinds blow westward across the tropical Pacific from a zone of upwelling water off the coast of Peru. The water warms up as it is transported westward to form a large warm-water pool above the thermocline in the western Pacific. The warm water causes the moist maritime air to rise and cool, bringing abundant rainfall to Indonesia. B. During an El Niño event, the tradewinds slacken and the pool of warm water moves eastward to the central Pacific. Descending cool, dry air brings drought conditions to Indonesia, while rising moist air above the warm-water pool greatly increases rainfall in the mid-Pacific. Surface waters in the eastern Pacific become warmer, and downwelling shuts off the supply of deep-water nutrients, adversely affecting the normally productive fishing ground off the coast of Peru.

A.

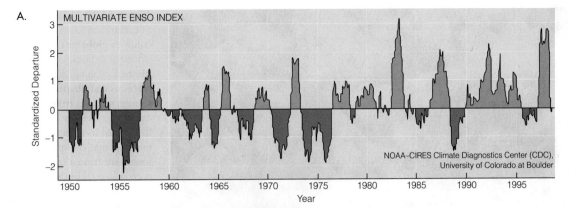

B.

B.

Figure 11.16 Geologic records of past El Niño events. A. A half-century record of ENSO events (1950 to 1998), showing the irregular frequency and amplitude (strength) of the events. The strongest recorded El Niños were those of 1982–1983 and 1997–1998. B. A slice through a living coral from the Galapagos Islands shows the annual layering (alternating dark and light bands rising from bottom to top of the section) of the calcium-carbonate skeleton. This layering preserves a record of changing surface water conditions, and therefore, of El Niño events.

annual growth layers, a record of historic and prehistoric El Niño events can be reconstructed for different oceanic sites and the dynamics of each cycle can be analyzed. Whereas the historical record of ENSO events extends back only about half a century (Fig. 11.16A), the coral studies can now extend the chronology much farther back in time. Ultimately, such data may make it possible to predict future El Niño events with considerable confidence.

BIOTIC ZONES AND SEDIMENTS OF THE OCEANS

Plants and animals living in the uppermost waters of the ocean occupy the *pelagic* zone and are called pelagic organisms. Animals that swim freely under their own locomotion include reptiles, squids, fish, and marine mammals. Benthic organisms live on the bottom or within bottom sediments (the *benthic* zone). Floating or drifting (planktic) organisms include phytoplankton, which are mainly single-celled plants, and zooplankton, which are tiny animals. Among the most important zooplankton are single-celled foraminifera and radiolaria. Foraminifera have a calcarous shell, whereas radiolarian remains consist of silica, an important distinction in determining the composition of deep-sea sediments (see below).

Plant life is restricted to the upper 200 m of the ocean, because in this zone (the photic zone) sufficient light energy is available for the process of photosynthesis. However, plants also require nutrients, and

to the central Pacific near the international date line, simultaneously bringing drought conditions to Indonesia (Fig. 11.15B). At the peak of an event, equatorial surface water moves from west to east and also poleward. This flow gradually reduces the equatorial pool of warm water, leading to intensification of the tradewinds and eventual return to normal conditions.

In an attempt to extend the detailed record of El Niño events farther back in time and improve our ability to predict future occurrence, scientists have examined the growth rings of living corals, because the rings record annual variations in seawater conditions (Fig. 11.16B). Corals are abundant and widespread throughout the region most strongly affected by El Niño, and individual colonies can live as long as 800 years. Their skeletal chemistry closely reflects surrounding environmental conditions, with the isotope or trace-metal composition of new skeletal material changing as water temperature and water chemistry change. By measuring the chemical composition of

Figure 11.17 Skeletons of calcareous foraminifera (smooth globular objects), siliceous radiolaria (delicate meshed objects), and siliceous rod-shaped sponge spicules from a deep-sea ooze, photographed by scanning electron microscope. The fossils are from a sediment core collected in the western Indian Ocean during a Deep Sea Drilling Project cruise.

these are available mainly along coasts and shallow continental margins. Over much of the deep ocean, plant life is limited because of a lack of nutrients, and therefore primary productivity (total organic material produced by photosynthesis) is low.

Over vast areas, the deep seafloor is mantled with sediment consisting largely of the skeletal remains of single-celled planktic and benthic animals and plants (Fig. 11.17). When more than 30 percent of the surface sediment consists of such remains, it is called a *calcareous ooze* or *siliceous ooze*, depending on the chemical composition of the major component (Fig. 11.18). Calcareous ooze covers broad areas of low to middle latitudes where warm surface waters favor the growth of carbonate-secreting organisms. Their tiny shells settle to the seafloor in vast numbers, but they accumulate at an average rate of only about 1–3 cm per thousand years.

Calcareous ooze, however, is not found at these same latitudes where the water is unusually deep. Cold, deep ocean water is under high pressure and contains more dissolved carbon dioxide than shallower waters. For this reason, these deep waters can readily dissolve away any carbonate particles that reach their level. In the Pacific Ocean this level lies at 4000–5000 m, whereas in the Atlantic it is somewhat shallower. This explains why over large portions of the deep north and south Pacific Ocean and some marginal parts of the Atlantic Ocean calcareous ooze is absent.

In broad belts across the equatorial and far northern Pacific Ocean, sectors of the Indian Ocean, and a belt around the Southern Ocean productivity is high. In these regions, siliceous organisms predominate and become the major component of the bottom sediments.

Lithic sediments consisting of rock fragments derived from continental sources mantle the continental shelves and slopes. Such sediments also are important where debris-laden glaciers generate icebergs that raft sediment seaward of glacier margins. Still other vast areas of the deep-sea floor, mostly far from land and in regions of low productivity, are dominated by very fine-grained reddish or brownish clay, generally called *red clay*. Much of the clay likely consists of fine wind-blown dust, the color of which is imparted by the oxidation of iron-rich minerals in the sediment.

OCEAN WAVES

Major currents and gyres are large-scale geographic features of the ocean surface. Finer-scale motions of surface ocean water are a response to the interaction of the atmosphere and ocean surface and, in special cases, to movements of the solid earth.

Surface ocean waves receive their energy from winds that blow across the water surface. The size of a wave depends on how fast, how far, and how long the wind blows. A gentle breeze blowing across a bay may ripple the water or form low waves less than a meter high. At the opposite extreme, storm waves produced by hurricane-force winds (>115 km/h, or 72 mi/h) blowing for days across hundreds or thousands of kilometers of open water may become so high that they tower over ships unfortunate enough to be caught in them.

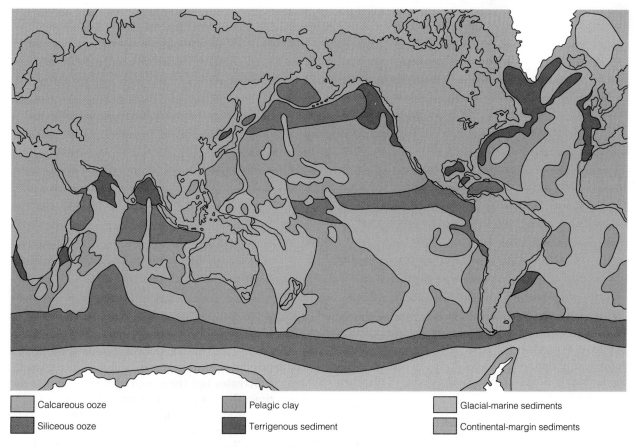

■ Calcareous ooze	■ Pelagic clay	■ Glacial-marine sediments
■ Siliceous ooze	■ Terrigenous sediment	■ Continental-margin sediments

Figure 11.18 Map of the World Ocean showing generalized distribution of the principal kinds of sediment on the ocean floor.

Wave Motion and Wave Base

Figure 11.19 shows the significant dimensions of a wave traveling in deep water, where it is unaffected by the bottom far below. As the wave moves forward, each small parcel of water revolves in a loop, returning very nearly to its former position once the wave has passed. Because wave form is created by a looplike motion of water parcels, the diameters of the loops at the water surface exactly equal wave height (H in Fig. 8.16). Below the water surface, a progressive loss of energy occurs with increasing depth, expressed as a decrease in loop diameter. At a depth equal to half the **wavelength** (the distance between successive wave crests or troughs; L in Fig. 11.19), the diameters of the loops have become so small that water motion is negligible.

The depth $L/2$ is the effective lower limit of wave motion and is generally referred to as the **wave base** (Figs. 11.19 and 11.20). In the Pacific Ocean, wavelengths as long as 600 m (1970 ft) have been measured. For them, $L/2$ equals 300 m, a depth half again as great as the average depth of the outer edge of the

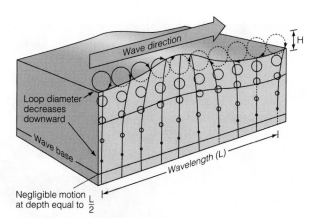

Figure 11.19 Looplike motion of water in a wave in deep water. To trace the motion of a small parcel of water at the surface, follow the arrows in the largest loops from right to left. The resultant motion is the same as watching the wave crest travel from left to right. Parcels of water in smaller loops beneath the surface have corresponding positions, marked by nearly vertical lines. Dashed lines represent wave form and parcel positions one-eighth of a period later.

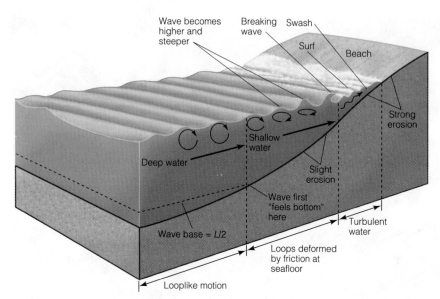

Figure 11.20 Waves change form as they travel from deep water through shallow water to shore. In the process, the circular motion of water parcels found in deep water changes to elliptical motion as the water shallows and the wave encounters frictional resistance to forward movement. Vertical scale is exaggerated, as is the size of loops relative to the scale of the waves.

continental shelves (about 200 m). Although the wavelengths of most ocean waves are far shorter than 600 m, it nevertheless is possible for very large waves approaching these dimensions to move seafloor sediment even on the outer parts of the continental shelves.

Breaking Waves

Toward the land, as water depth becomes less than $L/2$, the circular motion of the lowest water parcels is influenced by the increasingly shallow seafloor, which restricts movement in the vertical direction. As the water depth decreases, the loops of the water parcels become progressively flatter until, in the shallow water zone, the movement of water at the seafloor is limited to a rapid back and forth motion (Fig. 11.20). As depth decreases, the wave's shape is distorted; its height increases and the wavelength shortens. At the same time, the wave's front grows steeper. Eventually, the steep front is unable to support the advancing wave, and as the rear part continues to move forward, the wave collapses or *breaks* (Figs. 11.20 and 11.21).

When a wave breaks, the motion of its water instantly becomes turbulent **surf,** defined as wave activity between the line of breakers and the shore. In surf, each wave finally dashes against rock or rushes up a sloping beach until its energy is expended; then it

Figure 11.21 Orienting his board for the best ride, a surfer skims the inside of a breaking wave off the coast of Hawaii.

Figure 11.22 Waves arriving obliquely along a coast near Oceanside, California, change orientation as they encounter the bottom and begin to slow down. As a result, each wave front is refracted so that it more closely parallels the bottom contours. The arriving waves develop a longshore current that moves from right to left in this view.

flows back. Surf possesses most of the original energy of each wave that created it. This energy is quickly consumed in turbulence, in friction at the bottom, and in moving the sediment that is thrown violently into suspension from the bottom. Although fine sediment is transported seaward from the surf zone, most of the geologic work of waves is accomplished by surf shoreward of the line of breakers.

Ocean waves typically break at depths that range between wave height and 1.5 times wave height. Because waves are seldom more than 6 m (17 ft) high, the depth of vigorous erosion by surf should be limited to 6 m times 1.5, or 9 m (30 ft) below sea level. This theoretical limit is confirmed by observation of breakwaters and other coastal structures, which are only rarely affected by surf below a depth of 7 m (23 ft).

Wave Refraction and Longshore Currents

A wave approaching a coast generally does not encounter the bottom simultaneously all along its length. As any segment of the wave touches the seafloor, that part slows down, the wavelength begins to decrease, and the wave height increases. Gradually, the trend of the wave becomes realigned to parallel the bottom contours (Fig. 11.22). Known as **wave refraction**, this process changes the direction of a series of waves moving in shallow water at an angle to the shoreline. In this way, waves approaching the margin of a deep-water bay at an angle of 40° or 50° may, after refraction, reach the shore at an angle of 5° or less.

Waves passing over a submerged ridge off a headland will be refracted and converge on the headland

(Fig. 11.23). This convergence, as well as the increased wave height that accompanies it, concentrates wave energy on the headland, which is eroded vigorously. Conversely, refraction of waves approaching a bay will make them diverge, diffusing their energy at the shore. In the course of time, the net tendency of these contrasting effects is to make irregular coasts smoother and less indented.

The path of an incoming wave can be resolved into two directional components, one oriented perpendicular to the shore and the other parallel to it. Whereas the perpendicular component produces the crashing

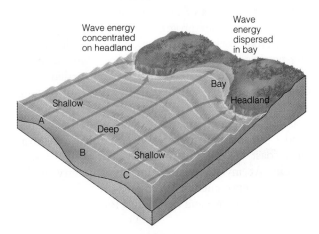

Figure 11.23 Refraction of waves concentrates wave energy on headlands and disperses it along bays. This oblique view shows how waves become progressively distorted as they approach the shore over a bottom that is deepest opposite the bay. The result is vigorous erosion on the exposed headland and sedimentation along the margin of the bay.

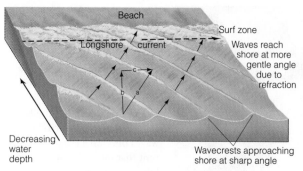

Figure 11.24 A longshore current develops parallel to the shore as waves approach a beach at a right angle and are refracted. A line drawn perpendicular to the front of each approaching wave (a) can be resolved into two components. The component oriented perpendicular to the shore (b) produces surf, whereas the component oriented parallel to the shore (c) is responsible for the longshore current. Such a current can transport considerable amounts of sediment along a coast.

surf, the parallel component sets up a **longshore current** within the surf zone, a current that flows parallel to the shore (Fig. 11.24). The direction of longshore currents may change seasonally if the prevailing wind directions change, thereby causing changes in the direction of arriving waves.

WHERE LAND AND OCEAN MEET

The oceans meet the land in a zone of dynamic activity marked by erosion and the creation, transport, and deposition of sediment. At a coast, waves that may have traveled unimpeded across hundreds or thousands of kilometers of open ocean encounter an obstruction to further progress. They dash against the shore, erode rock and sediment, and move the resulting particles about. In the surf zone, joint-bounded blocks of bedrock are plucked out and carried away, and the continuous rubbing and grinding of moving rock particles wears away solid rock. In effect, the surf acts like a knife or saw, cutting horizontally into the land. Over time, the net effect is substantial. The form of a coast changes, often slowly, but at times very rapidly. At any given moment, the geometry of the shoreline represents an approximate equilibrium between constructive and destructive forces.

Beaches and Other Coastal Deposits

Beaches are a primary landform of most coasts (Fig. 11.25); even coasts dominated by steep, rocky cliffs generally have beaches interspersed with rocky headlands. A **beach** consists of wave-washed sediment along a coast, including sediment in the surf zone that

is in constant motion. Although some beach sediment is derived from erosion of adjacent cliffs or older beach deposits, most sediment reaches beaches by rivers that enter the sea.

During storms, powerful surf erodes the exposed part of a beach and makes it narrower. In calm weather, the exposed beach is likely to receive more sediment than it loses and therefore becomes wider. Because storminess tends to be seasonal, beaches also change character seasonally.

Elongated ridges of sand or gravel, called *spits*, that project from land and end in open water are another conspicuous coastal landform. Most spits are merely seaward continuations of beaches (Fig. 11.26A). Many spits are built of sediment moved by longshore currents and dropped at the mouth of a bay, where the current encounters deeper water and its velocity decreases.

Barrier islands, which are long, narrow sandy islands lying parallel to a coast and separated from it by

Figure 11.25 A sandy beach along the shore of Bora Bora, a volcanic island in French Polynesia, consists of coral and shell debris carried landward by wave action and mixed with lava fragments from the eroding volcano.

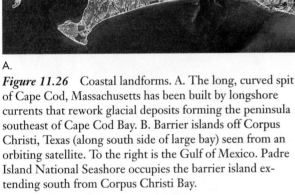

A.

B.

Figure 11.26 Coastal landforms. A. The long, curved spit of Cape Cod, Massachusetts has been built by longshore currents that rework glacial deposits forming the peninsula southeast of Cape Cod Bay. B. Barrier islands off Corpus Christi, Texas (along south side of large bay) seen from an orbiting satellite. To the right is the Gulf of Mexico. Padre Island National Seashore occupies the barrier island extending south from Corpus Christi Bay.

a lagoon, are found along most of the world's lowland coasts (Fig. 11.26B). Some barrier islands, like those off the North Carolina coast, occasionally receive the full fury of destructive hurricanes, which erode and reshape these ephemeral landforms.

Marine Deltas

Where surf and currents are inadequate to erode all new sediment carried to the sea by a large stream, the sediment builds outward as a *marine delta* (Fig. 11.27). The size and shape of a delta reflect the balance reached between sedimentation and erosion at the coast. Some deltas, such as that of the Mississippi River, consist of a complex of subdeltas of different ages, indicating a long and complicated history.

Some of the world's largest deltas constitute prime agricultural lands and, collectively, are home to millions of people (e.g., Egypt's Nile Delta, China's Huang He and Yangtze deltas, the Mississippi Delta on the Gulf coast of the United States, and the Ganges Delta of Bangladesh). Each of these deltas is a geologically young feature, having been built during the latest postglacial rise of world sea level. The surface of each also lies within a few meters of present sea level and is therefore especially vulnerable to destructive coastal storms that can inundate farmland, destroy towns and villages, and devastate local populations.

Estuaries and Coastal Wetlands

Estuaries and associated salty to brackish marine wetlands are among the most important ecological areas on the Earth. An **estuary** is a semi-enclosed marine embayment that is diluted with fresh water, normally entering it by one or more streams. Several types of estuaries are recognized, distinguished on the basis of their internal circulation and salinity distribution. The

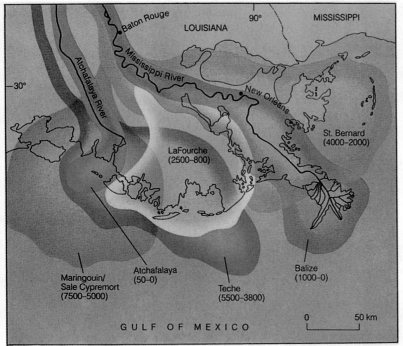

Figure 11.27 The Mississippi River has built a series of overlapping subdeltas as it has continually dumped sediment into the Gulf of Mexico. The ages of subdeltas are given in radiocarbon years before the present.

simplest type is found where a river enters directly into marine water. The fresh water, being less dense, flows directly out to sea over denser saltwater. The rapid flow of the river retards saltwater incursion, producing a boundary tilted downward in the upstream direction that is between a wedge of saltwater moving upstream and an overlying thinner wedge of stream water. The turbulent stream water entrains saltwater as it flows seaward, increasing its salinity. However, very little river water is mixed downward, and so such estuaries are poorly mixed. Examples are found at the mouth of the Mississippi River and also the Columbia River, where the saltwater incursion extends as much as 24 km (15 mi) upstream. Estuaries characterized by strong tidal turbulence and low river outflow tend to be well mixed, whereas those marked both by strong river and tidal inflow tend to be partially mixed.

Estuaries and marine wetlands offer important habitats for an array of plants and animals. They are spawning grounds for many species of commercial fish and shell fish, and often support large bird populations. Estuaries are also attractive to people. The establishment of industrial cities, commercial ports, and fisheries in estuarine settings has led to their increasing modification and degradation (See "A Closer Look: Human Modification of the San Francisco Bay Estuary"). Pollution of estuaries by human and indus-

trial wastes and loss of prime wetland habitat have alerted local governments to the need for establishing controls that will permit natural environments and people to coexist.

Reefs and Atolls

Many of the world's tropical coastlines consist of limestone reefs built by vast colonies of corals and other carbonate-secreting organisms. Three principal reef types are recognized: a *fringing reef* is either attached to or closely borders the adjacent land and lacks a lagoon (Fig. 11.28A); a *barrier reef* is a reef separated

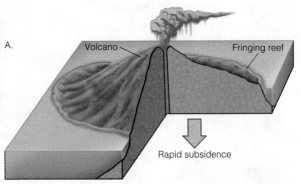

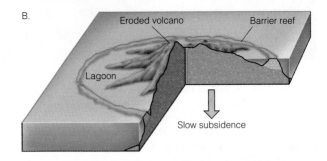

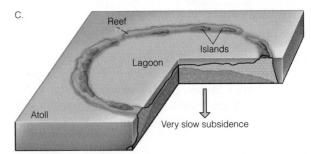

Figure 11.28 Evolution of an atoll from a subsiding oceanic volcano. A. Rapid extrusion of lava builds a shield volcano that begins to subside as the ocean crust is loaded by the growing volcanic pile; a fringing reef grows upward, keeping pace with subsidence. B. As subsidence continues, the fringing reef becomes a barrier reef, separated from the eroded volcano by a lagoon. C. With continuing subsidence and upward reef growth, the last remnants of volcanic rock are submerged, leaving an atoll reef surrounding a central lagoon.

A CLOSER LOOK

Human Modification of the San Francisco Bay Estuary

San Francisco Bay is one of the world's most famous estuaries (Fig. C11.1). As many as 20,000 Native Americans populated its shores when the first European explorers arrived. Today, the population surrounding the estuary numbers in the millions and continues to grow. Because of California's pressing need for fresh water to support human consumption as well as major agricultural and industrial enterprises, diversion of stream water has reduced freshwater inflow to the bay to less than half what it was in 1850. Simultaneously, increasing urbanization has led to the loss of 95 percent of the bordering marine wetlands due to filling and diking. The California gold rush also had an impact. Vast quantities of sediment produced by hydraulic mining in the foothills of the Sierra Nevada choked streams, destroyed fish spawning ground, obstructed navigation, and ultimately reached the bay where it reduced the area and volume of the estuary and modified tidal circulation.

As a result of this human tampering, most commercial fisheries have disappeared. Reduced freshwater flushing of the estuary has concentrated agricultural, domestic, and industrial wastes, contaminating both the sediments and the organisms that feed on them. Natural bird habitats have been extensively modified or destroyed.

Such changes are not unique to San Francisco Bay. Other large estuaries (e.g., the Susquehanna and Potomac rivers, the Rhine River in the Netherlands, and the Thames River in England) have similar problems. All are sensitive to human-induced changes and susceptible to steady deterioration. Because estuarine systems are complicated, the best hope for reversing their decline is through improved understanding of how human actions affect the natural physical, chemical, and biological processes and what measures are needed to curtail potentially destructive activities.

Figure C11.1 Vertical satellite image of the shrinking San Francisco Bay estuary. Filling and diking of tidal marshes to create farmland, evaporation ponds, and residential and industrial developments has reduced 2200 km² of wetland marshes that existed before 1850 to less than 130 km² today.

from the land by a lagoon and may be of considerable length and width (Fig. 11.28B); and an *atoll* is a roughly circular coral reef enclosing a shallow lagoon (Fig. 11.28C) that forms when a tropical volcanic island with a fringing reef slowly subsides. Charles Darwin was the first to deduce, during his voyage on the H.M.S. *Beagle* in the 1830s, that slow subsidence forces reef organisms to grow upward so that they can survive near sea level. As a volcanic island subsides, the fringing reef is transformed into an offshore barrier reef and eventually into an atoll. Atolls generally lie in deep water in the open ocean and are as large as 130 km (80 mi) in diameter. Darwin's hypothesis was confirmed a century after he proposed it by drill holes on atolls that reached volcanic rock after penetrating thick sections of ancient reef rock.

OCEAN TIDES

Tides, the rhythmic, twice-daily rise and fall of ocean waters, are caused by the gravitational attraction between the Moon (and to a lesser degree, the Sun) and the Earth. A sailor in the open sea may not detect tidal motion, but near coasts the effect of the tides is amplified and they become geologically important.

Tide-Raising Force

The gravitational pull that the Moon exerts on the solid Earth is balanced by an equal but opposite inertial force (which tends to maintain a body in uniform linear motion) created by the Earth's movement with respect to the center of mass of the Earth–Moon system (Fig. 11.29). At the center of the Earth, gravitational and inertial forces are balanced, but they are not balanced from place to place on the Earth's surface. A water particle in the ocean on the side facing the Moon is attracted more strongly by the Moon's gravitation than it would be if it were at the Earth's center, which lies at a greater distance. Although the attractive force is small, liquid water is easily deformed, and so each water particle on this side of the Earth is pulled toward a point directly beneath the Moon. This creates a bulge on the ocean surface.

Although the magnitude of the Moon's gravitational attraction on a particle of water at the surface of the ocean varies over the Earth's surface, the inertial force at any point on the surface is the same. On the side nearest the Moon, gravitational attraction and inertial force combine, and the excess gravitational force (or *tide-raising force*) is directed toward the Moon. On the opposite side of the Earth, the inertial force exceeds the Moon's gravitational attraction, and the

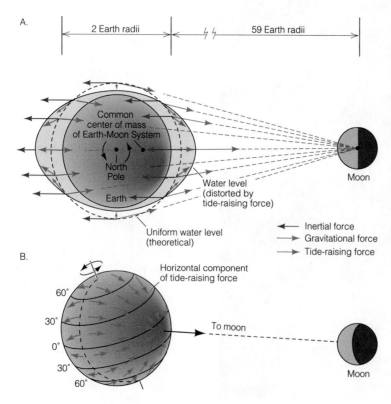

Figure 11.29 Tidal forces. A. Tide-raising forces are produced by the Moon's gravitational attraction and by inertial force. On the side toward the Moon, gravitational attraction exceeds the inertial force. The excess (tide-raising) force distorts the water level from that of a sphere, and raises a tidal bulge. On the opposite side of the Earth, where inertial force is greater than the gravitational force of the Moon, the excess inertial force also creates a tidal bulge. B. The horizontal component of the tide-raising force is shown by arrows on an oblique view of the Earth. The arrows are directed toward the point where a line connecting the Earth and Moon intersects the Earth's surface. This point shifts latitude with time as the relative positions of the Earth and Moon change.

tide-raising force is directed away from the Moon (Fig. 11.29). These unbalanced forces generate the daily ocean tides.

Tidal Bulges

The tidal bulges created by the tide-raising force on opposite sides of the Earth appear to move continually around the Earth as it rotates. In fact, the bulges remain essentially stationary beneath the tide-producing body (the Moon) while the Earth rotates. At most places on the ocean margins, two high tides and two low tides are observed each day as a coast encounters both tidal bulges. In effect, at every high tide, the coastline runs into a mass of water, which piles up against it. This water then flows back to the ocean basin as the coastline passes beyond each tidal bulge.

Earth–Sun gravitational forces also affect the tides, sometimes opposing the Moon by pulling at a right angle and sometimes aiding by pulling in the same direction. Twice during each lunar month, the Earth is directly aligned with the Sun and the Moon, whose gravitational effects are thereby reinforced, producing higher high tides and lower low tides (Fig. 11.30). At positions halfway between these extremes, the gravitational pull of the Sun partially cancels that of the Moon, thus reducing the tidal range. However, the Sun is only 46 percent as effective as the Moon in producing tides, so the two tidal effects never entirely cancel each other.

In the open sea, the effect of the tides is small (less than 1 m, or 3 ft), and along most coasts the tidal range commonly is no more than 2 m (6.5 ft). However, in bays, straits, estuaries, and other narrow places along coasts, tidal fluctuations are amplified and may reach 16 m (52 ft) or more (Fig. 11.31). Associated tidal currents are often rapid and may approach 25 km/h (16 mi/h).

CHANGING SEA LEVEL

On most coasts, the level of the sea is changing with respect to the land. Rapid changes can result from local or regional tectonic or isostatic movements of the crust. A much slower change, apparently worldwide and possibly related to global warming (Chapter 14), is causing sea levels to rise about 2.4 mm/year.

Far greater changes in world sea level occur over longer time intervals due to changes in water volume as continental glaciers wax and wane, and to changes in ocean-basin volume as lithospheric plates shift position. We refer to such changes as *eustatic*, meaning a rise or fall in sea level that affects all the world's

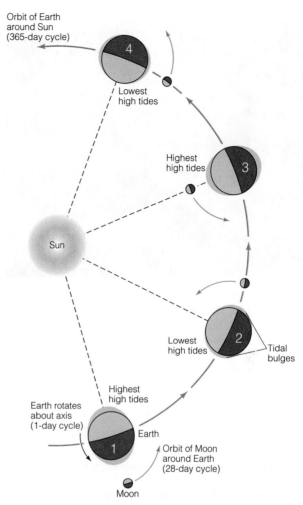

Figure 11.30 When the Earth, Moon, and Sun are aligned (positions 1 and 3), tides of highest amplitude are observed. When the Moon and Sun are pulling at right angles to each other (positions 2 and 4), tides of lowest amplitude are experienced.

oceans. Over the span of a human lifetime, these slow changes are imperceptible, but on geologic time scales they contribute in an important way to the evolution of the world's coasts.

Submergence and Emergence

Whatever their nature, nearly all coasts have experienced **submergence**, a rise of water level relative to the land. This is the result of a worldwide rise of sea level that occurred when glaciers melted away at the end of the last ice age (Fig. 11.32). Because of this submergence, evidence of lower glacial-age sea level is almost universally found seaward of the present coastlines and to depths of 100 m (330 ft) or more. By contrast, evidence of interglacial sea levels is often found

A.

B.

Figure 11.31 The tidal range in the Bay of Fundy, eastern Canada, is one of the largest in the world. A. Coastal harbor of Alma, New Brunswick, at high tide. B. Same view at low tide.

inland from and higher than present coastlines. Such evidence points to **emergence,** a lowering of water level relative to the land since the high-level coastal features formed. Repeated cycles of emergence and submergence along the world's coasts are related to the buildup and decay of vast ice-age glacier systems. Three components of the Earth system interact to produce such changes. The global climate system (atmosphere) controls the volume of glaciers (cryosphere) on land, which in turn determines the amount of water residing in the world ocean (hydrosphere). Changing ocean water volume, related to changing climate, thus will cause sea level to rise or fall.

Figure 11.32 Coastal submergence of eastern North America and western Europe resulted when meltwater from wasting ice sheets returned to the ocean basins at the close of the last glaciation. A. Area of northeastern North America covered by glacier ice during the last glacial maximum, the approximate position of the shoreline at the glacial maximum (18,000 years ago), and coastal areas submerged by the postglacial rise of sea level. B. Areas covered by ice sheets in Western Europe at the last glacial maximum and land areas that have been submerged during the postglacial rise of sea level.

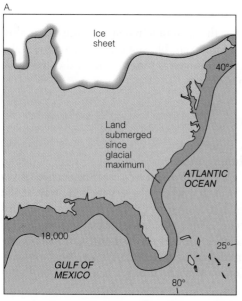

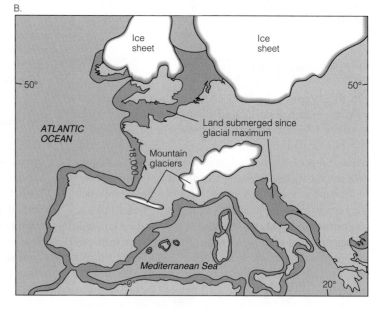

GUEST ESSAY

Where the Sun Never Rises, Seeing Is Believing

Victoria A. Kaharl, is the author of Water Baby: The Story of Alvin *(Oxford University Press, 1990).*

For most of the time humans have inhabited the Earth, the depths of the ocean have remained unknown and unknowable, a world cut off from the land, the sky, and the sun-lit top layers of the sea. For a long time, nobody believed there could be life in the frigid black depths. The nineteenth-century naturalist Edward Forbes probed the ocean by dangling hemp line attached to nets and dredges. Because he never brought up any life from deeper than 550 meters, Forbes concluded there was none. Beneath 550 meters, he wrote in 1841, lay the lifeless *azoic zone*.

At the time, however, there was some evidence to the contrary. The crews who repaired submarine telegraphic cables occasionally found animals, such as starfish, latched onto a deep-sea cable.

More evidence for life in the deep sea came during the research expedition of the H.M.S. *Challenger.* The British naval corvette left England three days before Christmas in 1872 for a round-the-world research cruise that would last three years and five months. Although research cruises had sailed before this, the *Challenger* cruise is often used to mark the beginning of oceanography as a truly organized and comprehensive science. The naturalists on the *Challenger* pulled up life, some marine worms, from 5500 meters.

The hemp rope used by the nineteenth-century naturalists eventually gave way to stronger piano wire for plumbing the depths. Modern oceanographers still use nets and dredges but lower them on thick steel cable. Whatever the line, however, this kind of selective sampling is done blindly, like hanging out of an airplane with a net at midnight to catch butterflies.

Because the ocean is extremely inhospitable to the explorer species *Homo sapiens,* anyone who probes this inner space is utterly dependent on technology. Oceanographic engineers are fond of saying the ocean is more hostile than the Moon. There is no saltwater on the Moon to corrode electronics, no pressure to implode instruments, no slimy animals looking for a nook to make their home

After World War II, some 70 years after the *Challenger* cruise, cameras and lights that could withstand the crushing pressures of the ocean and the corrosive effects of saltwater revealed more of the faraway deep ocean. Thousands of black and white photographs captured snatches of the deep sea. Almost all of the animals in the photographs were small; most were no bigger than flies. The photographic record showed that, although there was life in the deep ocean, there wasn't much of it. Oceanographers likened the deep ocean to a graveyard or a desert.

It would be decades before our perception and understanding of this world began to change. We had to await a new technology, one that could not only take humans to the deep sea but also allow us to move around, to actually explore and conduct experiments in that realm. This new tool was *Alvin,* a three-passenger research submarine commissioned in 1964 as a national oceanographic facility.

About the size of a UPS delivery truck, *Alvin* usually carries a pilot and two scientists on an average eight-hour dive, which is about the lifetime of its batteries. Passengers crouch beside one of the three small plastic portholes inside a 2-meter-wide titanium sphere. They breathe normally; oxygen is automatically bled into the sphere, and carbon-dioxide is removed to keep the atmosphere the same as room air. Because there is no heat inside the submarine, passengers dress warmly because it is always cold in the deep sea. With a generous lunch bag hanging from the center of the passenger sphere, there is no room to stand up. Most people don't mind; they're too busy looking. The world outside *Alvin's* small windows, each wide enough for only one pair of eyes, is still new and will continue to be for decades to come. We have explored only a fraction of the largest environment on Earth, but the little we have seen has changed once and for all a host of misconceptions about the deep sea.

Today, with more than 3000 dives to its credit, *Alvin* takes scientists to deep-sea sites with names like the Garden of Eden, Clam Acres, and Anemone Heaven. *Graveyard* and *desert* do not apply to any of these communities; each is a hydrothermal vent teeming with life.

Relative Movements of Land and Sea

A change of global sea level that causes coastal submergence or emergence affects all parts of the world ocean at the same time. On the other hand, uplift and subsidence of the land may cause local submergence or emergence. For example, vertical tectonic movements along the margins of converging plates have uplifted beaches and tropical reefs to positions far above sea level (Fig. 11.33). Because tectonic movements and eustatic sea-level changes may occur simultaneously, either in the same or opposite directions and at different rates, unraveling the history of sea-level fluctuations along a coast can be a challenging exercise.

B.

Figure 11.33   Coastal emergence of eastern New Guinea. A. The emergent coast of the Huon Peninsula in eastern Papua New Guinea is flanked by a series of ancient coral reefs that form flat terracelike benches parallel to the shoreline. Each reef formed at sea level and was subsequently uplifted along this active plate margin. The highest reefs lie several hundred meters above sea level and are hundreds of thousands of years old. B. Curve of sea-level fluctuations constructed by uranium-isotope dating of the uplifted reefs on the Huon Peninsula. Prior to a recent high stand, sea level remained lower than at present since the last interglaciation, about 120,000 years ago.

SUMMARY

1. Seawater covers nearly 71 percent of the Earth's surface and is concentrated in the Pacific, Atlantic, and Indian oceans. Each is connected to the Southern Ocean, which encircles Antarctica.

2. Although the greatest ocean depth is more than 11 km, the average depth is about 4.5 km. More than half the ocean water resides in the Pacific basin.

3. Evidence from sedimentary strata implies that the world has had an ocean since at least 3.95 billion years ago. Most likely, the ocean condensed from steam produced during primordial volcanic eruptions.

4. More than 99.9 percent of the saltiness of seawater is due to eight ions that are derived through chemical weathering of rocks on land and then transported by streams to the sea. Other sources of ions include airborne dust and human pollutants.

5. Sea-surface temperatures are strongly related to latitude, with the warmest temperatures measured in equatorial latitudes. Surface salinity is also strongly latitude-dependent and is related to both evaporation and precipitation.

6. Ocean water is stratified as a result of density differences that are related to salinity and temperature. A thin surface zone of relatively low density is separated from a deep zone of dense water by the pycnocline, a zone where density changes rapidly with increasing depth. The pycnocline coincides approximately with the halocline, a zone of rapid salinity change, and the thermocline, a zone of rapidly changing temperature.

7. Huge wind-driven surface ocean currents that circulate clockwise in the northern hemisphere and counterclockwise in the southern hemisphere carry warm equatorial water toward the polar regions.

8. The Coriolis effect deflects ocean currents to the right in the northern hemisphere and to the left in the southern hemisphere. The magnitude of the effect increases from the equator toward the poles.

9. El Niño/Southern Oscillation, an important element of year-to-year climatic variation, occurs when the tradewinds slacken and surface waters of the central and eastern Pacific become anomalously warm.

10. Ekman transport is the average flow velocity in the upper 100 m of the ocean, and moves at 90° to the prevailing wind direction. Downwelling occurs when Ekman transport is toward a coast, whereas upwelling occurs if it is away from a coast.

11. Sinking of cold and (or) saline high-latitude surface waters leads to oceanwide thermohaline circulation. Operating like a great conveyor belt and driven by density contrasts, this global circulation system replenishes the deep water of the world ocean, replacing it in about 1000 years.

12. The motion of wind-driven surface waves terminates downward at the wave base, a distance equal to half the wavelength. A wave breaks in shallowing water as interference with the bottom causes the wave to grow higher and steeper. Waves approaching shore are refracted as the wave base reaches the bottom, realigning the wave so that it reaches the shore at a gentler angle.

13. The remains of tiny pelagic and benthic organisms accumulate on the ocean floor to form widespread deep-sea oozes. In many regions where the water depth exceeds 4 km, carbonate shells are dissolved before reaching the bottom and instead the ocean floor is mantled with red clay.

14. Twice-daily ocean tides, resulting from the gravitational attraction of the Moon and Sun, are produced as the surface of the rotating Earth passes through tidal bulges on opposite sides of the planet.

15. Recently, nearly all coasts have experienced submergence as a result of the postglacial rise of sea level. Some coasts have experienced more complicated histories of relative sea-level change where tectonic movements have been superimposed over the worldwide sea-level rise.

IMPORTANT TERMS TO REMEMBER

beach *260*
Coriolis effect *248*
downwelling *251*
Ekman transport *250*
El Niño/Southern
   Oscillation (ENSO)
   254

emergence *266*
estuary *261*
gyre *249*
longshore current *260*
North Atlantic Deep Water
   (NADW) *252*

salinity *244*
submergence *265*
surf *258*
thermocline *248*
thermohaline circulation
   252

upwelling *251*
wave base *257*
wavelength *257*
wave refraction *259*

QUESTIONS FOR REVIEW

1. In terms of plate tectonics, explain why the Pacific Ocean basin is the largest of the three major ocean basins of the Earth.

2. What geologic evidence indicates that liquid water has been present at the Earth's surface for at least 3.95 billion years?

3. If the quantity of dissolved ions carried to the oceans by streams throughout geologic history far exceeds the known quantity of these substances in modern seawater, why is seawater not far more salty than it is?

4. How and why are the temperature and salinity of surface ocean water related to latitude?

5. What factors cause ocean water to be vertically stratified?

6. Explain why the seasonal temperature range is less at the western coast of North America than it is several hundred kilometers inland.

7. What geologic features would you look for to determine whether a coastal region had experienced emergence or submergence in the recent geologic past?

8. Explain the existence of large midocean gyres both north and south of the equator.

9. Describe the effects of an El Niño event at three sites in a transect across the equatorial Pacific Ocean: in Indonesia, at an island in the middle of the ocean, and at the coast of Peru.

10. What contrasting ocean conditions give rise to upwelling and downwelling along continental margins?

11. Explain why carbonate oozes are limited to oceanic regions where the water depth is less than about 4000 m.

12. Why does a ship tied to a dock experience two high tides and two low tides each day?

13. Describe how North Atlantic Deep Water is produced and explain its role in the global ocean conveyor system.

QUESTIONS FOR DISCUSSION

1. Imagine yourself a navigational assistant on the H.M.S. *Beagle* as it sets sail, with Charles Darwin aboard as naturalist, from England toward the southeastern coast of South America. Suggest a course that Captain Fitzroy could follow in order to take advantage of winds and currents as you travel from Portsmouth (50.8° N) to Montevideo (35° S).

2. Consider the following hypothesis: the thermohaline circulation system of the oceans has not operated continuously but from time to time has shut down. What might cause the system to shut down, and what evidence might you look for to see if this had happened in the past?

CHAPTER 12

Composition and Structure of the Atmosphere

● *The Air We Breathe*

One very important criterion must be met if a planet is to be habitable: the atmosphere must be breathable, and the most essential ingredient of a breathable atmosphere is oxygen.

Where oxygen is concerned, the human body must rely entirely on the atmosphere to supply its needs. Although the body has some capacity to adjust for a change in the amount of oxygen available, the range of adjustment is limited. A measure of the lower limit of this range is provided by people who live, work, or visit at high altitude. Miners in the Andes Mountains of Chile and Peru, for instance, work at elevations as great as 5300 m (17,400 ft). At this altitude, a lungful of air contains only 50 percent of the amount of oxygen contained in a lungful at sea level. Despite the small amount of oxygen available in each breath, miners who grow up in the mountains become acclimatized and lead active working lives.

The absolutely lowest intake of oxygen a body can handle varies from person to person, so what a Peruvian miner can do is not possible for everyone. However, for at least limited times, visitors to high altitudes can handle levels of oxygen even lower than the 50 percent value quoted above. Mountain climbers in the Himalaya, for example, have camped and climbed for weeks on end from bases as high as 7000 m (23,000 ft), at which height a lungful of air contains only 44 percent of the oxygen in a sea-level lungful. This amount of oxygen is probably very close to the lower human limit because balloonists who have attempted long-time flights at heights in excess of 7000 m have found it necessary to breathe bottled oxygen; some who attempted to fly without extra oxygen perished in the attempt.

By experiments on divers and in hospitals, the upper limit of oxygen has been found to be 55 percent above the amount found in sea-level air. Beyond this limit, oxygen becomes toxic because the body responds, in effect, by starting to burn up. For safety, the upper limit of oxygen used in hospitals for patients having breathing difficulties is a 40 percent increase above normal sea-level air.

We can conclude therefore that, to be habitable, a planet must have an oxygen level ranging from 40 percent above to 44 percent below the level found in today's air. As far as we can tell from the geological record, the oxygen content has varied, but it has not moved outside the habitable range for several hundred million years. This means that, if a time machine really were possible, we could turn the clock back and visit the dinosaurs and breathe their air

(opposite) A climber working near the limit of human capacity, on the southwest face of Mount Everest, the world's highest mountain. At the elevation of Mount Everest, the air pressure is so reduced a climber gets insufficient oxygen with each breath and has to supplement the air with bottled oxygen.

quite comfortably. However, the fact that the oxygen content of air has varied may have had major consequences for the biosphere in ages past. For example, samples of air trapped in amber (a fossil tree resin) suggest that, 100 million years ago, during the Cretaceous, the oxygen content of the atmosphere was 40 percent higher than in today's atmosphere. Although the dinosaurs were sent to extinction at the end of the Cretaceous, probably by a great meteorite impact, they had been declining in numbers for a long time before the impact. As the Cretaceous came to a close, the oxygen level started to decline. One hypothesis for the decline of the dinosaurs is that they had small lungs because the air contained so much oxygen. As the oxygen level dropped, their small lungs could not adjust, and so, like high-flying balloonists, they died out as a result of respiratory stress. ●

WEATHER AND CLIMATE

The weather is said to be the most popular topic of conversation for all cultures. The weather deserves this high ranking because it plays such an important role in our daily lives—from the clothes we wear to the activities we pursue. Even though we talk a lot about the weather there is sometimes a bit of confusion about what is actually being discussed. To avoid confusion, we follow the lead of *meteorologists* (weather scientists) who have a formal definition for **weather**; they define it as the state of the atmosphere at a given time and place. The five variables used to determine the state of the atmosphere are:

1. Temperature
2. Air pressure
3. Humidity
4. Cloudiness
5. Wind speed and direction

These five weather variables are also used to measure the climate, but there is an important difference between weather and climate. Weather is a short-term event, whereas climate is a long-term one. Weather can change over a short time span. For example, the weather can be cold and wet in the morning but warm and dry in the afternoon. Climate, on the other hand, must be measured over a period of years because **climate** is the *average* weather condition of a place. The climate of northern Canada is cold and wet, for example, even though there are many warm, dry days. Over a period of years, cold, wet days are more numerous than warm, dry days, and so the average weather (that is, the climate) of northern Canada is cold and wet regardless of what the weather may be on a given day or even during a given week or month. The opposite condition is found in the Sahara of northern Africa; there, hot, dry days are far more common than cold,

wet ones, and so the climate of the Sahara is classified as hot and dry.

Weather and climate are sensitive indicators of changes in the Earth system. This is so because a rapid change to the atmosphere can quickly change weather. For example, a volcanic eruption can affect weather around the world in a matter of days. Because the atmosphere is an important sensor in the Earth system, Chapters 12 and 13 are dedicated to the properties of the atmosphere and to how changes in these properties affect the weather. Chapter 12 focuses primarily on the composition, structure, and physical properties of the atmosphere, and Chapter 13 addresses motions in the atmosphere. Note that the topics in Chapters 12 and 13 are concerned primarily with weather rather than climate. Because climate can be affected by long-term changes in the geosphere, hydrosphere, and biosphere, as well as in the atmosphere, the controls on climate and climatic changes are addressed separately in Chapter 14, and potential future human influences on the climate are discussed in Chapter 20.

COMPOSITION AND STRUCTURE OF THE ATMOSPHERE

Two things energize the atmosphere: the Sun's heat and the Earth's rotation. As discussed in Chapter 3, energy from the Sun reaches the Earth in the form of electromagnetic radiation. Solar radiation is the energy source responsible for clouds, rain, snow storms, and much of the local weather. By contrast, the tilt of the Earth's axis of rotation is responsible for the annual seasons, as described in Chapter 2, and the rotation is also responsible mainly for large-scale effects in the atmosphere—effects such as the midlatitude west-to-east movement of weather patterns, the jet stream, and global wind systems. Although the two major energy sources, the Sun's heat and the Earth's rotation, operate in concert, they can nevertheless be separated for discussion, and that is the basis on which the topics in Chapters 12 and 13 are separated.

To appreciate how the two energizing sources play the roles they do, we first have to understand the structure of the atmosphere. It may seem strange to say "structure" when we discuss a gaseous medium, but measurements show that, with increasing altitude, there are distinct variations in such things as the composition, temperature, pressure, and humidity of the atmosphere. We humans live at the bottom of the atmosphere, and so most of the variations are far above our heads. What is known about the upper reaches of the atmosphere, like our knowledge of the geosphere beneath our feet, comes mainly from instrument probes of one kind or another.

Composition

There is a subtle difference between the words *atmosphere* and *air*. An *atmosphere* is the gaseous envelope that surrounds a planet or any other celestial body. **Air**, by contrast, is the invisible, odorless mixture of gases and suspended particles that surrounds one special planet, the Earth. In other words, air is the Earth's atmosphere; it is also our most precious commodity. Denied access to air, the human body dies within a few minutes.

Considering our total dependence on air, you might think that everyone would know and care a lot about the air we breathe. Unfortunately, that does not seem to be the case. All too often, we take this most precious commodity for granted and carelessly treat the air as if it had an endless capacity to absorb pollutants.

Air is a complex mixture of gases and tiny suspended particles. Because air pressure decreases with altitude, the amount of air per unit volume (that is, the density) also varies with altitude. In order to separate changes due to composition from those due to density, the composition of air is always discussed in terms of the *relative* rather than the absolute amounts of the different constituents present.

The relative composition, it turns out, varies somewhat from place to place on the surface of the Earth and even from time to time in the same place. There are two reasons for the variations, the presence of aerosols and the presence of water vapor, both of which vary widely in amount:

1. **Aerosols** are tiny liquid droplets or tiny solid particles that are so small they remain suspended in the air (Fig. 12.1). Common liquid aerosols, such as the water droplets in fogs, are familiar to all of us. Solid aerosols such as tiny ice crystals, smoke

particles from fires, sea-salt crystals from ocean spray, dust stirred up by winds, volcanic emissions, and pollutants from industrial activities are less familiar but nevertheless widespread. Aerosols are everywhere in the atmosphere, particularly in the air nearest the ground. We breathe them in all the time, but fortunately the human respiratory system is designed to prevent them from causing harm to our lungs.

2. Because the Earth has a hydrosphere from which water evaporates, there is always water vapor in the atmosphere. The amount of water vapor in the air, for which the term **humidity** is used, is quite variable. On a hot, humid day in the tropics, as much as 4 percent of the air by volume may be water vapor, whereas on a crisp, cold day, less than 0.3 percent water vapor may be present.

Because the water vapor and aerosol content of the air vary widely, the relative amounts of the remaining gases in the air are reported on a dry (meaning free of water vapor) and aerosol-free basis. Once these two

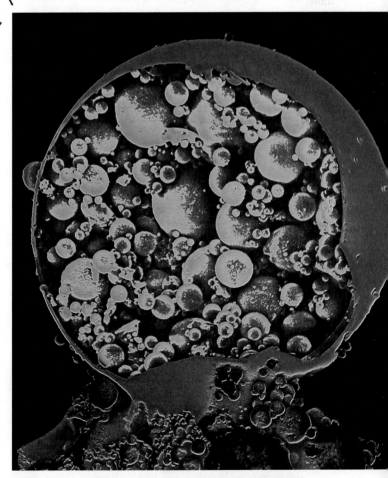

Figure 12.1 False-color electron micrograph image of fly ash, a common aerosol. Fly ash comes from power plants and other sources of combustion. Individual spheres range in size from about 0.001 to 0.01 in diameter.

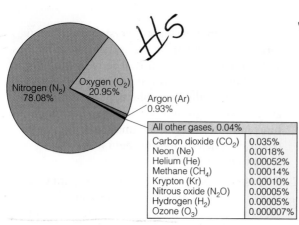

All other gases, 0.04%	
Carbon dioxide (CO_2)	0.035%
Neon (Ne)	0.0018%
Helium (He)	0.00052%
Methane (CH_4)	0.00014%
Krypton (Kr)	0.00010%
Nitrous oxide (N_2O)	0.00005%
Hydrogen (H_2)	0.00005%
Ozone (O_3)	0.000007%

Figure 12.2 Composition of dry, aerosol-free air in volume percent. Three gases—nitrogen, oxygen, and argon—make up 99.96 percent of the air.

variable constituents are removed, the relative proportions of the remaining gases in the air turn out to be essentially constant regardless of altitude. As shown in Figure 12.2, three gases—nitrogen, oxygen, and argon—make up 99.96 percent of dry air by volume. Even though the relative amounts of the remaining gases are very small, these minor gases are profoundly important for life on the Earth because they act both as a warming blanket and as a shield from deadly ultraviolet radiation.

Carbon dioxide, water vapor, methane, ozone, and nitrous oxide are the minor gases that create the Earth's life-maintaining blanket. These five gases are commonly called the *greenhouse gases* because, like a glass-covered greenhouse, they create a warm environment. A greenhouse is kept warm because air heated by the Sun's radiation is prevented from escaping (Fig. 12.3). The atmosphere works in a similar way. Like the glass of a greenhouse, the atmosphere is transparent to the Sun's short-wavelength radiation, which passes through and warms the Earth's surface. The warm surface radiates long-wavelength infrared radiation back into space. However, the greenhouse gases absorb some of the infrared radiation, and this absorbed radiation heats the atmosphere and keeps the air in contact with the Earth's surface in a comfortable temperature range. Changes in the amounts of greenhouse gases in the atmosphere lead to changes in the heat-absorbing capacity of the atmosphere, and therefore to changes in temperature, and eventually to changes in the climate. If you move ahead to Chapter 20, you will find a more detailed discussion of the radiation absorption effect and the role it plays in global climate changes.

The part of the solar radiation spectrum that is dangerous to humans and to many other living creatures is the ultraviolet (see Fig. C2.2). Most of this deadly radiation is prevented from reaching the Earth's surface as a result of absorption by three forms

of oxygen gas: O, O_2, and O_3 (ozone) (Fig. 12.4). The most important of the three shielding gases is ozone. Even though it occurs in only minute amounts high in the atmosphere, it is able to absorb the most lethal kind of ultraviolet rays.

At very great altitudes, 80 km (50 mi) and higher, the composition of dry, aerosol-free air changes a little from what it is at the Earth's surface; it is depleted in the heavier gases, such as argon and neon, and enriched in the lighter gases such as helium. For most purposes of discussion where the Earth system is concerned, however, we don't have to be worried about such high-altitude changes in the composition because weather and climate effects happen in the lower atmosphere, where the relative compositions do not vary.

Temperature

Temperature is the most important and most familiar variable used to define the state of the atmosphere. The human body is sensitive to changes in temperature as small as 1°C (1.8°F); as a result everyone is aware that temperature varies from hour to hour,

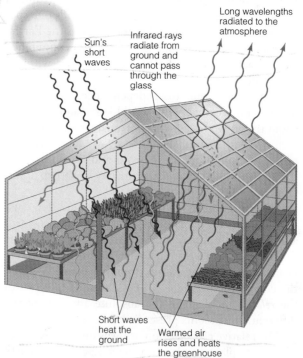

Figure 12.3 The way a greenhouse works. Short-wavelength radiation from the Sun passes through the glass roof and heats the ground. Some of the heat from the ground then warms the air in the greenhouse; the rest is re-radiated back as infrared radiation, which is then trapped by the glass roof, producing additional heating inside. The warmed air emits long-wavelength radiation which passes through the glass and escapes into the atmosphere. When a balance is reached, the incoming radiation equals the escaping radiation.

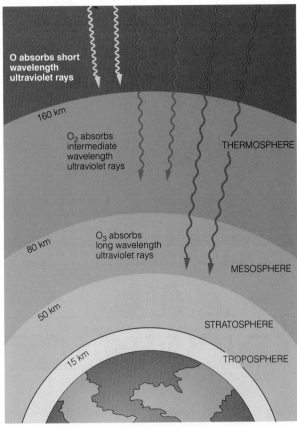

O absorbs short wavelength ultraviolet rays

160 km

O₂ absorbs intermediate wavelength ultraviolet rays

THERMOSPHERE

80 km

O₃ absorbs long wavelength ultraviolet rays

MESOSPHERE

50 km

STRATOSPHERE

15 km

TROPOSPHERE

Figure 12.4 Life-protecting layers of O, O₂, and O₃ in the atmosphere absorb lethal ultraviolet radiation.

from place to place, from day to night, and from season to season.

Temperature Versus Heat

Before we discuss temperature variations in the atmosphere, it is important to establish the difference between heat and temperature. The definition of **heat** and *heat energy* (the two terms mean exactly the same thing) is the total kinetic energy (energy of motion) of all the atoms in a substance. Not all the atoms in a given sample move with the same speed, however, and so there is a range of kinetic energies among them. **Temperature** is a measure of the *average* kinetic energy of all the atoms in a body. Note the difference—heat is the *total* energy, whereas temperature is a measure of the *average* energy. Even though two bodies of the same substance, such as a cup of water and a pail of water, have the same temperature, say 25°C (77°F), there are so many more water molecules in the pail than in the cup that the pail has far more heat energy. Furthermore, if the temperature of the cup of water is raised from 25°C (77°F) to boiling (100°C or 212°F), which means that the average speed of the atoms in the water molecules rises, the heat energy in the lower-temperature but much larger pail will probably

still exceed the heat energy in the small, higher-temperature cup. The reason is that the heat energy of the sum of a large number of slow-moving atoms may well exceed the heat energy of a small number of fast-moving atoms.

The atmosphere gets its heat energy from the Sun. As we discussed in Chapter 3, the flux of energy coming in from the Sun is 1370 W/m². This is the energy flux that would be measured by a satellite orbiting the Earth outside the atmosphere. The **insolation**, which is the energy from the Sun that actually reaches the Earth's surface, is considerably less than 1370 W/m². (See "A Closer Look: Sunlight and the Atmosphere.")

The insolation is less than 1370 W/m² for three principal reasons. The first two reasons concern the atmosphere, which reflects some of the incoming radiation back into space and absorbs some more of the radiation as it passes through. The third reason is the shape of the Earth. Even if the Earth were devoid of an atmosphere, there is only one place on it, as shown in Figure 2.5, that would receive 1370 W/m², and that is the spot where the Sun is directly overhead. At all other places, because of the Earth's curvature, 1370 W would be spread over an area larger than a square meter.

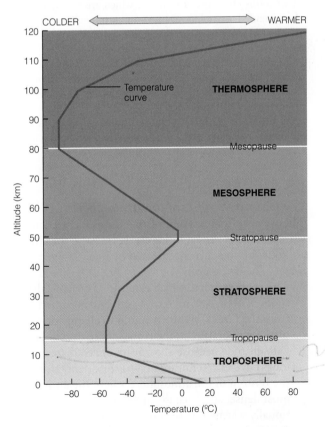

COLDER ⟷ WARMER

THERMOSPHERE

Mesopause

MESOSPHERE

Stratopause

STRATOSPHERE

Tropopause

TROPOSPHERE

Temperature curve

Altitude (km)

Temperature (°C)

Figure 12.5 The variation of temperature with altitude in the atmosphere. The atmosphere is divided into four temperature zones. The outermost zone, the thermosphere, continues to an altitude of about 700 km.

A CLOSER LOOK

Sunlight and the Atmosphere

Many things happen to sunlight when it passes through the atmosphere. Recall from Chapter 3 and specifically from Figure 3.11 that the spectrum of the Sun's radiation measured on the surface of the Earth differs from the spectrum measured in space.

When sunlight enters the atmosphere, four things can happen: it can pass through unchanged and be absorbed by the land or the sea; it can be reflected, unchanged, back into space; it can be scattered by particles in the air; or it can be absorbed by gases in the atmosphere. The last two of these four effects—scattering and absorption—explain why the sunlight spectrum at sea level differs from the spectrum in space (Fig.C12.1).

Scattering

Scattering is the dispersal of radiation in all directions, as Figure C12.2 shows. (It is often called spherical scattering to emphasize that radiation moving in a straight line is scattered equally in all directions.) Radiation comes in from the Sun in a straight line; if some of the radiation is scattered, an observer will obviously notice a reduction in the amount of radiation reaching the surface of the Earth. We are all familiar with this effect because clouds, which are simply masses of suspended particles, scatter sunlight in all directions. When a cloud passes overhead, the intensity of sunlight drops.

Aerosols and gas molecules also cause scattering, but there is an important difference: aerosols scatter all wavelengths of visible light; gas molecules do not. In general, if the diameters of the scattering particles are less than one-tenth the wavelength (λ) of the incoming radiation, as they are in gas molecules, scattering is governed by the Rayleigh Law. The Rayleigh scattering relationship, which was

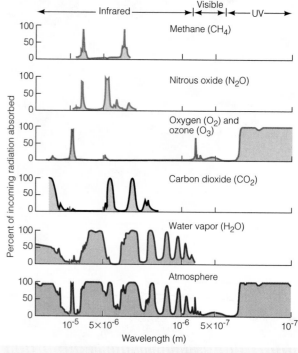

Figure C12.1 Absorption of incoming solar radiation by gases in the atmosphere. The percentage of a given wavelength range absorbed is indicated by the height of the peak, from 0 to 100 percent. The bottom panel is the sum of all the panels above. Note that, except for its longest wavelengths, ultraviolet radiation is almost fully absorbed.

Temperature Profile of the Atmosphere

The way the temperature of the atmosphere changes with altitude is shown in Figure 12.5. Note that there are four thermal layers separated by boundaries called *pauses*. The layers are:

1. The **troposphere**, which extends to variable altitudes of 10 to 16 km (6 to 10 mi) and is named from the Greek, *tropos*, meaning to change or to turn. Temperature in the troposphere decreases with altitude because absorption of reradiated long-wavelength infrared rays is most effective at the bottom of the atmosphere where the air is most dense, and because air at the bottom is continually warmed by the ground and the ocean. The troposphere is so named because it is endlessly convecting, with warm ground-level air rising upward and colder air from above sinking downward to take its place. Most of the weather is a conse-

quence of these thermal motions in the troposphere.

Because more heat per unit area reaches the Earth's surface in the tropics than at the poles, the top of the troposphere, the **tropopause**, is 16 km (6 mi) high at the equator but only 10 km or less at the poles. The tropopause height does not change smoothly from the equator to the poles. Rather, it declines very gently from the equator to a latitude of about 40°; then it decreases sharply to about 10 km and continues near that height to the poles (Fig.12.6). As will be discussed in Chapter 13, this sharp change in incline in the tropopause has important consequences for weather because it gives rise to the phenomenon known as the *jet stream*.

2. The **stratosphere** lies above the tropopause and is a region in which the temperature increases with

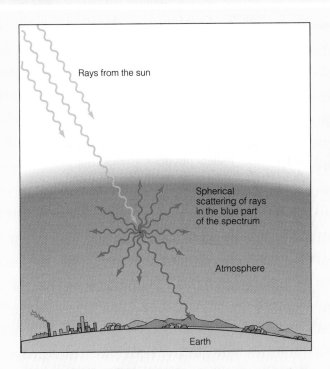

Figure C12.2 Light coming from the Sun is scattered. Air molecules are so small they scatter shorter blue wavelengths more easily than longer red and yellow wavelengths. The scattered blue wavelengths make the sky appear blue in all directions. If there were no scattering by air molecules, the sky would appear pitch black and stars would be visible all day.

discovered in 1881 by an English physicist, Lord Rayleigh, says that the amount of scattering is proportional to $\frac{1}{(\lambda)^4}$. The smaller λ is, the larger $\frac{1}{(\lambda)^4}$ will be, and therefore the greater the scattering will be for the short wavelength end of the spectrum.

In the visible portion of the solar spectrum, the predominant scattering is at the blue end because that is the short-wavelength end (see Chapter 3). The sky appears blue to us because white, unscattered light comes straight through, while blue radiation in sunlight is scattered in all directions. What we see when we look at the sky is this scattered blue radiation. Similarly, because the blue end of the spectrum is reduced in intensity by scattering, the Sun appears a little more yellow to an Earthbound observer than it does to an observer in space.

Absorption

Absorption can be of two kinds. In the first, certain specific wavelengths of solar radiation make atoms or molecules vibrate with the same frequency as the wavelength. In effect, such an atom or molecule absorbs the radiation and then re-emits it at the same wavelength. However, the radiation is re-emitted equally in all directions, and so the observer of the incoming radiation sees a diminution of that wavelength. Most of the absorption by H_2O, CO_2, N_2O and CH_4 shown in Figure C12.1 is of this kind.

In the second type of absorption, molecules absorb the radiation and break apart into atoms or smaller molecules as a result. This is the process by which ozone is formed in the stratosphere. Very-short-wavelength radiation, in the ultraviolet range (λ less than 3×10^{-7}m), is almost entirely absorbed by oxygen and ozone. When the molecules are re-formed, the trapped energy is released at a different and nonvisible wavelength.

altitude, reaching a maximum at about 50 km (30 mi). Strato means layer and is derived from *stratum*, meaning spread out, the same Latin word used to describe layers of sediment. The temperature in the stratosphere increases with altitude because ozone absorbs ultraviolet radiation coming from the Sun. Most of the ozone in the atmosphere is present in the stratosphere, and so that is the layer in which absorption occurs. Absorption converts the energy of ultraviolet rays into longer wavelength radiation, and this longer-wavelength radiation heats the air. The absorption of ultraviolet rays is a maximum at the top of the stratosphere, so this is where the highest temperatures are found. As the Sun's rays pass through the stratosphere, less and less ultraviolet radiation is left to be absorbed, so the lowest temperatures are found at the bottom of the layer.

The upper boundary of stratosphere is the **stratopause**.

3. The **mesosphere** (from the Greek *mesos*, for middle) is a region in which temperature again decreases with increasing altitude, reaching a minimum of about $-100°C$ ($-148°F$) at about 85 km (53 mi). The mesosphere is terminated by the **mesopause**.

4. The **thermosphere**, which reaches out to about 500 km (310 mi), is a region of increasing temperature. The temperature increase in the thermosphere arises partly from the absorption of solar radiation by gases in the atmosphere and partly from the bombardment of gas molecules by protons and electrons given off by the Sun. During periods of strong sunspot activity, when the flux of protons and electrons is at a maximum, the bom-

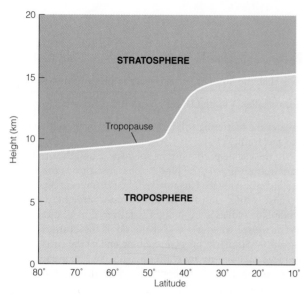

Figure 12.6 Schematic section through the tropopause showing the altitude variation with latitude. As discussed in Chapter 13, the sharp change in altitude in middle latitudes plays a major role in the polar jet stream.

bardment is so great that the temperature at the top of the thermosphere may reach as high as 1500°C (2732°F).

Despite the high temperatures reached in the thermosphere, very few molecules of gas are present and so there is very little heat. Strange as it may seem, we would feel very cold if we were exposed to a 1500°C atmosphere as thin as that in the thermosphere.

One of the most spectacular sights on the Earth, an aurora, occurs in the thermosphere (see Fig.3.12C). When radiation from the Sun is absorbed by molecules of gas in the thermosphere, some of the molecules are broken apart to form electrically charged ions. The region of ionized gases, from 100 to 400 km (160 to 250 mi) in altitude, is called the *ionosphere*. Auroras occur when electrons streaming in from the Sun combine with the ionized gases, form neutral atoms, and give off light rays in the process.

Air pressure

Everyone knows that the higher you go above sea level, the less oxygen there is and the harder it becomes to breathe. That is why planes have emergency oxygen masks in case cabin pressures should fail and why mountain climbers often carry tanks of oxygen.

The oxygen supply becomes short not because of a change in the composition of the atmosphere but rather because of a reduction in the air pressure and therefore a reduction in the air density. Because the size of a human lung is fixed, we get less oxygen when we breathe less dense air. Air at sea level and air at

9000 m (5.6 mi), near the summit of Mount Everest, have the same relative amount of oxygen—20.9 percent by volume in each case. However, a lungful of air at the top of Mount Everest has only 38 percent of the pressure that a lungful of air has at sea level and therefore only 38 percent of the amount of oxygen.

Measuring Air Pressure

Air pressure is measured with a device called a **barometer.** The two kinds of barometers are mercury and aneroid. The mercury barometer was invented in 1644 by the Italian physicist Evangelista Torricelli (1608–1647), who performed the following experiment: he sealed a 1-m-long glass tube at one end and then filled it with mercury. Then, with his finger over the open end in order to prevent the mercury from running out, he inverted the tube and put the open end into a bowl of mercury (Fig.12.7). When he removed his finger, some of the mercury flowed into the bowl, but most of it stayed in the tube. Torricelli reasoned that the air pressing on mercury in the bowl must be holding up the column of mercury in the glass tube.

Scientists were quick to exploit Torricelli's great discovery. Day-to-day measurements soon showed that the height of the mercury column fluctuated

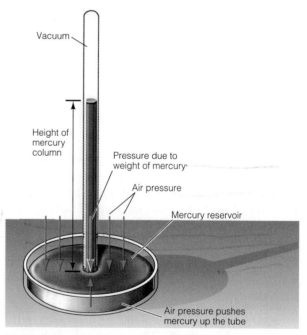

Figure 12.7 Sketch of a simple mercury barometer. Air pressure on the surface of the open bowl holds up the column of mercury in the glass tube. The downward pressure exerted by the air exactly balances the downward pressure exerted by the column of mercury on the bowl. When the air pressure changes, the height of the column adjusts in response.

slightly and therefore that the air pressure must vary from time to time.

Meteorologists still measure air pressure. When TV weather forecasters talk about "highs" and "lows," they are referring to air pressure that is higher or lower than average. Similarly, when a weather-watcher says "the glass" is falling or rising, the reference is to the mercury in a glass barometer that is rising (air pressure is increasing) or falling (air pressure decreasing).

The aneroid barometer (from the Greek *a* and *neros*, meaning no liquid) employs a sealed metal bellows that expands and contracts as the air pressure changes.

Air Pressure Variation with Altitude

Blaise Pascal (1623–1662), a young French scientist, carried out a very important experiment in 1658. He arranged for rock climbers to ascend a prominent volcanic rock in France, Puy-de-Dôme (Fig.12.8), and measure the air pressure at several places during the ascent. Despite the inconvenience of carrying a meter-long glass tube and a flask of mercury up the steep slope of the Puy, the climbers successfully performed the task, and from their measurements Pascal demonstrated that air pressure decreases with altitude.

Today, meteorologists use balloons filled with helium to carry pressure-recording instruments aloft.

Figure 12.9 A radiosonde, a helium-filled balloon used to carry measuring instruments into the upper atmosphere. This large, modern balloon has just been filled with helium and is about to be released. The size of the balloon can be judged from the scientists on the ground.

Such balloons, which are called *radiosondes* (Fig.12.9), can reach an altitude of 30 km (19 mi), at which point they burst and the recording instruments are parachuted back to the ground. To make measurements above 30 km, meteorologists resort to rockets and, most recently, to orbiting weather satellites.

At any given altitude, the air pressure is caused by the weight of air above. The average air pressure at sea level is about 10^5 Pa[1], equivalent to the pressure produced by a column of water about 10 m (33 ft) high. Why doesn't this pressure crush us and the houses we live in? It doesn't do so because air pressure is the same in all directions—up, down, and sideways, inside and outside. Because the outward-pointing air pressure inside a house is exactly the same as the inward-pointing pressure outside, the net pressure on the house is zero.

Figure 12.8 Puy-en-Velay, an ancient volcanic neck in France, where Blaise Pascal proved that air pressure decreases with altitude. Mountaineers carried a barometer up the Puy, making air pressure measurements at several places along the way.

[1] Because the earliest pressure measurements were made with mercury barometers, air pressures are still sometimes reported as the height of a mercury column. A model, or average pressure, called the **standard atmosphere** at sea level, is 760 mm or 29.9 in. of mercury. Another commonly employed unit is the bar, which is a pressure of 1 kg/cm². The standard atmosphere is 1013.25 millibars. In this book SI units are used throughout. The SI pressure unit is the pascal (Pa); the standard atmosphere is 101,325 Pa or 101.325 kilopascals (kPa).

As shown in Figure 12.10, the air pressure decreases smoothly with altitude. The air pressure curve is not a straight line, however, because gases are highly compressible. This means that the air near the ground is compressed by the weight of air above (Fig. 12.11). If air were not compressible (if it were more like water, for example), the pressure-versus-altitude curve would be a straight line.

As a result of the compressibility of air, half of the mass of the atmosphere lies below an altitude of 5.5 km (3.4 mi) and 99 percent lies below 32 km (20 mi). At a height of 32 km, the air is so thin that it is like a laboratory vacuum. The 1 percent of the atmosphere that lies above 32 km continues out to an altitude of about 500 km (310 mi), with the air getting thinner and thinner until it simply merges into the vacuum of space. The few gas atoms present at the outermost fringe of the atmosphere are mostly atoms of hydrogen and helium that have reached the Earth from the Sun.

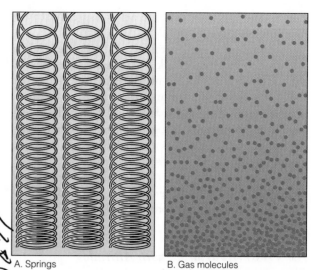

A. Springs   B. Gas molecules

Figure 12.11 Air pressure decreases with altitude because air is compressible and behaves like a pile of springs. A. The springs near the base are compressed by the weight of the springs above. B. Air, like the springs, is compressed by the weight of the air above. Molecules of the gases nearest the ground are squeezed closer together than molecules higher up. Compression is the explanation for the shape of the curve in Figure 12.10.

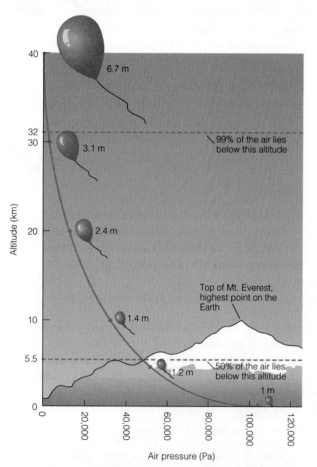

Figure 12.10 Air pressure decreases smoothly with altitude. If a helium balloon 1 m in diameter is released at sea level, it expands as it floats upward because of the pressure decrease. If the balloon did not burst, it would be 6.7 m in diameter at a height of 40 km.

MOISTURE IN THE ATMOSPHERE

The compound H_2O is so familiar that we sometimes forget what an unusual substance it is. In truth, H_2O is the most remarkable substance around, and in no small measure the Earth system works the way it does because H_2O has the properties it does.

One of the most important properties of H_2O is its existence in three physical states at the Earth's surface—as a solid (ice), a liquid (water), and a gas (water vapor). Under some conditions of temperature and pressure, all compounds can form solids, liquids, and gases, but among naturally occurring compounds only H_2O can do so under the conditions that exist at the Earth's surface. To establish why this property of H_2O is so important we first have to explore what happens when H_2O changes from one state to another.

Changes of State

Whenever matter changes from one state to another, energy is either absorbed by it or released (Fig. 12.12). In going from a more ordered state (a solid) to a less ordered one (a liquid) or to a fully disordered one (a gas), energy is absorbed. The reverse process occurs, and heat is released when the change is from a less ordered to a more ordered state. The amount of heat *released* or absorbed per gram during a change of state is known as the **latent heat** (from the Latin *latens*, meaning hidden, hence hidden heat). For example, the latent heat of condensation (less ordered water vapor condensing to more ordered liquid water) is 2260 J/g, while the latent heat of freezing (again less ordered to more ordered) is 330 J/g. The latent heat of evaporation (more ordered liquid water vaporizing to less ordered water vapor) is 2260 J/g, while the latent heat of melting (more ordered solid ice melting to less ordered fluid water) is 330 J/g.

One familiar phenomenon involving a change of state is evaporation. The 2260 J needed to evaporate a gram of water has to come from somewhere. The reason you feel cool after you wet yourself down on a hot day is that some of the heat needed for evaporation is absorbed from your skin and as a result your body temperature drops. Before the invention of ice chests and refrigerators, the best way to keep food cool in hot weather was a "cool safe" in which the evaporation of water kept the temperature low.

The six changes of state shown in Figure 12.12 (the six arrows) all play a role in weather, but evaporation and condensation are far more important than the other changes. Evaporation and condensation play vitally important roles in the weather (1) because they give rise to clouds, fogs, and rain, and (2) because they are the means by which huge amounts of heat are moved from equatorial regions toward the poles. To explore how these phenomena occur, it is necessary to discuss humidity, which, as mentioned previously, is the amount of water vapor in the atmosphere.

Relative Humidity

Water vapor gets into the air by evaporation. Evaporation is a process by which fast-moving liquid molecules manage to escape from the liquid and pass into the vapor above. Because molecules in a vapor move randomly in all directions, some of the gas molecules in the vapor will also move back into the liquid. When the number of molecules that evaporate (going from liquid to gas) equals the number condensing (going from gas to liquid), the vapor is said to be *saturated.* This is the maximum concentration of H_2O molecules in the vapor phase at any specified temperature.

It is common practice to report vapor properties in terms of the vapor pressure. One of the important properties of gases is that pressures in a mixture of gases are additive; this property is known as *Dalton's law of partial pressures*, and it means that the total pressure of a mixture of gases is the sum of the partial pressures exerted by all the individual gases present. Partial pressure, in turn, is a measure of the volume percent of a gas in a mixture. For example, because the content of oxygen in dry air is 20.9 percent by volume, the fraction of the pressure of standard air (101.325 kPa) due to oxygen is 101.325 × 0.209 = 21.2 kPa. The additivity of gas pressures is the reason

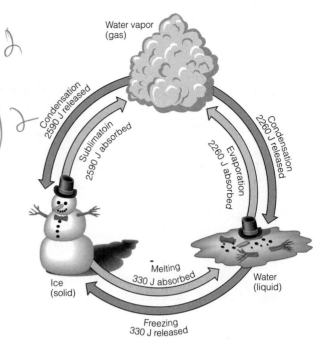

Figure 12.12 Amount of heat added to or released from a gram of H_2O during a change of state.

that the water vapor content of air is just as often reported as a pressure rather than a percentage. The saturation vapor pressure of water at various temperatures is shown in Figure 12.13.

The *saturation vapor pressure*, which is also known as the *water vapor capacity* of air at any given temperature, cannot be exceeded. If the vapor pressure exceeds the capacity, condensation occurs in order to reduce it to the saturation vapor pressure. (**Condensation** is the formation of a more ordered liquid from a less ordered gas.) The vapor pressure can, however, be lower than the saturation value. For example, if saturated air is removed from contact with water and is then heated, the vapor pressure will fall below saturation level and the air will then be undersaturated.

Meteorologists prefer to use the term *relative humidity* rather than undersaturation when they are discussing the amount of water vapor in undersaturated air. The **relative humidity** is the ratio of the vapor pressure in a sample of air to the saturation vapor pressure at the same temperature, expressed as a percentage. For example, saturated air at 20°C has a water vapor pressure of 2.338 kPa. Air at 20°C with a water vapor pressure of 1.403 kPa will therefore have a relative humidity of $\frac{1.403}{2.338} \times 100\%$, or 60 percent.

Note that the relative humidity does not refer to a specific amount of water vapor in the air; it refers to the ratio of what is present at a given temperature to the maximum possible amount that the air could hold at the same temperature. The fact that relative humidity is a ratio sometimes confuses people and leads to

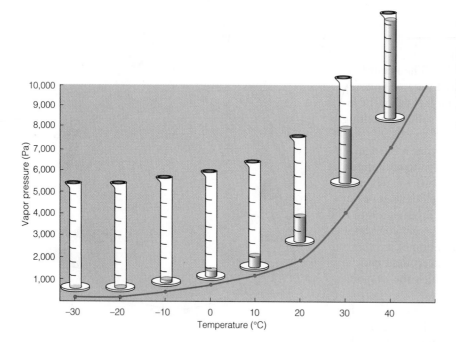

Figure 12.13 Saturation pressure of water vapor as a function of temperature. The amount of H_2O present is shown by the water in a measuring cylinder if all the vapor in a kilogram of saturated air at sea level were condensed and put in the cylinder.

misconceptions. One misconception is that if air feels damp and humid, it must contain more H_2O than air that feels dry. The confusion arises because temperature exerts a strong control on the water vapor capacity of air. For example, desert air at 30°C and relative humidity of 25 percent feels very dry even though it contains 6.62 g of H_2O per kg of air, while air at 10°C and relative humidity of 80 percent, which feels damp and humid, contains only 5.60 g of H_2O per kg of air.

Relative humidity can be changed in two ways—by addition of water vapor or by change of temperature. When the relative humidity is below 100 percent and air is in contact with water, evaporation will raise the relative humidity. This is why air in contact with the ocean usually has a high relative humidity and why so much of the water vapor that enters the atmosphere does so over the oceans (Fig.12.14).

Temperature changes can also affect the relative humidity whether or not H_2O is added. If the amount of water vapor in the air is kept constant and the temperature drops, the relative humidity will rise. The temperature at which the relative humidity reaches 100 percent and condensation starts is called the **dew point**. When the ground is cold and the air warm, the layer of air in contact with the ground may cool sufficiently for the dew point to be reached, so that *dew*, which is a film of water coating the ground, is formed. If the ground temperature is below freezing, frost forms instead of dew. In contrast to condensation, if the temperature rises, the relative humidity drops. People who live in centrally heated houses are very familiar with the problem of declines in relative humidity. In winter, as air is drawn in from outside and heated by a furnace, the relative humidity drops. For example, consider what happens when the outside air

temperature is −10°C (14°F) and the relative humidity is a comfortable 60 percent. If the air is now drawn into the house and heated to 25°C (77°F) without the addition of any water vapor, the relative humidity drops to 6 percent, a level at which many people feel discomfort. To counteract the effects of low humidity, a humidifier must be used to add water vapor to the heated atmosphere.

CONDENSATION AND THE FORMATION OF CLOUDS

If you have ever pumped up a bicycle tire, you will have noticed that the pump becomes hot when the air was compressed. Similarly, if the compressed air in a tire is allowed to escape, the air is noticeably cool as it expands. These two effects, compressional warming and expansional cooling, are examples of what are called **adiabatic processes**, from the Greek *adiabatos*, meaning no passage. Adiabatic processes are so named because they are processes that occur without the addition or subtraction of heat from an external source. When air is compressed, the mechanical energy of pumping is converted to heat and the temperature rises as a result. When compressed air expands the energy required comes from the heat energy of the gas, and consequently a temperature drop follows.

Warm air is less dense than cold air and therefore rises, creating a convection cell in the process. Because it is warmest at ground level, air in the troposphere is continually rising or, after cooling, falling. However, because air pressure decreases with increasing altitude, the rising air expands, and because there

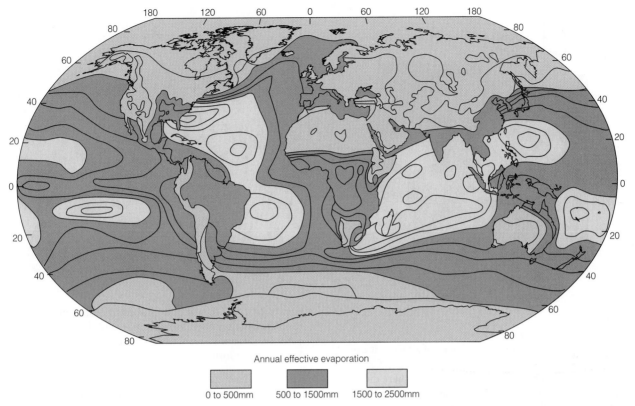

Annual effective evaporation

| 0 to 500mm | 500 to 1500mm | 1500 to 2500mm |

Figure 12.14 The annual addition of water vapor to the atmosphere as a function of geography. The amount evaporated per year is measured in millimeters of water. Note that areas of highest evaporation are over the oceans in equatorial and midlatitudes. Evaporation is low in the deserts because deserts have so little water to evaporate.

is no heat source in the troposphere, the rising air expands adiabatically and so its temperature falls. In the case of sinking air, the reverse happens—the air is compressed and so its temperature rises.

Adiabatic Lapse Rate

When a parcel of unsaturated air rises and expands adiabatically, the temperature drops at a constant rate of 1°C/100 m (1°F/183 ft). Conversely, if cool, unsaturated air sinks toward the Earth's surface, it is compressed and the temperature rises at a rate of 1°C/100 m (Fig. 12.15). The way temperature changes with altitude in rising or falling unsaturated air is called the **dry adiabatic lapse rate**. Eventually, when air rises far enough, it will cool sufficiently to become saturated and condensation will start. The latent heat of condensation that is released as the vapor condenses works against the adiabatic cooling process; in other words, the release of latent heat slows the cooling rate. The greater the amount of latent heat released (the greater the condensation), the less the temperature increase with altitude. The way temperature drops in a rising mass of saturated air is called the **moist adiabatic lapse rate** and it ranges from 0.4° C/100 m (0.4°F/183 ft) m to 0.9°C/100 m (0.9°F/183 ft), with an average of 0.6°C/100 m (0.6°F/183 ft) (Fig. 12.15).

Note that the moist adiabatic lapse rate is always less than the dry rate because of the addition of latent heat to the rising air above the level of condensation. Note, too, that when clouds start to form because a parcel of air has reached saturation and condensation has begun, the dry adiabatic lapse rate changes immediately to the moist rate.

Condensation

When the air becomes saturated with water vapor (that is, the dew point is reached), one of two things happens: either water condenses or, if the temperature is low enough, ice crystals precipitate. The processes seem simple, but in fact they are quite complex. In order for a droplet of water or an ice crystal to form, energy is needed. This process is called *nucleation*, and energy is required because a new surface (the surface of the drop or crystal) is formed. The amount of energy is small if nucleation occurs on a preexisting surface but large if no surface is available. When saturated air is in contact with the ground, for example, the ground itself serves as a nucleation surface and the result, depending on the temperature, is either dew or frost.

When condensation occurs in a rising parcel of air, aerosols provide the nucleation surfaces. Most

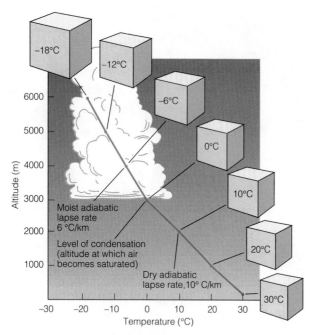

Figure 12.15 As an unsaturated mass of air rises, it expands and cools at the dry adiabatic lapse rate (10°C/km). When the air temperature falls to the point where the air is saturated, condensation commences and latent heat is released. Further altitude increase causes more condensation and the release of more latent heat; the air temperature now decreases at the moist adiabatic lapse rate (6°C/km). Note that the *speed* with which the air rises does not necessarily change; what changes is the temperature drop with altitude. Shown beside the curve is the volume of a mass of rising air that starts as a cube 1 km on an edge.

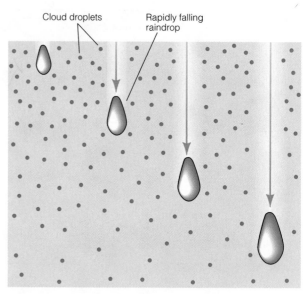

Figure 12.16 Growth of raindrops by coalescence. When droplets fall, they combine with other droplets in their path. Coalescence occurs when clouds are warm enough to keep ice from forming.

aerosols are tiny—no more than 1 micrometer in diameter. When condensation on aerosols commences, it happens very rapidly and water droplets reach a diameter of 20 to 25 micrometers in about a minute. Thereafter, droplet growth rate slows because the remaining water vapor must be spread over billions of droplets as it condenses.

Cloud droplets are so small that air turbulence within the cloud keeps them suspended. A density of about 1000 droplets/cm³ (or 1 drop/mm³) is sufficient to keep the drops apart. When the density of droplets increases above this value, they start to coalesce, and eventually a few drops become too big to remain suspended. As the drops fall, they bump into, and coalesce with, more and more droplets until finally a raindrop has formed (Fig.12.16). A single raindrop contains about 1 million cloud droplets.

When the cloud temperature is below 0°C, the process of precipitation is more complex than simple condensation. Complexity arises because water droplets can be supercooled below 0°C without freezing to ice. The person who first recognized the role of supercooled water in cloud precipitation was a Scandi-

navian scientist, Tor Bergeron, and the process is now named the **Bergeron process** in honor of his discovery. What Bergeron discovered is that ice crystals grow at the expense of supercooled water drops.

Clouds with temperatures between 0°C and −9°C (32°F and 16°F) contain only supercooled water droplets. When the temperature is between −10°C and −20°C (14°F and −4°F), ice crystals also nucleate so that the cloud becomes a mixture of supercooled water drops and ice crystals. Below −20°C, water drops disappear and clouds contain only ice crystals. As air cools and a cloud grows, therefore, the mixture of supercooled water droplets and ice crystals becomes increasingly dominated by ice crystals. Bergeron discovered that, in a mixture of supercooled water droplets and ice crystals, the water droplets slowly evaporate and release water vapor that is then deposited on the ice crystals, making them grow larger. Eventually, the ice crystals become so large that they start to fall (Fig. 12.17). If the temperature all the way to the ground is everywhere below 0°C, the result is snowflakes. If the temperature near the ground is above 0°C, the ice crystals melt and rain drops hit the ground. (In other words, most rain starts out as ice crystals.) If raindrops fall through a layer of air near the ground where the temperature is below 0°C, the drops freeze and *sleet*—frozen raindrops—is the result.

Clouds

Clouds form when air rises and becomes saturated in response to adiabatic cooling. There are four principal

Figure 12.17 Growth of ice crystals by the Bergeron process. Supercooled water droplets evaporate, and ice crystals grow by incorporating the newly formed water vapor.

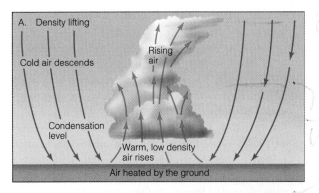

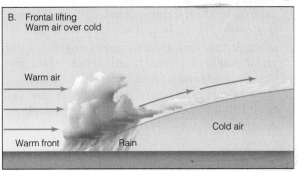

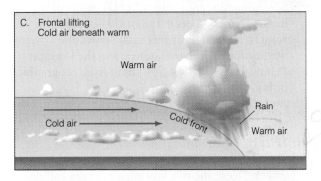

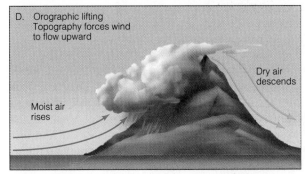

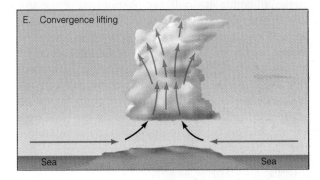

reasons for the upward movement of air. While it is possible to separate the reasons for discussion, most individual circumstances involve more than one lifting force. The four lifting forces are:

1. Density lifting, which occurs when warm, low-density air rises convectively and displaces cooler, dense air (Fig. 12.18A).

Figure 12.18 The lifting forces that lead to cloud formation. Note that under most circumstances two or more lifting forces operate at the same time. A. Density lifting causes a convection cell as warm, low-density air rises and cold, higher-density air sinks. B. Frontal lifting. A warm front occurs when flowing warm air overrides cold air and is forced upward. C. Frontal lifting. A cold front occurs when a wedge of forward-moving cold air slides under warm air and forces it upward. D. Orographic lifting occurs when flowing air is forced upward by mountains or other sloping ground. E. Convergence lifting occurs when masses of air collide and are forced upward.

2. Frontal lifting, which occurs when two flowing air masses of different density meet. The boundaries between air masses of different temperature and humidity, and therefore different density, are called fronts. The boundaries are between 10 and 150 km (6 and 93 mi) in width and mark the advance of one air mass into another. The name "front" is used because Norwegian meteorologists of World War I likened the clash of two air masses to battle lines, or fronts, between armies. When warm, humid air advances over cold air (an advancing **warm front**), the warm air rises up and over the cold air, forming clouds and possibly rain as a result (Fig. 12.18B). A similar process occurs when denser, cold, air flows in and displaces warm air by pushing it upward (a **cold front**), again producing clouds and possibly rain (Fig. 12.18C). When a cold front overtakes a warm front and two cooler air masses meet, the result is an **occluded front**.

3. Orographic lifting occurs when flowing air is forced upward as a result of a sloping terrain, such as a mountain range (Fig. 12.18D). Some of the highest rainfall spots in the world—such as the western coast of Tasmania in Australia, the Owen Stanley Range in New Guinea, and the Olympic Peninsula in Washington—result from orographic lifting.

4. Convergence lifting occurs when flowing air masses converge and are forced upward (Fig. 12.18E). The Florida peninsula provides an example; air flows landward off the ocean from both east and west; the two flowing air masses collide and force some of the air to rise. Clouds therefore form, and the result is the familiar frequent afternoon thunderstorms.

Clouds are visible aggregations of minute water droplets, tiny ice crystals, or both. They are such prominent and beautiful features of the sky that poets from time immemorial have written about them, painters beyond number have painted them, and meteorologists spend a great deal of time studying them.

Because clouds form by condensation of water vapor, all common clouds are phenomena of the troposphere. They are classified on the basis of shape, appearance, and height into three families: cumulus, stratus, and cirrus (Table 12.1, Figs. 12.19 and 12.20).

Cumulus clouds are puffy, globular, individual clouds that form when hot, humid air rises convectively, reaches a level of condensation where cloud formation starts, but continues to rise. These are the flat-based cauliflower-shaped clouds that children like to draw—the flat base marks the level of condensation. When

cumulus clouds coalesce to form a puffy layer, the term *stratocumulus* is applied.

When large cumulus clouds rise to the top of the troposphere, they expand horizontally and form *cumulonimbus* clouds. These are the familiar thunderstorm clouds or "thunderheads" of summer. There is a great deal of energy and turbulence in a cumulonimbus cloud, and some of the energy causes thunder and lightning within a cloud, between adjacent clouds, and between clouds and the ground.

Cumulus clouds that form at altitudes between 2 and 6 km (1 and 4 mi) are given the modifying name *altocumulus* clouds; those that form between 6 and 15 km (4 and 10 mi) are called *cirrocumulus* clouds. Frequently, altocumulus and stratocumulus clouds are arranged in regular rows or clumps separated by clear sky. Convection cells up to several hundred meters across give rise to the patterns.

Stratus clouds are sheets of cloud cover that form at altitudes from 2 km to about 15 km (1 to 10 mi) and cover the entire sky. Stratus clouds form when air rises as a result of frontal lifting, reaches its level of condensation, and then spreads laterally but not vertically. If the cloud blanket is several kilometers thick, the day is dark and dreary and the cloud is called *nimbostratus*.

Depending principally on the altitude, stratus clouds, analogous to cumulus clouds, are given modifying names; between 6 and 8 kilometers (4 and 5 mi), *altostratus*; between 8 and 12 km (5 and 8 mi), *cirrostratus*.

Cirrus clouds are the highest of the clouds in the troposphere. Looking like fine, wispy filaments, or feathers, cirrus clouds form only above 6 km (4 mi) in altitude and are composed entirely of ice crystals.

Two rare kinds of cloud are known to form in the stratosphere. *Nacreous clouds* are beautiful translucent sheets (meaning light gets through, like a frosted window pane) of minute ice crystals that form at altitudes between 20 and 30 km (12 and 19 mi). Even less common are *noctilucent clouds*, which are so thin they look like gossamer veils. They are composed entirely of minute ice crystals and form at altitudes as high as 90 km (56 mi). No clouds have ever been reported above 90 km (56 mi), and none are likely because water vapor is too scarce to form clouds at such altitudes.

A HABITABLE PLANET

We learned in the opening essay of this chapter that oxygen must be present within well-defined limits in an atmosphere if a planet is to be habitable. The reason that no other planet or moon in the solar system is habitable is the absence of oxygen.

Table 12.1   Classification of Clouds in the Troposphere by Altitude

Height	Name	Shape and Appearance
High-level clouds Cloud base 6 to 15 km above sea level	Cirrus	Feathery streaks
	Cirrocumulus	Small ripples and delicate puffs
	Cirrostratus	Translucent to transparent sheet, like a veil across the sky
Middle-level clouds Cloud base 2 to 6 km above sea level	Altocumulus	White to dark gray puffs and elongate ripples
	Altostratus	Uniform white to gray sheet covering the sky
Low-level clouds Cloud base below 2 km above sea level	Stratus	Uniform dull gray cover over the sky
	Nimbostratus	Uniform gray cover, rain generally falling
	Stratocumulus	Patches of soft gray; in places patches coalescing to a layer
Clouds with great vertical development Cloud base below 3 km above sea level	Cumulus	Puffy cauliflower shape with flat base
	Cumulonimbus	Large, puffy; white, gray and black; great vertical extent, often with anvil-shaped head

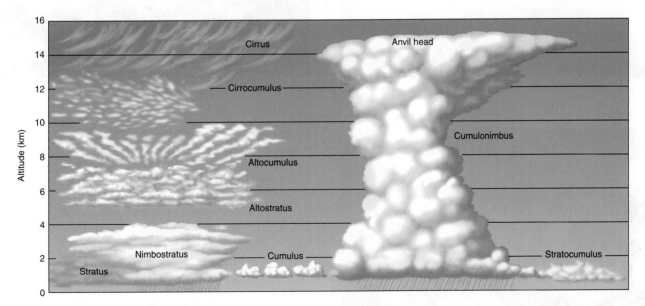

Figure 12.19   The altitudes of clouds. An anvil head is the flattened top of a cumulonimbus cloud that spreads across the top of the troposphere.

In addition to the presence of oxygen in the atmosphere, two other essential criteria make a planet habitable: water vapor must be present, and the temperature must be neither too high nor too low.

Humans cannot live for long where the air is completely dry because water vapor is needed for our lungs to work properly. A habitable planet must therefore have a hydrosphere so that there is some way of getting water vapor into the atmosphere.

Where temperature is concerned, the limits of a habitable planet are made clear by the way we humans live. The parts of the body that must be protected from temperature fluctuation are the core organs—the brain, heart, lung, liver and digestive system. Although the temperature of the skin can vary widely with no harmful consequences, the temperature of the core organs, which is 37°C (98.6°F), cannot vary safely by more than ± 2°C (3.6°F). To maintain a stable core-organ temperature, the body has two cooling mechanisms—perspiration and dilation of blood vessels in the skin, both of which cool because they get rid of heat—and two heating mechanisms—shivering and contraction of blood vessels in the skin. In addition to these natural mechanisms, humans can put on clothes and take other measures to avoid exposure to extreme temperatures.

The most effective way to handle the body's temperature requirements, of course, is to live where the annual mean temperature is comfortable. Over 90 percent of the global population lives in places where the annual mean temperature is between 6°C (43°F) and 27°C (81°F). A habitable planet must therefore have regions that fall in this temperature range. Much of the Earth enjoys annual mean temperatures in the equitable 6°C to 27°C range largely because of the atmosphere, which acts as a warming blanket.

A.

B.

C.

Figure 12.20 Principal types of clouds. A. Cumulus
B. Cumulonimbus. Note the plume spreading sideways.
C. Altocumulus D. Cirrostratus E. Stratocumulus F. Stratus
G. Altostratus H. Nimbostratus I. Cirrus.

D.

E.

F.

G.

H.

I.

SUMMARY

1. Weather is the state of the atmosphere at a given time and place; the variables that define weather are temperature, air pressure, humidity, cloudiness, and wind speed and direction.

2. Climate is the average weather condition at a given place. In order to determine the climate, the weather must be averaged over a period of years.

3. Two energy sources drive the atmosphere: the Sun's heat and the Earth's rotation.

4. The relative composition of the atmosphere is constant on a dry and aerosol-free basis. Both the humidity and the amount of aerosols vary from place to place and time to time, but the gases of the atmosphere have fixed relative proportions.

5. The three major gases, nitrogen, oxygen, and argon, account for 99.96 percent of dry, aerosol-free air.

6. The minor gases, water vapor, carbon dioxide, methane, nitrous oxide and ozone, are important for their roles in trapping the Sun's heat and shielding the Earth from ultraviolet rays.

7. The temperature profile of the atmosphere is highly structured. In the troposphere, to an altitude between 10 and 16 km, temperature decreases with altitude. In the stratosphere, 16 to 50 km, the temperature increases with altitude; temperature then decreases with altitude in the mesosphere (50 to 85 km), and finally increases again with altitude in the thermosphere (85 to 500 km).

8. The atmosphere extends out to an altitude of about 500 km, at which point it blends into the vacuum of space.

9. Air pressure decreases with altitude, but because air is highly compressible, the decrease is not linear. Fifty percent of the atmosphere lies below an altitude of 5.5 km and 99 percent below 32 km.

10. The troposphere is where most of the Earth's weather is generated, where clouds form, and where rain and snow develop.

11. When H_2O changes from one state to another, heat is absorbed or released. The most important changes of state as far as the weather is concerned are condensation, precipitation, and evaporation. Condensation and precipitation release latent heat. Evaporation absorbs heat.

12. The amount of water vapor in the atmosphere cannot exceed the saturation capacity. Air that has reached the saturation capacity has 100 percent relative humidity.

13. In air that contains less water vapor than the saturation value, the water vapor content is measured by the relative humidity (the ratio of the vapor in a given sample of air to the saturation vapor pressure at the same temperature, expressed as a percentage).

14. Rising air cools adiabatically, which means without losing or gaining heat energy. When adiabatically cooled air becomes saturated with water vapor, condensation commences and clouds start to form.

15. Rising or falling air masses cool or warm, respectively, as a result of adiabatic expansion or compression. If the air is unsaturated, the temperature changes at the adiabatic lapse rate; if unsaturated air becomes saturated so that latent heat is added, the temperature changes at the moist adiabatic lapse rate.

16. Aerosols serve as the nuclei on which water droplets and ice crystals nucleate.

17. In clouds that contain a mixture of supercooled water droplets and ice crystals, the water droplets evaporate and the ice crystals grow larger. This is known as the Bergeron process.

18. Droplets of water coalesce in warm clouds, and when the coalesced drops are large enough, they fall as rain. Most clouds are so cold that ice particles eventually predominate over water droplets. As ice particles fall into warmer air below the clouds, they melt and reach the ground as rain.

19. Clouds form when air rises adiabatically, becomes saturated, and condensation commences as a result of four kinds of lifting forces: density lifting, frontal lifting, orographic lifting, and convergence.

20. Clouds are classified by shape, appearance, and height. There are three cloud families, based on shape: cumulus, stratus, and cirrus clouds. Modifying prefixes are used to designate the altitude of the clouds; for example, cirrocumulus and cirrostratus are high-altitude cumulus and stratus clouds, respectively.

21. Two more kinds of clouds form in the stratosphere, nacreous and noctilucent clouds.

IMPORTANT TERMS TO REMEMBER

QUESTIONS FOR REVIEW

1. What are the five variables used to define weather and climate?

2. How does weather differ from climate, and how is climate measured from weather variables?

3. What are the sources of energy that drive activities in the atmosphere?

4. Mars and Venus have atmospheres, but are their atmospheres air? Why or why not? *no not*

5. The main ingredient of air is nitrogen. What is the second most common ingredient? What percentage of the air is made up by this second most abundant ingredient?

6. What is the difference between humidity and relative humidity?

7. Why is it possible for dry air in a desert to contain more water vapor than moist air in the Arctic?

8. Which minor gases in the atmosphere are known as the greenhouse gases, and why?

9. Ultraviolet radiation from the Sun is lethal to many forms of life, including humans. Explain how it is possible for humans to live on the Earth's surface without being harmed by these lethal rays.

10. Explain why, since the composition of the air doesn't change, there is less oxygen to breathe at the top of Mount Everest than at sea level.

11. The way air pressure decreases with increasing altitude is nonlinear. Explain why this is so.

12. How does heat differ from temperature?

13. The energy flux from the Sun that reaches the outer edge of the atmosphere is 1370 W/m²; the energy flux that reaches the Earth's surface (the insolation) is considerably less than 1370 W/m². Cite three effects that cause the reduction.

14. Name the four temperature regions of the atmosphere in order of altitude, starting with the lowest region. Give the approximate altitudes where one temperature zone passes into another.

15. Why does temperature decrease with altitude in the troposphere but increase with altitude in the stratosphere?

16. What is latent heat? Give two examples of a change of state in which latent heat is released. Give two examples of a change in which latent heat is absorbed.

17. When air is rapidly compressed, as in a bicycle pump, it becomes heated. Explain why this is so. What is the name given to processes such as the heating or cooling of a gas as a result of compression or expansion?

18. When a mass of air rises, the rate at which temperature changes with altitude above cloud level (the level of condensation) is different from that rate below cloud level. Explain why this is so.

19. What is the Bergeron process, and what role does it play in the formation of raindrops?

20. Describe four ways by which a mass of air can be lifted. Why are these lifting processes important for the formation of clouds?

21. What is the difference between a cold front and a warm front? Rain is commonly associated with both kinds of front. Why should that be so?

22. Name the three major families of clouds and describe their general differences.

Questions for A Closer Look

1. Why does scattering reduce the amount of incoming solar radiation that reaches the Earth's surface?

2. Why is the sky blue?

3. Which of the gases in the atmosphere removes the most wavelengths from the solar radiation?

QUESTIONS FOR DISCUSSION

1. When Venus is viewed through a telescope, all that can be seen is a thick cover of clouds. We know that Venus does not have a hydrosphere. What explanations can you offer for the formation of Venusian clouds?

2. Would you expect the Earth's cloud cover 25,000 years ago, during the most recent ice age, to have been more extensive, less extensive, or about the same as it is today? Did precipitation during the ice age have to be different from what it is today, or could precipitation have been the same and temperature the only thing that changed?

3. When atom bombs are exploded, a huge cloud rises into the atmosphere. At some height in the troposphere, the cloud starts to spread sideways and takes on a mushroom shape. The shape is similar to that of a great thunderhead. Why do thunderheads and atomic clouds spread sideways?

4. Air pressure is usually the same in all directions. However, during the passage of a tornado or hurricane all the windows in a house may break if they are closed. Why might this be so?

CHAPTER 13

Winds, the Weather, and Deserts

● *Harvesting the Wind*

A hard-working human can generate just enough power to keep a 75-watt light bulb burning. Since 75 watts is only one-tenth of a horsepower, it is hardly surprising that our ancestors sought other sources of power. They solved the problem by harnessing draft animals and by "harvesting" the wind. Long before written history, animals were being harnessed to plows, and by 8000 years ago wind-filled sails were pushing boats down rivers and around the shores of the eastern Mediterranean Sea. Many millennia were to pass, however, before the next step in the use of wind power. When the step came, it was a giant one: the invention of windmills provided a way for wind energy to drive machines.

The earliest written accounts of windmills, dating to about the middle of the seventh century, record their use for grinding grain in Persia (today's Iran). Knowledge of windmills appears to have reached Europe in about A.D. 1100, and from that time onward there is a reliable record of their use. By the nineteenth century, Dutch windmills had reached such a peak of technological excellence that individual mills could generate power at a rate of 30,000 watts.

The availability of inexpensive gasoline and the spread of electricity into rural communities sent windmills into decline from the 1930s onward. However, when the price of petroleum started to rise in the 1970s, attention returned to wind power. This time the attention was directed principally to the use of windmills to generate electricity. The first windmill designed specifically to produce electricity is believed to have been built in Denmark about 1895. In the United States, a 1,250,000-watt wind-driven turbine was installed at Grandpa's Knob in Vermont in the 1940s; this system eventually failed when one of its two propeller blades broke. Now, in the 1990s, the approach is to place hundreds of smaller turbines in wind farms located where the wind blows for long periods at high speeds. Large banks of wind-driven turbines have been installed in California at Altamont Pass and San Gorgonio Pass, in New Hampshire, in Denmark, and elsewhere.

What is the likely future for wind power? Three problems are obvious: first, winds blow only intermittently and so wind-driven generators need to be used in conjunction with other power sources. The second problem is location. Electricity is needed in cities, and land near cities is expensive. Also no one wants to live close to a wind farm because wind-driven turbines are noisy; wind farms therefore have to be in isolated places. The third problem concerns birds, especially birds during migration, which are killed by the fast-moving blades. Despite all the problems, many experts believe wind power has a great future. Many locations for wind farms do exist (offshore, for example), their efficiency is growing rapidly, and all developed countries have electrical networks that can hook wind farms with other generators of electricity. Some optimistic experts foresee a future in which as much as 10 percent of the world's electricity might be supplied by the wind. ●

(opposite) A wind farm at San Gorgonio Pass, near Palm Springs, California. Each windmill drives an electrical generator.

WHY AIR MOVES

Wind is horizontal air movement arising from differences in air pressure. Nature always moves to eliminate a pressure difference, and wind is the result when air flows from a place of high pressure to one of low pressure. Since air pressure is related to density—high pressure means the air is more compressed and therefore more dense, low pressure means less compression and lower density—horizontal movement is always associated to some degree with vertical movement.

Wind Speed

When discussing wind speed, it is necessary to separate momentary gusts from steady flow. For that reason, wind speed measurements are averaged over a specific period, commonly five minutes.

A wind speed of 20 km/h[1] (12 mi/h) is a pleasant breeze that rustles and moves all the leaves in a tree

[1]Because wind was the power source for sailors for so many centuries, wind speeds are sometimes still reported in nautical units. We do not follow the practice in this book, but if the weather reports in your newspaper or on the TV use nautical units, here is how to convert them: The unit of distance at sea is the nautical mile, and the unit of speed is the knot; a knot is 1 nautical mile/hour. A nautical mile is equal to 1.1508 land miles. To convert knots to miles/h, multiply by 1.1508; to convert knots to km/h, multiply by 1.852. Thus, 30 knots is 34.5 mile/h or 55.6 km/h.

and produces small, white-capped waves on a lake. At 45 km/h (28 mi/h), all the branches in a tree start to sway and spray forms on open water. When wind speeds reach 65 km/h (40 mi/h), twigs and small branches break off trees, waves are high, and foam forms on wave crests. At 90 km/h (56 mi/h), trees are uprooted and the wind will knock you down; at 180 km/h (112 mi/h), it can pick you up. Fortunately, such high-speed winds are very rare. When they do occur, they tend to be associated with tornadoes or hurricanes, so that the damage is localized.

The greatest wind speed ever recorded on the surface of the Earth is 372 km/h (231 mi/h) on Mount Washington, New Hampshire, in April 1934. Speeds of 325 km/h (202 mi/h) have been recorded in hurricanes, and winds up to 335 km/h (208 mi/h) have been reported during severe storms in Greenland. Such high-speed winds can do remarkable things (Fig.13.1). High-speed tornado winds, for example, have been reported to kill chickens and geese and also to pluck off all their feathers.

Most places around the world have wind speeds that average between 10 and 30 km/h (6 and 19 mi/h). The windiest place on the Earth is at Cape Dennison, in Antarctica, where the average speed is 70 km/h (43 mi/h). Mount Washington, winner of the crown for the highest speed, averages only 55 km/h (34 mi/h) year round.

Figure 13.1 All that remains of a town in Florida after Hurricane Andrew passed through in 1992. All the damage seen here was caused by high-speed winds.

The Windchill Factor

In places where temperatures drop below freezing, it has become customary for weather forecasters to report a **windchill factor**. This variable measures the heat loss from exposed skin as a result of the combined effects of low temperature and wind speed. Here's how it's calculated.

Immediately adjacent to the human body (and also to any other solid surface) is a thin layer of still air called the *boundary layer* (Fig.13.2). This layer is still because friction prevents movement. Heat escaping from the body must pass through the boundary layer by conduction. Because air is a poor conductor, the boundary layer serves as an effective insulator. The key to the windchill factor is that as wind speed increases, the thickness of the boundary layer decreases, thereby reducing its effectiveness as an insulator and increasing the rate at which heat is lost from the body.

Note that the air temperature does not drop as a result of the wind. What happens is that the higher the wind speed, the faster heat is lost from the skin and therefore the faster the skin temperature approaches the temperature of the air. If the skin reaches freezing temperature, frostbite ensues.

The windchill factor should correctly be called the windchill equivalent temperature because, for a given air temperature and given wind speed, the windchill factor is the air temperature at which exposed parts of the body would lose heat at the same rate if there were no wind. For example, if the air temperature is $-3°C$ (about $27°F$) and the wind speed is a sprightly 32 km/h (20 mi/h), then the windchill equivalent temperature is $-18°C$ (equal to $0°F$). Windchill is not the only factor that affects our comfort in cold weather, but it is certainly one of the most important as far as safety is concerned.

Factors Affecting Wind Speed and Direction

If the Earth did not rotate, wind would blow in a straight line; if the Earth had a frictionless surface, the wind would flow longer and harder than it does. Neither of these ifs applies, and wind is therefore controlled by the following factors:

1. The **air pressure gradient**, which is the drop in air pressure/unit distance.

2. The **Coriolis effect**, which, as discussed in Chapter 11, is the deviation from a straight line in the path of a moving body due to the Earth's rotation.

3. **Friction**, which is the resistance to movement when two bodies are in contact.

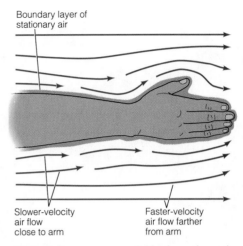

Boundary layer of
stationary air

Slower-velocity
air flow
close to arm

Faster-velocity
air flow farther
from arm

Figure 13.2 Adjacent to any solid body, such as a human arm, there is a thin layer of air held stationary by friction. Away from the body, wind speed, indicated by the length of the arrows, increases as the effects of friction become weaker and weaker.

Air Pressure Gradient

A pressure gradient is determined from the isobars on a weather map. **Isobars** are lines on a map connecting places of equal air pressure[2] (Fig.13.3), and the spacing between isobars determines the air pressure gradient. When isobars are close together, the gradient is steep; when they are far apart the gradient is low. When isobars are close together, air flows rapidly down the pressure gradient and a high-speed wind is the result (Fig.13.4).

As was mentioned earlier in this chapter, air pressure differences develop both horizontally and vertically in the atmosphere. In order to assess the effects caused by horizontal pressure gradients, vertical pressure differences must be avoided, and doing that requires measurements at constant altitude. Mean sea level is the elevation generally chosen. Since air pressure decreases with altitude, a vertical pressure gradient is everywhere present. In order to draw sea-level isobars on maps, air pressure measurements made at elevations higher than sea level must be corrected to the sea-level value. Once the corrections are made, isobars are drawn, generally at a contour interval of 0.4 kPa.

Coriolis Effect

The Coriolis effect, as discussed in Chapter 11 in the section on ocean currents, influences all moving bodies. Because wind is moving air, the directions of all

---

[2]Although the SI pressure unit is the pascal, weather maps usually report air pressure in an older unit, the millibar (Mb). A millibar is a pressure of 0.001 kg.cm². To convert Mb to kilopascals (kPa), divide the Mb by 10.

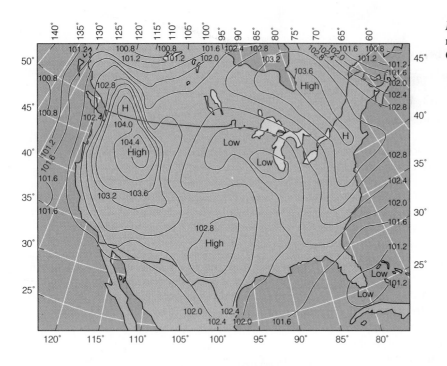

Figure 13.3 A typical air pressure map of the United States and part of Canada for November 26, 1993.

winds are subject to the Coriolis effect—that is, a deflection toward the right in the northern hemisphere and to the left in the southern hemisphere (see Fig.11.8).

The speed of a moving object influences the magnitude of the Coriolis effect because a fast-moving object covers a greater distance in a given time than a slow-moving object. The longer the trajectory, the greater the change in angular velocity and therefore the greater the Coriolis deflection. Where air flow is concerned, the Coriolis effect is of greatest importance in large-scale wind systems such as the tradewinds so loved by mariners, but of only minor importance in small-scale, local wind systems such as thunderstorms.

Friction

When wind blows across the ground, through trees, or over solid objects of any kind, friction slows its speed. Remember that the magnitude of the Coriolis effect is proportional to the speed of a moving body. A *reduction* in speed due to friction will therefore reduce the Coriolis deflection, in effect causing northern hemisphere winds to turn a little to the *left* and southern hemisphere winds a little to the *right*.

Friction is important for small-scale air motions and for the layer of air in contact with the surface, which means for winds within 1 km (0.6 mi) of the Earth's surface. It is much less important for winds higher than 1 km above the Earth's surface, such as the winds of the jet stream.

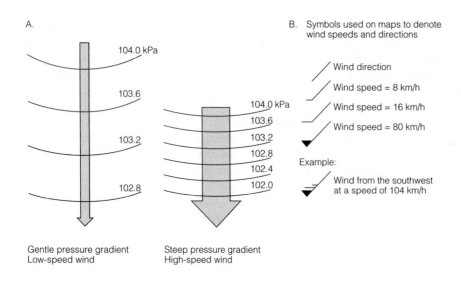

Figure 13.4 Winds and pressure gradients. A. Widely spaced isobars indicate a slow pressure drop over a long distance and thus a low-pressure gradient and low-speed winds. Closely spaced isobars indicate a steep pressure gradient; high-speed winds are the result. B. Symbols used on weather maps to indicate wind direction and speed. The orientation of the stem indicates wind direction, and the barbs indicate speed. If more than one barb is on the wind stem, add the barbs together to get the wind speed.

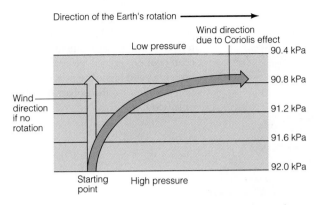

Figure 13.5 A geostrophic wind. A high-altitude wind is deflected by the Coriolis effect until a balance is reached between the direction of flow due to the pressure gradient and the direction due to the Coriolis deflection, at which point flow is parallel to the isobars.

Combinations of Factors

Winds are always subject to more than one factor. Consider the least complicated example, a high-altitude wind that is not in contact with the ground and therefore is not affected by friction. Such a wind starts to flow because an air pressure gradient exists, and the direction of flow is down the gradient perpendicular to the isobars—the steeper the gradient the greater the wind speed. Once flow starts, the Coriolis effect becomes important, and so the flowing air is deflected. Deflection means that the wind direction is no longer perpendicular to the isobars; instead, the wind crosses them at an oblique angle (Fig.13.5). Eventu-

ally, when the pressure-gradient flow and the Coriolis deflection are in balance, the wind flows parallel to the isobars; in the northern hemisphere, the low-pressure air is to the left and the high-pressure air is to the right.

Geostrophic Winds

Winds that result from a balance between pressure-gradient flow and the Coriolis deflection are called **geostrophic winds**. (Recall from Chapter 11, especially Figures 11.11A and 11.13, that geostrophic flow also occurs in the ocean.) The daily weather maps published by the National Weather Service reveal that geostrophic winds are almost always blowing in the upper part of the troposphere. Figure 13.6 is a map of North America at a height above sea level of 5.5 km (3.4 mi), well above any frictional effects. The contours are of pressures in millibars. The highest contour, 588 Mb, is on the right-hand side of the diagram at the center of a high-pressure air mass; the lowest contour, 480 Mb, is at the top of the diagram, associated with a low-pressure zone over the Canadian Arctic islands. Note that wind directions are parallel (or nearly so) to the isobars.

Within 1 km (0.6 mi) of the Earth's surface, friction complicates air flow by upsetting the balance between pressure-gradient flow and Coriolis deflection. Friction slows the wind and thereby reduces the Coriolis deflection. Now a balance must be reached between pressure-gradient flow, Coriolis deflection, and frictional slowing. As a result, winds near the surface flow at oblique angles to the isobars. The angle size is a

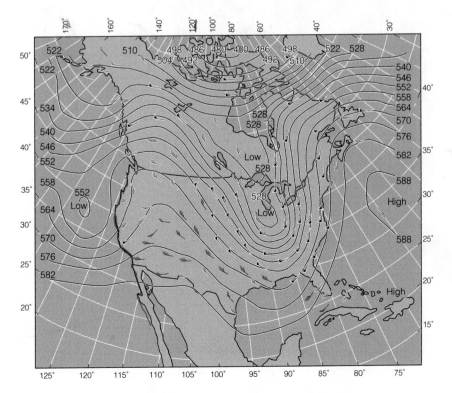

Figure 13.6 Map of North America showing upper-atmosphere winds at 7.00 A.M., November 28, 1993. The lines represent the pressure contours in millibars, at a height above sealevel of 5.5 km. Note that winds are nearly all parallel to the isobars and therefore are geostrophic. Map compiled by National Weather Service.

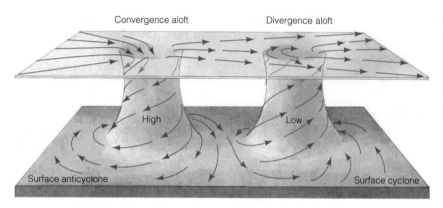

Convergence aloft   Divergence aloft

High   Low

Surface anticyclone   Surface cyclone

Figure 13.7 Air spirals into a low and out from a high. Lows are centers of convergence, while highs are centers of divergence. Note that in both lows and highs the flow direction is oblique to the isobars because of friction.

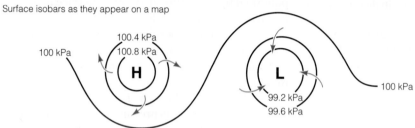

Surface isobars as they appear on a map

100 kPa   100.4 kPa   100.8 kPa   **H**

L   99.2 kPa   99.6 kPa   100 kPa

function of the roughness of the terrain. If the surface is very rough and the friction effect large, the angle between the air flow and the isobars can be as great as 50°. If the surface is smooth, such as the surface of the sea, the angle will be closer to 10 or 20°.

Convergent and Divergent Flow

As air near the ground flows inward from all directions toward a low-pressure center, frictional drag causes the flow direction to be across the isobars at an oblique angle. As a consequence, winds around a low-pressure center develop an *inward* spiral motion

Figure 13.8 A low-pressure center (cyclone) centered over Ireland and moving eastward over Europe. The counterclockwise winds of a northern hemisphere low are clearly shown by the spiral cloud pattern.

A.

B.

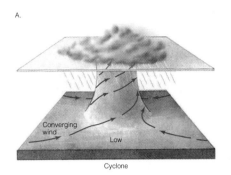
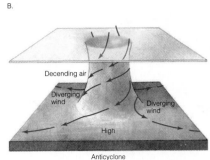

Cyclone

Anticyclone

Figure 13.9 Convergence in a cyclone causes a rising updraft of air and with it clouds and probably precipitation. Divergence in an anticyclone draws in high-altitude air, creating a downdraft; clear skies and fair weather are the result.

(Fig.13.7). By the same process, air flow spirals *outward* from a high-pressure area. In the northern hemisphere, the inward-flowing low pressure spirals rotate counterclockwise, and the high-pressure spirals rotate clockwise. In the southern hemisphere, the reverse is true.

Spiral flow was first explained by a Swedish scientist, Valfrid Ekman (1867–1954), and for this reason the spirals are sometimes called *Ekman spirals*. Ekman actually explained the spirals from his study of oceanography, as mentioned in Chapter 11, but the phenomenon, which arises from three counterbalancing effects—the pressure-gradient flow, Coriolis effect, and friction—is the same in the atmosphere as it is in the ocean.

The spiral pattern of air flow in the lower atmosphere can be seen almost daily on the weather map and is dramatically seen in the satellite images of cloud patterns shown on TV (Fig.13.8). Air spiraling inward around a low-pressure center, which is designated **L** for **Low** on the weather map, is called a **cyclone**. Air spiraling outward, away from a high pressure center designated **H** for **High** on the map, is called an **anticyclone**.

The inward spiral flow in a cyclone causes **convergence**, which leads to an upward flow of air at the center of the low. This upward flow leads to cloud cover and rain (Fig.13.9A). The outward spiral flow in an anticyclone causes **divergence**, which leads to an outward flow of air from the center. This outward flow means that a high must draw high-altitude air downward into the center (Fig.13.9B). Remember from Chapter 12 that cold air drawn downward is compressed and heated adiabatically, thus dropping the relative humidity and leading to clear, cloudless skies.

Lows tend to be associated with cloudy, unsettled weather, and highs with clear, dry weather. This is why weather forecasters always emphasize the location and movement of high- and low-pressure zones. It also explains, why barometers are good predictors of weather changes. When the barometer is falling,

the air pressure is dropping and a low is approaching, so cloudy weather can be forecast. When the barometer is rising, air pressure is increasing because a high is approaching, and dry, sunny weather is on the way.

GLOBAL AIR CIRCULATION

Mariners have long known about and used global-scale wind systems. Early Polynesian navigators, for example, were intrepid sailors who used the northeast and southeast tradewinds (the word *trade* once meant a direction or course), to discover and then settle the Hawaiian islands more than 1300 years ago. Christopher Columbus also relied on his knowledge of global winds when he set forth in 1492. He had previously sailed to the Azores islands, which lie at latitude 37° N about one-third of the way from Spain to America. Persistent westerly winds slowed the trip to the Azores and prevented Columbus (and earlier sailors as well) from getting any farther west. Columbus knew, however, from reports of Portuguese sailors that, if he sailed south down the coast of Africa, he would find easterly winds. When he left Spain on August 3, 1492, he therefore sailed south as far as the Canary islands (Fig.13.10), picked up the easterly-blowing tradewinds, and the rest is history. With the tradewind behind him, he crossed the Atlantic Ocean and blazed the trail used by Europeans to invade the Americas. On his return voyage to Europe, Columbus sailed northward and picked up the westerly-blowing winds.

Hadley Cell Circulation

The person who first offered an explanation for the persistent easterly and westerly winds reported by mariners was George Hadley (1685–1768), an English mathematician. Hadley pointed out in 1735 that the underlying cause of global winds is that more of the Sun's heat reaches the surface at the equator than at the poles. The reason for the disparity of heat reach-

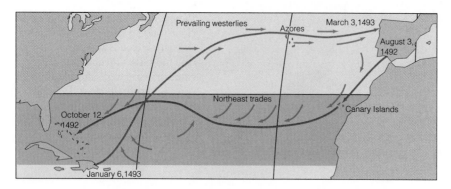

Figure 13.10 The winds used by Columbus on his first voyage to America. Outward bound after visiting the Canary Islands from August 12 to September 8, 1492, he sailed west with the northeast trades behind him. On his return voyage, Columbus sailed north to pick up the prevailing westerlies that had prevented previous European mariners from sailing any further west than the Azores. Columbus stayed in the Azores from February 15 to 24, 1493, and reached Europe on March 3, 1493.

ing the surface, as explained in Chapter 3, is that the Earth is round. The solar heat imbalance, Hadley pointed out, means that warm equatorial air must flow toward the pole and cold air must flow toward the equator, creating huge convection cells.

If the Earth were a nonrotating sphere, one convection cell would carry heat from the equator all the way to the poles (Fig. 13.11). Warm, low-density air

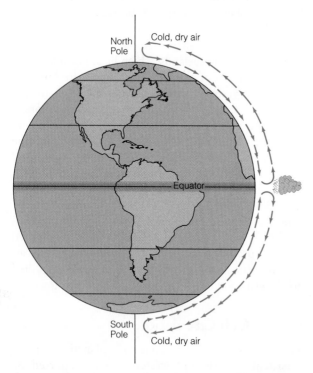

Figure 13.11 Global circulation as it would happen on a nonrotating Earth. Huge convection cells would transfer heat from equatorial regions, where the solar energy/unit area is greatest, to the poles, where the solar input is least. The equatorial region would be a zone of low pressure, while the poles would be high-pressure zones.

rising above the equator would flow poleward, and cool polar air would flow back across the surface toward the equator. Thus, the equatorial region would be a zone of convergence and therefore a low surface-pressure region, while the two polar regions would be zones of divergence and hence high surface-pressure zones.

Since the Earth is not stationary, the poleward air flow and the equatorward return flow are deflected as a result of the Coriolis effect. Convection does operate, but the flow is not as simple as the case described for a nonrotating Earth. On a rotating Earth, as on a non-rotating one, warm air rises in the tropics and creates a low-pressure zone of convergence called the **intertropical convergence zone**[3] (Fig.13.12). By the time the poleward-flowing air, high in the troposphere, reaches latitudes of 30° N or 30° S, it has been deflected by the Coriolis effect and is a westerly geostrophic wind. Remember that a westerly wind flows to the east. Obviously, a wind that flows due east cannot reach the poles, and so air tends to pile up at 30N and 30S, creating two belts of high-pressure air around the world centered approximately on those latitudes. Air in these high-pressure belts sinks back toward the surface, creating a zone of divergence. Some of the divergent air flows toward the poles, but most flows back toward the equator, creating convection cells on both sides of the equator that dominate the winds in tropical and equatorial regions. The cells, which are labeled in Figure 13.12, are called **Hadley cells** in honor of the man who explained their existence. The exact position of the intertropical convergence zone and of the two high-pressure belts varies

[3]Sea-level air pressure readings in equatorial regions are generally below the standard sea-level pressure of 101.3 kPa. They fall in the range 100.0 to 101.1 kPa, while pressures over the poles can be as high as 103.0 kPa.

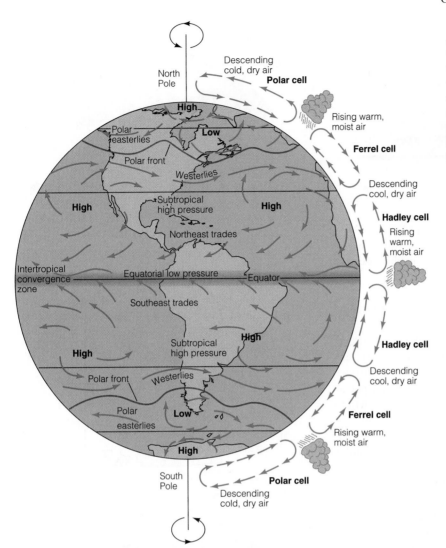

Figure 13.12 The Earth's global wind system. Moist air, heated in the warm equatorial zone, rises convectively and forms clouds that produce abundant rain. Cool, dry air descending at latitudes 20–30° N and S produces a belt of subtropical high pressure in which lie many of the world's great deserts.

with the seasons because the place where the Sun is directly overhead, and therefore where the Earth is receiving the greatest amount of heat, moves with the seasons.

In the Hadley cells, the high-level winds are westerlies, and the low-level winds bringing the return air toward the tropics are almost easterlies. The "almost" is necessary because friction comes into play. In the northern hemisphere the lower-level winds are northeasterly winds, called the *northeast trades*, and in the southern hemisphere, they are the *southeast trades* (Fig.13.12).

The Polar Front, Rossby Waves, and Jet Streams

In each hemisphere, poleward of the Hadley cells, a second, middle-latitude circulation occurs. The midlatitude cells are called *Ferrel cells* after the American meteorologist, William Ferrel (1817–1891). In Ferrel cells, the surface winds are westerlies because they are created, in part, by poleward flows of air from the high-pressure divergence regions at 30N and 30S.

These westerlies were the winds that prevented mariners before Columbus from sailing any farther west than the Azores islands.

A third region of high-latitude circulation, called *polar cells*, lies over the polar regions. In each polar cell, cold, dry, upper air descends near the pole, creating a high-pressure area of divergence. Then air from this area moves equatorward in a surface wind system called the polar easterlies. As this air moves slowly equatorward, it encounters the middle-latitude belt of surface westerlies in the Ferrel cells. The two wind systems meet along a zone called the **polar front** and create a low-pressure zone of convergence that is analogous to the intertropical convergence zone. The polar front is a region of unstable air along which severe atmospheric disturbances occur.

The high-level winds in the polar cells are westerly. Indeed, because some of the high-level air in the Hadley cells spills over into the midlatitude Ferrel cells, the prevailing high-level winds poleward of 30N and 30S are all westerlies. Flow in the upper atmosphere is not uniform, however. Rather, the winds flow in great undulating streams called *Rossby waves*;

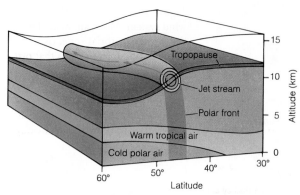

Figure 13.13 The jet stream is a high-speed westerly geostrophic wind that occurs at the top of the troposphere over the polar front where a steep pressure gradient exists between cold polar air and warm subtropical air.

these undulations resemble the meanders of streams and rivers.

Recall from Chapter 12 that the unequal heating of the Earth's surface causes the top of the troposphere (the tropopause) to be much lower at high latitudes than at low latitudes (16 km or 10 mi at the equator and 10 km or 10 mi at the poles). The region where the height of the tropopause changes most rapidly is over the polar front (Fig.13.13). A large body of cold, polar air fills the troposphere poleward of the polar front, while warmer, subtropical air fills the troposphere on the equatorial side, but the top of the troposphere is everywhere at the same pressure. In other words, the tropopause is an isobar. This means that, in the stratosphere, there is a very steep pressure gradient over the polar front; high pressure is on the poleward side, and low pressure on the equatorial side of the stratosphere. Steep pressure gradients mean high-speed winds, and because friction is not involved, the winds are geostrophic. Upper-atmosphere westerlies associated with this steep pressure gradient, called the *polar front jet stream*, can develop exceptionally high speeds. As high as 460 km/h (286 mi/h) have been reported by high-flying planes.

Rossby waves distort the polar-front jet stream into great undulations (Fig.13.14). As the jet stream undulates, it pushes and pulls the polar front with it and thus plays a major role in weather patterns between 45° and 60° north and south latitudes.

A second jet stream, also a geostrophic westerly, called the *subtropical jet stream*, forms above the tropopause over the Hadley cell between latitudes 20° and 30° north and south. Speeds of westerly winds in the subtropical jet stream reach 380 km/h (236 mi/h), but because the troposphere is higher at low latitudes, the subtropical jet is at a higher altitude than the polar-front jet and does not play the dominant weather role exerted by the polar-front jet stream.

A. Jet stream with small undulations

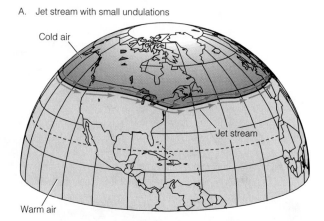

B. Rossby waves cause giant meanders to form

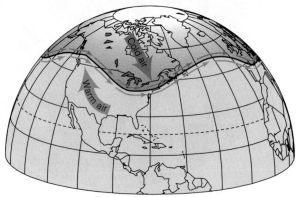

C. Strongly developed undulations pull a trough of cold air south

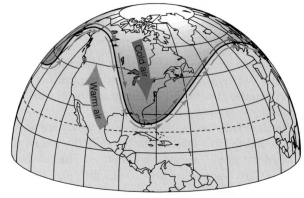

Figure 13.14 Rossby waves in the jet stream pull masses of cold air south as meanders form. (a) The axis of the jet stream starts out flowing to the east in a nearly straight line. (b) and (c) Undulations grow into gigantic meanders that pull masses of cold polar air down over the United States.

This discussion has said little about how land and sea are not distributed evenly around the world. In fact, both land distribution and land elevation play important roles in local wind patterns and in what are called monsoon systems. (See "A Closer Look: Monsoons.")

A CLOSER LOOK

Monsoons

Monsoonal circulation is characteristic of regions where local conditions bring about a seasonal reversal of the direction of surface winds. The places on the Earth where this phenomenon is most distinct are Asia and Africa, although a weak monsoon develops over eastern and central North America, too.

The Asian monsoon is the most distinct. Because the equator lies just south of the tip of India, the normal surface-wind pattern is a northeast trade blowing offshore from India into the Indian Ocean. For half a year during the winter months, the expected northeasterly wind pattern is observed because a high-pressure anticyclone sits over the high, cold plateau of central Asia while the low-pressure intertropical convergence zone lies south of the equator, where the Sun is overhead (Fig.C13.1). The winter months are therefore a time of cool, dry, cloudless days and northeast winds. For more than 2000 years, Arab sailors have used these northeasterly winds to sail home from India.

During the summer months, the Asian wind pattern is reversed. With the Sun overhead in the northern hemisphere, the intertropical convergence zone shifts north of the equator, and so the landmass of Asia heats up and is covered by low-pressure cyclones. The winds now blow southwesterly, from the Indian Ocean on to the land. Summer months are therefore a time of hot, humid weather and torrential rains. The summer monsoons start in southern India and Sri Lanka in late May, progress to central India by mid-June, and reach China by late July. During this period, the Arabs sailed from Arabia to India. The sailors referred to the change in wind direction as *mausim*, Arabic for change, and from this comes our word *monsoon*.

As seen in Figure C13.1. the monsoon system that affects India also occurs in North and West Africa. There are local differences, of course, but the main controlling factor in both cases is the seasonal movement of the intertropical convergence zone.

A weak monsoon system occurs in North America. During the summer months, there is a tendency for surface winds to bring warm, moisture-laden air from the Gulf of Mexico into the central and eastern United States. Humid weather and summer rains are the result. In the winter months, the winds reverse and there is a tendency for cold air to move southward from Canada into the Gulf.

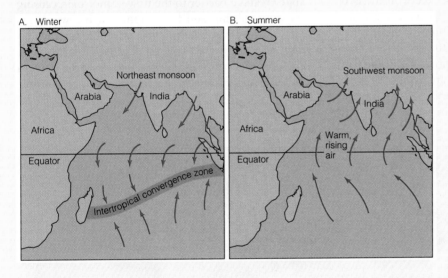

Figure C13.1 The reversing winds of the Indian monsoon. A. During the winter months when the Sun is overhead in the southern hemisphere, winds flow offshore from the northeast toward the intertropical convergence zone. Note how the winds curve toward the east as they cross the equator. B. During the summer months, the land heats up and winds flow from the southwest across Asia. When the Sun is overhead on land, the intertropical convergence zone is not a distinct band of low pressure.

GLOBAL PRECIPITATION AND THE DISTRIBUTION OF DESERTS

There are three global belts of high rainfall and four of low rainfall. The high rainfall belts are the three regions of global convergence—the intertropical convergence zone and the two polar fronts. The four belts of low rainfall are the regions of divergence—the two belts of subtropical highs (see Fig.13.12) centered on latitudes 30N and 30S, and the two polar regions.

The effect of the global air-circulation system is most clearly demonstrated by the distribution of deserts.

Deserts

Desert lands in which the annual rainfall is less than 250 mm (10 in) make up about 25 percent of the land area of the world outside the polar regions. In addition, a smaller, though still large, percentage of semi-desert land exists in which the annual rainfall ranges

Table 13.1 Main Types of Deserts and Their Origins

Desert Type	Origin	Examples
Subtropical	Centered in belts of descending dry air at 20–30° north and south latitude	Sahara, Sind, Kalahari, Great Australian
Continental	In continental interiors, far from moisture sources	Gobi, Takla Makan
Rainshadow	On the sheltered side of mountain barriers that trap moist air flowing from oceans	Deserts on the sheltered sides of Sierra Nevada, Cascades, and Andes
Coastal	Continental margins where cold, upwelling marine water cools maritime air flowing onshore	Coastal Peru and southwestern Africa
Polar	In regions where cold, dry air descends, creating very little precipitation	Northern Greenland, ice-free areas of Antarctica

between 250 and 500 mm (10 and 20 in). Together these desert and semidesert regions form a distinctive pattern on the world map (Fig.13.15). The regions are not randomly scattered across the globe but instead are related to the global atmospheric circulation and to local features of the Earth's geography. In all, five types of desert are recognized (Table 13.1).

When we compare Figure 13.15 with Figure 13.12, a relationship is immediately apparent. The most extensive deserts, the Sahara, Kalahari, Great Australian, and Rub-al-Khali, are associated with the two circumglobal belts of divergence, where dry air descends on the downward-flowing limbs of the Hadley cells, centered between latitudes 20° and 30°. These and other subtropical deserts comprise one of the five recognized types of desert. They are associated with anticyclonic regions of high pressure.

A second type of desert is found in continental interiors, far from sources of moisture, where hot summers and cold winters prevail (that is, a continental-type climate). The Gobi and Takla Makan deserts of central Asia fall into this category. These deserts form because wind that travels a very long distance over land, especially land that rises up to high plateaus, eventually contains so little water vapor that hardly any is left for precipitation.

A third kind of desert is found where a mountain range creates a barrier to the flow of moist air, causing orographic lifting and heavy rains on the windward side along with a zone of low precipitation called a *rainshadow* on the downwind side (see Fig.12.19D). The Cascade Range and Sierra Nevada of the western United States form such barriers and are responsible

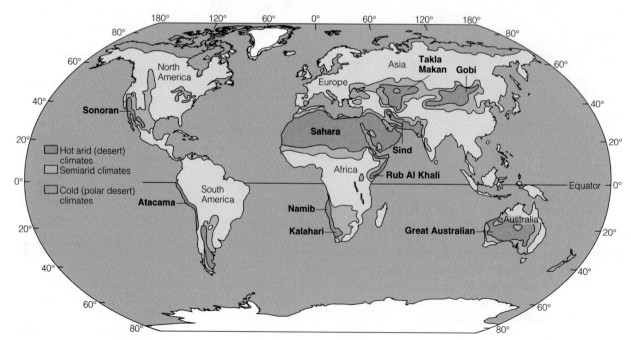

Figure 13.15 Arid and semiarid climates of the world and the major deserts associated with them. The very dry areas of the polar regions are polar deserts.

for desert regions lying immediately east of these mountains.

Coastal deserts, which constitute a fourth category, occur locally along western margins of certain continents. The flows of surface ocean currents can cause the local upwelling of cold bottom waters. The cold, upwelling seawater cools maritime air flowing onshore, thereby decreasing its ability to hold moisture. As the air encounters the land, the small amount of moisture it holds condenses, giving rise to coastal fogs. Nevertheless, in spite of the fog, the air contains too little moisture to generate much precipitation, and so the coastal region remains a desert. Coastal deserts of this type in Peru and southwestern Africa are among the driest places on the Earth.

The four kinds of desert mentioned thus far are all hot deserts, where rainfall is low and summer temperatures are high. In the fifth category are vast deserts of the polar regions where precipitation is also extremely low due to the sinking of cold, dry air. Remember that polar regions are high-pressure areas where cold, high-altitude air descends from the upper troposphere. However, cold deserts differ from hot deserts in one important respect: the surface of a polar desert, unlike the surfaces of warmer latitudes, is often underlain by abundant H_2O, nearly all in the form of ice. This ice accumulates, even though precipitation is exceedingly low, because the precipitation is always as snow and the snow doesn't melt. Even in midsummer, with the Sun above the horizon 24 hours a day, the air temperature may remain below freezing. Polar deserts are found in Greenland, arctic Canada, and Antarctica. Such deserts are considered to be the closest earthly analogs to the surface of Mars, where temperatures also remain below freezing and the rarefied atmosphere is extremely dry.

Dust Storms

One of the most striking features of deserts is dust storms. As discussed earlier in this chapter in the section on the windchill factor, at the surface of an object across which wind is flowing there is a boundary layer of still air less than 1 mm thick. Above the boundary layer is a thin zone in which air flow is laminar, and above the laminar air the flow is turbulent (Fig. 13.16). Large grains that protrude above the quiet, laminar-flow zone can be rolled and bounced along or may be swept aloft by rising turbulent winds. The larger grains that roll and bounce along mobilize fine sediment, which is then carried upward by the turbulent air. In this manner dust storms start.

Once in the air, dust constitutes the wind's suspended load. The grains of dust are continuously tossed about by eddies, like particles in a stream of turbulent water, while gravity tends to pull them toward the ground. Meanwhile, the wind carries them forward. Although in most cases grains suspended in the wind are deposited fairly near the place of origin, strong winds associated with large dust storms are known to carry very fine dust into the upper troposphere where it can be transported horizontally by geostrophic winds for thousands of kilometers.

In a dust storm, the visibility at eye level is reduced to 1000 m (0.6 mi) or less. Such storms are most frequent in the vast arid and semiarid regions of central Australia, western China, Soviet Central Asia, the Middle East, and North Africa (Fig.13.17). In the United States, blowing dust is especially common in the southern Great Plains and in the desert regions of California and Arizona.

The frequency of dust storms is commonly related to cycles of drought, with a marked rise in atmospheric dust concentration coinciding with severe drought. The frequency also has risen with increasing agricultural activity, especially in semiarid lands. An example of how human activities can contribute to an increase in dustiness is seen in records from the western desert of Egypt in the 1930s and 1940s. The number of dust storms rose from three or four per year before the Second World War to more than 40 between 1939 and 1941, when wartime tank action and artillery bombardment were at a peak, and then declined to four per year after military activity ceased.

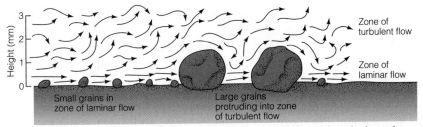

Figure 13.16 Particles of fine sand and silt at the ground lie within the boundary layer where wind speed is extremely low. As a result, it is difficult for the wind to dislodge and erode these small grains. Larger grains protrude into a zone of faster moving, turbulent air. The turbulence, which exerts a greater push on the top of the grains than does the still boundary-layer air at their base, starts the grains moving.

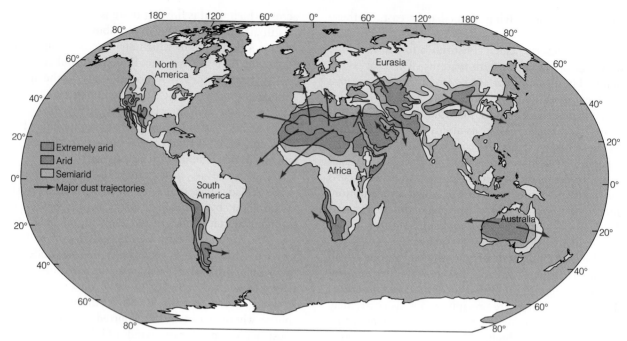

Figure 13.17 Major dust storms are most frequent in arid and semiarid regions that are concentrated in the subtropical high-pressure belts north and south of the equatorial zone. Arrows show the most common trajectories of dust transported during major storms.

LOCAL WIND SYSTEMS

In many localities, local winds are more important than global winds. Local winds, which may flow for tens or hundreds of kilometers rather than the thousands of kilometers involved in global winds, are the result of the local terrain.

Sea and Land Breezes

The least complicated example of a local wind system is the coupled land breeze and sea breeze that is familiar to anyone who lives on or near a coast. The origin of these breezes is illustrated in Figure 13.18. During the day the land heats up more rapidly than the sea, and the heated land causes the air in contact with it to heat up and expand. A pressure gradient develops, and the lower air layer flows toward the land, creating a *sea breeze*. Higher in the atmosphere, an upper-level reverse flow sets in; the coupled flows—rising air over the land and sinking air over the sea—form a convection cell.

During the night, heat is radiated more rapidly from the land than from the sea, and consequently the situation reverses. The sea is now warmer than the land, and air moves from the land to the sea, creating a *land breeze*.

Mountain and Valley Winds

Mountain winds and valley winds have a daily alternation of air flow in the same way that land and sea breezes do. During the day, the mountain slopes are heated by the Sun, and so air flows from the valley upward over the slopes. At night, the mountain slopes cool quickly, and so the flow reverses with air flowing from the mountain sides down into the valleys. Just as in the case of the land and sea breezes, mountain and valley winds respond to localized pressure gradients set up by heating and cooling of the lower air layer.

Katabatic Winds

The flow of cold, dense air under the influence of gravity is called a *katabatic wind*. Such winds occur in places where a mass of cold air accumulates over a high plateau or in a high valley in the interior of a mountain range. As the cold air accumulates, some eventually spills over a low pass or divide and flows down valleys on to the adjacent lowlands as a high-speed, cold wind.

Katabatic winds occur in most mountainous regions around the world and commonly have local names. The *mistral* is a notable example: it is a cold, dry wind that flows down the Rhone Valley in France, past Marseilles, and out onto the Mediterranean Sea. Another notable example is the *bora*, a northeasterly that rushes down from the cold highlands of Yugoslavia to the Adriatic Sea near Trieste. Wind gusts in Trieste during a bora can reach speeds of 150 km/h (93 mi/h).

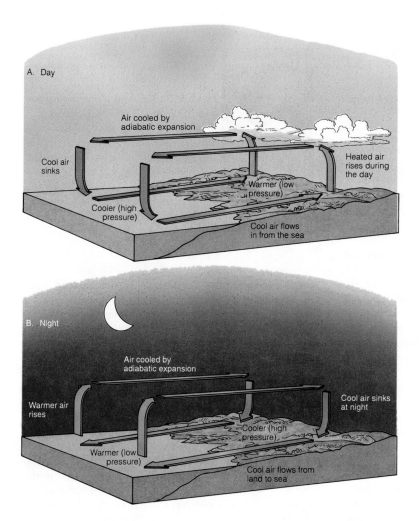

Figure 13.18 Land and sea breezes. A. During the day, the land heats up more rapidly than does the sea. Air rises over the land, creating a low-pressure area. Cooler air flows in to this area from the sea, creating a sea breeze. B. During the night, the land cools more rapidly than the sea, and the reverse flow, a land breeze, occurs.

The most striking examples of katabatic winds are those that occur around the edges of Greenland and Antarctica, where the frigid, high-pressure air masses that accumulate above the continental ice sheets pour down the sloping margins of the ice and out onto the adjacent ocean waters. When the ice slope is steep, the katabatic wind speed can be terrifyingly high. It is because of a katabatic wind that Cape Dennison in Antarctica has a higher annual average wind speed than any other place on Earth.

Chinooks

Related to katabatic winds is another class of downslope land winds known by various local names—*chinook* along the eastern slopes of the Rocky Mountains, *föhn* in Germany, *Santa Ana* in southern California. For simplicity, we speak now only of chinooks, but of course all we say is true of these winds regardless of the name being used. Chinooks are warm, dry winds. Because warm, dry air has a low density and so does not sink naturally, a chinook must be forced downward by large-scale wind and air pressure patterns. This forcing occurs when strong regional winds, commonly associated with anticyclones, rise and compress

higher level air masses as they pass over a mountain range and then are forced to flow down on the downwind side by the pressure of higher level air. The result is that the downward-flowing air is adiabatically heated and therefore dry—in short, a chinook.

AIR-MASS TYPES

People who dwell in the middle latitudes know that weather patterns generally last several days. The reason is that weather is controlled by huge air masses up to 2000 km across and several kilometers high. Such an air mass requires several days to cross a continent.

Within an air mass there are only small contrasts of temperature and humidity because any given mass forms over a surface that has roughly uniform properties. Four variables may affect air masses: whether a mass forms over a continent (c) or over a maritime region (m), and whether it forms in the tropics (T) or in polar regions (P). The characteristics of the four basic air-mass types are listed in Table 13.2. Warm fronts tend to be associated with mT air masses and anticyclones. Cold fronts are generally associated with cP or mP air masses and with cyclones. The kinds of air

GUEST ESSAY

Tracking Tornadoes Through the Southern Plains of the United States

Howard B. Bluestein is from the Boston area. He holds B.S. and M.S. degrees in electrical engineering, and M.S. and Ph.D. degrees in meteorology from the Massachusetts Institute of Technology. He is currently professor of meteorology at the University of Oklahoma, where he teaches both under-graduate and graduate courses, and does research on severe storms, tornadoes, mesoscale and synoptic meteorology, and tropical cyclones.

Tornadoes have been described as one of the last frontiers of meteorology. Relatively little is known about them because they are so difficult to study. Since they are only 100 m in diameter and usually last for less than half an hour, tornadoes affect very small areas of the Earth for very short periods of time and are extremely difficult to predict. The impact of their damage, however, can be enormous.

Prior to the 1970s, what we knew about tornadoes came mainly from serendipitous observations by nonmeteorologists. In the early 1970s, my predecessors at the University of Oklahoma and the National Severe Storms Laboratory in Norman, Oklahoma, began to increase our knowledge of tornado structure and behavior by setting out to intercept ("chase") tornadoes and tornado-producing thunderstorms, significantly increasing the number of observations.

Tornado interception works as follows. A forecast is made of where tornadoes are expected to occur, and meteorologists drive to within a 300-km radius of home base to the general area where the parent storms might form. If storms do form, the meteorologists decide which storm has the most tornado-producing potential and position themselves at a safe distance from the portion of the storm where tornadoes typically occur. At this distance (3 to 6 km), cloud features can still be seen clearly. In supercell storms (long-lived, rotating solitary storms), meteorologists look for tornadoes near the wall cloud, a rotating, lowered cloud base that is located near the rear (with respect to storm mo-

tion) of the storm. In nonsupercell storms, meteorologists look near the cloud base of rapidly building cloud towers in growing thunderstorms. We guess which storm has the most tornado-producing potential by combining theory with observational experience: it is both a science and an art to "pick" the tornadic storm from the zoo of storms out in the field.

I arrived in Norman in 1976, when meteorologists were beginning to evaluate the usefulness of Doppler radar for issuing severe weather warnings to the public. Conventional radar can assess only the intensity of the precipitation in a storm by measuring the backscattered radiation from rain-drops, ice crystals, and hail. Doppler radar, however, can also reveal features of the storm's wind field by measuring the shift in frequency of the backscattered radiation. At Norman we would observe a storm visually and correlate the wind "signature" seen by the Doppler radar with what we saw. After many observations, it became possible to de-

---

masses of greatest importance as far as the weather of North America is concerned are cP and mT. The cP air masses originate in Canada, in the Arctic, and to a lesser extent in Alaska; the mT air masses originate in the Gulf of Mexico, the Atlantic Ocean, the Caribbean Sea, and the Pacific Ocean (Fig.13.19).

As discussed in Chapter 12, the boundaries between air masses of different temperature and humidity, and therefore different density, are called fronts. The boundaries are between 10 and 150 km (6 and 93 mi) in width and mark the active advance of one air mass into another.

Table 13.2 Characteristics of Air Masses

Origin of Air	Temperature	Humidity
Continental polar (cP)	Cold	Low
Maritime polar (mP)	Cool	High
Continental tropical (cT)	Hot	High
Maritime tropical (mT)	Warm	High

SEVERE WEATHER

"Severe weather" and "storm" mean the same thing—a violent disturbance of the atmosphere attended by strong winds and commonly rain, snow, hail, sleet, thunder, and lightning. Severe weather can have many causes but most occurs along cold fronts. Three kinds of severe weather—thunderstorms, tornadoes, and hurricanes—cause so much damage, and even loss of life, that they deserve special attention in our coverage.

Thunderstorms

Thunderstorms develop when an updraft of warm, humid air releases a lot of latent heat very quickly and becomes unstable. Most thunderstorms in North America form along cold fronts and are associated with mT air masses formed over the Gulf of Mexico. The released heat causes stronger updrafts, which pull in more warm, moist air which in turn releases more latent heat, and so the process grows and the updraft intensifies. Cumulonimbus clouds form, and heavy

vise the technique currently being used by the National Weather Service to issue tornado warnings up to 20 or 30 minutes prior to the touchdown of a tornado within a supercell storm.

Our visual observations also proved useful for training storm spotters. These individuals watch approaching storms and provide the National Weather Service with information about whether a tornado is actually observed and whether a rotating wall cloud, a precursor to tornado development, is present.

As we became more proficient at intercepting tornadic storms, it became apparent to me that we should attempt to obtain more quantitative measurements. The wind, pressure, temperature, and moisture distributions within tornadoes were not known. From 1980 to 1983, we attempted to make these measurements, using TOTO (the Totable Tornado Observatory, not Dorothy's pet), an instrumented device designed at the Wave Propagation Laboratory in Boulder, Colorado. Our goal was to place TOTO directly in the path of a tornado, and then retrieve it and analyze the recorded data. Although we obtained data under wall clouds and near tornados, we found that it was too difficult to place TOTO directly in the path of a tornado: usually the tornado dissipated or changed direction before hitting TOTO, or there were no roads leading to the path of the tornado. We were also concerned that a strong tornado might damage TOTO.

From 1984 to 1989, we used a commercially available portable radiosonde (instrumented weather balloon) to learn about the vertical distribution of temperature and moisture inside and outside of severe storms. This information was especially useful for scientists who simulate the life history of severe storms on a computer and need to know atmospheric conditions that precede the development of the storm.

From 1987 to 1994 we used a portable 3-cm wavelength Doppler radar designed at the Los Alamos National Laboratory. Using the radar, we positioned ourselves at a safe distance from tornadoes and made wind measurements, without having to get directly in their paths. Not until 1990 and 1991 did we successfully collect data. On April 26, 1991, we made measurements of wind speeds as high as 120 to 125 m/s in a large tornado in north-central Oklahoma. Prior to this, wind speeds this high were only inferred from damage assessments or from photogrammetric analysis of debris and cloud tags in tornado movies.

In 1993 and 1994 we experimented with a 3-mm wavelength Doppler radar from the University of Massachusetts. In 1998 we used an improved version of the radar, capable of resolving volumes of air only 10m on a side.

Tracking tornadoes and making scientific measurements in them is an exhilarating, challenging, and often frustrating endeavor. On the average, we intercept tornadoes on only one out of nine chases. We drive great distances and often see nothing, or we miss a tornado that occurred just minutes earlier, or we intercept a tornado and have an instrument fail. However, when we do have a successful chase, we have the opportunity to witness one of nature's most awesome displays of power and to unravel its mysteries.

rainfall, commonly hail, thunder, and lightning are the result.

The towering masses of cumulonimbus clouds associated with thunderstorms can reach as high as 18 km (11 mi) (Fig.13.20), and winds in a thunderhead can exceed 100 km/h (60 mi/h). Updrafts in a thunderhead can be so strong that large hailstones can form by coalescence of tiny ice particles and be held aloft until a sudden downdraft deposits them on the ground.

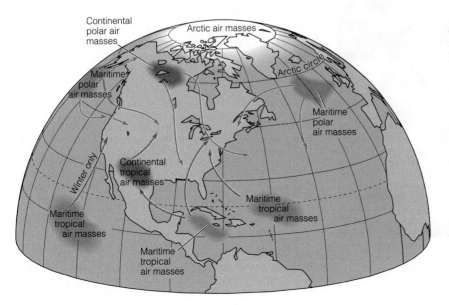

Figure 13.19 Sources of the air masses that control the weather of North America.

Lightning and *thunder* accompany each other and are due to electrical charges. The electrical charges form during the growth of a cumulonimbus cloud. The turbulent movement of precipitation inside the cloud causes particles in the upper part to become positively charged and particles in the lower part to become negatively charged. Exactly how the charge builds up is not clearly understood, but the buildup can reach hundreds of millions of volts. The charges can be released by a lightning strike either to the ground or to another cloud. As the lightning strike passes, it heats the surrounding air so rapidly that the air expands explosively and we hear the effect as thunder.

Tornadoes

Tornadoes are violent windstorms produced by a spiraling column of air that extends downward from a cumulonimbus cloud (Fig.13.21). Tornadoes are approximately funnel-shaped, and they are made visible by clouds, dust, and debris sucked into the funnel. By convention, a tornado funnel is called a *funnel cloud* if it stays aloft and a *tornado* if it reaches the ground.

Tornadoes are small features relative to the thunderstorms with which they are associated. Because they are so violent, many details of their formation are still unresolved. The funnel develops as a result of a spiraling updraft in a thunderstorm. Such updrafts are commonly 10 to 20 km (6 to 12 mi) in diameter. For reasons not clearly understood, a spiraling updraft in certain thunderstorms will narrow and spiral down to

Figure 13.20 A thunderstorm over Tucson, Arizona. Note the dark cumulonimbus clouds, the dense rain, and the lightning in the clouds.

Figure 13.21 Hail falling from a thunderstorm, Santa Cruz County, Arizona.

a tornado funnel from 0.1 to 1.5 km (0.06 to 0.9 mi) in diameter.

Fortunately, most tornadoes are not especially strong and do not cause much damage. At the other end of the strength scale, there are some tornadoes that completely destroy any object in their path. The strength of a tornado can be estimated from the damage it causes by referring to the F-scale, named for Professor T. Theodore Fujita of the University of Chicago, who devised it (Table 13.3).

An average of 780 tornadoes take place each year in the United States, and they are known to occur in all states and during any month of the year. However, there is a distinctly intense period of tornado activity from April to August, with a peak in May. Because the severe thunderstorms that are the parents of tornadoes form along cold fronts and because the most violent cold fronts are those associated with cP air from the Canadian Arctic and mT air from the Gulf of Mexico, most tornadoes occur in the midcontinent states because that is where the two air masses are most likely to meet.

Hurricanes

Hurricanes are violent, oceanic cyclones that, by definition, have maximum wind speeds in excess of 119 km/h (74 mi/h) (Fig. 13.22). Hurricanes are particularly devastating to island settlements and coastal regions. Once a hurricane leaves the ocean and moves onshore, wind speeds diminish and the hurricane quickly dies down. For this reason, most hurricane damage occurs within 250 km (155 mi) of the coast.

Besides wind damage, two other hurricane effects can be devastating. The first is a *storm surge*, a local, exceptional flood of ocean water. The center of a hur-

Table 13.3 Scale for Tornado Intensity

F-Scale	Category	Estimated Wind Speed, km/h	Damage
0	)	65–118	Minor, break twigs.
	) weak		
1	)	119–181	Down trees, move mobile homes off foundation.
2	)	182–253	Demolish mobile homes; roof off frame houses.
	) strong		
3	)	254–332	Lift motor vehicles. Destroy well-constructed buildings.
4	)	333–419	Level buildings, toss automobiles around.
	) violent		
5	)	420–513	Lift and toss around houses.

ricane is a region of very low air pressure—below 92kPa in the greatest hurricanes—and a drop in air pressure raises local sea level. In the eye of a great hurricane, sea level may be 8 or 9 m (27 or 30 ft) above normal, and when hurricane-force winds drive such high seas onshore, extensive flooding results. The second effect associated with hurricanes is rain. Torrential rains and consequential flooding accompany most hurricanes—falls of 25 cm (10 in) are not uncommon—and even after wind speeds have dropped below hurricane force, violent rainstorms can continue.

Hurricanes start as cyclones over warm ocean water. Experience has shown that they require a sea-surface temperature of at least 26.5°C (80°F) because the energy source that sustains a hurricane is warm water; the water evaporates and subsequently condenses in the hurricane, releasing latent heat. Hurricanes die over land or over bodies of cold water because they no longer have a warm-water energy source to sustain them. Hurricanes develop from cyclones, and so they can be spawned only in latitudes where the Coriolis effect is strong enough for cyclonic circulation to develop—that is, higher than about latitude 5°. Hurricanes cannot form along the equator. The necessary conditions for hurricane breeding happen in only a few places around the world (Fig.13.23). Note, however, that although the phenomenon is everywhere the same, hurricanes are called typhoons in the western Pacific and cyclones in northern Australia.

Figure 13.22 A tornado crossing the plains of North Dakota.

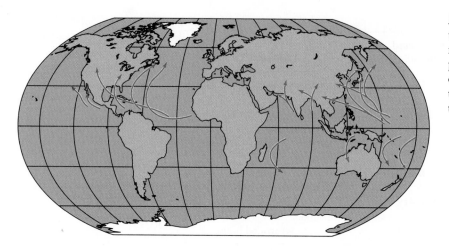

Figure 13.23 Hurricanes form in those places in the world where the right conditions of ocean water temperature and the Coriolis effect occur. Arrows show the usual directions followed by hurricanes once they form.

SUMMARY

1. Wind results from the horizontal movement of air in response to differences in air pressure.

2. The windchill factor results from wind reducing the insulating effect of the boundary layer of stationary air adjacent to the skin.

3. Wind speeds and wind directions are controlled by air pressure gradients, the Coriolis effect, and friction.

4. Air pressure gradients can be determined from a weather map by measuring the distance between isobars, which are lines connecting places of equal air pressure at sea level. When isobars are close together, the pressure gradient is steep and winds are strong.

5. The Coriolis effect, which arises as a result of the Earth's rotation, deflects wind toward the right in the northern hemisphere and to the left in the southern hemisphere.

6. The magnitude of the Coriolis effect is a function of latitude and wind speed. The effect is zero at the equator and a maximum at the poles. Wind speed contributes to the Coriolis deflection because, at high speed, a body moves a long distance in a short time. The longer the trajectory, the greater the deflection.

7. Friction between air and the ground slows winds and therefore reduces the Coriolis effect.

8. High-altitude geostrophic winds are Coriolis deflected and eventually flow parallel to isobars, with the low pressure on the left and the high pressure to the right.

9. Friction causes spiral air flow directed inward toward a low-pressure area. Such lows are called cyclones and rotate clockwise in the northern hemisphere and counterclockwise in the southern hemisphere. The opposite flow occurs as air spirals outward from a high-pressure area called an anticyclone. The spiral direction of an anticyclone is counterclockwise in the northern hemisphere and clockwise in the southern hemisphere.

10. Inward air flow in a cyclone produces a low air pressure zone of convergence; outward air flow in an anticyclone produces high air pressure and is a zone of divergence.

11. The global air circulation pattern arises from a combination of two factors—the flow of air from the equator toward the poles in response to a thermal imbalance, and the Coriolis effect.

12. At the top of the troposphere, along steep pressure gradients formed above the polar fronts, there are westerly geostrophic winds called the polar front jet streams. Similar subtropical jet streams occur above the descending limbs of the Hadley cells.

13. Around the equator is a region of low pressure caused by rising currents of warm, humid air. Centered on latitude 30° N and S are two belts of high pressure due to descending, low-humidity air. The world's major deserts are located in these belts. The cells of rising moist air and descending dry air are called Hadley cells.

14. Deserts form in four ways: as a result of the global air circulation, in continental interiors far from sources of moisture, in rainshadows, and along coasts adjacent to cold-upwelling seawater.

15. Local wind systems, such as land and sea breezes, mountain and valley winds, katabatic winds, and chinooks arise from local terrain effects. Such winds are often of much greater importance locally than global winds.

16. Thunderstorms form along cold fronts as a result of updrafts of warm, humid air; they are maintained by the latent heat of condensation from the humid air.

17. Tornadoes, which are violent, upward-spiraling columns of air associated with cumulonimbus clouds, form as a result of spiral updrafts in certain thunderstorms. Many aspects of their formation remain uncertain.

18. Hurricanes are violent oceanic cyclones in which maximum wind speeds exceed 119 km/h. Because hurricanes are oceanic phenomena, they cause their greatest damage to island and coastal regions.

IMPORTANT TERMS TO REMEMBER

air pressure gradient   *295*   divergence   *299*   high (h)   *299*   low (l)   *299*
anticyclone   *299*   friction   *295*   intertropical convergence   polar front   *301*
convergence   *299*   geostrophic wind   *297*   zone   *300*   wind   *294*
cyclone   *299*   Hadley cell   *300*   isobar   *295*   windchill factor   *295*

QUESTIONS FOR REVIEW

1. Explain why air density and air pressure are related.

2. Why do weather reporters give the windchill factor during cold weather but not during warm weather? Can windchill cause you harm if the air temperature is 15° C?

3. Name the three factors that control the speed and direction in which wind flows and briefly explain how each factor works.

4. On a weather map on which the isobar contour interval is 0.4 kPa, one region has the contours 20 km apart while another has them 200 km apart. In which region would you experience the stronger winds?

5. Explain why the Coriolis deflection of wind direction is always to the right in the northern hemisphere and always to the left in the southern hemisphere.

6. What are geostrophic winds and how do they arise? Name a well-known geostrophic wind.

7. How do cyclones and anticyclones form? What is the relationship between the highs and lows marked on a weather map and cyclones and anticyclones?

8. What kind of weather tends to be associated with cyclones?

9. What is a Hadley cell and how does it form? Describe the relationship between the intertropical convergence zone and the Hadley cells.

10. How do the tradewinds arise?

11. Use a drawing to illustrate the surface wind systems of the globe.

12. What is the polar front? Is it a zone of convergence or divergence? How does the polar front jet stream form?

13. Why are the world's major desert regions centered between latitudes 20° and 30°?

14. List three ways deserts can form other than as a result of the global air circulation.

15. What is the boundary layer and what role does it play in the formation of dust storms?

16. How and why do land and sea breezes occur?

17. What are katabatic winds? Give an example of a well-known katabatic wind.

18. List the four major categories of air masses. Which kinds of air masses are of greatest importance in weather development in North America?

19. What is a cold front and how does it differ from a warm front? Describe the kind of weather you might expect as a warm front advances into the area in which you live.

20. Briefly describe how a thunderstorm forms. Where does its energy come from?

21. What is the relationship between thunderstorms and tornadoes? Why are tornadoes most frequent in the central part of the United States?

QUESTIONS FOR DISCUSSION

1. Discuss in general terms the criteria needed for successful wind farms. Are the criteria most likely to be met by local or by global wind conditions?

2. About 250 million years ago, the continents of today were grouped together in a supercontinent called Pangaea. The site of New York City at that time was in the center of the supercontinent, thousands of kilometers from the sea. The latitude of the future New York City was about 15° N. What was the climate like?

3. What distributions of continents and oceans would effectively stop the formation of hurricanes? What distribution would make their formation rate even more frequent than it is today? Do some research and see if there have been times in the past when your predicted positions have occurred.

CHAPTER 14

The Earth's Changing Climate

● *The Tyrolean Iceman*

In the late summer of 1991, a remarkable discovery was made by a pair of German trekkers high in the Tyrolean Alps. The mummified body of a prehistoric man was seen protruding from slowly melting ice near the margin of Similaun Glacier at 3200 m (10,500 ft) altitude. With the corpse were a fur robe, woven grass cape, leather shoes, flint dagger, copper ax, wooden bow, and 14 arrows. Radiocarbon dating of the man's skin and bone indicated that he died about 5300 years ago. His antiquity, together with the associated artifacts and his bodily characteristics, showed him to be a member of the Late Neolithic and Bronze Age population of south-central Europe. The man was judged to be between 25 and 35 years old and about 1.6 m (5.2 ft) tall. He apparently had remained frozen more than five millennia until progressive thinning of the glacier since the middle of the nineteenth century eventually led to his exhumation.

The frozen corpse and the implements and clothing found with it proved to be a treasure trove for archaeologists. Prior to this discovery, scientists could only speculate about many aspects of Neolithic life on the basis of limited artifacts found at scattered sites. Now they had an actual person, complete with the tools of everyday life.

This discovery also was important to scientists studying climatic change, for it lent support to the interpretation of climate history in central Europe that had been pieced together from various lines of evidence. In the high Alpine valleys, receding glaciers have disgorged the remains of trees and bog vegetation that once flourished at sites subsequently covered by ice for many millennia. Dated by radiocarbon, the fossil trees and other plant remains tell us of an interval of mild climate during the early to middle Holocene Epoch (ca. 10,500–6000 years ago). This was a time, following the last ice age, when glaciers retreated high up in the mountains and plants invaded the upper slopes of alpine valleys. At the time the iceman lived, the climate was becoming cooler and glaciers were growing larger. The next 6000 years witnessed a succession of cool intervals, the latest of which is commonly referred to as the Little Ice Age. None of the intervening milder climatic periods apparently achieved temperatures equal to those of the first half of the Holocene. Thus, when the Tyrolean man died near the margin of Similaun Glacier, his remains became entombed in an alpine deep-freeze that kept him perfectly preserved for more than 5 millennia until the recent warming trend exposed him to view. ●

(opposite) The mummified body of a prehistoric man, exposed by retreat of a glacier high in the Tyrolean Alps, provides important clues about changing climate during the past 5000 years.

THE CLIMATE SYSTEM

In Chapter 12 we learned that climate is a measure of the average weather conditions of any place on the Earth and commonly is expressed in terms of mean temperature and mean precipitation. However, other parameters, including humidity, windiness, and cloudiness, also are important in characterizing climate, even though measurements of them are not routinely recorded in all places.

As shown in Figure 14.1, the Earth's climate system is complex and consists of several subsystems—atmosphere, hydrosphere (mainly oceans), cryosphere, lithosphere, and biosphere. The subsystems interact so closely that a change in one of these subsystems can lead to changes in one or more of the others. All but the lithosphere are driven by solar energy. Some of the incoming radiative energy is reflected back into space by clouds, atmospheric pollutants, ice, snow, and other reflective surfaces. The remainder is absorbed by the air, ocean, and land. Of these three energy reservoirs, the atmosphere responds most rapidly to outside influences, commonly within a month or less. The ocean surface responds more slowly (generally over months or years), whereas changes involving the deep ocean may take centuries. Although the land may respond rapidly or slowly to changes in the other components of the climate system, it has special significance on long time scales affecting the distribution of continents and ocean basins and the location and height of mountain ranges. The distribution and topography of the land directly influence the location and extent of glaciers and sea ice, as well as the character and extent of vegetation. Vegetation is important in the climate equation because it helps determine the reflectivity of the land surface. It also influences the composition of the air by absorbing carbon dioxide, and it affects humidity and therefore local cloud cover. Its absence can increase wind erosion, which in turn can influence climate by affecting the dustiness of the atmosphere.

Because of its ability to absorb and retain heat, the ocean serves as a great reservoir of heat energy that helps moderate climate. The ocean's effect is illustrated by the contrast between a coastal region having a maritime climate (little contrast between seasons) and an area farther inland having a continental climate (strong seasonal contrast). The world ocean also is ex-

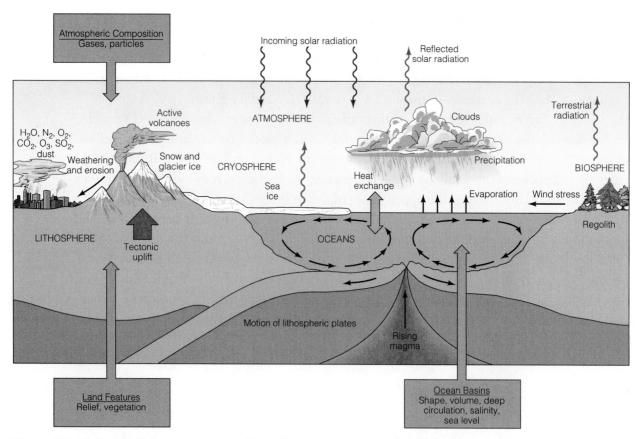

Figure 14.1 A diagrammatic representation of the Earth's climate system showing its five interacting components: lithosphere, atmosphere, hydrosphere (oceans), cryosphere, and biosphere.

tremely important in controlling atmospheric composition, for the ocean contains a large volume of dissolved carbon dioxide. If the balance between oceanic and atmospheric carbon dioxide reservoirs were to change by even a small amount, the radiation balance of the atmosphere would be affected, thereby bringing about a change in world climates (Chapter 20).

Understanding how the Earth's climate system works is a challenging task, and we are far from having all the answers. Important insights have been gained through the study of past climates, evidence of which is preserved in the geologic record. Such evidence offers important clues that can help tell us what causes climate to change, and how the different physical and biological systems of the Earth respond to changes of climate on different time scales.

In this chapter, we will investigate some of the evidence demonstrating that climates have changed during Earth history, and we will see how this evidence provides clues about why climates change. In Chapter 20, we will look more closely at how human activities are changing the atmosphere in ways that might lead to a significant change of climate during our lifetime.

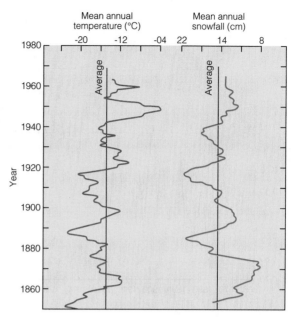

Figure 14.2 Variations of mean annual temperature and snowfall recorded at Great St. Bernard Pass on the Swiss-Italian border since the mid-nineteenth century. The vertical line through each graph shows the long-term average value.

EVIDENCE THAT CLIMATES CHANGE

Last winter may have been colder than the winter before, and last summer may have been wetter than the previous summer, but such observations do not mean that the climate is changing. The identification of a climatic change must be based on a shift in average conditions over a span of years. Several years of abnormal weather may not mean that a change is occurring, but trends that persist for a decade or more could signal a shift to a new climatic regime.

Historical Records of Climate

Our experience tells us that weather changes from year to year, but because climate is based on average conditions over many years, we may not find it easy to tell if the climate is changing. Your grandparents may recall that winters seemed colder half a century ago, but do such recollections actually point to a change of climate? Fortunately, weather records are kept throughout the world, and in some places they have been maintained for a century or more, long enough to see if average conditions have shifted in the past century or two.

One of the longest continuous climatic records available to us comes from Great St. Bernard Pass at the crest of the Alps, where the Augustinian friars have recorded temperatures since the 1820s and

snowfall since the 1850s (Fig. 14.2). Between 1860 and 1960, temperature and snowfall fluctuated approximately in phase, with times of cool temperature corresponding to times of above-average snowfall. Short-term trends persist for only about a decade or two, but over the entire period of the record there has been a general trend toward warmer temperatures.

The temperature pattern in the Alps is representative of that in other parts of the northern hemisphere, where average temperature experienced a fluctuating rise after the 1880s to reach a peak in the 1940s (Fig. 14.3). Thereafter, average temperatures declined slightly until the 1970s when they again began to rise, and in the early 1990s they reached the highest values yet recorded.

The amplitude of this recent long-term temperature increase, amounting to less than 1°C (1.8°F), seems small, yet its effects were seen widely, especially in high latitudes. During the six decades between 1880 and 1940, for example, mountain glaciers in most parts of the world shrank, some conspicuously (Fig. 14.4), and arctic sea ice was observed less frequently off the coast of Iceland. The biosphere also responded during this interval. The latitudinal limits of some plants and animals expanded slightly toward the poles, and an increase in the length of the summer growing season led to a general improvement in crop yields.

That the world's climates can change detectably within a human lifetime is a relatively new realization.

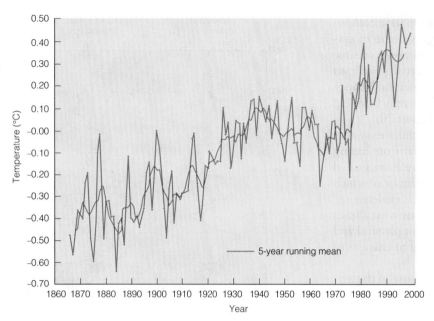

Figure 14.3 Annual mean air temperatures from 1866 to 1997 for the world's land areas. Because data from many different places are included, the annual mean temperatures are expressed as a variation from the average annual mean temperature for the 30 years from 1951 through 1980. From the late 1860s to the early 1940s, the annual mean temperature rose about 0.6°C. From the early 1940s to 1965 the temperature declined about 0.2°C and since 1965 the trend has been upward. The overall rise during the century and a quarter of record is ca. 0.85°C. The decade of the 1990s has been the warmest on record.

With this realization has come increasing concern about the impact of such changes on nature and on society, as well as the possible impact of human activities on the Earth's climate (Chapter 20).

The Geologic Record of Climatic Change

The evidence of climatic change on the Earth comes largely from the geologic record. Scientists have long puzzled over the occurrence of geologic features that seem out of place in their present climatic environment. For example, abundant fossil bones and teeth of hippopotamus—the same kind that lives in East Africa today—have been recovered from sediments in southeastern England. These sediments were deposited about 100,000 years ago under conditions that may have been like those in parts of modern tropical Africa. Another example involves many sites beyond the margin of the Great Lakes in north-central United States where plant remains show that this region formerly resembled arctic landscapes like those

Figure 14.4 In the late nineteenth century, Findelen Glacier in the Swiss Alps covered all the bare, rocky terrain seen here in the lower part of its valley. Since that time, the glacier terminus has retreated far upvalley in response to a general warming of the climate.

now seen in far northern Canada. In each case, a significant change in local climate apparently has taken place, so that the biota living in these areas today is very different from the fossil forms we see preserved in the geologic record.

Besides fossils, other anomalous features tell us that the climate has changed: glacial features in temperate lands, desert sand dunes now covered with stabilizing vegetation, beaches of extensive former lakes in dry desert basins, channel systems of now-dry streams, remains of dead trees above the present upper treeline in mountainous areas, and surface soils with profiles that are incompatible with the present climate.

Climate Proxy Records

Scientists attempting to reconstruct former climates can use instrumental records only for the very recent past. To extend the reconstruction back in time, they must rely on records of natural events that are controlled by, and closely mimic, climate. We call these *climate proxy records*, and although lacking the precision of instrumental data, they often can provide us not only with an indication of the year-to-year variability of weather but also with a good general picture of climatic trends. Four of the longest and most informative series are shown in Figure 14.5. Others include

A.

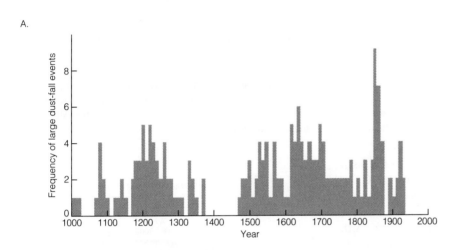

Figure 14.5 Climate proxy records spanning all or part of the last 1000 years: A. Frequency of major dust-fall events in China (*Source*: After Zhang, 1982); B. Severity of winters in England, recorded as the frequency of mild or severe months (*Source*: After Lamb, 1977); C. Number of weeks per year during which sea ice reached the coast of Iceland (*Source*: After Lamb, 1977); D. Freezing date of Lake Suwa in Japan relative to the long-term average (*Source*: After Lamb, 1966).

B.

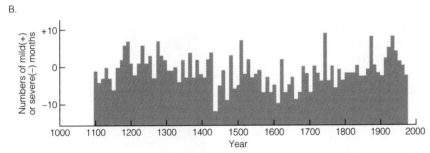

C.

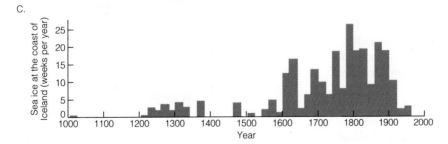

D.

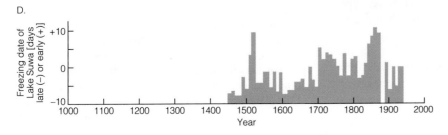

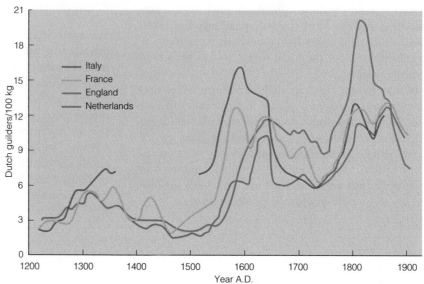

Figure 14.6 Fluctuations in the price of wheat in western Europe from the thirteenth to nineteenth century, expressed in Dutch guilders, track the course of climate. Intervals of cool, wet climate were unfavorable for wheat pro- duction, causing the price to rise. The two largest peaks, in the early seventeenth and early nineteenth centuries, coincide with the greatest advances of glaciers in the Alps during the Little Ice Age.

the number of severe winters in China since the sixth century A.D., the height of the Nile River at Cairo since A.D. 622, the quality of wine harvests in Germany since the ninth century A.D., dates for the blooming of cherry trees in Kyoto, Japan, since A.D. 812, and wheat prices (a reflection of climatic adversity) in England, France, the Netherlands, and northern Italy since A.D. 1200 (Fig. 14.6). Each of these phenomena bears a relationship to prevailing climate and therefore is regarded as a useful proxy for climatic variability.

As we learned in Chapter 10, another source of paleoclimate (i.e., past climate) information comes from ice cores collected from polar glaciers. Measurements of the ratio of two isotopes of oxygen (^{18}O and ^{16}O) in glacier ice enable us to estimate air temperature when the snow that later was transformed into that ice accumulated at the glacier surface. Cores obtained from the Greenland and Antarctic ice sheets, as well as from several smaller ice caps at lower latitudes, provide continuous records of fluctuating temperatures near the surface of these glaciers. In some cases, they extend back many tens of thousands of years (Fig. 14.7).

Trees offer additional important information about past climates. A tree living in middle latitudes typically adds a growth ring each year, the width and density of which reflect the local climate (Fig. 14.8). Many species live for hundreds of years; a few, like the Giant Sequoia and Bristlecone pine of the California mountains, can live thousands of years. Specialists in tree-ring analysis are able to reconstruct temperature

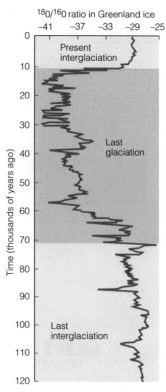

Figure 14.7 Variations in the oxygen-isotope ratio through the Greenland Ice Sheet. The zone of strongly negative values beginning about 70,000 years ago and ending about 10,000 years represents the last glaciation. The pronounced shift in values about 10,000 years ago marks an abrupt change from glacial to interglacial climate at the end of the glaciation.

A.

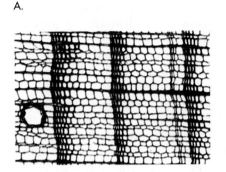

B.

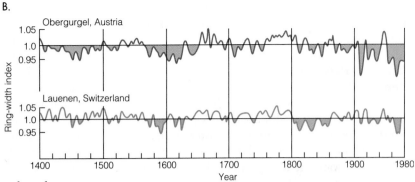

Figure 14.8 Climate and tree rings. A. An enlarged cross section of a 1500-year-old fossil larch tree found in the moraine of a Swiss glacier showing annual growth rings. Early wood of each year consists of large, well-formed cells. Late wood contains smaller, closely spaced cells. B. Tree-ring chronologies based on density measurements of spruce trees at two sites in the Alps that are 200 km apart. A general similarity can be seen among periods of low growth (shaded) in the two records that correspond with times of cold climate and glacier expansion.

and precipitation patterns from tree rings over broad geographic areas for any specific year in the past. These reconstructions provide pictures of both changing regional weather patterns and long-term climatic trends.

In Chapter 11 we saw how scientists use the growth rings of living corals to reconstruct water temperature oscillations related to the cyclic El Niño/Southern Oscillation. Because the rings are laid down annually (Fig. 11.16A), we can obtain a high-resolution record of changes in surface water temperature spanning hundreds of years.

Climate of the Last Millennium

A wealth of historical and climate proxy records provide us with an unusually comprehensive picture of climatic variations during the last thousand years. The varied evidence from the northern hemisphere shows that an episode of relatively mild climate during the Middle Ages gave way about 700 years ago to a colder period when temperatures in Western Europe averaged 1 to 2°C (2 to 4°F) lower. This cooler climate caused a lowering of the snowline by about 100 m (330 ft) in the world's high mountains, thereby causing glaciers to advance. Geologists refer to this interval of cooler climate and glacier advance as the Little Ice Age. Throughout much of western Europe and adjacent islands, the Little Ice Age climate was punctuated by unusually harsh conditions marked by snowy winters and cool, wet summers, expansion of sea ice in the North Atlantic, and an increase in the

frequency of violent wind storms and sea floods in mainland Europe. As summers became cooler and wetter, grain failed to ripen, wheat prices rose (Fig. 14.6), and famine became pervasive. In England the life expectancy fell by 10 years within a century.

By the early seventeenth century, advancing glaciers were overrunning farms in the Alps, Iceland, and Scandinavia. During the worst years of that century, sea ice completely surrounded Iceland, and the cod fishery in the Faeroe Islands failed because of increasing ice cover.

The 1810–1819 decade, the coldest in Europe since the seventeenth century, witnessed renewed advances of glaciers in the Alps and many other mountain ranges. Erratic weather in the nineteenth century led to further crop failures, rising grain prices, epidemics, and famines that resulted in large-scale emigrations of Europeans, especially to North America. Thus, many Canadians and Americans owe their present nationality to the vagaries of Little Ice Age climate.

Little Ice Age conditions persisted until the middle of the nineteenth century when a general warming trend caused mountain glaciers to retreat and the edge of the North Atlantic sea ice to retreat northward. Although minor fluctuations of climate have continued to take place (Fig. 14.3), the overall trend of increasing warmth in middle latitudes brought conditions that were increasingly favorable for crop production at a time when the human population was expanding rapidly and entering the industrial age.

The Last Glaciation

The last time the Earth's climate was dramatically different from what it is now was during the last **glaciation**, an interval when the Earth's global ice cover greatly exceeded that of today. The last glaciation, which culminated about 20,000 years ago, was the most recent of a long succession of glaciations, or ice ages, that characterized the Pleistocene Epoch. To reconstruct the climate of this latest ice age, earth scientists rely largely on sediments and ancient glacier ice that contain fossil and isotopic evidence of ice-age conditions.

Glaciers, Permafrost, and Sea Ice

During the last glaciation, the climate of northern middle and high latitudes became so cold that a vast ice sheet formed over central and eastern Canada and expanded southward toward the United States and westward toward the Rocky Mountains (Fig. 14.9). As it moved across the Great Lakes region, the glacier overwhelmed spruce trees growing in scattered groves beyond the ice margin. Ancient logs of that period, now exposed in the sides of stream valleys, are bent and twisted, indicating that they were alive when the glacier destroyed them. Some retain their bark, and some lie pointing in the direction of ice flow, like large aligned arrows. A radiocarbon date for the outermost wood or bark of such a log tells us the approximate time when the ice arrived and the tree was killed. The ages of buried trees discovered near the southern limit of the ice sheet indicate that the ice reached its greatest extent about 24,000 years ago. Still older wood found farther north pinpoints ages for the ice margin during its southward advance. Dividing the distance between two sample localities along a north-south transect by the difference in age of the logs at these sites yields an average rate of advance of the ice margin across that distance. Results from such pairs of dates suggest that the ice was advancing at an average rate of 25 to 100 m (82 to 330 ft) per year, a rate that is comparable to that of some large existing glaciers.

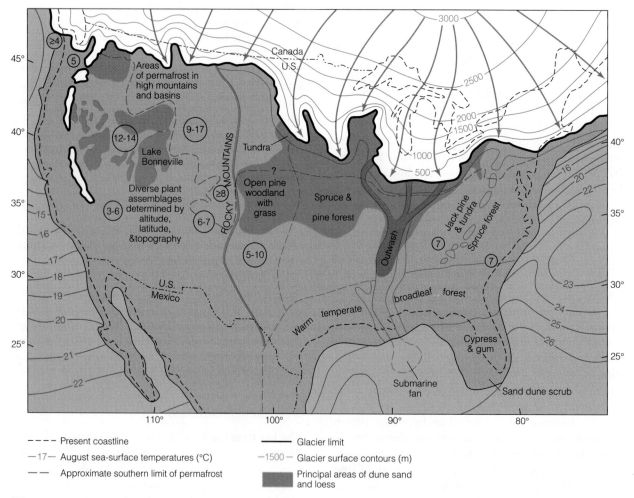

Figure 14.9 Geography of central North America about 20,000 years ago, during the last glaciation. Coastlines lie farther seaward owing to a fall of sea level of about 120 m. Sea-surface temperatures are based on analysis of microfossils in deep-sea cores. Circled numbers show estimated temperature lowering, relative to present temperatures, at selected sites based on various kinds of climate-proxy evidence.

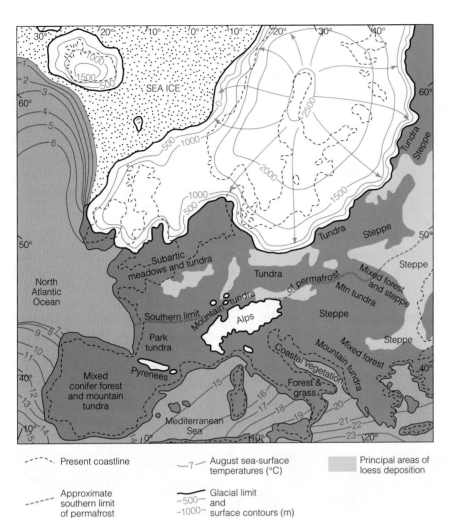

Figure 14.10 Geography of western Europe about 20,000 years ago during the last glaciation. At that time, a vast continental ice sheet which extended across northern Europe, was separated from the glacier-covered Alps by a frigid periglacial zone. Cold polar waters extended far south of their present limit in the North Atlantic Ocean. Because of lower sea level, areas now covered by waters of the English Channel and the North Sea were dry land.

- - - - - Present coastline

—7— August sea-surface temperatures (°C)

Principal areas of loess deposition

- - - - - Approximate southern limit of permafrost

—500—
—1000— Glacial limit and surface contours (m)

Simultaneously, other great ice sheets formed over the mountains of western Canada and over northern Europe (Fig. 14.10) and northwestern Asia. As ocean water was evaporated and then deposited as snow on these growing ice sheets, world sea level fell. The falling sea level allowed the great ice sheets of Greenland and Antarctica to grow larger as they spread across the adjacent, exposed continental shelves. Large glacier systems also formed in the Alps, Andes, Himalaya, and Rockies, and smaller glaciers developed on numerous other ranges and isolated peaks scattered widely through all latitudes.

We assume that ice shelves also existed under full-glacial conditions, but their size and distribution are not easy to determine. Some geologists postulate that an ice shelf may have covered all of the Arctic Ocean and extended south into the northern reaches of the Atlantic Ocean, thereby linking the major northern ice sheets into a continuous glacier system that covered nearly all of the arctic and much of the subarctic regions of the planet. Other geologists concede that ice shelves very likely were present in favorable places,

just as they are today around Antarctica, but suggest that the polar sea was largely covered by much thinner sea ice that extended far south of its present limit into the North Atlantic.

With the southward spread of ice sheets on the northern continents, periglacial zones were displaced to lower latitudes and lower altitudes. In Russia permafrost extended 1000 km (620 mi) or more south of the ice margin. However, in North America evidence of full-glacial permafrost is restricted largely to Alaska, to a narrow belt adjacent to the southernmost limit of the ice sheet in the northern Great Plains and Great Lakes regions, and to the high mountains of the American West, especially the Rockies. The contrast may largely reflect the fact that, whereas the massive Eurasian glacier lay north of 50° latitude, the ice sheet over central North America extended south of 40° into more temperate latitudes. The periglacial zone was therefore much narrower in the United States because the north-to-south gradient of climate there was far steeper.

The Dusty Ice-Age Atmosphere

At the height of the glacial age, the middle latitudes were both windier and dustier than they are today. We infer this from several lines of evidence. Glacial-age loess found south of the ice limit in the midwestern United States becomes both thinner and finer east of the floodplains of former meltwater streams, implying that the dust was picked up and distributed by strong westerly winds. The thick loess deposits of central China lie east of desert basins in central Asia that were swept by cold, dry winds during glacial times. Loess deposits in eastern Europe lie downwind from extensive meltwater sediments lying between the Alps and the southern limit of the great north European ice sheet. They contain fossil plants and animals consistent with cold, dry conditions, implying that dust deposition was characteristic of glacial times. In each of these regions, successive sheets of loess are separated by soils, each formed during an **interglaciation**, which is a time when both the climate and the global ice cover were similar to those of today.

That glacial times were both windy and dusty is also shown by studies of fine dust found in ice cores from the Greenland Ice Sheet. The percentage of wind-blown dust rises significantly in the part of the cores that corresponds to the last glaciation. Because Greenland and a large part of northern North America were ice-covered at that time, much of the dust likely originated in the deserts of central Asia and along the valleys of braided meltwater streams south of the North American ice sheets.

Sea-Level and Lake-Level Changes

The fall of world sea level that accompanied the buildup of glaciers on land changed the shape of the continents as broad areas of shallow continental shelf were exposed (Figs. 14.9 and 14.10). The fall in sea level also changed the gradients of the downstream segments of major streams, causing them to deepen their valleys as they reestablished equilibrium profiles. Stream sediments that had been dumped on the inner continental shelf wherever a river formerly entered

Figure 14.11 Horizontal benches at several levels above the surface of Great Salt Lake, Utah, mark shorelines of Lake Bonneville, a vast Pleistocene lake. At its maximum extent and depth during the last glaciation, the surface of Lake Bonneville stood more than 300 m above that of the present lake.

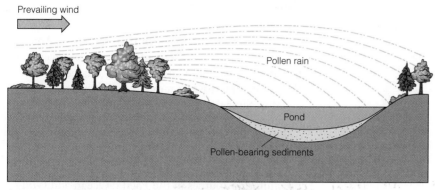

A.

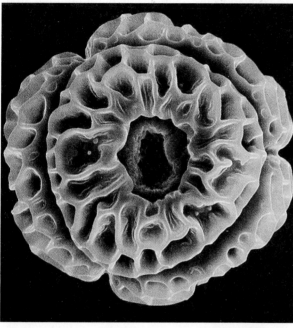

B.

Figure 14.12 Fossil pollen used to reconstruct past vegetation and climate. A. Windborne pollen grains from trees and shrubs fall into a nearby pond where they are incorporated as part of the accumulating sedimentary strata. B. Scanning electron microscope photograph of a grain of *Drymis winterii* pollen, having a diameter of 42 microns.

the ocean were now transported across the exposed shelf and deposited at the shelf margin. From there, the accumulating detritus could be carried by density currents swiftly down the continental slope to the deep sea.

In many arid and semiarid regions of the world, including the Sahara, the Middle East, southern Australia, and the American Southwest, the shift to glacial-age climates resulted either in the enlargement of existing lakes or the creation of new ones. For example, during the last glaciation the Great Salt Lake

basin in the western United States was occupied by a gigantic water body that geologists refer to as Lake Bonneville (Fig. 14.9). More than 300 m (985 ft) deeper than Great Salt Lake, Lake Bonneville had a volume comparable to that of modern Lake Michigan. Beaches, deltas of tributary streams, and lake-bottom sediments provide the evidence (Fig. 14.11). Although we might guess that expansion of the lake was caused by increased precipitation, evidence in fact points to *reduced* precipitation during glacial time. Lake Bonneville and other lakes of the American Southwest formed in a region where present-day evaporation rates are very high, and so an alternative explanation is that lake expansion may have resulted primarily from lower glacial age temperatures that led to reduced water loss by evaporation.

Ice-Age Vegetation

The glacial ages witnessed major changes not only in the cryosphere, hydrosphere, and atmosphere, but also in the biosphere. In fact, much of our knowledge of climatic conditions outside the great ice sheets during glacial times is based on interpretation of plant fossils. Large plant fragments permit identification of individual species, but they are far less numerous than fossil pollen grains, which possess a hard, waxy coating that resists destruction by chemical weathering. Most pollen is transported by the wind and settles into lakes, ponds, and bogs, where, protected from destructive oxidation in the wet environment, it slowly accumulates (Fig. 14.12). A sample of bog or lake sediment yields a vast number of pollen grains that can be identified by type, counted, and analyzed statistically. At any given level in a core, the pollen grains reveal the assemblage of plants that flourished near the site when the enclosing sediment layers were deposited (Fig. 14.13). If a modern vegetation assemblage can be found that has a composition like that implied by the fossil pollen, then the precipitation and temperature at the site of the modern assemblage can be used to

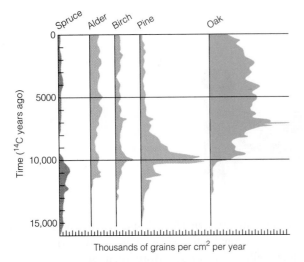

Figure 14.13 Simplified pollen diagram prepared from data collected at Rogers Lake, Connecticut. Variations in pollen influx are plotted as a function of time, and show progressive changes in forest composition. A major change occurred about 10,000 years ago at the end of the last glaciation when spruce/pine forest was replaced by a forest dominated by pine and deciduous trees.

estimate climatic conditions represented by the fossil assemblage.

The vegetation pattern in eastern North America prior to European settlement consisted of several approximately parallel belts that were related mainly to the gradual increase of temperature from pole toward equator. Superimposed on this latitudinal pattern was a change from moist forest in the east to dry grassland in the west. In the Far West, a complex mosaic of vegetation assemblages existed, with patterns determined by latitude, altitude, topography, and distance from precipitation sources in the Pacific Ocean and Gulf of Mexico. Pollen studies show us that in glacial times the vegetation distribution was quite different from this recent distribution (Fig. 14.9). About 20,000 years ago, a belt of tundra existed immediately south of the glacier margin, implying a much colder climate. Today's grassland country of the Great Plains was then mostly open pine woodland. It was once supposed that as the ice sheets advanced and then retreated, vegetation zones seen on today's map crept gradually southward and then back northward, each maintaining its own character. However, pollen stud-

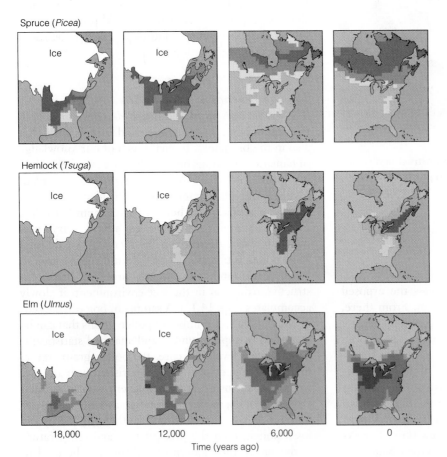

Figure 14.14 Changing distribution of spruce, hemlock, and elm trees in eastern North America at 6000-year intervals between 18,000 years ago and the present day based on fossil pollen data. The color intensities indicate relative abundance for each species, with the darkest shade of green being the highest and the lightest shade the lowest.

ies tell us that vegetation changes accompanying the advance and retreat of the great ice sheets were dynamic and far more complicated. Species were displaced in various directions, forming new plant communities that are unknown on the present landscape (Fig. 14.14).

In Europe the vegetational response to glaciation was similar (Fig. 14.10) but with one major difference. In North America plant species forced southward by the advancing ice could inhabit relatively warm lowlands that extended to the Gulf of Mexico. But in Europe, the glacier-clad Alps, 800 km (500 mi) long and 150 km (93 mi) wide, constituted a high, cold barrier north of the Mediterranean Sea. Many species were trapped between the large ice sheet to the north and the Alpine glaciers to the south and were driven to extinction. Thus, western Europe, which before the glacial ages had an abundance of tree types, now has only 30 naturally occurring species. By contrast, North America, with no mountain barrier standing between the Great Lakes and the Gulf of Mexico, has 130 species.

Changes in Temperature and Precipitation

In the popular imagination, glacial ages were times when temperatures were very cold, perhaps rivaling those in the middle of Antarctica today. Although such extreme cold did exist in some regions, in other places *average* temperatures at the culmination of the last glaciation were not very different from what they are now. The evidence for glacial-age temperatures is varied, and it comes from both the land and the ocean basins.

Estimates of temperature lowering on the land span a range of values. In midlatitude coastal regions, temperatures were generally reduced by about 5 to 8°C (9 to 14°F), whereas in continental interiors reductions of 10 to 15°C (18 to 27°F) occurred (Fig. 14.9). Some of the ways such estimates are derived include the following:

1. By comparing the snowlines of ice-age glaciers with those of modern glaciers, a value for snowline lowering can be obtained (Fig. 14.15). An estimate of temperature lowering can then be determined by assuming that present and past average rates for the upward decrease of temperature with altitude are similar (6°C/km). When this average rate is applied to the calculated snowline difference, the resulting depression in temperature can be found. However such estimates disregard the effects of precipitation, which also control snowline altitude (Chapter 10).

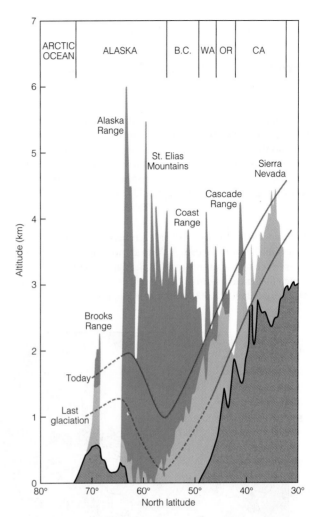

Figure 14.15 Transect along the coastal mountains of western North America showing the relationship of the present snowline to existing glaciers (blue) and of the ice age snowline to expanded glaciers during the last glaciation (light blue). The difference between present and ice age snowlines was about 900–1000 m along the southern part of the transect and about 600 m in northern Alaska. The change in slope of the two snowlines at about 55° latitude occurs where the transect passes inland across the Alaska Range and then northward across the Brooks Range to the Arctic Ocean.

2. By studying fossil pollen grains, ice-age vegetation assemblages can be reconstructed. Using the contemporary range of temperatures for these assemblages, past temperatures can be inferred. This approach works only where an assemblage reconstructed from fossil pollen matches a modern assemblage. A similar approach can be taken using animal fossils, such as beetles, which give comparable results.

A CLOSER LOOK

Reconstructing Ice-Age Ocean-Surface Conditions

The upper parts of drill cores taken from the seafloor throughout the world's oceans consist of soft sediments that commonly contain multitudes of tiny fossils. Most of the fossils are of microorganisms that live in the surface waters and whose shells rain down on the seafloor in vast numbers to form deep-sea oozes (Chapter 11). The rate of sedimen-

tation is extremely slow, however, so that it may take more than a thousand years for a single centimeter of sediment to accumulate. Because the assemblage of organisms that live in the surface waters is closely related to water temperature, the fossil remains in the sediment provide a record of changing conditions at the ocean surface.

In many deep-sea sediment cores, the fossil content changes downward, typically shifting back and forth from predominantly warm-water (interglacial) to cold-water (glacial) forms. By identifying the species present at any

Figure C14.1 Present and past temperatures of surface waters in the world ocean. A. Map showing modern August sea-surface temperatures (in °C). B. Map showing reconstructed August sea-surface temperatures during the last glaciation, about 18,000 years ago. Cold polar water ex-

tended far south of its present limit in the North Atlantic, and plumes of cool water extended westward from South America in the equatorial Pacific and from Africa in the Atlantic.

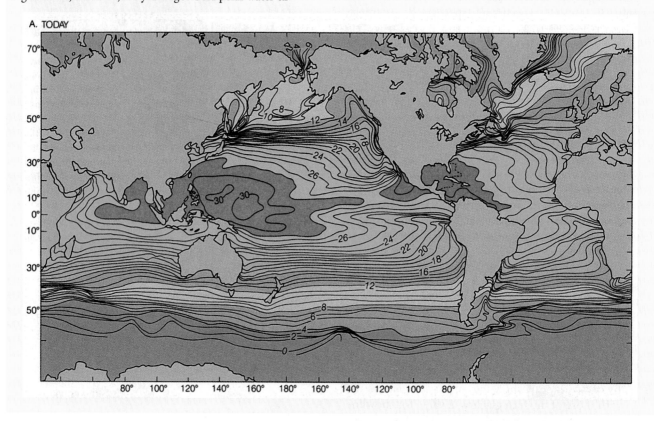

3. By obtaining measurements of the oxygen-isotope ratio in ice cores that penetrate ice of the last glacial age (Fig. 14.7), former surface air temperature can be estimated. The measurements show a marked change in isotope values at a level coinciding with the transition from mild (Holocene) interglacial climate recorded in the

upper parts of the cores to cold ice-age temperatures below.

4. By noting the distribution of certain periglacial features that indicate former permafrost conditions, an estimate can be made of the minimum temperature change that has taken place. At present, permafrost exists mainly in areas where the

level in a sediment core and comparing that assemblage with modern ones, it is possible to infer what the surface ocean temperature must have been when the shells were settling to the seafloor. In practice, geologists can select a level in a core that represents the peak of the last glaciation and determine, from the contained fossils, the surface water temperature at that time. Information from hundreds of cores scattered widely over the oceans has been used to derive a global map of sea-surface temperature for the last glacial maximum (Fig. C14.1B).

Surprisingly, the *average* global difference between present and ice-age sea-surface temperatures is only about 2.3°C (4°F), but this figure is somewhat misleading. In some large regions, like the subtropics, little or no change in temperature is detected. In others, such as the North Atlantic,

sea-surface temperatures were locally as much as 14°C (25°F) colder than now. In this region, cold polar water that is now found mainly north of latitude 60° descended far south at the peak of the glaciation to reach the shore of northeastern United States and the Iberian Peninsula in western Europe.

The greatest ocean temperature declines occurred in the North Atlantic around which large continental ice sheets were located, and in enclosed seas of the northwestern Pacific. They also were substantial near the equator, where cold water that welled up off the coasts of Africa and South America spread westward across the equatorial Atlantic and Pacific, respectively. However, over vast areas within the North and South Pacific midocean gyres, sea-surface temperatures apparently changed very little .

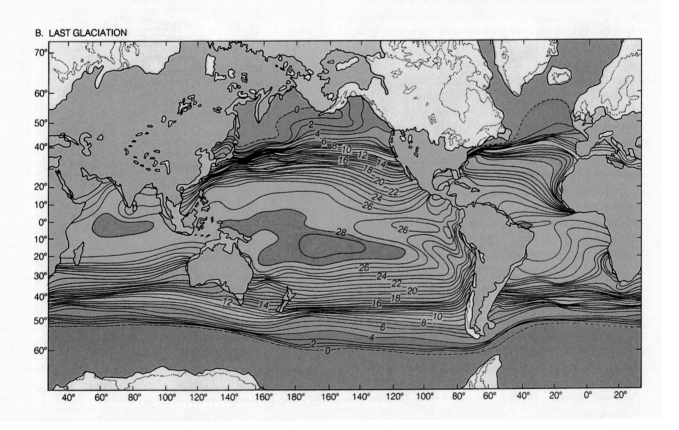

B. LAST GLACIATION

mean annual air temperature is below −5°C (23°F). If, for example, evidence of former permafrost is found at a place where the annual temperature is now 4°C (39°F), then the former periglacial climate is inferred to have been at least 9°C (16°F) colder.

5. By sampling deep-sea sediment cores for fossils of

microorganisms that lived near the ocean surface and by comparing the fossil assemblages with those now living in surface waters at various latitudes, sea-surface temperatures during the last glaciation can be reconstructed (see "A Closer Look: Reconstructing Ice-Age Ocean-Surface Conditions").

From these and other types of evidence obtained on land and from the oceans, we have learned an important fact: the changes accompanying a shift from interglacial to glacial conditions did not affect the whole world equally. The environments of some regions apparently changed little if at all, whereas others experienced profound changes.

Successive Pleistocene Glacial and Interglacial Ages

As recently as a few decades ago, it was thought that the Earth had experienced only four glacial ages during the Pleistocene Epoch. This assumption was based on studies of ice sheet and mountain glacier deposits, and it had its roots in early studies of the Alps where geologists identified stream terraces they thought were related to four ice advances. This traditional view was discarded when studies of deep-sea sediments disclosed a long succession of glaciations during the Pleistocene, the most recent of which was shown by radiocarbon dating to equate with deposits of the last glaciation on the continents. Paleomagnetic dating (Chapter 6) of deep-sea cores shows that the most recent glacial-interglacial cycles recorded in the sediments average about 100,000 years long and that during the last 800,000 years alone there have been about eight such episodes. For the Pleistocene Epoch as a whole (the last 1.8 million years), about 30 glacial ages are recorded rather than the traditional four. The implications are clear: whereas seafloor sediments provide a continuous historical record of climatic change, evidence of glaciation on land generally is incomplete and interrupted by many unconformities.

The seafloor evidence is of three kinds. First, with increasing depth in a core, the biologic component of the sediments shows repeated shifts from warm interglacial biota to cold glacial biota. Second, the percentage of calcium carbonate in cores from some ocean regions fluctuates in much the same manner. Third, the ^{18}O to ^{16}O isotope ratio fluctuates with a pattern similar to that shown in the biologic and mineral fractions of the sediments. Whereas the isotopic variations in ice cores are believed to represent fluctuations in air temperature near the glacier surface, in Pleistocene marine sediments they are thought primarily to reflect changes in global ice volume. During glacial ages, when water is evaporated from the oceans and precipitated on land to form glaciers, water containing the light isotope ^{16}O is more easily evaporated than water containing the heavier ^{18}O. As a result, Pleistocene glaciers contained more of the light isotope, whereas the oceans became enriched in the heavy isotope. Isotope curves derived from the sediments therefore give

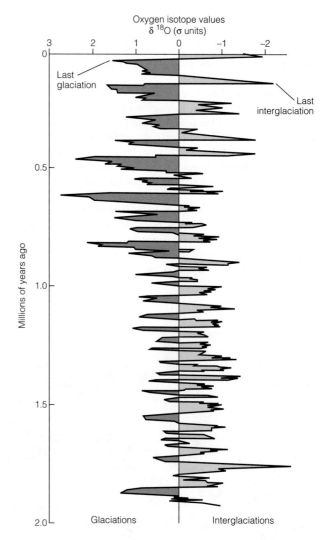

Figure 14.16 Curve of average oxygen-isotope variations during the last 2 million years based on analyses of deep-sea sediment cores. The curve illustrates changing global ice volume during successive glacial-interglacial cycles of the Quaternary Period.

us a continuous reading of changing ice volume on the planet (Fig. 14.16). Because glaciers wax and wane in response to changes of climate, the isotopes also give a generalized view of global climatic change.

For most of the last 800,000 years, peaks in the isotope curve that represent times of high global ice volume have a somewhat similar amplitude, suggesting that during each glaciation the amount of ice on land was about the same (Fig. 14.16). During this long interval, the average length of a glacial-interglacial cycle was about 100,000 years. Prior to that time, however, the amplitude of the cycles was smaller and their duration averaged only about 40,000 years. Why the cycle length changed is not yet known with certainty, but it clearly represents a fundamental shift in the Earth's climate system.

A record of ocean-surface temperatures, based on oxygen isotope values in deep-sea cores that penetrate Cenozoic sediment, shows that the oceans have grown colder over the last 50 million years (Fig. 14.17). During one pronounced cooling event about 35 million years ago, surface ocean temperatures declined by nearly 5°C (9°F) within only about 100,000 years. In concert with the long-term cooling trend, glaciers spread from highlands in Antarctica and reached the sea. About 12 to 10 million years ago, ice volume increased, and an ice sheet formed over Antarctica as temperatures continued to fall. The presence of such a large polar ice mass reduced average temperatures on the Earth still further and caused a substantial drop in sea level. From that time onward, large glaciers occupied mountain valleys of Alaska and the southern Andes. Although the evidence is still sketchy, it appears that large ice sheets did not form in northern middle latitudes until about 2.5 million years ago. If this inferred history is correct, glaciation has gradually affected more and more of the Earth's land surface during the Cenozoic: first the Antarctic, then high-latitude mountain systems, and more recently the northern middle latitudes.

Ancient glaciations, identified mainly by rocks of glacial origin and associated polished and striated rock surfaces, are known from pre-Cenozoic times as well. The earliest recorded glacial episode dates to about 2.3 billion years ago, in the middle Precambrian. Evidence of other glacial episodes has been found in rocks of late Precambrian, early Paleozoic, and late Paleozoic age (Fig. 8.3). During the latest of these intervals, 50 or more glaciations are believed to have occurred. The geologic record is fragmentary and not always easy to interpret, but evidence from such low-latitude regions as South America, Africa, and India, as well as from Antarctica, suggests that the Earth's land areas must have had a very different relationship to one another during the late Paleozoic glaciation than they do today. In the Mesozoic Era, glaciation of similar magnitude apparently did not occur, consistent with geologic evidence that points to a long interval of mild temperatures both on land and in the oceans.

The Warm Middle Cretaceous

It's probably a good thing we did not live 100 million years ago during the Middle Cretaceous Period. Not only was the world inhabited by huge carnivorous dinosaurs, but also the climate was one of the warmest in the Earth's history. Evidence that the world was much warmer in that period than it is today is compelling (Fig. 14.18). Warm-water marine faunas were widespread, coral reefs grew 5° to 15° closer to the poles than they do now, and vegetation zones were displaced about 15° poleward of their present positions. Peat deposits that would give rise to widespread coal formations formed at high latitudes, and dinosaurs, which are generally thought to have preferred warm climates, ranged north of the Arctic Circle. Sea level was 100 to 200 m (330 to 650 ft) higher than today, implying the absence of polar ice sheets, and isotopic measurements of deep-sea deposits indicate that intermediate and deep waters in the oceans

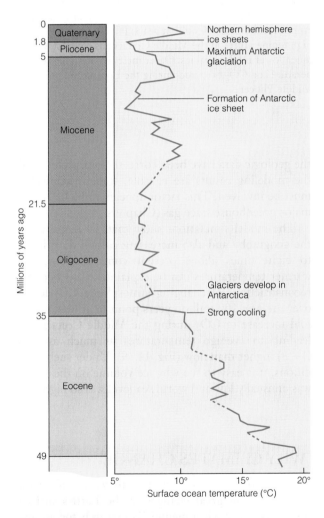

Figure 14.17 A long record of surface ocean temperatures based on oxygen-isotope ratios measured in a sediment core from the western Pacific Ocean. Relatively warm surface waters cooled abruptly about 35 million years ago, reflecting a climatic change that led to the buildup of glaciers in Antarctica. With further cooling, an ice sheet developed over Antarctica, and by 2.5 million years ago, northern hemisphere ice sheets had formed.

were 15 to 20°C (27 to 36°F) warmer than now. Based on such evidence, average global temperature is estimated to have been at least 6°C (11°F) milder than today and possibly as much as 14°C (25°F), with the greatest difference being in the polar regions. Whereas today the difference in temperature between the poles and the equator is 41°C (74°F), during the Middle Cretaceous it may have been no more than 26°C (47°F) and possibly as little as 17°C (31°F).

Computer simulations of past climates provide insights into the Middle Cretaceous world and suggest that several factors were likely involved in producing such warm conditions: geography, ocean circulation, and atmospheric composition. The simulations show that the Middle Cretaceous arrangement of continents and oceans (Fig. 14.18), which influenced ocean circulation and planetary albedo, could account for nearly 5°C (9°F) of warming. Of this 5°, about a third is attributable to the absence of polar ice sheets. However, geography alone is inadequate to explain warmer year-round temperatures at high latitudes. Could the poleward transfer of heat be the answer? The oceans now account for about a third of the present poleward heat transfer, but modeling shows that even with the geography and ocean circulation rearranged as they were in the Middle Cretaceous, oceanic heat transfer cannot explain the greater high-latitude warmth. If

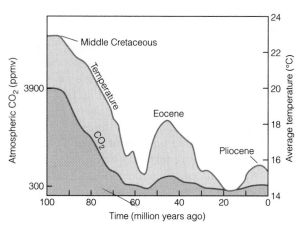

Figure 14.19 A geochemical reconstruction of changing atmospheric CO_2 concentration and average global temperature over the past 100 million years. High CO_2 values and high temperatures in the Middle Cretaceous contrast with much lower modern values. Other intervals of higher temperature and CO_2 occurred during the Eocene and the Middle Pliocene.

the geologic data have been correctly interpreted, and the modeling results are reliable, some other factor must be involved. This factor appears to be CO_2, the major greenhouse trace gas (Chapters 19 and 20).

The model simulations show that, by rearranging the geography and also increasing carbon dioxide six to eight times above present concentrations, the warmer temperatures can be explained. Geochemical reconstructions of changing atmospheric CO_2 levels over the past 100 million years point to at least a tenfold increase in CO_2 during the Middle Cretaceous, leading to average temperatures as much as 8°C (14°F) higher than now (Fig. 14.19). Under such conditions, it is easy to see why ice volume on the Earth was unusually low and world sea level was so high.

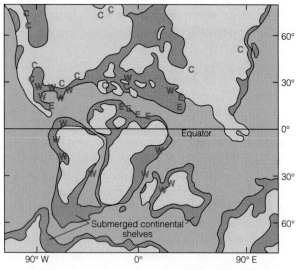

Figure 14.18 During the Middle Cretaceous Period, sea level was 100 to 200 m higher than now and ocean waters flooded large areas of the continents, producing shallow seas. Warm-water animal assemblages (W) and evaporite deposits (E) were present at low to middle latitudes, and coal deposits (C) developed in northern latitudes, implying warm year-round temperatures

WHY CLIMATES CHANGE

What factors cause the climate to warm and cool, bringing about great changes in the Earth's surface processes and environments? The search for an answer has proved difficult because climate fluctuates on different time scales, ranging from decades to many millions of years, and several quite different mechanisms appear to be responsible for these changes. Furthermore, these mechanisms involve not only the atmosphere, but also the lithosphere, the oceans, the biosphere, and extraterrestrial factors, all interacting in a complex way. The search for causes of climatic variability is therefore a challenging one.

Glacial Eras and Shifting Continents

The only reasonable explanation for a succession of glacial episodes during the last 2.3 billion years seems to be the slow but important geographic changes that affect the Earth's crust. These changes include the movement of continents as they are carried along with shifting plates of lithosphere, the creation of high mountain chains and plateaus where plates collide, and the opening or closing of ocean basins and seaways between moving landmasses.

How such movements affect climate is illustrated by the fact that low temperatures occur, and glaciers tend to form and persist, in high latitudes and at high altitudes, and especially in places where winds can supply abundant moisture evaporated from a nearby ocean. The Earth's largest existing glacier is centered on the South Pole, where temperatures are constantly below freezing and the land is surrounded by ocean. The only glaciers found at or close to the equator lie at extremely high altitudes.

Abundant evidence now leads us to conclude that the positions, shapes, and altitudes of landmasses have changed with time (Chapter 4). In the process, the paths of ocean currents and atmospheric circulation have been altered. As landmasses and ocean basins have shifted position, occasionally they have assumed an arrangement that was optimal for widespread glaciation in high latitudes. Where evidence of ancient ice sheet glaciation is now found in low latitudes, we invariably find evidence that such lands formerly were located in higher latitudes. Although this explanation appears adequate to explain the pattern of glaciation during and since the late Paleozoic, information about earlier glacial intervals is very fragmentary and more difficult to evaluate.

Why Was the Middle Cretaceous Climate So Warm?

Interspersed with ancient glacial intervals were episodes of exceptionally warm climate, like that of the Middle Cretaceous. If CO_2 was an important factor in Middle Cretaceous warming, as suggested earlier, we still are faced with explaining how this gas increased so substantially. A likely explanation is volcanic activity, which today constitutes a major natural source of CO_2 entering the atmosphere. Most of this CO_2 is generated by slow, noneruptive degassing of magmas in the upper crust.

Geologic evidence points to an unusually high rate of volcanic activity in the Middle Cretaceous. Rates of continental drift were then about three times as great as now, implying increased extrusion rates at spreading ridges. In addition, vast outpourings of lava created a succession of great undersea volcanic plateaus across the southern Pacific Ocean between 135 and 115 million years ago, the time of maximum Cretaceous warmth. One of these—the Ontong-Java Plateau in the southwestern Pacific—has more than twice the area of Alaska and reaches a thickness of 40 km (25 mi). Such a massive outpouring of lava likely released a huge volume of CO_2. Could this gas emission have been sufficient to warm the climate to unprecedented levels? By one calculation, the eruptions could have released enough CO_2 to raise the atmospheric concentration to 20 times its natural value at the beginning of the Industrial Revolution (ca. A.D. 1760), in the process raising average global temperature as much as 10°C (18°F). Other estimates range from 8 to 12 times the A.D. 1760 value.

Recently, geologists have proposed that each such vast lava outpouring is associated with a *superplume*, which is conceived of as a plumelike mass of unusually hot rock that rises from the base of the mantle. Moving upward at a rate of 10 to 20 cm/yr (4 to 8 in/yr), the hot rock spreads out in a mushroom shape as it reaches shallower depths where confining pressures are lower (Fig. 14.20). Such a superplume would be an efficient mechanism for allowing heat to escape from the Earth's core. If this hypothesis is correct, then the plate tectonic cycle cools the mantle both by heat loss at spreading ridges and by the downward plunge of plates of cool lithosphere, while superplumes cool the core. By this reasoning, the core and atmosphere are linked dynamically, and the warm Middle Cretaceous climate was a direct consequence of the cooling of the Earth's deep interior.

Ice-Age Periodicity and the Astronomical Theory

Determining the cause of the cyclic pattern of glacial and interglacial ages has long been a fundamental challenge to the development of a comprehensive theory of climate. A preliminary answer was provided by Scottish geologist John Croll, in the mid-nineteenth century, and later elaborated by Milutin Milankovitch, a Serbian astronomer of the early twentieth century.

Croll and Milankovitch recognized that minor variations in the Earth's orbit around the Sun and in the tilt of the Earth's axis cause slight but important variations in the amount of radiant energy reaching

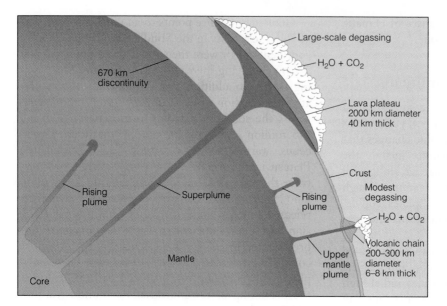

Figure 14.20 New evidence has led to the hypothesis that *superplumes*, rising slowly from the core-mantle boundary, build huge lava plateaus when they reach the top of the lithosphere. Simultaneous large-scale degassing of CO_2 could greatly enhance the atmospheric greenhouse effect. Smaller plumes rising from the base of the upper mantle at 670 km would produce much more-restricted hot spots that generate volcanoes like those of the Hawaiian chain.

any given latitude. Three movements are involved (Fig. 14.21).

First, the axis of rotation, which now points in the direction of the North Star, wobbles like the axis of a spinning top (Fig. 14.21A). The wobbling movement causes the North Pole to trace a cone in space, completing one full revolution every 26,000 years. At the same time, the axis of the Earth's elliptical orbit is also rotating, but much more slowly, in the opposite direction. These two motions together cause a progressive shift in the position of the four cardinal points of the Earth's orbit (spring and autumn equinoxes and winter and summer solstices). As the equinoxes move slowly around the orbital path, a motion called *precession of the equinoxes*, they complete one full cycle in about 23,000 years.

Second, the *tilt* of the axis, which now averages 23.5°, shifts about 1.5° to either side during a span of about 41,000 years (Fig. 14.21B).

Finally, the *eccentricity* of the orbit, which is a measure of its circularity, changes over a period of 100,000 years. About 50,000 years ago, the orbit was more circular (lower eccentricity) than it has been for the last 10,000 years (Fig. 14.21C).

The slow but predictable changes in precession, tilt, and eccentricity cause long-term variations of as much as 10 percent in the amount of radiant energy that reaches any particular latitude on the Earth's surface in a given season (Fig. 14.22). By reconstructing and dating the history of climatic variations over hundreds of thousands of years, geologists and oceanographers have shown that fluctuations of climate on

glacial-interglacial time scales match the predictable cyclic changes in the Earth's orbit and axial tilt. This persuasive evidence supports the theory that these astronomical factors control the *timing* of the glacial–interglacial cycles.

Amplification of Temperature Changes

Although orbital factors can explain the timing of the glacial-interglacial cycles, the variations in solar radiation reaching the Earth's surface are too small to account for the average global temperature changes of 4 to 10°C (7 to 18°F) implied by paleoclimatic evidence. Somehow, the slight temperature decreases caused by orbital changes must have been amplified into temperature changes sufficiently large to generate and maintain the huge Pleistocene ice sheets. We do not yet know how this amplification was accomplished, but some of the factors involved are likely to be changes in the chemical composition and dustiness of the atmosphere and changes in the reflectivity of the Earth's surface.

The chemical composition of air bubbles trapped in polar glaciers indicates that during glacial times the atmosphere contained less carbon dioxide and methane than it does today (Fig. 14.23). These two gases are important greenhouse gases (Chapter 20). If their concentration in the atmosphere is high, they trap radiant energy emitted from the Earth's surface that would otherwise escape to outer space. As a result, the lower atmosphere heats up and the Earth's climate becomes warmer. If the concentration of these

A.   Precession of the equinoxes (period = 23,000 years)

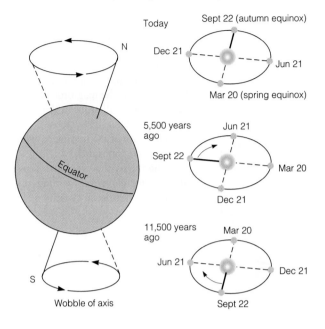

B.   Tilt of the axis (period = 41,000 years)

C.   Eccentricity (dominant period =100,000 years)

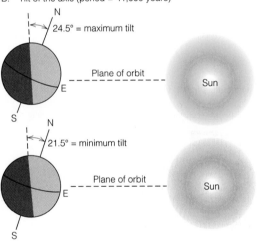

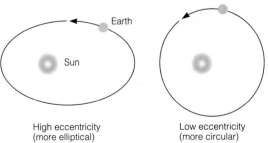

High eccentricity   Low eccentricity
(more elliptical)   (more circular)

gases is low, as it was during glacial times, surface air temperatures are reduced. Calculations suggest that the low levels of these two important atmospheric gases during glacial times can account for nearly half of the total ice-age temperature lowering. Therefore, the greenhouse gases likely played an important role in explaining the *magnitude* of past global temperature changes. Although we know that the atmospheric concentration of these gases fell during glacial times, we do not yet know for certain what caused them to fall.

As we learned earlier, ice core studies have shown that the amount of dust in the atmosphere was unusually high during glacial times when mid-latitude climates were generally drier and windier. The fine atmospheric dust scattered incoming radiation back into space, which would have further cooled the Earth's surface.

Whenever the world enters a glacial age, large areas of land are progressively covered by snow and glacier ice, and the extent of high-latitude sea ice increases. The highly reflective surfaces of snow and ice scatter incoming radiation back into space, further cooling the lower atmosphere. Together with lower greenhouse gas concentrations and increased atmospheric dust, this additional cooling would favor the expansion of glaciers.

Changes in Ocean Circulation

As we discussed in Chapter 11, the circulation of the world ocean plays an important role in global climate. The thermohaline circulation system links the atmosphere with the deep ocean. Warm surface water moving northward into the North Atlantic evaporates, and the remaining water becomes more saline and cools.

Figure 14.21 Geometry of the Earth's orbit and axial tilt. A. *Precession.* The Earth wobbles on its axis like a spinning top, making one revolution every 26,000 years. The axis of the Earth's elliptical orbit also rotates, though more slowly, in the opposite direction. These motions together cause a progressive shift, or precession, of the spring and autumn equinoxes, with each cycle lasting about 23,000 years. B. *Tilt.* The tilt of the Earth's axis, which now is about 23.5°, ranges from 21.5 to 24.5°. Each cycle lasts about 41,000 years. Increasing the tilt means a greater difference, for each hemisphere, between the amount of solar radiation received in summer and that received in winter. C. *Eccentricity.* The Earth's orbit is an ellipse with the Sun at one focus. Over 100,000 years, the shape of the orbit changes from almost circular (low eccentricity) to more elliptical (high eccentricity). The higher the eccentricity, the greater the seasonal variation in radiation received at any point on the Earth's surface.

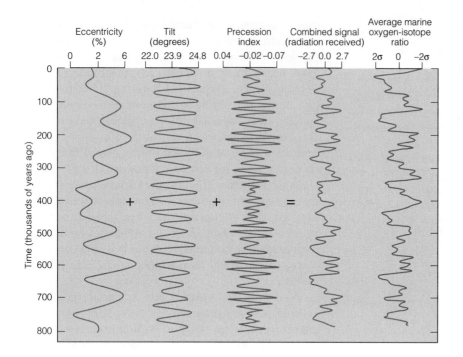

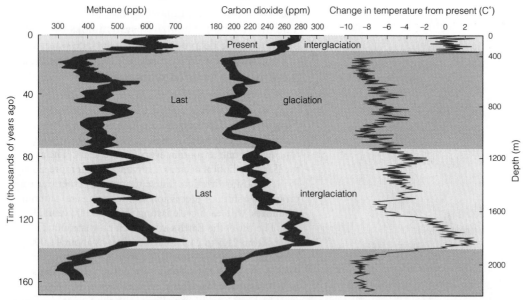

Figure 14.22 Curves showing variations in eccentricity, tilt, and precession during the last 800,000 years. Summing these factors produces a combined signal that shows the amount of radiation received on the Earth at a particular latitude through time. The magnitude and frequency of oscillations in the combined orbital signal closely matches those of the marine oxygen isotope curve (on right), which supports the theory that the Earth's orbital changes control the timing of the glacial-interglacial cycles.

Figure 14.23 Curves comparing changes in carbon dioxide and methane with temperature changes based on oxygen-isotope values in samples from a deep ice core drilled at Vostok Station, Antarctica. Concentrations of these greenhouse gases were high during the early part of the last interglaciation, just as they are during the present interglaciation, but they were lower during glacial times. The curves are consistent with the hypothesis that the atmospheric concentration of these gases contributed to warm interglacial climates and cold glacial climates.

The resulting cold, saline water is dense and sinks to produce cold North Atlantic Deep Water. Heat released to the atmosphere by evaporation maintains a relatively mild interglacial climate in northwestern Europe. Consider what would happen, however, any time this system closed down.

The rate of thermohaline circulation is sensitive to surface salinity at sites where deep water forms. Studies have shown that during times of reduced salinity, thermohaline circulation is also reduced. We therefore can postulate that as summer radiation decreased at the onset of a glaciation, the high latitude ocean and atmosphere cooled, decreasing evaporation and leading to expansion of sea ice. The resulting decreased saltiness of the high-latitude surface waters would have halted the production of dense saline water, thereby shutting off thermohaline circulation. Reduction of high-latitude evaporation, significantly reducing the release of heat to the atmosphere, would have maintained cold air masses moving eastward across the North Atlantic. Further cooled by an expanding sea-ice cover in the North Atlantic and extensive ice sheets on the adjacent continents, the climate of Europe became increasingly cold, causing permafrost to form in a broad zone beyond the ice sheet margin (Fig. 14.10).

Thus, a change in the ocean's thermohaline circulation system provides a means of further amplifying the relatively small climatic effect attributable to astronomical changes. Furthermore, it may help explain why the Earth's climate system appears to fluctuate between two relatively stable modes, one in which the ocean conveyor system is operational (interglaciation) and one in which it has shut down (glaciation).

Millennial-Scale Changes of Climate

Over the past decade, considerable interest has been focused on fluctuations of climate that have recurred at intervals of several thousand rather than tens of thousands of years. Such changes are clearly displayed in isotope records from deep-sea cores, and they also have been detected in pollen, loess, lake-sediment, ice-core, and glacial records on land. The climatic changes are of interest because they are relatively brief (commonly lasting only a few hundred to a thousand years) and their onset and termination are often rapid, even abrupt. Detailed studies of ice cores have shown that the shift from one climate state to another may take no more than a few decades, or less than a human lifespan. Furthermore, some changes have been so large as to suggest that the climate oscillated in and out of full-glacial conditions during much of the last glacial age. Clearly, we need to learn the causes of

these changes if we are to understand the global climate system.

An important clue came with the recognition of discrete layers of ice-rafted sediment layers in the upper parts of deep-sea cores from the North Atlantic Ocean. Referred to as **Heinrich layers**, after the marine geologist who first described them, they record sudden discharges of icebergs into the North Atlantic from the surrounding continental ice sheets during the last glaciation. Six main layers, designated H1 through H6, occurred within the interval spanning the last 70,000 years (Fig. 14.24). The discharges of icebergs led to deposition of coarse sediment on the seafloor as the bergs melted, producing the distinct layers of ice-rafted detritus (IRD). Decreased concentrations of the skeletal remains of surface-dwelling foraminifera at the level of the IRD layers in these cores show that the temperature of the ocean waters fell during the Heinrich events. Mineralogical studies have demonstrated that most of the debris layers originated in geologic regions then covered by the northeastern part of the largest North American ice sheet, and that debris in one layer (H3) likely originated in the Iceland-Scandinavian sector of the North Atlantic.

The climatic effects of the Heinrich events must have been profound and must have reverberated through key elements of the Earth's interacting atmosphere, hydrosphere, cryosphere, and biosphere systems. Air passing across the vast field of icebergs and cold low-salinity surface water associated with them was cooled as it swept eastward, causing temperatures to plummet in western Europe. There, the cold events are clearly recorded in lake sediments that show a shift to colder-climate vegetation assemblages during each event. The fact that similar climate shifts at these same times are found in pollen records from Florida, in loess deposits of China, and in alpine glacial deposits of western North America and South America implies that the Heinrich events affected climates worldwide. However, this raises the important question: how could icebergs in the North Atlantic affect climate globally?

The answer to this question most likely lies in the role of the Atlantic in the ocean's thermohaline circulation system. During a Heinrich event, a capping lid of cold meltwater laden with icebergs would have interrupted the generation of saline North Atlantic Deep Water, thereby shutting down the ocean conveyor system. Because that system affects the oceans worldwide (Fig. 11.13), world climate is also affected.

The large swings in climate associated with Heinrich events, and corresponding cooling events detected in the Greenland ice cores, appear to be a phe-

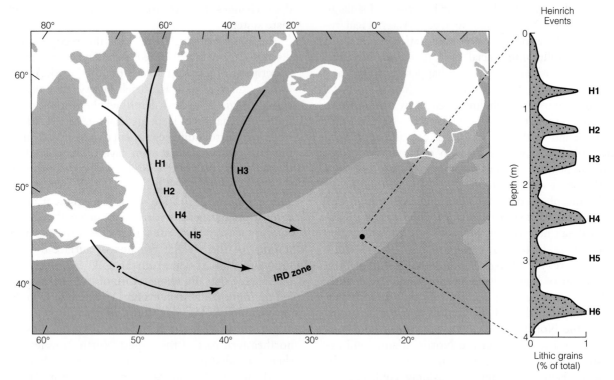

Figure 14.24 Layers of ice-rafted detritus (IRD) in a sediment core spanning the last 80,000 years from the eastern North Atlantic Ocean (labeled H1-H6) record massive outbursts of icebergs from nearby ice sheets. Sediment in five of the layers likely originated in glacier-covered northeastern Canada, whereas one (H3) apparently was derived mainly from the Iceland-Scandinavia sector of the North Atlantic. Air masses passing across the floating ice and frigid surface ocean water were cooled as they swept toward western Europe, bringing cold temperatures to the region south of the glaciers that covered Scandinavia and the British Isles.

nomenon of glacial times — times when large ice sheets encircled the North Atlantic. Events of comparable magnitude are not seen in records postdating the last glaciation. Smaller-scale ice-rafting events, recurring every 2000 to 3000 years, also punctuate the record of the last glaciation. These events may represent iceberg discharges of smaller magnitude from local sources in Iceland and eastern Canada. Possibly they are related to changes in sea level that destabilized the marine margins of the circum-North Atlantic ice sheets.

Yet another cycle of climate variation has recently been recognized in the marine record, one that recurs about every 1500 years. These climate shifts occur during both glacial and interglacial times and therefore operate independently of the astronomically controlled glacial-interglacial cycles. Their cause is not yet known.

Solar Variations and Volcanic Activity

Climatic fluctuations measured in centuries or decades were responsible for the Little Ice Age and similar episodes of glacier expansion. However, such fluctuations are too brief to be caused either by movements of continents or variations in the Earth's orbit, and so require us to seek other explanations for their cause. Two explanations have received special attention.

One hypothesis regarding the cause of short-lived glacial events like the Little Ice Age is based on the concept that the energy output of the Sun fluctuates over time. The idea is appealing because it might explain climatic variations on several time scales. However, although correlations have been proposed between weather patterns and rhythmic fluctuations in the number of sunspots appearing on the surface of the Sun, as yet there has been no clear demonstration that solar variations are responsible for climatic changes on the scale of the Little Ice Age.

Large explosive volcanic eruptions can eject huge quantities of fine ash into the atmosphere to create a veil of fine dust that circles the globe (Chapter 7 opener). Like other types of dust, the fine ash particles tend to scatter incoming solar radiation, resulting in a slight cooling at the Earth's surface. Although the dust

GUEST ESSAY

Is Our Climate Changing? Tree-ring Records Can Put Extreme Weather into Context

Lisa J. Graumlich studied botany (B.S.) and geography (M.S.) at the University of Wisconsin and received her Ph.D. in forestry at the University of Washington. She is currently Deputy Director of Biosphere 2 Center, Columbia University, where she is responsible for research and education programs. She maintains her affiliation with the Laboratory of Tree-Ring Research at the University of Arizona in order to continue her study of mountain climate and vegetation.

In the last several years, weather has moved from the back pages of our daily newspapers to the front-page headlines. During the winter of 1998, the most severe El Niño event of the twentieth century caused millions of dollars of storm-related damage in North America as well as drought-related crop failures in northeastern Brazil. Such events caused the "greenhouse effect" (see Chapter 18) to become a household term as citizens around the world debated whether human-induced changes in the composition of the atmosphere had significantly altered the Earth's climate. In the public's mind, the question emerges: What is happening to our weather? To the scientific community, a related question has gained widespread attention: Are these weather events a harbinger of large-scale climatic variation or simply a part of natural climatic variability?

These questions, though urgent, are not new. When I was in graduate school in the late 1970s, weather was also in the news. At that time, the west coast of the United States was experiencing its worst drought episode since the dust bowl years of the 1930s and the droughts were causing widespread famine in the Sahel of Africa. In my climatology classes, professors expressed dismay and articulated a question that has simultaneously haunted and intrigued me ever since: Are our weather records, which for most of the world extend back only to the early twentieth century, adequate to understand the complex dynamics of the climate system? As I puzzled over this question, one professor suggested that I might find an answer to that question by looking at the record of climate recorded in tree rings, a record that extends for several hundred, or, in some cases, several thousand years back in time. Acting on that suggestion, I began a life-long odyssey that has taken me to some of the most beautiful mountain ranges of the world in search of the old trees that are providing me the answers to current questions about climate change.

The field of tree-ring research, or dendrochronology, is based on several key concepts. During the growing season, trees produce xylem cells, specialized for the upward transport of water and nutrients, along the outermost circumference of the bole. Early in the growing season, the tree produces cells that are large, have thin walls, and appear light in color. Later in the growing season, the tree produces cells that are small, have thick walls, and appear dark in color. The distinctive alteration of light and dark cells marks a single annual growth ring, often visible to the naked eye when we examine a freshly cut stump. Luckily, tree-ring scientists do not have to cut down trees to study the patterns of the rights. We use a coring device that removes a core about the size of a pencil from the tree, a process that causes no damage to the tree. Back in the laboratory, we use a light microscope, interfaced with a computer-aided measuring device, to measure the right widths (see Fig. 14.8).

What do the tree rings tell us about the history of climate? If the tree is growing at a site characterized by cold temperatures and a short growing season (e.g., high-latitude or high-elevation sites), year-to-year variation in summer temperature will cause variation in the width of the right. If the tree is growing at a semiarid site, variations in precipitation will cause variation in the width of the ring. In my own research, I use these different types of sites to obtain different types of climatic histories. When I sample spruce trees at the northernmost limit of tree growth in the Brooks Range of Alaska, the samples indicate the history of summer temperature fluctuations in the Arctic. Alternatively, my cores from juniper trees at the margins of the Gobi Desert, high on the Tibetan Plateau of China, allow me to reconstruct a 1600-year history of the fluctuating rainfall. Working with Dr. Lonnie Thompson and our Chinese colleague Dr. Yao Tandong, we are comparing the ice core and tree-ring records to understand the complex dynamics of Asian monsoons.

settles out rather quickly, generally within a few months to a year, tiny droplets of sulfuric acid, produced by the interaction of volcanically emitted SO_2 gas and water vapor, also scatter the Sun's rays, and such droplets remain in the upper atmosphere for several years. The major eruptions of Krakatau (A.D. 1883) and Tambora (A.D. 1815) in the East Indies, and Pinatubo (1991) in the Philippines, lowered average surface temperatures in the northern hemisphere between 0.3 and 0.7°C (0.5 and 1.3°F). A far greater eruption of Toba volcano about 74,000 years ago, the largest known prehistoric explosive eruption, may have lowered surface temperatures in the northern hemisphere by 3 to 5°C (5 to 9°F)

SUMMARY

1. Climate, the average weather conditions over a period of years, is determined by such factors as temperature, precipitation, cloudiness, and windiness. Former climates are determined mainly from fossil plants and animals, sedimentary deposits, and isotopic studies.

2. Changes in climate within the last 100 to 200 years are recognized in instrumental records. Although decadal-scale fluctuations characterize the climatic record, the overall recent trend of average temperature has been upward, with values in the early 1990s reaching record levels.

3. Climatic changes during the last 1000 years are established (with less accuracy) by proxy records. In northern middle latitudes, an interval of mild climate during the Middle Ages was followed by cooler conditions during the Little Ice Age.

4. During the last glaciation, land-surface temperatures were 5 to 15° C lower than today's. Sea-surface temperatures locally fell as much as 14°C, with the greatest changes occurring in the North Atlantic, the North Pacific, and the equatorial zone. Temperature changes are determined from fossils on land and in deep-sea cores, from isotopic measurements of ice cores, from evidence of lowered snowlines, and from the distribution of periglacial features.

5. Glacial ages have alternated with interglacial ages, in which temperatures approximated those of today. Studies of marine cores indicate that during the last 800,000 years there have been eight glacial-interglacial cycles and that during the entire Pleistocene Epoch there have been at least 30 such cycles.

6. Glacial ages are discerned in many parts of the geologic column; their record extends back at least 2.3 billion years.

7. During the Middle Cretaceous Period, world temperatures are estimated to have been 6 to 14°C warmer than now and sea level was 100 to 200 m higher, indicating greatly reduced global ice volume. Computer modeling suggests that the increased warmth can be explained by a six- to eightfold increase in atmospheric carbon dioxide, possibly released during an episode of unusually intense volcanic activity.

8. Long intervals in Earth history that were marked by repeated glacial-interglacial cycles probably are related to favorable positioning of continents and ocean basins, brought about by movements of lithospheric plates. The timing of the glacial-interglacial cycles appears to be closely controlled by changes in the orbital path and axial tilt of the Earth, which affect the distribution of solar radiation received at the surface.

9. Millennial-scale fluctuations of climate during glacial ages are associated with vast discharges of icebergs in the North Atlantic that likely shut down the ocean's thermohaline circulation system. The effects were global, influencing climate and biota over intervals of hundreds to a thousand years or more.

10. Climatic variations on the scale of centuries and decades may be related to fluctuations in energy output from the Sun, from injections of volcanic dust and gases into the upper atmosphere, or from a combination of these factors.

IMPORTANT TERMS TO REMEMBER

glaciation   *321*   Heinrich layers   *337*   interglaciation   *324*

QUESTIONS FOR REVIEW

1. How did discovery of the frozen body of a Neolithic man in the Alps provide evidence of Holocene climatic change?

2. Describe two types of climate proxy records and explain how they are related to changing climate.

3. Evidence indicates that climates during glacial times were drier than they are today. If true, how can you explain the expansion of lakes in closed basins in such regions as the American Southwest and North Africa during glacial times?

4. How might we try to obtain an estimate of iceage land-surface temperature by studying fossil pollen grains? isotopes in glacier ice? evidence of permafrost distribution?

5. What evidence gained from the study of deep-sea cores indicates that glacial-interglacial cycles occurred repeatedly during the Pleistocene Epoch?

6. Why do oxygen-isotope measurements of deep-sea sediments provide evidence of changing global ice volume?

7. What factors may have contributed to making the Middle Cretaceous climate so much warmer than the present climate?

8. Describe the three orbital motions of the Earth that contribute to variations in the distribution of solar radiation reaching any point on the land surface over the course of a glacial-interglacial cycle.

9. How might global millennial-scale climate changes like the Heinrich events be related to glacial events in the North Atlantic Ocean?

10. What effect can large volcanic eruptions have on climate?

Questions for A Closer Look

1. How did sea-surface temperatures at the peak of the last glaciation differ from those of the present? How might differences in the degree of change among different oceanic regions be explained?

2. If the average sedimentation rate at a site in the Atlantic Ocean is 1 cm/1000 years, how many meters of deep-sea sediment must be extracted to obtain a record spanning the entire Quaternary Period (the last ca. 1.8 million years)?

3. How can sediments accumulating at the *floor* of the deep sea provide information on *surface* water conditions?

QUESTIONS FOR DISCUSSION

1. Name three pieces of evidence that point to a possible change of climate within your lifetime.

2. At the peak of the last glaciation, about 20,000 years ago, the average annual air temperature in middle latitudes was about 5 to 8°C lower than now. Describe what environmental effects such a temperature reduction produced in the region where you live.

3. Obtain or compile a record of annual temperature for your community (or a nearby location) for the past 50 to 100 years and compare it with the global temperature record plotted in Figure 14.3. How are the two records similar and how do they differ? What might explain any differences between the two records?

At our web site you will be able to learn more about global climate change through an interactive case study.

CHAPTER 15

A Planetary Perspective on Life

● Life in a Closed System

Biosphere-II

In the Arizona desert northeast of Tucson, an exotic, futuristic structure rises above the landscape (Fig. 15.1). It cost $120 million to build and had the best material seal ever produced. Eight people—four men and four women—lived in that closed system for two years, along with 3800 species of plants, nonhuman animals, and an unknown number of microorganisms. Their goal was to demonstrate that eight people could produce their own food, recycle all their own wastes, and maintain a breathable atmosphere and healthy environment in a system completely closed to the exchange of matter with the outside world.

The inhabitants came close to achieving their goal. But during the second year the oxygen concentration in the structure, called Biosphere-II, began to decline. When it reached the oxygen pressure equivalent to that found at 15,000 feet above sea level, a medical decision was made to open the air locks and provide some more oxygen, for the health of the inhabitants. Throughout this experiment—when it was proposed, while it was being planned, and during its early stages of operations—many scientists guessed that the system was bound to fail, and each gave a reason. Some suggested that the carbon dioxide level would get too high and become unhealthy for the people. Others suggested that nitrates would not be removed rapidly enough by bacteria and would pollute the waters. Still others suggested that some crop disease or an insect outbreak would make it impossible for the people to feed themselves.

But nobody forecast an oxygen decline. The cause of that decline was found to be certain bacteria which, in an atmosphere with elevated carbon dioxide, could break down the cement walls and in doing so would use up oxygen as they respired. Although in terms of its stated goals, Biosphere-II failed, it provided important insights, and it is the first example we have of a large-scale human-supporting, materially closed system. Another example of a closed system is found in Folsom's flasks, small, closed bottles of air, ocean water, marine sediments, algae, and bacteria that have sustained life for 40 years.

The following questions emerge from this discussion: What is required for life to *persist*? What are the characteristics of a system that contains *and* supports life? What kinds of changes are necessary or beneficial to life? ●

(opposite) Life's impact on the Earth is shown in this satellite image of biological productivity. The green shows land vegetation and marine algae. The image is a three-year composite from NASA's Coastal Zone Color System converted with a vegetation index.

Figure 15.1 Biosphere-II was an attempt to create a closed life-sustaining system, like that found of the whole Earth. It failed because of unanticipated problems. A closed system is closed to matter but open to energy; an isolated system is closed to both matter and energy; and an open system is open to both matter and energy.

A PLANETARY PHENOMENON

The examples of Biosphere-II and Folsom's flasks illustrate that natural ecological systems are incredibly complex and difficult to make predictions about. Simply put, nature doesn't always play by the rules of the game as we have believed them to be. This is an important lesson for us. People have always been fascinated by life—by its complexities, by the marvelous adaptations that plants and creatures make to meet their needs. We have learned a tremendous amount about life. We understand the genetic code by which the shape, form, and function of organisms are transmitted from one generation to the next. We know a great deal about how individual cells function, about the complex chemicals that make them up, and about the function of these chemicals. We can create hybrids of crop plants that are designed for specific environmental conditions. We have learned to cure many diseases. But when it comes to understanding what is necessary for life to persist, we are still perplexed. This understanding is one of the fundamental concerns of Earth system science. In this and the next two chapters, we will explore what we know about life and its environment, provide some answers as we understand them today, and discuss some of the principal unanswered questions.

We do know, at least in some fundamental ways, that *life is a planetary phenomenon*, and therefore the Earth system is an appropriate, and important, level at which to consider fundamental questions about life. The recent discovery of a meteorite known to have originated on Mars and observed to contain structures that may be fossils of bacteria-like life forms makes this question all the more intriguing. Could life have once existed on Mars but died out as water and oxygen became scarce? Could it be that life is not just a phenomenon of Earth but a more general phenomenon that requires just the right size and complexity of a planet, as well as the planet's right kind of environment, in order to persist? Many complex issues are bound up in this question, and the first and most fundamental is the question "What is life?"

What Is Life?

What do we mean when we say that something is alive? What are the *essential* differences between living and nonliving matter? Most of us seldom stop to ask this question. A dog runs about and barks, while a stone lies still and silent. What about a dog and a potato? The differences must obviously be something else besides the ability to run and bark. How, then, do we know when something is alive?

The four essential properties of living organisms are metabolism, growth, reproduction, and evolution. *Metabolism* is the sum of all the chemical reactions that take place in a living entity by which energy is provided and used for life processes. *Growth* involves the ordering and organizing of atoms and small molecules to make larger molecules. *Reproduction* occurs in single-celled organisms when cells divide. Reproduction in multicelled organisms occurs either through sexual or asexual reproduction. An example of asexual reproduction is live twigs of some trees that, when stuck in the ground, take root and produce a new tree. **Evolu-**

tion is the process by which certain individuals in a population leave more offspring than others. The inherited characteristics of those who leave more offspring come to dominate the entire population, changing the genetic characteristics of the population.

Living things are composed of complex, highly ordered, chemical compounds. At a microscopic level, the ordering of atoms and molecules can happen in two ways: polymerization and crystallization. *Polymerization* is the stringing together of small molecules, like beads on a necklace, to make large chain- or sheet-like molecules. *Crystallization*, as discussed in Chapter 4, is the packing of atoms or molecules in ordered geometric arrays. Crystallization is what happens when the randomly ordered H_2O molecules in water vapor form ice crystals on a cold window pane.

In both cases, polymerization and crystallization, an ordered pattern of atoms or molecules is replicated throughout the structure. There is an important difference between the two systems, however. Polymerization *absorbs* energy, whereas crystallization *releases* energy. (This is the latent heat of crystallization discussed in Chapter 12.) Growth in all living matter involves the polymerization of small organic molecules to form large organic molecules (*biopolymers*), and so growth requires a source of energy (Fig. 15.2).

Crystals have the power of growth, but unlike living organisms they have no metabolism; nor do they have the power to reproduce or to evolve, creating new kinds of crystals. Viruses, those nasty things that cause colds and flu and many other diseases, fall in a shadowy area somewhere between living and nonliving matter. Atoms in viruses are ordered in biopolymers, and viruses can reproduce by using the replication machinery of some other organism. But viruses cannot reproduce by themselves nor can they metabolize. Viruses are "almost" living; thus, the distinction between living and nonliving is not a sharp one.

Evolution gives a quality to life that sets it apart from the nonliving world—the living world is in a continual state of change. A crystal of quartz that grows today is the same as a crystal that grew 3 billion years ago. Living organisms, however, change over time. Today's organisms have arisen as a result of change. This change affects not only the individual organism in its life cycle of growth, reproduction, and death, but also the population to which it belongs, which is always responding to environmental fluctuations and occasionally undergoing more radical change in the generation of new species. Evolution will be discussed more fully in Chapter 17.

The Hierarchical Character of Life

The basic structural unit of all living organisms is the cell (Fig. 15.3A). A **cell** is a complex grouping of chemical compounds enclosed in a membrane, with a high degree of structure. The membrane separates the materials inside from the environment outside and facilitates the controlled exchange of materials and energy between the cell and its environment.

Life is organized as a hierarchy of many kinds of systems, from a single cell to the entire biosphere (Fig. 15.4). Cells form organs that form individual organisms (although some organisms are single-celled). A group of individual organisms of the same kind form a population that interbreeds. Populations make up **species**, which are groups of individuals capable of interbreeding and occasionally exchanging genes. Several species sharing genetic (inherited) characteristics are of the same **genus** (Table 15.1). **Ecological communities** are sets of interacting species. An

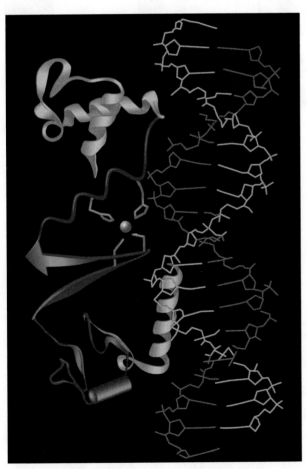

Figure 15.2 Biopolymers are large organic molecules formed by polymerization—the stringing together of small molecules, like beads on a necklace, to make large chain- or sheet-like molecules.

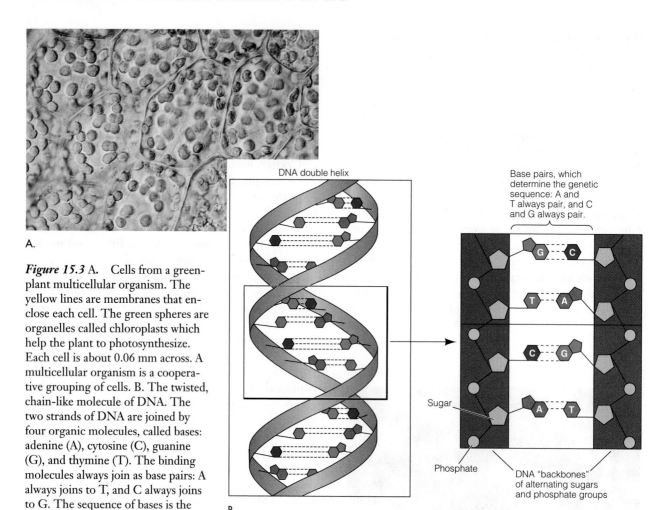

Figure 15.3 A. Cells from a green-plant multicellular organism. The yellow lines are membranes that enclose each cell. The green spheres are organelles called chloroplasts which help the plant to photosynthesize. Each cell is about 0.06 mm across. A multicellular organism is a cooperative grouping of cells. B. The twisted, chain-like molecule of DNA. The two strands of DNA are joined by four organic molecules, called bases: adenine (A), cytosine (C), guanine (G), and thymine (T). The binding molecules always join as base pairs: A always joins to T, and C always joins to G. The sequence of bases is the code that directs the activities of a cell.

Table 15.1 The Taxonomic Hierarchy

The systematic arrangement of plant and animal organisms according to their similarities. The example is a tiger.

Taxona[a]	Example	General Characteristics and Selected Representatives
Phylum (a group of classes)	Chordata	Bilaterally symmetrical animals with a dorsal nerve chord. Elephant, eel, human, tiger.
Class (a group of orders)	Mammalia	Warm-blooded animals with hair and mammary glands. Elephant, mouse, tiger, human.
Order (a group of families)	Carnivora	Flesh-eating placental mammals. Dogs, foxes, lions, tigers.
Family (a group of genera)	Felidae	Rounded-headed carnivores walking on toes; retractable claws. Lions, tigers, bobcats.
Genus (a group of species)	Panthera	Large cats, 50 kg (110 lb) or more. Lions, tigers, leopards, jaguars.
Species	Panthera	Striped coat, usually over 140 kg (310 lb). Bengal tiger, tiger, Siberian tiger.

[a]A convenient way to remember the taxonomic order is to make up a mnemonic device using the first letters of each word: for example, Philip cares only for gifted singers: phylum, class, order, family, genus, species.

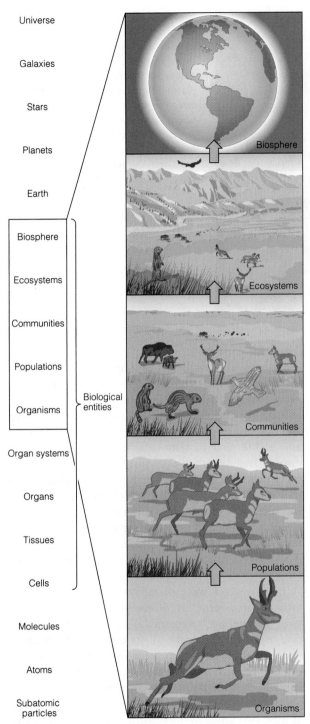

Figure 15.4 The hierarchical structure of life, from cells to the biosphere.

Together, they make up the tropical rainforest biome. Biomes make up the living part of the biosphere. The biosphere and the global environment form the Earth system. Each level within this hierarchy affects levels above and below it. From a planetary perspective, this hierarchical character is an important feature of life.

In this chapter, we focus on several levels in this hierarchy. First, we discuss the cell as the basic biological building block—the basic living unit. Then we discuss the ecosystem, because it is the basic unit that sustains life. Finally, we discuss populations, because these grow and can evolve and have effects on all the levels above them in the ecological hierarchy.

The Cell as the Basic Structure of Life

Many bacteria are *unicellular* (that is, they consist of one cell), but most organisms are *multicellular* and consist of hundreds to trillions of cells—the larger the organism, the more cells it contains. The growth of new living matter involves the construction of new cells. Cell growth follows specific plans, and each kind of cell has its own special plans. Full details of a growth plan are passed from cell to cell, generation after generation. The information is stored in *deoxyribonucleic acid (DNA)*, a biopolymer consisting of two twisted chain-like molecules held together by organic molecules called *bases*. There are four bases in DNA: guanine (G), cytosine (C), thymine (T), and adenine (A). The bases hook together to form *base pairs;* G always pairs with C and T with A (Fig. 15.3B). Each of the two strands of DNA thus carries four kinds of base pairs, and the sequence of base pairs along the chain can be varied almost infinitely. Like the barcodes used in supermarkets, the sequence of bases in DNA is the code that stores the genetic information. The stored information provides a cell with a reference library of how to carry out the activities of life, such as the details of reproduction, growth, and maintenance, including the polymerization of protein. The DNA codes are read and executed by *ribonucleic acid (RNA)*, a single-strand molecule similar to one-half of a double DNA chain.

Procaryotes and Eucaryotes

Cells may be smaller (0.01 to 0.02 mm, or 0.0004 to 0.0008 in) or larger (0.05 mm to a few centimeters, or 0.002 to 1.5 in or larger, in rare cases), but large or small, all cells are of two kinds, procaryotic or eucaryotic. *Procaryotic cells* (from the Greek *pro* = before and *karyote* = nucleus, hence before a nucleus) are generally small and comparatively simple in structure.

ecosystem is an ecological community of animals, plants, fungi and microorganisms and its local, non-biological environment (air, water, soil, etc.). Ecosystems make up **biomes**, which are classes of ecosystems. For example, tropical rainforests occur in Asia, Africa, Central and South America, and elsewhere.

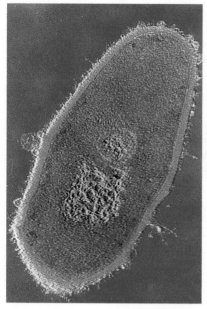
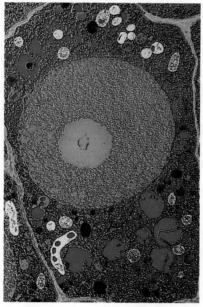

A.

B.

Figure 15.5 A. Procaryotic cell. A bacterial cell devoid of visible organelles and with the DNA concentrated in a poorly defined nucleoid that is not separated from the cytoplasm by a membrane. B. Eucaryotic cell from a plant root with a well-defined, membrane-bound nucleus and varied cytoplasmic organelles. Note that the cells are colored because they have been stained.

These cells house their DNA with all its genetic information in a poorly demarcated part of the cell (Fig. 15.5A). The main body of the cell, which is called the *cytoplasm*, lacks distinctly defined areas in which the various cell functions are carried out. Most importantly, the portion of the cell that houses the genetic information is not separated from the cytoplasm by a membrane. **Procaryotes** occur as bacteria and related organisms called archaebacteria and mycoplasmas.

Eucaryotic cells (from *eu* = true, hence with a true nucleus) are larger and more complex than procaryotic cells. Their genetic information is housed in a well-defined nucleus that is separated from the cytoplasm by a membrane (Figure 15.5B). The cytoplasm in a eucaryotic cell contains a variety of well-defined cell parts, called **organelles,** each having a particular function in the operation of the cell. We humans, and most other living things, including all animals, plants, and fungi, are **eucaryotes.**

THE ECOSYSTEM: THE MINIMUM SYSTEM THAT CAN SUPPORT LIFE

Now that we have discussed the characteristics that pertain to being alive, we can consider the characteristics of a system that maintains life and allows it to persist. When you think about Folsom's flasks, you realize that each contains representatives of the four subsystems of the Earth discussed in Chapter 1: the sediments represent the solid Earth; the water the hydrosphere; the air the atmosphere; and the algae and bacteria the biosphere.

Also consider again the experiment that took place in Biosphere-II. Within that sealed container, the eight humans had to obtain food from the plants and animals they farmed. They depended on plants, algae, and bacteria to give off oxygen, which is necessary for animal life, and to remove from the air the carbon dioxide they breathed out. They had to recycle water for drinking, washing, and irrigation, and all materials including their own wastes, which had to be cleaned for reuse. In short, their survival within the sealed structure depended on two primary processes: *a continual recycling of chemical elements* and *a flow of energy*. These are the two fundamental requirements for any long-term life-support system.

In Chapter 1 we explained why a flow of energy, rather than a recycling of energy, is necessary. A cycling of chemical elements is necessary because no one organism can make all its own food from inorganic chemicals and also break down all the complex organic chemicals, returning them to the same inorganic form that allows them to be reused. A set of organisms is necessary. At the minimum there must be at least two kinds of organisms: one that produces its own organic compounds from inorganic chemicals and another that decomposes the complex organic compounds so that chemical elements can be recycled. Organisms that use a source of energy to make their own chemical compounds from inorganic chemicals are called **autotrophs** from the Greek words for self and feed, hence "self-feeders." Organisms that decompose complex chemical compounds in order to obtain a source of energy are called *decomposers*; these are mainly bacteria and fungi. There are also organisms that feed on other living organisms. Organisms that feed on live or dead organic material are called **heterotrophs.**

Most autotrophs use sunlight as an energy source to combine carbon dioxide and water to form carbohydrate by a process called **photosynthesis**, according to the following reaction:

$$CO_2 + H_2O + energy \rightarrow CH_2O + O_2$$

Carbohydrates are a group of chemical compounds, including sugars, starches, and cellulose carbohydrates are found in all organisms, and have the general formula $Cx (H_2O)_y$. Green plants, algae, and photosynthetic bacteria are autotrophs. A second, smaller group of autotrophs obtain energy from chemical compounds and are known as *chemoautotrophs.*

When a heterotroph eats an autotroph or another heterotroph, the energy stored in the organic compounds of the consumed organism is released by either fermentation or respiration. *Respiration* is a process by which organisms use oxygen to oxidize carbohydrate to carbon dioxide and water, and release energy. Fermentation is a process whereby carbohydrate molecules are decomposed partially to form an alcohol plus carbon dioxide and water and release energy. Fermentation is carried out by anaerobic organisms, meaning that they do not use oxygen. Respiration is more efficient than fermentation because it releases more of the available energy.

To our knowledge, no single species of bacteria or fungus can decompose all the compounds in all autotrophs. Therefore, complete decomposition requires more than one species of bacteria or fungi. In addition to these kinds of organisms, there must be a fluid medium—a gas or liquid or both—to transmit the chemical elements among the different kinds of organisms. And so we can conceive of a minimum system that can support and sustain life. It is an open system with a source of energy and a sink of heat energy (as discussed in the Introduction). It has at least one autotroph and one heterotroph (a decomposer), but in fact will have a number of each as well as other heterotrophs—herbivores that feed on autotrophs and predators that feed on other heterotrophs. And it will have a nonliving environment consisting at least of fluids—water and air. Looking back at the definition of an ecosystem, we can see that what we have just described is a *minimal* ecosystem. Thus, we learn that *the minimum system that can sustain life is an ecosystem.*

Here we obtain a fundamental insight about life. Typically, we associate life with individual organisms, like ourselves, or a tree, a mushroom, some one-celled bacteria, because these are alive. But no individual is capable of life support. *The capacity to support life is the characteristic of an ecosystem, not of individual organisms or populations.* Life persists only in systems of a number of kinds of organisms and a nonliving environment.

Food Chains and Trophic Levels

When organisms feed on one another, energy and chemical elements are transferred from creature to creature along **food chains** (the linkage of who feeds on whom), which in more complex cases are called **food webs** (Fig. 15.6).

One of the simplest kinds of natural ecosystems are those found in salt lakes, such as the Aral Sea, which lies east of the Caspian Sea, in Asia, the Great Salt Lake in Utah, and Mono Lake in California. Mono Lake covers 40,000 acres in the high desert just east of the Sierra Nevada mountain range (Fig. 15.7). Mono Lake is famous for its beauty, for its strange salty waters—twice as salty as the ocean—and for more than one million birds that feed there and depend on small animals that grow in the lake. Figure 15.8 shows the food web at Mono Lake. A number of autotrophs—species of algae and photosynthetic bacteria—grow in the salty, alkaline waters. But there are only two heterotrophs—a brine shrimp and the larvae of the brine fly that can eat the algae and live in the salty water. Although the lake's chemistry creates an environment of great stress, this chemistry also provides an abundance of the chemical elements required by these species, which grow in incredible abundance in the lake.

Five species of birds are carnivorous and feed on the brine shrimp and brine fly. An unknown number of bacteria act to decompose the fecal material of the shrimp, fly, and the birds, and the dead bodies of these organisms. Notice in Figure 15.8 that each organism feeds a certain number of food web steps from the original source of energy. Ecologists group species that are the same number of food web steps from that source of energy. Each group is called a **trophic level**. For example, the autotrophs—algae and photosynthetic bacteria—in Mono Lake form the first trophic level; the two heterotrophs—organisms that feed on autotrophs—a brine shrimp and the larvae of the brine fly—form the second; and the birds form the third. These levels form a **trophic structure** (sometimes referred to as a *trophic pyramid*).

In Mono Lake, the food web is simple enough that each organism functions on just one trophic level. But in most ecosystems, the feeding structure is more complicated. Consider, for example, the food web of the harp seal, which feeds on fish (Fig. 15.9). In this food web, there are several carnivore trophic levels: carnivorous fish that feed on herbivores; carnivorous fish that feed on the fish that feed on herbivores; and

Figure 15.6 Plants (autotrophs) capture and lock up energy from the Sun in food molecules. Plants form the base of the food chain. Heterotrophs eat autotrophs to get energy. Zebras (herbivorous heterotrophs) feed on plants while lions (carnivorous heterotrophs) feed on the zebras.

Figure 15.7 Mono Lake, an example of a salt lake that has few species but high biological productivity. Mono Lake covers 40,000 acres east of the Sierra Nevada mountains in California. The rock structures are tufa formed when volcanic gases bubble through the salty and alkaline water of the lake.

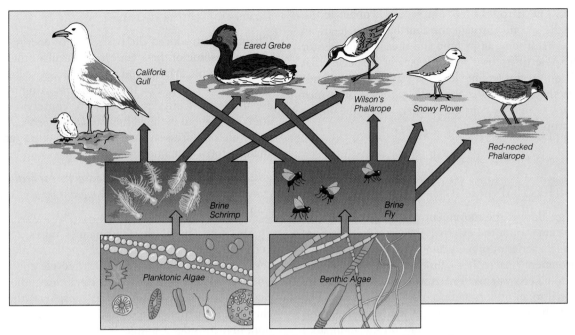

Figure 15.8 This diagram shows who feeds on whom in the Mono Lake ecosystem. Compared to most ecosystems, Mono Lake has few species and the food web is relatively simple.

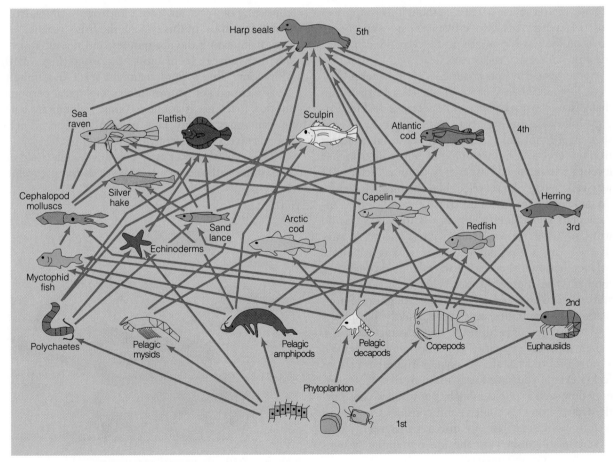

Figure 15.9 Food web of the harp seal. The arrows show who feeds on whom. The harp seal feeds on more than one tropical level.

so on. The harp seal feeds at several trophic levels, from the second through the fourth, and it feeds on predators of some of its prey and thus is a competitor with some of its own food. Typically, a species that feeds on several trophic levels is classified as belonging to the trophic level above the highest from which it feeds, so that we consider the harp seal to be on the fifth trophic level.

ENERGY AND LIFE

Energy flow is the movement of energy through an ecosystem—from the external environment (the Sun) through a series of organisms and back to the external environment (space). It is a fundamental process common to all ecosystems. Two aspects of this process are essential to an understanding of energy and life. The first is the function of energy in living systems; this can tell us about the ultimate limits on the abundance of life. The second is the pathways through which energy flows, as well as the efficiency with which energy is used, and the involvement of energy in the production and growth of organisms, populations, and the organic matter in ecosystems. Although life requires a flow of energy, life also influences the way energy flows on the Earth's surface. In this way life can affect the atmosphere and the oceans.

Some basic concepts about energy and energy flow were introduced in Chapter 1. Here we expand on that discussion to explain its importance to life. Energy follows a one-way path through an ecosystem such as Mono Lake. Energy flows from a lower trophic level to a higher one, but not from a higher level to a lower one. This energy is not recycled. But why can't energy be recycled, like chemical elements? To understand this, we must consider some of the fundamentals about energy. Energy is the ability to do work, to move matter. All life requires energy. As anyone who has dieted knows, our weight is a delicate balance between the energy we take in through our food and the energy we use. What we don't use, we store. This is true not only of people, but also of populations, communities, ecosystems, and the biosphere.

This energy is a difficult and abstract concept. When we buy electricity by the kilowatt-hour (1kWh = 3.6 x 103kJ), what are we buying? You can't see it or feel it, even if you have to pay for it. At first glance energy flow seems simple enough. But when we dig a little deeper into this subject, and consider the role of energy in an ecosystem, we discover that energy flow is what distinguishes life and life-containing systems from the rest of the universe.

Laws of Conservation

Chapter 1 introduced the basic laws of energy. Here we repeat some of these concepts to explain their significance to life. Matter and energy are both subject to *laws of conservation.* The laws we are referring to here are the laws of physics; the kind of conservation we are talking about is called the *first law of thermodynamics*, which addresses the observation that energy changes form, not amount. It states that

- *In any physical or chemical change, matter is neither created nor destroyed, but merely changed from one form to another.*

This first law of thermodynamics raises a seemingly simple question: if the total amount of energy is always constant, why can't we just recycle energy inside our bodies? Similarly, why can't energy be recycled in ecosystems and in the biosphere the way chemical elements are recycled? To answer this question, first consider a simple, hypothetical ecosystem.

Let us imagine how this might work, say, with frogs and mosquitoes. Frogs eat insects, including mosquitoes. Mosquitoes suck blood from vertebrates, including frogs. Consider an imaginary closed ecosystem consisting of water, air, a rock for frogs to sit on, frogs, and mosquitoes. In this system, the frogs get their energy from eating the mosquitoes, and the mosquitoes get their energy from biting the frogs (Fig. 15.10). Why can't this system maintain itself indefinitely? Such a closed system would be a biological perpetual motion machine: it could continue indefinitely without an input of any new material or energy. This sounds nice, but unfortunately it is impossible.

The general answer as to why this system could not persist lies with the *second law of thermodynamics*, which

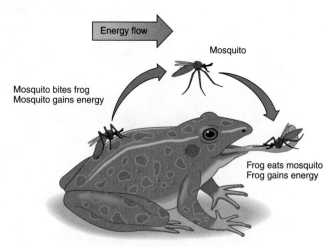

Figure 15.10 An impossible ecosystem.

A CLOSER LOOK

The Second Law of Thermodynamics

To better understand why we cannot recycle energy, imagine a spacecraft carrying an astronaut to Mars. Suppose the spacecraft is a completely isolated system (i.e., it will receive no external input of either energy or matter) containing a fuel cell that makes electricity from oxygen and hydrogen and produces water as a byproduct, a tank of water, air, electronic equipment including a computer with a hard disk and a small stereo system, as well as the other equipment the astronaut needs. Suppose the astronaut uses his computer and stereo. The electric energy is gradually converted to heat, some of it from the friction of the spinning hard drive. The electronic components also heat up as they operate. The heat from the devices gradually warms the entire system. When all the oxygen and hydrogen in the fuel cell are completely combined, the astronaut will have no more electricity and will no longer be able to use his computer and stereo. Because the system is completely closed, the energy is still inside the spacecraft, and the average temperature of the system is now higher than the starting temperature.

Why can't the astronaut recover all that energy, use it to dissociate the hydrogen and oxygen, put them back into the tanks for the fuel cell, run the fuel cell again, and be able to use his electronic equipment? Or why can't he set up a system that takes the electric energy after it had been used by his electronic equipment and dissociate the hydrogen and oxygen—all in one process? This is where the second law of thermodynamics applies. No real process that involves the use of energy can ever be 100 percent efficient. Whenever useful work is done, some energy is inevitably converted to heat. Heat energy is the random motion of molecules. The process of collecting all the energy dispersed as heat in this closed system must require more energy than can be recovered. The astronaut might be able to recharge his fuel cell, but he will always be in a losing situation. Eventually he must run out electricity.

Our imaginary spacecraft system began in a highly organized state, with energy compacted in the tanks of the fuel cell. It ends in a less organized state, with the energy dispersed throughout the system as heat. Physicists say that the energy has been degraded, and the system is said to have undergone a decrease in order. The measure of the decrease in order (the disorganization of energy) is called *entropy*. All energy of all systems tends to flow toward states of increasing entropy.

---

addresses the kind of change in form that energy undergoes:

- *Energy always changes from a more useful, more highly organized form to a less useful, unorganized form.* Energy cannot be completely recycled to its original state of organized, high-quality usefulness.

For this reason, the mosquito-frog system will eventually stop; there will not be enough useful energy left.

The second law of thermodynamics tells us that nature's way is to crumble its way to disorder (see "A Closer Look: The Second Law of Thermodynamics"). Unmaintained buildings fall to ruins, your desk and sock drawer get messier and messier unless you spend the time and energy to keep things in order. In fact, it takes energy to maintain order in any natural system. Life is no exception. Life makes orderly carbohydrate molecules out of disorderly simpler molecules, and this process takes energy.

Energy Flow

From what we have just discussed, it is clear that an ecosystem must have a source of usable energy and a sink for degraded (heat) energy. The ecosystem is said to be intermediate between the energy source and the energy sink. The energy source, ecosystem, and energy sink form a thermodynamic system. The ecosystem can only undergo an increase in order (called a local increase) if the entire system undergoes a decrease in order (called a global decrease).

The Earth is a closed system, meaning that matter is neither added nor lost, but that energy flows freely in and out of the system. As we saw in Chapter 1, all closed and open systems respond to inputs and have outputs. The same principle holds for the Earth system that holds for a single ecosystem: in order to support life, the Earth must be an intermediate system between a source of usable energy and a sink for heat. The usable energy source is principally the Sun, as well as a small additional amount of heat provided by radioactive decay in the deep Earth, emitted through deep-sea vents. The sink for degraded heat energy is the vastness of space; degraded heat is radiated out from the Earth's surface.

PRODUCTION AND GROWTH

The amount of usable energy provides an upper limit on rates of production of organic matter and on the amount of life that can be sustained by our Earth system. In reality, organisms are not even close to being

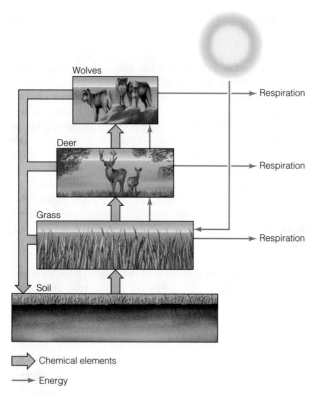

Chemical elements

Energy

Figure 15.11 Energy flows one-way through an ecosystem while chemical elements can recycle.

100 percent efficient in converting sunlight or chemical energy to energy stored in organic matter.

Biological Production

Carbohydrate production by photosynthetic organisms (autotrophs) is called *primary production*. The production of body mass by heterotrophs is called *secondary production*. The total amount of organic matter on Earth or in any particular ecosystem or area is called the **biomass** of that ecosystem. This includes all living things and all products of living things. Biomass is usually measured as the amount per unit surface area of the Earth (for example, as grams per square meter or pounds per square yard). Biomass is increased through biological production (growth); the change in biomass over a given period of time is called *net production*. Biological production involves the storage of usable energy through the production of organic compounds in which the energy is stored. There are three measures of production, which we can think of as the "currency" of production: biomass, energy stored, and carbon stored.

Pathways of Energy Flow

Consider again the Mono Lake food web. Energy flows upward in the diagram and enters the Mono Lake ecosystem as sunlight. Algae and certain photo-

synthetic bacteria are able to carry out photosynthesis; they are the primary producers. The brine shrimp and brine fly larvae are hetrotrophs, or secondary producers, that obtain energy by feeding on Mono Lake's algae and bacteria. The birds are also secondary producers because they obtain energy by feeding on the shrimp and fly.

Energy flows from a lower trophic level to a higher one (Fig. 15.11). As a rule of thumb, photosynthetic organisms (primary producers) are able to store about 1 percent of the energy available in sunlight. *A small fraction of the energy available to each trophic level is converted to net production of new organic matter. A large fraction of the energy available to each trophic level is used in respiration.* Basically, *respiration is the use of biomass to release energy that can be used to do work.* Respiration returns to the environment the carbon dioxide that has been removed by photosynthesis.

Gross and Net Production

There are three steps in the production of biomass and its use as a storage and source of energy by autotrophs. First, an organism produces organic matter within its body; then it uses some of this new organic matter as a fuel in respiration; and finally, some of the newly produced organic matter is stored for future use. The first step, production of organic matter before any use, is called **gross production**. This suggests another way to think about net production. Net production is what is left from gross production after use. In these terms,

Net Production = gross production - respiration

This is a fundamental production relationship. The difference between gross and net production is like that between a person's gross and net income. Gross income is the dollars you are paid, whereas net income is what you have left after money is deducted for taxes and other fixed costs. Respiration is equivalent to taxes and other fixed costs that must be paid in order for you to do your work.

POPULATION DYNAMICS AND THE BIOSPHERE

At any time on the Earth, some species are very abundant and others are very rare. Viewed from space, the land surface of the Earth and the upper waters of the ocean show that life is spread widely and that great geographic variation exists in the amount of biological production (Fig. 15.12). Over geological periods of time, the kinds of organisms that have been abundant have changed greatly, as we will see in Chapter 17.

A **population** is a group of individuals of the same

Figure 15.12 Satellite remote sensing images of the entire Earth. (from the National Geographic special centennial edition, *The Endangered Earth*.)

species living in the same area or interbreeding and sharing genetic information. A species is all individuals that are capable of interbreeding and is made up of populations. The study of population growth and change is known as *population dynamics*. The basic question that we are trying to answer in considering population dynamics is: what limits the abundance of life on the Earth? We ask this question at many different levels of organization, from a single population within a species to all the species in an ecosystem, to all of life in a major region of the Earth, and to all the life on the Earth. This question is important because a rapidly growing population can occupy large amounts of space, alter the physical and chemical environment, and therefore affect the global environment.

Classically, ecologists have suggested three possible processes that limit the abundance of life: (1) a population is self-limiting; (2) interactions among species lead to a control on the abundance on any one species; and (3) the nonbiological environment sets the limit. In the last case, the limit might be set by physical factors, such as climate or the amount of energy from sunlight; or it might be set by chemical factors, such as the availability of a crucial chemical element in a soil.

What accounts for differences in growth rates and abundance among species? Given enough energy and nutrients, life forms are capable of incredible rates of growth. For short periods, populations can undergo *exponential growth*, which means that they can increase at a fixed percentage every year, just the way com-

pound interest increases your money in a bank account. Populations of bacteria in a laboratory culture can grow this way. The elephant seal, once believed to be extinct but actually reduced to about 12 individuals at the turn of the twentieth century, grew to approximately 60,000 by the 1970s, a growth rate of 9 percent per year (Fig. 15.13.) However, populations cannot grow exponentially forever. In a finite world, nec-

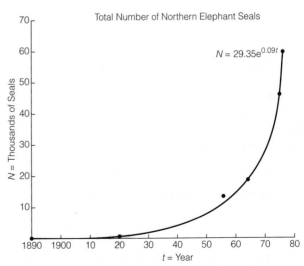

Figure 15.13 History of the Elephant seal population. This species was brought close to extinction about 1890 but has increased exponentially since then, increasing 9 percent per year. This illustrates that populations are at least temporarily capable of rapid growth.

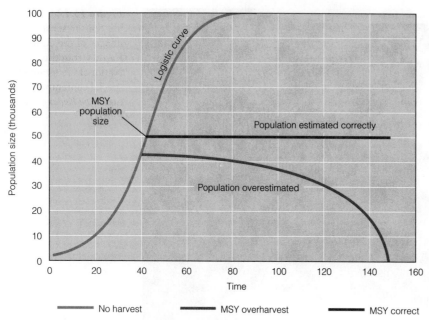

Figure 15.14   The logistic growth curve showing the carrying capacity and the Maximum Sustainable Yield (MSY). This curve is based on a mathematical equation that assumes that individuals in a population compete for resources. The more individuals, the less resources. As the population increases, the reproductive rate (newborn per adult) decreases and the death rate increases. Eventually a point is reached where the births equal the deaths and the population remains constant. The figure shows what happens to a population when we assume it is at MSY and it is not. Suppose that a population grows according to the logistic curve from a small number to a carrying capacity of 100,000 with an annual growth rate of 5 percent. The correct MSY population size would be 50,000. When the population reaches exactly that amount and we harvest at exactly the calculated maximum sustainable yield, the population continues to be constant. But if we make a mistake in estimating the size of the population and believe that it is 50,000 when, for example, it is really only about 42,000, then the harvest will always be too large and we will drive the population to extinction. This is a hypothetical curve, not found in nature, but useful to understand the principles of population growth.

essary resources must eventually limit a population. How rapidly a population changes depends on the *growth rate*, which is the difference between the *birth rate* and the *death rate*. A population with a constant growth rate grows exponentially. But if real populations cannot grow exponentially indefinitely, how do they grow?

One of the earliest proposals of the modern scientific era was that a population was self-regulating and that, under constant environmental conditions including a constant food supply, a population would grow smoothly following a curve that has become known as the *logistic* (Fig. 15.14). The idea behind the logistic is simple. When a population is small in comparison to its resources, every individual has plenty to eat and an abundance of all resources. The population grows almost at an exponential rate. But as the numbers in-

crease, the food and other resources per individual decline. As this decline takes place, individual growth and reproduction also decline, and death rates can increase. Eventually, a population level is reached at which there is just enough food for each individual. At this population size, birth and death rates are equal, and the population ceases to grow. In other words, it has reached a carrying capacity. In the case of the logistic, this carrying capacity is a stable abundance: if the population is reduced below it, births exceed deaths and the population will return to the carrying capacity. If the population reaches a level above the carrying capacity, deaths exceed births and the population declines to the carrying capacity.

Carrying capacity is an important concept for populations. There are two commonly used definitions of this term: (1) the constant abundance reached by a hypothetical population growing according to the logistic curve; and, a more general definition, (2) the maximum number of individuals of a species (or population) that can persist in an area without affecting the future ability of the population to maintain that same abundance.

A population that remains at carrying capacity is in a steady state and is called a *stable population*. A stable population in this sense has two attributes: first, it has a constant abundance unless disturbed; and second, when disturbed and released from the disturbing factor, it returns to the same abundance. A longstanding question in ecology has been whether populations exhibit this kind of stability.

The evidence available today suggests that populations undergo much more complex patterns of change over time; they do not grow according to the logistic curve. Instead, they are affected by environmental change, interactions with other species, random events, effects of individuals within the population on one another, and random time lags. For example, the population of adult salmon shows great variation over time, as shown in the catch on the Rogue River (see Fig. 15.15). Variation, rather than constancy, is likely to be the rule in nature. The question then is how to determine what factors cause the variation, and from these methods to learn to make forecasts about popu-

lation change. In regard to global change, forecasting the fate of populations, especially endangered ones, is important to understanding how change in the environment might affect the diversity of life and because of this change, affect changes in chemical cycling, energy flow, the surface coloration of the Earth, and other global factors.

The rapid growth of a population is generally viewed as good for that population and is often used as a sign that a population is "healthy" in some sense. But rapid growth for a long period can only be detrimental to a population. This is illustrated by the potential of our own population for growth, which is discussed at the end of this chapter. One of the often-stated concerns about the growth of the human population is that, as it gets larger and larger, it faces the danger of running out of resources and fouling its own nest—polluting its habitat—to an extent that might lead to its extinction. Thus, a population or species that is likely to persist for a long time may not be one that grows in a rapid, uncontrolled way for any time.

LIFE AND THE GLOBAL ENVIRONMENT

One of the important new ideas that has developed during the late twentieth century is that life not only depends on the global environment, but also affects that environment. It affects the physics and chemistry of the environment at a global level in a variety of ways. Photosynthetic organisms—especially land plants, but also algae and photosynthetic bacteria in the oceans—change the color of the Earth's surface and thereby change the reflectance properties and the amount of energy that is stored and reflected. The structure of vegetation on the land affects what is called the "roughness index," the roughness of the surface, which in turn reduces the flow of wind and thereby affects weather and climate, as discussed in Chapter 13.

Life on the land—especially vegetation and bacteria—also affects the atmosphere through the exchange of gases, altering the chemical cycles, as we will see in Chapter 16. Among these effects is the exchange of water vapor between plants and the atmosphere. Plants take up water from the soil and release it in their leaves, a process necessary for plants. A result is that plants increase evaporation and therefore decrease runoff. This decrease, together with the ability of the roots of plants to stabilize soil and take up water, affects rates of erosion. Life on the land therefore affects the rate of flux of dissolved and suspended particulate matter from the land to the oceans, via

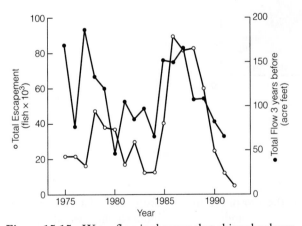

Figure 15.15 Water flow in the year that chinook salmon are born correlates strongly with the number of adults returning for the ocean to spawn three years later. The adult salmon are counted as they swim up a ladder at a dam on the Rogue River in Oregon. A high water year is followed three years later by a high number of adults returning to spawn, while a low water year is followed three years later by a smaller number of adults returning. (*Source:* Sobel, M.J., and D.B. Botkin, 1994, *Status and Future of Salmon of Western Oregon and Northern California, Forecasting Spring Chinook Runs,* Center for the Study of the Environment, Santa Barbara, CA, 42 pp.)

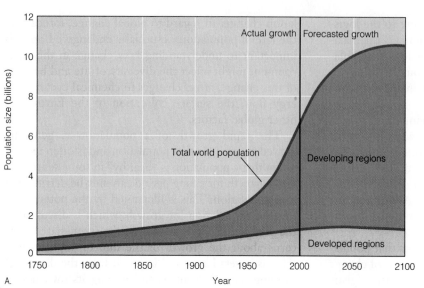

A.

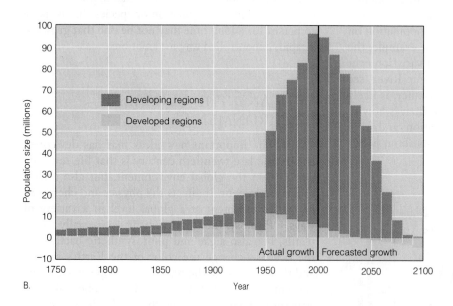

B.

Figure 15.16 World population shown as total numbers (*a*) and growth per decade (*b*), by development status, 1800–2100. For example, during the decade from 1980 to 1990, on average 82 million people were added each year, for a total addition of 820 million people. (*Source:* M. M. Kent and K. A. Crews, *World Populations: Fundamentals of Growth*, 1990, Population Reference Bureau, Washington, D.C.)

streams and rivers. In turn, life in the oceans, especially algae and bacteria, affects chemical transfers between the ocean and the atmosphere, as well as between the ocean and the solid sediments. There are many important sediments that are biological products or influenced indirectly by life.

The Gaia Hypothesis

Several intriguing questions arise concerning the extent to which life modifies the global environment. For example, are these modifications necessary for the persistence of life? If not necessary, are they beneficial to the biosphere in that life can persist longer or in greater abundance if life-induced changes take place? Answers to these questions have been phrased as the **Gaia Hypothesis** (from the Greek word "Gaia" for Mother Earth) proposed by the chemist James Lovelock and the biologist Lynn Margules. This hypothesis has three parts: (1) life has altered the environment at a global scale throughout life's history on the Earth and continues to do so; (2) life tends to stabilize the environment in the sense of reducing the variability of the physical and chemical aspects of the environment; and (3) these alterations benefit life in the sense that they increase the probability of the persistence of life.

Evidence related to the first two scientific projections of the Gaia Hypothesis can be found throughout this chapter as well as in the next two chapters.

Much evidence exists that life affects the global environment, and so the first part of the Gaia Hypothesis is generally accepted. The other parts remain controversial. Well-known writer Lewis Thomas describes Lovelock's fundamental contribution as a view that the Earth system is "a coherent system of life, self-regulating, self-changing, a sort of immense organism." This "super-organism" idea has made the Gaia Hypothesis controversial.

The Gaia Hypothesis has received much attention in recent years: not only was it the subject of a symposium of the American Geophysical Union, but also it entered into popular culture as a basis for one kind of global environmentalism. Some of these popularizations extend the hypothesis to a third idea, which is nonscientific: the belief that life altered the global environment on purpose, or with purposefulness. This last popular idea is a metaphysical statement beyond the realm of science and so is not a topic addressed directly in this book.

Human beings change the biosphere in many ways. The signs of change are everywhere around us—in the building of cities, the spread of pollutants, the cutting of forests. Since the discovery of fires, people have had large-scale effects on the land surface.

THE HUMAN POPULATION PROBLEM

The great capacity of populations to grow is a definite feature of our species. The total human population is approaching 6 billion, with a growth rate of approximately 1.8 percent per year (Fig. 15.16). At this rate, 95 million people are added to the Earth's population in a single year—a number more than equal to the 1990 population of Mexico! If this rate were to continue, then in slightly more than 1700 years the human population would weigh more than the mass of the entire Earth—5.98×10^{27}g or 5.98×10^{21}MT (a thousand billion billion metric tons), assuming an average body weight of 50 kg. Perhaps even more striking, consider the growth rate of the human population at what seems to be a small rate of increase. It is estimated that there were 10 million people at the time of Christ. If the human population had grown consistently at 1.8 percent a year after that time, by the time of the American Revolution in 1776 that human population of 10 million would have grown to a mass exceeding that of the entire present Earth system! Clearly it was not possible to maintain such a growth rate for a long period of time. Populations can grow exponentially for a brief period, but they cannot grow exponentially forever. Today, the human population influences the Earth system in many indirect ways, simply because of the rapid growth of our numbers. Assuming a constant per capita use of resources, the total human population will increasingly strain our ability to attain and sustain most resources, especially biological resources and mineral ores. Thus, the rapid growth of the human population underlies all environmental problems.

From our human perspective another set of questions arises. When people dominate an environment, things seem to go awry. Species become extinct; the environment becomes polluted. How do the algae and bacteria in a Folsom bottle avoid these disasters? Or, to put the question in a more scientific form, why do these kinds of disasters not take place in the Folsom bottles? Are there lessons we can learn from our study of life on the Blue Planet that will guide us to avoid environmental disasters that many have suggested lie in our future?

SUMMARY

1. Life is a planetary phenomenon, and life, as we know it on Earth, exists within a closed system.

2. The four essential properties of living organisms are metabolism, growth, reproduction, and biological evolution. Growth can occur through polymerization or crystallization; growth in living things is a result of polymerization, which requires a source of energy.

3 Life is organized as a hierarchy of many kinds of systems, from a single cell to the entire biosphere.

4. Cells may be procaryotic (the cells of bacteria—generally small and comparatively simple in structure), or cells can be eucaryotic (the cells of animals, plants, and fungi). These cells are larger and more complex with a well-defined nucleus and cell parts called organelles, each having a particular function of operation within the cell. Procaryotic life exists as bacteria. Many bacteria are unicellular, meaning they consist of one cell, but most other organisms are multicellular, consisting of hundreds to trillions of cells.

5. While individual organisms are alive, the persistence of life is a property of ecosystems. An ecosystem is a set of interacting species (called an ecological community) and the local, nonliving environment. Ecosystems can be small, like a small pond, or large, like a watershed covering hundreds of square kilometers in a wetland.

6. The flow of energy within an ecosystem begins with autotrophs, which get their energy from inorganic sources; they form the bottom of the food web and are on the first trophic level. Energy always flows from a lower trophic level to a higher one.

7. Autotrophs release energy either through fermentation or respiration; all aerobic organisms use energy in respiration, which is more efficient than fermentation because it releases more of the available energy.

8. Heterotrophs feed on autotrophs or other heterotrophs, thereby creating a trophic structure in which energy is moved upward from level to level via the food chain or food web.

9. Energy flow and the cycling of chemical elements are two fundamental processes required to sustain life.

10. The first law of thermodynamics tells us that matter is neither created nor destroyed, but merely changed from one form to another; the second law of thermodynamics tells us that energy always changes from a more useful form to a less useful, disorganized form. These laws have important implications for life. Specifically, life is sustained on Earth through a process wherein Earth acts as an intermediate system between a source of usable energy and a sink for heat; the Sun provides most of the energy, and space is the sink for heat.

11. The global ecosystem, also known as the biosphere, is the sum of all the smaller ecosystems on the Earth.

12. There are two kinds of production: primary production, which is a characteristic of autotrophs, and secondary production, which is characteristic of heterotrophs. Net production is a measure of the change (or growth) of biomass.

13. Organisms use a large amount of energy in respiration, and very little is left to convert to production of new tissue. Respiration returns to the environment the carbon dioxide removed by photosynthesis.

14. A population is a group of individuals of the same species sharing genetic information that inhabit an area or interbreed. Population dynamics are a function of the growth rate, birth rate, and death rate of a population.

15. When a population grows at a fixed rate, the growth is called exponential.

16. The logistic curve previously was understood as a method to chart population growth to what was considered a carrying capacity, which was a stable abundance. Evidence today suggests that populations do not exhibit this kind of stability.

17. The Gaia Hypothesis proposes that life has altered the environment at a global scale throughout history, and that these changes benefit life and have improved the probability of its persistence on Earth. A third tenet, much more controversial, is that the Earth system is a coherent system, self-regulating, and self-changing.

18. The biosphere influences and interacts with all of the other spheres, affecting the physics and chemistry of the environmental at a global level.

IMPORTANT TERMS TO REMEMBER

autotroph *348*	energy flow *352*	gross production *354*	trophic level *349*
biomass *354*	eucaryotes *348*	heterotroph *349*	trophic structure *349*
biomes *347*	evolution *344*	organelles *348*	
carbohydrate *349*	food chains *349*	photosynthesis *349*	
cell *345*	food web *349*	population *355*	
ecological community *345*	Gaia Hypothesis *358*	procaryotes *348*	
ecosystem *347*	genus *345*	species *345*	

QUESTIONS FOR REVIEW

1. Explain why life is a planetary phenomenon.

2. What examples do we have that one characteristic of life is its continual state of change?

3. What are the main differences between procaryotic cells and eucaryotic cells?

4. What are the two fundamental requirements for any long-term life-support system?

5. What is the difference between autotrophs and heterotrophs? Name two different classes of heterotrophs.

6. What is meant by an ecosystem? Describe an ecosystem you are familiar with in the area in which you live.

7. Why can't we recycle energy within our bodies or between species?

8. What do organisms need energy for, besides moving around?

9. When is exponential growth not good for a population?

10. Describe some examples found in this chapter that would support the Gaia Hypothesis; can you think of others that do not support the hypothesis?

QUESTIONS FOR DISCUSSION

1. Why is an ecosystem the smallest system that can support life? Discuss the flow of energy within an ecosystem.

2. Where are you on the food chain? Which trophic level?

3. Discuss the importance of production of biomass to the Earth system in terms of the hydrosphere, the solid Earth, and the atmosphere. How would a severe reduction of biomass production affect each? Consider Biosphere II.

4. How would you improve on the design of a closed system to support people for long term space travel.

5. How does life differ from other phenomena in the universe?

If you visit our web site, you will find several case studies about unique ecosystems.

CHAPTER 16

Geochemistry
and Life

● *Phosphorus and the Global Environment*

Phosphorus is one of the essential building blocks of life. Unlike the other essential building blocks—carbon, hydrogen, nitrogen, oxygen, and sulfur, all of which occur as gases in the atmosphere—phosphorus is difficult to obtain in nature. The original source of phosphorus is the rocks of the solid Earth, from which phosphorus containing soils are derived and from which in turn plants get their phosphorus, and animals, by eating plants, satisfy their needs. But agriculture removes phosphorus from the soil, and for good crop production phosphorus fertilizers are often needed. When the Indians showed the Pilgrims how to plant crop seeds with fish bones, they were teaching them a lesson in fertilizing; fish bones are rich in phosphorus.

Some soil phosphorus is carried away by groundwater and is eventually deposited in the sea. There, under rare and special circumstances, it may be concentrated into deposits that can be mined for the needed fertilizers. One of the most unusual circumstances of phosphorus concentration involves seabirds.

For centuries, phosphorus fertilizers were gathered on small coastal islands along the western coasts of South America, (Fig. 16.1). The "ores" were layer upon layer of whitish bird droppings—guano—left by millions of nesting sea birds over many years. When guano began to be exported to the northern hemisphere the harvesting of guano exceeded its creation. In addition, mining practices destroyed nests and bird populations on the islands was diminished.

Birds that nest on the guano islands obtain phosphorus by eating fish. The fish in turn feed on plankton, the floating "soup" of microscopic animals and algae in the upper layer of the ocean. The plankton get their phosphorus from seawater, and the places where plankton are most numerous are along coasts where deep, phosphorus-rich waters well up to the surface such as along the western coasts of South America and Africa.

Upwelling currents, plankton, fish, and birds are only part of the story. Guano is soluble and washes away in the rain. Guano can only accumulate in large amounts in dry climates. Thus, formation of a major source of guano requires a particular climate, a particular geography, and a particular set of fish and bird species. Occasionally, the relationships are disrupted; sometimes the upwellings fail because of the El Niño events (discussed in Chapter 8), the plankton disappear, fish die or swim away, and birds die in vast numbers.

(opposite) Phosphorous fertilizers are mined from deposits that have biological origins. This aerial view of the CF Industries Bonnie Mine in Bartow Florida shows extraction of phosphate rock produced by ancient marine biological activity. Such mines are the major modern source of phosphorous fertilizers.

There are other sources of phosphorus fertilizer, such as the marine deposits in Florida. The deposits there are in limestone formed by marine organisms. As with guano, the phosphorus is concentrated in the limestone as a result of the actions of marine animals. From guano islands off Peru to limestone mines in Florida, all of our important sources of phosphorus fertilizer are biological in origin.

The cycling of phosphorus, from rocks to soil, via groundwater to the ocean, from seawater through plankton and other sea creatures to guano and marine deposits, illustrates the intricate interrelationships between life and the global environment. Later in this chapter we will discuss the phosphorus cycle in more detail, but first we will consider the basic concepts of biogeochemical cycles. ●

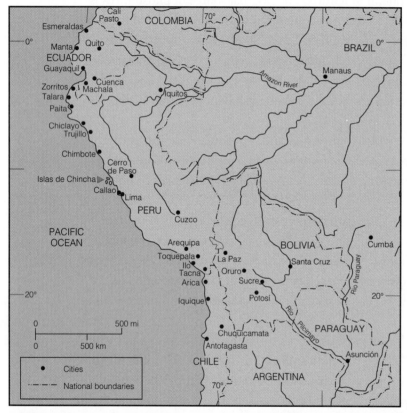

A.

Figure 16.1 A. Map showing location of Guano Islands off of Peru; B. Miners collecting guano to be used as phosphorous fertilizer.

B.

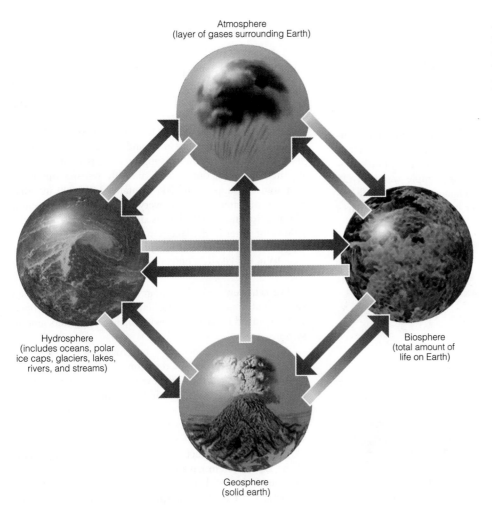

Atmosphere
(layer of gases surrounding Earth)

Hydrosphere
(includes oceans, polar
ice caps, glaciers, lakes,
rivers, and streams)

Biosphere
(total amount of
life on Earth)

Geosphere
(solid earth)

Figure 16.2 Cyclic transfer of chemical elements among Earth's four reservoirs (geosphere, hydrosphere, atmosphere, and biosphere.)

BIOGEOCHEMICAL CYCLES AND BIOLOGICAL EVOLUTION

Because organisms and populations evolve, the biosphere itself must also have changed through time and must still be changing today. This means that the Earth system must have undergone changes. One way the changes in the Earth system can be seen is in the cyclic transfer of chemical elements among the Earth's four reservoirs—the geosphere, hydrosphere, atmosphere, and biosphere (Figure 16.2). The cycling patterns of some elements have changed dramatically throughout the Earth's history. But the most marked changes have been in the cycling rates of biologically important elements such as carbon, oxygen, nitrogen, and phosphorus. Elements essential to the biosphere—and whose cycles are strongly influenced by it—are said to have *biogeochemical cycles*. (See "A Closer Look: A Biogeochemical Cycle").

The Basic Principles of Biogeochemical Cycling

The study of biogeochemical cycles is an attempt to answer two questions: What limits the abundance of life? And what role does life play in the chemistry of the Earth system? Often a specific chemical element sets an upper bound on the abundance of life, as in the case of phosphorus and the birds of the guano islands. We will also discover that life has profoundly altered the chemistry of the Earth systems, sometimes in ways that were not understood until recently. The chemical limits on life occur at each level in the ecological hierarchy discussed in Chapter 15—from an individual to populations, species, ecological communities, on up to the entire biosphere.

Less than 24 chemical elements are known to be required for life (Fig. 16.3). These are divided into the **macronutrients**, elements required in large amounts by all life; and **micronutrients**, elements required either in small amounts by all life or in moderate

A CLOSER LOOK

A Biogeochemical Cycle

The simplest way to view a biogeochemical cycle is as a box and arrow diagram, where the boxes represent places where a chemical element is stored and the arrows represent pathways of transfer (Fig. C16.1A). A biogeochemical cycle is generally drawn for a single chemical element, but sometimes it is drawn for a compound, for example, water (H_2O). Figure C16.1B shows these basic elements of a biogeochemical cycle applied to water. Water is stored temporarily in a

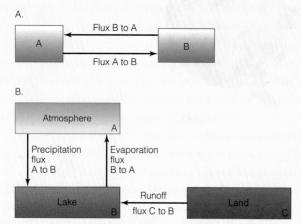

Figure C16.1 How the biogeochemical cycle is illustrated. A. A and B are reservoirs. Materials flow from each reservoir to the other. B. Some components of the hydrologic cycle.

lake (compartment B). It enters the lake from the atmosphere (compartment A) as precipitation or from the land around the lake as runoff (compartment C). It leaves the lake through evaporation to the atmosphere, or by runoff to a surface stream or to subsurface flows. There is a *flux* and *rate of transfer* for each element and a *residence time* in each compartment. For example, suppose there is a salt lake with no surface stream outlet, and the only transfer out is by evaporation. Suppose the lake contains 3,000,000 m^3, that 3000 m^3 of water evaporate every day, and that the amount of water flowing into the lake is also 3000 m^3/day so that the volume of water in the lake remains constant. Then the average residence time (the average time a molecule of water spends) in the lake is 3,000,000/3000 m^3/day, which is 1000 days (or 2.7 years).

We can view a single storage compartment as an open system and the entire biogeochemical cycle as a set of open systems linked by arrows showing material transfers among systems.

Another crucial aspect of a biogeochemical cycle is the set of factors or processes that control the flow from one compartment to another—for example, how air temperature or wind velocity across a lake influence the evaporation rate of water in the lake. To understand a biogeochemical cycle, these factors and processes should be quantified and understood from basic principles.

By their basic characteristics, the four major components of the Earth system have different average rates of storage. In general, the residence time of chemical elements is long in rocks, short in the atmosphere, and intermediate in the hydrosphere and the biosphere.

amounts by some forms of life and not others. The macronutrients include the "Big Six," the elements that form the fundamental building blocks of life. These are carbon, hydrogen, nitrogen, oxygen, phosphorus, and sulfur. Each plays a special role in organisms. Carbon is the basic building block of organic compounds. Along with oxygen and hydrogen, it forms carbohydrates. Nitrogen, along with the remaining three, makes proteins. Phosphorus is the "energy element" occurring in the compounds called ATP and ADP, which are important in the transfer and use of energy within cells. Other macronutrients also have important roles. For example, calcium is the "structure element" that occurs in bones of vertebrates, shells of shellfish, and the wood-forming cell walls of vegetation. Sodium and potassium are important to nerve signal transmission.

For any form of life to persist, chemical elements must be available at the right times, in the right amounts, and in the right relative concentrations to each other. When this does not happen, an element

can become a *limiting factor*, preventing the growth of an individual, population, or species, or even causing its local extinction.

Other elements are toxic. Some, like mercury, are toxic even in low concentrations. Others, like copper, are required in low concentrations but can be toxic if they are present in high concentrations. Paracelsus, a fifteenth-century German-Swiss physician who established the role of chemistry as a part of medicine, wrote that "everything is poisonous, yet nothing is poisonous." By this he meant that essentially any substance in too great amounts can be dangerous, yet anything in extremely small amounts can be relatively harmless. Every chemical element has a spectrum of possible effects on a particular organism. For example, selenium is required in small amounts by living things but may be toxic or cause cancer in cattle and wildlife when it is in high concentrations in soil. Finally, some elements are neutral for life—either they are chemically inert, such as the "noble gases" like Neon, which do not react with other elements, or they are present

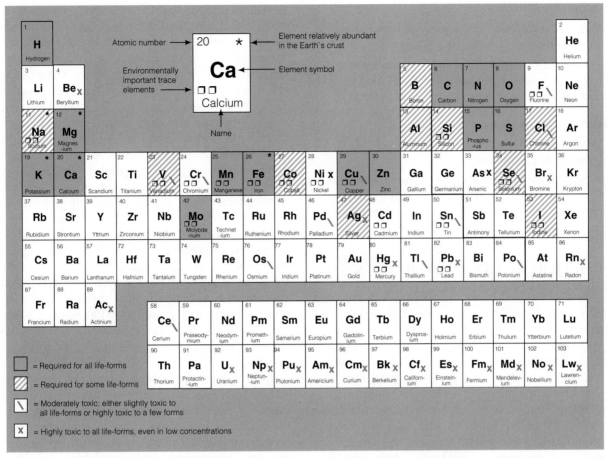

Figure 16.3   The periodic table of elements showing which are required for life and which are toxic to living things (Botkin & Keller, 1995).

on the Earth's surface in such low concentrations that they do not play a role in life processes; gold and platinum are examples.

HOW CHEMICAL ELEMENTS CYCLE

In its most general form, a **biogeochemical cycle** is the complete pathway that a chemical element follows through the Earth system—from biosphere to the atmosphere, to oceans, to sediments, soils, and rocks, and from rocks back to the atmosphere, ocean, sediments, soils, and biosphere. It is a *chemical* cycle because chemical elements are the form that we consider. It is *bio-* because these are the cycles that involve life. It is *geo-* because these cycles include rocks and soils. Although there are as many biogeochemical cycles as there are elements, certain general concepts hold true for these cycles. The key that unifies biogeochemical cycles is the involvement of the four

principal components on the Earth system: rocks, air, water, and life.

Many kinds of questions can only be answered through an understanding of biogeochemical cycles. Some examples follow. Although this chapter does not answer all of these questions, it does provide the background that will allow you to begin to find the answers.

Biological Questions

- What factors, including chemical elements, might be limiting abundance and growth?
- What are the sources of chemical elements required for life, and how might we make them more readily available?
- What problems occur when an element is too abundant?

Geological Questions

- What processes create desired mineral resources? Using this understanding, how can we better locate such resources?

Atmospheric Questions

- What factors determine the concentrations of elements and compounds in the atmosphere?
- Where the atmosphere is polluted as the result of human activities, how might we adjust a biogeochemical cycle in order to lower the pollution?

Hydrological Questions

- What determines whether a body of water will be biologically productive?
- When a body of water becomes polluted, how can we adjust the biogeochemical cycle that involves the pollutant in order to reduce its level and its effects?

BIOLOGICAL CONCENTRATION OF ELEMENTS

Organisms are highly selective in their uptake of chemical elements, allowing for the synthesis of highly specific compounds in their cells. And organisms concentrate certain chemical elements at much higher levels than they are found in the local environment. This is an important, fundamental ability of living things. For example, marine algae concentrate iron 100,000 times; that is, the algae have an internal concentration of iron 100,000 times or more than that found in ocean water (Fig. 16.4A). Nitrogen and phosphorus have concentration factors of 10,000 to 100,000. Manganese is concentrated by plankton 10,000 to 100,000 times above concentrations found in the ocean; zinc, nickel, copper, cadmium, and aluminum are concentrated 1000 to 10,000 times above the amount in the ocean water. In fact, marine algae concentrate *all* elements from ocean water except chlorine and sodium (the two most common dissolved elements in ocean water), and fluorine and magnesium (whose concentration factor is about 1, or the same as the water). If an element is concentrated at levels higher than the concentration in the environment, then energy must be expended. Organisms have *active metabolic* mechanisms to accomplish this. Through such active uptakes, living things change the chemical composition of their environment.

Land plants take up most required chemical elements through their roots (Fig. 16.4B). Land plants also show strong concentration factors, which differ for different elements. For example, the vegetable kale concentrates phosphorus 1600 times that in a soil water solution and potassium 1100 times. Calcium and magnesium have low concentration factors (less than 20), and some plants have been shown to actively reject sodium.

In conclusion, organisms can control the concentration of chemical elements in their local environment. By extension, organisms can affect concentrations globally. In this way, life is an active entity in an ecosystem; it is not merely a passive receptor, like a pail into which elements are poured. Instead, the biota form a community that functions as a highly selective chemical pump.

A.

B.

Figure 16.4 Photograph of A. a marine algae that concentrates chemicals in its environment; B. a land plant in a laboratory culture showing the roots that take up chemical elements.

The Right Ratios at the Right Times

The active concentration of chemical elements by living things has several consequences. Evidence suggests that marine organisms keep the chemical composition of ocean water essentially constant. Organic matter has a fairly constant composition, and therefore the ratios of chemical elements in living things are fairly constant. For example, in the oceans, small floating algae and animals that make up the marine plankton have an average chemical makeup in which there are 106 atoms of carbon and 16 atoms of nitrogen for every atom of phosphorus. The ratio of carbon:nitrogen:phosphorus for marine plankton is therefore 106:16:1. This is called the *Redfield ratio* after Alfred Redfield who first discovered this general property of living things. Through processes of active uptake, the plankton tend to bring the concentration of carbon, nitrogen, and phosphorus into this same ratio in the ocean water. If there is excess phosphorus, then bacteria can fix additional nitrogen from the atmosphere and algae and bacteria can add carbon through photosynthesis. The organisms will then take up the excess phosphorus. Only when the concentration of carbon:phosphorus ratio approaches 106:1 and the nitrogen:phosphorus ratio approaches 16:1 will the additional growth stop. The Redfield ratio is an important example of the mechanisms by which life affects the chemical environment. In its tendency to stabilize the chemistry of the environment, the Redfield ratio process is also an example of the Gaia Hypothesis discussed in Chapter 15.

We can determine what chemical element is limiting in an ecosystem by comparing the ratios of the concentration of elements in the environment with the ratios of the elements found in living tissue. If the ratio is much lower in the environment than in the organism, then the element with the low ratio will be limiting to growth.

Dose-Response Curves and Pollution Responses

Although life affects the chemistry of the environment, this chemistry affects life. Two basic concepts that help us understand the effects of chemistry of the environment on life are the dose-response curve and food chain concentration. The **dose-response curve** is a graph of the percentage of a population killed or injured versus the concentration or dose received of a toxin in the environment. The effect of an environmental pollutant is often described by the toxic dose-response curve. This shows a negative response, either death or injury, plotted against the increasing intensity of exposure to a pollutant. The upper limit of this curve represents 100 percent of the population affected (Fig. 16.5).

Food-chain concentration means that the concentration of a compound in living tissue increases as the compound is transferred up a food chain from autotroph to herbivore to predator. Both are illustrated by a famous and strange epidemic that occurred in Minamata, Japan as discussed in "A Closer Look: Analysis of a Disaster."

Threshold Effects

A controversy exists over whether and when thresholds can be found for environmental toxins. A **threshold** is a level below which no effect occurs and above which effects begin to occur. If a threshold exists, then any level in the environment below that threshold will be safe. Alternatively, it is possible that even the smallest amount of toxin has some negative effect (Fig. 16.5). Whether there is a threshold effect for environmental toxins is important to understanding biogeochemical cycles. Whether or not there is a threshold effect has important implications, often with heavy financial consequences. For example, asbestos, which was used for many years as an insulating material in homes, schools, and other buildings, has been found to cause certain kinds of cancer. The assumption that there is no threshold effect for this response—that even a single molecule of asbestos absorbed by a person can cause cancer—leads to expensive removal of this insulation, which might not have been necessary

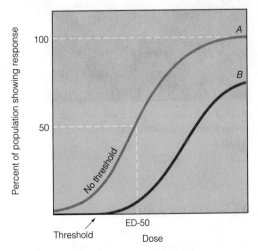

Figure 16.5 In this hypothetical toxic dose-response curve, toxin *A* has no threshold; even the smallest amount has some measurable effect on the population. The ED-50 for toxin *A* is the effective dose, or dose that causes an effect in 50 percent of the population. It is related to the onset of specific symptoms, such as loss of hearing, nausea, or slurred speech.

A CLOSER LOOK

Analysis of a Disaster

In 1956, doctors described curious symptoms in residents of Minamata Bay, Japan. The symptoms were indicative of severe toxification of the central nervous system. It was soon discovered that patients afflicted with the disease were all fishermen and their families, and that they regularly ate fish caught in Minamata Bay (Fig. C16.2). Further study showed that mercury was involved and that a chemical plant on Minamata Bay, which produced basic chemicals for the plastics industry, used mercury salts in the processing and was the source of the problem. The amounts used were tiny, however, so the question then became, how could the fishermen accumulate so much mercury in their bodies? The answer turned out to be a case of food-chain concentration.

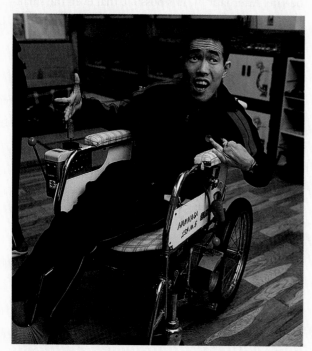

Figure C16.2   A victim of pollution. Physical deformity arising from mercury poisoning as a result of eating contaminated fish from Minamata Bay, Japan.

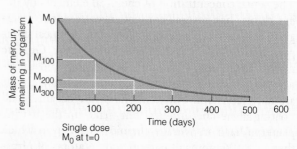

Figure C16.3   Decrease of mercury in the body of an organism that eliminates half of dose present each 100 days.

Organisms eliminate poisons at a rate that depends on the concentration of the poison in their tissues. If M is the amount of poison in the organism, M' the rate of elimination, and k a rate constant, then

$$M' = -k \cdot M$$

If an organism receives a single dose M_0 of the poison, a graph of M against time looks like Figure C16.3. M decreases in a regular manner so that after 100 days one-half of the initial dose $M_{0/2}$ remains; after another 100 days M has again been halved (to $M_{0/4}$), and so on. The halving period of 100 days is called the half-life of the poison in the organism. The rate constant (k) is related to the half-life by the following equation:

$$k = \frac{\text{natural log of 2}}{\text{half-life}} = \frac{0.7}{100}, \text{ or } 0.007 \text{ per day}$$

Suppose now that instead of receiving a single dose, the organism has a diet in which a constant amount of poison is ingested every day. At the start, the organism's poison content, M, will rise fairly steeply. As it builds up, however, the rate of elimination will increase until a steady state is reached where M' (rate of elimination) becomes equal to the rate of intake (see Fig. C16.4). Strictly speaking, the steady state is reached only after an infinite time, but it is near enough for practical purposes after about six half-lives.

had a threshold been established.

That there is a threshold for toxic effects of a pollutant can be treated as a scientific hypothesis and tested by experiments. Few experiments, however, have been directed toward testing this hypothesis. Some data suggest the existence of a threshold and of a toxic dose-response curve. For example, Figure 16.6 shows the dose-response curves for three species of trees exposed to chronic (long-term) radioactivity. These

trees were in a forest that was part of an experiment at Brookhaven National Laboratory on Long Island, New York, to test the effects of radioactivity on a natural ecosystem. In this experiment, a source of gamma radiation, cesium-134, was placed in the center of an oak-pine forest inside a vertical shaft so that the source could be lowered into a lead shield. The forest was irradiated for 20 hours every day for many years; during the 4 hours per day that the source was

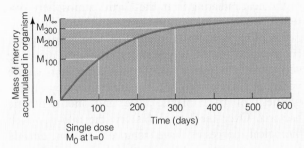

Figure C16.4 Buildup of mercury in the body of an organism that receives a steady diet of mercury each day. As the amount in the body builds up, the rate of elimination is increased. A balance is reached when the intake equals the elimination rate.

The poison content of an organism at steady state (assuming the poison has not killed it first) is the rate of intake divided by the rate constant k:

$$M = \text{daily intake}/0.007$$

Imagine an ecosystem with the trophic levels and diets given by the first two columns (mass and daily diet of individual) in Table C16.1. Now suppose that a small amount of mercury is introduced into the system and is taken up by the algae, producing a mercury concentration in them of 10

parts per billion (1 part in 10^8). The daily diet of a herbivore, 10 grams of algae, contains 10×10^{-8} or 10^{-7} grams of mercury. At steady state, the herbivore will contain $10^{-7}/0.007$ or 1.4×10^{-5} grams mercury, and the mercury concentration in its tissues will be $1.4 \times 10^{-5}/100$, or 1.4×10^{-7} (140 parts per billion). We can now complete Table C16.1 by calculating the mercury concentrations in the species at successive trophic levels in this food web.

Thus, in moving three trophic levels up the food web from the algae, the mercury gets concentrated by a factor of nearly 3000 ($2.9 \times 10^{-5}/10^{-8}$), reaching a concentration of 29 parts per million in the secondary carnivores: more than 500 times the acceptable limit of 0.05 ppm for food in the United States.

There are some weaknesses in this oversimplified model. For example, the half-life varies from one kind of tissue to another. Furthermore, it is unlikely that all of the prey species would survive the six half-lives necessary to reach the steady state because some would get killed and be eaten with lower concentrations of mercury in them. Finally, such an ecosystem might not be closed; fish that swam in after having fed elsewhere could be mercury-free. Nevertheless, in Minamata Bay top-level carnivores were found with as much as 50 ppm of mercury in them. A 60 kilogram Japanese eating 200 grams of such fish a day would accumulate 24 ppm of mercury in her tissue.

Table C16.1 Increase in Mercury Concentration Up the Food Web

Trophic Level	Mass of Individual	Daily Diet of Individual	Daily Mercury Intake	Steady-State Mercury Content	Mercury Concen./ Tissues (parts per billion)
Algae	—	—	—	—	10
Herbivores	100 g	10g algae	10^{-7} g	1.4×10^{-5} g	140
Primary carnivores	1000 g	1 herbivore	1.4×10^{-5} g	2×10^{-3} g	2000
Secondary carnivores	10,000 g	1 primary carnivore	2×10^{-3} g	0.29 g	29,000

shielded, scientists could enter the forest to examine the site.

The damage to the forest by the gamma radiation varied with the species and among individuals within species. There were several clearly distinguishable zones of effects. Nearest the source, where radiation was the most intense, a completely devastated zone was produced in which all woody plants were killed.

The dose-response curve shows the percentage of

tree mortality as a function of the amount of radiation they received. Essentially 100 percent of pitch pines were killed at an exposure of 10 roentgens/day, the level at which mortality is first measurable for white oak. Both scarlet oak and white oak can withstand ten times more radiation (100 roentgens/day) than the pine before suffering 100 percent mortality. This suggests that oaks have a higher threshold in regard to radiation than pitch pine. This example shows that at

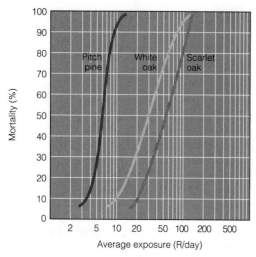

Figure 16.6 Toxic dose-response curve for trees subjected to radioactivity in an experiment at Brookhaven National Laboratory. (*Source:* G. M. Woodwell and A. L. Rebuck, 1967, "Effects of Chronic Gamma Radiation on the Structure and Diversity of an Oak-Pine Forest," *Ecological Monographs*, 37, pp. 53–69. Reprinted by permission.)

least in one case, there is a clear threshold effect.

SOME MAJOR BIOGEOCHEMICAL CYCLES

Now that we have considered some of the basic concepts of biogeochemical cycles, we can turn to specific cycles. Each element has a different cycle, with various different pathways. By comparing different chemical cycles, we can see in greater detail how life affects the chemistry of the environment and how the chemical environment affects life.

The Carbon Cycle

Carbon is the building block of life. It is the element that anchors all organic substances, from coal and oil to DNA, the compound that carries genetic information. Although of central importance to life, carbon is not one of the most abundant elements in the Earth's crust. It is fourteenth by weight, contributing only 0.032 percent of the weight of the crust, ranking far behind oxygen (45.2%), silicon (27.2%), aluminum (8.0%), iron (5.8%), calcium (5.1%), and magnesium (2.8%).

The major pathways and storage reservoirs of the carbon cycle are shown in Figure 16.7. Carbon is one of the elements that has a gaseous phase as part of its cycle, occurring in the Earth's atmosphere as carbon dioxide (CO_2) and methane (CH_4), both greenhouse

gases. Greenhouse gases, along with water vapor, warm the Earth's atmosphere because they trap some of the heat radiating from the Earth's atmosphere system. Carbon enters the atmosphere through the respiration of living things, through fires that burn organic compounds, through volcanoes, through decaying vegetation in wetlands, and through diffusion from the ocean. It is removed from the atmosphere by photosynthesis of green plants, algae, and photosynthetic bacteria. Over the Earth's history, the rate at which biological processes have removed carbon dioxide from the atmosphere has exceeded the rate of addition. As a result, the Earth's atmosphere now has far less carbon dioxide than would occur on a lifeless Earth and much less than occurs in the atmospheres of Venus and Mars, where carbon dioxide is the primary gas in the atmosphere. In the Earth's atmosphere, carbon dioxide can dissolve in water droplets to form a mild acid called carbonic acid (H_2CO_3) and return to the Earth's surface in precipitation. As mentioned earlier, this mild acid is important in weathering rocks at and near the surface of the land.

Carbon enters the ocean from the atmosphere as a simple solution of carbon dioxide. The carbon dioxide dissolves and is converted to carbonate and bicarbonate. (See Chapter 11 for a discussion of these processes.) Marine algae and photosynthetic bacteria obtain the carbon dioxide they use from the water, in one of these forms. Carbon is also transferred from the land to the ocean in rivers and streams as dissolved carbon, including organic compounds, and as organic particulates (fine particles of organic matter). Winds also blow small organic particulates from the land to the ocean. The transfer via rivers and streams makes up a comparatively small fraction of the total carbon flux to the ocean, but locally and regionally this flux is of great importance, influencing near-shore areas that are often highly productive biologically.

When an organism dies, most of its organic material decomposes into inorganic compounds including carbon dioxide, but some may be buried where there is no oxygen to make this conversion possible or where the temperatures are too cold for such decomposition, and so the organic matter is stored. Over years, decades, and centuries, such storage occurs in wetlands including parts of floodplains, lake basins, bogs, swamps, deep-sea sediments, and near polar regions. Over longer periods, some of this material may be buried under other sediments and form part of sedimentary rocks, and become fossil fuels—coal, oil, and gas. Large amounts of carbon are also stored in an oxidized form as limestone—produced primarily by biological activity—in the shells of marine organisms. In fact nearly all of the carbon stored in the lithosphere

A.

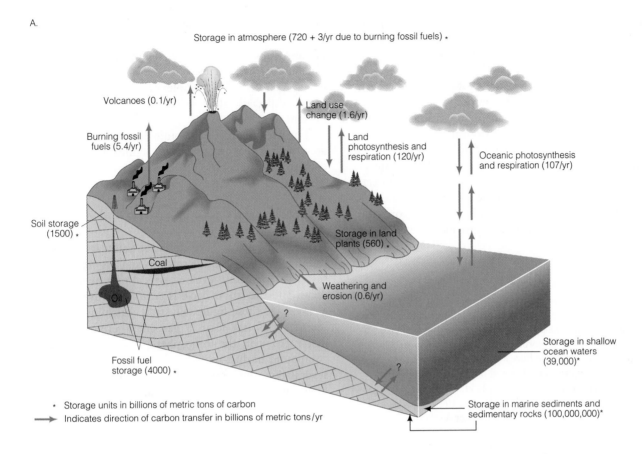

Storage in atmosphere (720 + 3/yr due to burning fossil fuels) *

Volcanoes (0.1/yr)

Burning fossil fuels (5.4/yr)

Land use change (1.6/yr)

Land photosynthesis and respiration (120/yr)

Oceanic photosynthesis and respiration (107/yr)

Soil storage (1500) *

Coal

Storage in land plants (560) *

Oil

Weathering and erosion (0.6/yr)

Fossil fuel storage (4000) *

Storage in shallow ocean waters (39,000)*

? ?

* Storage units in billions of metric tons of carbon

→ Indicates direction of carbon transfer in billions of metric tons / yr

Storage in marine sediments and sedimentary rocks (100,000,000)*

B.

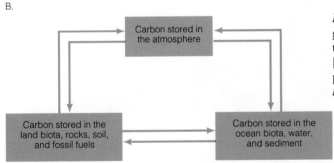

Carbon stored in the atmosphere

Carbon stored in the land biota, rocks, soil, and fossil fuels

Carbon stored in the ocean biota, water, and sediment

Figure 16.7 The global carbon cycle. A. Generalized global carbon cycle. B. Parts of the carbon cycle simplified to illustrate the cyclic nature of the movement of carbon. [*Source:* Modified after G. Lambert, 1987, *La Recherche*, 18, pp. 782–783, with some data from R. Houghton, 1993, *Bulletin of the Ecological Society of America*, 74(4), pp. 355–356.]

exists as sedimentary rocks in the form of carbonates such as limestone.

In addition to occurring in the solid Earth as fossil fuels, carbon is also found in a few inorganic forms. Pure carbon is found in a few inorganic materials in rocks, including graphite and diamonds.

The carbon cycle can be understood in terms of time scales: for example, short-term cycling (photosynthesis from an individual plant), medium-term cycling (storage of organic chemicals in the woody tissue of trees and forest soils), and long-term cycling (production of carbonate rocks and eventual return of CO_2 to the atmosphere via weathering and volcanic emissions).

The cycling of carbon dioxide between land organisms and the atmosphere is a large flux, with approximately 15 percent of the total carbon in the atmosphere taken up by photosynthesis and released by respiration annually. In this way, life has a large effect on the chemistry of the atmosphere. Because carbon is the most important organic compound and because it forms two of the most important greenhouse gases, much research has been devoted to understanding the carbon cycle. However, at a global level some key issues remain unanswered. For example, notice in Figure 16.7 that our burning of fossil fuels releases roughly 5.4 units of carbon per year and that land-use changes such as deforestation contribute roughly 1.6

units per year. (Deforestation leads to the decomposition and burning of trees and soils, thus converting organic carbon to carbon dioxide.) Monitoring of atmospheric carbon dioxide levels over the past 35 years suggests that of the approximate total amount of seven units released per year by human activities into the atmosphere, approximately 3.2 units remain, increasing the carbon dioxide concentration in the atmosphere. It is estimated that approximately two units should diffuse into the ocean. This leaves 1.8 units unaccounted for, ending up somewhere unknown to science. Inorganic processes do not account for the fate of this "missing carbon." Either marine or land photosynthesis, or both, must provide the additional

flux for this estimated 1.8 units of "missing carbon." At this time, however, scientists do not agree on which processes dominate, or in what regions of the Earth this missing flux occurs. The "missing carbon" problem illustrates the complexity of biogeochemical cycles, especially those where the biota play an important role.

The carbon cycle will continue to be an important area of research because of its significance to global climate investigations, especially to global warming, which we will discuss in Chapter 19.

Biogeochemical Cycles of a Metal and a Nonmetal

Different elements have different pathways, as illustrated in Figure 16.8 for calcium and Figure 16.9 for sulfur. The calcium cycle is typical of a metallic element, and the sulfur cycle is typical of a nonmetallic element. As these figures illustrate, calcium, like most metals, does not form a gas and therefore has no major phase in the atmosphere; it occurs only as a compound in dust particles. In contrast, sulfur forms several gases, including sulfur dioxide (a major air pollutant and component of acid rain, discussed later in

Figure 16.8 Annual calcium cycle in a forest ecosystem. In the circles are the amounts transferred per unit time (the flux rates) (kilograms per hectare per year). The other numbers are the amounts stored (kilograms per hectare). Unlike sulfur, calcium does not have a gaseous phase. The information in this diagram was obtained from Hubbard Brook Ecosystem. (*Source:* G. E. Likens, F. H. Bormann, R. S. Pierce, J. S. Eaton, and N. M. Johnson, 1977, *The Biogeochemistry of a Forested Ecosystem*, Springer-Verlag, New York.)

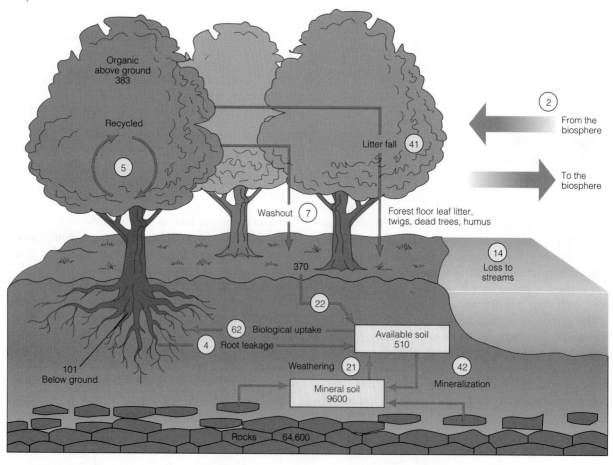

this chapter) and hydrogen sulfide (swamp or rotten egg gas, usually produced biologically).

Because sulfur has gas forms, it can be returned to the biosphere from other Earth spheres much more rapidly than can calcium. The annual input of sulfur from the atmosphere to a forest ecosystem has been measured to be ten times that of calcium. For this reason, elements without a gas phase are more likely to become limiting factors to the growth of organisms.

The Nitrogen Cycle

Oxygen can be thought of as life's gift to itself, but the atmosphere's main gift to life has been nitrogen, a relatively unreactive gas that seems to have been a minor

constituent of the initial atmospheres of all the planets. Nitrogen is essential to life because it is a necessary component of proteins including DNA.

The key to the nitrogen cycle is understanding how reduction (also called fixation) and oxidation (also called denitrification) of nitrogen take place and how nitrogen moves between the four major reservoirs—the atmosphere, the ocean, soil and sediment, and the biosphere (Fig. 16.10). Free nitrogen (N_2 uncombined with any other element) makes up approximately 80 percent of the Earth's atmosphere. However, except for a few bacteria, organisms cannot use free nitrogen directly. Plants, algae, and bacteria can take up nitrogen either as the nitrate ion ($NO_3^=$) or the ammonium ion (NH_4^+). Animals can take in nitrogen in the form of an organic compound made by the primary producers (autotrophs). In contrast to hydrogen, oxygen, and carbon, nitrogen is a relatively unreactive element, and there are few processes that convert molecular nitrogen to one of these com-

Figure 16.9 Annual sulfur cycle in a forest ecosystem. The circles show the amounts transferred per unit time (the flux rates) (kilograms per hectare per year). The uncircled numbers are the amounts stored (kilograms per hectare). Sulfur has a gaseous phase as H_2S and SO_2. Based on studies of the Hubbard Brook Ecosystem. (*Source:* G. E. Likens, F. H. Bormann, R. S. Pierce, J. S. Eaton, and N. M. Johnson, 1977, *The Biogeochemistry of a Forested Ecosystem*, Springer-Verlag, New York.)

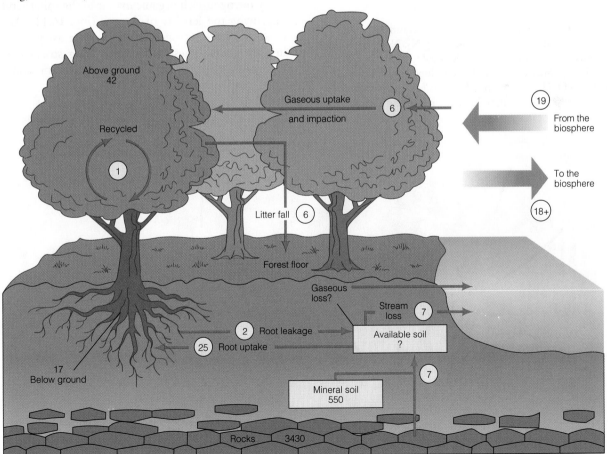

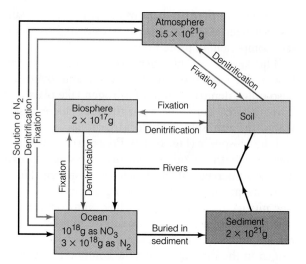

Figure 16.10 Basic processes in the nitrogen cycle. For more detail, see Figure 16.11.

pounds. Lightning oxidizes nitrogen, producing nitrate. This process is called **nitrogen fixation**. In nature, essentially 90 percent of the conversion of molecular nitrogen to biologically useful forms are conducted by bacteria.

Nitrogen is one of the most important and most complex global cycles (Fig.16.11). Basically, inorganic

Figure 16.11 Some details of the global nitrogen cycle. Pools (▢) and annual (→) flux in 10^{12} g N$_2$. Note that industrial fixation of nitrogen is nearly equal to global biological fixation. (*Source:* Data from R. Söderlund and T. Rosswall, 1982, in O. Hutzinger (ed.), *The Handbook of Environmental Chemistry*, Vol. 1, Pt. B, Springer-Verlag, New York.)

nitrogen in the atmosphere is transferred by lightning or bacterial uptake to nitrate or ammonia. Some bacteria can convert molecular nitrogen to ammonium ion. Once in either of these forms, it can be taken up on the land by plants and in the oceans by algae. Bacteria, plants, and algae then convert these inorganic nitrogen compounds into organic ones, and the nitrogen becomes available, through ecological food chains, as organic compounds. When organisms die, other bacteria are able to convert the organic compounds containing nitrogen back to nitrate, ammonia, or, by a series of chemical reactions, to molecular nitrogen, when it is then returned to the atmosphere. The process of releasing fixed nitrogen back to molecular nitrogen is called **denitrification.**

Symbiotic Relationships

It follows that ultimately all other organisms depend on nitrogen-converting bacteria; some organisms have evolved symbiotic relationships with these bacteria. For example, plants of the pea family have nodules in their roots that provide a habitat for such bacteria. The bacteria obtain organic compounds for food from the plants, and the plants obtain usable nitrogen. Such plants can grow in otherwise nitrogen-poor environments. When these plants die, they contribute relatively nitrogen-rich organic matter to the soil, thereby improving the fertility of the soil (Fig. 16.12). Alder trees also have nitrogen-fixing bacteria as symbionts in their roots. These trees grow along streams, and their nitrogen-rich leaves fall into the streams and in-

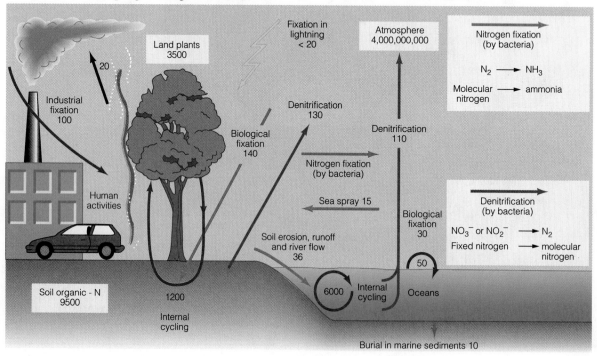

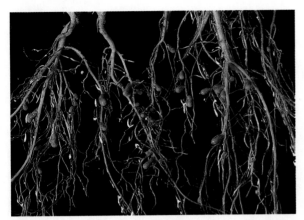

Figure 16.12 Root nodules on white clover produced by colonies of nitrogen-fixing bacteria.

crease the supply of the element in a biologically usable form to freshwater animals.

Nitrogen—Essential to Life, Driven by Life

In terms of availability for life, nitrogen is somewhere between carbon and phosphorus. Like carbon, nitrogen has a gaseous phase and is a major component of the Earth's atmosphere. However, unlike carbon, it is not very reactive, and its conversion depends heavily on biological activity. Thus, the nitrogen cycle is not only essential to life, but also primarily driven by life.

The nitrogen cycle is interesting because of its complexity and because parts of the cycle have had to evolve as the atmosphere became oxygenated. Because organisms cannot use N_2 directly, either some reduced nitrogen must have been available when life arose or the earliest organisms had the ability to reduce N_2. Anaerobic nitrogen-fixing bacteria are very ancient, and the fixation chemistry that evolved with them will not work in the presence of oxygen. Such bacteria must have evolved before the atmosphere contained oxygen. Today these bacteria live only in oxygen-free environments. A few nitrogen-fixing bacteria have developed an oxygen tolerance, even though they still use the old, anaerobic fixation chemistry. They perform this trick by making sure that the sites in their cells where fixation occurs are carefully guarded from oxygen.

As the oxygen content of the atmosphere increased, the amount of nitrate that rained into the soil also increased. This opened new niches that were soon occupied by organisms that learned to reduce NO_3 to NH_3. Many of the higher plants have this ability.

Once reduced, nitrogen tends to stay reduced, remain in the biosphere, and be either reused by other organisms or oxidized back to N_2 and returned to the atmosphere. The main route by which nitrogen returns to the atmosphere, however, is the reduction of nitrate. This route is kept open by bacteria that use the oxygen in nitrate in order to oxidize carbon compounds during metabolism. Denitrifying bacteria must therefore have evolved quite late in the history of the biosphere, after oxygen started to accumulate in the atmosphere. Thus, the simple nitrogen cycle of the early Earth has evolved into today's complex cycle in response to a changing atmosphere.

In the early part of the twentieth century, scientists discovered that electric sparks produced by industrial processes could convert molecular nitrogen into compounds usable by plants. This greatly increased the availability of nitrogen in fertilizers. Today industrial fixation is a major source of commercial nitrogen fertilizer and a large source of fixed nitrogen in the nitrogen cycle.

Nitrogen combines with oxygen in a high-temperature atmosphere. This occurs wherever the temperature and pressure conditions are appropriate. As a result, many modern industrial combustion processes produce oxides of nitrogen. This includes the burning of fossil fuels within gasoline and diesel engines. Thus, oxides of nitrogen, one kind of air pollution, are an indirect result of modern industrial activity and modern technology, and these oxides play a significant role in urban smog (see Chapter 19).

In summary, nitrogen compounds are a bane and a boon for society and for the environment. Nitrogen is required for all life, and its compounds are used in many technological processes and in modern agriculture. It is also a source of pollution of air and waters.

The Phosphorus Cycle

Unlike oxygen and nitrogen, which are gases in the Earth's atmosphere, phosphorus lacks a gaseous phase and exists in the atmosphere only in dust particles. Phosphorus in rocks on the continents is slowly eroded, used temporarily by life on the land, and then slowly washed to the oceans via streams and rivers (Fig. 16.13). In the oceans, phosphorus is temporarily available to the plankton but is eventually deposited in the deep oceans or in marine sediments. There is no short-term nonbiological return of phosphorus to the surface. As discussed in the opening case study, without life, the return of phosphorus takes place only through the very long-term geological process of the uplift of continents. At times when you think that solving global environmental problems is simple, remember the miners of bird guano on the dry islands off of South America. Only by understanding the intricate connections among life and its environment will we be able to solve the problems of our global environment.

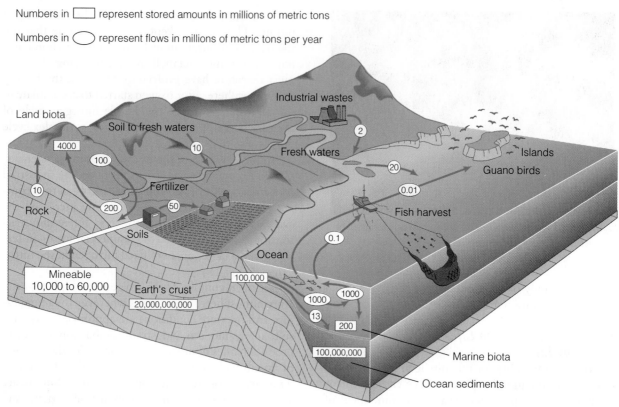

Numbers in ☐ represent stored amounts in millions of metric tons

Numbers in ◯ represent flows in millions of metric tons per year

Figure 16.13 The global phosphorus cycle. Phosphorus is recycled to soil and land biota by geologic processes that uplift the land and erode rocks, by birds that produce guano, and by human beings. Although Earth's crust contains a very large amount of phosphorus, only a small fraction of it is mineable by conventional techniques. Phosphorus is therefore one of our most precious resources. Values of the amount of phosphorus stored or in flux are compiled from various sources. Estimates are approximate to the order of magnitude. (*Source:* Based primarily on C. C. Delwiche and G. E. Likens, 1977, "Biological Response to Fossil Fuel Combustion Products," in W. Stumm, ed., *Global Chemical Cycles and Their Alterations by Man*, Abakon Verlagsgesellschaft, Berlin, pp. 73–88, and U. Pierrou, 1976, "The Global Phosphorus Cycle," in B. H. Svensson and R. Soderlund, eds, "Nitrogen, Phosphorus and Sulfur—Global Cycles," *Ecological Bulletin*, Stockholm, pp. 75–88.)

Figure 16.14 Experimental rice plots at the International Rice Research Institute, Phillipines. Genetic hybridization of rice vastly increased rice production per acre but also required greater use of fertilizer, four to seven times the water, and in some cases produced a rice that was undesirable to eat.

Phosphorus is often a limiting element for plant and algal growth as we see in two fields, one in which phosphorus fertilizers were used and another in which they were not (Fig. 16.14). However, if phosphorus is too abundant, it can cause environmental problems in bodies of water by promoting rapid growth of algae and blue-green photosynthetic bacteria, resulting in *eutrophication* (which is an increase in the concentration of elements required for living things). In addition, phosphorus tends to form compounds that are relatively insoluble in water and therefore are not readily eroded as part of the hydrologic cycle. Phosphorus commonly occurs in its oxidized state as phosphate, which in turn combines with calcium, potassium, magnesium, and iron to form minerals found in soils and in waters. As a result, the rate of transfer of phosphorus tends to be slow in comparison to carbon and nitrogen.

Phosphorus enters the biota through uptake by plants, algae, and bacteria—primarily through autotrophic organisms. Plants can take up phosphorus in its oxidized form, as phosphate, a common ion. Phosphorus slowly becomes available through the weathering of rocks or rock particles in the soil. In a relatively stable ecosystem, much of the phosphorus that is taken up by vegetation is returned to the soil. Some of the phosphorus, however, is inevitably lost to wind and water erosion. It is transported out of the soil in a water-soluble form or as suspended particles and is transported by rivers and streams to the oceans. The return of phosphorus to the land from the ocean is slow.

BIOGEOCHEMICAL LINKS AMONG THE SPHERES

Based on the discussion in this chapter, we can abstract some of the primary ways that life affects each of the other major components of the Earth system through biogeochemical cycling. This section summarizes some of those influences.

The Atmosphere

Among the Earth's spheres, (Fig. 16.2) the atmosphere has had the greatest influence on the biosphere and has in turn been most affected by the biosphere. Here are some examples.

Oxygen

The Archean atmosphere (atmosphere during the time of early life) provided the anaerobic (oxygen-free) conditions necessary for starting life. When life became autotrophic, the oxygen it produced created the oxygenated atmosphere that we know today. This step had the most profound significance for life. Without respiratory metabolism, none of the attributes of "higher" life could likely have developed. And without the ozone (O_3) of the upper atmosphere, there would have been insufficient protection from short-wave radiation for life to get a foothold on the land.

Then why is the atmospheric concentration of oxygen approximately 20 percent? Is this just happenstance, or does a feedback mechanism maintain the atmosphere at this level? James Lovelock suggests that this is an example of feedback mechanisms that are part of the Gaia Hypothesis. His argument is as follows: photosynthesis acts as an oxygen pump, continuously adding oxygen to the atmosphere. However, when oxygen concentration becomes very high, combustion with organic compounds, such as dead wood in forests, becomes very likely. Lovelock states that at oxygen concentrations not very much above 20 percent the probability of combustion becomes very high, and eventually a level is reached where spontaneous combustion (burning) of oxygen and organic compounds occurs. Thus, combustion acts as a *negative feedback*, with the stable point between input from photosynthesis and removal by combustion at around 20 percent. Since both free oxygen and combustible organic compounds are products of life, Lovelock suggests that this is an example of life controlling the chemistry of the atmosphere.

Carbon Dioxide

Another atmospheric gas that is vital to the biosphere is carbon dioxide, CO_2, the source of the carbon used by autotrophs to make carbohydrates by photosynthesis. It is present in today's atmosphere in a concentration of about 350 parts per million, and it is gradually increasing because of the burning of fossil fuels. The primary source of all CO_2 was the Earth's mantle, from which it issued through volcanoes. How much of the CO_2 coming from volcanoes today is recycled from subducted surface materials and how much is new is uncertain, but probably most volcanic CO_2 today is recycled. Of the total carbon that has been cycled through the biosphere, most is locked up in carbonate rocks (limestone and dolostone) and as fossil carbon in detrital sediments. (If you look ahead at Fig. 19.7, you will see a fine example of the kind of limestone in which CO_2 is locked up.) The little that remains in the atmosphere is vital to the autotrophs at the base of the global food web. Together with water vapor, it is also a climate-moderating influence as a greenhouse gas.

Figure 16.15 An extraordinary ecosystem around a black smoker at a depth of 2500 m in the Pacific Ocean. Bacteria that derive their energy inorganically through the oxidation of H_2S brought up by the smoker are the autotrophs. The heterotrophs that live directly or indirectly on the autotrophs include worms, clams, starfish, crabs, and skates.

The Hydrosphere

Water is indispensable for the biosphere. The hydrosphere is the water source for terrestrial ecosystems and the habitat of aquatic ecosystems. It mediates most of the gas exchange between aquatic ecosystems and the atmosphere, and it supplies to aquatic systems the essential elements that terrestrial systems derive from the soil. As with terrestrial systems, a good part of the essential elements used by aquatic organisms is recycled from dead organisms. In shallow environments (rivers, most lakes, and the continental shelves), resources are recycled fairly rapidly; in the deep-sea environment, the cycles are longer and slower.

The Geosphere

Ultimately, the geosphere (solid Earth) contributes all of the minor elements necessary to life and some of the major ones. The most important element it contributes, however, is phosphorus. This element plays two essential roles: in the form of sugar-phosphate units, phosphorus forms the helical framework of the DNA molecule, and it serves as the currency for all of life's energy transactions. The biosphere's phosphorus supply is released from the lithosphere by weathering. Since the growth of most ecosystems is limited by phosphorus availability, weathering is an important regulator of total biomass. The lithosphere gives up its phosphorus rather grudgingly from relatively insoluble minerals such as apatite (calcium phosphate).

Soil is the base of nearly all terrestrial ecosystems. It is often viewed as itself an ecosystem of many species and great complexity. Without the lithosphere there would be no soil. The inorganic constituents of soil supply a habitat to the organisms of a soil ecosystem. In addition, these inorganic constituents act as a substrate, together with water, nutrients, and essential trace elements, to the larger ecosystem (forest, prairie, or whatever it may be).

The Mantle

Through volcanism the mantle is the source of the carbon that (as CO_2) is the starting point for synthesis of organic compounds in the biosphere. Although today most of this carbon has probably been recycled, the mantle itself must be given credit for the initial supply that made the pre-biotic surface environment rich in CO_2.

Another interesting contribution of the mantle to life on Earth, discovered not very long ago, is chemical energy for a small group of autotrophs. In an expedition to the Galapagos rift in 1977, geologists in the small research submarine *Alvin* discovered submarine hot springs that are now known as black smokers—chimney-like structures of anhydrite ($CaSO_4$) and sulfide minerals from which issue plumes of hot water (350°C, or 662°F) darkened by a suspension of sulfide minerals of iron, zinc, and copper. Around these vents (and others like them discovered subsequently on other segments of the midocean ridge), at depths up to 3000 m (1.9 mi), lives a community of beard worms

(*Pogonophora*), clams, starfish, crabs, and other invertebrates, as well as deep-water skates (Fig. 16.15). All these animals are chalk white, except for the beard worms, which are reddish pink. Because the deep sea is devoid of sunlight, autotrophs cannot get their energy from sunlight through photosynthesis. Instead the energy comes from chemical compounds in the hot water. Bacteria are the primary producers in the unusual and diverse ecosystem around submarine hot springs. Living in complete darkness, they are chemoautotrophs, deriving energy from the oxidation of hydrogen sulfide (H_2S) in the water discharged from the smokers.

LINKS WITH HUMAN ACTIVITY

Humans have affected the geochemistry of life in many ways. Following are some examples.

Chemicals in the Human Food Web

The famous nineteenth-century English biologist Thomas Huxley (1825–1895) portrayed the gardener as perpetually at war with nature. The moment the gardener's back is turned, weeds begin to grow in the garden and stifle what is planted there. So the gardener's job is to be constantly vigilant and to defend the garden against nature's incursions, protecting pampered flowers and vegetables from competition with the wild species that are better adapted to the prevailing natural conditions. On a larger scale, the same is true of the farmer. For the past 50 years, nature has been increasingly kept at bay with herbicides and pesticides for the benefit of food crops and the people who eat them.

There are difficulties with herbicides and pesticides, however. In moving up the trophic structure in a food web, some chemical elements become concentrated in the tissues of organisms. Because some of the elements are toxic, what may appear to be harmless quantities of toxic substances can lead to serious consequences (see "A Closer Look: Analysis of a Disaster").

Soil Erosion

The soil of the continents serves as the base from which nutrition in all the terrestrial ecosystems (including ones with people) is derived. "Primitive" human cultures have generally known how to conserve this essential resource (there have been exceptions); modern peoples, going for short-term gain, are wasting it. The soil is the part of the lithosphere that is most vulnerable to erosion. Before humans began intensive farming and grazing, the average rate of soil erosion was about 10 billion tons a year for all the continents, and soil production and loss rates were in balance.

The present rate of erosion is about 25 billion tons a year (Fig. 16.16). With the average rate of soil formation being 10 billion tons a year, the current erosion rate could remove most of the world's topsoil in less than a century. This calculation is crude, but even if it's off by a factor of 5, a century from now there will be much less freedom to choose where to grow crops

Figure 16.16 Erosion as a result of overgrazing and poor farming practice in Ethiopia.

Figure 16.17 A large, open-pit phosphate mine in Florida, with piles of waste material. The land in the upper part of the photograph has been reclaimed following mining and is being used for pasture. These mines provide more than one-third of the entire world production of phosphate.

in the world's productive agricultural zones (the humid, midlatitude prairies) than there is today.

Mining

The subterranean resources of the Earth (metals, coal) exist in an oxygen-free environment. When this environment is exposed up to the atmosphere by mining, some compounds (mainly pyrite) are oxidized to acids (such as sulfuric acid). These acids then acidify the surrounding environment, especially streams, whose ecosystems are traumatized by the high acidity. In addition, the more acidic water of these streams mobilizes toxic elements, such as lead and cadmium, held in the rocks.

The mining of phosphorus provides another example. Some experts believe that, at current mining costs, the total U.S. reserves of phosphorus are about 2.2 billion metric tons, a quantity estimated to supply our needs for several decades. However, for a higher price per ton, more phosphorus is available. Florida is thought to have 8.1 billion metric tons of phosphorus recoverable with existing methods, and there are large deposits elsewhere. However, the mining processes have negative effects on the landscape. For example, at Bone Valley Florida, huge mining pits and slurry

ponds scar the landscape (Fig 16.17). Balancing the need for phosphorus with the environmental impacts of mining is a major environmental issue.

Acid Rain

Coal, formed mostly on ancient floodplains near sea level, contains varying amounts of sulfur (as pyrite derived from marine sulfate), which, on burning, is converted to sulfur dioxide. This latter compound forms sulfurous acid by combining with water, and eventually the sulfurous acid is oxidized by the atmosphere to sulfuric acid, which gets rained out onto the land downwind of the coal-burning installations. Protection of neighboring land is secured by building high smokestacks, and the result is often the export of pollution to distant environments. Thus, the United States exports acid rain to Canada, and Western Europe (especially Britain) exports it to Scandinavia.

Like acid-mine drainage, acid rain traumatizes aquatic ecosystems by acidifying the water. The long-term effects depend on geological conditions. In limestone country, the acid is quickly neutralized by the reactive carbonate rocks, and little harm may result. Unfortunately, Canada and Norway, two of the principal victims of acid rain, have much granitic bedrock, where neutralization is slow and aluminum (from feldspars and other aluminosilicate minerals) is released into the water. There it reaches concentrations that are toxic to wildlife.

Acid rain has been reported to have deleterious effects on terrestrial ecosystems as well. Conifers may have died as a result of acid rain in New England, and agricultural crops have been damaged in a variety of places. Further studies are needed to verify these reports.

Eutrophication of Surface Waters

Big cities produce large amounts of sewage. Whether raw or bacterially oxidized, this waste water usually contains fairly high concentrations of phosphorus (mostly from detergents) and nitrogen, in the approximate proportions that are optimal for plant growth. All too often, especially in the lakes or seaways where the cities are located, the result is **eutrophication**. Eutrophic bodies of water have a high level of key nutrients and consequently vigorous growth of algae (Fig. 16.18A), in contrast to water bodies with naturally occurring levels of phosphorus and nitrogen (Fig. 16.18B). As the algae die, they sink to the bottom, where their decay creates an oxygen demand that quickly makes the environment anoxic and asphyxiates all aerobic organisms living in it. A new generation of

A.

B.

Figure 16.18 A. Algal bloom due to eutrophication on a pond in western New Jersey. B. In contrast, a lake with a low rate of supply of nutrients shows no algae bloom, its waters are blue as seen in photo of Twin Lakes in Beartooth Mountains in Wyoming.

algae follows on the first, and so on, while masses of dead organic matter pile up on the bottom, sometimes thickly enough (as in the case of the Chesapeake Bay in the late 1960s) to obstruct shipping lanes.

Overturn of the water column in winter can cause fish kills as the anoxic water comes to the surface, and so the resources (fish, oysters, shrimp) that the ecosystem once offered to the human community are reduced to nil. In lakes, the rotting vegetation builds up until it reaches the surface and serves as the foundation for a bog ecosystem. In this case, human activity is merely speeding up nature's own ways.

SUMMARY

1. Chemical elements cycle throughout the Earth system. Those that are essential to life, toxic to life, or affected by life move through what are called biogeochemical cycles.

2. Biogeochemical cycles are one of the primary ways that life affects the atmosphere, hydrosphere, and the solid Earth.

3. Fewer than 24 chemical elements are known to be required by life. These are divided into macronutrients and micronutrients. Macronutrients are required in large amounts by all forms of life. Micronutrients are either required in small amounts by all forms of life or in moderate or large amounts by some but not all life forms. Macronutrients include the "Big Six": carbon, nitrogen, oxygen, hydrogen, phosphorus, and sulfur, the major building blocks of life.

4. A limiting factor is a chemical element that is in shortest supply and therefore limits the production and abundance of life.

5. Living things require chemical elements in certain amounts but also in certain ratios to one another. Thus, organisms are selective in taking up chemical elements from the environment. This selectivity can alter the concentrations and ratios of elements in nonliving parts of the environment. In this way, life can cause major alterations of the chemistry in the atmosphere, oceans, solid sediments, and, in the long term, the composition of rocks.

6. Many kinds of rocks are of biological origin, and others are influenced by biological activity because life altered the chemical composition of the material that later formed into rocks.

7. The curve showing the response of a population of organisms to the concentration of a chemical element or compound is called a dose-response curve. An important question about the dose-response curve is whether a threshold effect exists. A threshold is a level below which there is no detectable effect of the chemical on the population. If a threshold exists, then levels of a toxic substance below the threshold will have no measurable effect. Thus, it is important to determine whether thresholds exist in nature.

8. Food-chain concentration is a process through which chemicals can be concentrated as they move up trophic levels. For example, fat soluble chemicals such as the pesticide DDT are absorbed by fatty tissue and concentrated as one organism feeds on another. By this process, a toxic chemical that is in low concentration in the environment can become highly concentrated in certain organisms. Thus, the environmental effects of certain toxins can be amplified.

9. Although certain features are common to all biogeochemical cycles, these cycles can be divided into categories: cycles that include an atmospheric phase and those that do not; cycles of metals and nonmetals. Cycles that include an atmospheric phase are more rapid, and the chemical elements are therefore less likely to be limiting to life.

10. Humans continuously change and perturb the biosphere. Examples of change are soil erosion, the release of acid-mine drainage, concentration of herbicides and pesticides in the food chain, acid rain, and eutrophication of lakes and streams.

IMPORTANT TERMS TO REMEMBER

biogeochemical cycle   *367*
denitrification   *376*
dose-response curve   *369*

eutrophication   *382*
food-chain concentration
   369

macronutrient   *365*
micronutrient   *365*
nitrogen fixation   *376*

threshold   *369*

QUESTIONS FOR REVIEW

1. How is the evolution of the Earth system related to the cycling of chemical elements?

2. What is the main difference between a macronutrient and a micronutrient in terms of how it is used by biota?

3. What are the "Big Six"?

4. Why is phosphorus often a limiting factor?

5. Why is the residence time of chemical elements long in rocks and short in the atmosphere?

6. What function does concentration of elements play in a biogeochemical cycle?

7. What is the Redfield ratio? Is it an example of the Gaia Hypothesis; if so, how?

8. What process of carbon storage produces fossil fuels?

9. Where is most carbon stored in the lithosphere?

10. Why is accounting for the "missing carbon" considered important by those who study environmental issues?

11. Why are bacteria so important to the nitrogen cycle?

12. Cite one organism that has evolved a symbiotic relationship with nitrogen-converting bacteria.

13. What element is called the "energy" element? Why?

14. When can too much phosphorus be a problem?

15. What part did the development of respiration in organisms play in the evolution of the Earth?

16. Why does oxygen remain at a relatively constant concentration in the atmosphere, according to the Gaia Hypothesis?

17. What are the main ways the hydrosphere cycles chemical elements?

18. What is the most important element the lithosphere provides to biogeochemical cycling?

Questions for A Closer Look

1. Why do some toxic elements (such as mercury) become more concentrated in animal tissues as they move up trophic levels?

2. If the half-life of a toxin is one year and the daily intake of aquatic algae leads to a toxin concentration of 100 parts per billion, what will be the toxin level in a herbivore that eats 20 grams of algae a day?

QUESTIONS FOR DISCUSSION

1. Take one of the "Big Six" and describe its biogeochemical cycle. Describe how evolution of the Earth system would be affected if this element were a limiting factor.

2. Describe how the carbon cycle has such a strong effect on the chemistry of the atmosphere. Why doesn't the phosphorus cycle have a similarly strong effect?

3. What element is essential to life and also primarily driven by life processes? Describe why this is so.

4. Describe how "black smokers" survive and compare their life support system to those of an autotroph and a heterotroph.

Then links between geochemical processes and biological processes are explored in several of the case studies on our web pages.

CHAPTER 17

Biological Evolution and the History of the Biosphere

● Life—Abundance and Persistence

In the early 1960s, the view from the shore of Iceland was made up of an amazing display of smoke and rock hurled up by volcanic activity just below the surface—about 100 meters down—caused by molten lava making contact with cold ocean water. Large pieces of the seabed were thrown as high as 300 meters into the air. The cold water rapidly cooled the lava, turning it into fine-grained crystal particles called tephra. These explosions continued for a number of years. They were great enough to create a new island, which was named Surtsey, about 2 kilometers in diameter and about 100 meters high (see Chapter Opener Photo). The eruptions stopped in 1967. A few weeks afterward, a group of botanists visited the island, which was so newly formed that molten lava could still be seen flowing below the surface of the island where occasional cracks in the surface made the depths visible. Even though the island was only a few weeks free of the violence that created it, the botanists found that green plants had already appeared on the island. The first plant they found was a sea rocket, a small plant less than 5 cm high, already in flower. It had migrated to the new island as a seed, landed in the newly formed inorganic soil, sprouted, and grown to maturity (Fig.17.1).[1] That it had flowered so soon after the lava stopped flowing illustrates the speed with which life can invade new lands. Once there and given enough time, the life forms can evolve and adapt to the specific environmental conditions. The invasion of Surtsey Island by green plants encapsulates much of key features of the story of the history of life on the Earth. As life originated and evolved, it not only spread throughout the Earth system but greatly altered that system. How this happened is the subject of this chapter. ●

[1]These observations about Surtsey Island and its new plant life were made by David Challinor and are provided with his permission. They appear in: D. B. Botkin and D. Challinor, *Biological Invasions* (Geneva: Le Temps Stratigique, 1998).

(opposite) Photograph of the Isle of Surtsey, Iceland soon after it was formed.

A PLANETARY PERSPECTIVE ON SPECIES

In this chapter we address four tantalizing but still unresolved issues: How did life originate and evolve from its humble, microscopic beginnings to the complex biosphere that exists today? Why is life so diverse? What is the importance of biological diversity to the Earth system? And why and how did large groups of animals suddenly die out and new ones arise to take their places?

The answers to all of these questions begin with an understanding of the concepts of **biological diversity**. Interest in the great diversity of life on the Earth predates modern science. Ever since people have written, they have described the wonders of nature, especially the great diversity of life and the apparent exacting adaptation of each creature for its needs.

There are several kinds of biological diversity (see Table 17.1). Biologists group organisms according to their evolutionary relationships (see Table 15.1), and one could consider the diversity of any of these groups. In the present discussion, we will focus on species diversity. So many kinds of organisms exist that even today no one knows the exact number of species on the Earth. New species are being discovered all the time, especially in little-explored areas such as tropical rainforests and the deep ocean. Approximately 1.5 million species have been identified and named,[2,3] but estimates of the total number that may be present on the Earth are much higher, ranging from 3 million to 30 million.

Insects and plants make up most of the species: there are approximately 500,000 insect species and 400,000 plant species. Many of the insects are tropical beetles that inhabit local areas in rainforests. Mammals, the animal group to which people belong, comprise a comparatively small number of species—slightly more than 4000. As more explorations are carried out, especially in tropical areas, the number of identified invertebrates and plants will increase. There may be a few mammals yet undiscovered—a new mammalian species, the smallest monkey known, was discovered in 1997 in Brazilian forests (Fig. 17.3). However, it is not likely that the number of named mammals will increase much in the future.

---

[2]E. O. Wilson, ed. Biodiversity (Washington D. C.: National Academy Press, 1988).

[3]E. O. Olson, "The Biological Diversity Crisis" *BioScience* 35: 700–706.

Table 17.1 The Kinds of Biological Diversity

Biological diversity involves three different concepts:

(1) *Genetic diversity*—the total number of genetic characteristics, sometimes of a specific species, subspecies, or group of species.

(2) *Habitat diversity*—the diversity of habitats in a given unit area.

(3) *Species diversity*—which in turn involves three ideas:

    (a) *species richness*—the total number of species;

    (b) *species evenness*—the relative abundance of species;

    (c) *species dominance*—the most abundant species.

17.1 Photograph of a sea rocket in flower

Competition and the Competitive Exclusion Principle

The large number of species on the Earth is a curious phenomenon when we consider the **competitive exclusion principle**, which states that *two species that have exactly the same requirements cannot coexist in exactly the same habitat.* One species will always win out over the other. This can be demonstrated in laboratory experiments (Fig 17.2). This suggests that the number of species would be small.

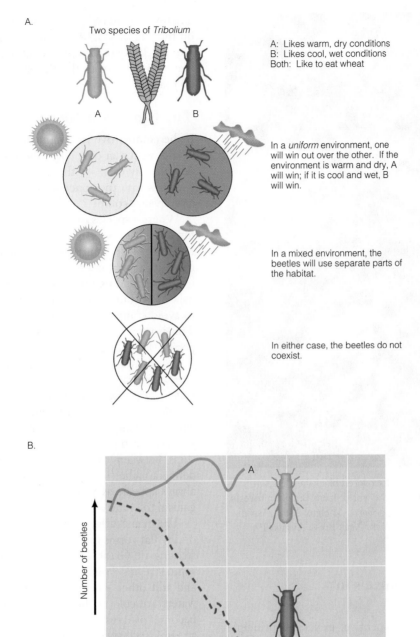

Figure 17.2   An experiment with flour beetles. A. The general process illustrating competitive exclusion; B. results of a specific, typical experiment under warm, dry conditions.

Figure 17.3 The newly discovered black-headed sagui dwarf, smallest monkey in the world, from Brazilian forests. An adult is only 10cm long. (*Source*: Michael Astor, Associated Press, in the *Santa Barbara News* Press, August 19, 1997, p. B5.)

Grey and Red Squirrels in Great Britain

Here is how the competitive exclusion principle works: The American grey squirrel was introduced into Great Britain because some people thought it was attractive and would add to the scenery in parks and towns. About a dozen different attempts were made at this introduction, the first perhaps as early as 1830, another in 1876. By the 1920s, the American grey squirrel was well established in Great Britain, and it underwent a major expansion in the 1940s and 1950s. Today the American grey squirrel is a problem; it competes with the native British red squirrel and is winning the competition (Fig. 17.4). As the American squirrel increased in numbers and advanced, the native red squirrel of Great Britain disappeared locally, losing in competition. At present, the red squirrel occurs in only a few parts of its former range—it is common only in Cumbria, Northumberland and Scotland.

There are scattered populations in East Anglia, Wales, on the Isle of Wight and in islands in Poole Harbor, Dorset. If present trends continue, the red squirrel may disappear from the British mainland in the next 20 years. [4]

The Ecological Niche

If the competitive exclusion principle is true, then it would seem that only a few species would exist on the Earth. Why then are there so many species? The answer lies with the concept of the ecological niche and how species compete and evolve to avoid competition.

Where a species lives is its **habitat**, but what it does for a living—its profession—is its **ecological niche**. For example, a squirrel's profession is eating seeds of trees. Its habitat is a forest or woodland. Suppose your neighbor is a bus driver. His niche is bus driving; his habitat is your hometown. Over time, species evolve to avoid competition by adapting to different niches. This is what allows the two species of flour beetles to persist (Fig. 17.2).

A quantitative definition of the niche is that *the niche represents the set of all environmental conditions under which a species can persist*. Multidimensional space can be represented only mathematically and is difficult to visualize, but two- or three-dimensional abstractions can often be used to illustrate the relationships between two or three selected variables that define a niche (Fig. 17.5). In such diagrams the niche always appears as a space, not a point, because the organism tolerates a *range* of values of each variable.

The quantitative niche is illustrated by the example of tiny flatworms that live on the bottom of freshwater streams in Great Britain, where it was found that some streams contained one species, some the other, and still other streams contained both.[5] The stream waters are cold at their source in the mountains and become progressively warmer as they flow downstream. Each species of flatworm occurs within a specific range of water temperatures. In streams where species A occurs alone, it is found from 6°C to 17°C (42.8°F to 62.6°F) (Fig. 17.6A). Where species B occurs alone, it is found from 6°C to 23°C (42.8°F to 73.4°F) (Fig. 17.6B). When they occur in the same stream, their temperature ranges are much reduced: A occurs in the upstream sections where the temperature ranges from 6°C to 14°C (42.8°F to 57.2°C), whereas B occurs in downstream areas where temperatures are warmer, from 14°C to 23°C (57.2°F to 73.4°C) (Fig. 17.6C).

[4]Rogers, C., "Red Squirrel: sciurus vulgaris." The Wild Screen Trust (1996).

[5]R. S. Miller, "Pattern and Process in Competition." *Advances in Ecol. Research* 4 (1967): 1–74.

Figure 17.4 American grey squirrel and British red squirrel. The introduced American grey squirrel is causing the

decline and possible extinction of the native red squirrel, an example of the competitive exclusion principle in action.

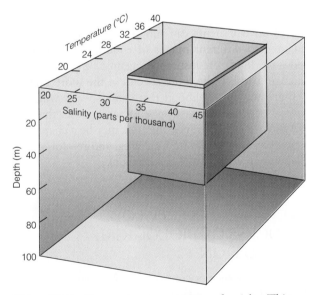

Figure 17.5 Geometric representation of a niche. This example illustrates the temperature-salinity-depth tolerances that bound the niche of a reef-forming coral. The graph tells us that the water temperature reaches from 16° to 40°C, for instance, but coral thrives only between 20° and 36°C. The water salinity ranges from 20 to 45 parts per thousand, but the coral lives only when the salinity is between 29 and 41 parts per thousand.

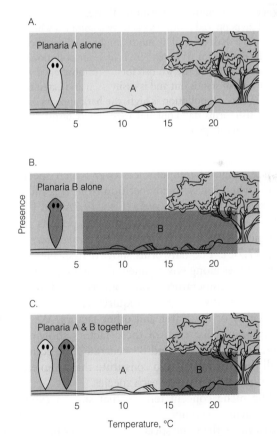

Figure 17.6 The occurrence of freshwater flatworms in cold mountain streams in Great Britain. A. The presence of species A in relation to temperature in streams where it occurs alone. B. The presence of species B in relation to temperature in streams where it occurs alone. C. The temperature range of both species in streams where they occur together. A. and B. show the fundamental niche; C. shows the realized niche of the two species.

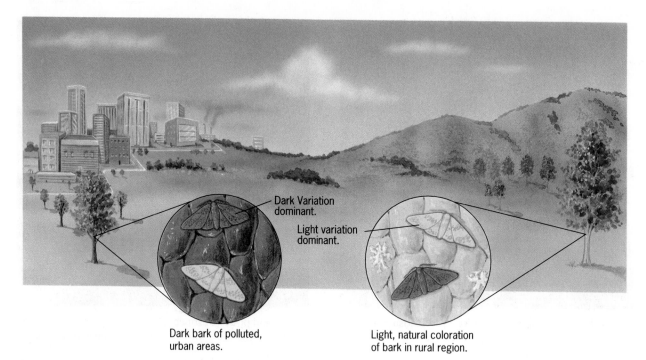

Dark Variation dominant.

Light variation dominant.

Dark bark of polluted, urban areas.

Light, natural coloration of bark in rural region.

Figure 17.7 Industrial melanism: Process of natural selection. The peppered moth occurs in two colors, a whitish color and a dark color. The moths alight on trees where birds feed on them. Where there is little air pollution, the tree bark is grayish; the whitish form is camouflaged while the dark form stands out and is easily seen and eaten by birds. Where air pollution is high, the bark appears dark; the whitish forms stand out and are eaten by birds; the dark form is camouflaged.

The range of temperatures over which species A occurs when it has no competition from B is called its *fundamental temperature niche* (Figs. 17.7A and 17.7B). The set of conditions under which it persists in the presence of B is called its *realized temperature niche* (Fig. 17.7C). The flatworms show that species divide up resources along environmental gradients in nature. Of course, temperature is only one aspect of the environment. Flatworms have requirements in terms of the acidity of the water and other factors. We could create graphs for each of these factors showing the range under which A and B occurred. The collection of all those graphs would constitute the complete description of the niche of each species.

The niche concept tells us that species evolve to avoid having to compete with one another. It also tells us that more than one species—sometimes many—can do the same job, have the same ecological profession, under slightly different environmental conditions (Fig. 17.2). Competitive exclusion and the niche concept tell us how there can be many species on the Earth. Another consequence of these ideas is that the greater the complexity of the environment, the more niche opportunities there may be. Thus, a planet with a great variety of environments can support many species, each adapted to different ranges of conditions. This provides a kind of insurance against environmental change, because the more species in the more niches, the greater the likelihood that some species will survive a large-scale catastrophe. This suggests that the persistence of life over a long time may be a characteristic of a planet, not of a local area, and reinforces the planetary perspective of this textbook.

NEW SPECIES FROM OLD: THE PROCESS OF BIOLOGICAL EVOLUTION

How do new species come about? New species evolve from old ones as the environment changes and creates new constraints and new opportunities. Environmental change can either create new niches or shift the balance from characteristics favoring one population within a species to another population. Biological evolution takes places as the competitive exclusion principle operates and there are niches available to be filled.

In trying to manage species, or predict their future condition and probable persistence, we sometimes assume that evolution will follow simple rules. However, we find that species "play tricks" on us—they adapt or fail to adapt over time in ways that we did not expect. The ecologist, Lawrence Slobodkin, has called evolution a game in which the only rule is to stay in the game. (You're winning if you, as a species, are still there; you lose by going extinct.)

Industrial Melanism: A Case Study in Variation and Natural Selection

The study of the peppered moth illustrates the principles of biological evolution. The peppered moth, a medium-sized insect, lives in Great Britain, where it often alights on tree trunks. Insect-eating birds try to catch the moths as they rest on the trees. Before the rise of industry, most of the moths were whitish, a color that melded with light tree bark and with lichens that covered much of the tree bark so that the moths were camouflaged from predatory birds. A few of the moths were black, but this color was easy to see against the tree trunks and lichens, and black individuals were readily caught by birds (Fig. 17.7).[6]

With the rise of industry, black soot from burning coal and other fuels became common near cities in Great Britain. The soot fell on the trees, darkening the bark. Lichens that grow on trees are very sensitive to air pollution, especially sulfur oxides; these may have declined in abundance as well, so that the darker natural color of some tree trunks would have become more apparent. Against the darker background, the whitish moths were readily visible, while the black ones were camouflaged. In a study of the moths, the number of each color taken by birds was recorded over a certain time period in two different locations, rural and urban. In the rural woodlands where the bark surface appeared light, more than 80 percent of the moths caught by birds were of the dark form. In the urban woodlands, polluted by soot and with bark surface appearing dark, more than 70 percent of the moths caught were of the light form. Pollution reversed the relative advantage of color. In a short time, most of the peppered moths living near cities were black, whereas those living in rural areas were mainly whitish. In recent years, modern air pollution controls have improved the air quality of British cities. Gas is burned instead of coal, and there is much less soot. Tree trunks are no longer covered with soot, and lichens can again grow where they had been killed by air pollution. The whitish forms are again becoming common in British cities.

Both forms of the moth have the potential for a high rate of reproduction, and the two forms compete for resources. Differential reproduction occurs. White forms survived better and had more offspring in clean, unpolluted air; black forms survived better and had more offspring in polluted, ash-filled air. This increase in the relative abundance of the dark form of the peppered moth is an example of **industrial melanism**. (Melanism refers to the dark form).

The peppered moths illustrate the process of natural selection as developed by Charles Darwin. It has four primary characteristics: (1) inheritance of traits from one generation to the next, and some variation in these traits; (2) environmental variability; (3) differential reproduction that varies with the environment; and (4) an influence of the environment on survival and reproduction. But how do we get from a change in gene frequency (i.e., percent of population with genes for whitish moths compared with dark moths) to a new species? Usually, this involves the change in a number of genes and geographic isolation, or some mechanism that makes it no longer possible for two genetically different populations to reproduce and exchange genetic information.

For example, if natural selection were to take place over a very long time, a number of characteristics could change in the peppered moths. Suppose the country and city populations of peppered moths became isolated completely and they no longer reproduced together for a long time. It is possible that some other changes in the two populations would alter them so much that they could no longer reproduce even when they were brought back into contact. In this case, two new species would have evolved from the original species. Thus natural selection and geographical separation of populations are mechanisms that make the evolution of new species possible.

INHERITANCE, MUTATION, MIGRATION, AND GENETIC DRIFT

Biological evolution is a key concept in understanding how life differs from everything else on our planet and in the universe. Biological evolution means the change in the inherited (genetic) characteristics of a population from generation to generation. **Natural selection** is a process by which organisms whose biological characteristics better fit them to the environment are better represented by descendants in future generations than those whose characteristics are less fit for the environment.

The inheritance of traits from one generation to the next is the result of the existence of genes, contained in chromosomes within cells. The genes are inherited from one generation to another. Genes are made up of a complex chemical compound called deoxyribonucleic acid (DNA), discussed in Chapter 15. Biologists now understand the chemistry of these compounds and are beginning to unravel the DNA code—what messages each set of chemical compounds transmits—for some species.

Mutation

When cells divide, the DNA is reproduced so that each daughter cell gets a copy. Sometimes an error or a failure occurs in the reproduction of the DNA, resulting in a change in the DNA and therefore in inherited characteristics. Other times an external environmental agent, such as radiation or certain toxic organic chemicals, comes in contact with DNA and alters the molecule. When DNA changes in any of these ways, the DNA is said to have undergone **mutation**.

In extreme cases, cells or offspring with a mutation cannot survive. In less extreme cases, individuals with the mutation are so different from their parents that they cannot reproduce with the "normal" offspring. In this case, a new species has been created. Mutation can result in a new species whether or not that species is better adapted to the environment than its parental species. In even less severe cases, the mutation can add variability to the inherited characteristics. For example, at some time in the past the dark form of the peppered moth might have originated from one or more such mutations.

In our discussion of the peppered moth, we discussed one visible feature. Sometimes, such a feature is determined by a single gene (one set of chemical codes in a DNA molecule), but most characteristics are affected by several genes. Organisms have many chromosomes, each of which consists of many molecules of DNA, and, therefore, inheritance of characteristics is a complex process.

Migration and Biological Invasions

Another process that can lead to changes in gene frequency is the **migration** or invasion of one population of a species into a habitat previously occupied by another population. Even organisms that do not move themselves have a reproductive structure that can migrate, as the case study about the small plant, the sea rocket, introducing this chapter illustrated.

Genetic Drift

Genetic drift refers to changes in the frequency of a gene in a population as a result of chance rather than selection, mutation, or migration. Chance may determine which individuals become isolated in a small group from a larger population and thus determine which genetic characteristics are most common in that isolated population. In such a case, the individuals may be poorly adapted to the environment.

Suppose, for example, that the only seeds of the sea rocket that reached Surtsey Island were from plants adapted to relatively warm conditions. Then, even though a few plants might sprout, grow, and flower in the locations warmed by the still-hot lava flows, soon after these sea rocket plants would be unable to cope with the cold climate of Iceland. The genetic flexibility would be lacking. The plant, the first known to reach the island, might be one of its first losses.

Changes in the environment that isolate formerly connected areas can lead to genetic drift, such as when ice-age glaciers isolate populations, or when changes in stream channels make migration of animals difficult where it was formerly possible, or where rising sea level creates islands in which there were once peninsulas. In this way, the solid Earth and hydrosphere affect biological evolution.

Diversity in the Global Ecosystem

Species with a wide range of tolerance (generalists) occupy large (i.e., broadly defined) niches, as in the case of Planaria (B) in Figure 17.6; those with a narrow range (specialists) occupy small (narrowly defined) ones, as in planaria (A) in Figure 17.6. Biologists speak of an "evolutionary strategy," a term meaning that, if organisms were sentient and could make decisions, they might be able to choose a strategy. By examining species and their niches, we can see that evolution takes a species along a pathway that, even without consciousness, leads to a certain strategy. As with the planaria, one species can continue to exist as the generalist while another can continue to exist as the specialist.

An environment that varies rapidly and over a wide range does not allow specialist species to survive. For example, in central Alaska near Fairbanks, the temperatures range in midsummer from the high 20°s C (high 80°s F) to the minus 40°s C (minus 50°s F) in the winter. A species that can survive only when temperatures are between 20° and 25° C could not persist there. As a result, the forests of the far north, called the boreal forests, have few species of trees—about 20 worldwide. Some occupy wetlands and some dry sandy soils, but all are generalists in terms of temperature tolerance. In the far north, then, we can refer to the niches as "large," meaning that any niche that has

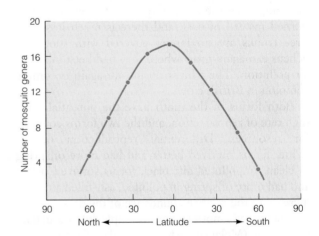

Figure 17.8 The number of mosquito genera as a function of latitude. Comparable patterns are observed in clams, turtles, parrots, foraminifera, termites, snails, frogs, snakes, lizards, crocodiles, reef-forming corals, amphibians, butterflies and palm trees.

Figure 17.9 The Great Barrier Reef on the continental shelf of northeastern Australia is one of the world's most diverse marine ecosystems

a wide range of an environmental variable. The larger the niches, the fewer the species and lower the diversity in that ecosystem; the smaller the niches, the more species and higher diversity of the ecosystem.

In low latitudes, where climatic conditions are more or less constant the year round, one might expect narrow niches and high diversity. In fact, such is the case both on land and in the oceans. The rainforest is the most diverse terrestrial ecosystem. In the marine environment, the continental shelves of the intertropical region are the most diverse ecosystems. In a general way, diversity decreases from a maximum at the equator to a minimum at the poles (Fig. 17.8).

Even near the equator, there are substantial variations in diversity, especially among marine organisms of the continental shelves. Proximity to landmasses can produce marked differences in the "climate" of the ocean. Examples are upwellings of cold, deep ocean water; the influx of fresh water and sediment from rivers draining the continents; and the frequency and severity of monsoons. For instance, there are no coral reefs at the mouth of the Amazon, which lies on the equator: the temperature is right, but with all the freshwater from the river, the salinity is too low for coral growth. The Australian shelf (latitude 20° S), however, is remote from continental influences and

has a remarkable reef system with the highest diversity in the equatorial belt (Fig. 17.9). Some of this diversity even spills over onto the shelves of the Indian Ocean, in spite of an unstable climate (monsoons) and a heavy influx of detrital sediment.

The Influence of Provinciality

It might be imagined that, if the equatorial climate extended over the whole Earth, global diversity would be greater than it is today. The high latitudes would be lush with subtropical vegetation (as they were in the Late Cretaceous Period). But this reasoning does not take into account the possibility that, in expanding poleward, the equatorial habitats might support not new species but simply more of the same species that live there at present. Species that now live in colder climates, such as polar bears, penguins, skuas, moose, caribou, and the rest, would not exist, and global diversity would be impoverished by their absence.

Here, then, another factor in global diversity, as important as climatic stability, comes into play. It is **provinciality**, which is the extent to which the Earth system is divided into subsystems by barriers to the migration of organisms. These barriers can take the form of (1) climatic gradients, (2) seaways or mountains between landmasses, or (3) land between seas. Provinciality is high at present as a result of these barrier types, and so is global diversity. For instance, several species of "anteater" (they really eat termites) are living in South America, South Africa, southwestern Asia, and Australia because they are kept apart by geographical barriers and cannot compete with each other (Fig. 17.10). Thus, the more the Earth's ecosystems are separated from each other, the greater will be the global diversity.

In 1876, A. R. Wallace suggested the idea of provinciality, which he referred to as "realms." Wallace noted that the world could be divided into six biogeographic regions on the basis of fundamental features of the animals found in them.[7] He named them Nearctic (North America), Neotropical (Central and South America), Palaearctic (Europe, northern

[7] A. R. Wallace, The Geographical Distribution of Animals, Vol. 1, reprint (New York: Hafner, 1962).

A.

B.

C.

D.

Figure 17.10 So-called anteaters (usually they eat termites) from different parts of the world are different species: they fill the same niches but are not competitive because they do not come in contact. A. Short-beaked echidna (Australia) B. Tamandua (Central America) C. Pangolin (Malaysia) D. Giant anteater (Venezuela)

Asia, and northern Africa), Ethiopian (central and southern Africa), Oriental (the Indian subcontinent and Malaysia), and Australian. These have become known as Wallace's realms. These realms should not be confused with the biomes that were discussed in Chapter 15. An example of a biome is a tropical forest, inhabited by many species having different ancestral heritages. In contrast, a realm is made up of many kinds of ecosystems (many biomes) with animals, plants, and fungi that have a common ancestral heritage. For example, the native mammals of Australia are marsupials; they exist within a variety of biomes but have a common heritage and are considered part of the Australian realm.

LIFE'S EVOLUTION ON THE EARTH

With this background in the process of biological evolution and biological diversity, we can now consider other major questions about the diversity of life on the Earth: How did life originate and evolve to the complex biosphere that exists today? And over geologic time what effect has the changing biosphere had on the evolution of the Earth system?

In this discussion, we once again rely on the concept of uniformitarianism, assuming that the processes of biological evolution that exist today existed in the past, and that therefore the changes in environment and the new constraints and new opportunities they gave rise to existed throughout Earth history. (Uniformitarianism was first discussed in the Introduction.) However, there is one thing that does not appear to fit with the idea of uniformitarianism: the origin of life. To our knowledge, new life does not spring spontaneously from inorganic matter. The theory of spontaneous generation of life from inorganic material was rejected in the nineteenth century. However, the fundamental laws of physics and chemistry applied then as now. Thus, the origin of life must lie with phenomena that exist today, even if the conditions that make this origin possible are no longer available on the Earth.

In trying to determine the origin of life, several principles can help us. One is the universality of biochemistry (see Chapter 16), as stated by the biophysicist Harold Morowitz: "There is a universal set of small organic molecules that constitutes a large portion of the total mass of all cellular systems." Furthermore, all proteins in all forms of life are made from the same group of amino acids. Among the many numbers of organic compounds, for most organisms more than 90 percent of a cell is composed of less than 50 compounds and polymers of the compounds.[8] The

[8]Harold Morowitz, 1992, *Source*, p 44.

membrane structure is of a universal type for all organisms. In addition, there is a network of reactions that make up metabolism that are common to all organisms. So we see that the principle of uniformitarianism still holds. From this we can only conclude that the conditions that allowed life to originate do not appear to exist on the Earth today.

Among the many remarkable aspects of the origin and history of life on the Earth is that life originated about 3.5 billion years ago, very early after the formation of the solid crust of the Earth, and about 1 billion years after the formation of the Earth.

HOW DID LIFE BEGIN?

A number of scientific theories about the origin of life have been proposed, but none of them has been proven. Could either a virus or a bacterium, two of the first known manifestations of life, have been the first living life form? The answer seems to be "no." Life cannot have begun as a virus because a virus requires a living host. Nor is it likely that life began as a bacterium because, even though bacteria are the simplest known organisms, they are nevertheless very complex—too complex, in all probability, to have been the first living organism (Fig. 17.11). The operating complexity of this simplest organism can be grasped by recognizing that a bacterium sustains life by hundreds of chemical reactions, and these reactions are all essential.

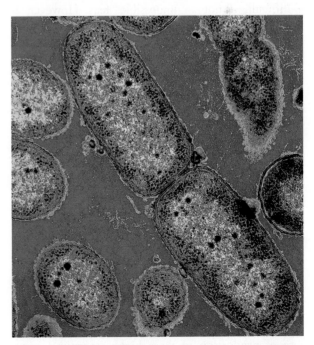

Figure 17.11 Although bacteria-like forms are the oldest known fossil organisms, it is unlikely that bacteria could have been the first living things. Their cells appear too complex to have originated intact as the beginning of life.

So, although today the boundary between the living and the nonliving is quite a sharp one, there must have once been some prebacterial form of life that has (so far as we can tell) left no record of itself. Did it metabolize? If so, what and how? If not, was it really alive? The search for evidence about how life began is one of the most intriguing and challenging quests facing scientists.

Whatever may have been the initial state of matter destined to become alive, we can specify three steps that must have been accomplished on the way to the complex life forms we know today: (1) **chemosynthesis**, the synthesis of small organic molecules such as amino acids; (2) **biosynthesis**, the polymerization of small organic molecules to form biopolymers, in particular proteins; and (3) the development of all the complex chemical machinery, including DNA and RNA, needed for replication. Connecting these processes is the requirement for metabolism, first discussed in Chapter 15.

One well-known theory of the origin of life is that the products of chemosynthesis collected as an organic "soup" in the surface waters of the primitive Earth. Most scientists agree with at least some aspects of the "soup" hypothesis. According to the theory, some of the less soluble organic compounds in the "soup" would clump together as droplets like those of butterfat in milk.

One of the main problems with the soup scenario is that when amino acids polymerize to make proteins, water is eliminated: in an aqueous environment, this is a difficult, if not an impossible, feat. However, if amino acids are dehydrated and heated, polymerization does happen. Protenoids may have formed on the early Earth by the drying out of some "soup" along ancient shorelines and subsequent polymerization by solar or volcanic heat. Polymerization might also have occurred if organic molecules had adsorbed onto the surfaces of clay minerals and the reactions had taken place there. Regardless of how polymerization may have occurred, protenoids have no metabolism. Whether such protenoids were capable, under some conditions, of further evolution and of becoming "alive" is not known.

Another major problem associated with the organic soup theory is that it appears to violate the second law of thermodynamics, which states that order tends to lead to disorder. The organic compounds in the primordial "soup" are compounds that are unlikely to come about by random collisions of atoms, and are molecules that would tend to have a relatively fast decay rate. For the organic soup theory to work, the rate of production of these highly organized compounds would have to exceed the rate of decay. Without enzymes to make the rate of production possible, this seems highly improbable under the temperature and pressure conditions that existed on the surface of the primitive Earth.

Black Smokers

Since the discovery of submarine hot springs called "black smokers" in 1977 (see Figure 16.14), a number of scientists have suggested these as possible sites for the origin of life. In this hypothesis, the first organic molecules were formed on the surfaces of pyrite grains that form around the vents. In this way, the pyrite rather than clay served as a concentrating mechanism. Prebiotic evolution could have begun with reactions that took place in organic layers as thin as one molecule thick. Reactions in the organic layer were akin to metabolism, and the products were the first cells. When completed, the cells became detached from the mineral substrate. This "black smoker" origin is simply another hypothesis with no firm evidence. These supposed earliest cells are not to be confused with the unicellular, chemoautotrophic bacteria that live around black smokers today and that now serve as the base of the food chain in this strange ecosystem. One strength of this theory is that the hot sulfur-containing gases provide an energy source that would allow a rapid rate of production of organic compounds in a highly localized area. Following from this theory, life could have originated in a comparatively small volume of the ocean where an energy source was concentrated, thereby avoiding some of the theoretical problems of the organic soup theory.

Panspermia

Arguing that the basic organic molecules of life would have been hard (but not impossible) to synthesize on the Earth, some scientists hypothesize that organic molecules may have arrived, ready made, from some other part of the solar system or even from the galaxy beyond the solar system. This hypothesis has attracted attention for two reasons. First, astronomers have demonstrated that many small organic molecules may be found in interstellar space. Second, among the many kinds of meteorites that fall on the Earth there is one class, called carbonaceous chrondrites, that contains a liberal amount of small organic molecules. If interstellar dust, carbonaceous chrondrites, or both, fell on the early Earth, perhaps they provided the starting molecules for life. The big problem is the next step: how did the early molecules polymerize to become "life" molecules?

It is hypothesized that biopolymers could perhaps have grown in space from small molecules and formed organisms there. Such organisms, moving on the interstellar winds of fortune, could have reached other parts of the galaxy, including planets. Most would have fallen on stony ground and perished, but conditions on the Earth were hospitable and allowed the further evolution of life. The hypothesis of the supposed origin of life in space, followed by a diaspora (migration) to various parts of the galaxy (including the Earth), is known as **panspermia**. The panspermia hypothesis is attractive because it extends by a few billion years the time available for biosynthesis of the earliest biopolymers, for life could have originated before the solar system formed. It is hard to imagine, however, that conditions in interstellar space could have favored biosynthesis. This theory presents the following problem: how could highly unlikely compounds form in high enough concentration over a short enough time for life to originate in space? What would be the energy source? Any location with a suitable temperature range would have been subject to concentrated short-wave electromagnetic radiation

unless it was on a planet with an atmosphere that filtered out the radiation. Such radiation, though it favors the formation of small organic molecules, is deadly to biopolymers. Furthermore, the dispersion of matter in outer space is so thin that the formation of polymers is unlikely in any case. Therefore, this theory seems highly speculative and unlikely.

THE EARLY HISTORY OF LIFE ON THE EARTH

Whatever its origin, early life was anaerobic (i.e., lived in the absence of oxygen). The geologic record shows that the early atmosphere contained little or no free oxygen. The general history of life on Earth is illustrated in Figure 17.12. The most ancient fossils that

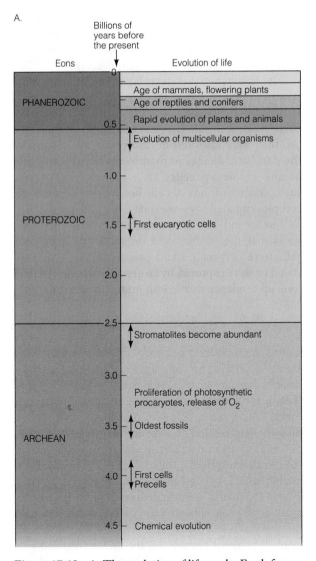

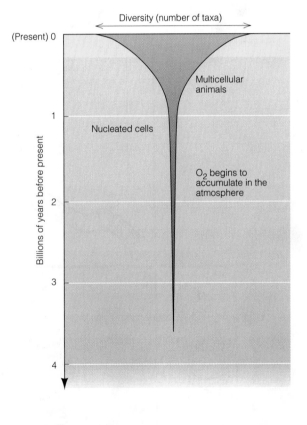

Figure 17.12 A. The evolution of life on the Earth from 4.6 billion years ago to the present. The rates at which new organisms appear and of biological diversity both increase with time. B. A simplified representation of global diversity through geologic time.

have been found to date are about 3.5 billion years old. Some of the ancient fossils are the remains of microscopic procaryotes (Fig. 17.13); others are structures made up, layer upon layer, of thin sheets of calcium carbonate that were precipitated as a result of blue-green bacteria (also procaryotes) influencing the chemistry of seawater (Fig. 17.14A). The layered structures, which are called stromatolites, are not fossils of the actual organisms, but they provide clear evidence of their presence because we can see and study similar structures being formed today (Fig. 17.14B).

For at least 2 billion years the only life on the Earth was procaryotic. Several different kinds of procaryotes evolved over their 2-billion-year supremacy; then, about 1.4 billion years ago, a profound change occurred, and eucaryotes appeared. How and where the first simple eucaryotes came into being is a matter for speculation. We can, however, be reasonably sure that eucaryotes arose from procaryotes. The chemical pathways in the two classes of cells are so similar that it is clear they must be related. Furthermore, the organelles in the eucaryotes so closely resemble some of the smaller procaryotes that most authorities believe that organelles were once procaryotic bacteria and that eucaryotes somehow arose by larger procaryotes enclosing, and thereby being able to use, the chemical products of smaller cells. Two hypotheses have been proposed about how this might have happened: by membrane invagination or by symbiosis (Fig. 17.15).

The first hypothesis, membrane invagination, proposes that the ancestral procaryotic cell folded in on itself, forming pockets in which particular enzymes could be concentrated. When the folds were pinched off, they were enclosed by and protected in some of the outer membranes of the procaryotic cell and thus became simple organelles. Over a long period of time, such pinched-off bits of cells became the increasingly complex organelles we see today.

The second, and more popular, symbiosis hypothesis is that the nucleus and organelles of eucaryotic cells were originally small procaryotic cells that invaded or were captured by larger procaryotic cells and took up residence there, with mutual benefits to both.

The Rise of Oxygen

There is abundant evidence showing that before about 1.5 billion years ago the Earth's atmosphere was deficient in oxygen. Water-worn grains of pyrite (FeS_2), for example, turn up in ancient sedimentary rocks. Today, when grains of pyrite get into streams, they are oxidized long before they accumulate in sediments. The most convincing evidence of an oxygen-deficient atmosphere is found in ancient chemical sediments called banded-iron formations (Fig. 17.16). These sediments were laid down in the sea, which must have been able to carry dissolved iron (something it can't do now because oxygen precipitates the iron). It is hypothesized that the ancient banded-iron formations were precipitated in places where photosynthetic bacteria were producing oxygen locally.

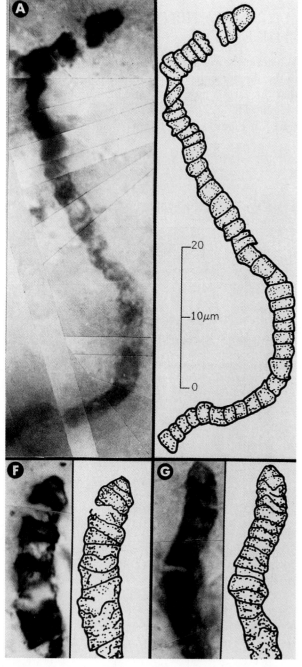

Figure 17.13 Examples of the most ancient fossils procaryotes ever found. 3.5 billion-year-old microfossils in chert from Western Australia. Adjacent to each photograph is a sketch. Magnification is indicated by the scale; 10μm is equal to 0.1 mm.

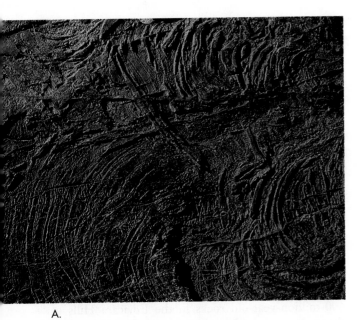

A.

B.

Oxygen was toxic to these bacteria and was released as a waste product. Scientists calculate that before oxygen started to accumulate in the air, 25 times the present-day amount of atmospheric oxygen had been neutralized by reducing agents such as dissolved iron.

During the anaerobic phase of life's history, photosynthesis became well established, but until sufficient free oxygen was in the atmosphere, living organisms could use energy only by fermentation. The energy yield from CO_2 fermentation is low, and the products include CO_2 and alcohol. Alcohol is a high-energy compound and, for the cell, a waste product that must be eliminated. The inability to use energy contained in alcohol puts limitations on the anaerobic cell:

1. A large surface-to-volume ratio is required to allow rapid diffusion of food in and waste out. Anaerobic cells must therefore be small.

2. Anaerobic cells have enough trouble keeping themselves supplied with energy. They cannot afford to deploy energy resources on the maintenance of specialized organelles, including the nucleus.

3. Procaryotes need free space around them; crowding interferes with the movement of nutrients and water into and out of the cell. Therefore, they live singly or are strung end-to-end in chains. They cannot form three-dimensional structures.

It took about 2 billion years for the Earth's oxygen sinks to be filled, and during this entire time the procaryotes had the world to themselves. Eventually, an oxygenated atmosphere started to form, and when it did, the organisms seem to have wasted little time in turning the lethal waste product, oxygen, to an advantage. This happened through the appearance of eucaryotic cells. Some of the characteristics of aerobic eucaryotic cells are as follows.

1. They use oxygen for respiration, and because oxidative respiration is much more efficient than fermentation, they do not require as large a surface-to-volume ratio as anaerobic cells do. Such cells are larger.

2. Aerobic cells, because of their superior metabolic efficiency, can maintain a nucleus and organelles.

Figure 17.14 Evidence of the antiquity of life. Stromatolites are layered growths that form in warm, shallow seas when photosynthetic bacteria cause dissolved salts to precipitate. A. Fossil stromatolites greater than 1.5 billion years old from the northern Flinders Range, South Australia. B. Modern stromatolites forming in the Intertidal zone, Shark's Bay, Western Australia.

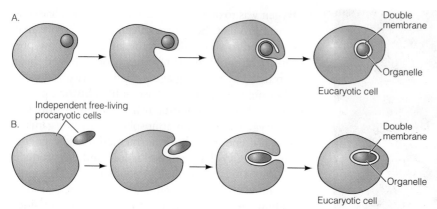

A.

B.

Independent free-living procaryotic cells

Double membrane

Organelle

Eucaryotic cell

Double membrane

Organelle

Eucaryotic cell

Figure 17.15 Hypotheses for the origin of eucaryotic cells. A. The invagination hypothesis in which a procaryotic cell encloses a portion of itself, which then becomes an organelle. B. The symbiosis hypothesis in which an independent, free-living procaryotic cell invades and lives inside another procaryotic cell. The invader evolves into an organelle.

3. Aerobic eucaryotes are not inhibited by crowding, so, unlike procaryotes, eucaryotes can form three-dimensional colonies of cells.

With the appearance of eucaryotes and the growth of an oxygenated atmosphere, the biosphere started to change rapidly and to influence more processes on the Earth, such as those that occurred with the occupation of the land.

Figure 17.16 Banded-iron formation of the Hamersley Range, Western Australia, formed during the Lower Proterozoic Eon. Banded-iron formations are chemical sediments and are thought to have formed when iron in solution in seawater was precipitated as a result of photosynthetic bacteria releasing oxygen. The woman in the foreground is Dr. Janet Watson, a distinguished English geologist.

Ediacaran Fauna

The earliest animal fossils were first discovered in 600-million-year-old rocks in the Ediacara Hills of South Australia and are known as **Ediacaran fauna**. Nearly identical fossils have subsequently been discovered in similar-aged rocks in other parts of the world. The Ediacaran fauna were animals that lived in quiet marine bays and lacked any hard parts. They seem to have been jelly-like animals without any external physical armor or defense.

Even if the Ediacaran fauna were the first, or at least among the first, animals to evolve, they nevertheless represent a huge jump in complexity from the first unicelled eucaryotes that appeared 800 million years earlier. Scientists have not yet been able to dis-

cover much about what went on during those 800 million years.

The Ediacaran animals are of three main kinds: (1) disc-like, resembling today's jellyfish (Fig. 17.17A); (2) pen-like, resembling today's sea-pens or soft corals, and (3) worm-like, resembling broad flat worms (Fig. 17.17B). The animals have some odd features. For example, the disc-like fossils are not really jellyfish because they lack the central radial structure and peripheral concentric structure of true jellyfish. Furthermore, although it may be a bit too much to read from such an ancient fossil, the "worms" don't seem to have guts.

The Cambrian Expansion

The Cambrian Period, beginning 570 million years ago, was the time of the introduction of internal and external skeletons. A famous fossil assemblage in the Burgess Shale (Fig. 17.18) preserves many of the soft-bodied forms and a few skeletonized ones. The Burgess Shale assemblage shows something else, too: a richness of forms that we no longer see in the contemporary bottom-dwelling marine fauna. The Cambrian was a time of almost unbridled growth and diversity in the marine environment: thousands of niches waited to be filled, and there was no lack of candidates for their occupancy. Most had to fall by the wayside, yielding to the nine surviving phyla that we know today.

Why did biological diversity explode in the Cambrian Period? That is another of the great unanswered questions about the biosphere. Many hypotheses have been offered, but none, as yet, is backed by hard evidence. One hypothesis is that sex and reproduction allowed more rapid evolution and made the explosion possible. Procaryotes are asexual and reproduce by a process called mitosis in which the cell splits its DNA into two equal halves in order to create a new cell. The new cell is identical to the parent. Eucaryotes reproduce sexually, and two individuals contribute their genetic information (their genes) equally; thus, any differences that exist are quickly spread among the growing population. Another hypothesis is that the rise of oxygen content of the atmosphere allowed calcium phosphate and calcium carbonate skeletons to develop, making more complex forms possible.

Whatever the reason, many changes occurred in the early Cambrian Period. Compact animals with skeletons evolved to replace the floppy ones of Ediacaran times: trilobites (Fig. 17.19), mollusks (clams), echinoderms (sea-urchins), and sea-snails—all types (except trilobites) that have persisted up to the present, and all equipped with gills, filters, efficient guts, a circulatory system, and other space-saving devices.

A.

B.

Figure 17.17 Two members of the Ediacara fauna from South Australia. These are the most ancient multicelled animals that have ever been found. A. *Mawsonia spriggi*, a discoid shape, possibly a floating animal like a jellyfish. B. *Dickinsonia costata*, a curious worm-like creature.

Figure 17.18 Most of the fossils in the Burgess Shale of British Columbia are soft-bodied. Of the 25 different species depicted in the drawing, only the five circled had hard parts. The Burgess Shale is the most complete assemblage of Cambrian fauna ever found, and it is presumed that the abundance of soft-bodied animals reflected the situation elsewhere in the ocean. The most ancient chordate, *Pikaia*, is seen as a small fish. (See Fig. 16.20 for a photo of a *Pikaia* fossil.)

Figure 17.19 Fossil trilobite from the Cambrian Period. Trilobites were one of the first animals to develop a hard. chitin covering, presumably as a defense against predators. This sample was collected in Utah.

LIFE ON LAND

Geologists divide the history of the Earth since the beginning of the Cambrian Period into three eras: Paleozoic, Mesozoic, and Cenozoic. The Cambrian Period was the first of the six periods in the Paleozoic Era, which lasted about 330 million years. The Cambrian Period was followed by the Ordovician, Silurian, Devonian, Carboniferous, and Permian. The five periods of the Mesozoic and Cenozoic Eras follow. Each period is characterized by the appearance of major groups of plants and animals, as summarized in Figure 17.20.

The great proliferation of life in the Cambrian Period was entirely confined to the sea, but by 500 million years ago, the main plans for animal life had been established. The one big step that remained for organisms was to leave the sea and occupy the land. Eventually, plants, insects and other animals, bacteria, and fungi all took the step.

Here are some of the requirements for multicellular organisms on land.

1. Structural support needed because, while aquatic organisms are buoyed up by the water, on land gravity becomes a real force with which to contend.

2. An internal aquatic environment, with a plumbing system giving it access to all parts of the organism, and devices for conserving the water against losses to the surrounding atmosphere.

3. Means for exchanging gases with air instead of with water.

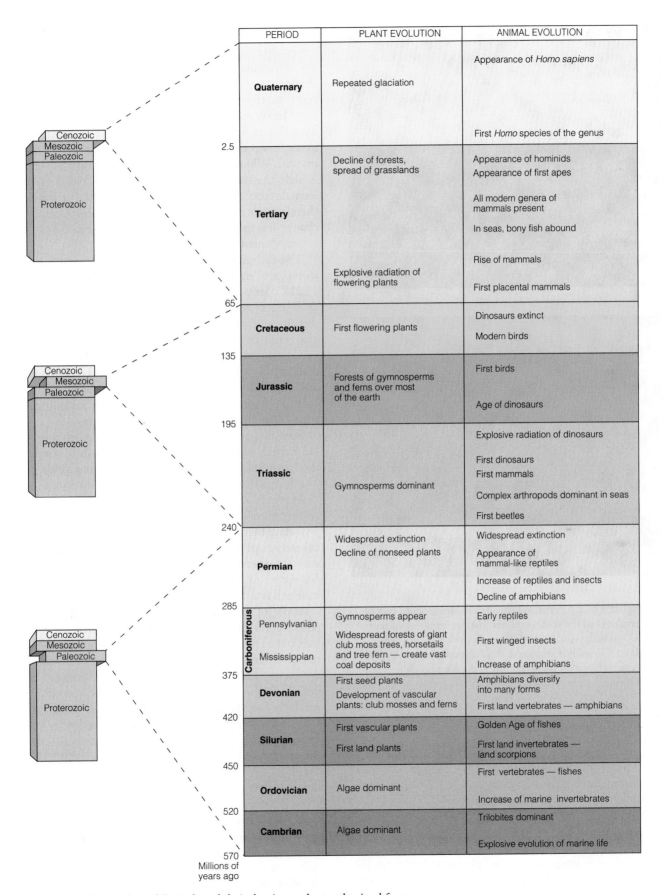

PERIOD		PLANT EVOLUTION	ANIMAL EVOLUTION
Quaternary		Repeated glaciation	Appearance of *Homo sapiens* First *Homo* species of the genus
Tertiary		Decline of forests, spread of grasslands	Appearance of hominids Appearance of first apes
			All modern genera of mammals present
			In seas, bony fish abound
		Explosive radiation of flowering plants	Rise of mammals
			First placental mammals
Cretaceous		First flowering plants	Dinosaurs extinct Modern birds
Jurassic		Forests of gymnosperms and ferns over most of the earth	First birds Age of dinosaurs
Triassic		Gymnosperms dominant	Explosive radiation of dinosaurs First dinosaurs First mammals Complex arthropods dominant in seas First beetles
Permian		Widespread extinction Decline of nonseed plants	Widespread extinction Appearance of mammal-like reptiles Increase of reptiles and insects Decline of amphibians
Carboniferous	Pennsylvanian	Gymnosperms appear	Early reptiles
	Mississippian	Widespread forests of giant club moss trees, horsetails and tree fern — create vast coal deposits	First winged insects Increase of amphibians
Devonian		First seed plants Development of vascular plants: club mosses and ferns	Amphibians diversify into many forms First land vertebrates — amphibians
Silurian		First vascular plants First land plants	Golden Age of fishes First land invertebrates — land scorpions
Ordovician		Algae dominant	First vertebrates — fishes Increase of marine invertebrates
Cambrian		Algae dominant	Trilobites dominant Explosive evolution of marine life

Figure 17.20 Geological Periods and their dominant plant and animal forms.

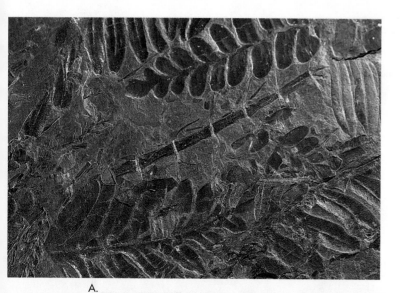

Figure 17.21 Ferns and club mosses are modern representatives of the seedless plants that first established themselves on the land in the Silurian Period. A. Fossil fern about 350 million years old. B. *Thelypteris phegopteris*, a modern fern that is also known as the long beech fern, showing spores on the undersides of the frond.

4. A moist environment for the reproductive system, essential for all sexually reproducing organisms.

Selection for these necessities is largely what has shaped terrestrial organisms into the familiar forms we know today.

Plants

It has been assumed that land plants evolved from the earliest known eucaryotes, green algae. Eventually, vascular plants evolved that had woody tissue for structural support for stems and limbs (requirement 1) and a vascular system (requirement 2), a system of channelways, by which water is transferred from the roots to the leaves and products of photosynthesis from leaves to roots. Requirement 3 (gas exchange) is by diffusion and is controlled by adjustable openings in the leaves called stomata (singular stoma, Greek for mouth). When carbon dioxide pressure inside the leaf is high, the stomata open; when low, they close. (The stomata also close when the plant is short of water, thereby protecting it against desiccation.) Gas exchange was managed by Devonian plants much as it is in today's plants.

The earliest plants, of which mosses and leafy liverworts are modern examples, were seedless (Fig. 17.21). Many of the seedless plants can tolerate some drought; mosses survive dry spells, releasing spores that lie dormant until moisture returns. But these plants rely on moisture for the sexual phase of the reproductive cycles; without it the sex cells have no medium in which to reach each other and fuse. Consequently, the seedless plants have never been able to colonize habitats where moisture is not unfailingly present for at least part of the growing season.

Seedless plants reached their peak in the Carboniferous Period, when they dominated vast forests that were then growing on the tropical floodplains and deltas of North America, Europe, and Asia, producing coal.

By the Middle Devonian Period, a few plants were already on the way to meeting requirement 4—that is, providing their own moist environment in order to facilitate sexual reproduction. The plants that evolved were the **gymnosperms**, or naked-seed plants, such as *Glossopteris* of Gondwana fame (Fig. 17.22A). The female cell is attached to the vascular system and thus has a supply of moisture. The male cell is carried in a pollen grain that has a hard coating. When the two fuse, a seed results. The seed is simply a supply of moisture and nutrients that will sustain the early growth of the young plant until it becomes self-supporting by photosynthesis. Naked-seed plants survive today; examples are the gingkos (Fig. 17.22B) and the conifers.

Gymnosperms had a huge success. Freed from the swampy habitat, they did not have to compete with the great seedless trees of the coal forests, but they could occupy the drier uplands of the newly forming supercontinent, Pangaea. By the end of the Carboniferous Period, they had spread over most of the world, and by the Triassic Period they were rivaling in size their former cousins of the swamps.

Life has one drawback for gymnosperms. The male cell-carrier, the pollen, is spread through the air. What chance does a pollen grain loose in the air have

A.

B.

Figure 17.22 Naked-seed plants developed from seedless plants late in the Devonian Period. A. A leaf of *Glossopteris*, a family of seed-fern plants that spread through Southern Gondwana. B. Leaves of modern fossil gingkos. The fossil is from North Dakota. Gingkos are long-lived relics of the ancient family of naked-seed plants.

of finding a female cell? To ensure success, gymnosperms have to make huge amounts of pollen. Flowering plants (**angiosperms**, or enclosed-seed plants) solved the problem of the random distribution of pollen. For a small incentive (nectar or a share of the pollen), insects will deliver the pollen (Fig. 17.23). It took longer for the angiosperms to evolve than it did for the gymnosperms, but by the end of the Cretaceous Period angiosperms had become the dominant

land plants. Their life cycle is not significantly different from that of gymnosperms, but they have specialized in symbiosis with animals: insects for pollination, birds and quadrupeds for seed dispersal.

The last frontier for plants—so far, at least—has been the dry steppes, savannas, and prairies. These were not colonized until the Tertiary Period, when grasses evolved. In arriving at this period, we have also reached the culmination of animal life on land, the

Figure 17.23 Pollen from a hollyhock (an angiosperm) coats a bumble bee collecting nectar. As the bumble bee moves from plant to plant, the pollen is efficiently distributed.

great grazing faunas of the high plains of all continents except Antarctica.

Insects

Among the numerous creatures in the Cambrian seas, many belonged to the phylum Arthropoda (trilobites of the Cambrian Period), so-called because of the presence of jointed legs. They include crabs, spiders, centipedes, and insects; these remain the most diverse phylum on the Earth. These animals, not the fish from which we humans are descended, were the first to make the change from sea to land.

Arthropods, with a few exceptions, were quite small, had lightweight structures, and were covered with a shell of chitin to provide structure support. Thus, they were admirably preadapted for life on land in regard to structural support and water conservation. The earliest to go on land were probably Silurian centipedes and millipedes. By the Carboniferous Period, insects were abundant and included dragon-

flies with a wing span of up to 60 cm (24 in). For all their success as land creatures, the arthropods have always had very primitive respiratory and vascular systems. They breathe through tiny tubes that penetrate the outer coating. Because the respiring mass of an aerobic organism increases as the cube of its length, while the area available for gas exchange increases only as the square, it is clear that this mode of respiration must severely limit the size of an organism. This is one reason that most insects are small.

Arthropods have an open vascular system. That is, their "blood" does not circulate in closed vessels, but is simply body fluid bathing the internal organs and generally kept in sluggish motion by a "heart" that is little more than a contractile tube. At first, it seems odd that the arthropods, with such a primitive arrangement, should have diversified into more than a million terrestrial species. The arthropods' vascular system isn't great, but it obviously works. And it's close to indestructible; whoever heard of a cockroach having a heart attack?

Animals with Backbones

Inconspicuous among the fossils of the Burgess Shale is a small fossil called *Pikaia* (Fig. 17.24). Pikaia is a chordate, a member of the phylum to which we humans belong, by virtue of possessing a notochord, a cartilaginous rod running along the back of the body. (We and other vertebrates have a notochord as embryos, later replaced by the backbone.) *Pikaia* (which may well be one of our ancestors) and other Cambrian fish were jawless, probably feeding on organic matter dredged from the seafloor. Jawed fish evolved afterward.

With the evolution of jawed fish, a great burst of diversification took place, giving access to a whole array of ecological niches that until then had been vacant. The original jawless fish, a few as long as many centimeters, were quickly joined by larger armored fish, including 9 m (30 ft) carnivores, as well as sharks and other boneless (cartilaginous) fish, and the huge order of ray-finned fish that are familiar as today's game and food fish.

The first fish to venture on land, an obscure group called the crossopterygians, did so in the Devonian Period (about 400 million years ago). They gave rise to the amphibians. The crossopterygians had several features that served to make the transition possible.

Figure 17.24 *Pikaia*, a soft-bodied animal from the Burgess Shale in British Columbia, is the earliest known chordate. *Pikaia* is the most ancient member of the group that became the vertebrates and to which humans belong.

Their lobe-like fins, for example, were preadapted as limbs because the lobes contain (much foreshortened) the elements of a quadruped limb, complete with small bones to form the extremity. They also had internal nostrils characteristic of air-breathing animals. Being fish, the crossopterygians already had a serviceable vascular system that was adequate to make a start on land. Water conservation, however, never became a strong point with amphibians: they retain permeable skins to this day, which is one reason they have never become independent of the aquatic environment.

Despite their limitations, the amphibians ruled the land for many millions of years during the Devonian Period. They had one difficulty that limited their expansion into many niches: they never met the reproductive requirement for life on land. In most species, the female amphibian lays her eggs in the water, the male fertilizes them there after a courtship ritual, and the young hatch as aquatic organisms. Like the seedless plants, the amphibians, with one foot on the land, so to speak, have remained tied to the water for breeding. Although some became quite large (2 to 3 m, or 2 to 3 yds long), amphibians are not highly diverse. One branch evolved to become reptiles; the rest that survive are frogs, toads, newts, salamanders, and limbless water "snakes".

The reptiles freed themselves from the water by evolving a water-tight skin and an egg with a shell that could be incubated outside of the water. These two

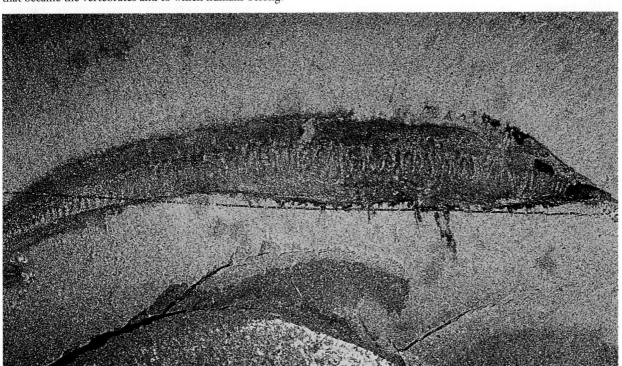

"inventions" gave them the versatility to occupy terrestrial niches that the amphibians had missed because of their bondage to the water. The amniotic egg did for reptilian diversity what jaws did for diversity in fishes. Reptiles originated in the Carboniferous Period coal swamps about 375 million years ago. By the Jurassic Period, some 185 million years later, the reptiles had moved onto the land, up in the air, and back to the water (as veritable sea monsters). This resulted in the production of the two orders of dinosaurs (Fig. 17.25) (the largest quadrupeds ever to walk the Earth) and gave rise to two new vertebrate classes—mammals and birds.

Mammals are in many ways better equipped to occupy terrestrial niches than were the great reptiles. It is difficult to pick out a single mammalian "invention" comparable to the jaws of fish or the reptilian egg, for the mammal is a fine-tuned quadruped, better adapted to a faster and more versatile life than the reptiles. The mammalian "invention" is perhaps just that: a set of interdependent improvements managed by a more capable brain and supported by a faster metabolism. The placental uterus is sometimes regarded as the key to mammalian success. It is mandated by the delicate intricacy of the fetus that lives in it, especially the brain.

Niches and the Evolution of Biological Diversity

The history of life's biological diversity can be viewed as the history of niche expansion, occupancy and elimination, over time, as the environment has changed. This history of biological diversity involves the availability of niches on the one hand and evolutionary innovations by successful mutations on the other. Organisms not only respond to environmental change but also create it. The oxygenation of the atmosphere by early photosynthetic organisms is an example of such a major biological change. An example at a smaller spatial scale is the vertical structural complexity provided by trees that create many niches. In a tropical rainforest, the top of the tree provides a sunlit environment that receives much rainfall, while down near the base of branches there can be shaded and comparatively dry habitats.

Availability of Niches

Rapid diversification results when a new "invention" (jaws in fish, the amniotic egg in reptiles) has adaptive advantages for a new way of life. However, a modification that adapts a species to a niche that is already occupied may confer little immediate advantage, even though the new species may be adaptively superior to the incumbent one.

Such was the case with the mammals. Originating in the Triassic Period, mammals were from the start more attuned to the niches they now occupy than were the established occupants, the dinosaurs. With more capable brains, faster metabolism, a uterus to shelter the fetus, milk for postnatal nourishment, and parent-offspring bonding, the early mammals had greater potential to exploit dinosaur niches than had the dinosaurs themselves. The great dinosaurs had

Figure 17.25 Gasosaurus, one of the huge flesh-eating Jurassic dinosaurs. Gasosaurus was up to 2 m high and 4 m long. Mammals that lived at the same time as Gasosaurus were small, about the size of mice or rabbits, and probably of little interest to the large dinosaur.

command of the food supply, however, and the mammals, instead of growing larger than their therapsid ancestors, became smaller, perhaps small enough not to interest carnivorous dinosaurs (Fig. 17.25). And so they remained for 150 million years, until the end of the Cretaceous Period. The mammals occupied vacated niches at the end of the dinosaur's rule in a burst of diversification that began in the Paleocene Epoch and continued through the Eocene. By comparing brain-to-body weight ratios in archaic and modern reptiles and mammals, it can be shown that increase in mammalian brain size is a continuing process, whereas in reptiles increase has not occurred: the ratios in modern reptiles do not differ significantly from those in archaic ones.

EXTINCTIONS AND THE BIOSPHERE

Few of the species alive at the beginning of the Cambrian Period had living descendants at the end. Throughout the Cambrian, species died out and became extinct; new species evolved to occupy the vacated niches. Why and how did great groups of animals suddenly die out and new ones arise to take their places? In the nineteenth century, when **paleontologists** (scientists who study extinct organisms) started to study the fossil record in a systematic way, they recognized that evidence of extinction is widespread and that most of the species that have *ever* lived on the Earth are now extinct.

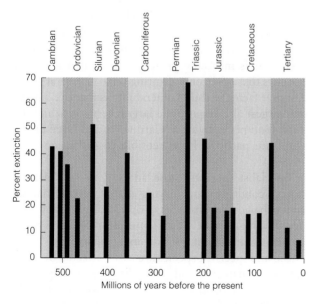

Figure 17.26 The extraordinary frequency of great extinction events that have occurred during the Phanerozoic Eon. The percentage of extinction was determined from the disappearance of genera of well-skeletonized animals.

Nowhere is the evidence of extinction more detailed or more striking than in the fossil record of marine animals. From that record we can estimate that the average time span of an ocean-dwelling species is about 4 million years. The record is equally clear, however, that organisms do not steadily disappear and new organisms do not steadily appear. Rather, the evidence suggests that a series of massive extinction events have occurred, followed by the rapid appearance of new species that occupied vacant niches.

Paleontologists David Raup and J. John Sepkoski, Jr., are two of the leading researchers of the extinction record. Using the appearance and disappearance of genera rather than species, Sepkoski has analyzed the fossil record of 34,000 genera of marine fossils from the Cambrian to the present. Figure 17.26 is a diagrammatic representation of 19 extinction events he has identified. Two of these major events occurred at the Cretaceous-Tertiary boundary, at the end of the Mesozoic Era, and at the Permian-Triassic boundary, at the end of the Paleozoic Era.

What causes these major extinction events? A popular theory is that the impact of a gigantic meteor at the end of the Cretaceous caused such an event. How important was this event in causing the extinction of the dinosaurs? A distinctive iridium-rich sediment is thought to have been formed as a result of a great meteorite impact that coincides with the Cretaceous-Tertiary boundary. A huge impact crater of exactly the right age has been found in Mexico, and the devastation caused by the impact must have been massive. As a result, many scientists today accept such an event as the probable cause of the massive extinction that coincides with the Cretaceous-Tertiary boundary.

Plate Tectonics and Extinctions

The greatest of all extinctions in the Earth's history, when some 95 percent of known fossil species were lost, marks the Permian-Triassic boundary at the end of the Paleozoic Era. There is no convincing evidence that the Permian-Triassic boundary extinction had an extraterrestrial cause. Although the Cretaceous-Tertiary extinction severely affected both terrestrial and marine life, it was mainly marine species that suffered in the Permian-Triassic one. This earlier extinction is associated with the assembly of Pangaea, the supercontinent that existed briefly from Late Permian to Late Triassic time. Because the assembly of Pangaea came about as a result of plate tectonic movement, the Permian-Triassic extinction is sometimes referred to as a plate tectonic extinction.

The Permian-Triassic extinction was associated with a marked drop in sea level caused by the dwindling of the ocean ridges that had driven the Paleozoic

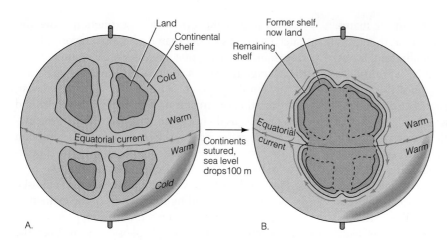

Figure 17.27 Idealized diagram of a hypothetical Earth with A. four separate continents and an equatorial seaway; and B. the four continents sutured together plus a 100 m drop in sea level. The resulting supercontinent lies across the equator, forcing the equatorial current poleward. The area of the continental shelf is drastically reduced, and the climatic gradient is weakened.

continents together. Consider the effect of these developments—continents massed together and sea level lowered—on the continental shelves. A simplified model (Fig. 17.27) shows that, if four continents of equal size are stuck together to make a single supercontinent, there is a 50 percent loss of coastline and, consequently, of continental shelf. A 100-meter (110 yd) drop in sea level will approximately halve the remaining shelf area, so that the total shelf area is reduced to about 25 percent of what was there before—a drastic reduction in living space for shelf organisms. Furthermore, with four separate continents there must be at least four ecological shelf provinces, and more if the climatic gradient is taken into account. (Remember: scientific evidence suggests that the south polar region was glaciated in the late Carboniferous Period.) When the four continents were joined together, the number of niches decreased because the shelves that remained became contiguous. Pangaea lay across the equator, forcing the warm, westward-flowing equatorial current into high latitudes and bringing heat from equator to poles. The latitudinal climatic gradient was much less steep than had been the case before or has existed since. It is therefore reasonable to suppose that supercontinent formation must have made a major contribution to the extinction.

Another line of thought connects plate tectonics with diversity in reptiles and mammals. Reptiles began diversifying in the late Carboniferous Period and had achieved most of their full diversity—some 9 of the 13 or so orders usually recognized—by the end of the Triassic Period 90 million years later. This amounts to the evolution of about 1 new order every 10 million years.

As noted earlier, the mammals originated in the Triassic Period but did not get a chance to diversify until the Cenozoic Era, producing about 30 orders by the end of the Eocene Epoch, a mere 30 million years later: 1 order per million years, or ten times the reptilian rate. Some of the mammals' diversity can be ascribed to the different circumstances in which reptiles and mammals diversified. Reptiles diversified on Pangaea, a single continent with a weak climatic gradient and therefore low diversity. Mammals diversified on seven separate continents following the breakup of Pangaea and faced a strong climatic gradient as the late Cenozoic ice age approached.

Common Traits of Endangered Species

Species likely to become endangered through human activities tend to share certain traits. Knowledge of these common traits helps us protect and manage such species. Easily endangered species, particularly vertebrates, are generally long-lived and large. Such species tend to have low reproductive rates and recover slowly from lowered population levels (Fig. 17.28A). It makes sense that the potential growth rate of a pair of rats that can have several litters of multiple offspring each year is markedly greater than that of a pair of condors that raises one offspring once in several years. Another factor in their vulnerability is that the biggest and largest also require the largest territories and the most food per individual. Carnivores and large herbivores are particularly susceptible to extinction (Fig. 17.28B).

Specialist species—those adapted to very narrow sets of conditions and having highly specific habitat and behavior requirements—are also especially vulnerable (Fig. 17.28C). As people clear land and modify the environment, the diversity of habitats is reduced and specialist species are easily affected by these changes. Adaptable and generalist species, especially those that share foods and habitats with people, do well. Rats, sparrows, cockroaches, cats, and dogs do well in a habitat dominated by humans, where there is an omnivore food base and a simplified environment.

A.

B.

C.

Figure 17.28   Certain characteristics tend to make some species more susceptible to extinction than others, especially from human activities. A. Large, long-lived animals, like the African black rhinoceros, often are desirable for some commercial product. The horn of the rhinoceros is used as a medicinal and can be sold for a high value. B. Species that are high on the food chain, such as the Osprey, a fisheating predatory bird, concentrate chemical toxins such as DDT, and are more susceptible to these chemicals than herbivores. C. Species with highly specialized diets, like the Giant Panda of China, which feeds only on certain kinds of bamboo, are at risk due to human land clearing and modification. Large animals that require large habitats, like the panda, are among the most vulnerable to this effect.

A key lesson to be learned from this discussion is relevant to the conservation of endangered species: a species does better with an ecosystem in good condition and a small population than an abundant population in an ecosystem in poor condition. Another lesson of this discussion points to an irony in evolution. When the environment is constant, a way to win at evolution is to become more and more specialized. But the more specialized the species is, the more vulnerable it is to environmental changes or changes in its competitors, predators, or prey.

LINKS WITH HUMAN ACTIVITY

Although extinction is a natural process in a world with a finite limit on its existence and the ultimate fate of all Earth-restricted species is extinction, human beings have greatly accelerated the rate of extinction, especially since the Industrial Revolution. This rate of extinction may have global impact. Ever since people learned how to use fire, they have had major effects on large areas of the land surface and have influenced the abundance of many species of animals and plants. Preindustrial, hunter-gatherer people also affected biological diversity through hunting. Extinction of some of the largest mammals in North America occurred near the end of the ice age when people migrated from Siberia to North America. Some anthropologists believe that these extinctions may have been caused by hunting. Similarly, the Polynesian settlers of New Zealand caused the extinction of some large flightless birds that had not been exposed to human beings before. As preindustrial peoples spread around

GUEST ESSAY

The Maturation of Earth and Life

In terms of its cosmological life expectancy, Earth is a middle-aged planet. In maturity, our planet contains large and widely dispersed continents, is glaciated at the poles, and is bathed in an oxygen-rich atmosphere. Perhaps most conspicuous, it supports an exuberance of life ranging from bacteria to redwoods. However, available portraits of the youthful Earth suggest a different countenance—a watery surface punctuated by volcanoes and limited areas of continental crust, an atmosphere rich in carbon dioxide and nearly devoid of oxygen, and no sign of biological activity.

The Earth's surface has changed markedly during the past 4 billion years. Given the close relationship between organisms and environment on the present-day Earth, it should not be surprising to learn that the two have evolved together through time, each influencing the other in a complex system of biogeochemical feedbacks.

The interplay between evolution and environment was particularly strong during our planet's early history. It is no overstatement to suggest that the evolutionary history of metabolism, explicit in evolutionary trees of both bacteria and organisms with nuclei, mirrors the chemical evolution of the oceans and atmosphere. One facet of this coevolution that has particularly interested me concerns the explosive evolution of animals near the Proterozoic-Cambrian boundary (ca. 543 million years before the present). Anyone who has ever walked through a stratigraphic section spanning this boundary knows that the mineralized skeletons of invertebrates increase abruptly as one strides into the Cambrian, as do the abundance and diversity of animal tracks, trails, and burrows. All the extant phyla of "higher" animals—that is, animals characterized by complex organ systems and a shape that is symmetric about a central axis—diversified within the first 20 million years of the Cambrian Period. Exquisitely well-preserved fossils from the Burgess Shale and similar deposits suggest that this remarkable radiation also included types of animals that no longer exist. Simple, mostly unskeletonized fossils of sponges, cnidarians (sea anemones, jellyfish, and their relatives), and ancestral bilaterians appear only slightly earlier, perhaps 570 million years ago. Although this seems like a long time ago, fossils provide unambiguous evidence of microbial life as early as 3500 million years before the present. Why did animals radiate so late in the evolutionary day?

Among the several theories advanced to account for this pattern, one really fascinates me: it holds that prior to 580 million years ago, the Earth's atmosphere contained too little oxygen to support the biology of large animals. If true, this suggests that both the Earth and its biota leapt toward maturity near the end of the Proterozoic Eon. It also provides a most dramatic example of coevolution between life and environment.

A wealth of sedimentological data support the hypothesis

Andrew H. Knoll is professor of organismic and evolutionary biology, and also earth and planetary sciences, at Harvard University. An avid field geologist, Knoll has traversed the globe, spending nights in such remote areas as the high Arctic, the Australian Outback, and the Namib Desert of Namibia. In 1991 he was elected to the National Academy of Sciences.

of late Proterozoic environmental change. Extensive glaciogenic rocks and sedimentary iron deposits document large-scale changes in the atmosphere and oceans at this time. Knowledge of the present-day Earth leads us to expect that such variations should relate to changes in the systems of biogeochemical cycling that control our environment. My colleagues and I have confirmed this hypothesis by documenting strong variations through late Proterozoic time in the ratio of the two stable isotopes of carbon (^{13}C and ^{12}C) in marine carbonate minerals and organic matter. Because this ratio reflects the relative rates of carbonate and organic carbon burial of the sea floor, we have been able to infer remarkable variations in the late Proterozoic carbon cycle—with high rates of organic carbon burial characterizing the period just prior to the initial appearance of large animals. Other colleagues have documented the intricate tectonic dance of the continents during this period, with a late Proterozoic supercontinent splintering and reamalgamating, only to break apart once more as the Cambrian began. These tectonic events provide clues to understanding the dynamic behavior of late Proterozoic climate and biogeochemistry.

As we learn more, the interval first singled out because of its biological importance turns out to be a time of remarkable tectonic, climactic, and biogeochemical change as well. There is as yet no direct means of measuring oxygen levels on the latest Proterozoic and Cambrian Earth, but geochemical models strongly support the hypothesis that atmospheric oxygen levels did indeed rise sharply 590 to 580 million years ago, just prior to the initial radiation of large animals. Thus, our planet may have come upon both biological and environmental maturity rather late in its development. Further biological innovations amplified by ecological interactions may have triggered the subsequent radiation that marks the Proterozoic-Cambrian boundary.

The end-Proterozoic example is striking, but it is hardly unique. It may even be general. Environmental dynamics have exerted a profound influence on evolution, and *vice versa*. As the Earth sciences become more fully integrated, we gain the prospect of new insights into long-standing problems of Earth history. Equally exciting, we will be able to generate new questions that will occupy our imaginations for decades to come.

the world, they altered biological diversity through hunting, changes in habitat, and the introduction of exotic species, such as dogs, and new crops. These effects are especially evident in the Pacific islands because the native animals and plants were long isolated and had no prior exposure to human beings.

Although the extinction of many species over large areas of the Earth is not new with the evolution of human beings, the rate of human-caused extinctions has accelerated greatly since the Industrial Revolution. Modern influences on biological diversity result more from habitat disruption than from the introduction of exotic species. At present, 1672 species of animals are listed as threatened or endangered worldwide, and an especially large number of these are fish and mollusks. Approximately 20,000 species of plants are believed to be endangered or threatened, according to the International Union for the Conservation of Nature. In terms of human-induced effects on biological diversity, the primary ones that have affected the Earth system are (1) deforestation; (2) desertification, which can also affect the atmosphere by changing reflectance and evaporation, and (3) the introduction of new species in exotic habitats.

The introduction of exotic species has had major effects worldwide on biological diversity. One of the major impacts of exotic species is the spread of disease organisms, including human diseases such as small pox, which was a major factor in the decline of American Indians following European settlement of North America. Diseases of plants, such as the so-called Dutch-elm disease that attacks the American elm, and the chestnut blight that eliminated chestnut as a mature tree through its range in the eastern forests of the United States, are other examples of such major effects.

Most major commercial fisheries have undergone large declines owing to overfishing, and these can have impacts on food chains by influencing competition and predation. A disproportionately large number of fish are threatened or endangered in part because people have introduced fish from one part of the world into lakes and rivers in another. Often these introductions took place for what were believed to be useful motivations—to provide new sources of food and sport fishing. But major lakes around the world have lost many of their native species, which have been unable to compete with introduced fish. This problem occurs locally, but it has become a global problem because fish have been introduced widely throughout the world.

An important connection exists between human effects on biological diversity and human effects on biogeochemical cycles: biological diversity can provide pollution control. Plants and bacteria in particular can remove toxic substances from the air, water, and soils. For example, carbon dioxide and sulfur dioxide are removed by vegetation, carbon monoxide is reduced and oxidized by soil fungi and bacteria, and nitric oxide is incorporated into the biological nitrogen cycle. Because different species have different capabilities for this removal, a diversity of species can provide the best range of pollution control.

At a global level, the major concern among biologists is that a rapid, human-induced decrease in genetic diversity may mean that the Earth will lose many kinds of biological capabilities, only a few of which we understand at present, but which might be important to ecosystem dynamics or have practical uses to human beings

SUMMARY

1. This chapter addresses several unresolved issues: (1) How did life evolve from microscopic beginnings to the complex set of species, ecosystems, and the entire biosphere that exists today? (2) Why is life so diverse? (3) What is the importance of biological diversity to the Earth system? and (4) How and why did great groups of animals suddenly die out and new ones arise to take their places?

2. There are several kinds of biological diversity, including genetic, species, and habitat diversities. Most discussions of biological diversity focus on species diversity.

3. Biological diversity is influenced by many factors, among them climate, sexual reproduction, evolutionary innovations, provinciality, and niche availability.

4. Two keys to understanding how the Earth can support so many species are the competitive exclusion principle and the niche concept. The competitive exclusion principle states that complete competitors cannot coexist. Species evolve to avoid such direct competition; they take advantage of different niches.

5. In concept, an ecological niche is the profession of a species—what it does to make a living. Quantitatively, the niche can be viewed as the range of all environmental conditions within which a species is found, as well as what the species does to obtain essential nutrients.

6. Natural selection is the set of processes that led to biological evolution. It has four primary characteristics: (1) inheritance of traits from one generation to the next

and some variation in these traits; (2) environmental variability; (3) differential reproduction that varies with the environment; and (4) an influence of the environment on survival and reproduction.

7. Populations evolve in response to changing environments and opportunities. In turn, biological evolution leads to changes in the environment. This means that the biosphere has changed through geologic time and is still changing today.

8. No one knows how life began or what the first living cell was like, but it must have been an anaerobic procaryote. Fossil procaryotes have been found in rocks 3.5 billion years old, so life appeared early in the Earth's history.

9. Bacteria, which are the simplest living organisms, and viruses, which are "almost" alive, cannot have been the first life forms because they are too complicated.

10. The four essential steps to form life are metabolism, chemosynthesis, biosynthesis, and development of the complex cellular machinery needed for reproduction.

11. A variety of theories about the origin of life have been propounded, none yet proven. Some hypotheses suggest that life may have arisen in the sea, possibly near submarine hot springs. Another hypothesis, called panspermia, is that life arose in space and arrived on the Earth ready made.

12. Procaryotes cannot form three-dimensional, multicellular structures; eucaryotes can. Thus, eucaryotes can develop much more complex body structures, such as elephants and redwood trees. Eucaryotes arose from procaryotes after oxygen started accumulating in the atmosphere.

13. Since early life, oxygen has been produced by photosynthesizing organisms. At first, the free oxygen was neutralized by reducing agents such as dissolved iron in sea-water. When the neutralizing capacity was exceeded, oxygen began to accumulate in the atmosphere, a build-up that began at least 1.5 billion years ago.

14. The earliest multicellular organisms known are soft-bodied marine animals, called the Ediacaran animals after the place in Australia where they were first found.

15. The Cambrian Period was a time of a great increase in biological diversity among marine life. The first skeletons, both internal and external, evolved in the Cambrian Period.

16. For an organism to leave the sea and live on land, four requirements have to be met: a structural support system; an internal aquatic system; a means of exchanging gases with air instead of water; and the availability of a moist environment for the reproductive system.

17. Among the first eucaryotic organisms to occupy the land were plants and insects. They did so in the early Silurian, more than 430 million years ago. The first land vertebrates, amphibians, occurred some 30 million years later in the early Devonian Period.

18. The first plants were seedless. From these evolved the naked-seeded gymnosperms, and the gymnosperms in turn gave rise to the enclosed-seed angiosperms.

19. The amphibians were succeeded by the reptiles. Amphibians have porous skins and require water in which to lay their eggs. Reptiles developed water-tight skins and amniotic eggs and thus freed themselves from the water.

20. The fossil record reveals that the biosphere has been repeatedly disrupted by times of great extinctions. The causes of the extinctions are the subjects of research but appear to include giant meteorite impacts, plate tectonic rearrangements of continents, declines of sea level, prolonged volcanic eruptions, and severe climatic changes.

IMPORTANT TERMS TO REMEMBER

angiosperm *407*
biological diversity *388*
biosynthesis *398*
chemosynthesis *398*
competitive exclusion
  principle *390*

ecological niche *390*
ediacaran fauna *402*
genetic drift *394*
gymnosperms *406*

habitat *390*
industrial melanism *393*
migration *394*
mutation *394*

natural selection *393*
paleontologist *411*
panspermia *399*
provinciality *395*

QUESTIONS FOR REVIEW

1. What are the essential characteristics of life?

2. Could life have started as a virus? Could it have started as bacteria? Explain why or why not.

3. It is thought that the Earth system has evolved through time because of the changes in the biosphere. What kind of changes in the biosphere could bring changes in the Earth system?

4. What are the three essential steps that must have occurred in order for life to form?

5. What is the "soup" hypothesis? Where and how did chemosynthesis occur in the hypothesis, and why is this part of the hypothesis now in doubt?

6. Describe the ways by which biosynthesis may have happened on the primitive Earth.

7. What characteristic must the earliest cells have had? Which organisms living today are thought to be representative of the most ancient life forms?

8. What is the panspermia hypothesis, and why do some scientists consider it favorably?

9. Why can't anaerobic cells have a nucleus?

10. It is hypothesized that eucaryotes arose from procaryotes. Why is that conclusion held, and how might eucaryotes have developed?

11. Contrast the size, feeding properties, and aggregating abilities of procaryotic and eucaryotic cells. Why are eucaryotic cells more efficient?

12. How long ago did the first eucaryotic cells develop?

13. Life expanded dramatically in the Cambrian Period and occupied innumerable ecological niches. What organisms occupied all the ecological niches during the Proterozoic Eon?

14. Give two hypotheses explaining why biological diversity exploded in the Cambrian Period.

15. What lessons can be learned from the fossil assemblages of the Burgess Shale?

16. When did life start leaving the sea and inhabiting the land? In what order did animals, insects, and plants become established on the land?

17. What four requirements had to be met in order for organisms to leave the sea and live on land? Are the requirements the same for plants, insects, and animals?

18. In what ways have plants modified their reproductive cycle since moving on to the land?

19. Describe how symbiosis has helped the angiosperms to develop.

20. Judged by their diversity, insects are a very successful group. Can you offer any possible reason for their success?

21. Describe two great inventions that dramatically changed the ability of animals to occupy new ecological niches.

22. Which animals first made their way from the sea to the land, and what group of animals evolved from them? Name two kinds of animals still living today that are representatives of the first group of land dwellers.

23. Amphibians are tied to the water. What characteristics did the reptiles develop that allowed them to free themselves from a dependence on water?

24. What evidence in the fossil record suggests that rapid extinctions have occurred many times in the past? Describe two hypotheses for the extinctions.

QUESTIONS FOR DISCUSSION

1. Could some form of life as we know it on Earth have developed on any of the other bodies of the solar system? If you had an unmanned space craft available to visit other planets and moons, what tests would you carry out to see if life now exists or ever existed there?

2. There are many intriguing but still unanswered questions concerning the history of the biosphere. Discuss the following questions:

a. In the absence of fossils, what kinds of evidence might be used to determine when the biosphere came into being?

b. What kind of research would you carry out in order to explain why the great Cambrian explosion of biodiversity occurred and why animals started to develop skeletons?

3. Choose one of the great extinctions in the fossil record, research it, and discuss evidence bearing on the cause of the extinction.

4. Ediacaran fauna have been identified at several places around the world. Do some research on where these places are, and from descriptions of the rocks in which the fossils are found, determine the characteristics of the environment in which the animals lived.

5. The great animal herds of the savannahs and prairies, such as buffalo and antelope, appeared during the Tertiary Period and seem to be coincident with the appearance of grasses. Do research on where and when the herds appeared. Discuss what effect climate change may have played in the rise of grasses and grass-grazing animals.

Several of the case studies available at our web site focus on special ecological niches.

CHAPTER 18

Resources from the Earth

● *Natural Resources and Civilization*

Civilization and natural resources—the former would not have been possible without the latter. Scholars even mark the stages of civilization by the natural resources our ancestors learned to use: the Stone Age, Bronze Age, and Iron Age are examples. The first resources our ancestors used were the fruits, grain, animals, and fish they ate. These first-used materials are *renewable resources* because new supplies grow each season.

Millions of years ago our ancestors also started to use a different class of natural resources. When they picked up suitably shaped stones and used them as hunting aids, they were using *nonrenewable resources*, so-called because even if they are replenished by natural processes, growth times are measured in millions of years rather than in annual seasons. Because the most desired stones were found in only a few restricted places, trading started. Next, our ancestors started gathering and trading another nonrenewable resource, salt. Originally, dietary needs for salt were satisfied by eating the meat brought home by hunters. When farming started and diets became cereal based, extra salt was needed. We don't know when or where the mining of salt started, but long before recorded history salt routes crisscrossed the globe.

Metals were first used more than 17,000 years ago. Both copper and gold can be found in their metallic form, and these were the earliest metals to be used. But natural copper is rare, and so eventually other sources of copper were needed. By 6000 years ago, our ancestors had learned how to extract copper from certain minerals by a chemical process called smelting. Before another thousand years had passed, they had discovered how to smelt minerals of lead, tin, zinc, silver, and other metals. The technique of mixing metals to make alloys came next; bronze (copper and tin) and pewter (tin, lead, and copper) came into use. The smelting of iron is more difficult than the smelting of copper, and so development of an iron industry came much later—about 3300 years ago.

The first people to use oil (a nonrenewable resource) instead of wood (a renewable resource) for fuel were the Babylonians, about 4500 years ago. The Babylonians lived in what is now Iraq, and they used oil from natural seeps in the valleys of the Tigris and Euphrates rivers. The first people to mine and use coal were the Chinese, about 3100 years ago. At about the same time, the Chinese drilled the first wells for natural gas—some were nearly 100 m (330 ft) deep.

By the time that first the Greek and then the Roman empires came into existence about 2500 years ago, our ancestors had come to depend on a very wide range of nonrenewable re-

(opposite) Many cultures around the world discovered the distinctive properties of gold and developed the skills to work with it. This mask from Columbia, now in the Bogota Gold Museum, was made about 2000 years ago by indigenous peoples living in the region.

sources—not just metals and fuels, but also processed materials such as cements, plasters, glasses, and porcelains. The list of materials we mine, process, and use has grown steadily larger ever since. Today we have industrial uses for almost all of the naturally occurring chemical elements, and more than 200 kinds of minerals and fuels are mined and used. Of course, we still harvest renewable resources, but society is now totally dependent on supplies of nonrenewable resources. Whether or not we realize it, we are slowly changing the global environment as we dig up nonrenewable resources in one place, use them in another, and finally discard the used and worn out products in yet another place. ●

MINERAL RESOURCES

Can you imagine a world without machines? Our modern world with its 6 billion people couldn't operate without them. Nor could it operate without bricks and cement, fertilizers, or plastics. All of these products and many more are made from nonrenewable mineral resources. Each of us now relies directly or indirectly (meaning through industry and public works) on a very large annual input of nonrenewable mineral resources (Fig.18.1).

Machines produce our food, make our clothes, transport us, and help us communicate. Bricks and cement are used to build houses and roadways, salts are used for chemicals and fertilizers, and plastics made from coal and oil are used in myriad ways. Metals, fuels, fertilizers, chemicals, and building materials are all dug or pumped from deposits formed through the ages by geological processes. Geologi-

cal processes therefore control the well-being of all people.

In some of the previous chapters, it was pointed out that under suitable conditions, geological processes such as weathering, sedimentation, and volcanism can form concentrations of valuable minerals and rocks. The "suitable conditions" are not common, however, and for this reason mineral and energy deposits are few and hard to find. No geological challenge is more demanding than the search for, and discovery of, new resources of minerals and energy, and no societal problem is more pressing than the recovery of natural resources with minimal disruption to the environment. Finding resources and managing the consequences of using them are two of the greatest problems facing human beings today.

We turn first to the minerals we mine. Mineral resources are nonrenewable and can be divided into two groups by the way they are used.

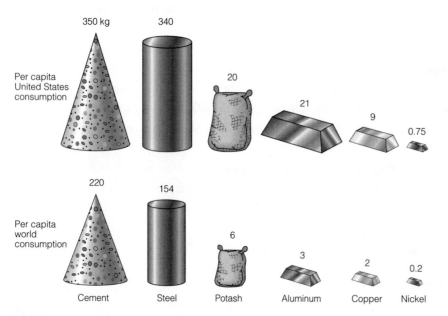

Figure 18.1 The average amount of material consumed per person per year (called the per capita consumption) is greater in an industrially advanced country such as the United States than it is for the world as a whole.

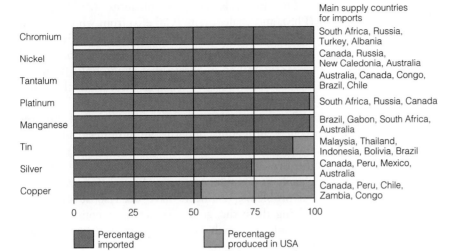

Main supply countries for imports

Mineral	Main supply countries for imports
Chromium	South Africa, Russia, Turkey, Albania
Nickel	Canada, Russia, New Caledonia, Australia
Tantalum	Australia, Canada, Congo, Brazil, Chile
Platinum	South Africa, Russia, Canada
Manganese	Brazil, Gabon, South Africa, Australia
Tin	Malaysia, Thailand, Indonesia, Bolivia, Brazil
Silver	Canada, Peru, Mexico, Australia
Copper	Canada, Peru, Chile, Zambia, Congo

■ Percentage imported ■ Percentage produced in USA

Figure 18.2 Selected mineral substances for which the United States' consumption exceeds production. The difference must be supplied by imports.

1. *Metallic minerals* are those from which metals such as copper, iron, gold, and zinc can be recovered by smelting.

2. *Nonmetallic minerals* are those used for their physical or chemical properties rather than for the chemical elements they contain; examples are sodium chloride, calcium sulfate, sodium carbonate, calcium fluoride, and clay for bricks.

Following a discussion of mineral resources, we turn to energy resources. Some energy resources, such as coal and uranium, are nonrenewable because they are rocks and minerals formed by the same geological processes that form metallic and nonmetallic minerals. Other energy sources, such as sunlight, wind, and running water, are renewable and have quite different properties than the nonrenewable.

Supplies of Minerals

Many industrialized nations possess rich **mineral deposits** (any volume of rock containing an enrichment of one or more minerals) which they are exploiting vigorously. Yet no nation is entirely self-sufficient in mineral supplies, and so each must trade with other nations to fulfill its needs (Fig. 18.2).

All mineral resources have three peculiarities that influence their use:

1. Usable minerals are limited in abundance and localized within the Earth's crust. This is the main reason why no nation is self-sufficient where mineral supplies are concerned.

2. The quantity of a resource available in any one country is never known with accuracy, because the likelihood that new deposits will be discovered is difficult to assess. A country that today can supply its needs for a given mineral substance may face a future in which it will become an importing nation. A century and a half ago, for example, Britain was a great mining nation, producing and exporting tin, copper, tungsten, lead, and iron. Today, the rich deposits have been worked out, and Britain imports all the metals it needs.

3. Unlike fruits and grains, which can be seasonally cropped and thus replenished, deposits of minerals are depleted by mining and are eventually exhausted. This disadvantage can be offset only by finding new occurrences or by using the same material repeatedly—that is, by recycling and making use of scrap.

Much ingenuity has been expended in bringing the production of minerals needed by society to its present state. Because known deposits are being rapidly exploited while demands for minerals continue to increase as the world's population grows ever larger, we can be sure that even more ingenuity will be needed in the future.

Ore

Minerals are sought in deposits from which the desired substances can be recovered least expensively. The more concentrated the desired minerals, the more valuable the deposit. In some deposits, the desired minerals are so highly concentrated that even rare substances such as gold and platinum can be seen with the naked eye. For every desired mineral substance, a *grade* (level of concentration) exists below which the deposit cannot be worked economically (Fig.18.3). To distinguish between profitable and unprofitable mineral deposits, the word **ore** is used, meaning an aggregate of minerals from which one or more minerals can be extracted profitably. It is not always possible to say exactly what the grade must be, or how much of a given mineral must be present, in order to constitute an ore. Two deposits may have the

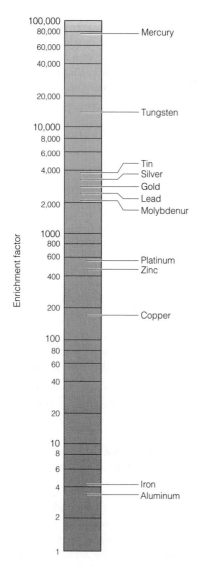

Figure 18.3 Before a mineral deposit can be worked profitably, the percentage of valuable metal in the deposit must be greatly enriched above its average percentage in the Earth's crust. The enrichment is greatest for metals that are least abundant in the crust, such as gold and mercury. As mining and mineral processing have become more efficient and less expensive, it has been possible to work leaner ore, and so there has been a historic decline in enrichment factors. Declines have ceased over the past 20 years, and for some metals, enrichment factors have increased slightly. Note that the scale is a magnitude (logarithmic) scale, in which the major divisions increase by multiples of ten.

same grade and be the same size, but one is ore and the other is not. There could be many reasons for the difference; for example, the uneconomic deposit could be too deeply buried or located in so remote an area that the costs of mining and transport would make the final product noncompetitive with the same product from other deposits. Furthermore, as costs and market prices fluctuate, a particular deposit may be an ore today but not tomorrow.

Ore minerals such as sphalerite (ZnS), galena (PbS), and chalcopyrite ($CuFeS_2$) from which desired metals are extracted are usually mixed with other minerals, collectively termed **gangue** (pronounced gang). Familiar minerals that commonly occur as gangue are quartz, feldspar, mica, calcite, and dolomite.

The ore challenge is twofold: (1) to find the ores (which altogether underlie an infinitesimally small proportion of the Earth's land area); and (2) to mine the ore and get rid of the gangue as cheaply and cleanly as possible. Mining and separation are technical problems; engineers have been so successful in solving them that some deposits now considered ore are only one-sixth as rich as were the lowest grade ores 100 years ago.

ORIGIN OF MINERAL DEPOSITS

All ores are mineral deposits because each of them is a local enrichment of one or more minerals or mineraloids. The reverse is not true, however: Not all mineral deposits are ores. "Ore" is an economic term, whereas "mineral deposit" is a scientific term. How, where, and why a mineral deposit forms is the result of one or more geological processes. Whether or not a given mineral deposit is an ore is determined by how much we human beings are prepared to pay for its content. Fascinating though the economics of ores and mining is, this topic cannot be explored in this volume. Instead, discussion is limited to the origin of mineral deposits without necessary regard to questions of economics.

In order for a deposit to form, some process or combination of processes must bring about a localized enrichment of one or more minerals. A convenient way to classify mineral deposits is through the principal concentrating process. Minerals become concentrated in five ways:

1. Concentration by hot, aqueous solutions flowing through fractures and pore spaces in crustal rock to form **hydrothermal mineral deposits**.

2. Concentration by magmatic processes within a body of igneous rock to form **magmatic mineral deposits**.

3. Concentration by precipitation from lake water or seawater to form **sedimentary mineral deposits**.

4. Concentration by flowing surface water in streams or along the shore to form **placers**.

5. Concentration by weathering processes to form **residual mineral deposits**.

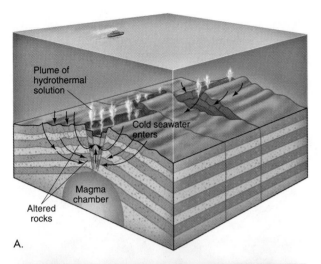

Figure 18.4 Hydrothermal solutions form mineral deposits on the seafloor. A. Seawater penetrates volcanic rocks on the seafloor at a spreading center. Heated by a magma chamber, seawater becomes a hydrothermal solution, alters rocks it passes through and extracts metals in the process, and rises at a midocean ridge as a hydrothermal plume. B. A so-called "black smoker" photographed at a depth of 2500 m below sea level on the East Pacific Rise at 21°N latitude. The "smoker" has a temperature of 320°C. The rising hydrothermal solution is actually clear; the black color is due to fine particles of iron sulfide and other minerals precipitated from solution as the plume is cooled through contact with cold seawater. The chimneylike structure is composed of pyrite, chalcopyrite and other ore minerals deposited by the hydrothermal solution.

Hydrothermal Mineral Deposits

Many of the most famous mines in the world contain ores that were formed when their ore minerals were deposited from hydrothermal solutions. More mineral deposits have probably been formed by deposition from hydrothermal solutions than by any other mechanism. However, the origins of hydrothermal solutions are often difficult to decipher. Some solutions originate when water dissolved in a magma is released as the magma rises and cools. Other solutions are formed from rainwater or seawater that circulates deep in the crust. (For a discussion of what is known about modern deposit-forming solutions, see "A Closer Look: Modern Hydrothermal Mineral Deposits.")

The heat source for seawater hydrothermal solutions of the kind illustrated in Figure 18.4 is spreading center volcanism. Because the ore minerals deposited are always sulfides, mineral deposits formed from such solutions are called *volcanogenic massive sulfide deposits.*

The ore–mineral constituents in volcanogenic massive sulfide deposits originate from the igneous rocks of the oceanic crust. Heated seawater reacts with the rocks it is in contact with, causing changes in both mineral composition and solution composition. For example, feldspars are changed to clays and epidote, and pyroxenes are changed to chlorites. As the minerals are transformed, trace metals such as copper and zinc, present by atomic substitution, are released and become concentrated in the slowly evolving hydrothermal solution.

Causes of Precipitation

When a hydrothermal solution moves slowly upward, as with groundwater percolating through an aquifer, the solution cools very slowly. If dissolved minerals were precipitated from such a slow-moving solution, they would be spread over great distances and would not be sufficiently concentrated to form an ore. But when a solution flows rapidly, as in an open fracture, through a mass of shattered rock, or through a layer of porous volcanic rock where flow is less restricted, cooling can be sudden and happen over short distances. Rapid precipitation and a concentrated mineral deposit are the result. Other effects—such as boiling, a rapid decrease in pressure, composition changes of the solution caused by reactions with adjacent rock, and cooling as a result of mixing with seawater—can also cause rapid precipitation and form concentrated deposits. When valuable minerals are present, an ore can be the result.

Examples of Precipitation

Veins form when hydrothermal solutions deposit minerals in open fractures, and many such veins are found in regions of volcanic activity (Fig. 18.5). The famous

A CLOSER LOOK

Modern Hydrothermal Mineral Deposits

Three extraordinary discoveries over a 15-year period changed our thinking about hydrothermal mineral deposits. The first discovery, in 1962, was accidental. Until that year no one was sure where to look for modern hydrothermal solutions or even how to recognize one when it was found. Drillers seeking oil and gas in the Imperial Valley of southern California were astonished when they struck a 320°C (608°F) brine at a depth of 1.5 km (0.9 mi). As the brine flowed upward, it cooled and precipitated minerals carried in solution. Over three months, the well deposited 8 tons of siliceous scale containing 20 percent copper and 8 percent silver by weight. The drillers had found a hydrothermal solution that could, under suitable flow conditions, form a rich mineral deposit.

The Imperial Valley is a sediment-filled graben covering the join between the Pacific and North American plates, where the plate spreading center called the East Pacific Rise passes under North America (Fig. C 18.1). Volcanism is the source of heat for the brine solution discovered in 1962. These brines provided the first unambiguous evidence that hydrothermal solutions can leach metals such as copper and silver from ordinary sediments.

The Imperial Valley brines also answered some important questions about the composition of hydrothermal solutions. The solubilities in water of ore minerals such as sphalerite (ZnS) and galena (PbS) are so incredibly low that geologists have long puzzled over how both metals and sulfur could be transported in the same solution. The answer is that in brines of appropriate composition, Cl^{1-} ions form anionic complexes such as $(ZnCl_4)^{2-}$ which shield the metal ions from S^{2-} ions in solution and thus effectively raise the solubility. Chloride complexing, as the process is called, had long been suspected, but it was the Imperial Valley brines that proved their existence.

Before geologists had a chance to fully absorb the significance of the Imperial Valley discovery, a second remarkable

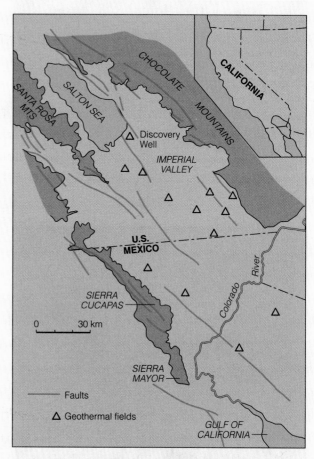

Figure C18.1 The Imperial Valley graben (also known as the Salton Trough). The graben is bounded by the Chocolate Mountains on the east and the Santa Rosa Mountains on the west. Hydrothermal solutions were discovered in a well drilled on the southern end of the Salton Sea. Places where geothermal activity is known, and where other hydrothermal solutions may be present at depth, are marked with triangles.

gold deposits at Cripple Creek, Colorado, were formed in fractures associated with a small caldera, and the huge tin and silver deposits in Bolivia are in fractures that are localized in and around volcanoes like those shown in Figure 4.23. In each case, the fractures formed as a result of volcanic activity, together with the magma chambers that fed the volcanoes, served as the sources of the hydrothermal solutions that rose up and formed the mineralized veins.

A cooling granitic body is a source of heat just as the magma chamber beneath a volcano is—and it can also be a source of hydrothermal solutions. Such solutions move outward from a cooling granite and will flow through any fracture or channel, altering the surrounding rock in the process and commonly depositing valuable minerals. Many famous ore bodies are as-

sociated with granites. The tin deposits of Cornwall, England, and the copper deposits at Butte, Montana; Bingham, Utah; and Bisbee, Arizona, are examples.

Magmatic Mineral Deposits

The processes of melting and crystallization discussed in Chapter 7 are two ways of separating some minerals from others. Fractional crystallization, in particular, can lead to the creation of valuable mineral deposits such as the chromite layers shown in Figure 7.16B. The processes involved are entirely magmatic, and such deposits are referred to as magmatic mineral deposits. Chromium, iron, platinum, nickel, vanadium, and titanium are the main resources concentrated by fractional crystallization.

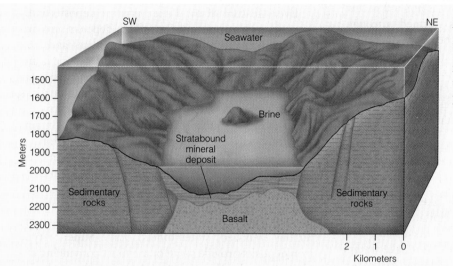

Figure C18.2 Topography of the Red Sea graben near the Atlantis II brine pool. Hot, dense brines rise up normal faults, pond on the floor of the graben, and form stratabound deposits rich in copper and zinc.

find was announced. In 1964, oceanographers discovered a series of hot, dense, brine pools at the bottom of the Red Sea. The brines were trapped in the graben formed by the spreading center between the Arabian and African plates (Fig. C 18.2). They were so much saltier, and therefore more dense, than seawater that they remained ponded in the graben even though they were as hot as 60°C (140°F). Many such brine pools have now been discovered.

The Red Sea brines rise up the normal faults associated with the central rift of a spreading center and, like the Imperial Valley brines, have evolved to their present compositions through reactions with the enclosing rocks. The Red Sea brine discovery was surprising, but even more surprising was the discovery that sediments at the bottom of the pools contained ore minerals such as chalcopyrite, galena, and sphalerite. In other words, the oceanographers had discovered modern stratabound mineral deposits in the process of formation.

The third remarkable discovery—really a series of discoveries—commenced in 1978. Scientists using deep-diving submarines made a series of dives on the East Pacific Rise at 21° N latitude. To their amazement, they found 300°C (572°F) hot springs emerging from the seafloor 2500 m (8100 ft) below sea level. Around the hot springs lay a blanket of sulfide minerals. The submariners watched a modern volcanogenic massive sulfide deposit forming before their eyes. From 1978 to the present, many more deposit-forming hot springs have been discovered along spreading centers in the ocean.

Each of the discovery sites—Imperial Valley, Red Sea, and 21° N—is on a spreading center, so there is no doubt that the deposits are forming as a result of plate tectonics. Soon the hunt was on to see if seafloor deposits could be found above subduction zones. In 1989, a joint German-Japanese oceanographic expedition to the western Pacific discovered the first modern subduction-related deposits, and two years later Canadian and Australian scientists discovered another modern subduction-related deposit in the Manus basin just north of Papua-New Guinea. No longer are geologists limited to speculating about how certain mineral deposits *might* have formed. Today they can be studied as they grow.

Pegmatites, which are coarse-grained igneous rocks formed by fractional crystallization of granitic magma, contain rich concentrations of elements such as lithium, beryllium, cesium, and niobium. Much of the world's lithium is mined from pegmatites such as those at King's Mountain, North Carolina, and Bikita in Zimbabwe. The great Tanco pegmatite in Manitoba, Canada, produces much of the world's cesium, and pegmatites in many countries yield beryl, one of the main ore minerals of beryllium.

Figure 18.5 A rich vein in Potosi, Bolivia, containing chalcopyrite, sphalerite, and galena cutting volcanic rocks. The rocks have been altered by the hydrothermal solution that deposited the ore minerals.

Sedimentary Mineral Deposits

The term *sedimentary mineral deposit* is applied to any local concentration of minerals formed through processes of sedimentation. Any process of sedimentation can form localized concentrations of minerals, but it has become common practice to restrict use of the term *sedimentary* to those mineral deposits formed through precipitation of substances carried in solution—that is, to chemical sedimentary deposits rather than clastic sedimentary deposits.

Evaporite Deposits

The most direct way in which sedimentary mineral deposits form is by evaporation of lake water or seawater. The layers of salts that precipitate as a consequence of evaporation are called *evaporite deposits*.

Examples of salts that precipitate from lake waters of suitable composition are sodium carbonate (Na_2CO_3), sodium sulfate (Na_2SO_4), and borax ($Na_2B_4O_7 \cdot 10H_2O$). Huge lake-water evaporite deposits of sodium carbonate were laid down in the Green River basin of Wyoming during the Eocene Epoch. Borax and other boron-containing minerals are mined from evaporite lake deposits in Death Valley and in Searles and Borax lakes, all in California, and in Argentina, Bolivia, Turkey, and China.

Much more common and important than lake-water evaporites are those formed by evaporation of seawater. The most important salts that precipitate from seawater are gypsum ($CaSO_4 \cdot 2H_2O$), halite (NaCl), and carnallite ($KMgCl_3 \cdot 6H_2O$). Low-grade metamorphism of marine evaporite deposits causes another important mineral, sylvite (KCl), to form from carnallite. Marine evaporite deposits are widespread; in North America, for example, strata of marine evaporites underlie as much as 30 percent of the entire land area (Fig. 18.6). Most of the salt that we use, as well as the gypsum used for plaster and the potassium used in plant fertilizers, is recovered from marine evaporites.

Iron Deposits

Sedimentary deposits of iron minerals are widespread, but the amount of iron in average seawater is so small that such deposits cannot have formed from seawater that is the same as today's seawater.

All sedimentary iron deposits are tiny by comparison with the class of deposits characterized by the *Lake Superior-type iron deposits*. These remarkable deposits, mined principally in Michigan and Minnesota, were long the mainstay of the United States steel industry but are declining in importance today as imported ores replace them. The deposits are of early Proterozoic age (about 2 billion years or older) and are found in sedimentary basins on every craton, par-ticularly in Labrador, Venezuela, Brazil, Russia, India, South Africa, and Australia.

Every feature of the Lake Superior deposits indicates chemical precipitation. The deposits are interbedded layers of chert (a form of silica) and iron minerals. Because the deposits are so large, it is inferred that the iron and silica must have been precipitated from seawater, but many aspects of the deposits remain conjectural. One puzzling aspect concerns the seawater origin of the deposits. Today's ocean contains tiny traces of dissolved iron, far too little to form a great deposit. This is so because today's oxygen-rich atmosphere keeps iron in seawater in the essentially insoluble ferric (oxidized) state. Two billion years ago there was apparently very little oxygen in the atmosphere, so iron could be dissolved in the more soluble ferrous (reduced) state. Indeed, Lake Superior-type iron deposits provide some of the most convincing evidence that the Earth's atmosphere has changed with time. Another puzzling aspect of the deposits concerns the cause of precipitation. Many experts suspect that Lake Superior-type deposits may be ancient marine evaporites from seawater that was not as salty as today's seawater. Other experts favor a hypothesis that tiny, floating organisms produced oxygen by photosynthesis and thereby caused precipitation.

Lake Superior-type iron deposits are not ores. The grades of the deposits range from 15 to 30 percent Fe by weight, and the deposits are so fine-grained that the iron minerals cannot be easily separated from the gangue. Two additional processes can form ore. First, leaching of silica during weathering can lead to a local enrichment of iron and produce ores containing as much as 66 percent Fe. Compare Figure 18.7A, which is a Lake Superior-type iron deposit in the Hamersley Range, Western Australia, with Figure 18.7B, a sample of ore developed by secondary enrichment in the Hamersley Range. The rocks in Figure 18.7A contain about 25 percent Fe, whereas those in Figure 18.7B have had most of the silica leached out by weathering and contain about 60 percent Fe.

The second way a Lake Superior-type iron deposit can become an ore is through metamorphism. Two changes occur as a result of metamorphism. First, grain sizes increase so that separating ore minerals from the gangue becomes easier and cheaper. Second, new mineral assemblages form, and the iron silicate and iron carbonate minerals that were originally present are replaced by magnetite or hematite, both of which are desirable ore minerals. The grade is not increased by metamorphism. It is the increase in grain size and the change in mineralogy that turn the sedimentary rock into an ore. Iron ores formed as a result of metamorphism are called *taconites*, and they are now the main kind of ore mined in the Lake Superior region.

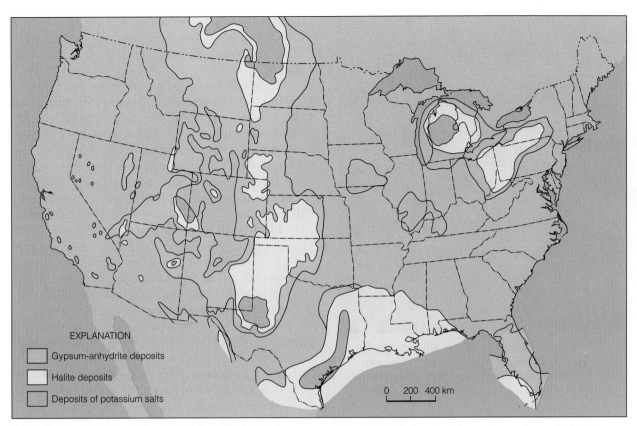

Figure 18.6 Portions of the United States known to be underlain by marine evaporite deposits. The areas underlain by gypsum and anhydrite do not contain halite. The areas underlain by potassium salts are also underlain by halite and by gypsum and anhydrite.

EXPLANATION

☐ Gypsum-anhydrite deposits

☐ Halite deposits

☐ Deposits of potassium salts

0   200   400 km

Stratabound Deposits

Some of the world's most important ores of lead, zinc, and copper occur in sedimentary rocks. The ore minerals—galena, sphalerite, chalcopyrite and pyrite—occur in such regular, fine layers that they look like sediments (Fig. 18.8). The sulfide mineral layers are enclosed by and parallel to the sedimentary strata in which they occur, and for this reason such deposits are called *stratabound mineral deposits*. They look like sediments but are not sediments in the truest sense of the term.

A.

B.

Figure 18.7 Sedimentary iron deposit of the Lake Superior type. A. Unaltered iron-rich sediments of the Brockman Iron Formation in Hamersley Range of Western Australia. The white layers are largely chert, while the darker bluish and reddish layers consist mainly of iron-rich silicate, oxide, and carbonate minerals. The grade is about 25% iron. B. Altered iron-rich sediment from the same formation shown in A. Leaching of silica during weathering has formed a secondarily enriched mass of iron minerals that is rich enough to be an ore. The grade is about 60% iron.

Figure 18.8 Stratabound ore of lead and zinc from Kimberley, British Columbia. The layers of pyrite (yellow), sphalerite (brown), and galena (grey) are parallel to the layering of the sedimentary rock in which they occur. The specimen is 4 cm across.

Stratabound deposits are formed when a hydrothermal solution invades and reacts with a muddy sediment. Chemical reactions between sediment grains and the solution cause deposition of the ore minerals. Deposition commonly occurs before the sediment has become a sedimentary rock.

The famous copper deposits of Zambia, in central Africa, are stratabound ores, as are the great Kupferschiefer deposits of Germany and Poland. Three of the world's largest and richest lead and zinc deposits ever discovered, at Broken Hill and Mount Isa in Australia, and at Kimberley in British Columbia, are also stratabound ores.

Placers

The way minerals and rock particles become sorted by flowing water was mentioned in Chapter 8. Differences in density are especially effective ways to effect sorting—more dense particles remain, while less dense minerals are washed away. Deposits of high-density minerals are called *placers*. The most important minerals concentrated in placers are gold, platinum, cassiterite (SnO_2), rutile (TiO_2), zircon ($ZrSiO_4$), and diamond. Typical locations of placers are illustrated in Figure 18.9.

Gold is the most important mineral recovered from placers; more than half of the gold recovered through-

out all of human history has come from placers. This is the result of the huge gold production from South Africa and Russia, almost all of which has come from placers.

The South African gold deposits are fossil placers, and they have many unusual features. Most placers are found in stream gravels that are geologically young. The South African fossil placers are a series of gold-bearing conglomerates (Fig.18.10) that were laid down about 2.7 billion years ago as gravels in the shallow marginal waters of a marine basin. Associated with the gold are grains of pyrite and uranium minerals. As far as size and richness are concerned, nothing like the deposits in the Witwatersrand basin has been discovered anywhere else. Nor has the original source of all the placer gold been discovered. It is therefore not possible to say, with any degree of certainty, why so much of the world's mineable gold should be concentrated in this one sedimentary basin.

Mining in the Witwatersrand basin has reached a depth of 3600 m (11,800 ft). This is the deepest mining in the world, and even though mining may continue to even greater depths, the heyday of gold mining in South Africa has probably passed because the deposits are running out of ore.

Through the middle years of the 1980s, the price of gold fluctuated at a high level, between about $14 and $16 a gram. This led to a boom in gold prospecting and to the discovery of a large number of new ore deposits in the United States, Canada, Australia, the Pacific islands, and elsewhere. Most of the new discoveries were hydrothermal deposits and world gold production rose continually for about ten years. Not surprisingly, by the middle of 1998 the price of gold had dropped to $9 a gram. Despite price drops and all the new discoveries, South Africa with its huge fossil placers continues to dominate the world's gold production. In 1997 South Africa supplied almost 30 percent of all the gold produced in the world, but the production rate is dropping steadily.

Residual Mineral Deposits

As we discussed in Chapter 6, all rocks are slowly altered or dissolved by naturally acid rainwater. Such chemical weathering processes remove the most soluble materials first and leave the less soluble minerals as a residual deposit. The two most important ore minerals found in residual deposits are limonite ($FeO \cdot OH$) and gibbsite ($Al(OH)_3$).

Limonite is among the least soluble of the minerals concentrated by chemical weathering. Under conditions of high rainfall in a warm, tropical climate, other minerals are slowly leached out of a soil, leaving an iron-rich limonitic residue called a lateritic soil or, if limonite is the only mineral present, *laterite*

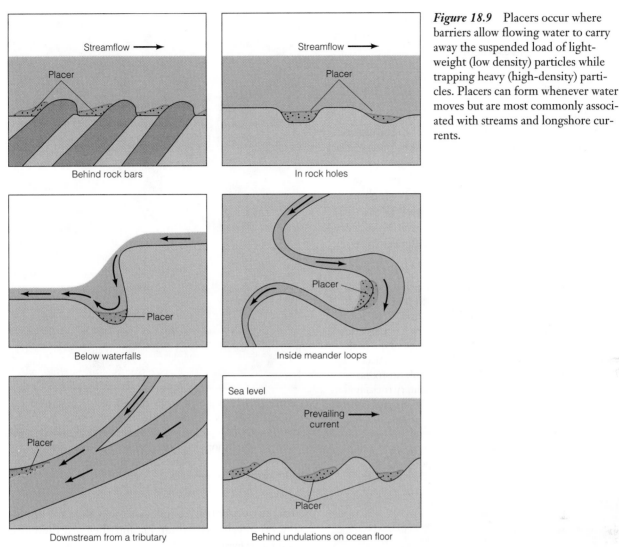

Behind rock bars

In rock holes

Below waterfalls

Inside meander loops

Downstream from a tributary

Behind undulations on ocean floor

Figure 18.9 Placers occur where barriers allow flowing water to carry away the suspended load of light-weight (low density) particles while trapping heavy (high-density) particles. Placers can form whenever water moves but are most commonly associated with streams and longshore currents.

(Fig. 18.11). In a few places, laterites can even be mined for iron.

Although iron-rich laterite is by far the most common kind of residual mineral deposit, the most important economic deposits as far as human exploitation is concerned are gibbsite-rich laterites called *bauxites*. Bauxites are the source of the world's aluminum.

Bauxites are widespread, but not nearly so abundant as iron-rich laterites, and they are concentrated in the tropics because that is where lateritic weathering occurs. Where bauxites are found in present-day temperate conditions, such as France, China, Hungary, and Arkansas, it is clear that the climate was tropical when the bauxites formed.

All bauxites and iron-rich laterites are vulnerable to erosion. They are not found in glaciated regions, for example, because overriding glaciers scrape off the soft surface materials. The vulnerability of bauxites and laterites means that most deposits are geologically young. More than 90 percent of all known deposits of

Figure 18.10 Gold is recovered from fossil placers in the Witwatersrand Basin, South Africa. The gold is found at the base of conglomerate layers interbedded with finer-grained sandstone, here seen in weathered outcrop at the site where gold was first discovered in 1886.

bauxites, for example, formed during the last 60 million years, and all of the very large deposits formed less than 25 million years ago.

Metallogenic Provinces

Many kinds of mineral deposits tend to occur in groups and to form what geologists call **metallogenic provinces**. These are defined as limited regions of the crust within which mineral deposits occur in unusually large numbers. A striking example is the metallogenic province shown in Figure 18.12 which runs along the western side of the Americas. Within the province is the world's greatest concentration of large hydrothermal copper deposits. These deposits are as-

Figure 18.11 Residual mineral deposits rich in iron and aluminum are typically formed under tropical or semitropical conditions. A. Red laterite enriched in iron, near Djenné, Mali. Laterites can sometimes be rich enough to be residual iron ores. Such ores have been mined in the past, but no large mining activity of residual iron ore is occurring today. B. Bauxite from Weipa in Queensland, Australia. Long-continued leaching of clastic sedimentary rocks under tropical conditions has removed most of the original constituents, such as silica, calcium, and magnesium, leaving a rich bauxite consisting largely of the mineral gibbsite $(Al(OH)_3)$. Nodules of gibbsite form by repeated solution and redeposition. The Weipa bauxite deposits are among the largest and richest in the world.

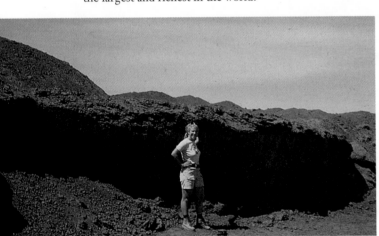

A.

B.

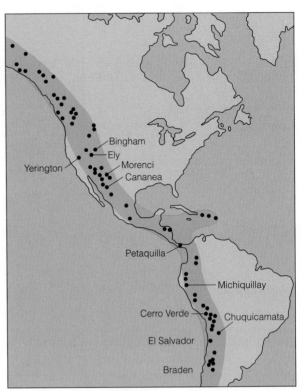

Figure 18.12 A metallogenic province of rich porphyry-copper deposits occurs along the western edge of the Americas. These chalcopyrite-rich deposits were formed by hydrothermal solutions associated with volcanoes; the volcanoes formed above the subduction edges of the South and North American plates.

sociated with intrusive igneous rocks that are invariably porphyritic (Chapter 7), and they are therefore called *porphyry copper deposits*. The intrusive igneous rocks, and therefore the deposits themselves, were formed as a consequence of subduction because they are in, or adjacent to, old stratovolcanoes.

Metallogenic provinces form as a result of either climatic control (as in the formation of bauxite deposits in the tropics) or plate tectonics. Magmatic, hydrothermal, and stratabound deposits all form near present or past plate boundaries (Fig. 18.13). This is hardly surprising as the deposits are related directly or indirectly to igneous activity, and most igneous activity we now know is related to plate tectonics.

ENERGY RESOURCES

A healthy person, working hard and continuously all day, can do just enough muscle work so that if the work were converted to electricity, the electricity would keep a single 75-watt light bulb alight. It costs about 10 cents to purchase the same amount of electrical energy from the local electrical utility. Viewed strictly as machines, humans aren't worth much. By

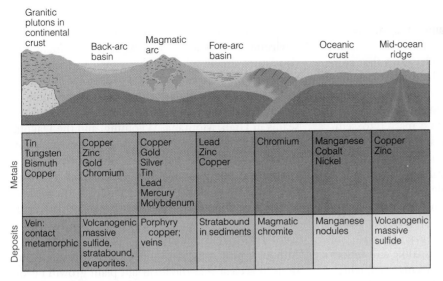

Figure 18.13 Locations of certain kinds of mineral deposits in terms of plate structures.

Metals	Tin Tungsten Bismuth Copper	Copper Zinc Gold Chromium	Copper Gold Silver Tin Lead Mercury Molybdenum	Lead Zinc Copper	Chromium	Manganese Cobalt Nickel	Copper Zinc
Deposits	Vein: contact metamorphic	Volcanogenic massive sulfide, stratabound, evaporites.	Porphyry copper; veins	Stratabound in sediments	Magmatic chromite	Manganese nodules	Volcanogenic massive sulfide

comparison, the amount of mechanical and electrical energy used each working day in North America could keep four hundred 75-watt bulbs burning for every person living there.

To see where all the energy is used, it is necessary to sum up all the energy employed to grow and transport food, make clothes, cut lumber for new homes, light streets, heat and cool office buildings, and do myriad other things. The uses can be grouped into three categories: transportation, home and commerce, and industry (meaning all manufacturing and raw material processing plus the growing of foodstuffs). The present-day uses of energy in the United States are summarized in Figure 18.14.

How much energy do all the people of the world use? The total is enormous. The energy drawn annually from the major fuels—coal, oil, and natural gas— as well as that from nuclear power plants, is 2.6×10^{20} J. Nobody keeps accurate accounts of all the wood and animal dung burned in the cooking fires of Africa and Asia, but the amount has been estimated to be so large that when it is added to the 2.6×10^{20} J figure, the world's total energy consumption rises to about 3.0×10^{20} J annually. This is equivalent to the burning of 2 metric tons of coal or 10 barrels of oil for every living man, woman, and child each year! Energy consumption around the world is very uneven, however. In less developed countries such as India and Tanzania, energy use is equivalent to burning only 2 or 3 barrels of oil per person per year, while in a developed country such as the United States and Canada, energy use is equivalent to burning more than 50 barrels of oil per person per year.

used for fuel. The kind of sedimentary rocks, the kind of organic matter trapped, and the changes in the organic matter as a result of burial determine the kind of fossil fuel that forms.

In the ocean, microscopic photosynthetic phytoplankton and bacteria are the principal sources of trapped organic matter. Fine-grained, clay-rich sediments that later become shales do most of the trapping. Once bacteria and phytoplankton are trapped in the sediment, the organic compounds they contain— *proteins*, *lipids*, and *carbohydrates*—become part of the

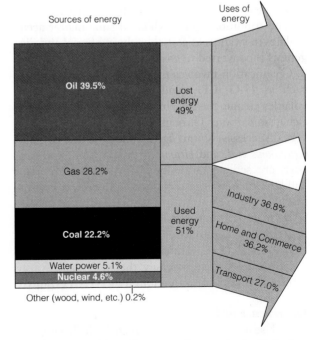

Figure 18.14 Uses and sources of energy in the United States. Lost energy arises both from inefficiencies of use and from the fact that the laws of thermodynamics impose a limit to the efficiency of any engine and therefore a limit on the fraction of available energy that can be usefully employed.

FOSSIL FUELS

The term **fossil fuel** refers to the remains of plants and animals trapped in sedimentary rocks that can be

sediment, and it is these compounds that are transformed (mainly by heat and burial) to oil and natural gas. The same burial and heating processes that convert organic compounds to oil and gas slowly change the enclosing sediment to a shale.

On land, it is trees, bushes, and grasses that contribute most of the trapped organic matter to sediments. These large land plants are rich in resins, waxes, and lignins, which tend to remain solid and form coals rather than oil or natural gas.

In many marine and lake shales, burial temperatures never reach the levels at which the original organic molecules are converted to the organic molecules found in oil and natural gas. Instead, an alteration process occurs in which waxlike substances with large molecules are formed. This material, which remains solid, is called *kerogen*, and it is the substance in so-called *oil shales*. Kerogen can be converted to oil and gas by mining the shale and heating it in a retort.

Coal

The black combustible sedimentary rock we call **coal** is the most abundant of the fossil fuels. Most of the coal mined is eventually burned either under boilers to make steam for electrical generators, or converted into coke, an essential ingredient in the smelting of iron ore and the making of steel. In addition to its use as a fuel, coal is a raw material for plastics such as nylon, as well as a multitude of other organic chemicals.

The conditions under which organic matter accumulates in swamps as *peat*, then during burial and diagenesis is converted to coal, was discussed in Chapter 8. Coalification involves the loss of volatile materials such as H_2O, CO_2, and CH_4 (methane). As the volatiles escape, the remaining coal is increasingly enriched in carbon. Through coalification, peat is converted successively into *lignite* (one type of coal), *subbituminous coal*, and *bituminous coal* (Fig. 8.9). These coals are sedimentary rocks. However, *anthracite*, a still later phase in the coalification process, is a metamorphic rock.

Because of its low volatile content, anthracite is difficult to ignite, but once alight it burns with almost no smoke. In contrast, lignite is rich in volatiles, burns smokily, and ignites so easily that it is dangerously subject to spontaneous ignition.

Occurrence of Coal

A coal seam is a flat, lens-shaped body having the same surface area as the swamp in which it originally accumulated. Most coal seams tend to occur in groups. In western Pennsylvania, for example, 60 seams of bituminous coal are found. This clustering indicates that the coal must have formed in a slowly subsiding site of sedimentation.

Coal swamps have formed in many sedimentary environments, of which two types predominate. One consists of slowly subsiding basins in continental interiors and the swampy margins of shallow inland seas formed at times of high sea level. This is the home environment of the bituminous and subbituminous coal seams in Utah, Montana, Wyoming, and the Dakotas. The second sedimentary environment consists of continental margins with wide continental shelves (that is, continental margins in plate interiors), that were flooded at times of high sea level. This is the environment of the bituminous coals of the Appalachian region.

Coal-forming Periods

Although peat can form under even subarctic conditions, it is clear that the luxuriant plant growth needed to form thick and extensive coal seams developed most readily under a tropical or semitropical climate. The Great Dismal Swamp in Virginia and North Carolina is one of the largest modern peat swamps. It contains an average thickness of 2 m (8 ft) of peat. However, unless the swamp lasts for millions of years, even that dense growth is insufficient to produce a coal seam as thick as some of the seams in Pennsylvania.

Peat formation has been widespread from the time land plants first appeared about 450 million years ago, during the Silurian Period. The size of peat swamps has varied greatly, however, and so, as a consequence, has the amount of coal formed. By far the greatest period of coal swamp formation occurred during the Carboniferous and Permian periods, when Pangaea existed. The great coal beds of Europe and the eastern United States formed at this time, when the plants of coal swamps were giant ferns and scale trees (gymnosperms). The second great period of coal deposition peaked during the Cretaceous Period, but commenced in the early Jurassic and continued until the mid-Tertiary. The plants of the coal swamps during this period were flowering plants (angiosperms), much like flowering plants today.

Petroleum: Oil and Natural Gas

Rock oil is one of the earliest resources our ancestors learned to use. However, the major use of oil really started in about 1847 when a merchant in Pittsburgh, Pennsylvania, started bottling and selling rock oil from natural seeps to be used as a lubricant. Five years later, in 1852, a Canadian chemist discovered that heating and distillation of rock oil yielded kerosene, a liquid that could be used in lamps. This discovery spelled doom for whale-oil lamps which had been widely used up to that time. Wells were soon being dug by hand near Oil Springs, Ontario, in order to produce oil. In Romania, by 1856, using the same hand-digging process, workers were producing 2000 barrels a year.[1] In 1859, the first successful oil well was drilled in Titusville, Pennsylvania. On August 27, 1859, at a depth of 21.2 m (70 ft), oil-bearing strata were encountered and up to 35 barrels of oil a day were pumped out. Oil was soon discovered in West Virginia (1860), Colorado (1862), Texas (1866), California (1875), and many other places.

The earliest known use of natural gas was about 3000 years ago in China, where gas seeping out of the ground was collected and transmitted through bamboo pipes to be used to evaporate saltwater in order to recover salt. It wasn't long before the Chinese were drilling wells to increase the flow of gas. Modern uses of gas started in the early seventeenth century in Europe, where gas made from wood and coal was used for illumination. Commercial gas companies were founded as early as 1812 in London and 1816 in Baltimore. The stage was set for the exploitation of an accidental discovery at Fredonia, New York, in 1821. A water well drilled in that year produced not only water, but also bubbles of a flammable gas. The gas was accidentally ignited and produced such a spectacular flame that a new well was drilled on the same site and wooden pipes were installed to carry the gas to a nearby hotel, where 66 gas lights were installed. By 1872, natural gas was being piped as far as 40 km (25 mi) from its source.

Origin of Petroleum

Petroleum is defined as gaseous, liquid, and semisolid naturally occurring substances that consist chiefly of *hydrocarbons* (chemical compounds of carbon and hydrogen). Petroleum is therefore a term that includes both oil and natural gas.

Petroleum is nearly always found in marine sedimentary rocks. In the ocean, microscopic phytoplankton (tiny floating plants) and bacteria (simple, single-celled organisms) are the principal sources of organic matter that is trapped and buried in sediment. Most of the organic matter is buried in clay that is slowly converted to shale. During this conversion, organic compounds are transformed to oil and natural gas.

Sampling on the continental shelves and along the base of the continental slopes has shown that fine muds beneath the seafloor contain up to 8 percent organic matter. From such observations geologists conclude that oil and gas originate primarily as organic matter trapped in sediment. Two additional kinds of evidence support the hypothesis that petroleum is a product of the decomposition of organic matter:

1. Oil possesses optical properties known only in hydrocarbons derived from organic matter.

2. Oil contains nitrogen and certain compounds believed to originate only in living matter.

A complex sequence of chemical reactions is involved in converting the original solid organic matter to oil and gas, and additional chemical changes may occur in the oil and gas even after they have formed. This explains why subtle chemical differences exist between the oil in one body of petroleum and another.

Once petroleum has formed in a shale, it is free to move. It is now well established that petroleum migrates through aquifers and can become trapped in reservoirs.

The migration of petroleum deserves further discussion. The sediment in which organic matter is accumulating today is rich in clay minerals, whereas most of the strata that constitute oil pools are sandstones (consisting of quartz grains), limestones and dolostones (consisting of carbonate minerals), and much-fractured rock of other kinds. Long ago, geologists realized that oil and gas form in one kind of material (shale) and at some later time migrate to another (sandstone or limestone).

Petroleum migration is analogous to groundwater migration. When oil and gas are squeezed out of the shale in which they originated and enter a body of sandstone or limestone somewhere above, they migrate readily because sandstones and limestones are much more permeable than any shale. The force of molecular attraction between oil and quartz or carbonate minerals is weaker than that between water and quartz or carbonate minerals. Hence, because oil and water do not mix, water remains fastened to the quartz or carbonate grains, while oil occupies the central parts of the larger openings in the porous sandstone or limestone. Because oil is lighter than water, the oil tends to glide upward past the carbonate- and quartz-held water. In this way, oil becomes segregated from the water; when it encounters a trap, it can form a pool (Fig. 18.15).

Most of the petroleum that forms in sediments does not find a suitable trap and eventually makes its way, along with groundwater, to the surface or the sea. It is estimated that no more than 0.1 percent of all the organic matter originally buried in a sediment is eventually trapped in an oil pool. It is not surprising, therefore, that the highest ratio of oil and gas pools to volume of sediment is found in rock no older than 2.5 million years—young enough so that little of the petroleum has leaked away—and that nearly 60 percent of all the oil and gas discovered so far has been found in strata of Cenozoic age (Fig. 18.16). This does not mean that older rocks produced less petroleum. It simply means that oil in older rocks has had a longer time in which to leak away.

[1]A barrel is equal to 42 U.S. gallons and is the volume generally used when commercial production of oil is discussed.

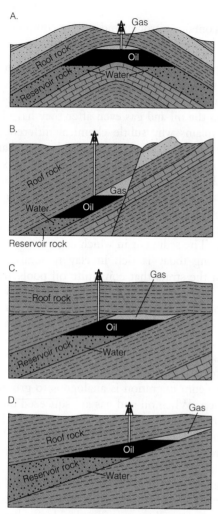

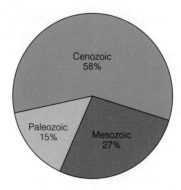

Figure 18.16 Percentage of world's total oil production from strata of different ages.

Figure 18.15 Four kinds of oil traps. A and B are structural traps formed as a result of folding A and fracturing B respectively. C and D are stratigraphic traps; in C an unconformity forms the seal of the trap; in D a porous stratum thins out and is overlain by an impermeable roof rock. Gas overlies oil, which floats on groundwater. Oil, gas and groundwater fill only the pore spaces in the reservoir rock.

Distribution of Petroleum

Petroleum deposits, like coal, are frequent but are distributed unevenly. The reasons for the uneven distribution are not as obvious as they are with coal. Suitable source sediments for petroleum are widespread and are as likely to form in subarctic waters as in tropical regions. The critical controls seem to be a supply of heat to effect the conversion of solid organic matter trapped in the sediment to oil and gas, and the formation of a suitable trap before the petroleum has leaked away.

Conversion of solid organic matter to oil and gas happens within a specific range of depth and temperature defined by the geothermal gradients shown in Figure 18.17. If a thermal gradient is too low—less than 1.8°C/100 m, (1°F/100 ft)—conversion does not occur to either oil or gas. If the gradient is above

5.5°C/100 m, (3°F/100 ft), conversion to gas starts at such shallow depths that leakage rates are high and little trapping occurs. The depth-temperature window within which oil and gas form and are trapped lies between the two thermal gradients. Once oil and gas have been formed, they will accumulate in pools only if suitable traps are present. Most oil and gas pools are found beneath anticlines; the timing of the folding event that forms an anticline is therefore a critical part of the trapping process. If folding occurs after petroleum has formed and migrated, pools cannot form. The great oil pools in the Middle East arose through the fortunate coincidence of the right thermal gradient and the development of anticlinal traps during the collision of Europe and Asia with Africa.

How much oil is there in the world? This is an extremely controversial question. Approximately 800 billion barrels of oil have already been pumped out of the ground. A lot of additional oil has been located by drilling but is still waiting to be pumped out. Possibly a great deal more oil remains to be found by drilling. Unlike coal, for which the volume of strata in a basin of sediment can be accurately estimated, the volume of undiscovered oil can only be guessed at. Guesses involve the use of accumulated experience from a century of drilling. Knowing how much oil has been found in an intensively drilled area, such as eastern Texas, experts make estimates of probable volumes in other regions where rock types and structures are similar to those in eastern Texas. Using this approach, and considering all the sedimentary basins of the world (Fig. 18.18), experts estimate that somewhere between 1500 and 3000 billion barrels of oil will eventually be discovered.

Tars

Oil that is exceedingly viscous and thick will not flow easily and cannot be pumped. Colloquially called **tar**, heavy, viscous oil acts as a cementing agent between mineral grains in an oil pool. The tar can be recovered only if the sandstone is mined and heated enough to

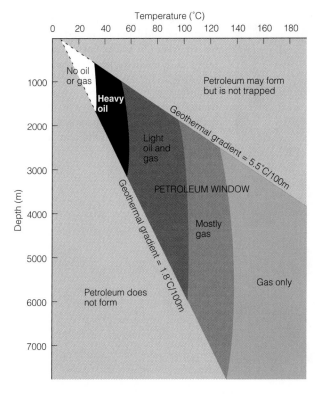

Figure 18.17 The petroleum window is that combination of depth and temperature within which oil and gas are generated and trapped.

Figure 18.18 Areas underlain by sedimentary rock and regions where large accumulations of oil and gas have been located. Where the ocean is deeper than 2000 m, sedimentary rock has yet to be tested for its oil and gas potential.

make the tar flow. The resulting tar must then be processed to recover the valuable gasoline fraction. Mining and treating "tar sands," as heavy, viscous oil deposits are called, carries a high cost, but it is technically possible. Tar sands are already an important source of fuel in Canada. The largest known occurrence is in Alberta, Canada, where the *Athabasca Tar Sand* covers an area of 5000 km² (1900 mi²) and reaches a thickness of 60 m (200 ft). Similar deposits almost as large are known in Venezuela and in Russia.

Oil Shale

Another potential source of petroleum is kerogen in shale. Kerogen is solid organic matter that has not been heated high enough to form petroleum, but if mined and heated, kerogen breaks down and forms liquid and gaseous hydrocarbons similar to those in oil and gas. All shales contain some kerogen, but to be considered an energy resource the kerogen must yield more energy than is required to mine and heat it. Only those shales that yield 40 or more liters (0.25 barrel) of distillate per ton can be considered because the energy needed to mine and process a ton of shale is equivalent to that created by burning 40 liters (0.25 barrel) of oil.

The world's largest deposit of rich oil shale is in the United States. During the Eocene Epoch, many large, shallow lakes existed in basins in Colorado, Wyoming, and Utah; in three of them a series of rich organic sediments was deposited that are now the Green River Oil Shales. The richest shales were deposited in the lake in Colorado, now called the Piceance basin. These shales are capable of producing as much as 240 liters (1.5 barrels) of oil per ton. Scientists of the U.S.

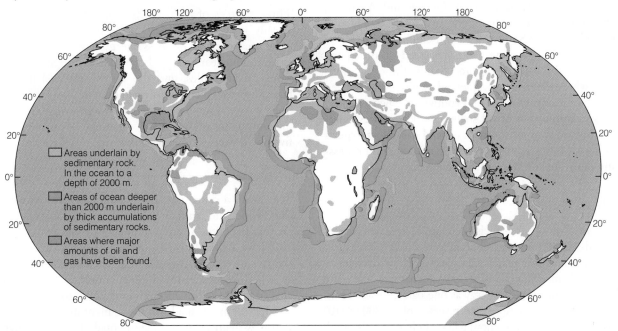

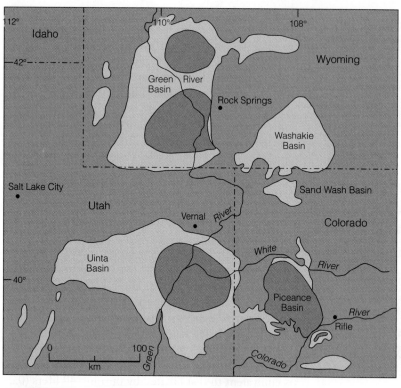

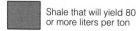

Figure 18.19 Large areas of Colorado, Wyoming, and Utah are underlain by the Green River Oil Shale. The extensive deposits of oil shale formed as organic-rich sediment that accumulated in ancient freshwater lakes and was buried, compacted, and cemented. If heated, the solid organic matter in the shale is converted to hydrocarbons similar to those in petroleum.

Geological Survey estimate that, in the Green River Oil Shale alone, oil-shale resources capable of producing 50 liters (0.3 barrels) or more of oil per ton of shale can ultimately yield about 2000 billion barrels of oil (Fig. 18.19).

Rich deposits of oil shale in other parts of the world have not been adequately explored, but there is a huge deposit in Brazil called the Irati Shale. Another very large deposit is known in Queensland, Australia, and others have been reported in such widely dispersed places as South Africa and China. Although oil shales have been mined and processed in an experimental fashion in the United States, the only countries where extensive commercial production has been tried are Russia and China. Production expenses today make exploitation of oil shales in all countries unattractive by comparison with oil and gas. Most experts believe, however, that large-scale mining and processing of oil shale will eventually happen.

How Much Fossil Fuel?

Are supplies of fossil fuels adequate to meet future demands? If we use a barrel of oil as our unit of measurement, we can compare quantities of all fossil fuels directly. Approximately 0.22 ton of coal produces the same amount of heat energy as one barrel of oil. Thus, the world's recoverable coal reserves of 13,800 × 10⁹ tons are equivalent to about 63,000 billion barrels of oil.

Considering the approximate world-use rate of barrels of oil (40 billion barrels a year), and comparing the estimated recoverable amounts of fossil fuels (Table 18.1), it is apparent that only coal seems to have the capacity to meet our long-term demands.

OTHER SOURCES OF ENERGY

Three sources of energy other than fossil fuels have already been developed to some extent: the Earth's plant life (so-called biomass energy), hydroelectric energy, and nuclear energy. Five others—the Sun's heat, winds, waves, tides, and the Earth's internal heat— have been tested and developed on a limited basis, but none has yet been developed on a large scale. The day may not be far off, however, when one or more of the five could become locally important.

Biomass Energy

Scientists working for the United Nations estimate that wood and animal dung used for cooking and heating fires now amount to energy production of 4×10^{19} J annually. This is approximately 14 percent of the world's total energy use. The greatest use of wood as a fuel occurs in developing countries, where the cost of fossil fuel is very high in relation to income.

Measurements made on living plant matter indicate that new plant growth on land equals 1.5×10^{11} met-

Table 18.1 Amounts of Fossil Fuels Possibly Recoverable Worldwide
(Unit of Comparison is a Barrel of Oil)

Fossil Fuel	Total Amount in Ground (billions of barrels)	Amount Possibly Recoverable (billions of barrels)
Coal	About 100,000	62,730[a]
Oil and gas (flowing)	1500–3000	1500–3000
Trapped oil in pumped-out pools	1500–3000	0–?
Viscous oil (tar sands)	3000–6000	500–?
Oil shale	Total unknown: much greater than coal	1000?

[a]0.22 tons of coal = 1 barrel of oil

ric tons of dry plant matter each year. If all of this were burned, or used in some other way as a biomass energy source, it would produce almost nine times more energy than the world uses each year. Obviously, this is a ridiculous suggestion because in order to do so all the forests would have to be destroyed, plants could not be eaten, and agricultural soils would be devastated. Nevertheless, controlled harvesting of fuel plants could probably increase the fraction of the biomass now used for fuel without serious disruption to forests or to food supplies. In several parts of the world, such as Brazil, China, and the United States, experiments are already under way to develop this obvious energy source.

Hydroelectric Power

Hydroelectric power is recovered from the potential energy of stream water as it flows to the sea. In order to convert the power of flowing water into electricity efficiently, it is necessary to dam streams. Unfortunately, as mentioned in Chapter 9, reservoirs behind dams fill with silt, and so even though water power is continuous, dams and reservoirs have limited lifetimes.

Water power has been used in small ways for thousands of years, but only in the twentieth century has it been used to any significant extent for generating electricity. All the water flowing in the streams of the world has a total recoverable energy estimated as 9.2×10^{19} J/yr, an amount equivalent to burning 15 billion barrels of oil per year. Thus, even if all the possible hydropower in the world were developed, we could satisfy only about one-third of the present world energy needs. We have to conclude that, for those fortunate countries with large rivers and suitable dam sites, hydropower is very important, but for most countries hydropower holds limited potential for development.

Nuclear Energy

Nuclear energy is the heat energy produced during controlled transformation of suitable radioactive iso-

topes (a process called **fission**). Three of the radioactive atoms that keep the Earth hot by spontaneous radioactive decay—^{238}U, ^{235}U, and ^{232}Th—can be mined and used in this way. Fission is accomplished by bombarding the radioactive atoms with neutrons, thus accelerating the rate of decay and the release of heat energy. The device in which this operation is carried out is called a **pile**.

When ^{235}U fissions, it not only releases heat and forms new elements but also ejects some neutrons from its nucleus. These neutrons can then be used to induce more ^{235}U atoms to fission, and a continuous chain reaction occurs. The function of a pile is to control the flux of neutrons so that the rate of fission can be controlled. When a chain reaction proceeds without control, an atomic explosion occurs. Controlled fission, therefore, is the method used by nuclear power plants, and a tremendous amount of energy can be obtained in the process. The fissioning of one gram of ^{235}U produces as much heat as the burning of 13.7 barrels of oil. Unfortunately, however, ^{235}U is the only naturally occurring radioactive isotope that will maintain a continuous chain reaction, and it is the least abundant of the three radioactive isotopes that are mined for nuclear energy. Only one atom of each 138.8 atoms of uranium in nature is ^{235}U. The remaining atoms are ^{238}U, which will not sustain a chain reaction. However, if ^{238}U is placed in a pile with ^{235}U that is undergoing a chain reaction, some of the neutrons will bombard the ^{238}U and convert it to plutonium–239 (^{239}Pu). This new isotope can, under suitable conditions, sustain a chain reaction of its own. The pile in which the conversion of ^{238}U takes place is called a **breeder reactor**. The same kind of device can be used to convert ^{232}Th into ^{233}U, which also will sustain a chain reaction. Unfortunately, breeder reactors and nuclear power plants based on them are more complex and less safe than ^{235}U plants; as a result, all the present nuclear power plants use ^{235}U.

There are many nuclear power plants operating around the world. They utilize the heat energy from fission to produce steam that drives turbines and gen-

GUEST ESSAY

The Pursuit of Undiscovered Fuel

As fossil fuel exploration opportunities diminish, scientists must put more emphasis on developing existing oil and gas resources. Such development, in turn, relies on improving techniques to locate these resources. While technical advancement has progressed using traditional geologic approaches, computer-aided analysis and modeling also have become increasingly important.

Several advances based on traditional geological studies help in the search for oil and gas. Plate tectonic reconstructions of regions are one of the advances: they are used to create a framework of paleogeography and paleoclimatology at the time sediments were being deposited in basins. This reconstruction tells earth scientists whether it is likely that petroleum has formed in a particular area. In addition, advances in organic geochemistry can now be used to evaluate the quality of source rock, its oil or gas yield, and its state of maturation. Some of the most important advances have come in our understanding of how sediments are deposited and how properties such as grain size and shape, cement, and porosity vary from place to place in a stratum. Using such information, reservoir rock properties can be evaluated for oil potential without relying entirely on drilling.

One of the areas where technological breakthroughs have occurred is in geophysical research. Geophysical research encompasses three different areas: petrophysics, seismic imaging, and borehole geophysics. All of these areas concentrate on directly recognizing the presence of gas and oil in the subsurface.

Petrophysics seeks to improve scientists' ability to evaluate sediments that could hold oil. This estimation involves analysis of sediment geometry, continuity, lithology, and fluid content through the combined use of surface seismic data and well logs. These logs involve lowering electronic instruments into a well borehole to determine resistivity, density, and porosity of sediments. Such electrical tools allow earth scientists to distinguish different sediments and types of fluids around the well bore. Advances in well log technologies emphasize new interpretation methods for rocks, such as shaly sands. In addition, they concentrate on methods which allow determination of type of fluid in reservoirs after a well has been drilled and has produced oil for some time. The well bore is also used for gathering seismic data closer to sediments of interest to locate structural and stratigraphic complexities.

Advances in seismic imaging provide explorationists with new tools to image the earth's subsurface. The goal of seismic imaging techniques is to create an accurate computer-

Pinar Oya Yilmaz holds a Ph.D. in structural geology from the University of Texas at Austin. At present she is External Technical Coordinator at Exxon Exploration Company in Houston, Texas. She has served on a number of professional committees, and currently is chairman of the International Liaison Committee of the AAPG (American Association of Petroleum Geologists).

generated "picture" of the subsurface in three dimensions to enable earth scientists to make the right decision on where to drill for oil. A scientist determines the area of interest and plans horizons which will be targeted for three dimensional (3-D) seismic imaging surveys. After the data is acquired and processed, the interpretation phase begins. High-powered computer workstations running GIS software (the same type of software used for remote sensing analysis discussed in the Chapter 1 Guest Essay) are used for complete geological and geophysical integration, including data from the surface and data provided from subsurface wells. Since seismic data lie in the time domain but geology must be interpreted in depth, well control is extremely important in understanding seismic data results. The interpreter has to be certain that proper depth conversions are used. Many dry holes have been drilled over seismic "highs" that were not checked for absolute depths.

Borehole geophysics integrates petrophysics and seismic analysis to better image the subsurface in structurally or stratigraphically complex areas, where there are abrupt vertical or lateral changes in sedimentary units. Borehole geophysics technologies place seismic tools downhole, at depths closer to the sediments of interest, in order to enhance resolution and to make it easier to image vertical features. Subsurface features are imaged around the well bore in the same manner as other seismic surveys. The images collected can be used to identify faults, locate salt-sediment interfaces, and characterize reservoirs according to rock type, porosity, and fluid content.

As the pursuit of undiscovered fuel intensifies, earth scientists must search for innovative and advanced methods to locate oil and gas. New tools and techniques enhance data-interpreters' predictive ability when searching for undiscovered oil. All of these advanced tools and techniques help earth scientists decide which parts of sedimentary basins housing oil fields show the greatest promise for new discoveries and the highest return on investment.

erates electricity. Approximately 8 percent of the world's electrical power is derived from nuclear power plants. In France, more than half of all the electrical power comes from nuclear plants; the fraction is quite high in some other European countries too, and is high and rising in Japan. The reason for the increase

is obvious. Japan and most European countries do not have adequate supplies of fossil fuels in order to be self-sufficient.

Many problems are associated with nuclear energy. The isotopes used in power plants are the same isotopes used in atomic weapons, so a security problem exists. The possibility of a power plant failing in some unexpected way creates a safety problem. The dreadful Chernobyl disaster in 1986 in the Ukraine is an example of such an event. Finally, the problem of safe burial of dangerous radioactive waste matter must be faced. Some of the waste matter will retain dangerous levels of radioactivity for thousands of years.

Geothermal Power

Geothermal power, as the Earth's internal heat flux is called, has been used for more than 50 years in New Zealand, Italy, and Iceland and more recently in other parts of the world, including the United States. The source of heat in most instances is a body of magma. The magma heats groundwater, and the steam that is formed can be used to drive turbines and make electricity. How this is done is illustrated in Figure 18.20.

Most of the world's geothermal stream reservoirs are close to plate margins because plate margins are where most recent volcanic activity has occurred. A depth of 3 km (1.9 mi) seems to be a rough lower limit for big geothermal steam and hot-water pools. It is estimated that the world's geothermal reservoirs can yield about 8×10^{19} J—equivalent to burning 13 billion barrels of oil. This estimate incorporates the observation that, in New Zealand and Italy, only about 1 percent of the energy in a geothermal reservoir is recoverable. If the recovery efficiency were to rise, the estimate of recoverable geothermal resources would also rise. But even if the efficiency increased to 50 percent, geothermal power, like hydropower, could only satisfy a small fraction of present human energy needs. For this reason, a good deal of attention is being given to creating artificial geothermal steam fields. So far experiments have been only partially successful.

Energy from Winds, Waves, Tides, and Sunlight

The most obvious source of energy is the Sun. The amount of energy reaching the Earth each year from the Sun is approximately 4×10^{24}J—that is, ten thousand times more than we humans use. We already put some of the Sun's energy to work in greenhouses and in solar homes, but the amount so used is tiny. The major challenge is to convert solar energy directly to electricity. Devices that effect such a conversion, called photovoltaic devices, have been invented. So far their costs are too high and their efficiencies too low for most uses, although they are already widely used in small calculators, radios, and other devices that use little power.

Winds and waves are both secondary expressions of solar energy. As discussed in Chapter 13, winds, in particular, have been used as an energy source for thousands of years through sails on ships and wind-

mills. Today, huge farms of windmills are being erected in suitably windy places. Although there are problems and high costs with windmills, by the year 2010 or sooner, windmills will likely be cost-competitive with coal-burning electrical power plants. Unfortunately, much of the wind energy is in high-altitude winds. Steady surface winds only have about 10 percent of the energy the human race now uses. As with hydro- and geothermal power, therefore, wind power may become locally significant but will probably not be globally important.

Waves, which arise from winds blowing over the ocean, contain an enormous amount of energy. We can see how powerful waves are along any coastline during a storm. Wave power has been used to ring

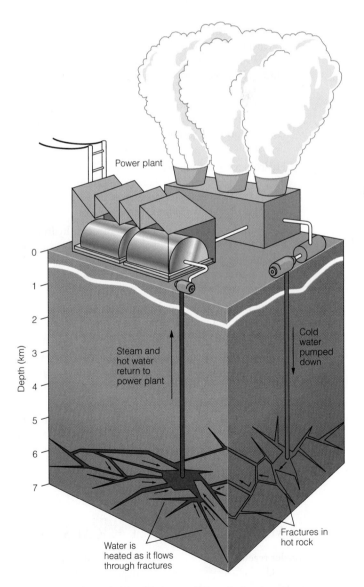

Figure 18.20 Geothermal energy. Water in fractures in hot rock forms steam which is brought to the surface and used to run a power plant. After use, waste water from condensed steam is pumped back underground.

bells and blow whistles as navigational aids for centuries, but so far no one has discovered how to tap wave energy on a large scale. Devices that have been designed to do this tend to fail because of corrosion or storm damage.

Tides arise from the gravitational forces exerted on the Earth by the Moon and the Sun. If a dam is put across the mouth of a bay so that water is trapped at high tide, the outward flowing water at low tide can drive a turbine. Unfortunately, the process is not efficient, and few places around the world have tides high enough to make tidal energy feasible.

It is clear that there are numerous sources of energy and that far more energy is available than we can use. What is not yet clear is when, or even whether, we will be clever enough to learn how to tap the different energy sources in nonhazardous ways that will not disrupt the environment.

SUMMARY

1. When a mineral deposit can be worked profitably, it is called an ore. The waste material mixed with ore minerals is gangue.

2. Mineral deposits form when minerals become concentrated in one of five ways: (1) precipitation from hydrothermal solutions to form hydrothermal mineral deposits; (2) concentration through crystallization to form magmatic mineral deposits; (3) concentration from lake water or seawater to form sedimentary mineral deposits; (4) concentration in flowing water to form placers; and (5) concentration through weathering to form residual deposits.

3. Hydrothermal solutions are brines, and they can either be given off by cooling magma or else form when either groundwater or seawater penetrates the crust, becomes heated, and reacts with the enclosing rocks.

4. Hydrothermal mineral deposits form when hydrothermal solutions deposit dissolved minerals because of cooling, boiling, pressure drop, mixing with cold groundwater or seawater, or through chemical reactions with enclosing rocks.

5. Chromite, the main ore mineral of chromium, is the most important mineral concentrated by fractional crystallization.

6. Sedimentary mineral deposits are varied. The largest and most important are evaporites. Marine evaporite deposits supply most of the world's gypsum, halite, and potassium minerals.

7. Gold, platinum, cassiterite, diamonds, and other minerals are commonly found mechanically concentrated in placers.

8. Bauxite, the main ore of aluminum, is the most important kind of residual mineral deposit. Bauxite forms as a result of tropical weathering.

9. The distribution of many kinds of mineral deposits is controlled by plate tectonics because most magmas and most sedimentary basins are where they are because of plate tectonics.

10. Nonmetallic substances are used mainly as chemicals, fertilizers, building materials, and ceramics and abrasives.

11. Coal originated as plant matter in ancient swamps, and is both abundant and widely distributed.

12. Oil and gas originated as solid organic matter trapped in shales and decomposed chemically due to heat and pressure following burial. Later, these fluids moved through reservoir rocks and were caught in geologic traps to form pools.

13. When heated, part of the solid organic matter found in shale—called kerogen—will convert to oil and gas. Kerogen in shales is the world's largest resource of fossil fuel. Unfortunately, most shale contains so little kerogen that more oil must be burned to heat the shale than is produced by the conversion process.

14. Nuclear energy is derived from atomic nuclei of radioactive isotopes, chiefly uranium. The nuclear energy available from naturally occurring radioactive elements is the single largest energy resource now available.

15. Other sources of energy currently used to some extent are geothermal heat, energy from flowing streams, winds, waves, tides, and the Sun's heat.

IMPORTANT TERMS TO REMEMBER

breeder reactor 437
coal 432
fission 437
fossil fuel 431
gangue 422

hydrothermal mineral
  deposit 422
magmatic mineral
  deposit 422
metallogenic province 430

mineral deposit 421
ore 421
petroleum 433
pile 437
placer 422

residual mineral
  deposit 422
sedimentary mineral
  deposit 422
tar 434

QUESTIONS FOR REVIEW

1. Describe the difference between renewable and nonrenewable resources. Name three things that you use daily that rely on renewable resources and three that rely on nonrenewable resources.

2. What are mineral deposits? Describe five ways by which a mineral deposit can form.

3. If there are any mineral deposits in the area where you live or study, what kind of deposits are they and how do such deposits form?

4. What factors determine whether or not a mineral deposit is ore?

5. How do hydrothermal solutions form, and how do they form mineral deposits?

6. Briefly describe the formation of three different kinds of sedimentary mineral deposit.

7. What factors control the concentration of minerals in placers? Name four minerals mined from placers.

8. How do residual mineral deposits form? What are the principal resources concentrated in residual deposits? If you were prospecting for such deposits, in what parts of the world would you concentrate your search?

9. What is a fossil fuel? Name four different kinds of fossil fuel.

10. Explain the steps that occur as organic matter becomes coal.

11. During what two periods in the Earth's history was most coal formed? Explain why coal formed when and where it did and why coal is now found where it is.

12. What kind of rocks serve as source rocks for petroleum? In what kinds of rocks does petroleum tend to be trapped? Why?

13. What is the source of the organic matter that forms petroleum? What observations lead geologists to conclude that organic matter is the source of petroleum?

14. Oil drillers find more petroleum per unit volume of rock in Cenozoic rocks than in Paleozoic rocks of the same kind. Explain.

15. Oil shales are rich in organic matter. Explain why such shales have not served as source rock for petroleum. How can oil shales be used as an energy resource?

16. Discuss the relative amounts of energy available from the different fossil fuels. What is your opinion about how fossil fuels will be used in the future?

17. What is nuclear energy? How is it used to make electricity and what possible dangers are there in developing nuclear energy?

18. What limitations are there to the development of hydroelectric power? of wave and wind power?

19. What is geothermal power and where is it found? Compare the magnitude of available geothermal power with the magnitude of petroleum resources.

20. Would it be possible to increase our use of biomass as an energy resource? What are the limitations to use of biomass energy?

QUESTIONS FOR DISCUSSION

1. Taking into consideration the world's population (6.0 billion in 1998) and the fact that it is growing by approximately 100 million a year, how do you think the world's energy needs will be met over the next century? If you decide that fossil fuel is to be the energy source of choice, what consequence might that have for the atmosphere?

2. Discuss the role recycling might play in making available metallic resources in the future. Can you imagine a scenario in which the mining of new metallic minerals might cease? Would your recycling arguments apply equally to metallic and nonmetallic minerals?

3. Considering how mineral deposits are known to form on the Earth, which planets or moons in the solar system would you recommend as places to prospect for mineral resources? Be sure to specify what kind of resources you would expect to be found.

To learn more about the geology, extraction, and environmental impacts of petroleum resources, please visit our web site.

CHAPTER 19

The Changing Face of the Land

● *Eroding Volcanoes in the Mid-Pacific Ocean*

We cannot directly observe the evolution of large-scale landscape features on the continents because of the immense time involved, but an ideal opportunity to visualize long-term landscape changes is provided by midocean volcanic island chains. Early geologists observed that the Hawaiian islands appear more dissected northwestward from the island of Hawaii, the site of active volcanism, and inferred that this change reflects the increasing age of the islands in that direction. K/Ar dating has now shown that the volcanoes indeed increase in age from less than half a million years on Hawaii to about 5 million years on the oldest major islands farther up the chain.

The island chain is believed to have evolved as the Pacific plate moved slowly northwestward across a midocean hot spot, above which frequent and voluminous eruptions built a succession of large volcanoes. Once formed, each volcanic island is carried slowly away from the hot spot in the direction of plate movement. Each of the emergent islands has approximately the same history. By examining the erosional landscapes on successively older volcanoes, one can obtain an understanding of how the oldest Hawaiian landscapes likely developed. During the initial constructional phase of a volcano, the rate of lava extrusion greatly exceeds the rate of erosion, and an island rises in the form of a broad shield volcano. As the volcanic pile accumulates, the localized added weight on the seafloor causes the underlying ocean crust to subside isostatically. This helps to limit the maximum altitude to which a volcanic island can rise (4-5 km) before the moving Pacific plate carries it beyond the magma source. High Hawaiian volcanoes intercept the moist tradewinds, resulting in abundant precipitation on windward slopes that enables streams to cut deep canyons, which indent the island flanks. Progressive dissection of the land, as major streams cut downward and backward, gradually eliminates the original constructional slopes of the volcanic shield. As extrusive activity slackens and finally ceases, continuing erosion ultimately reduces the dissected volcano to sea level. In the case of the Hawaiian chain, this entire evolutionary process takes about 10 million years. ●

(opposite) Vigorous erosion on the moist windward slope of the island of Hawaii has carved a deep valley through lava flows that are less than half a million years old. The flat landscape above the valley is the original constructional surface of the shield volcano.

THE EARTH'S VARIED LANDSCAPES

The major components of the Earth System meet and interact at the land surface. As a result, the evolution of the Earth's landscapes involves a complex of processes related to the solid Earth, the atmosphere, the hydrosphere, and the biosphere. Because of this complexity, gaining an understanding of landscape evolution involves taking an *interdisciplinary* approach; that is, scientists in a number of related Earth Science disciplines must collaborate to gain meaningful answers.

One of the most striking things about the Earth is its amazing variety of natural landscapes. Even a casual look at the land surface can raise some basic questions in our mind: How can the Earth's varied landscapes be explained? If landscapes have changed through time, what processes have controlled the changes? What clues do landscapes hold about the history of the Earth's mobile lithosphere and past climates?

The processes that produce changes in the land surface operate on many different time and spatial scales. Some act rapidly, even abruptly, and may cause only local changes. Sometimes these changes have a direct and adverse impact on people (for example, earthquakes, volcanic eruptions, storms, floods, landslides, and many other natural hazards). Other processes operate far more slowly—on geologic time scales—and are difficult or impossible to observe directly. Nevertheless, their effects can be seen both in surviving ancient landscapes and in the stratigraphic record, and computer models can be used to understand how they influence landscape evolution.

Not only can surface processes affect people going about their daily lives, but everywhere they live people, too, are changing the face of the land. The construction of dams across streams, excavations for buildings or waste disposal sites, and the building of highways, cities, and airports all modify the landscape. Although such individual actions may seem insignifi-

Figure 19.1 A The precipitous southern flank of Cerro Aconcagua (6959 m; 22,834 ft), the highest peak of the Andes and the southern Hemisphere, rises 3600 m (11,800 ft) above the glaciated valley at its base, making it among the areas of greatest local relief on the Earth. B. Subdued low-relief landscape of the Western Australia shield, with Tertiary laterite at surface.

A.

B.

A.

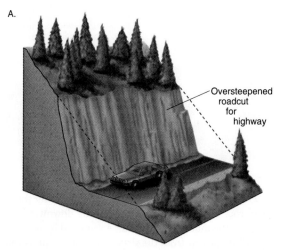

B.

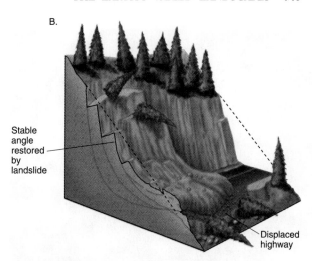

Figure 19.2 When a natural slope is oversteepened in building a road, failure can result. A. A highway cut exceeds the natural angle of the slope, producing an unstable situation. B. The oversteepened slope fails, and a landslide buries the road. The slope angle of the resulting deposit now is similar to the original natural one.

cant at the global scale, they can quickly add up until, over the span of several human generations, their effect on the landscape is substantial.

Competing Geologic Forces

The evidence is overwhelming that the Earth's surface is constantly changing, although the rate and magnitude of change vary considerably from region to region. The changes reflect an ongoing contest between forces that raise the lithosphere (tectonic and isostatic uplift, volcanism) and the force of gravity which, aided by physical and chemical weathering that break down rock, causes various erosional agents to transfer rock debris from high places to low places. The net result is the progressive sculpture of the land into a surface of varied topography and **relief** (the difference in altitude between the highest and lowest points on a landscape) (Fig. 19.1). Relief varies regionally because surface geologic processes operate not only under different climatic conditions, but also on rocks of different type and structure that have differing resistances to erosion. As we have seen in Chapters 9 and 10, streams and glaciers efficiently erode and transport rock debris and move large quantities of sediment downslope and toward the sea. However, another group of surface processes also is responsible for moving vast amounts of rock debris downslope under the influence of gravity.

Mass-Wasting

As any mountain climber will affirm, mountains can be dangerous, unstable places. Falling rocks, debris-

clad glaciers, and boulder-filled streams provide vivid evidence of the downslope movement of rock debris. Under the pull of gravity, debris falls, slides, and tumbles downslope where it is picked up by glaciers, streams, or wind and carried farther.

A smooth, vegetated hillslope may outwardly appear stable and show little obvious evidence of geologic activity. However, a time-lapse motion picture of almost any hillslope would show that the surface is moving and constantly changing. Most of the recorded motion would be the result of **mass-wasting**, the downslope movement of regolith under the pull of gravity. This definition implies that the motive force is gravity rather than a transporting medium such as water, wind, or ice. Although transport of debris by mass-wasting often involves movement of discrete particles, at times large volumes of debris move downslope en masse.

Landslide is a general term for a variety of mass-wasting processes that result in the downslope movement of a mass of rock or regolith, or a mixture of the two, under the influence of gravity. The composition and texture of the sediment involved, the amount of water and air mixed with the sediment, and the steepness of slope all influence the type and velocity of a landslide.

Although landslides may occur for no apparent reason, many are related to an unusual occurrence: (1) A major earthquake can trigger landslides throughout a large area. (2) Landslides often are associated with heavy or prolonged rains that saturate the ground and make it unstable. (3) A volcanic eruption, like that at Mount St. Helens in 1980, can trigger a variety of landslides, as well as mudflows that move rapidly into surrounding valleys. (4) Landslides often result when human activities modify natural slopes. Slides frequently occur where road construction significantly oversteepens natural slopes (Fig. 19.2). (5) A stream undercutting its bank can trigger a landslide, and pounding storm surf along a seacoast can also produce landslides when steep bluffs are undercut (Fig. 19.3).

Figure 19.3 Steep seacliffs of jointed basalt along the windward coast of Hawaii are undercut by pounding surf. When a cliff collapses, the resulting landslide debris is rapidly reworked by surf and currents, and the process begins anew.

Landslides and other types of mass movement contribute importantly to the production of landscapes that are dominantly controlled by streams or glaciers. A substantial fraction of the sediment transported by a mountain glacier reaches the glacier surface by rockfalls and rockslides from adjacent slopes. When a stream erodes into the bedrock floor of its valley, much of the rock ultimately removed reaches the stream channel by mass wasting (Fig. 19.4).

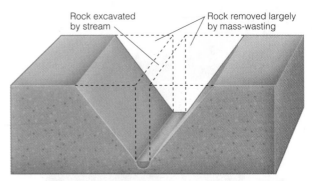

Figure 19.4 An idealized valley segment comparing the volume of rock excavated by a stream cutting downward in bedrock with the much greater volume removed by mass-wasting. All the debris from the slopes ultimately enters the stream and is transported out of the area.

Factors Controlling Landscape Development

We have already seen that distinctive landscape elements result from the activity of various surface processes. A sand dune has a form that is different from that of a moraine. We can also distinguish between an alluvial fan and a delta on the basis of their form. In each case, the active process and depositional environment lead to a unique end product, or landform. *Process*, then, is one factor that helps dictate the character of landforms.

Climate, in turn, helps determine which processes are active in any area. In humid climates, streams may be the primary agent that moves and deposits sediment, whereas in an arid region wind may locally assume the dominant role. Glaciers and periglacial phenomena are largely restricted to high latitudes and high altitudes where frigid climates prevail. Because

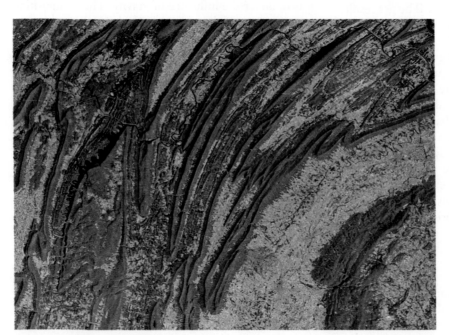

Figure 19.5 A satellite image of the region near Harrisburg, Pennsylvania, reveals a complicated series of northeast-trending ridges and valleys produced by differential erosion of sedimentary rocks. Ridges are underlain by resistant sandstones and conglomerates, whereas valleys are underlain by more-erodible shales. The folded structure of the rocks is clearly visible due to the pronounced topographic relief between the less-erodible and more-erodible strata.

climate also controls vegetation cover, it further controls the effectiveness of some important erosive processes; for example, a hillslope stabilized by plant roots that anchor the soil may become prone to mass-wasting if the plant cover is destroyed. Thus, hillslope erosion is linked to vegetation cover, which in turn is controlled by soil moisture, which is determined by climate—a complex linkage in which the atmosphere, the hydrosphere, and the biosphere all play a role in determining how surface processes affect the land surface.

In the same way that we can identify distinctive climatic regions of the Earth Appendix E, it is also possible to identify regions containing distinctive landforms produced by one or more surface processes. However, because climates have changed through time, the active surface processes in some regions also have changed. As a result, some modern landscapes largely reflect former conditions rather than those of the present.

Within any climatic zone, a certain surface process may interact with various exposed rocks differently, depending on their *lithology* (Fig. 19.5). Some rock types are less erodible than others and will produce greater relief and more prominent landforms than those more susceptible to erosion. Any single rock type, however, may behave differently under different climatic conditions. For example, limestone may underlie valleys in areas of moist climate where dissolution is very effective, but in dry desert regions the same kind of rock may form bold cliffs.

The ease with which a formation is eroded by streams depends chiefly on its composition and structure. Because of differential erosion, folded or faulted beds may stand in relief or control the drainage in such a way that they impart a distinctive pattern to the landscape, thereby disclosing the underlying structure (Fig. 19.5). Figure 19.6 shows some of the most common drainage patterns and the geologic structures that control them. An experienced geologist can often use drainage patterns to infer rock type, the orientation of a dipping rock unit, the manner in which the rocks are folded or offset, and the pattern and spacing of joints.

Relief of the land, another factor in landscape development, is related to tectonic environment. Tectonically active regions may have high rates of uplift, leading to high summit altitudes and steep slopes. Such landscapes tend to be extremely dynamic and generate high erosion rates (Fig. 19.7). In areas far away from active tectonism, relief typically is low, erosion rates are much lower, and landscape changes take place more gradually. Even in nontectonic areas, however, a rapid rise or fall of sea level, or regional isostatic movements related to changing ice or water loads, may produce significant changes in streams and the landscapes they influence.

Finally, the concept of landscape evolution necessarily involves the element of *time*. Although some landscape features can develop rapidly, even catastrophically, others develop only over long geologic intervals. We know this, or at least we infer this, from measurements of the present rates of surface processes and by dating deposits that place limits on the ages of specific landforms or land surfaces.

Dendritic Irregular branching of channels ("treelike") in many directions. Common in massive rock and in flat-lying strata. In such situations, differences in rock resistance are so slight that their control of the directions in which valleys grow headward is negligible.

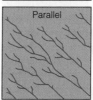

Parallel Parallel or subparallel channels that have formed on sloping surfaces underlain by homogeneous rocks. Parallel rills, gullies, or channels are often seen on freshly exposed highway cuts or excavations having gentle slopes.

Radial Channels radiate out, like the spokes of a wheel, from a topographically high area, such as a dome or a volcanic cone.

Rectangular Channel systems marked by right-angle bends. Generally results from the presence of joints and fractures in massive rocks or foliation in metamorphic rocks. Such structures, with their cross-cutting patterns, have guided the directions of valleys.

Trellised Rectangular arrangement of channels in which principal tributary streams are parallel and very long, like vines trained on a trellis. This pattern is common in areas where the outcropping edges of folded sedimentary rocks, both weak and resistant, form long, nearly parallel belts.

Annular Streams follow nearly circular or concentric paths along belts of weak rock that ring a dissected dome or basin where erosion has exposed successive belts of rock of varying degrees of erodibility.

Centripetal Streams converge toward a central depression, such as a volcanic crater or caldera, a structural basin, a breached dome, or a basin created by dissolution of carbonate rock.

Figure 19.6 From stream patterns, geologists can infer a good deal about the type and configuration of underlying rock and about the structural history of an area.

Figure 19.7 A stream cutting a deep gorge in the Hindu Kush of northern Pakistan transports a coarse gravel bedload, contributed largely by rockfalls and landslides from the steep adjacent cliffs.

LANDSCAPES PRODUCED BY RUNNING WATER

Most of the Earth's land areas, except those continuously covered by ice sheets, show the effects of running water. Even in extremely dry deserts, we can find evidence that ancient streams have had a hand in shaping the land.

Tectonic and Climatic Control of Continental Divides

All the continents except ice-covered Antarctica can be divided into large regions from which major through-flowing rivers enter one of the world's major oceans. The line separating any two such regions is a **continental divide**, one of the major landscape elements of our planet. In North America, continental divides lie at the head of major streams that drain into the Pacific, Atlantic, and Arctic oceans (Fig. 19.8). In South America, a single continental divide extends along the crest of the Andes and divides the continent into two regions of unequal size. Streams draining the western (Pacific) slope of the Andes are steep and short, whereas to the east the streams take much longer routes along more gentle gradients to reach the Atlantic shore.

Because continental divides often coincide with the crests of mountain ranges and because mountain ranges are the result of uplift associated with the interaction of tectonic plates, a close relationship must exist between plate tectonics and the location of primary stream divides and drainage basins.

When a divide lies close to a continental margin, and there is a strong climatic gradient across it, an unequal distribution of precipitation can lead to a marked landscape asymmetry across the divide.

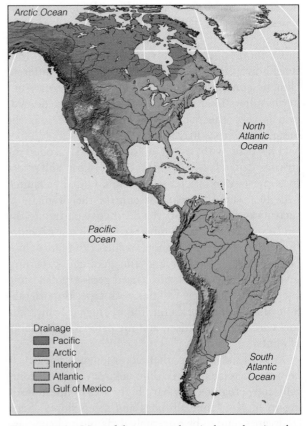

Figure 19.8 Map of the western hemisphere showing the location of major drainage divides. The continental divide separating streams draining to the Pacific, Arctic, and Gulf of Mexico in North America and to the Pacific and Atlantic in South America follows the crest of the high cordillera in both hemispheres. In eastern North America, the divide separating Atlantic and Gulf of Mexico drainage follows the Appalachian Mountains and the approximate limit of ice sheet glaciation south of the Great Lakes.

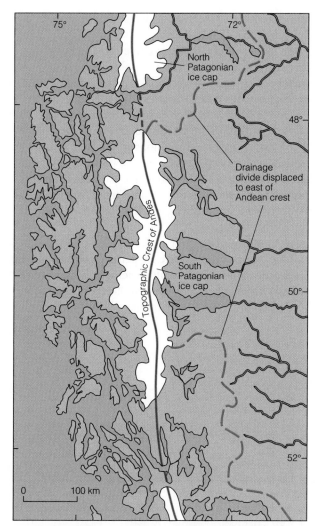

Figure 19.9 Map of southernmost Andes showing coastal fjords of Chile, Andean ice caps, and topographic crest of range (bold solid line). In two sectors, the continental drainage divide (dashed line) lies well east of the range crest and partly coincides with a belt of Pleistocene end moraines.

Streams draining the windward (wet) side commonly have steeper gradients than those draining the lee (dry) side, and erosion rates are higher. Over time, the net effect will be the headward growth of channels on the wet side, leading to a shift in the divide toward the drier side.

A striking example is seen in the case of the Andes in southern South America. Through nearly 60 degrees of latitude, from the northern tip of the continent to well south of Santiago, Chile, the drainage divide coincides with the topographic crest of the Andes. However, south of 45°, where easterly winds have given way to westerlies flowing off the Pacific, the continental divide has shifted eastward in several sectors. During glaciations, the mountain ice cap of the southern Andes apparently was thickest over the fjord region of southern Chile, which received heavy snowfall from the moist westerly winds flowing on-shore. Outlet glaciers of a large ice cap flowed eastward across saddles in the main Andean crest, progressively deepening and ultimately eliminating these topographic barriers. In several sectors, the continental divide between the Pacific and Atlantic oceans now lies along the crest of end moraines marking the limit of piedmont glaciers that terminated near or beyond the eastern front of the mountains (Fig. 19.9).

The Geographic Cycle of W. M. Davis

Mountain ranges and high plateaus are major landforms of the Earth's crust, and it is natural to ask: How long can such features persist? As soon as mountain building commences, the forces of erosion begin to attack the rising land. The average altitude of the land at any time, therefore, should reflect this contest between uplift and erosion.

One of the most influential theories of landscape evolution was proposed by the American geographer William Morris Davis in the late nineteenth century. Davis called his model the *geographic cycle*, implying that it had a beginning and an end. According to this concept, a cycle was initiated by rapid uplift of a landmass, with little accompanying erosion, so that the initial relief was large. Erosion then progressively sculptured the land and reduced its altitude until it was worn down close to sea level (Fig. 19.10). Davis deduced that a landscape passed through a series of stages. During the initial stage, streams cut down vigorously into the uplifted land mass and produce sharp V-shaped valleys, thereby increasing the local relief. Gradually, the original gentle upland surface is con-

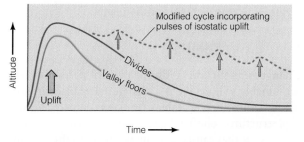

Figure 19.10 The Geographic Cycle of W. M. Davis begins with a pulse of uplift, during which the uplift rate increases, then diminishes. Initially, the landscape increases rapidly in altitude. As the uplift rate decreases, the land is lowered by erosion. Valleys deepen and local relief (the distance between divides and valley floors) reaches a maximum. During later stages of the cycle, the land is progressively lowered toward sea level and relief becomes lower and lower. If isostatic adjustments are also considered, a long time is required to erode the landscape to lower altitude (dashed line).

sumed as the drainage system expands and valleys become deeper and wider. During the next stage, the land achieves its greatest local relief. Streams now begin to meander in their valleys, and valley slopes are gradually worn down by mass-wasting and erosion. In the final stage, the landscape consists of broad valleys containing wide floodplains, stream divides are low and rounded, and the landscape is slowly worn down ever closer to sea level. Davis also envisaged interruptions of erosion cycles related to climatic fluctuations or to renewed uplift. Others later pointed out that isostatic response of the crust to the progressive transfer of sediment from the land to the ocean should lead to pulses of uplift, requiring that a cycle be longer than Davis envisaged (Fig. 19.10).

Davis's theory attracted wide attention and formed the basis for most interpretations of landscape evolution during the following decades. However, with renewed interest in surface process studies and the widespread acceptance of the theory of plate tectonics, it has become increasingly difficult to reconcile Davis's concept of an erosion cycle with what we know about global tectonics and Earth history. The major landscape features of the Earth have developed over long intervals of time as the lithosphere has evolved and continental arrangements have continually been reorganized. The lateral motions and resulting collisions of lithospheric plates, leading to the generation of orogenic belts, has provided much of the driving force for landscape change over hundreds of millions of years. The focus in landscape evolution has now shifted from the concept of landscape cycles to questions about landscape equilibrium, studies of the relative importance of uplift rates and erosion rates, and investigations of the complex relationships and feedbacks between the atmosphere, biosphere, and lithosphere in dictating the character and evolution of the major landscape elements of the planet.

Landscape Equilibrium

Change is implicit in landscape evolution. Landscapes presumably will evolve if a change takes place in any of the controlling variables (process, climate, lithology, structure, relief). Change may be started by a tectonic event that causes a landmass to be uplifted, or by a substantial fall of sea level that causes streams to assume new gradients. It can begin with a shift in climate that may modify the relative effectiveness of different surface processes. A change may also result as a stream, eroding downward through weak, erodible rock, suddenly encounters massive, hard rock beneath.

Over short intervals of time, rates of change may vary because of natural fluctuations in the magnitude and intensity of surface processes, but overall the landscape can be close to a steady state (Fig. 19.11a). Over longer intervals, the average rate of change may increase if the rate of tectonic uplift increases, or experience a gradual decrease as a land surface is progressively worn down toward sea level (Fig. 19.11b).

Does a landscape, then, ever achieve a state of complete equilibrium in which no change takes place? The answer apparently is no, for we have abundant evidence that the Earth's surface is now, and very likely always has been, a dynamic surface, constantly experiencing changes in response to the natural motions of the lithosphere, hydrosphere, and atmosphere. Nevertheless, it is apparent that conditions of near-equilibrium can be reached.

One reason why a state of perfect equilibrium in landscape systems is difficult to achieve is that the responses to change are often complex. Within a landscape unit such as a large drainage basin, a change in

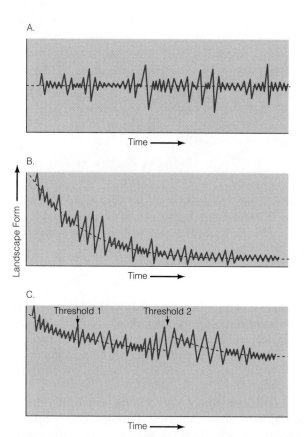

Figure 19.11 Three possible equilibrium conditions in landscape evolution. A. Steady-state equilibrium, with the landscape fluctuating about some average condition. B. Dynamic equilibrium, with the landscape oscillating about an average value that is declining with time. C. Dynamic equilibrium, with sudden changes to the landscape occurring whenever a critical threshold is reached.

one of the controlling factors may at first impact only a small part of the basin. A sudden vertical movement along a fault that crosses the basin near its mouth will affect the long profile of the stream at that point and begin a series of compensating adjustments both in the upstream and downstream directions. Areas in the headward part of the basin may initially be unaffected, but as the stream system adjusts, changes will progressively affect other parts of the basin, including tributary streams, valley slopes, and ultimately valley heads, until a near-equilibrium condition is once more achieved. The change may move through the system in a complex manner, with many lags and minor readjustments taking place.

Several changes may affect a stream system simultaneously. For example, a sudden intense rainstorm may be concentrated in one tributary and a massive landslide in another. The response of the stream to one of these events may lag its response to the other, resulting in a complex adjustment of the entire system. It is not surprising, therefore, that geologists working with the depositional and erosional products of such changes can have difficulty in sorting out the events that caused them.

Threshold Effects

In many natural systems, sudden changes can take place without any outside stimulus when a critical threshold condition is reached. Sand grains in a stream channel may remain at rest until a critical current velocity is reached, at which point they begin to move. Certain glaciers apparently start to surge when the buildup of water pressure at their base attains a critical threshold value that forces the ice to float off its bed (Chapter 10).

The concept of thresholds is also applicable to landscapes. It implies that the development of landscapes, rather than being progressive and steady, can be punctuated by occasional abrupt changes. A landscape in near-equilibrium may experience a sudden change in form if a process operating on it reaches some threshold level. The condition of equilibrium is thereby affected, and a change in the landscape takes place, as it moves toward a new equilibrium condition (Fig. 19.11C). Studies in the western United States have shown, for example, that the development of gullies on the floors of some alluvial valleys is related to valley slope. Where valley floor slopes reach a certain critical value, which depends on the area of the drainage basin, they become unstable and gullying begins. Where the slope is gentler, valley floors remain ungullied. The critical slope value therefore constitutes a threshold; when it is exceeded, erosion of the valley floor begins.

UPLIFT AND DENUDATION

Obtaining reliable geologic information about long-term rates of landscape change is not simple. First, we must determine how mobile a landmass has been through time. If it is rising tectonically, we must ask how rapidly it is rising and whether rates of uplift have changed through time. Second, we must be able to calculate changing rates of **denudation**, the combined destructive effects of weathering, mass-wasting, and erosion. Knowing uplift and denudation rates for a region, we can then infer something about how the landscape may have evolved and anticipate how it may evolve in the future.

Uplift Rates

One way geologists attack the problem of calculating uplift rates is to measure how much local uplift occurs during historic large earthquakes, try to determine the recurrence interval of such earthquakes, and then extrapolate the recent rates of uplift back in time. This approach assumes that the brief historic record is representative of longer intervals of geologic time, an assumption that may not always be valid.

A second approach is to measure the warping, or vertical dislocation, of originally horizontal geologic surfaces of known age. Examples are flood basalts that have been deformed since extrusion and uplifted coral reefs along a tectonically active coast (Fig. 11.33). In each case, we need to know both the difference between the present altitude of a basalt flow or coral reef and its altitude at the time of formation, as well as a radiometric age for the rock or coral.

Successive intervals of stream incision into a mountain range or plateau may produce a series of terraces that record uplift events. If the age of the terraces can be obtained, then uplift rates can be calculated. However, because terracing can also result from changes in stream activity caused by world sea-level fluctuations, changes of climate, or variations in the structure of rocks across which a stream flows, relating river terraces to tectonic events is not always straightforward.

Finally, rocks that formed deep within the crust and subsequently were exposed at the surface by uplift and erosion can provide uplift rates. When the mineral zircon crystallizes in a plutonic rock, the subsequent decay of radioactive isotopes trapped in the mineral damages the internal arrangement of atoms, leaving tiny tracks (called *fission tracks*) that can be detected under a microscope. At high temperatures, fission tracks can form but they quickly anneal and disappear, leaving no record of their former existence. Only when the cooling mineral falls below a certain critical temperature (its closure temperature) is the annealing rate so slow that fission tracks will be retained. The number of tracks forming in a mineral increases with time, and so track density can be used to date the time elapsed since a rock containing the mineral cooled below the closure temperature. In the example shown in Figure 19.12, we assume that the closure temperature for the mineral zircon is 240°C (464°F). By measuring the geothermal gradient (we'll assume in the example it is 40 C°/km), we then calculate that the 240°C isotherm lies at a depth of 6000 m (3.7 mi). The average rate of uplift can then be calculated because we know the fission track age of the rock and the total uplift since the rock first acquired fission tracks (i.e., the depth beneath the sample site at which the 240°C isotherm lies).

The current local uplift rate for one high region of the western Himalaya, based on fission-track measurements, is as high as 5 mm/year (0.2 in/yr). If sustained for only 2 million years, the total uplift would be 10 km (33,000 ft). Such a high rate of uplift is generally associated with steep slopes and high relief. This is certainly true in the Himalaya, where glacially eroded mountain slopes near the crest of the range typically exceed angles of 30° and local relief can exceed 5 km (16,500 ft). At lower altitudes, where streams are the dominant erosional force, slopes generally decline to between 15 and 20°. Mean erosion rates in these high mountains are estimated to be about 3 to 4 mm/year (0.1 to 0.2 in/yr), which is close to the average uplift rate.

Figure 19.12 Uplift rate across a mountain range calculated using fission-track ages. A. Two million years ago, a zircon crystal (A) in a cooling pluton passes the closure isotherm of 240°C and begins to acquire fission tracks. Another crystal (B) began acquiring tracks 2 million years earlier and since then has been uplifted 1200 m above the 240°C isotherm. B. Rock samples containing zircon crystals A and B collected from a stream valley eroded into the rising mountain range have fission track ages of 2 and 4 million years, respectively, and lie 6000 and 2400 m, respectively, above the closure isotherm of 240°C. C. By using these data, the average uplift of samples A and B are calculated as 3.0 and 0.6 mm/year, respectively.

Measured uplift rates in this and other tectonic belts are quite variable (ca. 1 to 10 mm/yr, averaged over intervals of several thousand to several million years). We can assume that for any region, the average values have changed through time as rates of seafloor spreading and plate convergence have varied. Each such change is likely to lead to compensating adjustment in landscapes as they begin to evolve toward a new condition of equilibrium.

Denudation Rates

The calculation of long-term denudation rates requires knowledge of how much rock debris has been removed from an area during a specified length of time. The total sediment removed from a mountain range, for example, will include sediment currently in transit, sediment temporarily stored on land on its way to the sea, and sediment deposited in the adjacent ocean basin, mostly on or near the continental shelf. If the volume of all this sediment can be measured, its solid rock equivalent and the average thickness of rock eroded from the source region can be calculated. Finally, if the duration of the erosional interval can be determined, the average denudation rate can be calculated.

For areas drained by major through-flowing streams, the volume of sediment reaching the ocean each year is a measure of the modern erosion rate. Most of this sediment ultimately is deposited in deltas, as a blanket across the continental shelves, and in vast submarine sediment fans. The volume of sediment deposited during a specific time interval can be estimated using drill-core and seismic records of the seafloor. Calculating the equivalent rock volume of this sediment and averaging it over the area of the drainage basin(s) from which it was derived then gives an average denudation rate for the source region.

The highest measured sediment yields are from the humid regions of southern Asia and Oceania, and from basins that drain steep, high-relief mountains of young orogenic belts, such as the Himalaya, the

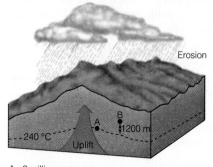

A. 2 million years ago

B. Today

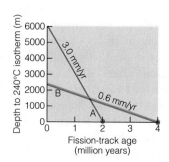

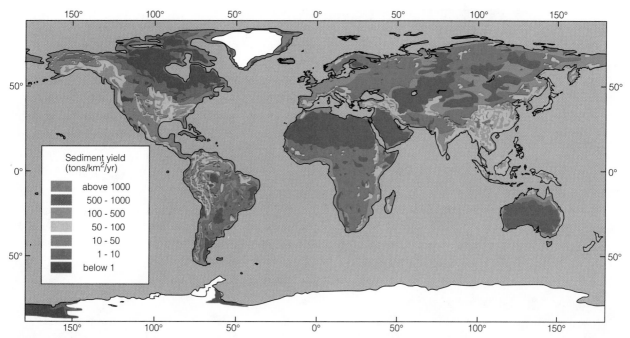

Figure 19.13 Estimated rates of sediment yield on the continents. The highest yields are from Southeast Asia, which receives high precipitation related to the summer monsoon climate, and major high-relief mountain belts, including the Himalaya, the Alps, the Andes, and the coastal ranges of Alaska and British Columbia.

Andes, and the Alps (Fig. 19.13). Low sediment yields characterize deserts and the polar and subpolar sectors of the northern continents. As one might expect, rates are high on steep slopes, and much higher in areas underlain by erodible clastic sediments or sedimentary rocks, or by low-grade metamorphic rocks, than in areas where crystalline or highly permeable carbonate rocks crop out. Structural factors also play a role, for rocks that are more highly jointed or fractured are more susceptible to erosion than massive ones. Denudation is surprisingly high in some dry climate regions, an important reason being that the surface often lacks a protective cover of vegetation. Soil protection by vegetation is likely a threshold phenomenon; below a certain vegetation density, erosion rates may increase.

The dominant erosional process can also strongly influence sediment yield. Measurements have shown that sediment yields generally increase with increasing glacier cover in a drainage basin. Values are unusually high in places like south-coastal Alaska, the most heavily glacierized temperate mountain region in the world, where denudation rates exceed those of comparable-sized basins in other regions that lack glaciers by an order of magnitude. Under climates favorable for the expansion of temperate valley glaciers, both chemical and physical weathering rates tend to be substantially higher than the global average.

The continent discharging the most sediment to the ocean is Asia, and it also is the continent on which the greatest average stream sediment loads have been measured. Asian rivers entering the sea between Korea and Pakistan are believed to contribute nearly half the total world sediment input to the oceans (Fig. 19.14). Second to Asia is the combined area of the large western Pacific islands of Indonesia, Japan, New Guinea, New Zealand, the Philippines, and Taiwan. Taiwan is especially remarkable because it produces only slightly less sediment than the entire contiguous United States!

One important factor that influences denudation rates is human activity, especially the clearing of forests, development of cultivated land, damming of streams, and construction of cities. Each of these activities has affected erosion rates and sediment yields in the drainage basins where they have occurred. Sometimes the results are dramatic: in parts of eastern United States, areas cleared for construction produce between 10 and 100 times more sediment than comparable rural areas or natural areas that are vegetated. On the other hand, in urbanized areas sediment yield tends to be low because the land is almost completely covered by buildings, sidewalks, and roads that protect the underlying rocks and sediments from erosion.

In many drainage basins, the measured and estimated sediment yields reflect conditions that are probably quite different from those of only a few decades ago because much of the sediment that formerly reached the sea is now being trapped in reservoirs behind large dams. The high Aswan Dam (Chapter 20) now intercepts most of the sediment that formerly was carried by the Nile to the Mediterranean Sea.

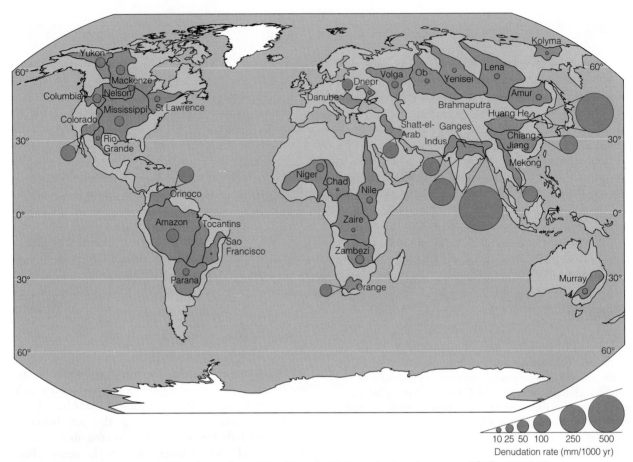

Figure 19.14 Present rates of denudation for the Earth's major drainage basins. Compare with Figure 19.13.

Figure 19.15 Ayers Rock, a massive reddish inselberg, rises about 360 m (nearly 1200 ft) above the surrounding flat plain in central Australia.

Ancient Landscapes of Low Relief

The ultimate reduction of a landmass to low altitude, as envisioned by Davis, is likely to occur only if changes in plate motion lead to cessation of tectonic uplift. Denudation can then gradually lower the relief. Examples of such landscapes can be found in the world's shield areas, where the roots of ancient mountain systems have been exposed as the crust has thinned through the action of long-continued erosion and compensating isostatic adjustment.

Australia is widely believed to have some of the oldest landscapes of any of the continents. In places these ancient landscapes are dominated by **inselbergs**, dome-like hills that rise above otherwise featureless plains (Fig. 19.15). Measurements of nuclides, produced by the bombardment of cosmic-rays, which are concentrated just beneath the surface of the granitic inselbergs, have shown that the tops of these domes are eroding at a rate of less than a meter per million years. A combination of a relatively arid climate, long-term tectonic stability, and low erodibility of the exposed granite contribute to the exceptional stability of these landforms. Based on these measurements, it seems likely that this landscape predates the Quaternary Period (Fig. 8.3) and may have originated well back in the Tertiary Period, or even earlier.

Widespread erosional landscapes that have low relief and low altitude are not common. This must either mean that the Earth's crust has been very active in the recent geologic past or that such landscapes take an extremely long time to develop. Estimates have been made of the time it would take to erode a landmass to or near sea level by extrapolating current denudation rates into the past. Such estimates must take into account two important factors. Studies have shown that rates of denudation are strongly related to altitude, implying that as the land is lowered the rate of denudation will decline. Furthermore, the eroding land will rise isostatically as the lithosphere adjusts to transfer of sediment from the land to the oceans. Both factors tend to increase the time it takes for the final reduction of a landmass to low altitude. Estimates of the time it would take to reduce a landmass about 1500 m high to near sea level, assuming no tectonic uplift, range from about 15 to 110 million years.

Although one might argue that much of the Earth's crust has been unusually active during the last 15 million years or more, could there have been earlier intervals of relative crustal quiet when low-relief surfaces did develop? Many ancient land surfaces are preserved in the geologic record as unconformities (Chapter 8). Some can be traced over thousands of square kilometers and can be shown to possess only slight relief (Fig. 19.16). Associated with such surfaces are weathering profiles that imply that the land surface was continuously exposed for long intervals at relatively low altitude. Such buried ancient landscapes are evidence of times when broad areas were eroded to low relief.

Figure 19.16 Nearly flat-lying sedimentary strata in the upper walls of the Grand Canyon of the Colorado River rest on tilted older strata. The angular unconformity separating the two groups of rocks is a surface of low relief that can be traced for considerable distances, and represents a subdued ancient land surface.

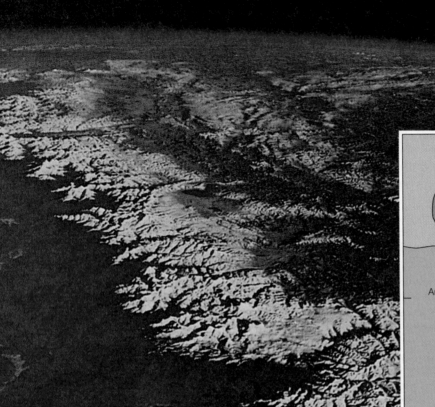

A.

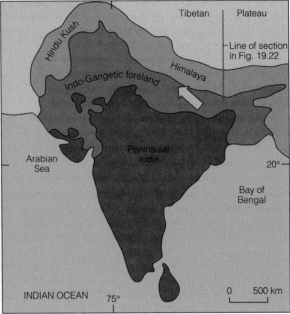

B.

Figure 19.17 A. An oblique view of the snow-covered Himalaya and the adjacent Tibetan Plateau rising above the Gangetic lowland (dark green) of India, as seen from an orbiting spacecraft. B. Map of India and adjacent parts of Asia showing relationship of the Himalaya and Hindu Kush mountain systems to the Tibetan Plateau and the Indian subcontinent. Arrow shows direction of view of Figure 19.17a.

CONTINENTAL UPLIFT AND CLIMATIC CHANGE

The height and form of the continents can strongly influence both regional and world climate. Of singular importance to the evolution of the world's present climate system has been the uplift of the Himalaya and the Tibetan Plateau in the lower middle latitudes of central Asia over the last 50 million years (since the early Eocene), caused by the collision of the Indian plate with Asia. The resulting high-altitude terrain may be the largest such feature produced on the Earth during Phanerozic time (the last 570 million years) (Fig. 19.17). This vast uplifted region has had a major effect on the climate of Asia, as well as global circulation patterns. Since the collision began, more than 2000 km of crustal convergence has been accommodated by both lateral and vertical tectonic movements within Asia. During the first 20 million years of collision, the convergence was accommodated mainly by movement of crustal blocks to the east and southeast along major strike-slip faults. Subsequently, the ac-

commodation involved major north-dipping faults in the Himalaya and southern Tibet, together with many minor faults, which led to uplift of the Himalaya and the adjacent plateau. A variety of evidence indicates that by about 8 million years ago (late Miocene) the Tibetan Plateau had reached an altitude high enough to intensify the Asian monsoon system and affect global climate.

Central Asia, though clearly the most important, is not the only continental region to have experienced major regional uplift in the recent geologic past. Other areas include the high plateaus and mountains of the American West, the high Andes and Altiplano of South America, and the extensive plateaus of eastern and southern Africa that have mean altitudes of more than 1000 m.

To explore the effects of large-scale uplift on the Earth's climate system, global climate models have been used to compare conditions with a Tibetan Plateau (i.e., today) and without (prior to uplift). The climate simulations, supported by geologic evidence, point to some major global changes:

- Uplift has led to cooling of the midlatitude high plateaus. Raising the plateaus has led to local cooling of more than 16°C over Tibet (Fig. 19.18) and more than 8°C across the American West. The average temperature lowering between 30 and 40° N exceeds 10°C, and the mean hemispheric change over land areas is 8°C.

- Under no-plateau conditions, the westerlies would display strong zonal (latitudinal) flow (Fig. 19.19). Uplift in western North America and central Asia has created obstacles to low-level atmospheric flow, diverting the westerlies northward around the uplifted regions, with a return southward flow to the lee (downwind) of the high land. In North America, enhanced meandering of the jet stream brings cold air farther south, a condition that would favor ice-sheet glaciation in northern middle latitudes.

- An uplifted plateau and the bordering Himalaya generates orographic precipitation (Chapter 12) on the windward flanks of the uplifted region as warm, moist summer air rises against the high topographic obstacle and cools. Increased rainfall in southeastern Asia, for example, contrasts with the formation of an arid rainshadow inland from the Himalaya, leading to desertification of the plateau surface (Fig. 19.19).

- Uplift has increased the seasonal contrast in precipitation related to an enhanced monsoon system. In southern Asia, summer heating of the high plateau causes warm air to rise, creating a low-pressure region that draws moist air inland from the adjacent ocean and produces the famous summer monsoon rains (Chapter 13; Fig. 19.19). In winter, the snow-covered surface of the plateau reflects solar radiation and cools down, creating a region of high pressure from which cold, dry air flows outward and downward. Periods of maximum precipitation and runoff occur during times of maximum solar insolation that are dominated by orbital precession (Chapter 14).

- Increased precipitation related to strengthening of the monsoon means more frequent large floods during the summer monsoon season.

- Uplift has caused regional drying of continental interiors at middle latitudes. As a result of restructured atmospheric circulation, hot, dry air from interior Asia moves southwestward over Arabia and northern Africa, leading to summer drying (Fig. 19.18).

- Changes in ocean salinity and temperature related to changed atmospheric circulation would cause changes to deep and intermediate ocean circulation in the North Pacific and North Atlantic (Chapter 11).

The variety and magnitude of environmental changes linked to uplift of the mountains and plateau of central

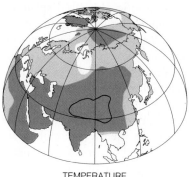

TEMPERATURE

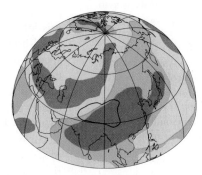

PRECIPITATION

Figure 19.18 Computer simulation of environmental changes in Asia resulting from uplift of the Tibetan Plateau. A. Changes in temperature. B. Changes in precipitation. C. Overall changes in major geographic regions.

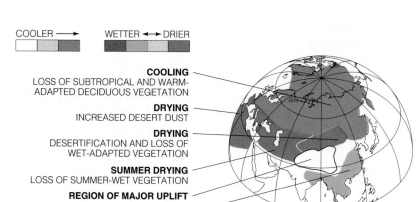

COOLER ——→   WETTER ←—→ DRIER

COOLING
LOSS OF SUBTROPICAL AND WARM-ADAPTED DECIDUOUS VEGETATION

DRYING
INCREASED DESERT DUST

DRYING
DESERTIFICATION AND LOSS OF WET-ADAPTED VEGETATION

SUMMER DRYING
LOSS OF SUMMER-WET VEGETATION

REGION OF MAJOR UPLIFT

CLIMATIC STABILITY
PERSISTENCE OF WARM- AND WET-ADAPTED VEGETATION

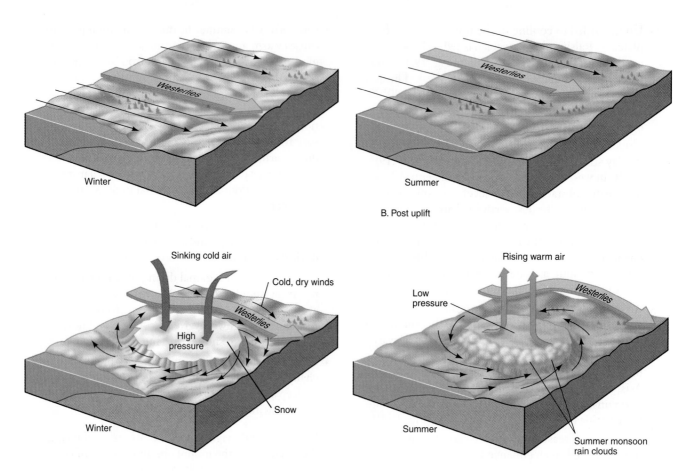

Figure 19.19 Development of a monsoon climate system. Before uplift, westerly winds pass directly across a landscape, the axis of flow shifting seasonally north and south. As a large plateau is uplifted, the westerly flow is diverted around the upland, a winter high-pressure system develops over the high, snow-covered plateau, and cold, dry winds blow clockwise off the upland region. In summer, as the plateau surface warms, the warm air rises, drawing moist marine air inland. The summer monsoon clouds move inland toward the plateau margin and lose their moisture as heavy rainfall.

Asia point to the complexity of the interlinked components of the Earth system. We see that a major tectonic event resulting from the collision of two continents strongly affected not only the lithosphere in and beyond the belt of collision, but also the global atmosphere, hydrosphere, cryosphere, and biosphere. This continental impact has been a major influence on the shaping of the Earth's surface over the last 50 million years and no doubt will continue to be far into the future.

UPLIFT, WEATHERING, AND THE CARBON CYCLE

In Chapter 14 we noted that the long-term trend of climate during the last 100 million years bears a general relationship to the reconstructed trend in atmos-

pheric CO_2 concentration (Fig. 14.19). Explaining this long-term trend is important in trying to understand how the Earth's climate system has evolved. Of special interest is explaining how the climate maintains a degree of equilibrium. Why doesn't atmospheric CO_2 rise to such high values that the Earth becomes a "hothouse" like Venus, or fall so low that the Earth cools down to become a frigid planet like Mars? According to one hypothesis, the average global rate of seafloor spreading controls the rate of CO_2 generation along midocean ridges and along zones of collision and subduction (Fig. 19.20). An additional and possibly larger source of atmospheric CO_2 than volcanic outgassing is metamorphism of carbon-rich pelagic sediments carried downward in subduction zones. Because subduction and seafloor spreading are fundamental aspects of plate tectonics, variations in plate motion necessarily must play a large role in controlling the rate of CO_2 buildup in the atmosphere. However, at the same time, CO_2 is removed from the atmosphere by the weathering of surface silicate rocks. As we have seen in Chapter 6, rainwater combines with CO_2 to form carbonic acid (H_2CO_3), and this acid weathers silicate rocks, as in the following reaction:

$$CaSiO_3 \; + \; H_2CO_3 \; \rightarrow \; CaCO_3 + SiO_2 \; + H_2O$$
(silicate rock) (carbonic acid) (weathering products) (water)

The weathering products are carried by streams to

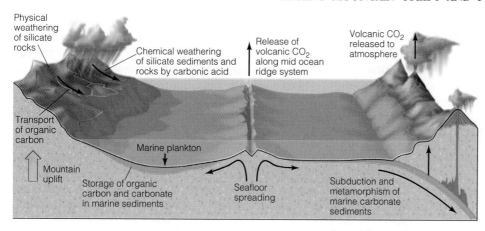

Figure 19.20 Diagram of carbon cycle showing primary natural sources of atmospheric CO_2 (volcanism and metamorphism), removal of CO_2 due to surface weathering, and transport and long-term storage of organic carbon in marine sediments.

the ocean. There the carbonate and silica are used by marine plankton, whose remains accumulate on the seafloor and are stored as sediment (Figs. 16.7 and 19.20). Weathering of silicate rocks, therefore, is a negative feedback in the climate system that can help move global climate toward an equilibrium condition. If seafloor spreading speeds up, more CO_2 enters the atmosphere, the Earth warms, the water vapor content of the atmosphere rises, and vegetation density increases. This speeds up the rate of chemical weathering of silicate rocks, which removes CO_2 from the atmosphere, thereby keeping the system in a more balanced state. If spreading slows, less CO_2 enters the atmosphere, the climate cools, weathering rates decrease, and the system adjusts to a near-balanced condition.

A further important negative feedback is burial of organic carbon. High erosion rates can lead to rapid sedimentation in sedimentary basins and in extensive submarine fans. Isolated from the weathering environment at the land surface, the carbon is quickly stored rather than being returned to the atmosphere by surface weathering. Recent studies suggest that carbon burial may be even more important than silicate weathering in long-term carbon storage.

An alternative hypothesis has been proposed in which tectonic uplift is the driving force behind changes in atmospheric CO_2 levels. Uplift could increase the rate of removal of atmospheric CO_2 because faulting exposes fresh, fractured rock; earthquakes promote landslides that dislodge rock from steep slopes; glacial and periglacial (physical) weathering is dominant on unvegetated high-altitude slopes; and runoff from high summer rainfall transfers physically weathered products from high and middle altitudes, where slopes are steepest, to lower altitudes, where they are chemically weathered in floodplains and deltas in warm, moist climates. Instead of being a relatively weak negative feedback, chemical weathering now becomes the major factor in controlling the long-term concentration of CO_2 in the atmosphere, and it focuses attention on areas of rapid uplift as primary sites influencing the Earth's carbon budget.

RECENT MOUNTAIN UPLIFT AND POSITIVE FEEDBACKS

The great altitude and relief of major mountain ranges like the Himalaya and Andes have led most geologists to assume that these impressive topographic features are the result of ongoing tectonic forces that have led to accelerated uplift during the late Cenozoic. But is tectonism the most likely or only cause? An alternative hypothesis questions this long-standing assumption and proposes that uplift may also be driven by important feedbacks in the Earth system.

If a mountain system is uplifted by tectonic forces (e.g., plate subduction or collision), erosion and mass-wasting will tear away at the rising landmass, transferring sediment to adjacent basins and to the ocean. Removal of this mass of rock creates an isostatic imbalance, the adjustment to which will cause the mountains to rise because of their reduced load on the underlying mantle (Fig. 19.21). Higher mountains provide an increased barrier to rain clouds, which will therefore lead to deep incision by rivers on the windward flank, further dissecting the mountains. The interesting consequence is that as the base of the crust rises due to a reduction in overlying load, and the mountains increase in altitude, the mean altitude of the mountains actually decreases. This somewhat surprising result is confirmed by recent studies of the topography of the Tibetan Plateau and the Himalaya. Although the mountain range rises to altitudes of more than 8000 m, its mean altitude of ca. 5000–5500 m is nearly identical to that of the adjacent relatively undissected plateau (Fig. 19.22).

An added factor in the equation is glaciation. If the isostatically rising mountains exceed the altitude of the snowline, glaciers will form and the rate of denudation may increase. This positive feedback link may lead to additional positive feedbacks which, as we have seen, include increased physical and chemical weathering, a reduction in atmospheric CO_2, a cooling of global climate, and a further lowering of the snowline. Repeated lowering of the snowline during dozens of Quaternary glaciations (Fig. 14.16), likely accelerated erosion at high altitudes and may have helped maintain a relatively constant mean altitude.

These contrasting hypotheses that attempt to explain landscape and climate evolution illustrate the complexity of the problem. The entire Earth system is involved, and not only the basic mechanisms but also the complex positive and negative feedbacks leave us with difficult questions to answer and challenging research problems to pursue. The answers will not come from any single field, but rather from collaborative investigations by scientists in many different fields of Earth science.

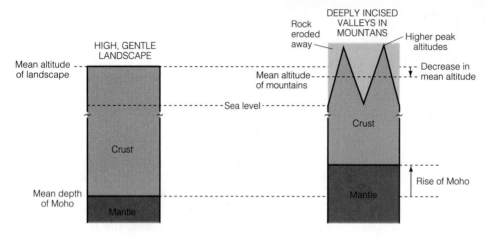

Figure 19.21 Simplified sections through the continental crust showing the effects of erosion on mean altitude, the depth of the Moho (crust/mantle boundary; see Chapter 5), and uplift of rock. If a change in climate causes a high, gentle landscape to be deeply incised by streams that erode valleys toward sea level, the remaining crustal rocks and the Moho would rise (due to isostasy), the mean altitude of the mountainous area would decrease slightly, and the highest peaks would be much higher than before.

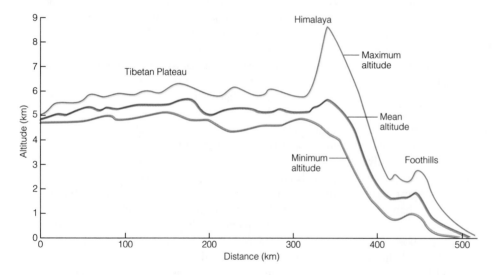

Figure 19.22 Topographic profile across the Tibetan Plateau and the Himalaya (see Figure 19.17b for orientation) showing that in spite of much higher altitudes, the average altitude of the Himalaya is nearly equal to that of the plateau.

SUMMARY

1. Natural processes are constantly changing the character of the landscape, but so, too, are human activities. Although generally operating at slow rates, natural processes can produce dramatic changes on geologic time scales.

2. Mass-wasting causes rock debris to move downslope under the pull of gravity without a transporting medium. The composition and texture of debris, the amount of entrapped air and water, and the steepness of the terrain influence the character and velocity of different types of slope movement.

3. The Earth's surface topography reflects sculpture by different erosional processes and deposition of resulting sediment. Water, ice, and wind are the principal agents that erode the land and transfer sediment toward the ocean basins. The principal factors influencing landscape development are process, climate, lithology, relief, and time.

4. A close relationship exists between rock structure resulting from the movement of crustal plates and the location of major drainage basins and continental divides.

5. Landscapes are constantly adjusting to changes in the factors that control their development and likely are never in a complete state of equilibrium.

6. Landscapes evolve through time as tectonic forces raise crustal rocks and erosional agents wear them away. Rates of denudation in some mountain areas appear to be approximately equal to rates of uplift.

7. Ancient landscapes with low relief imply long intervals of erosion that slowly lowered the land surface to low altitude. Although not common in modern landscapes at low altitudes, surfaces of low relief are preserved in some ancient rocks as widespread unconformities.

8. Late Cenozoic uplift of high mountains and plateaus has influenced world climate and led to local and hemispheric cooling, changed the path of the jet streams, created rainshadows that promoted desertification in continental interiors, and intensified monsoon circulation. Other effects include more frequent major summer monsoon floods and changes in the intermediate and deepwater circulation of the oceans.

9. Silicate weathering and burial of organic carbon remove CO_2 from the atmosphere and place it in long-term storage. Removal of CO_2 from the atmosphere reduces the greenhouse effect and influences world climate.

10. Mountain uplift may be the result both of tectonic activity and isostatic response to the erosion and transfer of rock debris away from mountain areas. In high, glaciated mountains rates of uplift and denudation may remain nearly balanced, resulting in relatively constant mean altitudes.

IMPORTANT TERMS TO REMEMBER

continental divide *448*   landslide *445*   relief *445*
denudation *451*   mass-wasting *445*

QUESTIONS FOR REVIEW

1. In what way does mass-wasting differ from stream erosion? Name three geologic factors that make high mountain regions especially prone to landslides.

2. How might one calculate the volume of sediment removed from the land surface during the last 2 million years by a major stream that has built a large delta where it enters the ocean?

3. How might drainage patterns provide clues to rock structure in an area of dense tropical forest?

4. How would you determine the total *relief* of an area appearing on a topographic map?

5. How do bedrock geology and structural history control the position of the continental divide in western North America?

6. What geologic evidence points to periods of relative landscape equilibrium at times in the past?

7. How does the concept of *thresholds* apply to landscape evolution?

8. Describe two ways that you might be able to determine the uplift rate in a coastal mountain system.

9. Do glaciers increase or decrease denudation rate? Why?

10. How can human activity influence sediment yield from a drainage basin?

11. How might biologic factors help control the denudation rate of a drainage basin?

12. How could mountain uplift be related to geologic factors other than tectonism?

QUESTIONS FOR DISCUSSION

1. Outline a plan for measuring the average present denudation rate in the region where you live. What erosional processes are active, and how might you measure their effectiveness?

2. During the first half of this century, erosional landscapes were generally believed to represent successive cycles of erosion, identifiable by stages of landscape evolution. This theory, developed by geographer William Morris Davis, predated the concept of plate tectonics. In your library, find and read some of Davis' work and discuss whether aspects of his theory are consistent with our present understanding of plate tectonics.

Through the case studies available at our web site, you can learn more about the geologic forces that shape our landscape.

CHAPTER 20

Global Change:
A Planet Under Stress

● Destroying the Ozone Shield

Ozone, a pale blue gas with a pungent odor, is present in the atmosphere in very small amounts—only 20 to 40 parts per billion by volume near the land surface. Without this gas, however, life on the Earth would be very different, for ozone provides living organisms with a protective shield against harmful ultraviolet radiation from the Sun. For humans, direct exposure to ultraviolet light damages the immune system, produces cataracts, substantially increases the frequency of skin cancer, and causes genetic mutations.

Maximum concentrations of ozone are found in a layer between 25 and 35 km (15.5 and 22 mi) above the Earth's surface, a region where ultraviolet radiation breaks down molecules of oxygen (O_2) into two oxygen atoms, which are then able to combine with other O_2 molecules to form molecules of ozone (O_3). The ozone is in turn broken down by ultraviolet radiation, thereby creating a balance among O, O_2, and O_3.

In 1985 British scientists working in Antarctica reported a startling discovery: a vast hole, about the size of Canada, had developed in the ozone layer above the Antarctic region. By 1987 measurements showed that, over Antarctica, concentrations of this life-protecting gas had dropped more than 50 percent since 1979 and, between altitudes of 15 and 20 km (9 and 12 mi), the depletion had reached 95 percent. Record low values were subsequently measured over Australia and New Zealand, and continuing surveys showed that ozone values at all latitudes south of 60° had decreased by 5 percent more since 1979.

What had happened to upset the natural atmospheric balance among the three gaseous forms of oxygen? A decade before the ozone hole was discovered, it was recognized that a group of synthetic industrial gases, the *chlorofluorocarbons* (*CFCs*), were entering the lower atmosphere and spreading rapidly around the world. As the CFCs ultimately rise into the upper atmosphere, ultraviolet radiation breaks them down, releasing chlorine. It is chlorine, in the form of chlorine monoxide (ClO), that does the damage: The chlorine atoms destroy the ozone, with each chlorine being capable of destroying as many as 100,000 ozone molecules before other chemical reactions remove the chlorine from the atmosphere. The sunlight and very cold springtime temperatures [−80°C (−176°F) or lower] in the upper atmosphere that are critical to ozone destruction are present in the south polar region, which is why the ozone hole is especially pronounced over Antarctica. In the Arctic, the period of the critical spring conditions is much shorter. With the documentation of ozone depletion in the upper atmosphere, scientists for the first time could show that human activity was having a detrimental global effect on one of the Earth's natural systems.

(opposite) Ozone in the upper atmosphere acts as a shield against harmful ultraviolet radiation. Record low concentrations of this gas in recent years, believed to be the result of human activity, places sunbathers at crowded beaches like this at increased risk of developing skin cancer.

Continuing measurements have shown that atmospheric ozone concentration is decreasing at all latitudes outside the tropics and that in the lower troposphere the rate of decrease is about 10 percent per decade. From ground-based and satellite observations, we know that ozone concentration during February 1993 was as much as 20 percent below normal over much of the northern hemisphere. Although the rate of increase of atmospheric CFC gas concentration has fallen markedly during the last five years with a cut in worldwide production, even if all production stops, recovery to natural conditions is likely to take a century or more. ●

THE CHANGING EARTH

Since earliest times, people have probably asked important questions about the Earth: How did the major features of our planet originate? Has the land always been the way we see it now? How can we explain the diversity of life on the Earth? What is the place of *Homo sapiens* in the multitude of living things that inhabit our planet? Have the Earth's climates always been as we find them today?

Study of the Earth has made it abundantly clear that our planet is a complex, dynamic system that is in a constant state of change. Furthermore, the solid, liquid, gaseous, and organic realms of the Earth are closely interlinked. A change in one part of the system is likely to affect other parts. For example, a massive earthquake might raise an extensive zone of coastal land, exposing and destroying nearshore marine habitats. A volcano might erupt lava that dams a river, thereby affecting other streams in the drainage system, while tephra and gases ejected into the atmosphere could lead to a hemisphere-wide drop in air temperatures. We can observe and measure such natural changes in progress or interpret them from the geologic record. However, only recently have we come to recognize that humans can also have a profound effect on the four major components of the Earth system—air, oceans, land, and biota.

Human-Induced Environmental Changes

In the 1840s, Henry David Thoreau traveled to Cape Cod and viewed the ocean from Provincetown at the very end of the Cape. At that time the Cape was little known to tourists and little visited. It was a rural countryside of fishermen and farmers. To Thoreau, the ocean seemed to be the wildest part of the Earth's surface and could never be affected by human actions. "Serpents, bears, hyenas, tigers, rapidly vanish as civilization advances," he wrote, "but the most populous and civilized city cannot scare a shark far from its wharves. It is no further advanced than Singapore, with its tigers, in this respect."[1] Little did he realize that events were already underway during his lifetime that would lead to human-induced global changes to the Blue Planet's life support system. Two kinds of human activities were underway in the ocean off his home state of Massachusetts: direct harvesting of marine populations and chemical alterations of the ocean water.

During one of his visits to Cape Cod, Thoreau saw fishermen driving small whales ashore where they were harvested for blubber. Thoreau reported this as an interesting economic activity, never suggesting the possibility that human actions might reduce the number of some whales to threaten them with extinction. But at that time, Yankee whaling ships were departing from New Bedford, Massachusetts, on a global hunt of Bowhead whales. By the beginning of World War I, the abundance of these huge mammals decreased from 20,000 to about 3,000, threatening them with extinction.

Since Thoreau's time, modern civilization has developed technologies that can and do affect the entire global environment. These effects arise from the sheer number of human beings and the growth of technology, which increases the power of each person to affect the environment.

Human impacts on the global environment can be divided into two major kinds, *direct* and *indirect*. Burning fossil fuels is an example of a direct global effect, because the combustion products are emitted into the air where they are mixed and directly affect the atmosphere at a global level. Pollution of water is indirect globally in the sense that the pollution occurs locally and regionally, but because the human population is widely distributed, water is polluted throughout the world. By summation, the effect is global.

OUR PLANET'S GROWING POPULATION

We had little direct knowledge about the size of the human population until rather recently. The first attempts to estimate global population were made near the end of the seventeenth century, when world population was probably no more than about 700 million (Fig. 20.1). National censuses helped improve estimates during the nineteenth century, but only after the Second World War did estimates of global population become reasonably reliable. Estimates of earlier human population size are based, in part, on the way people obtained food. Hunter-gatherers can live only at a low density without overusing their food supply. Primitive agriculture therefore allowed a major increase in human population density, but far less than modern agriculture permits. Based on this kind of information, and the fossil and historic record of the distribution of our species, estimates of human population growth have been made.

The long-term trend is clear: whereas it took countless millennia for the human species to number a billion (toward the beginning of the nineteenth century), that number has doubled twice since then and is now approaching 6 billion. At present, the human population is growing at the rate of approximately 1.5 percent per year, an increase of about 80 million people a year. This is more than the population of France (56,942,000), and about twice the population of Spain (40,092,000). The sheer growth in population has led to an ever-increasing interaction between people and their environment, so that now even the most remote parts of the planet are being affected.

Traditionally, the debate about the growing human population has been divided into two camps: the Malthusians and the anti-Malthusians. About 200 years ago, the English economist Thomas Malthus (1766–1834) eloquently stated that the human population problem is based on three premises: (1) food is necessary for people; (2) passion between the sexes is necessary and will remain nearly in its present state so that children will continue to be born; and (3) the power of population growth is much greater than the power of the Earth to produce subsistence. As a result, the human population will eventually grow beyond the Earth's capacity to provide food and other necessary resources. The anti-Malthusians argue that technology has always found a way, and will do so in the future, to make possible a greater density of people on the Earth. The global perspective of this book makes clear that the Earth system is finite and some upper bound must occur. The exact upper bound depends on the quality of life that people want to maintain, as well as the direct limit of resources.

Over many millennia prior to the Industrial Revolution, people slowly changed the Earth's natural landscapes as they built villages and cities, converted forests to agricultural land, and locally dammed and diverted streams. With the development of industrial technology, mineral and energy resources were needed to fuel an increasingly populous and demanding society. The exploitation of fossil fuels helped raise the standard of living for most people well beyond that of their forebears.

In spite of the obvious benefits involved, the exploitation of our planet's rich natural resources has not been without cost. In many parts of the world, environmental deterioration is epidemic. In addition to scarring and poisoning the Earth's land surface, we have also, unwittingly, polluted the oceans and groundwater and changed the composition of the atmosphere. Even in places long considered to be the most remote on the planet—the frigid ice sheets of Antarctica, the vast Amazon rainforest, the trackless desert of Saudi Arabia, the lofty summits of the Himalaya—the impact of human activities is being felt. Today, human and natural geologic activities are inextricably intertwined, and it is increasingly apparent that people have become a major factor—a global factor—in environmental change.

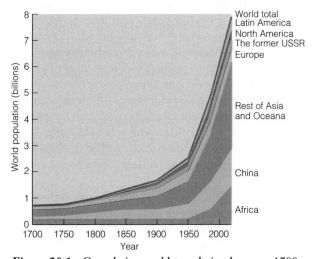

Figure 20.1 Cumulative world population between 1700 and the present obtained by summing values for each of the major inhabited regions of the world. Projected values extend the present estimates to the year 2020.

HUMAN IMPACTS ON FRESH WATERS

Rivers and People

As the human population has grown and become increasingly industrialized, people have altered the natural flow of rivers and communities have generated ever-larger amounts of human and industrial wastes. A good deal of these wastes have inevitably found their way into the very water that people must rely on for their existence. In many places, water is dwindling in both quantity and quality, creating important questions for the communities involved: Will there be enough clean water to sustain future needs? Is its quality adequate for the uses to which we put it? Is the water being used with a minimum of waste?

Few of the world's large rivers in the densely populated regions of the Earth now flow unrestricted to the sea, and most are contaminated, to various degrees, with human-generated wastes. The estuaries of almost every major river of the world are seriously altered by human actions, especially from the release of sewage and artificial chemicals. This human impact has produced serious problems for natural ecosystems, as well as for people who live beside rivers and use their water for personal and industrial purposes.

Hydroelectric Power and Artificial Dams

Hydroelectric power is recovered from the potential energy of water in streams as they flow downslope to the sea. Water (hydropower) is an expression of solar power because it is the Sun's heat energy that drives the water cycle. That cycle is continuous, and so energy obtained from flowing water is also continuous. Water power has been used in small ways for thousands of years, but only in the twentieth century has it been used widely for generating electricity. All the water flowing in the streams of the world has an estimated 9.2×10^{19} J/yr of recoverable energy, an amount equivalent to burning 15 billion barrels of oil per year. Hydropower appears to be greatly beneficial to people. Unlike coal and oil, hydropower cannot be used up; it is a renewable resource. Unlike fossil fuels, hydropower is "clean" in the sense that it does not contaminate the Earth's atmosphere or waters.

In the early twentieth century, hydropower was viewed as an environmentally benign and renewable resource, a kind of ideal energy source. Major hydropower projects are under construction in many developing nations. One of the most famous is in the Three Gorges Dam watershed in China, where scenery famous throughout the history of that populous nation will be buried under a series of large reservoirs in order to provide electricity and control floods. Today the ecological effects of dams and stream-channel control are recognized. The worldwide alteration of major rivers is having a widespread effect on biological diversity through the loss of habitat for fish of commercial and recreational interest, habitat for the food that sustains these fish and habitat for migrating waterfowl. A fundamental question facing people today is how best to meet the energy and irrigation requirements of human beings while sustaining river ecosystems.

One objective of constructing large artificial dams is flood control. A dam can help modulate the flow of water along a stream system, thereby reducing the impact of peak flow events. The results can be impressive, as illustrated by Egypt's Aswan Dam (Fig. 20.2). This huge dam, with its vast reservoir (Lake Nasser) backed up behind it, permits the controlled release of water. This markedly reduces the seasonal variability in discharge that formerly could produce devastating floods (Fig. 20.3).

Despite the success of the Aswan Dam project, it brought some unanticipated consequences. The Nile, like all other large streams, is a complex natural system, and its behavior reflects a delicate balance between water flow and sediment load. Ninety-eight percent of the Nile's sediment load is carried in suspension. Prior to construction of the dam, an average of 125 million metric tons (275 billion lbs) of sediment passed downstream each year, but the dam reduced this value to only 2.5 million metric tons (5.5 billion lbs) because nearly 98 percent of the suspended sediment is now deposited in Lake Nasser. It is estimated that the accumulating sediment will fill the reservoir in about 500 years, terminating its useful life. Under natural pre-dam conditions, floodwater carried this sediment downstream where much of it was deposited, thus adding to the rich agricultural soils in the Nile Valley and the delta at a rate of 6 to 15 cm/century (2 to 6 in/century). With this natural source of nourishment eliminated, farmers must now resort to artificial fertilizers and soil additives to keep the land productive. Some of the fertilizer seeps back into the river, causing pollution problems downstream.

Before the Aswan Dam was constructed, the annual Nile flood transported at least 90 million metric tons (200 billion lbs) of sediment to the Mediterranean Sea. The shoreline of the sea at that time reflected a balance between sediment supply and the attack of waves and currents that redistributed the sediment along the Mediterranean coast. Because the annual discharge of sediment has now been cut off, the coast has become increasingly vulnerable to erosion. Over time the shoreline will adjust to the reduction in Nile sediment until a new balance is reached.

Figure 20.2 View of Aswan High Dam, which impounds the Nile River (right) to form Lake Nasser (to south behind dam). Sediment formerly carried northward to the Mediterranean Sea and now settling out in the lake will eventually fill the reservoir and make it unusable.

Mining Groundwater

In the dry regions of western North America, where streams are few and average discharge is low, groundwater is a major source of water for human consumption. In many of these regions, withdrawal exceeds natural recharge, so that the volume of stored water is diminishing and the water table is falling, often at an increasing rate. In the same way that petroleum is being steadily withdrawn from the most accessible oil pools and minerals are being mined from accessible rocks of the upper crust, groundwater also is being mined. We regard petroleum, coal, and minerals as nonrenewable resources, for they form only over geologically long intervals of time. We don't often stop to think that groundwater can also be a nonrenewable resource. In some regions, natural recharge would take so long to replenish depleted aquifers that formerly vast underground water supplies have essentially been lost to future generations. Even where the problem has been recognized and measures have been taken to stem the loss, centuries or millennia of natural recharge will be required to return aquifers to their original state.

To halt the fall of the water table, methods have been developed to recharge groundwater artificially. For example, runoff from rainstorms in urban areas that normally would flow away in surface streams can

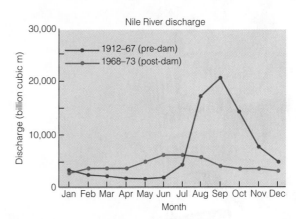

Figure 20.3 Prior to construction of the Aswan Dam, the discharge of the Nile River varied seasonally, with peak discharge coming during the late summer and early fall interval of flooding. Controlled release of water after the dam was built greatly reduced the seasonal variability in discharge.

A CLOSER LOOK

Toxic Groundwater in California

Selenium is a naturally occurring element and a necessary trace nutrient in our diet, as well as in the diet of livestock and many wild animals. However, if excessive amounts are ingested, it can prove toxic. In 1983, selenium attracted national attention when the U.S. Fish and Wildlife Service reported fish kills and high incidences of mortality, birth defects, and decreased hatching rates in nesting waterfowl at the Kesterson National Wildlife Refuge in California's San Joaquin Valley (Fig. C20.1). Laboratory studies showed high concentrations of selenium in fish from Kesterson Reservoir, located in the refuge, and birds using the reservoir were found to have high concentrations of selenium and obvious symptoms of selenium poisoning.

Geologists subsequently showed that the poisoning of wildlife at Kesterson resulted from a combination of geological, hydrological, and agricultural factors. The western San Joaquin Valley is a prime agricultural area, but because it lies in a zone of arid climate, the land is irrigated. The irrigation artificially raised the water table to such a degree that a system of subsurface drains was established to remove excess water and funnel it northward along the valley to Kesterson Reservoir.

Rainwater falling in the Coast Range immediately west of the San Joaquin Valley dissolves selenium-bearing salts from sedimentary rocks. Surface runoff then carries the dissolved salts to broad alluvial fans in the valley, where the water seeps into the ground and recharges a shallow regional aquifer. High evaporation in this arid region concentrates the salts in the soil. An irrigation system on the fan surfaces flushes salts out of the soil and into the drainage canal that leads to Kesterson Reservoir. Because the reservoir has no outlet, selenium became concentrated there until it reached toxic levels.

Groundwater conditions in the western San Joaquin Valley contribute to the Kesterson Reservoir problem. Most of the shallow groundwater in this area is alkaline and slightly to highly saline. Under these conditions, selenium is very soluble and is carried with the flowing groundwater. This mobility greatly increases the ease with which selenium moves from the soils and into the artificial drainage system.

A further condition is the presence of a clay layer 3 to 23 m beneath the land surface. Being impermeable, the clay layer restricts the downward percolation of selenium-bearing groundwater and produces a perched water body. It is this perched water body above the clay aquiclude that necessitates the drain system, which in turn creates the environmental hazard at Kesterson Reservoir.

Although we are constantly alerted to the environmental impact of pesticides and other manufactured poisons introduced into natural ecosystems, the Kesterson saga illustrates how natural substances that pose no special hazard under normal conditions can reach toxic levels through human intervention. The number of such problems is increasing as an expanding human population places ever greater demands on limited natural resources.

be channeled and collected in basins where it will seep into permeable strata below, thereby raising the water table. Groundwater withdrawn for nonpolluting industrial use can be pumped back into the ground through injection wells, recharging the saturated zone.

The water pressure in the pores of an aquifer helps support the weight of the overlying rocks or sediments. When groundwater is withdrawn, the pressure is reduced, and the particles of the aquifer shift and settle slightly. As a result, the land surface subsides. The famous Leaning Tower of Pisa, Italy, built on unstable fine-grained alluvial sediments, began to tilt when construction began in 1174 (Fig. 20.4). The tilting increased during the twentieth century as ground-

Figure 20.4 The Leaning Tower of Pisa, Italy, the tilting of which accelerated as groundwater was withdrawn from aquifers to supply the growing city.

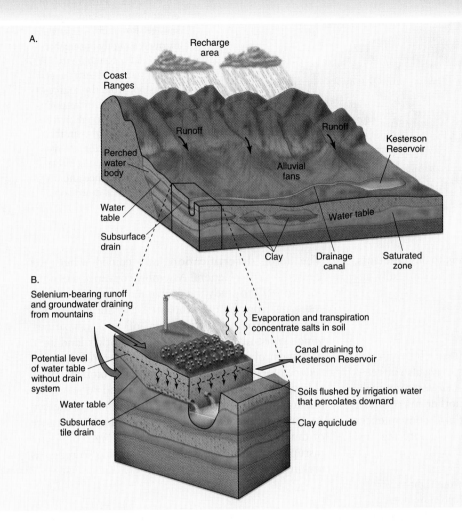

A.

Recharge area

Coast Ranges

Runoff

Runoff

Kesterson Reservoir

Perched water body

Alluvial fans

Water table

Subsurface drain

Water table

Clay

Drainage canal

Saturated zone

B.

Selenium-bearing runoff and groundwater draining from mountains

Evaporation and transpiration concentrate salts in soil

Canal draining to Kesterson Reservoir

Potential level of water table without drain system

Water table

Soils flushed by irrigation water that percolates downward

Subsurface tile drain

Clay aquiclude

Figure C20.1 Geologic setting of Kesterson Reservoir in the western San Joaquin Valley, California. Runoff carries dissolved selenium-bearing salts from the coastal mountains to alluvial fans where the salts are precipitated in the soil. Artificial irrigation water then flushes the salts into a drainage system that carries the selenium to Kesterson Reservoir where it becomes concentrated to toxic levels.

water was withdrawn from deep aquifers. Recent strengthening of the foundation is designed to keep the tower stable in the future, providing that groundwater withdrawal is strictly controlled.

Groundwater Contamination

Citizens of modern industrialized nations take it for granted that when they turn on a faucet, safe, drinkable water will flow from the tap. Throughout much of the world, however, the available water is often unfit for human consumption. Not only do natural dissolved substances make some water unpotable, but also many water supplies have become severely contaminated by human and industrial wastes (see "A Closer Look: Toxic Groundwater in California").

The most common source of water pollution in wells and springs is sewage. Drainage from septic tanks, broken sewers, privies, and barnyards contaminates groundwater. If water contaminated with sewage

bacteria passes through sediment or rock with large openings, such as very coarse gravel or cavernous limestone, it can travel long distances and remain polluted (Fig. 20.5). On the other hand, if the contaminated water seeps through sand or permeable sandstone, it can become purified within short distances, in some cases in less than about 30 m (33 yd). Sand is ideal for cleaning up polluted water; it promotes purification by (1) mechanically filtering out bacteria (water gets through, but most of the bacteria do not), (2) oxidizing bacteria so they are rendered harmless, and (3) containing other organisms that consume the bacteria. For this reason, purification plants that treat municipal water supplies and sewage pass these fluids through sand. Although natural purification is as effective, it occurs much more slowly than in a water treatment plant.

Groundwater discharging directly to the oceans carries dissolved chemicals that likely constitute a significant chemical input to coastal and nearshore

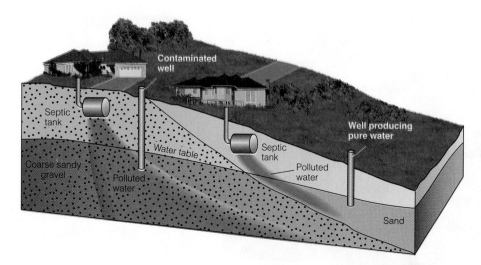

Figure 20.5 Purification of groundwater contaminated by sewage. Pollutants seeping through a highly permeable sandy gravel contaminate the groundwater and enter a well downslope from the source of contamination. Similar pollutants moving through permeable fine sand higher in the stratigraphic section are removed after traveling a relatively short distance and do not reach a well downslope.

zones. If such groundwater contains toxic pollutants, the effect on coastal ecosystems can be especially adverse.

Desertification

People have affected the environment by trying to make water more available, but they have also affected large regions by creating deserts—inadvertently making water less available. This is another major effect that is regional in extent but has global consequences because it is occurring over more and more regions.

The term **desertification** was coined when the United Nations General Assembly convened a conference in 1977 to study the problem of land degradation resulting from human impact. Desertification is the major environmental problem of arid landscapes, which constitute 40 percent of the world's land area. The most obvious symptoms include crop failures or reduced yields, reduction in rangeland biomass available to livestock, reduction in fuelwood supplies, reduction in water supplies resulting from decreased streamflow or a depressed groundwater table, advance of dune sand over agricultural lands (Fig. 20.6), and disruption of life-support systems leading to refugees seeking outside relief.

Figure 20.6 Barchan dunes advance from right to left across irrigated fields in the Danakil Depression, Egypt.

Africa's Sahel

Recently, significant land degradation and drylands expansion in the populated Sahel region of Africa, a semiarid belt bordering the Sahara, have attracted world attention because of widespread famine (Fig. 20.7). Although desert expansion can result from natural processes, it is widely thought that excessive human exploitation of drylands, generally linked to increasing human and livestock populations, can lead to progressive deterioration of the land and ultimately to desertification. Sometimes the process is triggered or exacerbated by natural drought. Although a strong scientific case can be made that human activity has caused desertification in some areas, opinions are not always consistent. Studies in the Sahel have led some researchers to conclude that arid land areas have increased by nearly 54 million hectares (133 million acres) since 1931 as a result of a 30 percent decline in rainfall during this interval. Other scientists found no evidence of persistent trends toward desert conditions between 1962 and 1984 and concluded that observed vegetation and crop changes could be explained by annual rainfall variations alone. A group studying the desert margin in western Africa, on the other hand, assembled evidence showing that the desert advanced an average of 10 km/yr (6 mi/yr) between 1961 and 1987.

The complex linkage between humans and their environment is illustrated by a hypothesis of progressive desertification that links grazing animals, the vegetation they consume, and the overlying atmosphere. As animals consume vegetation cover in drylands, the albedo (reflectivity) of the land surface increases. (Sand and bare rock have a higher albedo than grassland does.) This causes more of the incoming solar radiation to be reflected back into space, leading to a cooler ground surface. The cooler ground is associated with descending dry air and therefore with reduced precipitation. In this way, the degraded area becomes more and more desertlike. In other words, a *positive feedback* is set in motion that promotes increasing desertification once the process begins.

Forests and Deforestation

Over the last 3000 to 4000 years, humans have been the primary agent in changing the world's vegetation cover. As the human population expanded, forests were replaced with agricultural land. The increasing exploitation of forests for fuel and other economic

Figure 20.7 Overgrazing during years of drought killed much of the vegetation in this part of the Sahel in Senegal. Without vegetation, topsoil is eroded and the land becomes infertile.

Table 20.1   Food and Agriculture Organization (FAO) 1991 Estimates of Recent Forest Cover and Deforestation for the Tropical Regions

Continent	Forest Area in 1980 (10⁶ ha)	Forest Area in 1990 (10⁶ ha)	Annual Deforestation (1981–1990) (10⁶ ha)	Rate of Change (1981–1990) (%/year)
Africa	650	600	5.0	−0.8
Latin America and Caribbean	923	840	8.3	−0.9
Asia	321	275	3.6	−1.2
Total	1894	1715	16.9	

Source: From M. K. Tolba, and D. A. El-Kholy (eds.) *The World Environment 1972–1992.* (London: Chapman and Hall, 1992, p. 169.)

uses has had a dramatic impact in many regions. People have long had a love-hate relationship with forests. On the one hand, forests have always provided essential resources for civilization: fuel, material for construction, sources of medicinal plants. On the other, the dark interior of forests has long made these seem a fearful place about which many myths and stories have originated. Because forest lands often can be converted to productive agricultural fields, in many places forests have been removed not so much for their resources but to make room for crops. This was a prime motivation in the clearing of land in New England during colonial times and into the eighteenth century.

Over major parts of the world, the natural vegetation has all but disappeared, to be replaced by an artificial one dominated by agriculture and introduced species. It is generally estimated that forests covered a quarter of the Earth's land area in 1950. By 1980 the area had been reduced to only one fifth. Today there are approximately 2.5 billion ha (6.2 billion acres) of closed forest and 1.2 billion ha (3 billion acres) of open woodlands and savannas in the world.

The spread of European civilization since A.D. 1500 has been responsible for much of this **deforestation**, or forest removal. In the United States, an estimated 60,000 km² (23,000 mi²) had been cleared by 1850 and 660,000 km² (255,000 mi²) by 1910. In the tropics, forests covering an estimated 2,400,000 km² (930,000 mi²) were cleared between 1860 and 1978. Although much of the forest clearing carried out in North America and part of the tropics has been related to commercial exploitation of agricultural and forest products, in Africa nearly 60 percent is related to fuelwood production. At present, the tropical forests are receiving the greatest impact. Data for 87 countries in Africa, Latin America, and Asia show that between 1980 and 1990 forest clearing was eliminating an average of about 1 percent of the forest cover per year (Table 20.1).

Recent land clearing has been especially intense in the tropical Americas. In Costa Rica, for example, forest covered 67 percent of the country in 1940, but by 1983 the figure had been reduced to 17 percent (Fig. 20.8. Especially hard-hit has been the Amazon basin of Brazil, where the arrival of new settlers in the late 1970s reached 5000 per month. Massive land clearance adjacent to major access roads rapidly reduced the forest cover, but the unproductive soils made agriculture unprofitable, and the cleared lands have also proved unsuitable for sustainable cattle ranching.

Figure 20.8   Dramatic deforestation in Costa Rica between 1940 and 1983 reduced the percentage of forest as a proportion of the total area of the country from 67 percent to 17 percent.

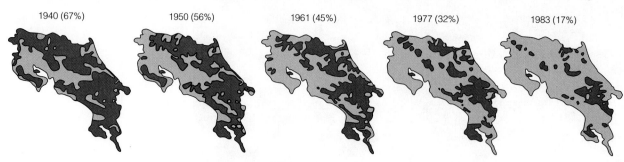

1940 (67%)   1950 (56%)   1961 (45%)   1977 (32%)   1983 (17%)

Deforestation and Climate

The environmental implications of deforestation in Amazonia reach well beyond the economic arena, however. Research in this region has shown that when the forest is cleared, the hydrological balance can be severely upset. In one study area, streams carry away 25 percent of the rainfall, whereas trees and other plants return 50 percent of the precipitation to the atmosphere by transpiration. Moist air masses moving inland from the Atlantic bring about half the water vapor that falls as rain; the other half is supplied by the forest as a result of of evapotranspiration. This means that the forest plays a key role in maintaining the precipitation balance, and thus the forest itself, in the Amazon basin.

Removal of forest changes the hydrological equation. Without a forest cover to promote infiltration, more of the rainfall runs off the land and far less is recycled through plants back to the atmosphere. As a result, the potential exists for a *negative feedback*, whereby destruction of forest leads to reduction in the rainfall that the forest requires for its very existence. Although the ultimate effects of continued deforestation are still difficult to predict, an estimated 12 percent of the Amazonian rainforest has already been cleared, and in some regions the rate of land clearance has been increasing exponentially. If continued unchecked, the repercussions could include a significant change in the regional hydrology, which might lead to widespread reduction in soil moisture, increased flood discharges, and worsening droughts.

Logging can also have a deleterious effect on natural streams. In the rugged coastal mountains of northwestern United States, logging operations have often involved **clearcutting**—the removal of all trees within an area, followed by replanting of seedlings to restore the forest. However, one effect of this practice has been a dramatic increase in sediment supplied to streams. Once the trees are felled, the natural protection from erosion provided by trees rooted on steep hillslopes is lost. Both hillslopes and logging roads then become sites of increased runoff and erosion. The end result is stream channels clogged with debris and disruption of natural ecosystems. This has led to new forest practices that are designed to permit both commercial forestry and fisheries to operate fairly and effectively, with minimum disruption of the environment.

Soil Erosion

With world population now approaching 6 billion, increasing competition for a finite amount of agricultural land is causing serious erosion of soils. Although soil erosion results from natural changes in topography, climate, or vegetation cover, the effects of human activities have overwhelmed natural systems in many parts of the world. Soil erosion is often closely related to deforestation. Widespread felling of trees has led to accelerated rates of surface runoff and destabilization of soils due to loss of anchoring roots. Soils in the humid tropics, when stripped of their natural vegetation cover and cultivated, quickly lose their fertility (Fig. 20.9). So widespread are the effects of soil ero-

Figure 20.9 Widespread deforestation in Rondonia, Brazil has devastated a formerly luxuriant rain forest and led to accelerated runoff and erosion. Soils on the landscape quickly lose their natural fertility when forest is converted to crops or grazing land, leaving a degraded landscape with little value.

sion and degradation that the problem has been described as epidemic. Because agriculture is the foundation of the world economy, progressive loss of soil signals a potential crisis that could undermine the economic stability of many countries.

The upper layers of a soil contain most of the organic matter and nutrients that support crops. When these layers are eroded away, not only the fertility but also the water-holding capacity of a soil diminishes. Because it generally takes between 80 and 400 years to form one centimeter of topsoil, soil erosion, for all practical purposes, is tantamount to mining the soil. It is estimated that farmers in the United States are now losing about 5 tons of soil for every ton of grain they produce, whereas in India the rate of soil erosion is estimated to be more than twice as high. Worldwide, the most productive soils are being depleted at the rate of 7 percent each decade. One recent estimate projected that, as a result of excessive soil erosion and increasing population, only two-thirds as much topsoil will be available to support each person at the end of the century as was available in 1984.

Although soil erosion and degradation are severely impacting many countries, effective control measures can substantially reduce these adverse trends. One method of reducing soil loss involves crop rotation. A study in Missouri showed that land which lost 49.25 tons of soil per hectare when planted continuously in corn lost only 6.75 tons per hectare when corn, wheat, and clover crops were rotated. In this case, the bare land exposed between rows of corn is far more susceptible to erosion than land planted with a more continuous cover of wheat or clover.

The most serious soil erosion problems occur on steep hillslopes. In Nigeria, land planted with cassava (a staple food source) and having a gentle 1 percent slope lost an average of 3 tons of soil per hectare each year, but on a 15 percent slope, the annual erosion rate increased to 221 tons per hectare, a rate that would remove all topsoil within a decade. Despite these grim statistics, steep slopes can be exploited through appropriate terracing.

Toxic Wastes and Agricultural Poisons

Vast quantities of human garbage and industrial wastes are deposited each year in open basins or excavations at the land surface. When a landfill site reaches its capacity, it generally is covered with earth and revegetated. Many of the waste products, now underground, are mobilized by precipitation that percolates through the site, carrying away soluble substances. In this way, harmful chemicals slowly leach into groundwater systems and contaminate them, potentially making them unfit for human use. The pollutants travel from landfill sites as plumes of contaminated water in directions that depend on the regional groundwater flow pattern, and they are dispersed at the same rates as the moving water (Fig. 20.10). The pollutants often are toxic not only to humans but also to plants and animals in the natural environment.

In the United States, pollution problems associated with landfill wastes have become so severe that the government has begun a major long-term effort (the Superfund Program) to clean up such sites and render them environmentally safe. However, the identified sites number in the tens of thousands, and it is difficult to judge how much time and money ultimately will be required to accomplish the task.

Another hazard is posed by toxic chemicals. Each year pesticides and herbicides are sprayed over agricultural fields to help improve quality and productivity. However, some of the chemicals have been linked with cancers and birth defects in humans, and others have led to major declines of wild animal populations. Because of the way they are spread, the toxic chemicals invade the groundwater system over wide areas as precipitation flushes them into the soil.

Underground Storage of Hazardous Wastes

A leading environmental concern of industrialized countries is the necessity of dealing with highly toxic or radioactive waste products. Experience has demonstrated that surface dumping quickly leads to contamination of surface and subsurface water supplies and thereby to the possibility of serious and potentially fatal health problems (Fig. 20.11). Countries with nuclear capacity have the special problem of disposing of high-level radioactive waste products, substances so highly toxic that even minute quantities can prove fatal if released to the surface environment.

Most studies concerning disposal of toxic and nuclear wastes have concluded that underground storage is appropriate, provided safe sites can be found. In the case of high-level nuclear wastes that can remain dangerous for tens or hundreds of thousands of years, a primary requirement is a site that will remain stable over long time periods. The only completely safe sites would be those where waste products and their containers would not be affected chemically by groundwater, physically by natural deformation such as earthquakes, or accidentally by people.

Impacts of Mining

People have been mining the Earth for mineral resources for millennia, but within the past two centuries mining activity has increased to the point where the impact on natural stream systems has sometimes

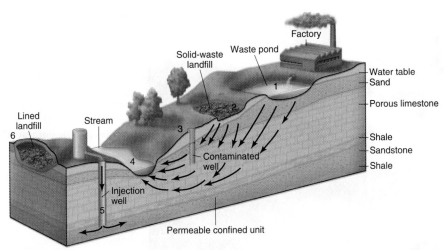

Figure 20.10 A groundwater system contaminated by toxic wastes. Toxic chemicals in an open waste pond (1) and an unlined landfill (2) percolate downward and contaminate an underlying aquifer. Also contaminated are a well downslope (3) and a stream (4) at the base of the hill. Safer, alternative approaches to waste management include injection of waste into a deep confined rock unit (5) that lies well below aquifers used for water supplies, and a carefully engineered surface landfill (6) that is fully lined to prevent downward seepage of wastes. Because neither of the latter approaches is completely foolproof, constant monitoring at both sites would be required.

Figure 20.11 Toxic substances in this New Jersey waste pond could constitute a serious health hazard for nearby residents if the fluid wastes leak into the groundwater system.

been dramatic. Mining can lead to increased soil erosion and to the release of toxic chemicals into the environment. The chemicals include those used in the mining process as well as those that are part of the mineral deposits.

Following the discovery, in the 1840s, of gold in stream channels draining California's Sierra Nevada, the rush for riches led to the development of hydraulic mining. This technique involves directing water under high pressure against unconsolidated gold-bearing sediments that accumulated along ancient stream channels. The sediments are washed into a wooden conduit where the heavy flakes and nuggets of gold are concentrated, while the remaining coarse alluvium is flushed away (Fig. 20.12). However, the unwanted alluvium was a load greater than the streams could handle, leading to widespread deposition downstream. The effects ultimately were felt as far away as San Francisco Bay where, in the years between 1850 and 1914, an estimated 1.28 billion cubic m (45 billion ft³) of hydraulically mined sediment was deposited in the northeastern arm of the bay. This is at least eight times the amount of sediment excavated during construction of the Panama Canal. The sediment destroyed fish spawning grounds, obstructed navigation, and ultimately reduced the area and volume of the bay, thereby modifying the tidal circulation.

Figure 20.12 Hydraulic mining in the foothills of California's Sierra Nevada. Gold-bearing alluvium is washed into wooden channels (sluices) where flakes and nuggets of gold are concentrated . Although this was an efficient way of mining in the years following the Gold Rush, it led to widespread environmental degradation of streams, and its impact was felt as far away as San Francisco Bay.

THE CHANGING ATMOSPHERE: THE INFLUENCE OF MODERN TECHNOLOGY

One of the direct ways that people are changing the global system—potentially one of the largest impacts people may have on the Earth system—is by changing the climate. As we learned in Chapter 14, the Earth's climate system consists of a number of interacting subsystems that involve the atmosphere, the hydrosphere, the solid Earth, and the biosphere (Fig. 14.1). The interactions are extremely complex and difficult to analyze. Consequently, only with the advent of supercomputers have we begun to answer some of the basic questions about how the climate system works.

The geologic record plays an important role in this enterprise, for it contains a history of the Earth's changing climates that extends into the remote past. Climatic change can be read from the stratigraphic record in many ways. For example, paleontologists infer past climates from the assemblages of fossil plants and animals they find in ancient strata. Sedimentologists and stratigraphers can infer many things about past climates from the nature of the sediments they study: the present distribution of these sediments, their mineralogy, the varied depositional environments represented, features that indicate the agencies of sediment transport, and the soils that represent former land surfaces. Isotope geologists can determine past surface temperatures from studies of sediments on land, in the oceans, and in polar ice sheets.

By reconstructing past climates, the range of climatic variability on different time scales can be determined, and the accuracy of computer models that try to simulate past climatic conditions can be tested.

These studies help us learn how the climate system behaves, what controls it, and how it is likely to change in the future. We know it will change, but we lack a clear view of how and at what rate. The answers are important not only scientifically, but sociologically and politically as well. In extracting and burning the Earth's immense supply of fossil fuels, people have unwittingly begun a great "geochemical experiment" that may well have a significant impact on our planet and its inhabitants.

The Carbon Cycle

A basic chemical substance involved in the climate "experiment" is carbon, an element that is essential to all forms of life. As explained in Chapter 16, carbon occurs in all four reservoirs of the Earth system (Fig. 20.13).

The key to the carbon cycle is the biosphere, where plants continuously extract CO_2 from the atmosphere and then break the CO_2 down by photosynthesis to form organic compounds. Animals consume plants and use these organic compounds in their metabolism. When plants and animals die, the organic compounds decay by combining with oxygen from the atmosphere to form CO_2 again. The passage of organic material through the biosphere is so rapid that the entire content of CO_2 in the atmosphere cycles every 4.5 years.

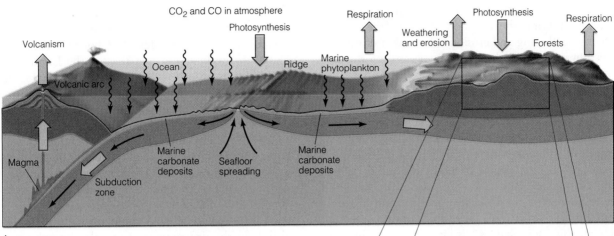

A.

Figure 20.13 The carbon cycle. A. Natural fluxes of carbon through the atmosphere, hydrosphere, biosphere, and lithosphere. Carbon enters the atmosphere through volcanism, weathering, biological respiration, and decay of organic matter in soils. Photosynthesis incorporates carbon in the biosphere, from which it can become part of the lithosphere, if buried with accumulating sediment. B. Human activities release carbon to the atmosphere through burning of forests and fossil fuels.

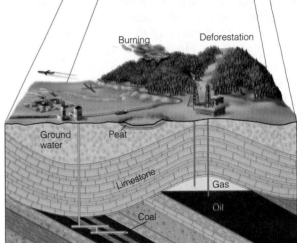

B.

Not all dead plant and animal matter in the biosphere immediately decays back to CO_2. A small fraction is transported and redeposited as sediment; some is then buried and incorporated in sedimentary rock where it locally forms deposits of coal and petroleum. The buried organic matter joins the slower moving rock cycle and can reenter the atmosphere naturally only after uplift and erosion have exposed the rock in which it is trapped.

Carbon dioxide from the atmosphere is also dissolved in the waters of the hydrosphere. There aquatic plants use it in the same way that land plants use CO_2 from the atmosphere. In addition, aquatic animals extract calcium and carbon dioxide from the water to make shells of $CaCO_3$. When the animals die, the shells accumulate on the seafloor, mixing with any $CaCO_3$ that may have been precipitated as chemical sediment. When compacted and cemented, the $CaCO_3$ forms limestone (Fig. 20.14). In this way, too,

Figure 20.14 Vast bodies of carbonate rocks, like the Dolomites of northern Italy, constitute reservoirs of carbon dioxide that have been temporarily removed from the carbon cycle. Once exposed at the surface, carbonate rocks are weathered and eroded, thereby freeing CO_2 which reenters the carbon cycle.

some carbon joins the rock cycle. Eventually, the rock cycle will bring the limestone back to the surface where weathering and erosion will break it down; the calcium returns in solution to the ocean, and the carbon escapes as CO_2 to the atmosphere.

Now, let's consider what happens when human activities change these four carbon reservoirs and influence the exchanges between them. While any individual action may appear insignificant, the cumulative effects of all human activities are now so great as to be measurable. The burning of fossil fuels and the clearing of forested land cause CO_2 to move from the lithosphere and the biosphere to the atmosphere at rates much faster than they would move naturally. Unless this additional CO_2 is dissolved in the hydrosphere or is buried in sediments as fast as it is generated, the CO_2 content of the atmosphere must inevitably increase. However, the rate at which these natural processes are removing CO_2 from the atmosphere is slower than the rate at which human activities are adding it, leading us to conclude that the CO_2 content of the atmosphere should be increasing.

The Greenhouse Effect

The atmosphere is the engine that drives the Earth's climate system, and the Sun provides the energy that allows the engine to work (Fig. 20.15). Some of the solar radiation that reaches the atmosphere is reflected off clouds and dust and bounces back into space. Of the radiation that reaches the Earth's surface, some is absorbed by the land and oceans, and some is reflected into space by water, snow, ice, and other highly reflective surfaces. This visible reflected solar radiation has a short wavelength.

The Earth also emits long-wave, infrared radiation. Some of the long-wave radiation does not escape into space, but instead is absorbed by atmospheric water vapor and CO_2. Because this radiant energy is retained in the lower atmosphere, the temperature at the Earth's surface rises. A comparable effect also explains why the air temperature in a glass greenhouse is warmer than the air outside: the glass, acting in much the same way as the atmospheric gases, prevents the escape of radiant energy. We refer to this phenomenon as the **greenhouse effect.**

It is the greenhouse effect that makes the Earth habitable. Without it, the surface of our planet would be as inhospitable as the surfaces of the other planets in the solar system. If the Earth lacked an atmosphere, its surface environment might be like that of the Moon: On the sunlit side the temperature is close to the boiling point of water, whereas on the dark side it is far below freezing. The nearly airless surface of

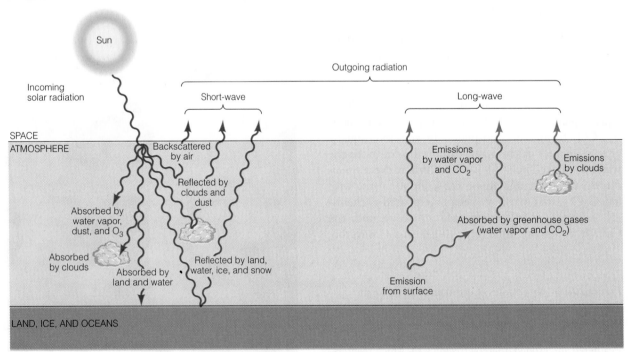

Figure 20.15 Some of the short-wave (visible) solar radiation reaching the Earth is absorbed by land, oceans, clouds, and atmospheric dust and gases, and some is reflected back into space by reflective surfaces that include snow, ice, clouds, and dust. The earth also radiates long-wave radiation back into space. Greenhouse gases trap some of the outgoing long-wave radiation, causing the air temperature of the lower atmosphere to rise.

Mars is a frigid landscape whose closest earthly analogs are the frozen polar deserts. By contrast, Venus is much closer to the Sun, and its atmosphere is so dense that the greenhouse effect generates surface temperatures hot enough to melt lead.

Greenhouse Gases

Dry air consists mainly of three gases: nitrogen (79%), oxygen (20%), and argon (1%). However, water vapor is usually present in the Earth's atmosphere in concentrations of up to several percent and accounts for about 80 percent of the natural greenhouse effect. The remaining 20 percent is due to other gases present in very small amounts. Despite their very low concentrations, measurable in parts per billion by volume (ppbv) of air, these *trace gases*, also called **greenhouse gases,** contribute significantly to the greenhouse effect (Table 20.2). Chief among the trace gases is carbon dioxide. Other significant gases—each a basic part of natural biogeochemical cycles and efficient in absorbing infrared radiation—are methane, nitrous oxide, and ozone. The commercially produced CFCs are an additional important group of greenhouse gases.

Trends in Greenhouse Gas Concentrations

During the past decade, the greenhouse gases have received increasing scientific and public attention as it has become clear that their atmospheric concentrations are rising.

Carbon Dioxide Beginning in 1958, measurements have been made of carbon dioxide concentration in the atmosphere near the top of Mauna Loa volcano, Hawaii. This site was chosen because of its altitude and remote location far from sources of atmospheric pollution. The measurements show two remarkable things. First, the amount of CO_2 fluctuates regularly with an annual rhythm (Fig. 20.16). In effect, the Earth or, more correctly, the biosphere, is breathing! During the growing season, CO_2 is absorbed by vegetation, and the atmospheric concentration falls; then, during the winter dormant period, more CO_2 enters the atmosphere than is removed by vegetation, and the concentration rises. Second, the long-term trend is an unmistakable rise in concentration. Since 1958, the atmospheric CO_2 concentration has risen from 315 to 365 ppmv, and the rise is not linear but exponential (i.e., the rate is increasing with time).

The rising curve of atmospheric CO_2 immediately raises two questions: (1) Is the observed rise unusual? and (2) How can it be explained? To answer the first question, we must turn to the geologic record. As discussed in Chapter 14, ice cores from the Antarctic and Greenland contain samples of ancient atmosphere.

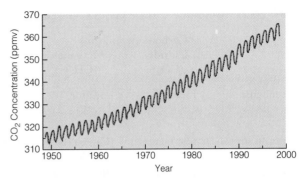

Figure 20.16 Concentration of carbon dioxide in dry air, since 1958, measured at the Mauna Loa Observatory, Hawaii (given in parts per million by volume = ppbv/1000). Annual fluctuations reflect seasonal changes in biologic uptake of CO_2, while the long-term trend shows a persistent increase in the atmospheric concentration of this greenhouse gas. *Source:* Dave Keeling and Tim Whorf (Scripps Institution of Oceanography).

The glacier records show that the preindustrial concentration of CO_2 was close to 280 ppmv, the typical value for an interglacial age. The subsequent rapid increase to 365 ppmv during the last 100 years is unprecedented in the ice-core record and implies that something very unusual is taking place.

A possible explanation for the extraordinary recent rise in atmospheric CO_2 is suggested if we examine the rate at which this gas has been added to the atmosphere since the beginning of the Industrial Revolution (Fig. 20.17). The curve tracking the increase in CO_2 closely resembles the curve showing the increase in carbon released to the atmosphere by the burning of fossil fuels. Because no known natural mechanism can explain such a rapid increase in CO_2, the inescapable conclusion is that the human burning of fossil fuels

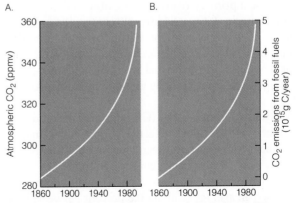

Figure 20.17 Since the beginning of the Industrial Revolution, about 1850, the atmospheric concentration of CO_2 has risen at an increasing rate (A). The increase matches the increasing rate at which CO_2 has been released through the burning of fossil fuels (B).

Table 20.2 Atmosphere Trace Gases Involved in the Greenhouse Effect

	Carbon Dioxide (CO_2)	Methane (CH_4)	Nitrous Oxide (N_2O)	Chlorofluoro-carbons (CFCs)	Tropospheric Ozone (O_3)	Water Vapor (H_2O)
Greenhouse	Heating	Heating	Heating	Heating	Heating	Heats in air; cools in clouds
Effect on stratospheric ozone	Can increase or decrease	Can increase or decrease	Can increase or decrease	Decrease	None	Decrease
Principal anthropogenic sources	Fossil fuels; deforestation	Rice culture; cattle; fossil fuels, biomass burning	Fertilizer; land-use conversion	Refrigerants; aerosols; industrial processes	Hydrocarbon (with NO_x); biomass burning	Land conversion; irrigation
Principal natural sources	Balanced in nature	Wetlands	Soils; tropical forests	None	Hydrocarbons	Evapotranspiration
Atmospheric lifetime	50–200 yr	10 yr	150 yr	60–100 yr	Weeks to months	Days
Present atmospheric concentration at land surface (ppbv)	355,000	1720	310	CFC-11: 0.28 CFC-12: 0.48	20–40	3000–6000
Preindustrial (1750–1800) concentration at land surface (ppbv)	280,000	790	288	0	10	Unknown
Present annual rate of increase	0.5%	1.1%	0.3%	5%	0.5–2.0%	Unknown
Relative contribution to the anthropenic greenhouse effect	60%	15%	5%	12%	8%	Unknown

Source: Earthquest 5, No. 1 (1991).

must be a primary reason for the observed increase in atmospheric CO_2. Additional contributing factors must be widespread deforestation, with its attendant burning and decay of cleared vegetation, and the use of wood as a primary fuel in many underdeveloped countries that have rapidly growing populations.

Methane Methane gas (CH_4) absorbs infrared radiation 25 times more effectively than CO_2, making methane an important greenhouse gas despite its relatively low atmospheric concentration (Table 20.2). The concentration of atmospheric methane has increased an average of about 0.8 percent per year; however, in northern high latitudes, the trend has been decreasing gradually over the past decade. Methane levels for earlier times obtained from ice-core studies show an increase that essentially parallels the rise in

the human population. This relationship is not surprising, for much of the methane now entering the atmosphere is generated (1) by biological activity related to rice cultivation and (2) as a byproduct of the digestive processes of domestic livestock, especially cattle. The global livestock population increased greatly in the past century, and the total acreage under rice cultivation has increased more than 40 percent since 1950. In prehistoric times, methane levels, like CO_2 levels, increased and decreased with the glacial/interglacial cycles.

Other Important Trace Gases CFC-12, used mostly as a refrigerant, has 20,000 times the capacity of carbon dioxide to trap ultraviolet radiation, whereas CFC-11, which is widely used in making plastic foams and as an aerosol propellant, has 17,500 times the capacity. Both compounds are increasing in the atmosphere at an annual rate of about 5 percent. As we have already seen, the observed increasing atmospheric concentration of CFCs in the 1980s produced worldwide concern because the scientific consensus is that these gases destroy ozone in the upper atmosphere, thereby leading to the formation of the Antarctic ozone hole as well as loss of ozone at the Arctic pole and in other parts of the atmosphere.

Tropospheric ozone and nitrous oxide are increasing annually at rates of 0.5 to 2 percent and 0.3 percent, respectively, and together account for about 13 percent of the greenhouse effect. Although ozone in the upper atmosphere is beneficial because it traps harmful infrared solar radiation, when this gas builds up in the troposphere it constitutes a greenhouse gas. Tropical forests are important in photosynthetically removing excess tropospheric ozone, which is largely produced by the combustion of fossil fuels. However, the wholesale destruction of these forests could lead to further concentration of ozone in the atmosphere. Nitrous oxide—released by microbial activity in soil, the burning of timber and fossil fuels, and the decay of agricultural residues—has a long lifetime in the atmosphere. Accordingly, atmospheric concentrations are likely to remain well above preindustrial levels even if emission rates stabilize.

GLOBAL WARMING

If the atmospheric concentration of the greenhouse gases is rising, what does this portend for future climate? Does it mean that the Earth's surface temperature is warming, and, if so, by how much and at what rate? To try and answer these questions, we first look at the historical record and then see how forecasts of the future can be made.

Historical Temperature Trends

Correctly assessing recent global changes in temperature is a very difficult task because few instrumental measurements were made before 1850, and the majority date to the time since the Second World War. The earliest records are from western Europe and eastern North America. Data for oceanic areas, which encompass 70 percent of the globe, are sparse and decrease significantly in number prior to 1945. Therefore, most "global" temperature curves are reconstructed primarily from land stations located mainly in the northern hemisphere.

Numerous curves of average annual temperature variations since the middle or late nineteenth century have been published, and although they differ in detail, they all show one characteristic feature: a long-term rise in temperature during the past century (Fig. 14.3). Although short-term departures from this trend are evident, the total temperature increase since 1860 when reliable hemispheric-wide records began is about 0.8°C (1.4°F).

Because the interval of rising temperatures coincides with the time of rapidly increasing greenhouse gas emissions, it is tempting to assume that the two phenomena are causally related. However, because the temperature reconstructions prior to 1950 are based on relatively few data that are unequally distributed across the globe, it is difficult to make a convincing case. Even if the temperature curves do approximate actual trends, it might be argued that the modest rise in global temperature during the past century falls within the natural variability of the Earth's climate system and would have occurred even if the greenhouse gas concentrations had not increased.

If the historical temperature record is judged to be inconclusive, we can still explore the linkage between greenhouse gas emissions and present and future climate by turning to models of the climate system.

Climate Models

Three-dimensional mathematical models of the Earth's climate system are an outgrowth of efforts to forecast the weather. The most sophisticated are **general circulation models** (GCMs) that attempt to link atmospheric, hydrospheric, and biospheric processes. The sheer complexity of these natural systems means that such models, of necessity, are greatly simplified representations of the real world. Furthermore, many of the linkages and processes in the climate system are still poorly understood and therefore difficult to model. For instance, the models do not yet adequately portray the dynamics of ocean circulation or cloud formation, two of the most important elements of the climate system. Also absent are many of the complex biogeochemical processes that link climate to the biosphere. Despite these limitations, GCMs have been successful in simulating the general character of present-day climates and have greatly improved weather forecasting. This success encourages us to use these models to gain a general global picture of future climates as the Earth's physical and chemical balance changes.

These models can generate a reasonable picture of global and hemispheric climatic conditions, but they are poor at resolving conditions at the scale of small countries, states, or counties. Until more powerful computers are built, the spatial resolution of GCMs is likely to remain relatively coarse.

A.

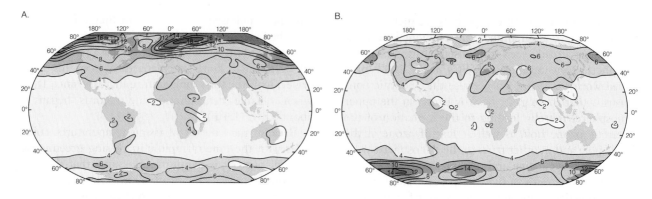

B.

C.

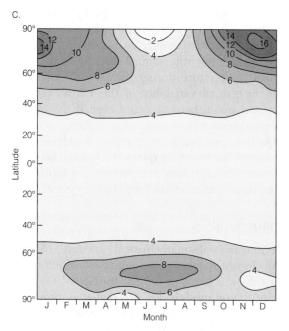

Figure 20.18 A forecast of future changes in surface air temperature (in °C) resulting from an effective doubling of atmospheric CO_2 concentration relative to that of the present. A. Temperature increases for Winter (December. January, February). For example, along the lines labeled 4, the projected temperature increase is everywhere 4°C. B. Temperature increases for Summer (June, July, August). C. A latitudinal cross section showing changes in zonal average air temperature through the year. This graph is a summary of the map patterns shown in A and B, but includes the spring and autumn months as well. Greenhouse warming is greatest at high latitudes, where temperature increases as great as 16 °C are forecast by the model for the northern hemisphere winter.

Estimates of Future Greenhouse Warming

Predictions of climatic change related to greenhouse warming are based mainly on the results of several climate models that differ in detail as well as in the assumptions they employ. Nevertheless, the models predict that the anthropogenically generated greenhouse gases already in the atmosphere would cause an average global temperature increase of 0.5 to 1.5°C (0.9 to 2.7°F) This prediction is consistent with the 0.5°C rise in temperature inferred from the instrumental record. The models further predict that if the greenhouse gases continue to build up until their combined effect is equivalent to a doubling of the preindustrial CO_2 concentration, then average global temperatures most likely will rise between 1.5 and 4.5°C (2.7 and 8.1°F). A best guess is an increase of 1.8°C (3.2°F) by the year 2030, with an eventual increase of 2.5°C (4.5°F) by the end of the twenty-first century. This does not mean that the temperature will increase uniformly all over the Earth. Rather, the projected temperature change varies geographically, with the greatest change occurring in the polar regions (Fig. 20.18; Table 20.3).

The rate of projected warming depends on a number of basic uncertainties: How rapidly will concentrations of the greenhouse gases increase? How, and how rapidly, will the oceans, a major reservoir of heat and a fundamental element in the climate system, respond to changing climate? How will changing climate affect ice sheets and cloud cover? What is the range of natural variations in the climate system on the century time scale? The potential complexity is well illustrated by clouds. If the temperature of the lower atmosphere increases, more water will evaporate from the oceans. The increased atmospheric moisture will create more clouds, but clouds reflect solar energy back into space, which will have a cooling effect on the surface air and consequently a result opposite that of the greenhouse effect.

Because of such uncertainties, scientists are reluctant to make firm forecasts and tend to be cautious in their predictions. They hedge their bets with qualifying adjectives like "possible," "probable," and "uncertain." Their understandable caution emphasizes the gap between what we know about the Earth and what

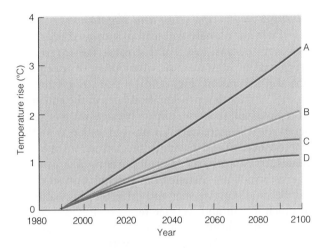

Figure 20.19 Estimates of future average global temperature rise due to the greenhouse effect based on different assumptions. Curve A represents "business as usual," with energy supply dominated by coal, continuing deforestation, and limited or no control of methane, nitrous oxide, and carbon monoxide emissions. Curve B assumes a shift toward lower-carbon fuels (e.g., natural gas), coupled with large increases in fuel efficiency, reversal of deforestation trends, and stringent carbon monoxide controls. Curve C assumes a shift toward renewable sources (solar, hydro-and wind power) and nuclear energy in the second half of the twenty-first century, a phaseout of CFC emissions, and limitations on agricultural methane and nitrous oxide emissions. Curve D assumes a shift to renewable and nuclear energy in the first half of the twenty-first century, thereby reducing carbon dioxide emissions to 50 percent of 1985 amounts by 2050, and stabilization of atmospheric concentrations of greenhouse gases through controls in industrialized countries and moderate growth in developing countries. In deriving all these estimates, world population was assumed to reach 10.5 billion in the second half of the next century, and economic growth was assumed to be modest (2-5%) in the next decade but decreasing thereafter.

we would like to know, and points to the many challenges that still face scientists studying global change.

Despite the uncertainties, the general consensus is that (1) human activities have led to increasing atmospheric concentrations of carbon dioxide and other trace gases that have enhanced the greenhouse effect; (2) global mean surface air temperature has increased by up to 0.8°C (1.4°F) during the last 100 years (Fig. 14.3), an increase that may be the direct result of the enhanced greenhouse effect; and (3) during the next century global average temperature will likely increase at about 0.3°C (0.5°F) per decade, assuming emission rates do not change. This projected increase may lead to an average global temperature about 1.5 to 1.8°C (2.7 to 3.2°F) warmer than at present by the year 2030 and as much as 2.5°C (4.5°F) warmer by the end of

the next century If governmental controls lead to lower emission rates, the per decade rise in temperature may be only 0.1 to 0.2°C (0.2 to 0.4°F) (Fig. 20.19). Nevertheless, the temperature increase related to the continued release of greenhouse gases will be larger and more rapid than any experienced in human history. Thus, we could be moving toward a "super interglaciation," warmer than any interglaciation of the past 2 million years (Fig. 20.20).

Table 20.3 Estimates of Potential Changes in Average Surface Air Temperature and Precipitation for Selected Regions, Preindustrial Times to 2030 A.D.

Region[a]	Temperature		Precipitation	
	Winter[b]	Summer[b]	Winter	Summer
Central North America	2 to 4°C warmer	2 to 3°C warmer	0 to 15% wetter	5 to 10% drier
Southeast Asia	1 to 3°C warmer	2 to 3°C warmer	5% drier to 15% wetter	5 to 15% wetter
Sahel, Africa	1 to 3°C warmer	1 to 2°C warmer	10% drier to no change	0 to 5%
Southern Europe	1 to 3°C warmer	2 to 3°C warmer	0 to 10% wetter	5 to 15% drier
Australia	1 to 3°C warmer	2°C warmer	5 to 15°% wetter	no change

[a]Within each region, considerable variation occurs. Confidence in these estimates is low, especially with regard to precipitation values.

[b]Winter = December, January, February; Summer = June, July, August.

Source: J. T. Houghton, G. J. Jenkins, and J. J. Ephraums (eds), *Climate Change, the IPCC Scientific Assessment.* (Cambridge: Cambridge University Press, 1990), Table 5.1.

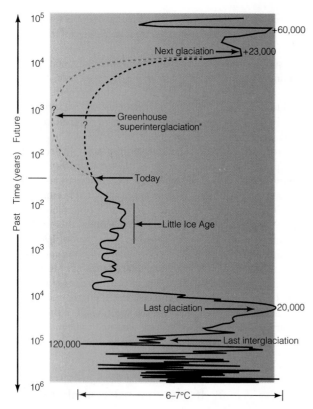

Figure 20.20 The course of average global temperature during the past million years and ca. 100,000 years into the future plotted as a function of estimated average annual temperature. The warmest part of the last interglaciation, the coldest part of the glaciation, and the next two glacial maxima are shown in years. The natural course of climate would likely be overall declining temperatures during the next several thousand years leading to the next glacial maximum (black dashed line) about 23,000 years from now. With greenhouse warming, a continuing rise of temperature may lead to a "super interglaciation" (dashed red line) within the next several centuries. The temperature may then be warmer than during the last interglaciation 120,000 years ago and warmer than at any time in human history. The decline toward the next glaciation would thereby be delayed a millennium or more.

Environmental Effects of Global Warming

An increase in global surface air temperature by a few degrees C does not sound like much. Surely, we can put up with this rather insignificant change. However, if we stop and consider that the difference in average global temperature between the present and the coldest part of the last ice age was only about 5°C (9°F), we can begin to see how a temperature change of even a degree or two could well have global repercussions.

Global warming is just one result of our great "geochemical experiment." There are many physical and biological side effects that are of considerable interest and concern. Among them are the following.

Changes in Precipitation and Vegetation

A warmer atmosphere will lead to increased evaporation from oceans, lakes, and streams and to greater precipitation. However, the distribution of precipitation will be uneven (Fig. 20.21; Table 20.3). Climate models suggest that precipitation rates in the equatorial regions will increase, in part because warmer temperatures will increase rates of evaporation over the tropical oceans and promote the formation of rainclouds. By contrast, the interior portions of large continents, which are distant from precipitation sources, will become both warmer and drier. Shifting patterns of precipitation and warmer temperatures will likely lead to significant local and regional changes in stream runoff and groundwater levels.

Shifting precipitation patterns are likely to upset ecosystems, causing vegetation communities and the animals dependent on them to adjust to new conditions. Forest boundaries may shift during coming centuries in response to altered temperature and precipitation patterns. Some prime midcontinental agricultural regions are likely to face increased droughts and substantially reduced soil moisture that will negatively impact crops. Higher-latitude regions with short, cool growing seasons may see increased agricultural production as summer temperatures increase.

Changes in the Global Ice Cover

Because warmer summers favor increased ablation, worldwide recession of low- and middle-latitude mountain glaciers is likely in a warmer world. On the other hand, warmer air in high latitudes can evaporate and transport more moisture from the oceans to adjacent ice sheets, which may cause them to grow larger.

The greatly enhanced heating projected for high northern latitudes (Fig. 20.18) favors the shrinkage of sea ice. A reduction in polar sea ice, which has a high albedo, should contribute to global warming by reducing the amount of short-wave solar radiation reflected back into space, thereby increasing the heat absorbed by the ocean. Models show much less heating in the high-latitude southern hemisphere, suggesting little change in sea–ice cover there.

Rising summer air temperatures will also begin to thaw vast regions of perennially frozen ground at high latitudes. The thawing will likely affect natural ecosystems as well as cities and engineering works built on frozen ground.

Worldwide Rise of Sea Level

As the temperature of ocean water rises, its volume will expand, causing world sea level to rise. This rise in sea level, supplemented by meltwater from shrinking glaciers, is likely to increase calving along the margins of tidewater glaciers and ice sheets, thereby leading to additional sea-level rise. The rising sea will inundate coastal regions where millions of people live and will make the tropical regions even more vulnerable to larger and more frequent cyclonic storms.

Decomposition of Organic Matter in Soils

As temperature rises, the rate of decomposition of organic matter in soil will increase. Soil decomposition releases CO_2 to the atmosphere, thereby further enhancing the greenhouse effect. If world temperature rises by 0.3°C (0.5°F) per decade, during the next 60 years soils will release an amount of CO_2 equal to nearly 20 percent of the projected CO_2 release due to combustion of fossil fuels, assuming the present rate of fuel consumption continues.

Breakdown of Gas Hydrates

Gas hydrates are icelike solids in which gas molecules, mainly methane, are locked up in water trapped in ocean sediments and beneath frozen ground. By one estimate, worldwide gas hydrates may hold 10,000 billion metric tons of carbon, twice the amount in all the known coal, gas, and oil reserves on land. They accumulate in ocean sediments beneath a water depth of 500 m (1640 ft), where the temperature is low enough and the pressure high enough to permit their formation. They also accumulate beneath permafrost, which acts as a seal to prevent upward migration and escape of the gas. When gas hydrates break down, they release methane. Global warming at high latitudes will thaw frozen ground, and this thawing may destabilize the hydrates there, releasing large volumes of methane, amplifying the greenhouse effect.

Feedback in the Climate System

A number of potential responses to global warming further enhance the warming trend and therefore constitute positive feedback (Chapter 15) i.e.; they move the system further in the direction of change. These include the reduction in global albedo as sea ice cover contracts in the polar oceans, release of additional CO_2 to the atmosphere as soil organic matter decomposes, and release of methane from gas hydrates as frozen ground thaws. At the same time, there are some potential negative feedbacks. For example, plants grow faster in air enriched with carbon dioxide. As the atmospheric concentration of CO_2 rises, plant growth may increase, thereby removing this greenhouse gas from the atmosphere at an increasing rate. The result could then be a reduction in the rate that CO_2 is added to the atmosphere (a negative feedback).

LESSONS FROM THE PAST

Examining changes to physical and biological systems that occurred when human influence was absent or minimal allows us to see how these systems responded to natural environmental change. Of particular interest are episodes of rapid climatic change—those occurring within a century or less—which may provide analogs of future climatic change. We can examine rapid shifts in climate that occurred at the end of the

Figure 20.21 A computer simulation showing possible changes in summer (June, July, August) precipitation resulting from a doubling of atmospheric CO_2. Many of the areas of projected decreased precipitation (large parts of central North America, eastern and southern South America, Western Europe, Africa, the Middle East and central Asia) are prime agricultural areas.

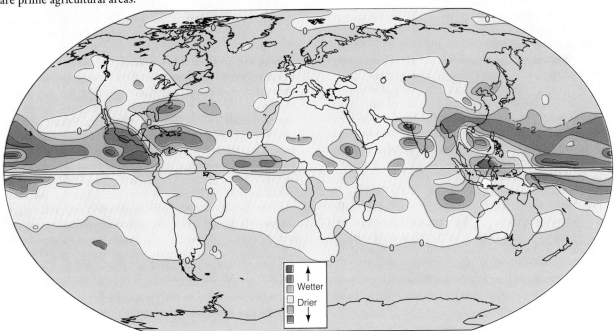

GUEST ESSAY

Assessing Projections for Future Climate Change

Thomas J. Crowley is a Professor of Oceanography at Texas A&M University and specializes in the study of past climates.

Much has been written about changes in climate that could result from continued emissions of carbon dioxide through deforestation and burning of fossil fuels. Other gases (e.g., methane, nitrous oxide, freons) also increase the so-called greenhouse effect—a term that refers to an absorption by atmospheric gases of upwelling radiation from the Earth's surface. A simple illustration of the greenhouse effect occurs on warm humid summer evenings , when water vapor in the atmosphere (another important greenhouse gas) warms the nighttime air much more than occurs in arid regions, where the surface radiation is emitted to space, thus leading to cooler nights.

The net effect of the anthropogenic (or man-made) disturbance is that the carbon dioxide concentration of the atmosphere is already at the highest level in several hundred thousand years, and we have only used about 5 percent of the available fossil fuel reservoir. Since the average "lifetime" of carbon dioxide in the atmosphere is about a century, changing the concentration now can affect atmospheric levels for generations to come. Such realizations have motivated the world community to critically evaluate the consequences of the human impact on the climate system and assess whether there is a need to impose emission restrictions on greenhouse gases.

Although in principle the greenhouse effect is straightforward, several key uncertainties complicate projections of its possible effects. The atmospheric and oceanic response to an initial greenhouse warming is uncertain, because we cannot accurately predict how clouds and the ocean will change in a warmer world. These responses, known as feedbacks, could further increase the warming. For example, snow and cloud cover could decrease, leading to more absorbed solar radiation and more warming. Feedbacks can be either positive or negative. A negative feedback might result in more clouds, leading to increased cooling that might partially offset the global warming trend. Climatologists have developed large computer models to try to assess the effects of these and other feedbacks in the climate system.

A comprehensive assessment of the greenhouse effect predicts that global temperatures could increase 1.5 to 4.5°C by sometime toward the middle to end of the next century. The range directly reflects the uncertainties in the atmospheric response to feedbacks. Sea level will also rise due to melting glaciers and the thermal expansion of seawater at higher temperatures. Flooding of coastal regions, especially during storm surges, is more likely in regions such as the Gulf Coast and eastern seaboard of the United States. Some models predict increased drying in the midwestern United States, which could severely affect agriculture. Recent observations suggest that extreme climate events (droughts, floods) have been more frequent in the last 20 years. But it is difficult to evaluate this problem from the statistical viewpoint, and computer models do not do a very good job of predicting extreme events. Therefore, caution is necessary with respect to evaluating the possibility of future changes in extreme events.

How significant is the temperature increase projected for the next century? The geological record of climate change affords a frame of reference against which such projections can be compared. Already global temperatures appear to be as high or higher than those at any time in the last 1000 years. Analysis of past climate change suggests that a 1.5°C warming is as large as (if not larger than) that at any time period in the last 200,000 years. A 4.5°C warming may be as large as that at any time in the last few tens of millions of years. Since the projected warming reflects the response to usage of only 20 percent of the available fossil fuel reservoir, climate change beyond a doubling of carbon dioxide will likely lead to a climate change that is large even on a geological scale.

The geologic record also indicates that there have sometimes been very rapid changes in climate. For example, ice cores on Greenland indicate that at the end of the last ice age the climate may have changed in less than a decade. Some of these changes may have involved an abrupt variation in the North Atlantic circulation that transports warm Gulf Stream waters to Europe, maintaining conditions there substantially warmer than occurs in Canada at the same latitude. Greenhouse model projections suggest that the North Atlantic could be affected in a similar manner, owing to "short circuiting" of the circulation. But there has been some concern as to whether such a prediction may be an artifact of the models. Similar occurrences of such events in the geologic record suggest that the model predictions should be taken seriously.

The geologic record can also be used to test other facets of greenhouse-climate projections. On time scales of thousands to millions of years, atmospheric carbon dioxide levels may have "naturally" changed due to, for example, variations in the ocean circulation during the ice age, degassing from midocean ridges, and changes in carbon storage due to the evolution of land plants. Comparison of the past record of climate change with reconstructions of past atmospheric carbon dioxide levels indicates a striking agreement between the two. Applying many of these same models to the geologic past (the last ice age, the Cretaceous warm interval, Pangaea) results in an impressive agreement between many of the features of observed climate change and those predicted by models. There are exceptions, but it is not clear whether the model-data differences reflect imperfections in the models or imprecise determination of geological evidence. Joint assessment of observations and models therefore suggests that overall the geologic record provides general support for the importance of greenhouse gas climate change and the ability of models to stimulate at the largest scales the responses to such gases. Future work will hopefully narrow these differences even more.

last glacial age and during the Little Ice Age. We can also search the geologic record for information about times when the climate was warmer than now. Because we may be on the brink of a warmer world, times of greatest interest are the early Holocene, from 10,000 to about 6000 years ago, when average temperatures were 0.5 to 1°C (0.9 to 1.8°F) warmer; the warmest part of the last interglaciation, about 120,000 years ago, when global temperatures were about 1 to 2°C (2 to 4°F) higher; and the Middle Pliocene, about 4.5 to 3 million years ago, when the Earth's climates may have been 3 to 4°C (5 to 7°F) warmer than at present. These periods do not provide perfect analogs for present global warming because the distribution of solar radiation reaching the Earth's surface was not always the same as now. Nevertheless, they enable us to see how plants and animals responded to climatic conditions that may have been broadly similar to those we may experience in the near future.

During the warmest parts of the Pliocene Epoch,

Figure 20.22 A. Acidity record from a Greenland ice core showing peaks in sulphuric-acid precipitation attributable to major volcanic eruptions. The largest acid peak dates to 1815–1816, the time of the huge Tambora eruption in the East Indies, which produced "the year without a summer" (1816). The volcanic dust and gases in the stratosphere reduced northern hemisphere temperatures at least 0.7 °C. Subsequent eruptions of Krakatau, Katmai, and Agung also produced detectable climatic effects of smaller magnitude. B. Observed atmospheric impact of the eruption of Mt. Pinatubo in June 1991. Vast amounts of sulfur dioxide gas were quickly transformed into sulfuric acid particles that blocked incoming radiation and caused a globally average drop in temperature of about 0.6°C (1.1°F) that lasted about two years. *Source:* From J. Hansen et al., in *National Geographic Research and Exploration*, vol 9, no 2, pp. 142–158,

for example, world sea level was tens of meters higher than now, pointing to reduced global ice cover. Isotopic measurements of North Atlantic deep-sea cores show that winter sea-surface temperatures were at least 3°C (5°F) warmer than at present. In response to warmer ocean waters, temperature-sensitive marine organisms had different distributions. The boundaries between tropical, subtropical, and temperate assemblages of ostracodes, a type of crustacean, shifted northward along the eastern coast of the United States during the Middle Pliocene in response to warmer coastal waters, whereas in China a warm, moist climate produced deep weathering profiles and tropical to subtropical soils.

PERSPECTIVES OF GLOBAL CHANGE

Despite considerable research, we do not yet have a clear vision of our climatic future. At present, the best we can conclude is that the force of scientific evidence and theory makes it very probable that the climate is warming and will continue to warm as we add greenhouse gases to the atmosphere. There also is a high probability that average global temperatures ultimately will increase by 2 to 4°C (4 to 7°F), leading to widespread environmental changes.

It is less probable that the temperature will increase steadily, for there are natural, and as yet largely unpredictable, modulations of the climate system on the time scale of years to decades. The huge eruption of Tambora in the East Indies in 1815 was followed by the "year without a summer," during which midsummer snow and frost caused severe hardships in Europe and New England (Fig. 20.22A). The eruption of

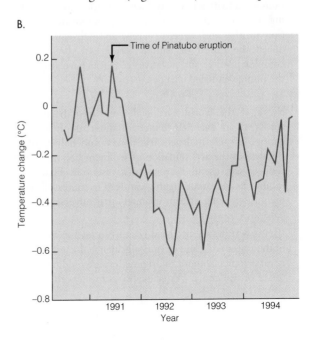

Pinatubo in June 1991 (Chapter 5) produced a veil of dust and gas that spread through the atmosphere, reflecting incoming solar radiation and reducing surface temperatures (Fig. 20.22B). Such a volcanic event may well reverse temporarily any upward trend in average global temperature attributable to continued emission of greenhouse gases.

Although the short-term prospect (on the scale of human generations) is for a warmer world, if we stand back and look at our great geochemical experiment from a geological perspective, we will perceive that it is only a brief, very rapid, yet nonrepeatable perturbation in the Earth's climatic history. It is nonrepeatable because once the Earth's store of easily extractable fossil fuels is used up, most likely within the next several hundred years, the human impact on the atmosphere will inevitably decline and the climate system should slowly return to its natural state. The greenhouse perturbation may well last a thousand years, and perhaps more, but ultimately the changing geometry of the Earth's orbit will propel the climate system into the next glacial age (Fig. 20.20).

SUMMARY

1. Although environmental change is natural, humans are creating new kinds of changes and altering the rates of many natural processes.

2. Humans have long had an effect on the environment, but their effect on the entire Earth system has accelerated greatly since the start of the Industrial Revolution.

3. Human effects on the global Earth system result from the sheer number of people (about 5.7 billion) and the rapid increase in technology. An expanding human population has led to global exploitation of natural resources and serious environmental deterioration. Human and natural geologic activities are intimately interlinked, and people have increasingly become a major agent of environmental change.

4. Humans cause environmental change both directly and indirectly. An example of a direct change is global warming, which results from the introduction of greenhouse gases into the atmosphere. Soil erosion is an example of an indirect change: it is accelerated by human activities at many specific locations, but the increase is so widely dispersed that overall the effect is global.

5. Human activities have produced unanticipated poisoning of natural ecosystems, helped promote desertification in arid and semiarid lands, and substantially reduced the extent of natural forests. In the process, complex positive and negative feedbacks have amplified the human-induced changes.

6. Modern civilization has had a great effect on the major rivers of the world. Artificial dams have been built across stream channels to increase water supply for irrigation, harness hydroelectric power, and help control floods, but the accumulation of sediment behind a dam reduces the useful lifetime of a reservoir. Hydraulic mining and clearcut logging can lead to increased sedimentation along stream channels that adversely affect natural ecosystems.

7. Excess withdrawal of groundwater can lead to lowering of the water table and to land subsidence. In some cases, natural replenishment to original levels would take centuries or millennia. Water quality is influenced by natural dissolved substances and human, agricultural, and industrial pollutants that seep into groundwater reservoirs.

8. Many human activities affect global biogeochemical cycles. Extremely toxic and radioactive wastes, which pose special problems, should be stored underground only if geologic conditions imply little or no change in groundwater systems over geologically long intervals of time.

9. Synthetic chlorofluorocarbon (CFC) gases entering the upper atmosphere break down to chlorine, which destroys the protective ozone layer. Discovery of a vast ozone hole over Antarctica has led to international efforts to eliminate CFC production by the end of the century.

10. Changes affecting the Earth's climate system operate on time scales ranging from decades to millions of years. The anthropogenic extraction and burning of fossil fuels perturb the natural carbon cycle and have led to an increase in atmospheric CO_2 since the start of the Industrial Revolution (about 1850).

11. The increase in atmospheric trace gases (CO_2, CH_4, O_3, N_2O, and the CFCs) due to human activities is projected to warm the lower atmosphere between 2 and 4°C by the end of the next century. A probable 0.5°C increase in average global temperature since the mid-nineteenth century may reflect the initial part of this warming. The rate of warming may reach 0.3°C per decade and could lead to a "super interglaciation," making the Earth warmer than at any time in human history.

12. The greenhouse effect, due to the trapping of longwave infrared radiation by trace gases in the atmosphere, makes the Earth a habitable planet.

13. Potential physical and biological consequences of global warming include global changes in precipitation and vegetation patterns; melting of glaciers, sea ice, and frozen ground; a worldwide rise of sea level; increased rates of organic decomposition in soils; and breakdown of gas hydrates trapped beneath high-latitude permafrost. Although the Earth's surface environments may

change substantially during the next several centuries in response to greenhouse warming, viewed from the geologic perspective, this interval will appear as only a brief perturbation in the Earth's climate history.

14. Using information from the geologic record, geologists can determine the magnitude and geographic extent of past climatic changes, determine the range of climatic variability on different time scales, and test the accuracy of computer models that simulate past climatic conditions, and provide insights into physical and biological responses to future global warming.

IMPORTANT TERMS TO REMEMBER

clearcutting *473*

deforestation *472*

desertification *470*

general circulation models *481*

global warming *484*

greenhouse effect *478*

greenhouse gases *479*

QUESTIONS FOR REVIEW

1. What is different about human-induced changes to the Earth system compared to natural changes?

2. Why does the human population pose the major future stress on the Earth system?

3. What are the net benefits and the negative environmental effects of hydroelectric power?

4. Why does chlorine have such an adverse affect on the ozone layer, despite the fact that it is released to the atmosphere in very small amounts?

5. Which global effect, decrease in the ozone layer or global warming, is likely to have the greatest effects on life? Give the reasons for your conclusion.

6. How might desertification brought on by overgrazing in a semiarid rangeland produce a positive feedback that enhances the shift to desert conditions?

7. Suggest ways in which widespread deforestation can affect (a) streams, (b) soils, and (c) local climate.

8. In what ways can mining affect life in (a) forests, (b) streams, (c) the ocean?

9. In what ways can carbon be trapped in the Earth and become part of the rock cycle? How can such stored carbon once again find its way into the atmosphere?

10. If atmospheric CO_2 can be dissolved in streams, lakes, groundwater, and the oceans, and also efficiently absorbed by vegetation, suggest why the burning of fossil fuels is causing the CO_2 content of the atmosphere to increase.

11. Why is the geologic record important in helping predict the environmental effects of greenhouse warming? Give two examples.

12. Describe the carbon cycle. Why do we regard it as one of the most important biogeochemical cycles? In what ways are you, as an individual human, involved in the carbon cycle?

13. Why does uncertainty exist about the extent to which average global temperature will rise in the next century as a result of greenhouse warming?

14. Based on what you have learned about potential future greenhouse warming, what are some of the possible changes that could affect your community during the next 50 years if the average world climate warms by as much as 2°C?

The environmental hazards of modern life are explored in some of the case studies available at our web site.

APPENDIX A
Units and Their Conversions

ABOUT SI UNITS

Regardless of the field of specialization, all scientists use the same units and scales of measurement. They do so to avoid confusion and the possibility that mistakes can creep in when data are converted from one system of units, or one scale, to another. By international agreement the SI units are used by all, and they are the units used in this text. SI is the abbreviation of Système International d'Unités (in English, the International System of Units).

Some of the SI units are likely to be familiar, some unfamiliar. The SI unit of length is the meter (m), of area the square meter (m^2), and of volume the cubic meter (m^3). The SI unit of mass is the kilogram (kg), and of time the second (s). The other SI units used in this book can be defined in terms of these basic units. Three important ones are:

1. The newton (N), a unit of force defined as that force needed to accelerate a mass of 1 kg by 1 m/s^2; hence 1 N = 1 kg·m/s^2. (The period between kg and m indicates multiplication.)

2. The joule (J), a unit of energy or work, defined as the work done when a force of 1 newton is displaced a distance of 1 meter; hence 1 J = 1 N·m. One important form of energy so far as the Earth is concerned is heat. The outward flow of the Earth's internal heat is measured in terms of the number of joules flowing outward from each square centimeter each second; thus, the unit of heat flow is J/cm^2/s.

3. The pascal (Pa), a unit of pressure defined as a force of 1 newton applied across an area of 1 square meter; hence 1 Pa = 1 N/m^2. The pascal is a numerically small unit. Atmospheric pressure, for example (15 lb/in^2), is 101,300 Pa. Pressure within the Earth reaches millions or billions of pascals. For convenience, earth scientists sometimes use 1 million pascals (megapascal, or MPa) as a unit.

Temperature is a measure of the internal kinetic energy (expressed as movement) of the atoms and molecules in a body. In the SI system, temperature is measured on the Kelvin scale (K). The temperature intervals on the Kelvin scale are arbitrary, and they are the same as the intervals on the more familiar Celsius scale (°C). The difference between the two scales is that the Celsius scale selects 100°C as the temperature at which water boils at sea level, and 0°C as the freezing temperature of water at sea level. Zero degrees Kelvin, on the other hand, is absolute zero, the temperature at which all atomic and molecular motions cease. Thus, 0°C is equal to 273.15 K, and 100°C is 373.15 K. The temperatures of processes on and within the Earth tend to be at or above 273.15 K. Despite the inconsistency, earth scientists still use the Celsius scale when geological processes are discussed.

Appendix A provides a table of conversion from older units to Standard International (SI) units.

PREFIXES FOR MULTIPLES AND SUBMULTIPLES

When very large or very small numbers have to be expressed, a standard set of prefixes is used in conjunction with the SI units. Some prefixes are probably already familiar; an example is the centimeter (which is one hundredth of a meter, or 10^{-2} m). The standard prefixes are

tera	1,000,000,000,000	= 10^{12}
giga	1,000,000,000	= 10^9
mega	1,000,000	= 10^6
kilo	1,000	= 10^3
hecto	100	= 10^2
deka	10	= 10
deci	0.1	= 10^{-1}
centi	0.01	= 10^{-2}
milli	0.001	= 10^{-3}
micro	0.000001	= 10^{-6}
nano	0.000000001	= 10^{-9}
pico	0.000000000001	= 10^{-12}

One measure used commonly in geology is the nanometer (nm), a unit by which the sizes of atoms are measured; 1 nanometer is equal to 10^{-9} meter.

COMMONLY USED UNITS OF MEASURE

LENGTH

Metric Measure

1 kilometer (km)	= 1000 meters (m)
1 meter (m)	= 100 centimeters (cm)
1 centimeter (cm)	= 10 millimeters (mm)
1 millimeter (mm)	= 1000 micrometers (μm) (formerly called microns)
1 micrometer (μm)	= 0.001 millimeter (mm)
1 angstrom (Å)	= 10^{-8} centimeters (cm)

Nonmetric Measure

1 mile (mi)	= 5280 feet (ft) = 1760 yards (yd)
1 yard (yd)	= 3 feet (ft)
1 fathom (fath)	= 6 feet (ft)

Conversions

1 kilometer (km)	= 0.6214 mile (mi)
1 meter (m)	= 1.094 yards (yd) = 3.281 feet (ft)
1 centimeter (cm)	= 0.3937 inch (in)
1 millimeter (mm)	= 0.0394 inch (in)
1 mile (mi)	= 1.609 kilometers (km)
1 yard (yd)	= 0.9144 meter (m)
1 foot (ft)	= 0.3048 meter (m)
1 inch (in)	= 2.54 centimeters (cm)
1 inch (in)	= 25.4 millimeters (mm)
1 fathom (fath)	= 1.8288 meters (m)

AREA

Metric Measure

1 square kilometer (km^2)	= 1,000,000 square meters (m^2)
	= 100 hectares (ha)
1 square meter (m^2)	= 10,000 square centimeters (cm^2)
1 hectare (ha)	= 10,000 square meters (m^2)

Nonmetric Measure

1 square mile (mi^2)	= 640 acres (ac)
1 acre (ac)	= 4840 square yards (yd^2)
1 square foot (ft^2)	= 144 square inches (in^2)

Conversions

1 square kilometer (km^2)	= 0.386 square mile (mi^2)
1 hectare (ha)	= 2.471 acres (ac)
1 square meter (m^2)	= 1.196 square yards (yd^2)
	= 10.764 square feet (ft^2)
1 square centimeter (cm^2)	= 0.155 square inch (in^2)
1 square mile (mi^2)	= 2.59 square kilometers (km^2)
1 acre (ac)	= 0.4047 hectare (ha)
1 square yard (yd^2)	= 0.836 square meter (m^2)
1 square foot (ft^2)	= 0.0929 square meter (m^2)
1 square inch (in^2)	= 6.4516 square centimeter (cm^2)

VOLUME

Metric Measure

1 cubic meter (m^3)	= 1,000,000 cubic centimeters (cm^3)
1 liter (l)	= 1000 milliliters (ml)
	= 0.001 cubic meter (m^3)
1 centiliter (cl)	= 10 milliliters (ml)
1 milliliter (ml)	= 1 cubic centimeter (cm^3)

Nonmetric Measure

1 cubic yard (yd^3)	= 27 cubic feet (ft^3)
1 cubic foot (ft^3)	= 1728 cubic inches (in^3)
1 barrel (oil) (bbl)	= 42 gallons (U.S.) (gal)

Conversions

1 cubic kilometer (km^3)	= 0.24 cubic miles (mi^3)
1 cubic meter (m^3)	= 264.2 gallons (U.S.) (gal)
	= 35.314 cubic feet (ft^3)
1 liter (l)	= 1.057 quarts (U.S.) (qt)
	= 33.815 ounces (U.S. fluid) (fl. oz.)
1 cubic centimeter (cm^3)	= 0.0610 cubic inch (in^3)
1 cubic mile (mi^3)	= 4.168 cubic kilometers (km^3)
1 acre-foot (ac-ft)	= 1233.46 cubic meters (m^3)
1 cubic yard (yd^3)	= 0.7646 cubic meter (m^3)
1 cubic foot (ft^3)	= 0.0283 cubic meter (m^3)
1 cubic inch (in^3)	= 16.39 cubic centimeters (cm^3)
1 gallon (gal)	= 3.784 liters (l)

MASS

Metric Measure

1000 kilograms (kg)	= 1 metric ton (also called a tonne) (m.t)
1 kilogram (kg)	= 1000 grams (g)

Nonmetric Measure

1 short ton (sh.t)	= 2000 pounds (lb)
1 long ton (l.t)	= 2240 pounds (lb)
1 pound (avoirdupois) (lb.)	= 16 ounces (avoirdupois) (oz) = 7000 grains (gr)
1 ounce (avoirdupois) (oz)	= 437.5 grains (gr)
1 pound (Troy) (Tr. lb)	= 12 ounces (Troy) (Tr. oz)
1 ounce (Troy) (Tr. oz)	= 20 pennyweight (dwt)

Conversions

1 metric ton (m.t)	= 2205 pounds (avoirdupois) (lb)
1 kilogram (kg)	= 2.205 pounds (avoirdupois) (lb)
1 gram (g)	= 0.03527 ounce (avoirdupois) (oz) 0.03215 ounce (Troy) (Tr. oz) = 15,432 grains (gr)
1 pound (lb)	= 0.4536 kilogram (kg)
1 ounce (avoirdupois) (oz)	= 28.35 grams (g)
1 ounce (avoirdupois) (oz)	= 1.097 ounces (Troy) (Tr. oz)

Pressure

1 pascal (Pa)	= 1 newton/square meter (N/m^2)
1 kilogram/square centimeter (kg/cm^2)	= 0.96784 atmosphere (atm) = 14.2233 pounds/square inch (lb/in^2) = 0.098067 bar
1 bar	= 0.98692 atmosphere (atm) = 10^5 pascals (Pa) = 1.02 kilograms/square centimeter (kg/cm^2)

ENERGY AND POWER

Energy

1 joule (J)	= 1 newton meter (N.m)
	= 2.390×10^{-1} calorie (cal)
	= 9.47×10^{-4} British thermal unit (Btu)
	= 2.78×10^{-7} kilowatt-hour (kWh)
1 calorie (cal)	= 4.184 joule (J)
	= 3.968×10^{-3} British thermal unit (Btu)
	= 1.16×10^{-6} kilowatt-hour (kWh)
1 British thermal unit (Btu)	= 1055.87 joules (J)
	= 252.19 calories (cal)
	= 2.928×10^{-4} kilowatt-hour (kWh)
1 kilowatt hour	= 3.6×10^{6} joules (J)
	= 8.60×10^{5} calories (cal)
	= 3.41×10^{3} British thermal units (Btu)

Power (energy per unit time)

1 watt (W)	= 1 joule per second (J/s)
	= 3.4129 Btu/h
	= 1.341×10^{-3} horsepower (hp)
	= 14.34 calories per minute (cal/min)
1 horsepower (hp)	= 7.46×10^{2} watts (W)

Temperature

To change from Fahrenheit (F) to Celsius (C)

$$°C = \frac{(°F - 32°)}{1.8}$$

To change from Celsius (C) to Fahrenheit (F)

$$°F = (°C \times 1.8) + 32°$$

To change from Celsius (C) to Kelvin (K)

$$K = °C + 273.15$$

To change from Fahrenheit (F) to Kelvin (K)

$$K = \frac{(°F - 32)}{1.8} + 273.15$$

Tables of the Chemical Elements and Naturally Occurring Isotopes

Table B.1 Alphabetical List of the Elements

Element	Symbol	Atomic Number	Crustal Abundance, Weight Percent	Element	Symbol	Atomic Number	Crustal Abundance, Weight Percent
Actinium	Ac	89	Human-made	Iron	Fe	26	5.80
Aluminum	Al	13	8.00	Krypton	Kr	36	Not known
Americium	Am	95	Human-made	Lanthanum	La	57	0.0050
Antimony	Sb	51	0.00002	Lawrencium	Lw	103	Human-made
Argon	Ar	18	Not known	Lead	Pb	82	0.0010
Arsenic	As	33	0.00020	Lithium	Li	3	0.0020
Astatine	At	85	Human-made	Lutetium	Lu	71	0.000080
Barium	Ba	56	0.0380	Magnesium	Mg	12	2.77
Berkelium	Bk	97	Human-made	Manganese	Mn	25	0.100
Beryllium	Be	4	0.00020	Meitnerium	Mt	109	Human-made
Bismuth	Bi	83	0.0000004	Mendelevium	Md	101	Human-made
Boron	B	5	0.0007	Mercury	Hg	80	0.000002
Bromine	Br	35	0.00040	Molybdenum	Mo	42	0.00012
Cadmium	Cd	48	0.000018	Neodymium	Nd	60	0.0044
Calcium	Ca	20	5.06	Neon	Ne	10	Not known
Californium	Cf	98	Human-made	Neptunium	Np	93	Human-made
Carbon[a]	C	6	0.02	Nickel	Ni	28	0.0072
Cerium	Ce	58	0.0083	Nielsbohrium	Ns	107	Human-made
Cesium	Cs	55	0.00016	Niobium	Nb	41	0.0020
Chlorine	Cl	17	0.0190	Nitrogen	N	7	0.0020
Chromium	Cr	24	0.0096	Nobelium	No	102	Human-made
Cobalt	Co	27	0.0028	Osmium	Os	76	0.00000002
Copper	Cu	29	0.0058	Oxygen[b]	O	8	45.2
Curium	Cm	96	Human-made	Palladium	Pd	46	0.0000003
Dysprosium	Dy	66	0.00085	Phosphorus	P	15	0.1010
Einsteinium	Es	99	Human-made	Platinum	Pt	78	0.0000005
Erbium	Er	68	0.00036	Plutonium	Pu	94	Human-made
Europium	Eu	63	0.00022	Polonium	Po	84	Footnote[d]
Fermium	Fm	100	Human-made	Potassium	K	19	1.68
Fluorine	F	9	0.0460	Praseodymium	Pr	59	0.0013
Francium	Fr	87	Human-made	Promethium	Pm	61	Human-made
Gadolinium	Gd	64	0.00063	Protactinium	Pa	91	Footnote[d]
Gallium	Ga	31	0.0017	Radium	Ra	88	Footnote[d]
Germanium	Ge	32	0.00013	Radon	Rn	86	Footnote[d]
Gold	Au	79	0.0000002	Rhenium	Re	75	0.00000004
Hafnium	Hf	72	0.004	Rhodium[c]	Rh	45	0.00000001
Hahnium	Ha	105	Human-made	Rubidium	Rb	37	0.0070
Hassium	Hs	108	Human-made	Ruthenium[c]	Ru	44	0.00000001
Helium	He	2	Not known	Samarium	Sm	62	0.00077
Holmium	Ho	67	0.00016	Scandium	Sc	21	0.0022
Hydrogen[b]	H	1	0.14	Seaborgium	Sg	106	Human-made
Indium	In	49	0.00002	Selenium	Se	34	0.000005
Iodine	I	53	0.00005	Silicon	Si	14	27.20
Iridium	Ir	77	0.00000002	Silver	Ag	47	0.000008

Table B.1 (*Continued*)

Element	Symbol	Atomic Number	Crustal Abundance, Weight Percent	Element	Symbol	Atomic Number	Crustal Abundance, Weight Percent
Sodium	Na	11	2.32	Tin	Sn	50	0.00015
Strontium	Sr	38	0.0450	Titanium	Ti	22	0.86
Sulfur	S	16	0.030	Tungsten	W	74	0.00010
Tantalum	Ta	73	0.00024	Uranium	U	92	0.00016
Technetium	Tc	43	Human-made	Vanadium	V	23	0.0170
Tellurium[c]	Te	52	0.000001	Xenon	Xe	54	Not known
Terbium	Tb	65	0.00010	Ytterbium	Yb	70	0.00034
Thallium	Tl	81	0.000047	Yttrium	Y	39	0.0035
Thorium	Th	90	0.00058	Zinc	Zn	30	0.0082
Thulium	Tm	69	0.000052	Zirconium	Zr	40	0.0140

Source: After K. K. Turekian, 1969.

[a]Estimate from S. R. Taylor (1964).

[b]Analyses of crustal rocks do not usually include separate determinations for hydrogen and oxygen. Both combine in essentially constant proportions with other elements, so abundances can be calculated.

[c]Estimates are uncertain and have a very low reliability.

[d]Elements formed by decay of uranium and thorium. The daughter products are radioactive with such short half-lives that crustal accumulations are too low to be measured accurately.

Table B.2 Naturally Occurring Elements Listed in Order of Atomic Numbers, Together with the Naturally Occurring Isotopes of Each Element, Listed in Order of Mass Numbers

Atomic Number[a]	Name	Symbol	Mass Numbers[b] of Natural Isotopes	Atomic Number[a]	Name	Symbol	Mass Numbers[b] of Natural Isotopes
1	Hydrogen	H	1, 2, ☐3 [c]	29	Copper	Cu	63, 65
2	Helium	He	3, 4	30	Zinc	Zn	64, 66, 67, 68, 70
3	Lithium	Li	6, 7	31	Gallium	Ga	69, 71
4	Beryllium	Be	9, ☐10	32	Germanium	Ge	70, 72, 73, 74, 76
5	Boron	B	10, 11	33	Arsenic	As	75
6	Carbon	C	12, 13, ☐14	34	Selenium	Se	74, 76, 77, 80, 82
7	Nitrogen	N	14, 15	35	Bromine	Br	79, 81
8	Oxygen	O	16, 17, 18	36	Krypton	Kr	78, 80, 82, 83, 84, 86
9	Fluorine	F	19				
10	Neon	Ne	20, 21, 22	37	Rubidium	Rb	85, ☐87
11	Sodium	Na	23	38	Strontium	Sr	84, 86, 87, 88
12	Magnesium	Mg	24, 25, 26	39	Yttrium	Y	89
13	Aluminum	Al	27	40	Zirconium	Zr	90, 91, 92, 94, 96
14	Silicon	Si	28, 29 30	41	Niobium	Nb	93
15	Phosphorus	P	31	42	Molybdenum	Mo	92, 94, 95, 96, 97, 98, 100
16	Sulfur	S	32, 33, 34, 36				
17	Chlorine	Cl	35, 37	44	Ruthenium	Ru	96, 98, 99, 100, 101, 102, 104
18	Argon	Ar	36, 38, 40				
19	Potassium	K	39, ☐40, 41	45	Rhodium	Rh	103
20	Calcium	Ca	40, 42, 43, 44, 46, ☐48	46	Palladium	Pd	102, 104, 105, 106, 108, 110
21	Scandium	Sc	45	47	Silver	Ag	107, 109
22	Titanium	Ti	46, 47, 48, 49 50	48	Cadmium	Cd	106, 108, 110, 111, 112, 113, 114, 116
23	Vanadium	V	☐50, 51				
24	Chromium	Cr	50, 52, 53, 54	49	Indium	In	113, ☐115
25	Manganese	Mn	55	50	Tin	Sn	112, 114, 115, 116, 117, 118, 119, 120, 122, 124
26	Iron	Fe	54, 56, 57, 58				
27	Cobalt	Co	59				
28	Nickel	Ni	58, 60, 61, 62, 64	51	Antimony	Sb	121, 123

Table B.2   *(Continued)*

Atomic Number[a]	Name	Symbol	Mass Numbers[b] of Natural Isotopes	Atomic Number[a]	Name	Symbol	Mass Numbers[b] of Natural Isotopes
52	Tellurium	Te	120, 122, 123, 124, 125, 126, 128, 130	70	Ytterbium	Yb	168, 170, 171, 172, 173, 174, 176
53	Iodine	I	127	71	Lutetium	Lu	175, 176̄
54	Xenon	Xe	124, 126, 128, 129, 130, 131, 132, 134, 136	72	Hafnium	Hf	174, 176, 177, 178, 179, 180
				73	Tantalum	Ta	180, 181
55	Cesium	Cs	133	74	Tungsten	W	180, 182, 183, 184, 186
56	Barium	Ba	130, 132, 134, 135, 137, 138	75	Rhenium	Re	185, 187̄
57	Lanthanum	La	138̄, 139	76	Osmium	Os	184, 186, 187, 188, 189, 190, 192
58	Cerium	Ce	136, 138, 140, 142̄				
59	Praseodymium	Pr	141	77	Iridium	Ir	191, 193
60	Neodymium	Nd	142, 143, 144̄, 145, 146, 148, 150	78	Platinum	Pt	190, 192, 195, 196, 198
62	Samarium	Sm	144, 147̄, 148̄, 149̄, 150, 152, 154	79	Gold	Au	197
				80	Mercury	Hg	196, 198, 199, 200, 201, 202, 204
63	Europium	Eu	151, 153				
64	Gadolinium	Gd	152̄, 154, 155, 156, 157, 158, 160	81	Thallium	Tl	203, 205
				82	Lead	Pb	204, 206, 207, 208
65	Terbium	Tb	159	83	Bismuth	Bi	209
66	Dysprosium	Dy	156, 158, 160, 161, 162, 163, 164	84	Polonium	Po	210̄
				86	Radon	Rn	222̄
67	Holmium	Ho	165	88	Radium	Ra	226̄
68	Erbium	Er	162, 166, 167, 168, 170	90	Thorium	Th	232̄
				91	Protactinium	Pa	231̄
69	Thulium	Tm	169	92	Uranium	U	234̄, 235̄, 238̄

[a]Atomic number = number of protons.
[b]Mass number = protons + neutrons.
[c]☐ indicates isotope is radioactive.

APPENDIX C
Tables of the Properties of Common Minerals

Table C.1 Properties of the Common Minerals with Metallic Luster

Mineral	Chemical Composition	Form and Habit	Cleavage	Hardness / Specific Gravity		Other Properties	Most Distinctive Properties
Bornite	Cu$_5$FeS$_4$	Massive. Crystals very rare.	None. Uneven fracture.	3	5	Brownish bronze on fresh surface. Tarnishes purple, blue, and black. Grayish-black streak.	Color, streak.
Chalcocite	Cu$_2$S	Massive. Crystals very rare.	None. Conchoidal fracture.	2.5	5.7	Steel-gray to black. Dark gray streak.	Streak.
Chalcopyrite	CuFeS$_2$	Massive or granular.	None. Uneven fracture.	3.5–4	4.2	Golden yellow to brassy yellow. Dark green to black streak.	Streak. Hardness distinguishes from pyrite.
Chromite	FeCr$_2$O$_4$	Massive or granular.	None. Uneven fracture.	5.5	4.6	Iron black to brownish black. Dark brown streak.	Streak and lack of magnetism distinguishes from ilmenite and magnetite.
Copper	Cu	Massive, twisted leaves and wires.	None. Can be cut with a knife.	2.5–3	9	Copper color but commonly stained green.	Color, specific gravity, malleable.
Galena	PbS	Cubic crystals, coarse or fine-grained granular masses.	Perfect in three directions at right angles.	2.5	7.6	Lead-gray color. Gray to gray-black streak.	Cleavage and streak.
Gold	Au	Small irregular grains.	None. Malleable.	2.5	19.3	Gold color. Can be flattened without breakage.	Color, specific gravity, malleability.
Hematite	Fe$_2$O$_3$	Massive, granular, micaceous.	Uneven fracture.	5–6	5	Reddish-brown, gray to black. Reddish-brown streak.	Streak, hardness.
Ilmenite	FeTiO$_3$	Massive or irregular grains.	Uneven fracture.	5.5–6	4.7	Iron-black. Brown-reddish streak differing from hematite.	Streak distinguishes hematite. Lack of magnetism distinguishes magnetite.
Limonite (*Goethite* is most common.)	A complex mixture of minerals, mainly hydrous iron oxides.	Massive, coatings, botryoidal crusts, earthy masses.	None.	1–5.5	3.5–4	Yellow, brown, black, yellowish-brown streak.	Streak.
Magnetite	Fe$_3$O$_4$	Massive, granular. Crystals have octahedral shape.	None. Uneven fracture.	5.5–6.5	5	Black. Black streak. Strongly attracted to a magnet.	Streak, magnetism.
Pyrite ("Fool's gold")	FeS$_2$	Cubic crystals with striated faces. Massive.	None. Uneven fracture.	6–6.5	5.2	Pale brass-yellow, darker if tarnished. Greenish-black streak.	Streak. Hardness distinguishes from chalcopyrite. Not malleable, which distinguishes from gold.

Table C.1 *(Continued)*

Mineral	Chemical Composition	Form and Habit	Cleavage	Hardness / Specific Gravity	Other Properties	Most Distinctive Properties
Pyrolusite	MnO_2	Crystals rare. Massive, coatings on fracture surfaces.	Crystals have a perfect cleavage. Massive, breaks unevenly.	2–6.5 / 5	Dark gray, black or bluish black. Black streak.	Color, streak.
Pyrrhotite	FeS	Crystals rare. Massive or granular.	None. Conchoidal fracture.	4 / 4.6	Brownish-bronze. Black streak. Magnetic.	Color and hardness distinguish from pyrite, magnetism from chalcopyrite.
Rutile	TiO_2	Slender, prismatic crystals or granular masses.	Good in one direction. Conchoidal fracture in others.	6–6.5 / 4.2	Reddish-brown (common) black (rare). Brownish streak. Adamantine luster.	Luster, habit, hardness.
Sphalerite (zinc blende)	ZnS	Fine to coarse granular masses. Tetrahedron shaped crystals.	Perfect in six directions.	3.5–4 / 4	Yellowish-brown to black. White to yellowish-brown streak. Resinous luster.	Cleavage, hardness, luster.
Uraninite	UO_2 to U_3O_8	Massive, with botryoidal forms. Rare crystals with cubic shapes.	None. Uneven fracture.	5–6 / 6.5–10	Black to dark brown. Streak black to dark brown. Dull luster.	Luster and specific gravity distinguish from magnetite. Streak distinguishes from ilmenite and hematite.

Table C.2 Properties of Rock-Forming Minerals with Nonmetallic Luster

Mineral	Chemical Composition	Form and Habit	Cleavage	Hardness / Specific Gravity	Other Properties	Most Distinctive Properties
Amphiboles (A complex family of minerals, *Hornblende* is most common.)	$X_2Y_5Si_8O_{22}(OH)_2$ where X = Ca, Na; Y = Mg, Fe, Al.	Long, six-sided crystals; also fibers and irregular grains.	Two; intersecting at 56° and 124°.	5–6 / 2.9–3.8	Common in metamorphic and igneous rocks. *Hornblende* is dark green to black; *actinolite*, green; *tremolite*, white.	Cleavage, habit.
Andalusite	Al_2SiO_5	Long crystals, often square in cross-section.	Weak, parallel to length of crystal.	7.5 / 3.2	Found in metamorphic rocks. Often flesh-colored.	Hardness, form.
Anhydrite	$CaSO_4$	Crystals are rare. Irregular grains or fibers.	Three, at right angles.	3 / 2.9	Alters to gypsum. Pearly luster, white or colorless.	Cleavage, hardness.
Apatite	$Ca_5(PO_4)_3(F, OH, Cl)$	Granular masses. Perfect six-sided crystals.	Poor. One direction.	5 / 3.2	Green, brown, blue, or white. Common in many kinds of rocks in small amounts.	Hardness, form.
Aragonite	$CaCO_3$	Massive, or slender, needle-like crystals.	Poor. Two directions.	3.5 / 2.9	Colorless or white. Effervesces with dilute HCl.	Effervescence with acid. Poor cleavage distinguishes from calcite.
Asbestos			See Serpentine			
Augite			See Pyroxene			
Biotite			See Mica			
Calcite	$CaCO_3$	Tapering crystals and granular masses.	Three perfect; at oblique angles to give a rhomb-shaped fragment.	3 / 2.7	Colorless or white. Effervesces with dilute. HCl.	Cleavage, effervescence with acid.

Table C.2   (Continued)

Mineral	Chemical Composition	Form and Habit	Cleavage	Hardness / Specific Gravity	Other Properties	Most Distinctive Properties
Chlorite	$(Mg, Fe)_5(Al, Fe)_2 Si_3O_{10}(OH)_8$	Flaky masses of minute scales.	One perfect; parallel to flakes.	2–2.5 / 2.6–2.9	Common in metamorphic rocks. Light to dark green. Greasy luster.	Cleavage—flakes not elastic, distinguishes from mica. Color.
Dolomite	$CaMg(CO_3)_2$	Crystals with rhomb-shaped faces. Granular masses.	Perfect in three directions as in calcite.	3.5 / 2.8	White or gray. Does not effervesce in cold, dilute HCl unless powdered. Pearly luster.	Cleavage. Lack of effervescence with acid.
Epidote	Complex silicate of Ca, Fe and Al	Small elongate crystals. Fibrous.	One perfect, one poor.	6–7 / 3.4	Yellowish-green to dark green. Common in metamorphic rocks.	Habit, color. Hardness distinguishes from chlorite.
Feldspars: *Potassium* feldspar (*orthoclase* is a common variety	$KAlSi_3O_8$	Prism-shaped crystals, granular masses.	Two perfect, at right angles.	6 / 2.6	Common mineral. Flesh-colored, pink, white, or gray.	Color, cleavage.
Plagioclase	$NaAlSi_3O_8$ (albite) and $CaAl_2Si_2O_8$ (anorthite) and all compositions between.	Irregular grains, cleavable masses. Tabular crystals.	Two perfect, not quite at right angles.	6–6.5 / 2.6–2.7	White to dark gray. Cleavage planes may show fine parallel striations.	Cleavage. Striations on cleavage planes will distinguish from potassium feldspar.
Fluorite	CaF_2	Cubic crystals, granular masses.	Perfect in four directions.	4 / 3.2	Colorless, bluish green. Always an accessory mineral.	Hardness, cleavage does not effervesce with acid.
Garnets	$X_3Y_2 (SiO_4)_3$; X = Ca, Mg, Fe, Mn; Y = Al, Fe, Ti, Cr.	Perfect crystals with 12 or 24 sides. Granular masses.	None. Uneven fracture.	6.5–7.5 / 3.5–4.3	Common in metamorphic rocks. Red, brown, yellowish-green, black.	Crystals, hardness, no cleavage.
Graphite	C	Scaly masses.	One, perfect. Forms slippery flakes.	1–2 / 2.2	Metamorphic rocks. Black with metallic to dull luster.	Cleavage, color. Marks paper.
Gypsum	$CaSO_4 \cdot 2H_2O$	Elongate or tabular crystals. Fibrous and earthy masses.	One, perfect. Flakes bend but are not elastic.	2 / 2.3	Vitreous to pearly luster. Colorless.	Hardness, cleavage.
Halite	$NaCl$	Cubic crystals.	Perfect to give cubes.	2.5 / 2.2	Tastes salty. Colorless, blue.	Taste, cleavage.
Hornblende			See Amphibole			
Kaolinite	$Al_2Si_2O_5(OH)_4$	Soft, earthy masses. Submicroscopic crystals.	One, perfect.	2–2.5 / 2.6	White, yellowish. Plastic when wet; emits clayey odor. Dull luster.	Feel, plasticity, odor.
Kyanite	Al_2SiO_5	Bladed crystals.	One perfect. One imperfect.	4.5 parallel to blade, 7 across blade / 3.6	Blue, white, gray. Common in metamorphic rocks.	Variable hardness, distinguishes from sillimanite. Color.
Mica: *Biotite*	$K(Mg, Fe)_3 AlSi_3O_{10} (OH)_2$	Irregular masses of flakes.	One, perfect.	2.5–3 / 2.8–3.2	Common in igneous and metamorphic rocks. Black, brown, dark green.	Cleavage, color. Flakes are elastic.
Muscovite	$KAl_3Si_3O_{10}(OH)_2$	Thin flakes.	One, perfect.	2–2.5 / 2.7	Common in igneous and metamorphic rocks. Colorless, pale green or brown.	Cleavage, color. Flakes are elastic.

Table C.2 (*Continued*)

Mineral	Chemical Composition	Form and Habit	Cleavage	Hardness / Specific Gravity	Other Properties	Most Distinctive Properties
Olivine	$(Mg, Fe)_2SiO_4$	Small grains, granular masses.	None. Conchoidal fracture.	6.5–7 / 3.2–4.3	Igneous rocks. Olive green to yellowish-green.	Color, fracture, habit.
Orthoclase			See Feldspar			
Plagioclase			See Feldspar			
Pyroxene (A complex family of minerals. *Augite* is most common.)	$XY(SiO_3)_2$ X = Y = Ca, Mg. Fe	8-sided stubby crystals. Granular masses.	Two, perfect, nearly at right angles.	5–6 / 3.2–3.9	Igneous and metamorphic rocks. *Augite*, dark green to black; other varieties white to green.	Cleavage.
Quartz	SiO_2	6-sided crystals, granular masses.	None. Conchoidal fractures.	7 / 2.6	Colorless, white, gray, but may have any color, depending on impurities. Vitreous to greasy luster.	Form, fracture, striations across crystal faces at right angles to long dimension.
Serpentine (Fibrous variety is *asbestos*)	$Mg_3Si_2O_5(OH)_4$	Platy or fibrous.	One, perfect.	2.5–5 / 2.2–2.6	Light to dark green. Smooth, greasy feel.	Habit, hardness.
Sillimanite	Al_2SiO_5	Long, needle-like crystals, fibers.	Breaks irregularly, except in fibrous variety.	6–7 / 3.2	White, gray. Metamorphic rocks.	Hardness distinguishes from kyanite. Habit.
Talc	$Mg_3Si_4O_{10}(OH)_2$	Small scales, compact masses.	One, perfect.	1 / 2.6–2.8	Feels slippery. Pearly luster. White to greenish.	Hardness, luster, feel cleavage.
Tourmaline	Complex silicate of B, Al, Na, Ca, Fe, Li and Mg.	Elongate crystals, commonly with triangular cross section.	None.	7–7.5 / 3–3.3	Black, brown, red, pink, green, blue, and yellow. An accessory mineral in many rocks.	Habit.
Wollastonite	$CaSiO_3$	Fibrous or bladed aggregates of crystals.	Two, perfect.	4.5–5 / 2.8–2.9	Colorless, white, yellowish. Metamorphic rocks. Soluble in HCl.	Habit. Solubility in HCl and hardness distinguish amphiboles, kyanite, sillimanite.

APPENDIX D

World Soils and Soil Classification System

Table D.1 Orders of the Soil Classification System Used in the United States

Soil Order (Meaning of Name)	Main Characteristics
Alfisol (Pedalfer[a] soil)	Thin A horizon over clay-rich B horizon, in places separated by light-gray E horizon. Typical of humid middle latitudes.
Aridisol (Arid soil)	Thin A horizon above relatively thin B horizon and often with carbonate accumulation in K horizon. Typical of dry climates.
Entisol (Recent soil)	Soil lacking well-developed horizons. Only a thin incipient A horizon may be present.
Histosol (Organic soil)	Peaty soil, rich in organic matter. Typical of cool, moist climates.
Inceptisol (Young soil)	Weakly developed soil, but with recognizable A horizon and incipient B horizon lacking clay or iron enrichment. Generally occurs under moist conditions.
Mollisol (Soft soil)	Grassland soil with thick dark A horizon, rich in organic matter. B horizon may be enriched in clay. E and K horizons may be present.
Oxisol (Oxide soil)	Relatively infertile soil with A horizon over oxidized and often thick B horizon.
Spodosol (Ashy soil)	Acidic soil marked by highly organic O and A horizons, an E horizon, and iron/aluminum-rich B horizon. Occurs in cool forest zones.
Ultisol (Ultimate soil)	Strongly weathered soil characterized by A and E horizons over highly weathered and clay-rich B horizon. Characteristic of tropical and subtropical climates.
Vertisol (Inverted soil)	Organic-rich soil having very high content of clays that shrink and expand as moisture varies seasonally.
Andisol (Dark soil)	Soil developed on pyroclastic deposits and characterized by low bulk density and high content of amorphous minerals.

[a]Soils rich in iron and aluminum.

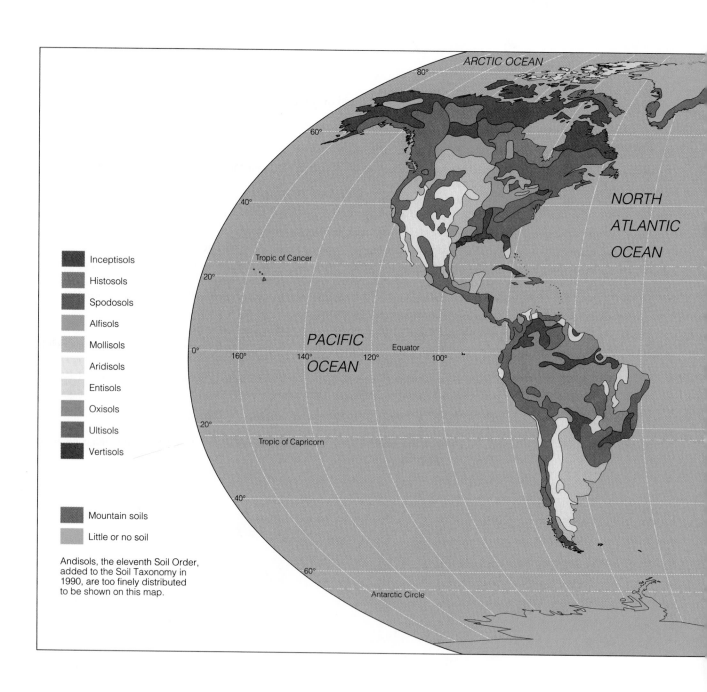

Inceptisols

Histosols

Spodosols

Alfisols

Mollisols

Aridisols

Entisols

Oxisols

Ultisols

Vertisols

Mountain soils

Little or no soil

Andisols, the eleventh Soil Order, added to the Soil Taxonomy in 1990, are too finely distributed to be shown on this map.

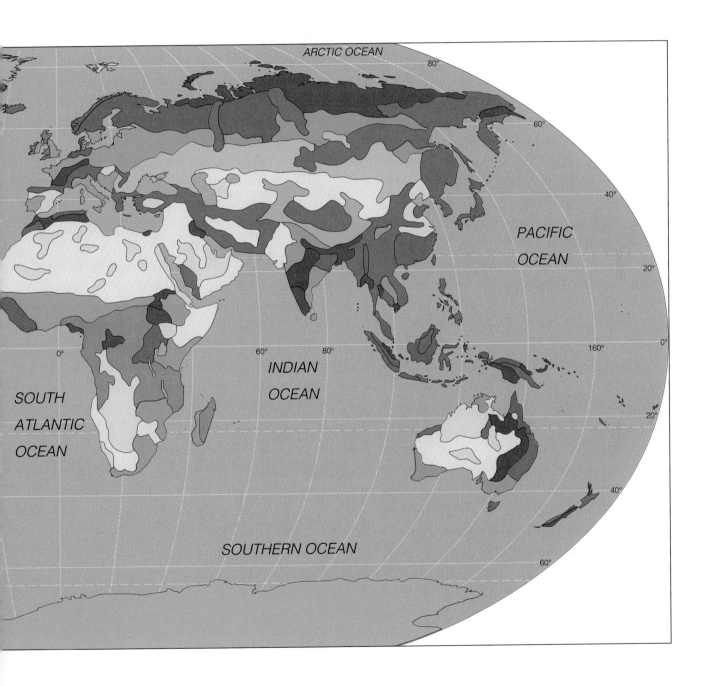

The Köppen Climatic Classification

There are five basic climatic categories in the Köppen system, symbolized as A, B, C, D, and E. These are further subdivided by adding lower-case letters to indicate lesser variations of temperature and moisture within the major groupings.

A. TROPICAL HUMID CLIMATES

Coolest month must be above 18°C (64.4°F).

Af—Tropical Rain Forest (f—feucht or moist) No dry season. Driest month must attain at least 6 cm (2.4 in) of rainfall.

Aw—Tropical Savanna (w—winter) Winter dry season. At least one month must attain less than 6 cm (2.4 in) of rainfall.

The following lower-case letters may be added for clarification in special situations:

m (monsoon)—despite a dry season, total rainfall is so heavy that rain forest vegetation is not impeded.

w'—autumn rainfall maximum.

w"—two dry seasons during a single year.

s (summer)—summer dry season.

i—annual temperature range must be less than 5°C (9°F).

g (Ganges)—hottest month occurs prior to summer solstice.

B. DRY CLIMATES

No specific amount of moisture makes a climate dry. Rather, the rate of evaporation (determined by temperature), relative to the amount of precipitation dictates how dry a climate is in terms of its ability to support plant growth. This is reckoned through the use of formulas that are not included here.

BW (W-wuste or wasteland)—desert.

BS (S—steppe)—semiarid.

The following lower-case letters may be added for clarification in special situations.

h (heiss or hot)—average annual temperature must be above 18°C (64.4°F).

k (kalt or cold)—average annual temperature must be under 18°C (64.4°F).

k'—temperature of warmest month must be under 18°C (64.4°F).

s—summer dry season. At least three times as much precipitation in the wettest month as in the driest.

w—winter dry season. At least ten times as much precipitation in the wettest month as in the driest.

n (nebel or fog)—frequent fog.

C. TEMPERATE HUMID CLIMATES

Coldest month average must be below 18°C (64.4°F), but above −3°C (26.6°F). Warmest month average must be above 10°C (50°F).

Cf—no dry season. Driest month must attain at least 3 cm (1.2 in) of precipitation.

CW—winter dry season. At least ten times as much rain in the wettest month as in the driest.

Cs—summer dry season. At least three times as much rain in the wettest month as in the driest. Driest month must receive less than 3 cm (1.2 in) of rainfall.

The following lower-case letters may be added for clarification in special situations:

a—hot summer. Warmest month must average above 22°C (71.6°F).

b—cool summer. Warmest month must average below 22°C (71.6°F). At least 4 months above 10°C (50°F).

c—short, cool summer. Less than four months over 10°C (50°F).

i—see A climate.

g—see A climate.

n—see B climate.

x—maximum precipitation in late spring or early summer.

D. COLD HUMID CLIMATES

Coldest month average must be below –3°C (26.6°F). Warmest month average must be above 10°C (50°F).

Df—no dry season.
Dw—winter dry season.

The following lower-case letters may be added for clarification in special situations:

a—see C climate.
b—see C climate.

c—see C climate.
d—coldest month average must be below –38°C (–36.4°F).
f—see A climate.
s—see A climate.
w—see A climate.

E. POLAR CLIMATES

Warmest month average must be below 10°C (50°F).

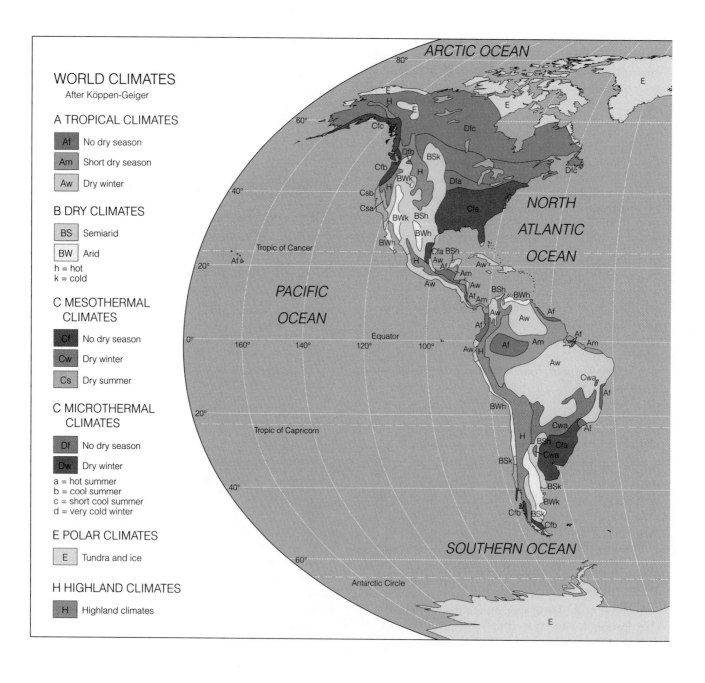

WORLD CLIMATES
After Köppen-Geiger

A TROPICAL CLIMATES

Af	No dry season
Am	Short dry season
Aw	Dry winter

B DRY CLIMATES

| BS | Semiarid |
| BW | Arid |

h = hot
k = cold

C MESOTHERMAL CLIMATES

Cf	No dry season
Cw	Dry winter
Cs	Dry summer

C MICROTHERMAL CLIMATES

| Df | No dry season |
| Dw | Dry winter |

a = hot summer
b = cool summer
c = short cool summer
d = very cold winter

E POLAR CLIMATES

| E | Tundra and ice |

H HIGHLAND CLIMATES

| H | Highland climates |

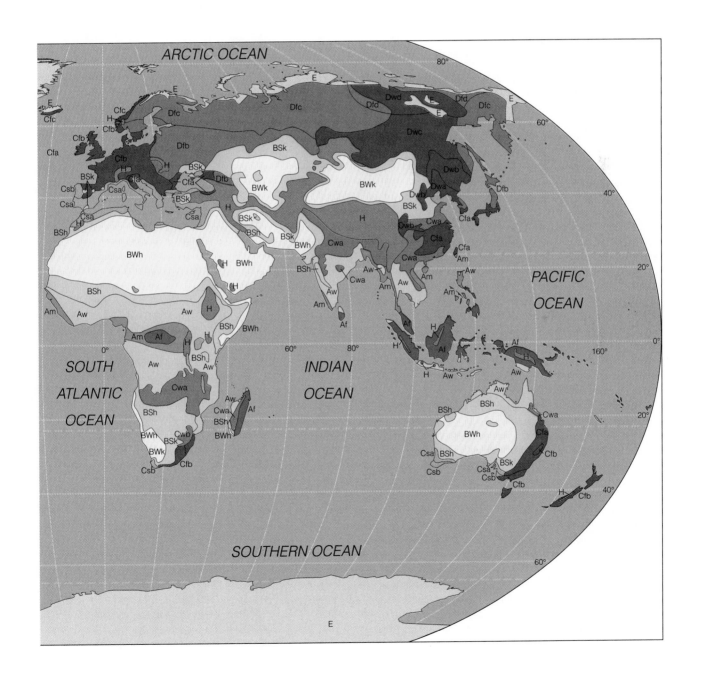

A P P E N D I X   F

Topographic Maps

USES OF MAPS

An important part of the accumulated information about the geology and morphology of the Earth's crust exists in the form of maps. Nearly everyone has used automobile road maps in planning a trip or in following an unmarked road. A road map of a state, province, or county does what all maps have done since their invention at some unknown time more than 5000 years ago; it reduces the pattern of part of the Earth's surface to a size small enough to be seen as a whole. Maps are especially important for an understanding of geologic relations because a continent, a mountain chain, or a major river valley are of such large size that they cannot be viewed as a whole unless represented on a map.

A map can be made to express much information within a small space by the use of various kinds of symbols. Just as some aspect of physics and chemistry use the symbolic language of mathematics to express significant relationships, so many aspects of geology use the simple symbolic language of maps to depict relationships too large to be observed within a single view. Maps made or used by geologists generally depict either of two sorts of things.

1. The shape of the Earth's surface, on which are shown hills, valleys, and other features. Maps of this kind are *topographic maps.*

2. The distribution and attitudes of bodies of rock or regolith. Maps showing such things are *geologic maps.* They are often plotted on a topographic base map.

BASE MAPS

Every map is made for some special purpose. Road maps, charts for sea or air navigation, and geologic maps are examples of three special purposes. However, whatever the purpose, all maps have two classes of data: base data and special-purpose data. As base data, most geologic maps show a latitude-longitude grid, streams, and inhabited places; many also show roads and railroads and details of topography. Geologists may take an existing base map that contains such data and plot geologic information on it, or they may start with blank paper and plot on it both base and ge-

ologic data—a much slower process if the map is made accurately.

Two-Dimensional Base Maps

Many base maps used for plotting geologic data are two-dimensional; that is, they represent length and breadth but not height. A point can be located only in terms of its horizontal distance, in a particular direction, from some other point. Hence, a base map always embodies the basic concepts of direction and distance. Two natural reference points on the Earth are the north and south poles. Using these two points, a grid is constructed by means of which any other point can be located. The grid we use consists of lines of *longitude* (half circles joining the poles) and *latitude* (parallel circles concentric to the poles and perpendicular to the axis connecting the poles) (Fig. F.1). The longitude lines (*meridians*) run exactly north–south, cross the east–west *parallels* of latitude at right angles. Since the circumference of the Earth at its equator (and the somewhat smaller circumference through its

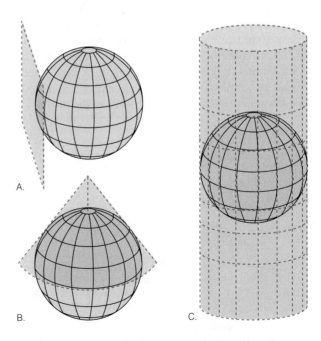

Figure F.1 The Earth's latitude–longitude grid can be projected onto a plane a. cylinder c. or core b. that theoretically can be cut and flattened out.

511

two poles) is known with fair accuracy, it is possible to define any point on the Earth in terms of direction and distance from either pole or from the point of intersection of any parallel with any meridian.

For convenience in reading, most maps are drawn so that the north direction is at the top or upper edge of the map. This is an arbitrary convention adopted mainly to save time. The north direction could just as well be placed elsewhere, provided its position is clearly indicated.

Map Projections

The Earth's surface is nearly spherical, whereas maps (other than globes) are two-dimensional planes, usually sheets of paper. It is geometrically impossible to represent any part of a spherical surface on a plane surface without distortion (Fig. F.2). The latitude-longitude grid has to be projected from the curved surface to the flat one. This can be done in various ways, each of which has advantages, but all of which represent a sacrifice of accuracy in that the resulting scale on the flat map will vary from one part of the map to another. The most famous of these is the Mercator projection, prepared by projecting all points radially onto a cylinder (Fig. F.1c), then unfolding the cylinder; although it distorts the polar regions very greatly, compass directions drawn on a Mercator projection are straight lines. Because this is of enormous value in navigation, the Mercator projection is widely used in navigator's charts.

Other kinds of projections are shown in Figure F.1. A commonly used projection for small regions of the Earth's surface is the conic projection (Fig. F.1b). Some commonly used varieties are polyconic, in which not one cone, as in Figure F.1b, but several cones are employed, each one tangent to the globe at a different latitude. This device reduces distortion.

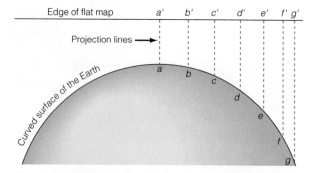

Figure F.2 Equally spaced points (*a, b, c,* ...) along a line in any direction on the Earth's surface become unequally spaced when projected onto a plane. That is why all flat maps are distorted.

Map Scales

The accuracy with which distance is represented determines the accuracy of the map. *The proportion between a unit of distance on a map and the unit it represents on the Earth's surface* is the *scale* of the map. It is expressed as a simple proportion, such as 1:1,000,000. This ratio means that 1 meter, 1 foot, or other unit on the map represents exactly 1,000,000 meters, feet, or other units on the Earth's surface. It is approximately the scale of many of the road maps widely used by motorists in North America. Scale is also expressed graphically by means of a numbered bar, as is done on most of the maps in this book. A map with a latitude-longitude grid needs no other indication of scale (except for convenience), because the lengths of a degree of longitude (varying from 110.7 km at the equator to 0 at the poles) and of latitude (varying from 109.9 km at the equator to 110.9 km at the poles) are known.

The most commonly used scale for both topographic and geologic maps prepared in the United States by the Geological Survey is 1:24,000. Many older maps employ a scale of 1:62,500, approximately equal to 1 in. = 1 mile. When maps of larger regions are prepared, scales of 1:100,000, 1:250,000, and 1:1,000,000 are employed. Use of scales at 1:24,000 and 1:62,500 arises from the practice of preparing maps that cover a quadrangular segment of the surface that is either 15 minutes (15') of longitude by 15' of latitude (scale, 1:62,500), or 7½' × 7½' (scale, 1:24,000).

Contours and Topographic Maps

A more complete kind of base map is three-dimensional; it represents not only length and breadth but also height. Therefore, it shows **relief** (*the difference in altitude between the high and low parts of a land surface*) and also **topography,** defined as *the relief and form of the land. A map that shows topography* is a **topographic map.** Topographic maps can give the form of the land in various ways. The maps most commonly used by geologists show it by contour lines.

A **contour line** (often called simply a **contour**) is *a line passing through points having the same altitude above sea level.* If we start at a certain altitude on an irregular surface and walk in such a way as to go neither uphill nor downhill, we will trace out a path that corresponds to a contour line. Such a path will curve around hills, bend upstream in valleys, and swing outward around ridges. Viewed broadly, every contour must be a closed line, just as the shoreline of an island or of a continent returns upon itself, however long it may be. Even on maps of small areas, many contours are closed lines, such as those at or near the tops of hills. Many, however, do not close within a given map area; they extend to the edges of the map and join the contours on adjacent maps.

Imagine an island in the sea crowned by two prominent isolated hills, with much steeper slopes on one side than on the other and with an irregular shoreline. The shoreline is a contour line (the zero contour) because the surface of the water is horizontal. If the island is pictured as submerged until only the two isolated peaks project above the sea, and then raised above the sea 5 m at a time, the successive new shorelines will form a series of contour lines separated by 5 m contour intervals. (A **contour interval** is *the vertical distance between two successive contour lines* and is commonly the same throughout any one map.) At first, two small islands will appear, each with its own shoreline, and the contours marking their shorelines will have the form of two closed lines. When the main mass of the island rises above the water, the remaining shorelines or contours will pass completely around the landmass. The final shoreline is represented by the zero contour, which now forms the lowest of a series of contours separated by vertical distances of 5 m.

The following rules apply to contours:

1. All points on a contour have the same *elevation*, (also called *altitude*), which is *the vertical distance above mean sea level.*

2. A contour separates all points of higher elevation from all points of lower elevation.

3. In order to facilitate reading the contours on a map, certain contours (usually every fifth line) are drawn as a bolder line. Contours are numbered at convenient intervals for ready identification. The numbers are always multiples of the contour interval. For example, contour intervals of 5 m mean that successive contours are drawn at 10, 15, 20 m, etc.

4. Contours do not split or cross over, but at vertical cliffs they merge.

5. Because the contours that represent a depression without an outlet resemble those of an isolated hill, it is necessary to give them a distinctive appearance. Therefore, depression contours are *hatched;* that is, they are marked on the downslope side with short transverse lines called *hatchures.* An example is shown on one contour in Figure F.3*b.* The contour interval employed is the same as in other contours on the same map.

6. Closely spaced contours indicate steep slopes, and widely spaced contours indicate gentle slopes.

7. Contours crossing a valley form a V-shape pointing *up* the valley, while contours on a ridge form a V-shape pointing *down* the ridge.

Idealized Example of a Topographic Map

Figures F.3*a* and *b* show the relation between the surface of the land and the contour map representing it. Figure F.3*a,* a perspective sketch, shows a stream valley between two hills, viewed from the south. In the foreground is the sea, with a bay sheltered by a curving spit. Terraces in which small streams have excavated gullies border the valley. The hill on the east has a rounded summit and sloping spurs. Some of the spurs are truncated at their lower ends by a wave-cut cliff, at the base of which is a beach. The hill on the west stands abruptly above the valley with a steep scarp and slopes gently westward, trenched by a few shallow gullies.

Each of the features on the map (Fig. F.3*b*) is represented by contours directly beneath its position in the sketch.

Topographic Profile

The *outline of the land surface along a given line* is called a **topographic profile.** A profile can be drawn along any line on a topographic map. Both the horizontal and the vertical scales must be designated for a profile. Most commonly, the map scale is chosen for the horizontal scale. The vertical scale is commonly made somewhat larger than the horizontal scale in order to exaggerate, or emphasize, the topography. The *ratio of the horizontal scale to the vertical scale in a topographic profile* is called the **vertical exaggeration.** If the horizontal scale is 1 cm = 1000 m, and the vertical scale is 1 cm = 200 m, the vertical exaggeration is 1000/200 = 5X. A topographic profile of Figure F.3*b* is shown at a vertical exaggeration of 5X in Figure F.4.

To prepare a topographic profile perform the following steps:

1. Select the vertical and horizontal scales.

2. On a sheet of paper, select one of the horizontal lines as a baseline. Choice of a base varies from profile to profile; it can be sea level or any convenient height, such as the contour below the lowest point on the profile. Then mark in elevations on the graph paper, choosing the spacing appropriate to the vertical scale.

3. Place the edge of the graph paper on the line of the profile marked on the map.

4. Wherever a contour crosses the line of the profile, make a line on the graph paper at the appropriate elevation.

5. Join the points on the graph paper with a smooth line.

A.

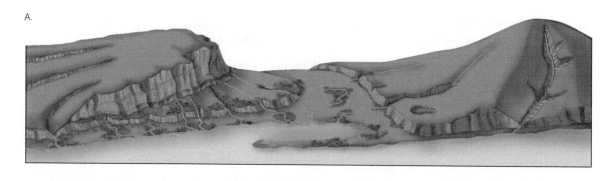

B.

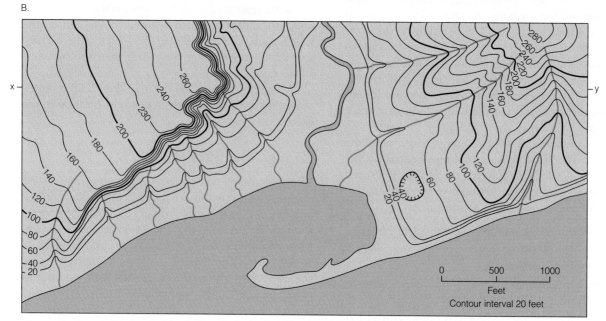

0   500   1000

Feet
Contour interval 20 feet

Figure F.3   (*a*) Perspective sketch of a landscape. (*Source:* Modified from U.S. Geological Survey.) (*b*) Topographic map of the area shown in Figure F.3*a*. Note that this map is scaled in feet and the contour interval is 20 ft. (*Source:* Modified from U.S. Geological Survey.)

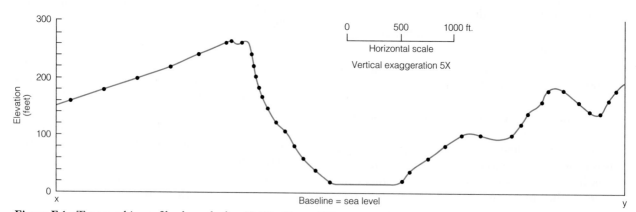

Figure F.4   Topographic profile along the line X–Y in Figure F.3*b*.

GLOSSARY

Ablation. The loss of mass from a glacier. (Ch. 10)

Ablation area. A region of net loss on a glacier characterized by a surface of bare ice and old snow from which the last winter's snowcover has melted away. (Ch. 10)

Abyssal plain. A large flat area of the deep seafloor having slopes less than about 1 m/km, and ranging in depth below sea level from 3 to 6 km. (Ch. 4)

Accreted terrane. Block of crust moved laterally by strike-slip faulting or by a combination of strike-slip faulting and subduction, then accreted to a larger mass of continental crust. Also called a *suspect terrane*. (Ch. 8)

Accumulation. The addition of mass to a glacier. (Ch. 10)

Accumulation area. An upper zone on a glacier, covered by remnants of the previous winter's snowfall and representing an area of net gain in mass. (Ch. 10)

Adiabatic lapse rate. The way temperature changes with altitude in rising or falling air. (Ch. 12)

Adiabatic process. A process that happens without the addition or subtraction of heat from an external source. (Ch. 12)

Aerosol. A tiny liquid droplet or tiny solid particle so small it remains suspended in air. (Ch. 12)

Agglomerate. A pyroclastic rock consisting of bomb-sized tephra, i.e., tephra in which the average particle diameter is greater than 64 mm. (Ch. 7)

Air. The invisible, odorless mixture of gases and suspended particles that surrounds the Earth. (Ch. 12)

Air-pressure gradient. The air pressure drop per unit distance. (Ch. 13)

Albedo. The reflectivity of the surface of a planet. (Ch. 10)

Alluvial fan. A fan-shaped body of alluvium typically built where a stream leaves a steep mountain valley. (Ch. 9)

Alluvium. Sediment deposited by streams in nonmarine environments. (Ch. 9)

Amphibolite. A metamorphic rock of intermediate grade, generally coarse-grained, containing abundant amphibole. (Ch. 8)

Amino acid. Organic molecule containing an amino (NH_2) group; the building block of proteins. (Ch. 16)

Anaerobic. Without oxygen. (Ch. 15)

Andesite. A fine-grained igneous rock with the composition of a diorite. (Ch. 7)

Angiosperm. A plant whose seeds are surrounded by fruit. (Ch. 17)

Anion. An ion with a negative electrical charge. (Ch. 6)

Anticyclone. Air spiraling outward away from a high-pressure center. (Ch. 13)

Aquiclude. A body of impermeable or distinctly less permeable rock adjacent to an *aquifer*. (Ch. 9)

Aquifer. A body of permeable rock or regolith saturated with water and through which groundwater moves. (Ch. 9)

Artesian aquifer. An aquifer in which water is under hydraulic pressure. (Ch. 9)

Asthenosphere. The region of the mantle where rocks become ductile, having little strength, and are easily deformed. It lies at a depth of 100 to 350 km below the surface. (Ch. 2)

Atmosphere. The mixture of gases, predominantly nitrogen, oxygen, carbon dioxide, and water vapor that surrounds the Earth. (Ch. 1)

Atom. The smallest individual particle that retains all the properties of a given chemical element. (Ch. 6)

Autotrophs. Organisms that can get energy directly from sunlight. (Ch. 15)

Barometer. A device that measures air pressure. (Ch. 12)

Basalt. A fine-grained igneous rock with the composition of a gabbro. (Ch. 7)

Batholith. The largest kind of pluton. A very large, igneous body of irregular shape that cuts across the layering of the rock it intrudes. (Ch. 7)

Beach. Wave-washed sediment along a coast, extending throughout the surf zone. (Ch. 11)

Bed. The smallest formal unit of a body of sediment or sedimentary rock. (Ch. 8)

Bedding. The layered arrangement of strata in a body of sediment or sedimentary rock. (Ch. 8)

Bed load. Coarse particles that move along the bottom of a stream channel. (Ch. 9)

Bergeron process. The evaporation of supercooled water droplets in a cloud to release water vapor that is then deposited on ice crystals within the cloud, leading to precipitation. (Ch. 12)

Biogeochemical cycle. A natural cycle describing the movements and interactions through the Earth's spheres of the chemicals essential to life. (Ch. 16)

Biological diversity. Used loosely to mean the variety of life on Earth, but technically this concept consists of three components: (1) genetic diversity—the total number of genetic characteristics; (2) species diversity; and (3) habitat or ecosystem diversity—the number of kinds of habitats or ecosystems in a given unit area. Species diversity in turn includes three concepts: *species richness*, *evenness*, and *dominance*. (Ch. 17)

Biomass. Usually used to mean the amount of living material, both as live and dead material, as in the leaves (live) and stem wood (dead) of trees. (Ch. 15)

Biome. A kind of ecosystem. The rain forest is an example of a biome; rain forests occur in many parts of the world but are not all connected with each other. (Ch. 15)

Biosphere. The totality of the Earth's

organisms and, in addition, organic matter that has not yet been completely decomposed. (Ch. 1)

Biosynthesis. The polymerization of small organic molecules within a living organism to form biopolymers, particularly proteins. (Ch. 17)

Black body radiator. A (hypothetical) perfect radiator of light that absorbs all light that strikes it and reflects none; its light output depends only on its temperature. (Ch. 2)

Body waves. Seismic waves that travel outward from an earthquake focus and pass through the Earth. (Ch. 5)

Boundary current. A current that flows generally poleward, parallel to a continental coastline. (Ch. 11)

Bowen's reaction series. A schematic description of the order in which different minerals crystallize during the cooling and progressive crystallization of a magma. (Ch. 7)

Braided stream. A channel system consisting of a tangled network of two or more smaller branching and reuniting channels that are separated by islands or bars. (Ch. 9)

Breeder reactor. A nuclear reactor in which fission takes place, specifically, the pile in which the conversion of ^{238}U takes place. (Ch. 18)

Burial metamorphism. Metamorphism caused solely by the burial of sedimentary or pyroclastic rocks. (Ch. 8)

Caldera. A roughly circular, steep-walled volcanic basin several kilometers or more in diameter. (Ch. 7)

Calving. The progressive breaking off of icebergs from a glacier that terminates in deep water. (Ch. 10)

Carbohydrates. Organic compounds composed of carbon, hydrogen and oxygen, of which sugars, starches and cellulose are examples. Carbohydrates are formed by all green plants and constitute a major source of food for animals. (Ch. 15)

Carnivores. Meat-eating heterotrophs. (Ch. 15)

Carrying capacity. The limit on the population that an ecosystem can carry, imposed by the limited resources of that ecosystem. (Ch. 15)

Catastrophism. The concept that all of the Earth's major features, such as mountains, valleys, and oceans, have been produced by a few great catastrophic events. (Ch. 1)

Cation. A positive ion. (Ch. 10)

Cell. The basic structural unit of all living organisms. (Ch. 15)

Channel. The passageway in which a stream flows. (Ch. 9)

Chemical sediment. Sediment formed by precipitation of minerals from solutions in water. (Ch. 8)

Chemical weathering. The decomposition of rocks through chemical reactions such as hydration and oxidation. (Ch. 6)

Chemoautotrophs. Organisms that derive energy from the oxidation of hydrogen sulfide in the water discharged from black smokers. (Ch. 15)

Chemosynthesis. The synthesis of small organic molecules such as amino acids. (Ch. 17)

Cirque. A bowl-shaped hollow on a mountainside, open downstream, bounded upstream by a steep slope (headwall), and excavated mainly by frost wedging and by glacial abrasion and plucking. (Ch. 10)

Clastic sediment. The loose fragmented debris produced by the mechanical breakdown of older rocks. (Ch. 8)

Clear cutting. A practice in lumbering by which all trees in a tract are felled. (Ch. 20)

Cleavage. The tendency of a mineral to break in preferred directions along bright, reflective plane surfaces. (Ch. 6)

Climate. The average weather conditions of a place or area over a period of years. (Ch. 12)

Cloud. Visible aggregations of minute water droplets, tiny ice crystals, or both. (Ch. 12)

Coal. A black, combustible, sedimentary or metamorphic rock consisting chiefly of decomposed plant matter and containing more than 50 percent organic matter. (Chs. 8 and 18)

Cold front. A front in which dense, cold air flows in and displaces warmer air by pushing it upward, producing clouds and possibly rain. (Ch. 12)

Collision margin. A convergent plate margin where two bodies of continental crust collide. (Ch. 4)

Competitive exclusion principle. The idea that two populations of different species with exactly the same requirements cannot persist indefinitely in the same habitat—one will always win out and the other will become extinct. Which one wins depends on the exact environmental conditions. Referred to as a principle, the idea has some basis in observation and experimentation. (Ch. 17)

Condensation. The formation of a more ordered liquid from a less ordered gas. (Ch. 12)

Conduction. The means by which heat is transmitted through solids without deforming the solid. (Ch. 4)

Cone of depression. A conical depression in the water table immediately surrounding a well. (Ch. 9)

Confined aquifer. An aquifer bounded by impermeable rock. (Ch. 9)

Conglomerate. A sedimentary rock composed of clasts of rounded gravel set in a finer-grained matrix. (Ch. 8)

Constellation. A distinctive star pattern in the sky, named mostly for animals and mythical characters. (Ch. 2)

Continental crust. The part of the Earth's crust that comprises the continents, which has an average thickness of 45 km. (Ch. 2)

Continental divide. A line that separates streams flowing towards opposite sides of a continent, usually into different oceans. (Ch. 19)

Continental rise. A region of gently changing slope where the floor of the ocean basin meets the margin of a continent. (Ch. 4)

Continental shelf. A submerged platform of variable width that forms a fringe around a continent. (Ch. 4)

Continental shield. An assemblage of cratons and orogens that has reached isostatic equilibrium. (Ch. 8)

Continental slope. A pronounced slope beyond the seaward margin of the continental shelf. (Ch. 4)

Convection. The process by which hot, less dense materials rise upward, being replaced by cold, dense, downward flowing material to create a convection current. (Ch. 4)

Convection current. The flow of material as a result of convection. (Ch. 4)

Convergence. The coming together of air masses, caused by the inward spiral flow in a cyclone and leading to an upward flow of air at the center of the low-pressure center. (Ch. 13)

Convergent margin. The zone where plates meet as they move toward each other. See *subduction zone*. (Ch. 4)

Core. The spherical mass, largely metallic iron, at the center of the Earth. (Ch. 2)

Coriolis effect. An effect that causes any body that moves freely with respect to the rotating solid Earth to veer toward the right in the northern hemisphere and toward the left in the southern hemisphere, regardless of the initial direction of the moving body. (Ch. 11)

Craton. A core of ancient rock in the continental crust that has attained tectonic and isostatic stability. (Ch. 8)

Crevasse. A deep, gaping fissure in the upper surface of a glacier. (Ch. 10)

Cross bedding. Beds that are inclined with respect to a thicker stratum within which they occur. (Ch. 8)

Crust. The outermost and thinnest of the Earth's compositional layers, which consists of rocky matter that is less dense than the rocks of the mantle below. (Ch. 2)

Cryosphere. The part of the Earth's surface that remains perennially frozen. (Ch. 10)

Crystal. A solid compound composed of ordered, three-dimensional arrays of atoms or ions chemically bonded together and displaying crystal form. (Ch. 6)

Crystal faces. The planar surfaces that bound a crystal. (Ch. 6)

Crystal form. The geometric arrangement of crystal faces. (Ch. 6)

Crystal structure. The geometric pattern that atoms assume in a solid. Any solid that has a crystal structure is said to be *crystalline*. (Ch. 6)

Curie point. A temperature above which permanent magnetism is not possible. (Ch. 4)

Cyclone. Air spiraling inward around a low-pressure center. (Ch. 13)

Cytoplasm. The main body of the cell, excluding the nucleus and the plasma membrane. (Ch. 16)

Deforestation. The process of forest clearing. (Ch. 20)

Delta. A body of sediment deposited by a stream where it flows into standing water. (Ch. 9)

Denitrification. The conversion of nitrate to molecular nitrogen by the action of bacteria—an important step in the nitrogen cycle. (Ch. 16)

Density. The average mass per unit volume. (Ch. 6)

Denudation. The sum of the weathering, mass-wasting, and erosional processes that result in the progressive lowering of the Earth's surface. (Ch. 19)

Deoxyribonucleic acid (DNA). A biopolymer consisting of two twisted, chainlike molecules held together by organic molecules called bases; the genetic material for all organisms except viruses, it stores the information on how to make proteins. (Ch. 16)

Desertification. The invasion of desert into nondesert areas. (Ch. 20)

Dew point. The temperature at which the relative humidity reaches 100 percent and condensation starts. (Ch. 12)

Differential stress. Stress in a solid that is not equal in all directions. (Ch. 8)

Dikes. Tabular, parallel-sided sheets of intrusive igneous rocks that cut across the layering of the intruded rock. (Ch. 7)

Diorite. A coarse-grained igneous rock consisting mainly of plagioclase and ferromagnesian minerals. Quartz is sparse or absent. (Ch. 7)

Discharge. The quantity of water that passes a given point in a stream channel per unit time. (Ch. 9)

Discharge area. Area where subsurface water is discharged to streams or to bodies of surface water. (Ch. 9)

Dissolution. The chemical weathering process whereby minerals and rock material pass directly into solution. (Ch. 9)

Dissolved load. Matter dissolved in stream water. (Ch. 9)

Divergence. The separation of air masses in different directions, caused by the outward spiral flow in an anticyclone and leading to an outward flow of air from the center of a high-pressure center. (Ch. 13)

Divergent margin (of a plate). A fracture in the lithosphere where two plates move apart. Also called a *spreading center*. (Ch. 4)

Divide. The line that separates adjacent drainage basins. (Ch. 9)

Dolostone. A sedimentary rock composed chiefly of the mineral dolomite. (Ch. 8)

Dose-response curve. A graph showing the effect on a population of a substance or energy as a function of the concentration of that substance or energy in the environment. The effect is usually expressed in terms of the percentage of the population injured or killed by the substance or energy. For example, there is usually interest in the concentration that kills one-half of the population, known as the "LD-50" dose (for "Lethal Dose 50%"). (Ch. 16)

Downwelling. The process by which surface water thickens and sinks. (Ch. 11)

Drainage basin. The total area that contributes water to a stream. (Ch. 9)

Dry adiabatic lapse rate. The way temperature changes with altitude in a rising mass of unsaturated air. (Ch. 12)

Dune. A mound or ridge of sand deposited by wind. (Ch. 11)

Earthquake focus. The point of the first release of energy that causes an earthquake. (Ch. 5)

Earth system science. The science that studies the whole Earth as a system of many interacting parts and focuses on the changes within and between these parts. (Ch. 1)

Ecliptic. Plane of the Earth's orbit around the Sun. (Ch. 2)

Ecological community. This term has two meanings. (1) A conceptual or functional meaning: a set of interacting species that occur in the same place (sometimes extended to mean a set that interacts in a way to sustain life). (2) An operational meaning: a set of species found in an area, whether or not they are interacting. (Ch. 15)

Ecological niche. The sum of the conditions (including habitat and resources) that allows an organism and its offspring to sustain themselves and breed. (Ch. 17)

Ecosystem. A trophic pyramid and its habitat. (Ch. 15)

Ediacaran animals. The earliest fossils of multicellular organisms discovered in 600-million-year-old rocks in the Ediacara Hills of South Australia. (Ch. 16)

Ediacaran fauna. A group of marine animals known only from fossil imprints found in uppermost strata of Proterozoic age immediately preceding Cambrian strata. (Ch. 17)

Ekman spiral. A spiraling current pattern from the water's surface to deeper layers, caused by the Coriolis effect, as each successive, slower moving layer of water is shifted to the right. (Ch. 8)

Ekman transport. The average flow of water in a current over the full depth of the Ekman spiral. (Ch. 11)

Elastic deformation. The reversible or nonpermanent deformation that occurs when an elastic solid is stretched and squeezed and the force is then removed. (Ch. 5)

Elastic rebound theory. The theory that earthquakes result from the release of stored elastic energy by slippage on faults. (Ch. 5)

El Niño/Southern oscillation (ENSO). A periodic climatic variation in which tradewinds slacken and surface waters of the central and eastern Pacific become anomalously warm. (Ch. 11)

Electromagnetic radiation. A self-propagating electric and magnetic wave, such as light, radio, ultraviolet, or infrared radiation; all types travel at the same speed and differ in wavelength or frequency, which relates to the energy. (Ch. 3)

Element (chemical). The most fundamental substance into which matter can be separated by chemical means. (Ch. 6)

Emergence. An increase in the area of land exposed above sea level resulting from uplift of the land and/or fall of sea level. (Ch. 11)

End moraine. A ridgelike accumulation of drift deposited along the margin of a glacier. (Ch. 10)

Energy flow. The movement of energy through an ecosystem from the external environment through a series of organisms and back to the external environment. It is one of the fundamental processes common to all ecosystems. (Ch. 15)

Energy-level shell. The specific energy level of electrons as they orbit the nucleus of an atom. (Ch. 6)

Enzyme. A protein that catalyzes a chemical reaction in an organism. (Ch. 17)

Epicenter. That point on the Earth's surface that lies vertically above the focus of an earthquake. (Ch. 5)

Equilibrium line. A line that marks the level on a glacier where net mass loss equals net gain. (Ch. 10)

Erosion. The complex group of related processes by which rock is broken down physically and chemically and the products are moved. (Ch. 1)

Estuary. A semienclosed body of coastal water within which seawater is diluted with fresh water. (Ch. 11)

Eucaryote. Organism whose cells have nuclei and certain other characteristics that separate it from the procaryotes. The eucaryotes include flowering plants, animals, and many single-cell organisms. (Ch. 15)

Eucaryotic cell (eucaryotes). A cell that includes a nucleus with a membrane, as well as other membrane-bound organelles. (Ch. 16)

Eutrophication. Bodies of water with a high level of plant nutrients and consequently high levels of algae growth. (Ch. 16)

Evaporite deposits. Layers of salts that precipitate as a consequence of evaporation. (Ch. 11)

Evolution. The changes that species undergo through time, eventually leading to the formation of new species. (Ch. 15)

Extrusive igneous rock. Rock formed by the solidification of magma poured out onto the Earth's surface. (Ch. 7)

Fault. A fracture in a rock along which movement occurs. (Ch. 5)

Fission. Controlled radioactive transformation. (Ch. 18)

Fjord. A deep, glacially carved valley submerged by the sea. Also spelled *fiord*. (Ch. 10)

Floodplain. The part of any stream valley that is inundated during floods. (Ch. 11)

Flux. The amount of energy flowing through a given area in a given time. (Ch. 3)

Foliation. The planar texture of mineral grains, principally micas, produced by metamorphism. (Ch. 8)

Food chains. The pathways by which energy (as food) is moved from one trophic level to another. (Chs. 15 and 16)

Food web. The map of all interconnections among food chains for an ecosystem. (Ch. 15)

Fossil. The naturally preserved remains or traces of an animal or a plant. (Ch. 8)

Fossil fuel. Remains of plants and animals trapped in sediment that may be used for fuel. (Ch. 18)

Friction. The resistance to movement when two bodies are in contact. (Ch. 13)

Front. The boundary between air masses of different temperature and humidity, and therefore different density. (Ch. 12)

Gabbro. A coarse-grained igneous rock in which olivine and pyroxene are the predominant minerals and plagioclase is the feldspar present. Quartz is absent. (Ch. 7)

Gaia. The Gaia hypothesis states that the surface environment of the Earth, with respect to such factors as the atmospheric composition of reactive gases (for example, oxygen, carbon dioxide, and methane), the acidity-alkalinity of waters, and the surface temperature are actively regulated by the sensing, growth, metabolism and other activities of the biota. Interaction between the physical and biological system on the Earth's surface has led to a planetwide physiology which began more than 3 billion years ago and the evolution of which can be detected in the fossil record. (Ch. 15)

Galaxy. A cluster of a billion or more stars, plus gas and dust, that is held together by gravity. (Ch. 3)

Gangue. The nonvaluable minerals of an ore. (Ch. 18)

Genus. A category of classifications of plants and animals that is intermediate between *family* and *species*. (Ch. 15)

General Circulation Model (GCM). A mathematical model used to simulate present and past climate conditions on the Earth. (Ch. 20)

Genetic drift. Changes in the frequency of a gene in a population as a result of chance rather than of mutation, selection, or migration. (Ch. 17)

Geocentric. A universe in which a stationary Earth is at the center and everything else revolves around it. (Ch. 2)

Geologic column. A composite diagram combining in chronological order the succession of known strata, fitted together on the basis of their fossils or other evidence of relative or actual age. (Ch. 8)

Geostrophic current. A flow of surface water around a gyre that is not deflected toward the center of the gyre. (Ch. 8)

Geostrophic wind. A wind that results from a balance between pressure-gradient flow and the Coriolis deflection. (Ch. 13)

Geothermal energy. Heat energy drawn from the Earth's internal heat. (Ch. 4)

Geothermal gradient. The rate of increase of temperature downward in the Earth. (Ch. 4)

Glaciation. The modification of the land surface by the action of glacier ice. (Ch. 11)

Glacier. A permanent body of ice, consisting largely of recrystallized snow, that shows evidence of downslope or outward movement, due to the stress of its own weight. (Ch. 10)

Glacier ice. Snow that gradually becomes denser and denser until it is no longer permeable to air. (Ch. 10)

Global change. The changes produced in the Earth system as a result of human activities. (Ch. 1)

Gneiss. A high-grade metamorphic rock, always coarse-grained and foliated, with marked compositional layering but with imperfect cleavage. (Ch. 8)

Gradient. A measure of the vertical drop over a given horizontal distance. (Ch. 9)

Granite. A coarse-grained igneous rock containing quartz and feldspar, with potassium feldspar being more abundant than plagioclase. (Ch. 7)

Gravitation (law of). Every body in the universe attracts every other body. (Ch. 2)

Gravity anomaly. Variations in the pull of gravity after correction for latitude and altitude. (Ch. 5)

Greenhouse effect. The property of the Earth's atmosphere by which long wavelength heat rays from the Earth's surface are trapped or reflected back by the atmosphere. (Ch. 20)

Greenhouse gases. The gases in the atmosphere, mainly H_2O, CO_2, CFCs and CH_4, that cause the greenhouse effect. (Ch. 20)

Greenschist. A low-grade metamorphic rock rich in chlorite. (Ch. 8)

Gross production (biology). Production before respiration losses are subtracted. (Ch. 15)

Groundwater. All the water contained in the spaces within bedrock and regolith. (Ch. 9)

Gymnosperms. Naked-seed plants. (Ch. 17)

Gyre. A large subcircular current system of which each major ocean current is a part. (Ch. 11)

Habitat. Where an individual, population, or species exists or can exist. For example, the habitat of the Joshua tree is the Mojave Desert of North America. (Ch. 17)

Hadley cell. Convection cells on both sides of the equator that dominate the winds in tropical and equatorial regions. (Ch. 13)

Halocline. A zone of the ocean, below the surface zone, which is marked by a substantial increase of salinity with depth. (Ch. 8)

Hardness. Relative resistance of a mineral to scratching. (Ch. 6)

Heat. The energy a body has due to the motions of its atoms. (Ch. 12)

Heat capacity. The amount of heat required to raise or lower the temperature of a material. (Ch. 11)

Heat energy. The energy of a hot body. (Ch. 12)

Heinrich layers. Layers of ice-rafted sediment identified in North Atlantic deep-sea cores, originating from massive discharges of icebergs during glacial ages. (Ch. 14)

Heliocentric. A universe in which a stationary Sun is at the center and everything else revolves around it. (Ch. 2)

Herbivores. Plant-eating heterotrophs. (Ch. 15)

Hertzsprung-Russell diagram (H-R diagram). A plot of a star's luminosity versus its temperature. (Ch. 3)

Heterotrophs. Organisms that are unable to use the energy from sunlight directly and so must get their energy by eating autotrophs or other heterotrophs. (Ch. 15)

High (H). An area of relatively high air pressure, characterized by diverging winds. (Ch. 13)

Homeostatis. The maintenance of fairly constant internal conditions; a balance within an ecosystem. (Ch. 15)

Humidity. The amount of water vapor in the air. (Ch. 12)

Hydrosphere. The totality of the Earth's water, including the oceans, lakes, streams, water underground, and all the snow and ice, including glaciers. (Ch. 1)

Hydrothermal mineral deposit. Any local concentration of minerals formed by deposition from a *hydrothermal solution*. (Ch. 18)

Hypothesis. An unproved explanation for the way things happen. (Ch. 1)

Igneous rock. Rock formed by the cooling and consolidation of magma. (Ch. 6)

Industrial melanism. A form of natural selection where the color of a living thing helps it to blend in with an urban, industrial environment. (Ch. 17)

Inner core. The central, solid portion of the Earth's core. (Ch. 2)

Insolation. The energy that reaches the surface of the Earth from the Sun. (Ch. 12)

Interglaciation. Period between glacial epochs. (Ch. 14)

Intertropical convergence zone. A low-pressure zone of convergent air masses caused by warm air rising in the tropics. (Ch. 13)

Intrusive igneous rock. Any igneous rock formed by solidification of magma below the Earth's surface. (Ch. 7)

Ion. An atom that has excess positive or negative charges caused by electron transfer. (Ch. 6)

Isobar. Lines on a map connecting places of equal air pressure. (Ch. 13)

Isostasy. The ideal property of flotational balance among segments of the lithosphere. (Ch. 5)

Isotopes. Atoms of an element having the same atopic number but differing mass numbers. (Ch. 6)

Jovian planets. Giant planets in the outer regions of the solar system that are characterized by great masses, low densities, and thick atmospheres consisting primarily of hydrogen and helium. (Ch. 2)

Karst. Topography formed on limestones due to solution by groundwater; characterized by sinkholes and caves. (Ch. 9)

Karst topography. An assemblage of topographic forms resulting from dissolution of carbonate bedrock and consisting primarily of closely spaced sinkholes. (Ch. 11)

Laccolith. A lenticular pluton intruded parallel to the layering of the intruded rock, above which the layers of the invaded country rock have been bent upward to form a dome. (Ch. 7)

Lake. Inland body of water in a depression on the Earth's surface; the water may be fresh or saline. (Ch. 9)

Landslide. Any perceptible downslope movement of a mass of bedrock or regolith, or a mixture of the two. (Ch. 19)

Latent heat. The amount of heat released or absorbed per gram during a change of state. (Ch. 12)

Lava. Magma that reaches the Earth's surface through a volcanic vent. (Ch. 7)

Law (scientific). A statement that some aspect of nature is always observed to happen in the same way and that no deviations have ever been seen. (Ch. 1)

Lead (in sea ice). A linear opening in thin ice cover caused by stresses resulting from the diverging movement of the ice cover. (Ch. 10)

Limestone. A sedimentary rock consisting chiefly of calcium carbonate, mainly in the form of the mineral calcite. (Ch. 8)

Lithosphere. The outer 100 km of the solid Earth, where rocks are harder and more rigid than those in the plastic asthenosphere. (Ch. 2)

Load. The material that is moved or carried by a natural transporting agent, such as a stream, the wind, a glacier, or waves, tides, and currents. (Ch. 9)

Loess. Wind-deposited silt, sometimes accompanied by some clay and fine sand. (Ch. 11)

Longshore current. A current, within the surf zone, that flows parallel to the coast. (Ch. 11)

Low (L). An area of relatively low air pressure, characterized by converging winds, ascending air, and precipitation. (Ch. 13)

Luminosity. The total amount of energy radiated outward each second by the Sun or any other star. (Ch. 3)

Luster. The quality and intensity of light reflected from a mineral. (Ch. 6)

Macronutrients. Elements required in large amounts by living things. These include the big six—carbon, hydrogen, oxygen, nitrogen, phosphorus, and sulfur. (Ch. 16)

Magma. Molten rock, together with any suspended mineral grains and dissolved gases, that forms when temperatures rise and melting occurs in the mantle or crust. (Ch. 7)

Magmatic mineral deposit. Any local concentration of minerals formed by magmatic process in an igneous rock. (Ch. 18)

Main sequence. The principal series of stars in the Hertzsprung-Russell diagram, which includes stars that are converting hydrogen to helium. (Ch. 3)

Mantle. The thick shell of dense, rocky matter that surrounds the core. (Ch. 2)

Marble. A metamorphic rock derived from limestone and consisting largely of calcite. (Ch. 8)

Mass balance (of a glacier). The sum of the accumulation and ablation on a glacier during a year. (Ch. 10)

Mass-wasting. The movement of regolith downslope by gravity without the aid of a transporting medium. (Ch. 19)

M-discontinuity. See *Mohorovičić discontinuity*. (Ch. 5)

Meander. A looplike bend of a stream channel. (Ch. 9)

Mesopause. The boundary between the mesosphere and the thermosphere. (Ch. 12)

Mesosphere (atmospheric science). One of the four thermal layers of the atmosphere, lying above the stratosphere. (Ch. 12)

Mesosphere (geology). The region between the base of the asthenosphere and the core/mantle boundary. (Chs. 1, 2, and 12)

Metabolism. The sum of all the chemical reactions in an organism, by which it grows and maintains itself. (Ch. 16)

Metallogenic provinces. Limited regions of the crust within which mineral deposits occur in unusually large numbers. (Ch. 18)

Metamorphic facies. Contrasting assemblages of minerals that reach equilibrium during a metamorphism within a specific range of physical conditions belonging to the same metamorphic facies. (Ch. 8)

Metamorphic rock. Rock whose original compounds or textures, or both, have been transformed to new compounds and new textures by reactions in the solid state as a result of high temperature, high pressure, or both. (Ch. 6)

Metamorphism. All changes in mineral assemblage and rock texture, or both, that take place in sedimentary and igneous rocks in the solid state within the Earth's crust as a result of changes in temperature and pressure. (Ch. 8)

Micronutrients. Chemical elements required in very small amounts by at least some forms of life. Boron, copper, and molybdenum are examples of micronutrients. (Ch. 16)

Midocean ridges. Continuous rocky ridges on the ocean floor, many hundreds to a few thousand kilometers wide with a relief of more than 0.6 km. Also called *oceanic ridges* and *oceanic rises*. (Ch. 6)

Migration. The movement of an individual, population, or species from one habitat to another or more simply from one geographic area to another. (Ch. 17)

Mineral. Any naturally formed, crystalline solid with a definite chemical composition and a characteristic crystal structure. (Ch. 6)

Mineral assemblage. The variety and abundance of minerals present in a rock. (Ch. 6)

Mineral deposit. Any volume of rock containing an enrichment of one or more minerals. (Ch. 18)

Modified Mercalli Scale. A scale used to compare earthquakes based on the intensity of damage caused by the quake. (Ch. 5)

Moho. See *Mohorovičić discontinuity*. (Ch. 5)

Mohorovičić discontinuity (also called *M-discontinuity* and *Moho*). The seismic discontinuity that marks the base of the crust. (Ch. 5)

Moist adiabatic lapse rate. The way temperature changes with altitude in a rising mass of saturated air. (Ch. 12)

Moraine. An accumulation of drift deposited beneath or at the margin of a glacier and having a surface form that is unrelated to the underlying bedrock. (Ch. 10)

Mutation. Stated most simply, a chemical change in a DNA molecule. It means that the DNA carries a different message than it did before, and this change can affect the expressed characteristics when cells or individual organisms reproduce. (Ch. 17)

Natural selection. A process by which organisms whose biological characteristics better fit them to the environment are better represented by descendants in future generations than those whose characteristics are less fit for the environment. (Ch. 17)

Negative feedback. The influence of a product on the process that produces it, such that production decreases with the growth of the product. (Ch. 15)

Nitrogen fixation. The process by which atmospheric nitrogen is converted to ammonia, nitrate ion, or amino acids. Microorganisms perform most of the conversion but a small amount is also converted by lightning. (Ch. 16)

North Atlantic Deep Water (NADW). A deep-ocean mass in the North Atlantic that extends from the intermediate water to the ocean floor; dense and cold, it originates at several sites near the surface of the North Atlantic, flows downward, and spreads southward into the South Atlantic. (Ch. 11)

Occluded front. A front resulting from a cold front overtaking a warm front. (Ch. 12)

Oceanic crust. The crust beneath the oceans. (Chs. 1 and 2)

Oceanic ridges. See *midocean ridges*. (Ch. 6)

Omnivores. Heterotrophs that eat both meat and plants. (Ch. 15)

Organelles. A well-defined cell part that has a particular function in the operation of the cell. (Ch. 15)

Ore. An aggregate of minerals from which one or more minerals can be extracted profitably. (Ch. 17)

Original horizontality (law of). Waterlaid sediments are deposited in strata that are horizontal, or nearly horizontal, and parallel, or nearly parallel, to the Earth's surface. (Ch. 8)

Orogens. Elongate regions of the crust that have been intensively folded, faulted, and thickened as a result of continental collisions. (Ch. 8)

Outer core. The outer portion of the Earth's core, which is molten. (Ch. 2)

Paleomagnetism. Remanent magnetism in ancient rock recording the di-

rection of the magnetic poles at some time in the past. (Ch. 4)

Paleontologist. A scientist who studies extinct organisms. (Ch. 17)

Panspermia. The hypothesis of the supposed origin of life in space, followed by a diaspora to various parts of the galaxy (including the Earth). (Ch. 17)

Peat. An unconsolidated deposit of plant remains that is the first stage in the conversion of plant matter to coal. (Ch. 8)

Periglacial. A land area beyond the limit of glaciers where low temperature and frost action are important factors in determining landscape characteristics. (Ch. 10)

Permafrost. Sediment, soil, or bedrock that remains continuously at a temperature below 0°C for an extended time. (Ch. 10)

Permeability. A measure of how easily a solid allows a fluid to pass through it. (Ch. 9)

Petroleum. Gaseous, liquid, and semi-solid substances occurring naturally and consisting chiefly of chemical compounds of carbon and hydrogen. (Ch. 18)

Photosynthesis. Synthesis of sugars from carbon dioxide and water by living organisms using light as energy. Oxygen is given off as a by-product. (Ch. 15)

Phyllite. A well-foliated metamorphic rock in which the component platy minerals are just visible. (Ch. 8)

Physical weathering. The disintegration (physical breakup) of rocks. (Ch. 6)

Pile. A device in which nuclear fission can be controlled. (Ch. 18)

Placer. A deposit of heavy minerals concentrated mechanically. (Ch. 18)

Planetary accretion. The process by which bits of condensed solid matter were gathered to form the planets. (Ch. 2)

Plate tectonics. The special branch of tectonics that deals with the processes by which the lithosphere is moved laterally over the asthenosphere. (Ch. 4)

Plate tectonic cycle. Cyclic process whereby magma formed in the mantle rises to form new oceanic crust along a spreading edge; the crust moves laterally away from the spreading edge and eventually sinks back into the mantle at a subduction edge. (Ch. 4)

Pluton. Any body of intrusive igneous rock, regardless of shape or size. (Ch. 7)

Polar front. The region where equatorward-moving polar easterlies meet poleward-moving westerlies. (Ch. 13)

Polar (cold) glacier. A glacier in which the ice is below the pressure melting point throughout, and the ice is frozen to its bed. (Ch. 10)

Population. A group of individuals of the same species living in the same area or interbreeding and sharing genetic information. (Ch. 1)

Porosity. The proportion (in percent) of the total volume of a given body of bedrock or regolith that consists of pore spaces. (Ch. 9)

Porphyry. Any igneous rock consisting of coarse mineral grains scattered through a mixture of fine material grains. (Ch. 7)

Positive feedback. The influence of a product on the process that produces it, such that production increases the growth of the product. (Ch. 15)

Pressure melting point. The temperature at which ice can melt at a given pressure. (Ch. 10)

Primary waves. See *P waves*. (Ch. 5)

Principle of Stratigraphic superposition. See *Stratigraphic superposition*. (Ch. 8)

Principle of Uniformitarianism. The same external and internal processes we recognize in action today have been operating unchanged, though at different rates, throughout most of the Earth's history. (Ch. 1)

Procaryote. A kind of organism that lacks a true cell nucleus and has other cellular characteristics that distinguish it from the *eucaryotes*. Bacteria and blue-green algae are procaryotes. (Ch. 15)

Procaryotic cell (procaryotes). Cells without a nucleus; refers to single-celled organisms that have no membrane separating their DNA from the cytoplasm. (Ch. 16)

Protein. Molecule formed through the polymerization of an amino acid. (Ch. 16)

Provinciality. The extent to which the global ecosystem is divided into subsystems by barriers to the migration of organisms. (Ch. 17)

P waves. Seismic body waves transmitted by alternating pulses of compression and expansion. *P* waves pass through solids, liquids, and gases. (Ch. 5)

Pycnocline. An ocean zone beneath the surface zone in which water density increases rapidly, as a result of a decrease in temperature, an increase in salinity, or both. (Ch. 8)

Pyroclast. A fragment of rock ejected during a volcanic eruption. (Ch. 7)

Pyroclastic rocks. Rocks formed from pyroclasts. (Ch. 7)

Quartzite. A metamorphic rock consisting largely of quartz, and derived from a sandstone. (Ch. 8)

Radiation. Transmission of heat energy through the passage of electromagnetic waves. (Ch. 4)

Recharge. The addition of water to the saturated zone of a groundwater system. (Ch. 9)

Recharge area. Area where water is added to the saturated zone. (Ch. 9)

Red giant. A large, cool star with a high luminosity and a low surface temperature (about 2500 K), which is largely convective and has fusion reactions going on in shells. (Ch. 3)

Reflection. The bouncing of a wave off the surface between two media. (Ch. 5)

Refraction. The change in velocity when a wave passes from one medium to another; the process by which the path of a beam of light is bent when the beam crosses from one transparent material to another. (Ch. 5)

Regional metamorphism. Metamorphism affecting large volumes of crust and involving both mechanical and chemical changes. (Ch. 8)

Regolith. The irregular blanket of loose, noncemented rock particles that covers the Earth. (Ch. 1)

Relative humidity. The ratio of the vapor pressure in a sample of air to the saturation vapor pressure at the same temperature, expressed as a percentage. (Ch. 12)

Relief (topographic). The range in altitude of a land surface. (Ch. 19)

Residual mineral deposit. Any local concentration of minerals formed as a result of weathering. (Ch. 18)

Rhyolite. A fine-grained igneous rock with the composition of a granite. (Ch. 7)

Ribonucleic acid (RNA). A single-stranded molecule similar to the DNA molecule, but with a slightly different chemical composition; it reads and executes the codes contained in the DNA. (Ch. 16)

Richter magnitude scale. A scale, based on the recorded amplitudes of seismic body waves, for comparing the amounts of energy released by earthquakes. (Ch. 5)

Rock. Any naturally formed, nonliving, firm, and coherent aggregate mass of mineral matter that constitutes part of a planet. (Ch. 1)

Rock cycle. The cyclic movement of rock material, in the course of which rock is created, destroyed, and altered through the operation of internal and external Earth processes. (Chs. 1 and 4)

Runoff. The fraction of precipitation that flows over the land surface. (Ch. 9)

Salinity. The measure of the sea's saltiness; expressed in parts per thousand. (Ch. 11)

Saltation. The progressive forward movement of a sediment particle in a series of short intermittent jumps along arcing paths. (Ch. 9)

Sand sea. Vast tract of shifting sand. (Ch. 11)

Sandstone. A medium-grained clastic sedimentary rock composed chiefly of sand-sized grains. (Ch. 8)

Saturated zone. The groundwater zone in which all openings are filled with water. (Ch. 9)

Schist. A well-foliated metamorphic rock in which the component platy minerals are clearly visible. (Ch. 8)

Scientific method. The use of evidence that can be seen and tested by anyone who has the means to do so, consisting often of observation, formation of a hypothesis, testing of that hypothesis and formation of a theory, formation of a law, and continued reexamination. (Ch. 1)

Seafloor spreading (theory of). A theory proposed during the early 1960s in which lateral movement of the oceanic crust away from midocean ridges was postulated. (Ch. 4)

Sea ice. A thin veneer of ice at the ocean surface in the polar latitudes; accounts for approximately two-thirds of the Earth's permanent ice cover. (Ch. 10)

Secondary waves. See *S waves*. (Ch. 6)

Sediment. Regolith that has been transported by any of the external processes. (Ch. 6)

Sedimentary mineral deposit. Any local concentration of minerals formed through processes of sedimentation. (Ch. 18)

Sedimentary rock. Any rock formed by chemical precipitation or by sedimentation and cementation of mineral grains transported to a site of deposition by water, wind, ice, or gravity. (Ch. 6)

Seismic sea waves (also called *tsunami*). Long wavelength ocean waves produced by sudden movement of the seafloor following an earthquake. Incorrectly called tidal waves. (Ch. 5)

Seismic waves. Elastic disturbances spreading outward from an earthquake focus. (Ch. 5)

Shale. A fine-grained, clastic sedimentary rock. (Ch. 8)

Shell fusion. The process of nuclear fusion in a star, in which the hydrogen in the shell around its core is converted into helium after the hydrogen in the core itself has already been depleted; such a star becomes a red giant. (Ch. 3)

Shield volcano. A volcano that emits fluid lava and builds up a broad dome-shaped edifice with a surface slope of only a few degrees. (Ch. 7)

Silicate (*-silicate mineral*). A mineral that contains the silicate anion. (Ch. 6)

Silicate anion. A complex ion $(SiO_4)^{-4}$, that is present in all silicate minerals. (Ch. 6)

Sills. Tabular, parallel-sided sheets of intrusive igneous rock that are parallel to the layering of the intruded rock. (Ch. 7)

Siltstone. A sedimentary rock composed mainly of silt-sized mineral fragments. (Ch. 8)

Sinkhole. A large solution cavity open to the sky. (Ch. 9)

Slate. A low-grade metamorphic rock with a pronounced slaty cleavage. (Ch. 8)

Snowline. The lower limit of perennial snow. (Ch. 10)

Soil. The part of the regolith that can support rooted plants. (Ch. 6)

Soil horizons. The subhorizontal weathered zones formed as a soil develops. (Ch. 6)

Soil profile. A vertical section through a soil that displays its component horizons. (Ch. 6)

Solar nebula. A flattened rotating disc of gas and dust surrounding the Sun. (Ch. 2)

Species. A population of individuals that can interbreed to produce offspring that are, in turn, interfertile with each other. (Ch. 15)

Specific gravity. A number stating the ratio of the weight of a substance to the weight of an equal volume of pure water. A dimensionless number numerically equal to the density. (Ch. 6)

Spectrum. A group of electromagnetic rays arranged in order of increasing or decreasing wavelength. (Ch. 3)

Spring. A flow of groundwater emerging naturally at the ground surface. (Ch. 9)

Standard atmosphere. The model or average air pressure at sea level: 760 mm or 29.9 inches of mercury. (Ch. 12)

Stock. A small, irregular body of intrusive igneous rock, smaller than a batholith, that cuts across the layering of the intruded rock. (Ch. 7)

Strata. See *stratum*.

Stratification. The layered arrangement of sediments, sedimentary rocks, or extrusive igneous rocks. (Ch. 8)

Stratigraphic superposition (principle of). In a sequence of strata, not later

overturned, the order in which they were deposited is from bottom to top. (Ch. 8)

Stratigraphy. The study of strata. (Ch. 8)

Stratopause. The boundary between the stratosphere and the mesosphere. (Ch. 12)

Stratosphere. One of the four thermal layers of the atmosphere, lying above the troposphere and reaching a maximum of about 50 km. (Ch. 12)

Stratovolcano(es). Volcanoes that emit both tephra and viscous lava, and that build up steep conical mounds. (Ch. 7)

Stratum (plural = *strata*). A distinct layer of sediment that accumulated at the Earth's surface. (Ch. 8)

Stratus clouds. Uniform dull grey clouds with a cloud base less than two km above sea level. (Ch. 12)

Streak. A thin layer of powdered mineral made by rubbing a specimen on a nonglazed porcelain plate. (Ch. 6)

Stream. A body of water that carries detrital particles and dissolved substances and flows down a slope in a definite channel. (Ch. 9)

Striations. Subparallel scratches inscribed on a clast or bedrock surface by rock debris embedded in the base of the glacier. (Ch. 10)

Subduction zone (also called a *convergent margin*). The linear zone along which a plate of lithosphere sinks down into the asthenosphere. (Chs. 4 and 6)

Submergence. A rise of water level relative to the land so that areas formerly dry are inundated. (Ch. 11)

Supernova. A stupendous explosion of a star, which increases its brightness hundreds of millions of times in a few days; a supernova releases heavy elements into space, and what remains of its core becomes a black hole. (Ch. 3)

Surf. Wave activity between the line of breakers and the shore. (Ch. 11)

Surface waves. Seismic waves that are guided by the Earth's surface and do not pass through the body of the Earth. (Ch. 5)

Surge. An unusually rapid movement of a glacier marked by dramatic changes in glacier flow and form. (Ch. 10)

Suspended load. Fine particles suspended in a stream. (Ch. 9)

S waves. Seismic body waves transmitted by an alternating series of sideways (shear) movements in a solid. *S* waves cause a change of shape and cannot be transmitted through liquids and gases. (Ch. 5)

Symbiotic. A close, long-term relationship between individuals of different species. (Ch. 15)

Tar (also called *asphalt*). An oil that is viscous and so thick it will not flow. (Ch. 18)

Temperate (warm) glacier. A glacier in which the ice is at the pressure-melting point and water and ice coexist in equilibrium. (Ch. 10)

Temperature. A measure of the average kinetic energy of all the atoms in a body. (Ch. 12)

Tephra. A loose assemblage of pyroclasts. (Ch. 7)

Tephra cone. A cone-shaped pile of tephra deposited around a volcanic vent. (Ch. 7)

Terminus. The outer, lower margin of a glacier. (Ch. 10)

Terrace. An abandoned floodplain formed when a stream flowed at a level above the level of its present channel and floodplain. (Ch. 11)

Terrestrial planets. The innermost planets of the solar system (Mercury, Venus, Earth, and Mars), which have high densities and rocky compositions. (Ch. 2)

Texture. The overall appearance that a rock has because of the size, shape, and arrangement of its constituent mineral grains. (Ch. 6)

Theory. A hypothesis that has been examined and found to withstand numerous tests. (Ch. 1)

Thermocline. A zone of ocean water lying beneath the surface zone, characterized by a marked decrease in temperature. (Ch. 11)

Thermohaline circulation. Global patterns of water circulation propelled by the sinking of dense cold and salty water. (Ch. 11)

Thermosphere. One of the four thermal layers of the atmosphere, reaching out to about 500 km. (Ch. 12)

Threshold. A point in the operation of a system at which a change occurs. With respect to toxicology it is a level below which effects are not observable and above which effects become apparent. (Ch. 16)

Topographic relief. The difference in altitude between the highest and lowest points on a landscape. (Ch. 11)

Transform fault margin (of a plate). A fracture in the lithosphere along which two plates slide past each other. (Ch. 4)

Tributary. A stream that joins a larger stream. (Ch. 9)

Trophic level. In an ecological community, all the organisms that are the same number of food-chain steps from the primary source of energy. For example, in a grassland the green grasses are on the first trophic level, grasshoppers are on the second, birds that feed on grasshoppers are on the third, and so forth. (Ch. 15)

Trophic pyramid. The hierarchy of organisms in which energy is moved from one level to the next. (Ch. 15)

Trophic structure. A hierarchy of organisms in which energy is moved from one level to the next. (Ch. 15)

Tropopause. The boundary between the troposphere and the stratosphere. (Ch. 12)

Troposphere. One of the four thermal layers of the atmosphere, which extends from the surface of the Earth to variable altitudes of 10 to 16 km. (Ch. 12)

Tsunami. See *seismic sea waves*. (Ch. 5)

Tuff. A pyroclastic rock consisting of ash- or lapilli-sized tephra, hence *ash tuff* and *lapilli tuff*. (Ch. 7)

Unconfined aquifer. An aquifer with an upper surface that coincides with the water table. (Ch. 9)

Unconformity. A substantial break or gap in a stratigraphic sequence that marks the absence of part of the rock record. (Ch. 8)

Uniform stress. Stress that is equal in all directions. Also called *confining stress* or *homogeneous stress*. (Ch. 8)

Uniformitarianism. See *Principle of Uniformitarianism*. (Ch. 1)

Upwelling. The process by which subsurface waters flow upward and replace the water moving away. (Ch. 11)

Varve. A pair of sedimentary layers deposited during the seasonal cycle of a single year. (Ch. 8)

Viscosity. The internal property of a substance that offers resistance to flow. (Ch. 7)

Volcanic neck. The approximately cylindrical conduit of igneous rock forming the feeder pipe of a volcanic vent that has been stripped of its surrounding rock by erosion. (Ch. 7)

Volcanic pipe. A cylindrical conduit of igneous rock below a volcanic vent. (Ch. 7)

Volcanic rock. Rock formed from the volcanic eruption of lava or tephra; often very fertile. (Ch. 7)

Volcano. The vent from which igneous matter, solid rock, debris, and gases are erupted. (Ch. 7)

Warm front. A front in which warm, humid aid advances over colder air, producing clouds and possibly rain. (Ch. 12)

Water table. The upper surface of the saturated zone of groundwater. (Ch. 9)

Wave base. The effective lower limit of wave motion, which is half of the wavelength. (Ch. 11)

Wavelength. The distance between the crests or troughs of adjacent waves. (Ch. 11)

Wave refraction. The process by which the direction of a series of waves, moving into shallow water at an angle to the shoreline, is changed. (Ch. 11)

Weather. The state of the atmosphere at a given time and place. (Ch. 12)

Weathering. The chemical alteration and mechanical breakdown of rock materials during exposure to air, moisture, and organic matter. (Ch. 6)

Welded tuff (also called *ignimbrite*). Pyroclastic rocks, the glassy fragments of which were plastic and so hot when deposited that they fused to form a glassy rock. (Ch. 7)

Western boundary current. A current that flows generally poleward, parallel to a continental coastline; the poleward direction is caused by the deflection of westward-flowing equatorial currents as they encounter land. (Ch. 8)

White dwarf. A small, dense star that has exhausted its nuclear fuel and shines from residual heat; it has a high surface temperature but low luminosity. (Ch. 3)

Wind. Horizontal air movement arising from differences in air pressure. (Ch. 13)

Windchill factor. The heat loss from exposed skin as a result of the combined effects of low temperature and wind speed. (Ch. 13)

Zodiac. The 12 constellations through which the Sun passes. (Ch. 3)

Zone of aeration. The groundwater zone in which open spaces in regolith or bedrock are filled mainly with air. (Ch. 9)

PHOTO CREDITS

CHAPTER 1

Opener: Tom Walker/Tony Stone Images/New York, Inc. **Figure 1.4:** Courtesy NASA. **Figure 1.11a:** © John S. Shelton. **Figure 1.11b:** William E. Ferguson. **Figure 1.12:** Courtesy K. Roy Gill. **Figure 1.13:** © John S. Shelton. **Figure 1.14:** W. Alvarez/Photo Researchers. **Figure 1.15:** Grant Heilman Photography. **Figure 1.16:** George Gerster/Comstock, Inc. **Figure 1.17:** Y. Arthis/Peter Arnold, Inc. **Essayist:** Courtesy Compton Tucker.

CHAPTER 2

Opener: Tony Stone Images/New York, Inc. **Figure 2.1:** ©Mary Evans Picture Library. **Figure 2.10:** Courtesy Royal Observatory, Edinburgh, and Anglo-Australian Telescope Board. **Figure 2.16:** Courtesy Space Photography Laboratory, Arizona State University.

CHAPTER 3

Opener: Tony Hallas/Science Photo Library/Photo Researchers. **Figure 3.7:** Courtesy NASA. **Figure 3.8:** François Gohier/Photo Researchers. **Figure 3.12a:** Courtesy NASA. **Figure 3.12b:** Science Photo Library/Photo Researchers. **Figure 3.12c:** Pekka Parviainen/Science Photo Library/Photo Researchers. **Figure 3.14:** Jon Sanford/Science Photo Library/Photo Researchers. **Figure C3.3a:** Richard Megna/Fundamental Photographs. **Figure C3.6a:** Galen Rowell/Peter Arnold, Inc.

CHAPTER 4

Opener: Barbara Cushman/DRK Photo. **Figure 4.5:** Bruce Heezen and Marie Tharp. **Figure 4.6a:** Earth Satellite Corp/Science Photo Library/Photo Researchers. **Figure 4.23:** Loren McIntrye/Woodfin Camp & Associates, Inc.

CHAPTER 5

Opener: Iwasa/Sipa Press. **Figure 5.10:** Joe Caravetta/Gamma Liaison. **Figure 5.11:** Boris Yuchenko/AP/Wide World Photos. **Figure 5.12:** Stan Waymen/Life Picture Service. **Essayist:** Courtesy Jeffrey Park.

CHAPTER 6

Opener: Mark Snyder/Tony Stone Images/New York, Inc. **Figure 6.7:** William Sacco. **Figure 6.9:** Brian J. Skinner. **Figure 6.10:** William E. Ferguson. **Figures 6.11 and**

6.12a: S.C. Porter. **Figure 6.12b:** Courtesy Fiorenzo Ugolini. **Figures C6.1, C6.3, C6.4, C6.5 and C6.7:** William Sacco. **Figure C6.2:** Brian J. Skinner. **Figure C6.6:** Breck P. Kent/Animals Animals. **Essayist:** Courtesy Jill Schneiderman.

CHAPTER 7

Opener: Roger Werth/Woodfin Camp & Associates. **Figure 7.2:** G. Brad Lewis/Gamma Liaison. **Figures 7.3 and 7.5:** Courtesy J.D. Griggs/USGS. **Figure 7.4:** Krafft Explorer/Photo Researchers. **Figure 7.6a:** William E. Ferguson. **Figure 7.6b:** Courtesy J.P. Lockwood/USGS. **Figure 7.6c:** Steven L. Nelson/Alaska Stock Images. **Figure 7.8:** S.C. Porter. **Figure 7.9a:** Krafft Explorer/Photo Researchers. **Figure 7.9b:** Tom Bean/DRK Photo. **Figure 7.10:** Steve Vidler/Leo de Wys, Inc. **Figure 7.11:** Greg Vaughn/Tom Stack & Associates. **Figure 7.13:** David Hiser/Photographers/Aspen/PNI. **Figure 7.14:** Ron Sanford/f/STOP Pictures. **Figure 7.16b:** Brian J. Skinner. **Figure C7.1a-d:** William Sacco. **Figures C7.2 and C7.3:** Brian J. Skinner.

CHAPTER 8

Opener: Brian Stablyk/Tony Stone Images/New York, Inc. **Figure 8.1:** Josef Muench. **Figures 8.6 and 8.7:** S.C. Porter. **Figure 8.8:** Jim Frazier. **Figure 8.11:** William Sacco. **Figure 8.13a:** Courtesy Schalk W. vand der Merwe. **Figure C8.3:** Brian J. Skinner. **Figure C8.5:** Courtesy Craig Johnson.

CHAPTER 9

Opener: Tom Bean/DRK Photo. **Figure 9.1:** Courtesy C. Clapperton. **Figure 9.2:** F. Stuart Westmorland/Photo Researchers. **Figure 9.5:** S.C. Porter. **Figure 9.8:** John Eastcott & Yva Momatiuk/DRK Photo. **Figure 9.11:** Hiroji Kubota/Magnum Photos, Inc. **Figure 9.15:** S.C. Porter. **Figure 9.16:** Martin Miller. **Figures 9.17 and 9.18:** ©Earth Satellite Corporation. **Figure 9.19:** S.C. Porter. **Figure 9.28:** Michael Nichols/Magnum Photos, Inc. **Figure 9.29:** Jim Tuten/Black Star. **Figure 9.30:** Hiroji Kubota/Magnum Photos, Inc. **Figure 9.31:** S.C. Porter. **Essayist:** Courtesy Waite Osterkamp.

CHAPTER 10

Opener: Ralph A. Clevenger/West Light. **Figure 10.1:** Courtesy NASA. **Figure 10.2:** S.C. Porter. **Figure 10.4:**

Breck P. Kent/Animals Animals. **Figure 10.5**: Sam Abell/National Geographic Society. **Figure 10.6**: Courtesy USGS. **Figure 10.7**: Courtesy John Price, USDA. **Figure 10.8**: Courtesy NASA. **Figure 10.11**: S.C. Porter. **Figure 10.18**: Courtesy Austin Post. **Figures 10.19 and 10.20**: S.C. Porter. **Figure 10.21**: Ole P. Rorvik/Aune Forlag. **Figure 10.22**: Tom Bean/AllStock/Tony Stone Images/New York, Inc. **Figure 10.23**: Gerald Osborn. **Figure 10.24**: S.C. Porter. **Figure 10.25**: Courtesy N. Unter-steiner. **Figure 10.28**: NASA/Science Photo Library/Photo Researchers. **Figure 10.31**: Steve McCutcheon. **Figure 10.32**: Charlie Ott/Photo Researchers. **Figure C10.1**: Courtesy Austin Post.

CHAPTER 11

Opener: ©Herb Kane. **Figure 11.2**: Courtesy Walter H.F. Smith and David T. Sandwell. **Figure 11.16a**: S.C. Porter. **Figure 11.17**: Courtesy Deep Sea Drilling Project, Scripps Institution of Oceanography. **Figure 11.21**: Jeff Divine/FPG International. **Figure 11.22**: © John S. Shelton. **Figure 11.25**: S.C. Porter. **Figure 11.26a**: Spaceshots, Inc. **Figure 11.26b**: Courtesy NASA. **Figure 11.27**: ©Earth Satellite Corporation. **Figure 11.31**: Greg Scott/Masterfile. **Figure 11.33**: Arthur Bloom. **Figure C11.1**: Courtesy Advanced Satellite Productions, Inc., and Earth Observation Satellite Co. **Essayist**: Courtesy Victoria Kaharl.

CHAPTER 12

Opener: Chris Noble/Tony Stone Images/New York, Inc. **Figure 12.1**: Dr. Gerald L. Fisher/Science Photo Library/Photo Researchers. **Figure 12.8**: Courtesy French Government Tourist Office. **Figure 12.9**: Courtesy NASA. **Figure 12.20**: Tom Bean/DRK Photo. **Figure 12.20c**: S. Nielsen/Bruce Coleman, Inc. **Figure 12.20d**: Joyce Photographers/Photo Researchers. **Figure 12.20e**: Steve Mc-Cutcheon. **Figure 12.20f**: Darryl Torckler/Tony Stone World Wide. **Figure 12.20g**: Keith Gunnar/Bruce Coleman, Inc. **Figure 12.20h**: ©Barrie Rokeach. **Figure 12.20i**: Stan Osolinski/Oxford Scientific Films Ltd.

CHAPTER 13

Opener: Stephen J. Krasemann/DRK Photo. **Figure 13.1**: Stephen J. Kraseman/DRK Photo. **Figure 13.8**: National Oceanic and Atmospheric Administration/DLR-FRG/Starlight. **Figure 13.20**: Warren Faidley/Weather-stock. **Figure 13.21**: John H. Hoffman/Bruce Coleman, Inc. **Figure 13.22**: Impact Photos. **Essayist**: Courtesy Howard Bluestein.

CHAPTER 14

Opener: Hanny Paul/Gamma Liaison. **Figure 14.4**: S.C. Porter. **Figure 14.7a**: Freidrich Rothlisberger. **Figure 14.8a**: S.C. Porter. **Figure 14.11**: Tom Bean. **Figure 14.12b**: Matsuo Tsukada. **Essayist**: Courtesy L. Graumlich.

CHAPTER 15

Opener: Courtesy NASA. **Figure 15.1**: Scott McMullen/Space Biospheres Ventures. **Figure 15.3a**: Manfred Kage/Peter Arnold, Inc. **Figure 15.5a**: Dr. Jeremy Burgess/Photo Researchers. **Figure 15.5b**: CNRI/Photo Researchers. **Figure 15.6**: Wendy Stone/Gamma Liaison. **Figure 15.7**: Joe Sohm/Photo Researchers. **Figure 15.12**: WorldSat International & J. Knighton/Science Source/Photo Researchers.

CHAPTER 16

Opener: Lynn M. Stone/Bruce Coleman, Inc. **Figure 16.1b**: Tui De Roy/Minden Pictures, Inc. **Figure 16.4a**: Randy Morse/Tom Stack & Associates. **Figure 16.4b**: Dwight R. Kuhn/DRK Photo. **Figure C16.2**: Fred Ward/Black Star. **Figure 16.12**: Phil Degginger/Bruce Coleman, Inc. **Figure 16.14**: Courtesy Jesse Victolero, International Rice Research Institute. **Figure 16.15**: Courtesy Robert R. Hessler, Scripps Institution of Oceanography. **Figure 16.16**: Mike Andrews/Animals Animals. **Figure 16.17**: William Felger/Grant Heilman Photography. **Figure 16.18a**: Michael P. Gadomski/Photo Researchers. **Figure 16.18b**: Barbara Gerlach/DRK Photo.

CHAPTER 17

Opener: Regina P. Simon/Tom Stack & Associates. **Figure 17.1**: Brother Eric Vogel F.S.C. **Figure 17.4**: AP/Wide World Photos. **Figure 17.5a**: Alvin E. Staffan/Photo Researchers. **Figure 17.5b**: H. Reinhard/Bruce Coleman, Inc. **Figure 17.10**: Valerie Taylor/Ardea London. **Figure 17.11a,b**: ©Jany Sauvent/Natural History Photographic Agency. **Figure 17.11c**: A. Cosmos Blank/Photo Researchers. **Figure 17.11d**: Karl Weidmann/Photo Researchers. **Figure 17.12**: John Cardmore/BPS/Tony Stone Images/ New York, Inc. **Figure 17.13**: Courtesy J. William Schopf, Dept. of Earth and Planetary Sciences, UCLA. **Figure 17.14a**: Ford Kristo/Planet Earth Pictures. **Figure 17.14b**: Francois Gohier/Ardea London. **Figure 17.16**: Brian J. Skinner. **Figure 17.17a**: D. Pauer and E. Paner-Cook/Auscape International Pty. Ltd. **Figure 17.17b**: Ken Lucas/Planet Earth Pictures. **Figure 17.19**: A. Kerstitch/Bruce Coleman, Inc. **Figure 17.21a**: Edward R. Degginger/Bruce Coleman, Inc. **Figure 17.21b**: John Shaw/Bruce Coleman, Inc. **Figure 17.22a**: Martin Land/Science Photo Library/Photo Researchers. **Figure 17.22b**: Breck P. Kent/Animals Animals. **Figure 17.23**: Patti Murray/Animals Animals. **Figure 17.24**: Courtesy S. Conway Morris. **Figure 17.25**: Courtesy The Natural History Museum, London. **Figure 17.28a**: Stephen J. Krasemann/Tony Stone Images/New York, Inc. **Figure 17.28b**: N. N. Birks/Auscape International Pty. Ltd. **Figure 17.28c**: Michael George/Bruce Coleman, Inc. **Essayist**: Courtesy Andrew Knoll.

CHAPTER 18

Opener: Victor Englebert/Photo Researchers. **Figure 18.4b**: Courtesy Dudley Foster, Woods Hole Oceanographic Institution. **Figures 18.5 and 18.7**: Brian J. Skinner. **Figure 18.8**: William Sacco. **Figures 18.10 and 18.11**: Brian J. Skinner. **Essayist**: Courtesy Pinar Yilmaz.

CHAPTER 19

Opener: Courtesy John Dohrenwend/USGS. **Figure 19.1a**: S.C. Porter. **Figure 19.1b**: Brian J. Skinner. **Figure 19.3**: S.C. Porter. **Figure 19.5**: ©Earth Satellite Corporation. **Figure 19.7**: S.C. Porter. **Figure 19.15**: Mannfred Gottschalk/Tom Stack & Associates. **Figure 19.16**: S.C. Porter. **Figure 19.17a**: Courtesy NASA.

CHAPTER 20

Opener: Vince Streano/Tony Stone Images/ New York, Inc. **Figure 20.2**: Lloyd Cluff. **Figure 20.4**: S.C. Porter. **Figure 20.6**: George Gerster/Comstock, Inc. **Figure 20.7**: Courtesy Agency for International Development. **Figure 20.9**: Nani Gois/Abril Imagens. **Figure 20.11**: Michael Melford/The Image Bank. **Figure 20.12**: Courtesy Bancroft Library, University of California. **Figure 20.14**: Giorgio Gualco/Bruce Coleman, Inc. **Essayist**: Courtesy Thomas J. Crowley.

FIGURE AND TABLE CREDITS

The following figures are reprinted or adapted from Barbara W. Murck, Brian J. Skinner, and Stephen C. Porter, *Environmental Geology* (New York: Wiley, 1996): 1.2 (p. 3), 1.3 (p. 4), 1.5 (p. 5), 1.6 (p. 6), 1.8 (p. 8), 1.9 (p. 10), 2.13 (p. 36). The following figures are based on Michael Zeilik (1991), *Astronomy: The evolving universe*, 6th ed. (New York, Wiley): 3.9 (p. 52), 3.10 (p. 53), 3.11 (p. 53), 3.15 (p. 56), 3.16 (p. 57), C3.4 (p. 58), C3.5 (p. 59). The following figures are reprinted or adapted from Daniel B. Botkin and Edward A. Keller, *Environmental Science: Earth as a Living Planet*, 2d ed. (New York: Wiley, 1998): 15.9 (p. 351), 15.10 (p. 352), 15.14 (p. 356), C16.1 (p. 366), 16.3 (p. 367), 16.5 (p. 369), 16.13 (p. 378), 17.2 A&B (p. 389), 17.6 A–C (p. 391).

Figure 3.6 (p. 50): Based on John Gribbin (1993, March 13), Inside the sun, *New Scientist*, p. 1; Edward J. Tarbuck and Frederick J. Lutgens (1994), *Earth science*, 7th ed. (New York, Macmillan), Fig. 19.18, p. 656; Michael Zeilik (1991), *Astronomy: The evolving universe*, 6th ed. (New York, Wiley), Fig. 13.31, p. 284; Cesare Emiliani (1992), *Planet earth: Cosmology, geology, and the evolution of life and environment* (Cambridge, Cambridge University Press), Fig. 6.3.

Figure 3.13 (p. 55): Adapted from Michael Zeilik (1991), *Astronomy: The evolving universe*, 6th ed. (New York, Wiley), Fig. 13.20, p. 278, copyright ©1991 by Michael Zeilik, reprinted by permission of John Wiley & Sons, Inc.; and reprinted by permission from H. Yosibmura, *Astrophysical Journal*, vol. 227, p. 1047.

Figure 3.17 (p. 60): Adapted from Michael Zeilik (1991), *Astronomy: The evolving universe*, 6th ed. (New York, Wiley), Fig. 14.15, p. 300, copyright © 1991 by Michael Zeilik, reprinted by permission of John Wiley & Sons, Inc.

Table 7.2 (p. 146): From a Report by the Task Group for the International Decade of Natural Disaster Reduction (1990), in *Bulletin of the Volcanological Society of Japan*, series 2, vol. 35, no. 1, pp. 80–95.

Figure 9.10 (p. 197): Adapted from P.E. Potter et al. (1988), Teaching and field guide to alluvial processes and sedimentation of the Mississippi River, Fulton County, Kentucky, and Lake County Tennessee, *Kentucky Geological Survey*, Fig. 8.

Figure 9.12A (p. 198): Adapted by permission from C.F. Nordin, Jr. et al (1980), Size distribution of Amazon River bed sediment, *Nature*, vol. 286, Fig. 2a. Copyright © 1980 Macmillan Magazines Limited.

Figure 9.12B (p. 198): Adapted from P.E. Potter et al. (1988), Teaching and field guide to alluvial processes and sedimentation of the Mississippi River, Fulton County, Kentucky, and Lake County Tennessee, *Kentucky Geological Survey*, Fig. 13.

Figure 9.13 A&B (p. 198): Adapted by permission

from A. Foucault and D.J. Stanley (1989), Late quaternary paleoclimatic oscillations in East Africa recorded by heavy minerals in the Nile delta, *Nature*, vol. 339, Figs. 1 and 2, p. 44. Copyright © 1989 Macmillan Magazines Limited.

Figure 10.3 (p. 217): Adapted by permission from M.F. Meier and A.S. Post (1962), Recent variations in mass net budgets of glaciers in western North America, in *Symposium of Obergurgl: Variations of the Regime of Existing Glaciers*, IAHS Pub. 58, pp. 63–77.

Figure 10.26 A&B (p. 233): Adapted from Central Intelligence Agency (1978), *Polar regions atlas*, pp. 12, 38).

Figure 10.29 (p. 236): Adapted by permission from T.L. Péwé (1983), The penglacial environment in North America during Wisconsin time, in H.E. Wright, Jr., and S.C. Porter (eds.), *Late quaternary environments of the United States, vol. 1: The late Pleistocene* (Minneapolis, University of Minnesota Press), Fig. 9.1. Copyright © 1983 by the University of Minnesota.

Figure 10.30 (p. 237): Adapted by permission from S.W. Muller (1947), *Permafrost of permanently frozen ground and related engineering problems* (Ann Arbor, Mich., J.W. Edwards), Fig. 3.

Figure 11.1 (p. 242): Map adapted from *National Geographic*, vol. 150 (October 1976), p. 517.

Figure 11.3 (p. 244): Based on maps from *National Geographic Atlas of the World*, 1963, p. 10.

Figure 11.5 (p. 246): Based on plate 3 in *The Times Atlas of the World*, Comprehensive edition (1967), published by Houghton Mifflin; © John Bartholomew & Son Ltd., Edinburgh.

Figure 11.6 (p. 247): Adapted by permission of Andrew McIntyre from CLIMAP Project Members (1981), Map and Chart Series MC-36, Map 6A, published by the Geological Society of America.

Figure 11.12 (p. 251): Adapted by permission from A.L. Gordon (1990/91), The role of thermohaline circulation in global climate change, *Lamont-Doherty Geological Observatory (Columbia University) Report 1990/91*, Fig. 3, p. 48.

Figure 11.13 (p. 253): Based on information from *Lamont-Doherty Geological Observatory (Columbia University) Report 1990/91*, Fig. 4, p. 50; and from J. Imbrie et al. (1992), On the structure and origin of major glaciation cycles, 1: Linear responses to Milankovitch forcing. *Paleoceanography* 7, Fig. 1b, p. 704. Published by American Geophysical Union, Washington, D.C.

Figure 11.14 (p. 253): Adapted by permission from S.J. Lehman and L.D. Keigwin (1992), Deep circulation revisited, *Nature*, vol. 358, pp. 197–198. Copyright © 1992 Macmillan Magazines Limited.

Figure 11.15 (p. 254): Drawing by Ian Worpole in *Scientific American*, June 1986, p. 76.

Figure 11.16A (p. 255): Based on National Oceanic and Atmospheric Administration (NOAA) graph at http://www.cdc.noaa.gov/ENSO.enso.mei.index.html.

Figure 11.33B (p. 268): Adapted from N.J. Shackleton (1987), Oxygen isotopes, ice volume, and sea level, *Quatenary Science Reviews*, vol. 6, Fig. 5, p. 187, copyright 1987, with permission from Elsevier Science.

Figure C12.1 (p. 276): Based on J.M. Moran and M.D. Morgan (1991), *Meteorology*, 3rd ed. (New York Macmillan); and R.G. Fleagle and J. Businger (1963), *An introduction to atmospheric physics* (New York, Academic Press), p. 153.

Figure 12.14 (p. 283): Annual addition of water vapor to atmosphere. *Source*: Reprinted by permission from S.D. Gedzelman (1980), *The source and wonders of the atmosphere* (Wiley, 1980), Fig. 10.6. Copyright © 1980 by John Wiley & Sons. Reprinted by permission of John Wiley & Sons, Inc.

Figure 13.1 (p. 296): From National Weather Service.

Figure 13.6 (p. 297): From National Weather Service.

Figure 13.13 (p. 302): Based on data from National Weather Service in Alan H. Strahler and Arthur N. Strahler (1992), *Modern physical geography*, 4th ed. (New York, Wiley), p. 85; and H. Riehl (1962), *Jet Streams of the Atmosphere* (Fort Collins, Colorado State University).

Figure C13.1 (p. 303): Reprinted by permission from S.D. Gedzelman (1980), *The source and wonders of the atmosphere* (New York, Wiley), Fig. 16.7. Copyright © 1980 by John Wiley & Sons. Reprinted by permission of John Wiley & Sons, Inc.

Figure 13.16 (p. 305): Reprinted by permission with kind permission from Kluwer Academic Publishers from K. Pye and H. Tsoar (1990), *Aeolian sand and sand dunes* (London, Unwin Hyman), Fig. 2.17.

Figure 13.17 (p. 306): Based on several maps in Deserts in *Encyclopaedia Britannica* (15th ed.), Figs. 4, 5, 6, 7, and 8.

Figure 13.19 (p. 309): From *Meteorology*, 4th ed., by Joseph Moran and Michael Morgan. © 1994. Reprinted by permission of Prentice-Hall, Inc., Upper Saddle River, NJ.

Figure 13.23 (p. 312): From *Meteorology*, 4th ed., by Joseph Moran and Michael Morgan. © 1994. Reprinted by permission of Prentice-Hall, Inc., Upper Saddle River, NJ.

Figure 14.2 (p. 317): Data from Bernard Janin (1970), *Le col du Grand-Saint-Bernard climat et variations climatiques* (Aoste, Italy, Marguerettaz-Musumeci), Tables 38 and 40.

Figure 14.3 (p. 318): Based on NCAR sea-level temperature records.

Figure 14.5A (p. 319): Adapted by permission from

Zhang De-er (1982), Analysis of dust rain in the historic times of China. *Kexue Tongbao*, vol. 27, pp. 294–297.

Figure 14.5B&C: Adapted by permission of Taylor & Francis (Routledge) from H.H. Lamb (1977), *Climate — Present, past, and future, vol. 2: Climatic history and the future* (London, Methuen).

Figure 14.5D (p. 319): Adapted by permission of Taylor & Francis (Routledge) from H.H. Lamb (1966), *The changing climate: Selected papers* (London, Methuen), app. 3, p. 223.

Figure 14.6 (p. 320): Reprinted by permission of Taylor & Francis (Routledge) from H.H. Lamb (1977), *Climate — Present, past, and future, vol. 2: Climatic history and future* (London, Methuen), Fig. 17.17, p. 462; derived from an unpublished paper by L.M. Libby, R and D. Associates, Santa Monica, CA, 1974.

Figure 14.7 (p. 320): Adapted by permission from W.S.B. Paterson et al. (1977), An oxygen-isotope climatic record from the Devon Island ice cap, arctic Canada, *Nature*, vol. 266, Fig. 6 (article on pp. 508–511). Copyright © 1977 Macmillan Magazines Limited.

Figure 14.14 (p. 326): Adapted by permission from Thompson Webb III et al. (1993), Vegetation, lake levels, and climate in eastern North America form the past 18,000 years in H.E. Wright, Jr., et al. (eds.), *Global climates since the last glacial maximum* (Minneapolis, University of Minnesota Press), Fig. 17.10. Copyright © 1993 by the Regents of the University of Minnesota.

Figure C14.1A&B (pp. 328–329): Adapted by permission of Andrew McIntyre from CLIMAP Project Members (1981), Seasonal reconstructions of the Earth's surface at the last glacial maximum, MC-36, maps 3A and 3B, published by the Geological Society of America.

Figure 14.17 (p. 331): Based in part on N.J. Shackleton and J.P. Kennett (1975), Paleotemperature history of the Cenozoic and the initiation of Antarctic glaciation: Oxygen and carbon isotope analyses in Deep Sea Drilling Project sites, 227, 279, and 281, in J.P. Kennett et al. (1975), *Initial reports of the Deep Sea Drilling Project*, vol. 29, National Ocean Sediment Coring Program (Washington, D.C., National Science Foundation), Fig. 7.

Figure 14.18 (p. 332): Adapted from Fig. 8.3 of *Paleoclimatology*, by Thomas J. Crowley and Gerald R. North. Copyright © 1991 by Oxford University Press, Inc. Used by permission of Oxford University Press, Inc. And adapted by permission of Elsevier Science Publishers and E.J. Barron from E.J. Barron and W.M. Washington (1982), Cretaceous climate: A comparison of atmospheric simulations with the geologic record. *Paleogeography, Paleoclimatology, Paleoecology*, vol. 40, Fig. 9.

Figure 14.19 (p. 332): Reprinted from A.C. Lasaga, R.A. Berner, and R.M. Garrels (1985), A geochemical model of atmospheric CO_2 fluctuations over the past 100 million years, in *The carbon cycle and atmospheric CO_2: Natural variations Archean to present* (AGU Monograph 32), Figs. 7 and 8. Copyright 1985 by the American Geophysical Union.

Figure 14.23 (p. 336): Based in part on C. Lorius et al. (1990), The ice-core record: Climate sensitivity and future greenhouse warming, *Nature*, vol. 347, pp. 139–145; and J.T. Houghton, G.J. Jenkins, and J.J.Ephraums (1990), *Climate change: The IPCC scientific assessment* (Cambridge, Cambridge University Press), Fig. 2.

Figure 15.3B (p. 346): Adapted by permission from Gil Brum, Larry McKane, and Gerry Karp (1994), *Biology: Explaining life*, 2d ed. (New York, Wiley), Fig. 14.4. Copyright © 1994 by John Wiley & Sons. Reprinted by permission of John Wiley & Sons, Inc.

Figure 15.8 (p. 351): Reprinted by permission from D.B. Botkin, W.S. Broecker, L.G. Everett, J. Shapiro, and J.A. Wiens (1988). *The future of Mono Lake*, Report #68 (Riverside, University of California, California Water Resources Center).

Figure 15.15 (p. 357): Reprinted by permission from M.J. Sobel and D.B. Botkin (1994), *Status and future of salmon of Western Oregon and Northern California: Forecasting spring chinook runs* (Santa Barbara, Calif., Center for the Study of the Environment), 42 pages.

Figure 15.16 (p. 358): Adapted by permission from M.M. Kent and K.A. Crews (1990), *World populations: Fundamentals of growth* (Washington, D.C., Population Reference Bureau).

Figure 16.6 (p. 372): From G.M. Woodwell and A.L. Rebuck (1967), Effects of chronic gamma radiation on the structure and diversity of an oak-pine forest, *Ecological Monographs*, 37, pp. 53–69.

Figure 16.7A&B (p. 373): Modified from G. Lambert (1987), *La recherche*, 18, pp. 782–783, with some data from R. Houghton (1993), *Bulletin of the Ecological Society of America*, 74(4), pp. 355–356.

Figure 16.8, 16.9 (p. 374): Reprinted by permission from G.E. Likens, F.H. Bormann, R.S. Pierce, J.S. Eaton, and N.M. Johnson (1995), *The biogeochemistry of a forested ecosystem*, 2d ed. (New York, Springer-Verlag).

Figure 16.11 (p. 376): Data from R. Söderlund and T. Rosswall (1982), in O. Hutzinger (ed.), *The handbook of environmental chemistry*, vol. 1, pt. B (New York, Springer-Verlag).

Figure 16.13 (p. 378): Based primarily on C.C. Delwiche and G.E. Likens (1977), Biological response to fossil fuel combustion products, in W. Stumm (ed.), *Global chemical cycles and their alterations by man* (Berlin, Abakon Verlagsgesellschaft), pp. 73–88; and U. Pierrou (1976), The global phosphorus cycle, in B.H. Svensson and R. Söderlund (eds.), Nitrogen, phosphorus, and sulfur — global cycles, *Ecological Bulletin*, Stockholm, pp. 75–88.

Figure 17.7 (p. 392): Drawn from data in H.B.D. Kettlewell (1959, March). Darwin's missing evidence. *Scientific American*, 200, pp. 51–52.

Figure 17.8 (p. 394): Reprinted by permission of Yale University Press from F.G. Stehli (1968), Taxonomic diversity gradients in pole location: The recent model, in E.T. Drake (ed.), *Evolution and environment*, Fig. 6, p. 170. Copyright © 1968 by Yale University Press.

Figure 17.12A (p. 399): Based on G. Brum, L. McKane, and G. Karp (1994), *Biology: Explaining life*, 2d ed. (New York, Wiley), Fig. 35.3.

Figure 17.12B (p. 399): Adapted by permission from E. Barghoorn, The oldest fossils, *Scientific American*, May 1971, p. 41. Bunji Tagawa illustration.

Figure 17.15 (p. 402): Adapted by permission from Gil Brum, Larry McKane, and Gerry Karp (1994), *Biology: Explaining life*, 2d ed. (New York, Wiley), Fig. 5.24 (a and b). Copyright © 1994 by John Wiley & Sons. Reprinted by permission of John Wiley & Sons, Inc.

Figure 17.18 (p. 404): Based on a drawing by Virge Kask in Derek E.G. Briggs (1991), Extraordinary fossils, *American Scientist*, vol. 79, no. 2, p. 133 (article on pp. 130–141).

Figure 17.20 (p. 405): Composite drawing adapted by permission from Gil Brum, Larry McKane, and Gerry Karp (1994), *Biology: Explaining life*, 2d ed. (New York, Wiley), Figs. 35.9, 35.10, and 35.12. Copyright © 1994 by John Wiley & Sons. Reprinted by permission of John Wiley & Sons, Inc.

Figure 17.26 (p. 411): Adapted by permission of the American Geological Institute from J. John Sepkoski, Jr., Extinction and the fossil record, *Geotimes*, March 1994, p. 17.

Figure 19.12 (p. 452): Based on P.F. Cerveny et al. (in press), History of uplift and relief of the Himalaya during the past 18 million years: Evidence from fission-track ages of detrital zircons from sandstones of the Siwalik Group, in Paola and Kleinspehn (eds.), *New perspectives in sedimentary Basin Analysis* (Springer Verlag), Fig. 3.4.

Figure 19.13 (p. 453): From W. Ludwig and J.-L. Probst (1998), River sediment discharge to the oceans: Present-day controls and global budgets, *American Journal of Science*, 298, Fig. 9, p. 287. Reprinted by permission of the *American Journal of Science*.

Figure 19.14 (p. 454): M.A. Summerfield and N.J. Hulton (1994), Natural controls on fluvial denudation rates in major world drainage basins, *Journal of Geophysical Research*, 99, Fig. 12, p. 13872.

Figure 19.18 A–C (p. 457): Reprinted by permission of Dr. Pat Behling from W.F. Ruddiman and J.E. Kutzbach (1991, March), Plateau uplift and climate change, *Scientific American*, 264, pp. 70–71.

Figure 19.19 (p. 458): Adapted by permission of the Royal Society of Edinburgh and W.F. Ruddiman from W.F. Ruddiman and J.E. Kutzbach (1990), Late Cenozoic plateau uplift and climate change, *Transactions of the Royal Society of Edinburgh: Earth Sciences*, vol. 81, pt. 4, Fig. 1, p. 302.

Figure 19.21 (p. 460): Adapted by permission from Peter Molnar and Philip England (1990), Late Cenozoic uplift of mountain ranges and global climate change: Chicken or egg? *Nature*, 346, Fig. 2, p. 30. Copyright 1990 Macmillan Magazines Ltd.

Figure 19.22 (p. 460): Adapted by permission from D.W. Burbank (1992), Characteristic size of relief, *Nature*, 359, Fig. 1, p. 483. Copyright 1992 Macmillan Magazines Ltd.

Figure 20.1 (p. 465): Adapted with the permission of Cambridge University Press from B.L. Turner et al. (eds.) (1990), *The Earth as transformed by human action*, Fig. 3.1, p. 43. *Source:* Adapted by permission of Cambridge University Press from B.L. Turner et al. (eds.) (1990), *The Earth as transformed by human action*, Fig. 3.1, p. 43.

Table 20.1 (p. 472): Reprinted by permission of Routledge, ITPS Ltd., and the U.N. Environment Programme in Kenya from M.K. Tolba and D.A. El-Kholy (eds.) (1992). *The world environment 1972–1992* (London, Chapman and Hall), p. 169.

Figure 20.8 (p. 472): Adapted by permission of Routledge, ITPS Ltd., and the U.N. Environment Programme in Kenya from M.K. Tolba and D.A. El-Kholy (eds.) (1992), *The World Environment 1972–1992* (London, Chapman and Hall), Fig. 3, p. 164; by permission of Oxford University Press from T.C. Whitmore (1990), *An introduction to tropical forests* (Oxford, Clarendon Press); and by permission of the Association for Tropical Biology from S. Sader and A. Joyce (1988), Deforestation rates and trends in Costa Rica, 1940–1983, *Biotropica*, 20, 11–19.

Figure 20.15 (p. 478): Adapted with permission from U.S. Committee from the Global Atmospheric Research Program, *Understanding climatic change: A program for action*, Fig. 3.2. Copyright 1975 by the National Academy of Sciences. Courtesy of the National Academy Press, Washington, D.C.

Figure 20.16 (p. 479): Adapted from C.D. Keeling et al. (1989), A three-dimensional model of atmospheric CO_2 transport based on observed winds: 1. Analysis of observational data. *Aspects of climate variability in the Pacific and the Western Americas* (AGU Monograph 55), Fig. 16, p. 179. Copyright 1989 by the American Geophysical Union. Data for the years 1989 on from Dave Keeling and Tim Whorf, Scripps Institution of Oceanography.

Figure 20.17A (p. 479): Adapted from A.M. Solomon et al. (1985), The global carbon cycle, in J.R. Trabalka (ed.), *Atmospheric carbon dioxide and the global carbon cycle*, U.S. Department of Energy Report DOE/ER-0239, Fig. 1.2, p. 7.

Figure 20.17B (p. 479): Adapted by permission of the Royal Swedish Academy of Sciences from International Geosphere Biosphere Programme (1990), Past global changes (PAGES), in *The International Geosphere-Biosphere Programme: A Study of Global Change*, IGBP Report 12, Fig. 1, p. 7.1–4 (including data from Stauffer, 1988, and Barnett, 1988).

Table 20.2 (p. 480): Reprinted by permission from Atmospheric trace gases that are relatively active and of significance to global change (1991), *EarthQuest* (UCAR/Office of Interdisciplinary Earth Studies), vol. 5, no. 1

Figure 20.18 (p. 482): Adapted from M. Schlesinger and J. Mitchell (1987), Climate model simulations of the equilibrium climate response to increased carbon dioxide, *Reviews of Geophysics*, vol. 25, Figs. 14, 15, and 16. Copyright 1987 by the American Geophysical Union.

Figure 20.19 (p. 483): Reprinted by permission of IPCC Scientific Assessment from J.T. Houghton, G.J. Jenkins, and J.J. Ephraums (eds.) (1990), *Climate change: The IPCC scientific assessment* (Cambridge, Cambridge University Press), Fig. 5.

Table 20.3 (p. 483): Reprinted by permission of IPCC Scientific Assessment from J.T. Houghton, G.J. Jenkins, and J.J. Ephraums (eds.) (1990), *CLimate change: The IPCC scientific assessment* (Cambridge, Cambridge University Press), Table 5.1

Figure 20.21 (p. 485): Reprinted by permission of IPCC Scientific Assessment from GCM output in J.T. Houghton, G.J. Jenkins, and J.J. Ephraums (eds.) (1990), *Climate change: The IPCC scientific assessment* (Cambridge, Cambridge University Press).

Figure 20.22A (p. 487): Based on data of K. Hammer, personal communication.

Figure 20.22B (p. 487): From J. Hansen et al. (1993), in *National Geographic Research and Exploration*, vol. 9, no. 2, pp. 142–158; from *Consequences*, vol. 3, no. 2 (1997), p. 25.

Table B.1 (pp. 495–496): After K.K. Turekian (1969). Also note (a): Estimate from S.R. Taylor (1964).

Figure F.3 A&B (p. 514): Modified from U.S. Geological Survey.

INDEX

Page references in *italics* indicate illustrations. Page references followed by italic *table* indicate material in tables. Page references followed by lowercase italic *n* indicate material in footnotes.

/

Praise for Debbie Macomber's
Twenty Wishes

"Even the most hard-hearted readers
will find themselves rooting for the women
in this hopeful story while surreptitiously wiping away tears
and making their own lists of wishes."
—*Booklist*

"It's impossible not to cheer for
Macomber's characters, and there's a story in this book for
women of every age. When it comes to creating a special
place and memorable, honorable characters,
nobody does it better than Macomber."
—*BookPage*

"Macomber's assured storytelling and affirming narrative
is as welcoming as your favorite easy chair."
—*Publishers Weekly*

A "heart-felt, thought-provoking return to Blossom Street."
—*Genre Go Round Reviews*

"*Twenty Wishes* is a heartwarming story of friends and a
second chance at love. Reading a Debbie Macomber book
is like wrapping yourself in a warm, cozy blanket."
—*BookLoons*

"*Twenty Wishes* is an emotionally gripping novel...
not a book to be missed."
—*Romance Junkies.com*

"Debbie committed...her heart to paper,
and the result is spectacular."
—*Books2Mention*

"Macomber's latest is completely delightful....
By turns humorous, sad and optimistic,
it's a wonderful book to curl up with at any time."
—*Romantic Times BOOKreviews*

DEBBIE MACOMBER
Twenty Wishes

MIRA®

MIRA®

Recycling programs
for this product may
not exist in your area.

ISBN-13: 978-0-7783-2631-1
ISBN-10: 0-7783-2631-4

TWENTY WISHES

Copyright © 2008 by Debbie Macomber.

All rights reserved. Except for use in any review, the reproduction or utilization of this work in whole or in part in any form by any electronic, mechanical or other means, now known or hereafter invented, including xerography, photocopying and recording, or in any information storage or retrieval system, is forbidden without the written permission of the publisher, MIRA Books, 225 Duncan Mill Road, Don Mills, Ontario, Canada M3B 3K9.

This is a work of fiction. Names, characters, places and incidents are either the product of the author's imagination or are used fictitiously, and any resemblance to actual persons, living or dead, business establishments, events or locales is entirely coincidental.

MIRA and the Star Colophon are trademarks used under license and registered in Australia, New Zealand, Philippines, United States Patent and Trademark Office and in other countries.

www.MIRABooks.com

Printed in U.S.A.

To
June Scobee Rodgers
My dear friend
An inspiration
And a joy

April 2009

Dear Friends,

When we were children, my cousins and I often lay on the grass during those warm summer nights, gazing up at the heavens and wishing upon a star. It seems the child in us never really goes away, does it? I was reminded of this some time ago, when I met a reader named Arliene Zeigler at an autographing and she told me about her list of wishes. They weren't resolutions, decisions or even goals. They were simply wishes. Some of them were places she wanted to go, people she longed to meet and experiences she hoped to have.

Don't we all have wishes in one form or another? Secret desires we rarely talk about because they might sound silly? As I started to write *Twenty Wishes*, I made up a completely new list of my own. I want to cuddle with my husband and reminisce about the years we've been together. I'd like to blow bubbles with my grandchildren and chase butterflies. I want to sing on Broadway. Okay, that's carrying it a bit far, but one can dream….

I hope you enjoy spending a few hours with Anne Marie, her friends (especially the widows) and everyone else on Blossom Street. Alix has the coffee brewing over at the French Café and Susannah's setting out flowers on the sidewalk outside Susannah's Garden. I see that Whiskers has curled up in the display window at A Good Yarn, and Lydia has turned over the Open sign. The door at Blossom Street Books is open, too, so come on in!

Hearing from my readers is one of my joys as an author. You can contact me through my Web site at www.DebbieMacomber.com or at P.O. Box 1458, Port Orchard, WA 98366.

Debbie Macomber

Chapter

1

It was six o'clock on Valentine's Day, an hour that should have marked the beginning of a celebration—the way it had when she and Robert were married. When Robert was alive. But tonight, on the most romantic day of the year, thirty-eight-year-old Anne Marie Roche was alone. Turning over the closed sign on the door of Blossom Street Books, she glanced at the Valentine's display with its cutout hearts and pink balloons and the collection of romance novels she didn't read anymore. Then she looked outside. Streetlights flickered on as evening settled over the Seattle neighborhood.

The truth was, Anne Marie hated her life. Well, okay, *hate* was putting it too strongly. After all, she was healthy, reasonably young and reasonably attractive, financially solvent, *and* she owned the most popular bookstore in the area. But she didn't have anyone to love, anyone who loved her. She was no longer part of something larger than herself. Every morning when she woke, she found the other side of the bed empty and she didn't think she'd ever get accustomed to that desolate feeling.

Her husband had died nine months ago. So, technically, she was a widow, although she and Robert had been separated. But they saw each other regularly and were working on a reconciliation.

Then, suddenly, it was all over, all hope gone. Just when they were on the verge of reuniting, her husband had a massive heart attack. He'd collapsed at the office and died even before the paramedics could arrive.

Anne Marie's mother had warned her about the risks of marrying an older man, but fifteen years wasn't *that* much older. Robert, charismatic and handsome, had been in his mid-forties when they met. They'd been happy together, well matched in every way but one.

Anne Marie wanted a baby.

Robert hadn't.

He'd had a family—two children—with his first wife, Pamela, and wasn't interested in starting a second one. When she'd married him, Anne Marie had agreed to his stipulation. At the time it hadn't seemed important. She was madly in love with Robert—and then two years ago it hit her. This longing, this need for a baby, grew more and more intense, and Robert's refusal became more adamant. His solution had been to buy her a dog she'd named Baxter. Much as she loved her Yorkie, her feelings hadn't changed. She'd still wanted a baby.

The situation wasn't helped by Melissa, Robert's twenty-four-year-old daughter, who disliked Anne Marie and always had. Over the years Anne Marie had made many attempts to ease the tension between them, all of which failed. Fortunately she had a good relationship with Brandon, Robert's son, who was five years older than his sister.

When problems arose in Anne Marie and Robert's marriage, Melissa hadn't been able to disguise her glee. Her stepdaughter seemed absolutely delighted when Robert moved out the autumn before last, seven months before his death.

Anne Marie didn't know what she'd done to warrant such passionate loathing, other than to fall in love with Melissa's father. She supposed the girl's ardent hope that her parents would one day remarry was responsible for her bitterness. Every child wanted his or her family intact. And Melissa was a young teen when Anne Marie married Robert—a hard age made harder by the family's circumstances. Anne Marie didn't blame Robert's daughter, but his marriage to Pamela had been dead long before she entered the picture. Still, try as she might, Anne Marie had never been able to find common ground with Melissa. In fact, she hadn't heard from her since the funeral.

Anne Marie opened the shop door as Elise Beaumont approached. Elise's husband, Maverick, had recently passed away after a lengthy battle with cancer. In her mid-sixties, she was a retired librarian who'd reconnected with her husband after nearly thirty years apart, only to lose him again after less than three. She was a slight, gray-haired woman who'd become almost gaunt, but the sternness of her features was softened by the sadness in her eyes. A frequent patron of the bookstore, she and Anne Marie had become friends during the months of Maverick's decline. In many ways his death was a release, yet Anne Marie understood how difficult it was to let go of someone you loved.

"I was hoping you'd come," Anne Marie told her with a quick hug. She'd closed the store two hours early, giving Steve Handley, her usual Thursday-night assistant, a free evening for his own Valentine celebration.

Elise slipped off her coat and draped it over the back of an overstuffed chair. "I didn't think I would and then I decided that being with the other widows was exactly what I needed tonight."

The widows.

They'd met in a book group Anne Marie had organized at the store. After Robert died, she'd suggested reading Lolly Winston's *Good Grief*, a novel about a young woman adjusting to widowhood. It was through the group that Anne Marie had met Lillie Higgins and Barbie Foster. Colette Blake had joined, too. She'd been a widow who'd rented the apartment above A Good Yarn, Lydia Goetz's yarn store. Colette had married again the previous year.

Although the larger group had read and discussed other books, the widows had gravitated together and begun to meet on their own. Their sessions were often informal gatherings over coffee at the nearby French Café or a glass of wine upstairs at Anne Marie's.

Lillie and Barbie were a unique pair of widows, mother and daughter. They'd lost their husbands in a private plane crash three years earlier. Anne Marie remembered reading about the Learjet incident in the paper; both pilots and their two passengers had been killed in a freak accident on landing in Seattle. Lillie's husband and son-in-law were executives at a perfume company and often took business trips together.

Lillie Higgins was close to Elise's age, but that was all they shared. Actually, it was difficult to tell exactly how old Lillie was. She looked barely fifty, but with a forty-year-old daughter, she had to be in her mid-sixties. Petite and delicate, she was one of those rare women who never seemed to age. Her wardrobe consisted of ultra-expensive knits and gold jewelry. Anne Marie had the impression that if Lillie wanted, she could purchase this bookstore ten times over.

Her daughter, Barbie Foster, was a lot like her mother and aptly named, at least as far as appearances went. She had long blond hair that never seemed to get mussed, gorgeous crystal-blue eyes, a flawless figure. It was hard to believe she had eighteen-year-old twin sons who were college freshmen; Anne Marie would bet that most people assumed she was their sister rather than their mother. If Anne Marie didn't like Barbie so much, it would be easy to resent her for being so…perfect.

"Thanks for closing early tonight. I'd much rather be here than spend another evening alone," Elise said, breaking into Anne Marie's thoughts.

There was that word again.

Alone.

Despite her own misgivings about Valentine's Day, Anne Marie tried to smile. She gestured toward the rear of the store. "I've got the bubble wrap and everything set up in the back room."

The previous month, as they discussed an Elizabeth Buchan novel, the subject of Valentine's Day had come up. Anne Marie learned from her friends that this was perhaps the most painful holiday for widows. That was when their small group decided to plan their own celebration. Only instead of romantic love and marriage, they'd celebrate friendship. They'd defy the world's pitying glances and toast each other's past loves and future hopes.

Elise managed a quivering smile as she peered into the back of the store. "Bubble wrap?"

"I have tons," Anne Marie informed her. "You can't imagine how many shippers use it."

"But why is it on the floor?"

"Well…" It seemed silly now that Anne Marie was trying to explain. "I always have this insatiable urge to pop it, so I thought we could do it together—by walking on it."

"You want us to step on bubble wrap?" Elise asked, sounding confused.

"Think of it as our own Valentine's dance and fireworks in one."

"But fireworks are for Independence Day or maybe New Year's."

"That's the point," Anne Marie said bracingly. "New beginnings."

"And we'll drink champagne, too?"

"You bet. I've got a couple bottles of the real stuff, Veuve Clicquot."

"*Veuve* means widow, you know. The widow Clicquot's bubbly—what else could we possibly drink?"

The door opened, and Lillie and Barbie entered in a cloud of some elegant scent. As soon as they were inside, Anne Marie locked the shop.

"Party time," Lillie said, handing Anne Marie a white box filled with pastries.

"I brought chocolate," Barbie announced, holding up a box of dark Belgian chocolates. She wore a red pantsuit with a wide black belt that emphasized her petite waist. Was there no justice in this world? The woman had the figure of a goddess and she ate *chocolate?*

"I read that dark chocolate and red wine have all kinds of natural benefits," Elise said.

Anne Marie had read that, too.

Lillie shook her head in mock astonishment. "First wine and now chocolate. Life is good."

Leading the way to the back room, Anne Marie dimmed the lights in the front of the shop. Beside the champagne and flutes, she'd arranged a crystal vase of red roses; they'd been a gift from Susannah's Garden, the flower shop next door. All the retailers on Blossom Street were friends. Hearing about the small party, Alix Turner from the French Café had dropped off a tray of cheese, crackers and seedless green grapes, which Anne Marie had placed on her work table, now covered with a lacy cloth. Lydia had insisted they use it for their celebration. It was so beautiful it reawakened Anne Marie's desire to learn to knit.

She wished she could see her friends' gifts as more than expressions of sympathy, but her state of mind made that impossible. Still, because of the other widows, for their sake as well as her own, she was determined to try.

"This is going to be fun," Elise said, telling them why Anne Marie had spread out the bubble wrap.

"What a wonderful idea!" Barbie exclaimed.

"Shall I pour?" Anne Marie asked, ignoring the sense of oppression she couldn't seem to escape. It had been present for months and she'd thought life would be better by now. Perhaps she needed counseling. One thing was certain; she needed *something*.

"By all means," Lillie said, motioning toward the champagne.

Anne Marie opened the bottle and filled the four glasses and then they toasted one another, clicking the rims of the flutes.

"To love," Elise said. "To Maverick." Her voice broke.

"To chocolate!" Barbie made a silly face, perhaps to draw attention away from Elise's tears.

"And the Widow's champagne," Lillie threw in.

Anne Marie remained silent.

Although it'd been nine months, her grief didn't seem to diminish or become any easier to bear. She worked too much, ate too little and grieved for all the might-have-beens. It was more than the fact that the man she'd loved was dead. With his death, she was forced to give up the dream of all she'd hoped her marriage would be. A true companionship—and the foundation of a family. Even if she were to fall in love again, which seemed unlikely, a pregnancy past the age of forty was risky. The dream of having her own child had died with Robert.

The four sipped their champagne in silence, each caught up in her own memories. Anne Marie saw the sorrow on Elise's face, the contemplative look on Lillie's, Barbie's half smile.

"Will we be removing our shoes in order to pop the bubble wrap?" Lillie asked a moment later.

"Mom has this thing about walking around in stocking feet," Barbie said, glancing at her mother. "She doesn't approve."

"It just wasn't done in our household," Lillie murmured.

"There's no reason to take our shoes off," Anne Marie said. "The whole idea is to have fun. Make a bit of noise, celebrate our friendship and our memories."

"Then I say, let 'er rip," Elise said. She raised her sensibly shod foot and stomped on a bubble. A popping sound exploded in the room.

Barbie went next, her step firm. Her high heels effectively demolished a series of bubbles.

Pop. Pop. Pop.

Pop.

Lillie followed. Her movements were tentative, almost apologetic.

Pop.

Anne Marie went last. It felt…good. Really good, and the noise only added to the unexpected sense of fun and exhilaration. For the first time since the party had begun, she smiled.

By then they were all flushed with excitement and champagne. The others were laughing giddily; Anne Marie couldn't quite manage that but she could *almost* laugh. The ability to express joy had left her when Robert died. That wasn't all she'd lost. She used to sing, freely and without self-consciousness. But after Robert's funeral Anne Marie discovered she couldn't sing anymore. She just couldn't. Her throat closed up whenever she tried. What came out were strangled sounds that barely resembled music, and after a while she gave up. It'd been months since she'd even attempted a song.

The popping continued as they paraded around on the bubble wrap, pausing now and then to sip champagne. They marched with all the pomp and ceremony of soldiers in procession, saluting one another with their champagne flutes.

Thanks to her friends, Anne Marie found that her mood had begun to lift.

Soon all the bubbles were popped. Bringing their champagne, they sat in the chairs where the reader groups met and toasted each other again in the dimly lit store.

Leaning back, Anne Marie tried to relax. Despite her earlier laughter, despite spending this evening with friends, her eyes filled with tears. She blinked them away, but new tears came, and it wasn't long before Barbie noticed. Her friend placed a reassuring hand on Anne Marie's knee.

"Does it ever hurt any less?" Anne Marie asked.

Searching for a tissue in her hip pocket, she blotted her eyes. She hated breaking down like this. She wanted to explain that she'd never been a weepy or sentimental woman. All her emotions had become more intense since Robert's death.

Lillie and Barbie exchanged knowing looks. They'd been widows the longest.

"It does," Lillie promised her, growing serious, too. "But it takes time."

"I feel so alone."

"That's to be expected," Barbie said, passing her the box of chocolates. "Here, have another one. You'll feel better."

"That's what my grandmother used to say," Elise added. "Eat, and everything will seem better."

"Mine always said I'd be good as new if I did something for someone else," Lillie said. "Grams swore that showing kindness to others was the cure for any kind of unhappiness."

"Exercise helps, too," Barbie put in. "I spent many, many hours at the gym."

"Can't I just buy something?" Anne Marie asked plaintively, and hiccuped a laugh as she made the suggestion.

The others smiled.

"I wish it was that easy," Elise said in a solemn voice.

Anne Marie's appetite had been nonexistent for months and she didn't really enjoy going to a gym—walking nowhere on a treadmill seemed rather pointless to her. She didn't feel like doing volunteer work, either, at least not right now—although helping another person might get her past this slump, this interval of self-absorption.

"We're all looking for a quick fix, aren't we?" Barbie said quietly.

"Maybe." Lillie settled back in her chair. "Of these different options, the one I could really sink my teeth into is buying something."

"So could I," Barbie said with a laugh.

"I realize you're joking—well, partly—but material things won't help," Elise cautioned, bringing them all back to reality. "Any relief a spending spree offers is bound to be temporary."

As tempting as the idea of buying herself a gift might be, Anne Marie supposed she was right.

"We all need to take care of ourselves physically. Eat right. Exercise," Elise said thoughtfully. "It's important we get our finances in order, too."

"I couldn't agree with you more on *that*," Lillie said.

"Let's make a list of our suggestions," Elise went on. Reaching for her purse, she took out a small spiral notebook.

"If I'm going to make a list," Lillie piped up, "it won't be about eating cauliflower and going jogging. Instead, I'd plan to do some of the things I've put off for years."

"Such as?" Anne Marie asked.

"Oh, something fun," Lillie said, "like traveling to Paris."

Anne Marie felt as if a bolt of lightning had struck her. When they were first married, Robert had promised her that one day he'd take her to Paris. They talked about it frequently, discussing every aspect of their trip to the City of Light. The museums they'd visit, the places they'd walk, the meals they'd eat…

"I want to go to Paris with someone I love," she whispered.

"I want to fall in love again," Barbie said decisively. "Head over heels in love like I was before. A love that'll change my life."

They all grew quiet for a long moment, considering her words.

Anne Marie couldn't believe Barbie would lack for male companionship. They'd never discussed the subject, but she was surprised that a woman as attractive as Barbie didn't have her choice of men. Maybe she did. Maybe she simply had high standards. If so, Anne Marie couldn't blame her.

"We all want to be loved," Lillie said. "It's a basic human need."

"I had love," Elise told them, her voice hoarse with pain. "I don't expect to find that kind of love again."

"I had it, too," Barbie said.

Another hush fell over them.

"Making a list is a good idea," Elise stated emphatically. "A list of things to do."

Anne Marie nodded, fingering one of the suspended Valentine's decorations as she did. The idea had caught her interest. She needed to revive her enthusiasm. She needed to find inspiration and motivation—and a list might just do that. She was a list-maker anyway, but this would be different. It wouldn't be the usual catalog of appointments and everyday obligations.

"Personally I don't need another to-do list," Lillie murmured, echoing Anne Marie's thought. "I have enough of those already."

"This wouldn't be like that," Anne Marie responded, glancing at Elise for verification. "This would be a…an inventory of wishes," she said, thinking out loud. She recognized that there were plenty of *shoulds* involved in widowhood; her friends were right about that. She did need to get her financial affairs in order and pay attention to her health.

"Twenty wishes," she said suddenly.

"Why twenty?" Elise asked, leaning forward, her interest obvious.

"I'm not sure. It sounds right." Anne Marie shrugged lightly. The number had leaped into her head, and she didn't know quite why. *Twenty.* Twenty wishes that would help her recapture her excitement about life. Twenty dreams written down. Twenty possibilities that would give her a reason to look toward the future instead of staying mired in her grief. She couldn't continue to drag from one day to the next, lost in pain and heartache because Robert was dead. She needed a new sense of purpose. She owed that to herself—and to him.

"Twenty wishes," Barbie repeated slowly. "I think that works. Twenty's a manageable number. Not like a hundred, say."

"And it's not too few—like two or three," her mother said.

Anne Marie could tell that her friends were taking the idea seriously, which only strengthened her own certainty about it. "Wishes and hopes for the future."

"Let's do it!" Lillie proclaimed.

Barbie sat up straighter in her chair. "You should learn French," she said, smiling at Anne Marie.

"French?"

"For when you're in Paris."

"I had two years of French in high school." However, about all she remembered was how to conjugate the verbs *être* and *avoir*.

"Take a refresher course." Barbie slid onto the edge of her cushion.

"Maybe I will."

"I might learn how to belly dance," Barbie said next.

The others looked at her with expressions of surprise; Anne Marie grinned in approval.

"Lillie mentioned this earlier, but I think it would do us all a world of good to be volunteers,"Elise said."I've become a Lunch Buddy at my grandson's school and I really look forward to my time with Malcolm."

"Lunch Buddy? What's that?"

"A program for children at risk," Elise explained. "Once a week I visit the school and have lunch with a little boy in third grade. Malcolm is a sweet-natured child, and he's flourished under my attention. The minute I walk into the school, he races toward me as if he's been waiting for my visit all week."

"So the two of you have lunch?"

"Well, yes, but he also likes to show me his schoolwork. He's struggling with reading. However, he's trying hard, and every once in a while he'll read to me or I'll read to him. I've introduced him to the Lemony Snicket books and he's loving those."

"You tutor him, then?"

"No, no, he has a reading tutor. It's not that kind of program. I'm his *friend*. Or more like an extra grandmother."

The idea appealed to Anne Marie, but she didn't know if this was the right program for her. She'd consider it. Her day off was Wednesday and every other Saturday when Theresa came into the store. She had to admit that volunteering at an elementary school would give her something to do other than feel sorry for herself.

It wasn't a *wish*, exactly. Still, Elise claimed she felt better because of it. Helping someone else—perhaps that was the key.

The party broke up around nine-thirty, and after she'd waved everyone off, Anne Marie locked the front door.

Then she climbed the stairs to her tiny apartment above the bookstore. Her ever-faithful Baxter was waiting for her, running circles around her legs until she bent down and lifted him up and lavished him with the attention he craved. After taking him out for a brief walk, she returned to the apartment, still thinking about the widows' new project.

She made a cup of tea and grabbed a notepad, sitting on the couch with Baxter curled up beside her. At the top of the page she wrote:

Twenty Wishes

It took her a long time to write down the first item.

1. Find one good thing about life

She felt almost embarrassed that all she could come up with was such a plaintive, pathetic desire, one that betrayed the sorry state of her mental health. Sitting back, she closed her eyes and tried to remember what she used to dream about, the half-expressed wishes of her younger years.

She added a second item, silly though it was.

2. Buy myself a pair of red cowboy boots

In her twenties, long before she married Robert, Anne Marie had seen a pair in a display window and they'd stopped her cold. She absolutely *had* to have those boots. When she'd gone into the store and tried them on, they were a perfect fit. Perfect. Unfortunately the price tag wasn't. No way could she afford $1500 for a pair of cowboy boots! With reluctance she'd walked out of the store, abandoning that small dream.

She couldn't have afforded such an extravagance working part-time at the university bookstore. But she still thought about those boots. She still wanted them, and the price no longer daunted her as it had all those years ago.

Somehow, she'd find herself a pair of decadent cowboy boots. Red ones.

Chewing on the end of her pen, she contemplated other wishes. Really, this shouldn't be so difficult....

It occurred to her that if she was going to buy red cowboy boots, she should think of something to do in them.

3. Learn how to line dance

She suspected line dancing might be a bit passé in Seattle—as opposed to, say, Dallas—but the good thing was that it didn't require a partner. She could show up and just have fun without worrying about being part of a couple. She wasn't ready for another relationship; perhaps in time, but definitely not yet. After a few minutes she crossed out the line-dancing wish. She didn't have the energy to be sociable. She read over her first wish and scratched that out, too. She didn't know how to gauge whether she'd actually found something good about life. It wasn't specific enough.

A host of possibilities bounced around in her head but she didn't bother to write any of them on her list.

Lillie was right; she needed to get her finances in order. She wrote that down on a second sheet of paper, along with getting her annual physical and—maybe—signing up for the gym. The only thing on the first sheet, her wish list, was those boots.

So now she had two separate lists—one for wishes and the second for the more practical aspects of life. Not that each wish wouldn't ultimately require its own to-do list, but that was a concern for another day. She closed her eyes and tried to figure out what she wanted most, what wish she hoped to fulfill. The next few ideas were all sensible ones, like scheduling appointments she'd postponed for months. It was a sad commentary that her one wish, the lone desire of her heart, was an outrageously priced pair of boots.

That was the problem; she no longer *knew* what she wanted. Shrouded in grief and lost dreams, her joy had vanished, the same way laughter and singing had.

So far, her second list outnumbered the wish list. It included booking appointments with an accountant, an attorney, the vet and a couple of doctors. Sad, sad, sad. She could well imagine what Lillie and Barbie's lists looked like. They'd have wonderful ideas. Places to go, experiences to savor, people to meet.

Anne Marie stared at her wish list with its one ridiculous statement, tempted to crumple it up.

She didn't. For reasons she couldn't explain, she left it sitting on her kitchen counter. Lists were important; she knew that. Over the years she'd read enough about goal-setting to realize the value of writing things down. In fact, the store carried a number of bestselling titles on that very topic.

Okay, this was a start. She wasn't going to abandon the idea. And at least she'd taken control of some immediate needs. She'd identified what she *had* to do.

Sometime later, she'd list what she *wanted* to do.

She ran her finger over the word *boots*. Foolish, impractical, ridiculous—but she didn't care. She was determined to have the things.

Already the thought of listing her wishes was making a difference; already she felt a tiny bit of hope, a whisper of excitement. The thawing had begun.

Eventually other desires, other wishes, would come to her. She had nineteen left. She felt as if the genie had finally escaped the lamp and was waiting to hear her greatest desires. All she had to do was listen to her own heart and as soon as she did, her wildest dreams would come true.

If only life could be that simple.

It wasn't, of course, but Anne Marie decided she was willing to pretend.

All that next week Anne Marie continued to look at her list. The sheet of paper with TWENTY WISHES written across the top became a patchwork of scribbles and scratched-out lines. She wrote *I want to sing again,* then changed her mind, deciding it was unnecessary to waste a wish on something she was convinced would return in its own time.

Eventually she transferred her list, such as it was, to a yellow legal pad, which somehow made her wishes seem more official. Then on Wednesday, her day off, she walked past a craft store on her way back from the accountant's and noticed the scrapbooking supplies in the window. She stared at the beautifully embellished pages displayed in the showcase. She used to possess a certain decorative flair. She wasn't sure she did anymore, but the idea of creating pages like that for her meager list of wishes appealed to her. A scrapbook to compile her wishes, make her plans and document her efforts. Those wishes would encourage her to look forward, to focus on the future with an optimism that had been lacking since her separation from Robert.

With that in mind, Anne Marie bought the necessary supplies, then lugged them home. As she passed A Good Yarn, the shop just two doors down from the bookstore, she impulsively stepped inside. First, she wanted to thank Lydia for the table covering and second…she'd ask about classes.

She'd add knitting to her wish list. Anne Marie wondered why she hadn't thought of that earlier. Elise was a consummate knitter and often encouraged the others to learn. She described the satisfactions of knitting in such a compelling way, Anne Marie had flirted more than once with the idea of taking a class. Lydia Goetz, who owned A Good Yarn, was a much-loved and admired member of the Blossom Street neighborhood. Anne Marie was friendly with her and had often gone inside the yarn store, but never with the serious intent of learning to knit. Now, the prospect of knitting filled her with unfamiliar enthusiasm.

Lydia was sitting at the table in the back of the shop with her sister, Margaret. Although Lydia was petite and graceful, her sister was rather big-boned, a little ungainly. At first glance it was hard to believe they were even related. Once the surprise of learning they were sisters wore off, the resemblance revealed itself in the shape of their eyes and the thrust of their chins.

When Anne Marie entered the store, the sisters were obviously involved in their conversation; as they spoke, Lydia was knitting, Margaret crocheting. The bell above the door jingled, startling them both.

A smile instantly broke out on Lydia's face. "Anne Marie, how nice to see you! I'm glad you stopped by."

Lydia had a natural warmth that made customers feel welcome.

"Good morning," Anne Marie said, smiling at the two women. "Lydia, I came to thank you again for the gorgeous tablecloth."

"Oh, you're welcome. You know, it's really a lace shawl I knit years ago. I hope you'll have occasion to use it again."

"Oh, I will."

"I've been meaning to visit the bookstore," Lydia told her. "I want to pick up a couple of new mysteries. By the way, how did the Valentine's party go?"

"It was wonderful," Anne Marie said, gazing around. Whenever she went into the yarn shop, she was astonished by the range of beautiful colors and inviting textures. She walked over to the blue, green and teal yarns that lined one area of the shelves. Putting down her packages, she reached out a hand to touch a skein of irresistibly soft wool.

"Can I help you find something?" Lydia asked.

Anne Marie nodded and, strangely, felt a bit hesitant. "I'd like to learn to knit." This was the first positive step she'd taken toward acting on her wish list. She'd been searching for somewhere to start, and knitting would do very well. "I...saw the notice in the window for a beginners' class last week, but there isn't a sign now. Do you have one scheduled anytime soon?"

"As it happens, Margaret and I were just discussing a beginners' class for Thursday afternoons."

Anne Marie shook her head. "I work all day on Thursdays."

"I'm also thinking about starting a new class for people who work. How about lunchtime on Tuesdays?" Lydia suggested next. "Would you like to sign up for that?"

Before Anne Marie could respond, Margaret was on her feet. "That's too many classes," she muttered. "Lydia's teaching far too many classes and it exhausts her."

"Margaret!" Lydia protested and cast a despairing look at her sister.

"Well, it's true. You need to get someone else in here who can teach. I do as much as I can," she said, "but there are times I've got more customers than I can handle and you're involved with all those classes."

Lydia ignored her sister. "Anne Marie, if you want to learn how to knit, I'll teach you myself."

It occurred to Anne Marie that what she really wanted was a class. She'd rejected line dancing because that had seemed like an overwhelming social occasion; a small knitting group was far less threatening. Other than the Valentine's event with the widows, she hadn't gone anywhere or done much of anything since Robert's funeral. Until now, the mere thought of making cheerful conversation with anyone outside the bookstore was beyond her. She decided she could ease into socializing with a knitting class. A few like-minded women, all focused on the same task…

"I appreciate the offer," Anne Marie told Lydia. "However, I think Margaret's probably right. You've got a lot on your plate. Let me know if that noontime beginner class pans out."

"Of course."

After they'd exchanged farewells, Anne Marie picked up her shopping bags and left the yarn store. As she strolled past the shop window she noticed Whiskers, Lydia's cat, curled up in a basket of red wool. When Anne Marie walked Baxter, he often stood on his hind legs, front paws against the window, fixated on Lydia's cat—who wanted nothing to do with him.

Hauling the scrapbooking supplies upstairs to her apartment, Anne Marie set her bags on the kitchen table, then scooped up her dog, stroking his silky fur. "Hey, Mr. Baxter. I just saw your friend Whiskers."

He wriggled excitedly and she put him down, collecting a biscuit from a box on the counter. "Here you go." She smiled as he loudly crunched his cookie, licking up each and every crumb. "Maybe I'll knit you a little coat sometime...and maybe I won't."

Now that a knitting class apparently wasn't a sure thing, Anne Marie was shocked at how discouraged she felt. One roadblock, and she was ready to pack it in. Less than a year ago, hardly anything seemed to defeat her, but these days even the most mundane problems were disheartening.

At least Baxter's needs were straightforward and easily met, and he viewed her with unwavering devotion. There was comfort in that.

Eager to start her scrapbook project, she got to work. The three-ring binder was black with a clear plastic cover. For the next thirty minutes she cut out letters, decorated them with glitter glue and pasted them on a bright pink sheet. Then she slipped it behind the cover so the front of the binder read TWENTY WISHES. In addition to the binder, Anne Marie had purchased twenty plastic folders, one for each wish.

She became so involved in her work that it was well past one before she realized she hadn't eaten lunch. She emptied a can of soup into a bowl, and it was heating in the microwave when her phone rang.

Startled, she picked up the receiver on the first ring. The beeper went off at the same time, indicating that her meal was ready.

"Hello," she said, cradling the phone against her shoulder as she opened the microwave. She rarely got calls at home anymore. In the weeks after Robert's funeral, she'd heard from a number of couples they'd been friends with, but those people had gradually drifted away. Anne Marie hadn't made the effort to keep in touch, either. It was easier to lose herself in her grief than to reach out to others.

"Anne Marie, it's Lillie. Guess what?" her friend said breathlessly.

"What?" Hearing the excitement in Lillie's voice lifted her own spirits.

"Remember what you said Valentine's night?"

Anne Marie frowned. "Not exactly. I said various things. Which one do you mean?"

"Oh, you know. Elise was talking about eating something to feel better and then someone else—me, I think—brought up volunteering and you said…" She giggled. "You asked why we couldn't just buy ourselves something."

Anne Marie smiled. She'd been joking at the time, but it appeared that Lillie had taken her seriously. "Are you about to tell me you bought yourself something?"

"I sure did," Lillie said gleefully.

"Well, don't leave me in suspense. What did you get?"

Lillie giggled again. "A brand-new shiny red convertible."

"No!" Anne Marie feigned shock.

"Yes. Can you imagine me at sixty-three buying myself a sports car?"

"What kind is it?" Anne Marie knew next to nothing about cars, which was why she belonged to Triple A. In truth, Robert had been pretty helpless, too.

"A BMW."

It must've been expensive; Anne Marie knew that much. Well, Lillie could afford it. The perfume company had been more than generous to her and Barbie after the plane crash, and they were both financially secure.

"Want to go for a ride?"

Anne Marie's first inclination was to decline. Almost immediately she changed her mind. Why not go? Lillie's excitement was so contagious, she couldn't resist joining in.

"I'd love to," she said warmly.

"Great. I'll meet you in front of the bookstore in twenty minutes."

"Uh, what about Jacqueline?" She knew Lillie had plenty of other friends and that she and Jacqueline Donovan were especially close. They'd raised their children together, belonged to the same country club and were active members of several charitable organizations. Jacqueline, too, was a frequent customer at Blossom Street Books, not to mention all the other neighborhood stores.

"Rest assured, she'll get her turn," Lillie told her. "So, do you want to go for a ride or not?"

"I do. I just thought…never mind. I'd love to ride in your shiny new red convertible."

Gulping down her soup and then grabbing her coat, Anne Marie waited outside by the curb. Lillie pulled up right on time. The car, a convertible, was certainly bright red, and it shone from fender to fender. Despite the overcast skies, her friend had the top down.

Anne Marie stepped forward, gawking at the vehicle. "Lillie, it's fabulous!"

The older woman grinned. "I think so, too."

"What did Barbie have to say?"

Lillie shook her head. "She doesn't know yet. No one does. I'd just driven it off the showroom floor when I called you."

"Why me?"

"You're the one who inspired the idea. So it's only fitting that you be the first one to ride in it."

Anne Marie remembered the "eat something," "do something" conversation, but she never would've guessed she'd end up riding in a brand-new BMW because of it.

"It's the first time in my life that I've purchased my own car. I negotiated the deal myself," Lillie announced proudly. "And I had all my facts straight before I even walked inside. Those salesmen take one look at me and see dollar signs. I needed to prove to them—and to myself—that I'm no pushover."

"I'm sure you did—and then some."

Lillie nodded. "I got on the Internet and found a Web site that showed the invoice price, and then broke out the dealer's typical overheads and advertising costs."

Anne Marie was more impressed by the minute. "You really did your research."

"My dear, you can find out just about anything on the Internet." She raised her eyebrows. "I also discovered that the dealer cost includes a holdback for profit." Lillie smiled roguishly as she continued her story. "The salesman was a charming fellow, I will say that. He expected to walk away with a substantial commission check, but I quickly disavowed him of *that* notion."

Anne Marie stared at her, astonished. "How did you do it?"

"We started negotiating and I had him at the point of accepting my offer when I remembered that dealers sometimes get incentives and rebates on cars sold."

"You mentioned that, too?"

"Darn right I did and he agreed to my terms."

"Lillie, congratulations." Anne Marie had no idea the older woman had such a head for business. As far as she was aware, Lillie hadn't worked a day in her life, or at least not outside the home. In many ways Barbie was a younger version of her mother. Both women had married young, and each had chosen a husband ten or so years her senior. That was something Anne Marie had in common with them; the fact that they were both mothers was not. They'd promptly delivered the requisite child, in Barbie's case, twin sons. If Anne Marie recalled correctly, the Foster boys, Eric and Kurt, were enrolled in separate East Coast schools—very elite ones, naturally.

"It feels so good to drive a vehicle I negotiated for myself," Lillie said. "And this came about because of you."

"Really, I just made an off hand comment."

"It's more than purchasing my own car," she said, as though Anne Marie hadn't spoken, "it was managing everything myself instead of handing the task over to someone else. I've always felt I could be a good businesswoman if I'd been given the opportunity." She rubbed her hand over the arc of the steering wheel. "No one seemed to consider me capable of running my own affairs. Ironically, the person I needed to convince most was me. Thanks to you, I did."

Anne Marie felt a bit uncomfortable; Lillie was giving her far more credit than she deserved.

"Come on," Lillie said. "Get in."

Swinging open the passenger door, Anne Marie climbed into the convertible and fastened her seat belt.

Lillie gripped the steering wheel tightly, throwing back her head. "I have to tell you, I'm really getting into this Twenty Wishes thing."

"I am, too," Anne Marie said. "When you phoned I was in the middle of making a scrapbook, a page for each wish. I'm going to cut out magazine pictures to visualize them and to document the various steps."

Lillie turned to smile at her. "What a great idea."

The praise encouraged her and Anne Marie quickly went on to describe the craft-store supplies she'd purchased. "I don't have much of a list as yet, but I'm working on it. How about you?"

Lillie was silent for a moment. "I've decided I want to fall in love." She spoke with a determination Anne Marie had never heard from her.

"Barbie said the same thing at our Valentine's party," Anne Marie pointed out.

"I know."

Anne Marie waited.

"I've had plenty of men ask me out," Lillie told her. "I don't mean to sound egotistical, but I'm not interested in most of them."

Anne Marie nodded, not surprised that "plenty of men" would find Lillie attractive.

"I've learned a thing or two in the last sixty-odd years," Lillie was saying, "and I'm not as impressed with riches or connections as I once was. When I fall in love, I want it to be with a man of integrity. Someone who's decent and kind and—" She paused as though searching for the right word. "Honorable. I want to fall in love with an honorable man." She seemed embarrassed at having spoken her wish aloud, and leaned forward to start the engine. "As you might've guessed, my marriage—unlike my daughter's—wasn't a particularly good one. I don't want to repeat the mistakes I made when I was younger." The car roared to life, then purred with the sound of a flawlessly tuned engine.

Checking behind her, Lillie backed out of the parking space on Blossom Street. From there they headed toward the freeway on-ramp. Lillie proposed a drive through the Kent Valley and along the Green River, and Anne Marie agreed.

Closing her eyes, Anne Marie let the cold February wind sweep past her. Lillie turned on the radio just as the DJ announced a hit from the late 1960s. Soon she was crooning along to The Lovin' Spoonful's "Did You Ever Have To Make Up Your Mind." Anne Marie remembered her mother singing that song as a girl. Perhaps it was unusual to find herself good friends with a woman who was her mother's contemporary. Sadly, although Anne Marie was an only child, she and her mother weren't close. Her parents had divorced when she was in sixth grade, and the bitterness, especially on her mother's part, had lingered through the years. It didn't help that Anne Marie resembled her father. She'd had little contact with him after the divorce, and he died in a boating accident on Lake Washington when she was twenty-five. Her mother had never remarried.

Because they had such an uneasy relationship, Anne Marie avoided frequent visits home. She made a point of calling her mother at least once a month. Even then, it seemed they didn't have much to discuss. Sad as it was to admit, Anne Marie had more in common with Lillie than she did with her own mother.

As Lillie's voice grew louder, Anne Marie stayed quiet, afraid that if she attempted to sing she'd embarrass herself. After about twenty minutes, Lillie exited the freeway and drove toward the road that ran beside the banks of the Green River.

This was about as perfect a moment as Anne Marie

could remember since Robert's death. They had the road to themselves. The sun was on her face and the wind tossed her hair in every direction and she couldn't have cared less.

Lillie, however, had wrapped a silk scarf over her elegantly arranged hair, which held it neatly in place.

Darting around the twisting country roads, Lillie revealed her skill as a driver. Then, in the middle of a sharp turn, she let out a small cry of alarm.

"What's wrong?" Anne Marie was instantly on edge. She grasped the passenger door as Lillie struggled to control the vehicle.

"The steering wheel," she gasped. She pulled the car over to the side of the road and cut the engine. She looked wide-eyed at Anne Marie. "There's something wrong with the steering."

"This is a brand-new car!"

"You don't need to remind me," Lillie said through clenched teeth. She opened the car door and got out, then reached behind the seat for her purse. Taking out her cell phone, she exhaled slowly. "Fortunately I have the dealership's number in my Calls Received." She wrapped one arm around her waist while she waited for someone to answer.

"Hello," she said, speaking without even a hint of irritation in her voice. "This is Lillie Higgins. I was in the dealership earlier this afternoon. Could I speak with Darryl Pierpont, please? He's the salesman who sold me this vehicle." She waited, and it seemed the salesman was unavailable because Lillie asked to speak with the manager, who was apparently out of the office, as well. Lillie then said, "All right, answer me this. Has the dealership deposited the check I wrote?" She turned to Anne Marie, eyes fierce. "I suggest you don't, as I'm about to put a stop payment order on it."

That quickly got her the attention she sought. After explaining what had happened and listening for a moment, then describing her location, Lillie closed the cell.

"The dealership's sending a tow truck for the car. The service manager is bringing me a replacement vehicle until they can determine what's wrong with mine."

"As they should."

"Until then we have to sit here and wait."

They climbed back into the car and chatted for half an hour or so until another BMW arrived, followed by a tow truck. A Hispanic man stepped out of the car. "Ms. Higgins?" he asked with a slight Mexican accent, looking at Lillie.

"Yes."

"I'm Hector Silva, manager of the service department. I would like to personally apologize for this inconvenience."

"I've owned this car for less than two hours!"

Hector shook his head. "I give you my word that we will find out what caused the problem and repair it properly. Until then, the dealership would like you to use this loaner car."

Anne Marie liked the man immediately. He was around Lillie's age, she guessed, with lovely tanned skin and salt-and-pepper hair. He handed Lillie some papers to sign and then the keys to the other car.

"Would you like a ride back to the dealership, Mr. Silva?" Lillie offered, surprising Anne Marie.

"No, thank you, I'll escort your convertible with the tow truck driver. I'll have your car back to you as soon as possible."

"Thank you."

He bowed his head. "It is my pleasure, Ms. Higgins."

While Hector Silva and the driver of the tow truck conferred, Lillie and Anne Marie slipped into the second car, a luxury sedan.

"He was so nice," Anne Marie commented. The service manager couldn't have been more accommodating or polite.

"I was looking forward to giving the dealership a piece of my mind," Lillie said with a sigh. "But how can I when everyone's being so wonderful? Well," she said, grinning, "after I threatened them."

"That had nothing to do with Mr. Silva, though."

"I agree," Lillie said. "He struck me as genuine."

They resumed their drive, except that this time Lillie headed straight back to the city, stopping in front of Blossom Street Books.

"Thank you, Lillie," Anne Marie said as she climbed out. "I've never enjoyed a car ride more."

"Bye." And with a smile that shone from her eyes and her heart, Lillie drove off.

Standing in front of Woodrow Wilson Elementary School, Anne Marie took a deep breath. Elise Beaumont had repeatedly encouraged her to become a volunteer and had recommended the Lunch Buddy program. Elise herself was a Lunch Buddy at a different school—her grandson's—but Woodrow Wilson was closer to Blossom Street. She'd sounded so positive about the experience that Anne Marie had felt inspired to make the initial call. Volunteering was now number three on her list of Twenty Wishes, after the red boots and learning to knit.

Lillie had bought her red BMW convertible and despite the problems that first day, she was thrilled with her purchase. Buoyed by that sense of exhilaration, Lillie had decided to look more closely into the financial matters she'd left in the hands of others. She, too, was working on her list, as were Barbie and Elise.

Last week Elise had said she was applying for a part-time job. For the last three years of her husband's illness, she'd been Maverick's primary caregiver. Now that her husband was gone, Elise needed some kind of activity to fill her time. Maverick wouldn't have wanted her to mope uselessly around the house, she insisted.

Although Anne Marie had only met Maverick Beaumont—a professional poker player—once or twice, she felt Elise was right. Maverick was obviously a man of action and he would've urged his wife to do something constructive and meaningful with her remaining years. The Lunch Buddy program was a worthwhile start, but Elise had extra time, lots of it, and energy to spare.

Anne Marie wasn't sure how Robert would react if he were to find out she'd volunteered as a Lunch Buddy—let alone that she'd begun a list of Twenty Wishes. Would he consider it frivolous? Self-involved? Or would he think it was a good idea, a good way of recapturing her enthusiasm for life? They'd been married almost eleven years and there were days Anne Marie felt she'd never really known her husband.

Robert was a private person who kept his feelings hidden from the world and sometimes even from her. When she first told him she wanted a child, Robert had simply left the room. Not until three days later was he willing to discuss the matter. He'd told her that a second family was out of the question; as far as he was concerned, they'd made that decision before their marriage. He was right. She'd agreed there'd be no children. What he didn't understand or seem capable of acknowledging was that she'd been at a very different point in her life when she'd married him. She'd been too young to realize how intense the desire for a baby would become as the years went on.

Robert said he already had his family, that it was time to think about grandchildren, not more children. She'd agreed to his terms and, according to him, that agreement was binding.

Anne Marie had tried to ignore her yearning for a child. With Robert's encouragement and support, she'd purchased Blossom Street Books with a small inheritance from her grandparents' estate, which she'd invested years before. That hadn't solved the problem, nor had Baxter, the Yorkie he'd surprised her with one evening. Much as she loved Robert, her bookstore and her dog, her need for a baby was still there, growing until she could no longer ignore it.

She wanted a baby. Robert's baby. The promise she'd made him had been more than eleven years ago. She'd changed her mind, but he refused to change his. She'd pleaded and cajoled, all to no avail.

To complicate everything, Robert had discussed this personal and private matter with his daughter, who'd naturally sided with her father. That made Anne Marie's relationship with Melissa—and with Robert—even more difficult.

Melissa had hated Anne Marie from the day she married Robert. Granted, the girl had only been thirteen at the time, but she'd rejected Anne Marie's overtures in no uncertain terms, and her attitude had become more adamant, more intolerant, with age. His daughter had always been Daddy's little girl and her resentment toward Anne Marie was unyielding. Melissa had done everything possible to make her feel like an outsider. Anne Marie hadn't been invited to graduations, birthdays or other family events. Brandon, her stepson, had accepted her from the beginning, and they'd held their own little celebrations. During the first few years, Robert had tried to build a bridge between her and his daughter, but that effort had fallen by the wayside. After a while both she and Robert had given up. His relationship with Melissa had become something completely separate from his marriage.

Still, Anne Marie felt deeply betrayed when her husband took a private matter between the two of them to his daughter. He'd been disloyal to her. Even worse was learning about it from Melissa, who'd taunted Anne Marie with what she knew. That had added humiliation to the pain.

Robert listened stoically as she wept and cried out her fury. Nothing she said seemed to affect him. He listened, his face impassive, and then a few days later, packed a bag and moved out. Just like that.

The shock of it had left Anne Marie reeling for weeks. After a month in which she refused to give him the satisfaction of calling, Robert had briefly returned to the house to suggest a legal separation.

Remaining as unemotional as possible, Anne Marie had agreed. Perhaps living apart would be best while they both considered their options. By then, Anne Marie had been angry. Okay, furious. She'd wondered if Robert had ever really loved her. How selfish, how unfair, how…male of him.

Anne Marie felt it was imperative that Robert know she was serious about a baby. He'd moved out of the house and, following his lead, she'd moved out, too, leaving the place to sit vacant. Fortunately she had the apartment above the bookstore, which had recently become available. She hoped such a drastic action would give Robert notice that she was more than able to support herself— more than capable of living her life without him. In his own fit of defiance, Robert had listed the house, which was in his name. Everyone was surprised when it sold the first week. Anne Marie's things, whatever she hadn't moved to the apartment, had been taken to a storage unit. It had all been so petty, so juvenile.

Their separation had become a battle of wills, each of them intent on showing how unnecessary and superfluous the other was. They were clearly destined for the divorce court, until Anne Marie decided enough was enough. After all, this was the man she loved. Despite everything—her disappointment, her anger toward Melissa—her feelings for her husband hadn't changed. The day she called Robert at the office had been a turning point. She admitted she missed him and was sorry the situation had deteriorated so far. He seemed surprised to hear from her and at the same time delighted. He said he was sorry, too, and they'd agreed to meet for dinner.

The one stipulation was that there be no talk about Anne Marie having a baby. Although she didn't like it, she'd promised. Dinner was wonderful and Robert had gone out of his way to make the evening as romantic as possible.

Robert Roche could certainly be charming when he put his mind to it, and that night he'd charmed himself right into her bed. Their lovemaking had always been powerful and it felt so wonderful to be with him again. Then, in the morning when she awoke, Anne Marie discovered he'd left during the night. That was like a slap in the face. It would serve him right if she ended up pregnant, she'd thought angrily.

Only she hadn't.

They'd continued to meet and to talk regularly but that was the last time they'd made love.

Shaking her head, trying to free herself from the memories, Anne Marie realized she'd been standing in front of the elementary school for ten minutes without moving. Making a determined effort, she walked into the building.

She had an appointment with the school counselor, Ms. Helen Mayer, at ten-thirty and she was already five minutes late.

As soon as Anne Marie entered the school, the hallway immediately filled with noisy youngsters, all of them trying to get past her and outside. But for the first time that day, the sun peeked out through dark clouds, and she took that as a favorable sign.

Eventually Anne Marie located the school office, which had a small waiting area, a large counter that stretched across the room and a number of offices behind it.

"May I help you?" the woman at the counter asked.

"I'm Anne Marie Roche. I have an appointment with Ms. Mayer."

"You're here for the Lunch Buddy program?"

"That's right." Anne Marie nervously brushed her hair away from her face. She wore it straight, shoulder-length, and had dressed in wool slacks and a white turtleneck sweater. Now that she was actually at the office, her uncertainty returned. She wasn't convinced this was the best project for her, wish list or not. She didn't know anything about children of elementary-school age, or any age for that matter. Her experience with Melissa hadn't exactly inspired confidence in her ability to relate to kids.

"Ms. Mayer is meeting with the other volunteers in Room 121," the woman told her. "There's an orientation first."

"Okay," Anne Marie said with a nod, figuring the

orientation would help her decide. "How do I find Room 121?"

"It's easy. Just go out the way you came in, take a left and follow the hallway to the end."

"Thank you."

"You're welcome," the secretary mumbled as she turned back to her computer screen.

Mentally repeating the directions, Anne Marie stepped out of the office. For a moment she hesitated, thinking she could just leave now, simply walk out. She didn't know any young children and couldn't imagine what they'd want to talk about. But her hesitation was brief. The prospect of confessing to Elise that she hadn't even tried compelled Anne Marie to go to Room 121.

Two other women and one man were already seated on metal folding chairs at a long conference table. There was a chalkboard behind them. Helen Mayer welcomed her with a gesture toward an empty seat.

"You must be Anne Marie," she said. "Meet Maggie, Lois and John."

Anne Marie nodded in the direction of the other volunteers and pulled out a chair. She still felt the urge to make an excuse and walk out. She couldn't, though. Not without at least going through the orientation.

"I believe that's everyone," Helen said, reaching for a piece of chalk. She walked over to the board and wrote each person's name.

During the next thirty minutes, Anne Marie learned that this was a four-month commitment. She must agree to meet faithfully with her lunch buddy once a week for that period of time.

"Every week?" one of the other women asked.

"Yes, the same day if possible but it's understandable

if you occasionally need to change days. It's best for the children to have a sense of routine and trust that you'll be here for them."

The others all nodded. A little belatedly, Anne Marie did, too.

"Next, we ask that you eat the food from the cafeteria. Lunch Buddy kids get their lunch free, thanks to a government subsidy, but you can buy yours at a minimal charge. If you must bring in food from outside, please check to be sure the child you're paired with doesn't have any food allergies."

That was reasonable, Anne Marie thought.

"After lunch you can let the child take you to his or her classroom. Or you can go outside for recess if you prefer. The idea is to spend the entire lunch period with your assigned child."

"Do they still jump rope?" Lois asked.

Ms. Mayer nodded. "With the same rhymes we used when I was a girl."

The women exchanged smiles.

"The important thing is to interact with the child," the school counselor continued. "Get to know him or her and forge a friendship."

"What about seeing the child outside school?" This question came from Maggie, who appeared to be in her early fifties.

"That'll have to be approved by the child's parent or guardian."

Anne Marie couldn't imagine seeing the child other than inside the protected walls of the school. She didn't want to get emotionally attached. Besides, that wasn't part of the deal. All that was required was to come in and have lunch with her young charge. If he or she wanted to show off school assignments, fine. But that was the limit of what Anne Marie could handle. She had enough to cope with; she didn't need to add anything else to the mix. Any relationship with an at-risk child would have to remain casual. Nothing beyond the most basic obligations.

The orientation meeting took the full half hour. Several additional questions were asked, but Anne Marie only half listened. While the others chatted, she struggled, asking herself over and over if this was the right volunteer program. She couldn't imagine why Elise seemed to think she'd be a perfect Lunch Buddy. Anne Marie didn't feel perfect. What she felt was…nothing. Nothing at all. Zoned out. Emotionally dead. Disinterested.

Ms. Mayer handed out the assignments, leaving Anne Marie for last. She must have sensed her doubts because she asked, "Do you have any further questions?"

Anne Marie shook her head. "Not really. I'm just wondering if I'm really a good candidate for this."

"Why not give it a try? I suspect you'll enjoy it. Almost everyone does."

The other woman's reassurance warmed her. "Okay, I will."

"The child I have in mind for you is named Ellen Falk," she went on to say. "Ellen is eight years old and in second grade. Because of the Right to Privacy Laws, I'm not allowed to reveal any details about her home background. However, I can tell you that Ellen is currently living with her maternal grandmother."

"Has she been in this school long?"

"Ellen's been a student here for the past two years."

"Okay."

Before Anne Marie could ask why the school counselor had decided to pair her with this particular child, Helen Mayer continued. "Ellen is an intense child. Very quiet.

Shy. She doesn't have a lot to say, but don't let that discourage you."

"Okay," Anne Marie said again.

"Talk to her and be patient. She'll speak to you when she's ready."

Oh, great. She'd have to carry the entire conversation for heaven only knew how many weeks. "Is there a reason you decided to match me up with this child?" she asked. Surely there was another one, another little girl who was more personable. Anne Marie wasn't much of a talker herself these days, and she wasn't sure that pairing her with an intense, reticent child would work.

"That's an excellent question," Helen Mayer said approvingly. "Ellen loves to read, and since you own Blossom Street Books...well, it seemed to be a good fit."

"Oh."

"Ellen is one of our top second-grade readers."

Rather than suggest being paired with a different child, Anne Marie decided to go ahead with this arrangement. "I look forward to meeting her," she said, wincing inwardly at the lie.

"Ellen has first lunch, which starts in a few minutes, so if you'll come with me, I'll introduce you."

Anne Marie still wasn't convinced she was ready for this. However, it was now or never. Once she walked out of Room 121, Anne Marie knew that unless she met the child immediately, she wouldn't be back.

Ms. Mayer led her down the hallway to a row of classrooms, each door marked with the grade and the teacher's name. Ellen was in Ms. Peterski's class. Helen Mayer waited until a young woman—obviously Ms. Peterski—and twenty or so children had filed out, then walked inside, Anne Marie a few steps behind her.

The first thing Anne Marie noticed was how impossibly small the desks were. The second was the child sitting in the far corner all alone. Her head was lowered, and her stick-straight hair fell forward, hiding her eyes.

"Ellen," the school counselor said, her voice full of enthusiasm. "I want you to meet your Lunch Buddy."

The little girl, dressed in dirty tennis shoes, jeans and a red T-shirt, slid out of her chair and moved toward them, her gaze on the floor.

"Anne Marie, meet Ellen."

"Hello, Ellen," Anne Marie said dutifully. She kept her voice soft and modulated.

Ellen didn't acknowledge the greeting.

After an awkward silence, Ms. Mayer spoke again. "Ellen, would you please escort your guest to the lunchroom?"

In response Ellen nodded and walked quickly out of the room. She stood outside the door until Anne Marie caught up.

"That's a nice T-shirt you're wearing," Anne Marie said, testing the waters. "Red is one of my favorite colors."

No response.

The noise from the cafeteria grew louder as they made their way down the hall. Ellen joined the other students in the lunch line and Anne Marie stood behind her.

"What's for lunch today?" Anne Marie asked.

Ellen pointed to one of the students at a nearby table, spooning macaroni and cheese into her mouth. "That."

At last! The eight-year-old actually had a voice.

The line started to move. "Macaroni and cheese used to be one of my favorite lunches," Anne Marie said. "Do you like it, too?"

Ellen shrugged.

"What's your favorite?"

She expected the universal response of pizza. Instead Ellen said, "Chili and corn bread."

"I like that, too." Well, she didn't hate it, but it wasn't one of Anne Marie's favorites. Thus far they didn't seem to have a lot in common.

Their lunch consisted of macaroni and cheese, a gelatin salad, carrot sticks, milk and an oatmeal cookie. Carrying her tray, Anne Marie followed the girl to a table near the back of the room. Ellen chose to sit at the far end, away from the other children.

Anne Marie set her tray across from Ellen, then pulled out her chair and sat down. Ellen bowed her head and folded her hands on her lap for a silent moment before she reached for her silverware. Apparently she was saying grace before eating her lunch.

Anne Marie took a sip of milk once Ellen had taken her first bite. "I understand you like to read," she said conversationally.

Ellen nodded.

"I own a bookstore. Have you read any of the Harry Potter books?"

Ellen shook her head. "My grandma said they're too advanced for me. She said I could read them in fourth grade."

"Your grandmother's probably right."

Ellen crunched down on a carrot stick.

"Who's your favorite author?" Anne Marie asked, encouraged by the girl's response.

Ellen swallowed. "I like lots of authors."

Again, this was progress. Of a sort. *And* the girl didn't talk with her mouth full, which meant she'd been taught some manners.

"When I was your age, books were my best friends." Anne Marie could recall reading in her bedroom with the door closed to drown out the sound of her parents arguing.

That comment didn't warrant a response. Anne Marie took another bite of her lunch as she mentally sorted through potential topics of conversation. It was hard to remember what she'd liked when she was eight. She didn't think Ellen would be interested in hearing about her widowed friends or her list of Twenty Wishes.

They continued to eat in silence until an idea struck Anne Marie. "Do you like dogs?"

Ellen nodded vigorously.

"I have a dog."

For the first time since they'd sat at the table, Ellen looked up. "A boy dog or a girl dog?"

"A boy. His name is Baxter."

"Baxter." A hint of a smile flashed in her eyes.

Anne Marie felt a surge of relief. She'd hit pay dirt. Ellen liked dogs. "He's a Yorkshire terrier. Do you know what kind of dog that is?"

Ellen shook her head.

"Baxter is small but he has the heart of a tiger. He's not afraid of anything."

Ellen's eyes brightened.

"Would you like to meet him one day?"

Ellen nodded again. "What color is he?"

"Mostly he's black but his face is sort of a tan, and he has funny-looking ears that stick straight up."

"My ears stick out, too," Ellen said in a solemn voice.

Anne Marie studied the child. She could see the faint outline of Ellen's ears beneath her straight hair, which hung just below her chin. "I had ears like that when I was your age," Anne Marie told her. "Then I grew up and my ears stayed the same size and everything else got bigger."

Ellen took another bite of her macaroni and cheese.

Anne Marie did, too. She finished the lunch period by telling the girl stories about Baxter. Ellen asked dozens of questions and even giggled once.

The other children gradually left the lunchroom, drifting out to the schoolyard. The muted sound of their play could be heard through the windows. Anne Marie looked out several times; when she asked if Ellen wanted to go outside, the youngster declined.

The bell finally rang, signaling the end of lunch. Ellen stood.

So did Anne Marie.

Ellen carried her dirty tray to the kitchen and showed Anne Marie where to place it.

"I guess you have to go back to class now," Anne Marie said.

Ellen nodded. Anne Marie walked her to the classroom door and just as she was about to leave, Ellen whispered something she couldn't quite hear.

"What did you say?" Anne Marie asked.

Ellen glanced up. "Thank you," she said more loudly.

"You're welcome, Ellen. I'll see you next Wednesday."

Ellen smiled, then quietly entered the room and walked to her desk.

As Anne Marie watched, her chest constricted with a sensation that felt alien to her. It was a good feeling, though—one that came from reaching out to someone else.

Elise was right; Anne Marie did feel better for volunteering. Little Ellen Falk needed a friend.

The ironic thing was that Anne Marie needed one even more.

Chapter
4

After leaving Woodrow Wilson Elementary, Anne Marie ran a few errands in the neighborhood. She bought groceries, went to the post office and picked up some dry cleaning. Her Wednesdays were generally crowded with appointments and chores.

When she brought the groceries up to her apartment, she noticed that the light on her answering machine was flashing. After greeting a sleepy Baxter and putting the perishables in the refrigerator, she grabbed a pen and pad and pushed the message button.

The first one was from the school counselor. "Anne Marie, this is Helen Mayer. I wanted to see how everything went with Ellen. If you have any questions, please feel free to contact me at the school." She then repeated the phone number. "See you next Wednesday."

The second message began. "Anne Marie—" Melissa Roche's voice stopped Anne Marie cold.

"Could you call me at your earliest convenience?" Her question was followed by a slight hesitation. "It's important."

The recording ended with Melissa reciting her phone number. "This is a new number. If I don't hear from you by the end of the day, I'll call the bookstore."

That sounded almost like a threat.

Anne Marie wondered about Melissa's request as she finished putting the groceries away. When she was done, she tentatively reached for the phone. If Melissa was seeking her out, it had to be something serious, although she couldn't imagine what. The call connected and the phone rang twice. Anne Marie was hoping for a reprieve. She didn't get one.

"Hello," Melissa answered. Her voice seemed clipped, defensive.

"This is Anne Marie," she said, trying to keep her own voice as unemotional as possible.

"I know who it is," Melissa said. "I have Caller ID."

"You left a message for me," Anne Marie reminded her. The enmity between them remained, despite the fact that Robert was gone.

"I need to talk to you," Melissa told her.

"I'm free now." Anne Marie would rather get this over with.

"I mean, I need to talk to you face-to-face."

That was exactly what Anne Marie had hoped to avoid. Naturally, she was suspicious of Melissa's sudden need for a meeting. "Why?"

"Anne Marie, please, I wouldn't ask if it wasn't necessary."

She exhaled slowly. "All right. When?"

"What about tomorrow night? We could meet for dinner...."

"I close the store on Thursday nights. It would have to be after eight."

"What about Friday night then?" Melissa suggested.

"Okay." Anne Marie knew her reluctance must be evident. She could think of a dozen ways she'd rather spend Friday evening than sitting across a table from her stepdaughter.

Melissa chose a restaurant and they set the time. The conversation ended shortly thereafter, and when she put the phone back, Anne Marie felt queasy. Everything about their short conversation had unnerved her. She hated going into this meeting with Melissa so unprepared, but then it occurred to her that perhaps Brandon knew what was going on. She hadn't spoken to her stepson in a few weeks, and this was a good excuse to catch up with him. She hoped he could clear up the mystery; if he had any idea why Melissa had contacted her after all these months, he'd certainly tell her.

Anne Marie opened a drawer in the kitchen and removed the telephone directory, then flipped through the pages until she found her stepson's work number.

Brandon answered immediately, obviously pleased to hear from her.

"Anne Marie! How are you doing?" he asked. Although Robert had been especially close to Melissa, the relationship between father and son was often strained.

"I'm fine. How about you?"

"Good. Good. What can I do for you?"

Brandon was a claims adjuster for an insurance company and she was well aware that he didn't have time to waste on idle chitchat.

"I heard from Melissa this afternoon."

"Melissa called you?" That was strange enough to instantly get his attention. "What did she want?" he asked curiously.

"To talk to me, or so she says. We're meeting for dinner. Can you tell me what that's about?"

"Melissa called you?" Brandon repeated. He seemed completely at a loss. "I couldn't begin to tell you what she wants."

Anne Marie sighed. "I can't figure it out, either. She insists we talk face-to-face."

"Would you like me to give her a call?" he asked.

"No, that's okay. I'll find out soon enough." Whatever it was didn't appear to involve Brandon.

"Let me know what's up, will you?"

"You haven't heard from her?" Brandon and Melissa had always been fairly close, even though he openly disapproved of his sister's attitude toward Anne Marie.

"Not in a couple of weeks, which isn't like her. After Dad died, I heard from her practically every day. Lately, though, she's been keeping to herself."

"You haven't called her?"

"I've left her a couple of messages. Apparently she's been spending all her time with that guy she's seeing. If I'm reading the situation right, it sounds like she and Michael are serious."

"Is that good news or bad?" Anne Marie asked.

"I think it's good. I like Michael and as far as I can tell, he really cares about Melissa."

"So you've met him?"

"Yeah, a couple of times. He came to Dad's funeral."

Anne Marie had been too grief-stricken to remember who'd been there; not only that, Michael would've been a stranger to her, one among many.

Was Melissa planning to confide in her about this young man? Hard to believe, but Anne Marie's curiosity was even more pronounced now.

She replaced the phone, staring out the kitchen window onto the alley behind Blossom Street. She'd just have to wait until Friday to learn the reason for Melissa's phone call.

On Friday, Anne Marie got to the restaurant shortly before the predetermined time of seven. Based on past experience, she expected Melissa to be late; that was usually the case, especially if the event happened to include Anne Marie—like dinner at her and Robert's house or a holiday get-together. It was yet another way she displayed her complete lack of regard for her stepmother. But when Anne Marie arrived Melissa was already there, pacing outside the restaurant. Anne Marie was shocked, to say the least.

Melissa had suggested a well-known seafood place on the waterfront close to Pike Place Market. Walking fast, it was about twenty minutes from the bookstore, and Anne Marie had worn an extra sweater against the cold wind coming off Elliot Bay.

Her stepdaughter abruptly stopped her pacing the moment she saw her. Because of their long, unfortunate history, Anne Marie didn't—couldn't—lower her guard. She'd been sucker punched too many times by some slyly cruel comment or unmistakable slight.

"Hello, Melissa," she said, maintaining a cool facade. "You're looking well." Her stepdaughter was an attractive woman, tall and willowy in stature. Her hair was dark and fell in soft natural curls about her face. She was wearing black jeans and an expensive three-quarter-length khaki raincoat. Even as a girl, she'd been almost obsessed with fashion and appearances, an obsession her father had indulged.

"You look good, too," Melissa said carelessly. "Are you dating anyone?"

Anne Marie bit her tongue. "No. If that's what you want to talk about, I think I should leave now."

"Calm down, would you?" Melissa snapped. "This doesn't have anything to do with you dating."

The derisive, scornful attitude was there in full display, and Anne Marie wondered why she still tried. Her stepdaughter seemed unreachable—by her, anyway—and had been from the day they met.

"I…I shouldn't have asked," Melissa murmured in what might have passed for an apology if her voice hadn't held the same level of hostility. "It isn't really any of my business."

"Shall we go inside?" Anne Marie said. The wind was growing stronger, and the rain seemed about to start any minute.

"Yes," Melissa agreed, moving quickly to the door.

Melissa had made a reservation, and they were soon seated at a table by the window. The water was as dark as the sky but Anne Marie gazed out at the lights, dimly visible in the fog. Then she turned to her menu. She and Melissa both seemed determined to make a thorough study of it. With her nerves on edge, Anne Marie didn't have much of an appetite. She decided on clam chowder in a bread bowl and when the server came, she was surprised to hear Melissa order the same thing.

"I'd like some coffee, too," Melissa told him.

"I would, as well."

Once the waiter had left, Melissa nervously reached for her linen napkin, which she spread carefully across her lap. Then she rearranged her silverware.

"Are you ready to tell me what this is about?" Anne Marie asked. Any exchange of pleasantries was pointless.

There was a pause. "It's probably unfair to come to you

about this," Melissa finally said, "but I didn't know what else to do."

Anne Marie closed her eyes briefly. "Rather than hint at what you want to say, why don't you just say it?"

Melissa placed her hands in her lap and lowered her head. "I…I haven't been doing well since Dad died."

Anne Marie nodded. "I haven't, either."

Melissa looked up and bit her lip. "I miss him so much."

Anne Marie tried to swallow the sudden lump in her throat. "Me, too."

"I thought if I went into his office and talked to his friends I'd feel better."

The waiter brought their coffee, and Anne Marie welcomed the distraction. She could feel tears welling up and she didn't want the embarrassment of crying in front of Melissa.

When they were alone again, Melissa dumped sugar in her coffee. "Like I said, I decided to stop by the office," she muttered, scooping up three tiny half-and-half cups and peeling away the tops. "Dad was always so proud of his role in the business."

Robert had every right to be proud. He'd worked for the data storage business almost from its inception and much of the company's success could be attributed to his efforts. He enjoyed his job, although the demands on his time had increased constantly. For three consecutive years, Robert had planned to take Anne Marie to Paris for their wedding anniversary. Each year he'd been forced to cancel their vacation plans because of business.

"Everyone must've been happy to see you," Anne Marie commented politely.

Melissa shrugged. "Even in this short amount of time, there've been a lot of changes."

That was understandable. Robert had died almost ten months ago, and life had a way of creeping forward, no matter what the circumstances.

"Do you remember Rebecca Gilroy?" Melissa asked.

"Of course." The young woman had been Robert's personal assistant. As Anne Marie recalled, Rebecca had started working for the company a year or so before Robert's heart attack.

"She had a baby."

"I didn't know she was pregnant." Had she learned of it, Anne Marie would've sent her a gift. She'd only met Rebecca on a few occasions, but she'd liked her.

"She isn't married." Melissa's gaze held hers.

Anne Marie didn't consider that significant. "It's hardly a prerequisite these days."

Melissa picked up her coffee and Anne Marie noticed that her hands were trembling.

"Do you remember exactly when you and my dad separated?"

Anne Marie expelled her breath. "It's not something I'm likely to forget, Melissa. Of course I remember. He…left on September 18th the year before last." She lifted her shoulders as she took in a deep breath, feeling raw and vulnerable. "I was miserable without your father. I still am." She wasn't sure where this conversation was leading and strained to hold on to her patience. Exhaling, she added, "Despite the fact that you dislike me, we've always had something very important in common. We both loved your father."

Melissa didn't acknowledge the comment; instead she stared down at the table. "One night a couple of months after you and Dad separated, I decided to treat him to dinner. He was working too hard and he often stayed late at the office."

That was a fairly typical occurrence throughout their marriage. As a company executive, Robert put in long hours.

"I picked up a couple of sandwiches and some of his favorite soup and went over there to surprise him."

Anne Marie nodded patiently, wondering when her stepdaughter would get to the point.

"The security guard let me in and when I walked into the office…"

The waiter approached the table with their order; Melissa stopped talking and even seemed grateful for the intrusion.

Anne Marie took her first taste, delicious despite her lack of appetite. Realizing Melissa hadn't continued, she gestured with her spoon. "Go on. You walked into the office and?"

Melissa nodded and reluctantly picked up her own spoon. "Rebecca was there, too."

"Mandatory overtime was one of the job requirements."

"She wasn't exactly…working."

Anne Marie frowned. "What do you mean?"

Melissa glared at her then. "Do I have to spell it out for you?" she demanded. "If you're going to make me say it, then fine. Rebecca and my father were…they were having sex."

Anne Marie's spoon clattered to the floor as the shock overwhelmed her. Her body felt mercifully numb, and her mind refused to accept what she'd heard. It was like the day the company president had come to the bookstore to personally tell her Robert had died. The same kind of dazed unbelief.

"I'm sorry, Anne Marie," Melissa whispered. "I…I shouldn't have been so straightforward, but I didn't know how else to say it."

Melissa's words had begun to fall together in her mind. Robert and Rebecca sexually involved. Rebecca pregnant and unmarried. *Rebecca had a child.*

Anne Marie could no longer breathe.

"Rebecca's baby…"

Melissa's eyes held hers. "I'm not positive…but I think so. You know her better than I do. I only saw her the one time…with Dad, and then when I stopped by the office recently. I…I had the impression that she isn't the type to sleep around. Oh, and she was at the funeral."

Anne Marie closed her eyes and shook her head. All of a sudden, the few spoonfuls of soup she'd managed to swallow came back up her throat. Grabbing her napkin, she held it over her mouth and leaped from her chair. She weaved unsteadily around the tables, then bolted for the ladies' room and made it inside just in time. Stumbling into a vacant stall, Anne Marie was violently ill. When she finished, she was so weak she couldn't immediately get up.

Melissa was waiting for her as she came out of the stall and handed her a dampened paper towel. Tears had forged wet trails down the younger woman's cheeks. "I'm so sorry…I shouldn't have told you. I…I had no idea what else to do."

Anne Marie held the cold, wet towel to her face with both hands. Shock, betrayal, outrage—all these emotions bombarded her with such force she didn't know which one to react to first.

"I should've talked to Brandon," Melissa whispered, leaning against the wall. She slid down until she was in a crouching position. "I shouldn't have told you...I shouldn't have told you."

A waitress came into the ladies' room. "Is everything all right?" she asked, looking concerned. "The manager asked me to make sure there wasn't anything wrong with your dinner."

As Melissa straightened, Anne Marie tried to reassure the woman that this had nothing to do with the food. "We're fine. It wasn't the soup...it's nothing to worry about."

"There'll be no charge for your dinners."

"No, please. I'll pay." The anger had begun to fortify her now, and she washed her hands with a grim determination that was sure to kill any potential germs.

Melissa waited for her by the washroom door, following her back to the table. Anne Marie scooped up her purse and slapped two twenty-dollar bills down on the table. That should more than cover their soup and coffee. Like a stray puppy, her stepdaughter trailed her outside, a foot or two behind.

The rain had begun in earnest by then and was falling so hard large drops bounced on the sidewalk. Anne Marie flattened herself against the side of the building while she struggled to comprehend what she'd heard. It seemed impossible. Unbelievable.

It *couldn't* be right. Robert would never risk getting Rebecca pregnant. Even the one night they'd spent together— She froze. They hadn't used protection. She'd told him she was off her birth control pills and it was as if it no longer mattered to him. His lack of concern had thrilled Anne Marie. She saw it as the first crack in his stubborn unwillingness to accept her need for a baby.

"Anne Marie…" Melissa choked out her name. The tears ran down her stepdaughter's face, mingling with the rain. Her hair hung in wet clumps but she didn't seem to notice. "Someone needs to talk to Rebecca—to ask her…"

"Not me."

"I can't," Melissa wailed.

"Why not?" she asked. "What difference does it make now?"

"If the baby's Dad's, then…then it's related to me. And if that baby really *is* Dad's, then…I have to know. I've got a right to know."

Anne Marie wondered if Robert's daughter would have been as tolerant toward a child she might have had. "Did Rebecca—did she have a boy or a girl?"

"A boy."

The pain was as searing as a hot poker against her skin. It took her a moment to find her voice. "If the child is Robert's, why hasn't Rebecca said anything?"

"I…I don't know," Melissa whispered. "I shouldn't have told you…."

"You wanted to hurt me," Anne Marie said coldly.

"No!" Melissa's denial was instantaneous.

"There's no love lost between us." Anne Marie had no illusions about her stepdaughter's motives. "You don't like me. You never have. All these years you've been try-ing to get back at me, to *punish* me, and now you have."

Not bothering to deny the accusation, Melissa buried her face in her hands and started to weep uncontrollably. "I'm sorry, so sorry."

Anne Marie wanted to turn her back on Robert's daughter and walk away. But she couldn't bear to hear Melissa weep. Even though *she* was the one Robert had betrayed, Anne Marie reached for his daughter and folded her arms around Melissa.

The two women clung together, hardly aware of the people scurrying by.

Anne Marie's reserve broke apart and the pain of Robert's betrayal came over her in an explosive, unstoppable rush. She wept as she never had before, even at Robert's funeral. Her shoulders heaved and the noisy, racking sobs consumed her.

Then it was Melissa who was holding her, comforting her. After all the years of looking for common ground with her stepdaughter, Anne Marie had finally found it.

In her husband's betrayal.

Barbie Foster stood in line at the movie theater multiplex, waiting to purchase a ticket, preferably for a comedy. She needed a reason to laugh. Her day had started early when she opened Barbie's, her dress shop, two blocks off Blossom Street. The shop was high-end, exclusive and *very* expensive. Her clientele were women who could easily afford to drop four figures on a dress. Barbie made sure they got their money's worth, providing advice, accessories and free alterations. She had a number of regular customers who counted on her for their entire wardrobes. Her own sense of style had served her well.

She didn't want to sound conceited, but Barbie was aware that she was an attractive woman. Since Gary's death, she'd received no shortage of attention from the opposite sex. Men wanted a woman like her on their arm— and, she suspected, they wanted her money. Barbie, however, wasn't easily swayed by flattery. She'd been happy in her marriage and had loved her husband. At this point in her life she wasn't willing to settle for mere companionship or, heaven forbid, no-strings sex. She wanted *love*. She longed for a man who'd treat her like a princess the way Gary had. Her friends told her that was a dated attitude; Barbie didn't care. Unfortunately there weren't many princes around these days.

She'd married young. In retrospect she recognized how fortunate she'd been in finding Gary. She'd had no real life experience, so the fact that she'd met a really wonderful man and fallen in love with him was pure luck. He was ten years her senior; at thirty, he'd had a wisdom beyond his years and a great capacity for love, for loyalty. He'd been working for her father at the time and came to the house often. She'd had a crush on him that developed into genuine love, although it took her a few years to recognize just how genuine it was. At nineteen, she always made sure she happened to be around whenever he stopped by, and enjoyed parading through the house to the pool—in her bikini, of course. She still smiled at the way Gary had looked in every direction except hers.

They'd married when she was twenty-one, with her father's blessing and, surprisingly perhaps, her mother's. She got pregnant the first week of their honeymoon. When she'd delivered identical twin sons, Gary had been over the moon. The pregnancy had been difficult, however, and he'd insisted the two boys were family enough.

The twins, Eric and Kurt, filled their lives and they were idyllically happy. Not that she and Gary didn't have their share of differences and arguments, but they forgave each other quickly and never confused disagreement with anger. Their household had been calm, orderly, contented. The plane crash ended all that. Barbie had always been close to her sons, but following the tragic deaths, of Gary and her father, the three of them were closer than ever.

They helped one another through their grief, and even now they talked almost every day.

Encouraged by her mother and sons, a year after Gary's death Barbie started her own business. The dress shop helped take her mind off her loneliness and gave her purpose. Her sons were eighteen and growing increasingly independent. They'd be on their own soon. As it happened, they were attending colleges on the opposite side of the country. Swallowing her natural instinct to hold on to her children, she flew out to Boston and New York with her sons, got them settled in their respective schools and then flew home. She'd wept like a baby throughout the entire five-hour flight back to the West Coast.

Her house seemed so empty without the boys—her house and her life. She'd never felt more alone than she had since last September when she'd accompanied Kurt and Eric to their East Coast schools. Thankfully, though, they'd both come home for Thanksgiving and Christmas.

She'd kept herself occupied with the shop, but the Valentine's get-together with the other widows had revealed a different kind of opportunity. Barbie had begun to compose her list of Twenty Wishes, hoping to discover a new objective, some new goal to pursue. Her mother had leaped at this idea with an enthusiasm she hadn't shown in years, and if for no other reason, Barbie had followed suit. They often did things together and, in fact, her mother was Barbie's best friend.

The line moved. Barbie approached the teenage cashier and handed her a ten-dollar bill.

"Which movie?"

Barbie smiled at her. "You decide. Preferably a comedy."

The girl searched her face. "There are three or four showing. You don't care which one?"

"Not really." All Barbie wanted to do was escape reality for the next two hours.

The teenager took her money and a single ticket shot up, which she gave Barbie, along with her change. "Theater number twelve," she instructed. "The movie starts at four twenty-five."

Although she wasn't hungry, the instant Barbie stepped into the lobby, the scent of popcorn made her mouth water. She purchased a small bag and a soft drink, then headed for theater number twelve.

The previews were underway, and Barbie quickly located a seat in a middle row. She settled down with her popcorn and drink, dropping her purse in the empty seat beside her.

Glancing about, Barbie saw nothing but couples, most of them older and presumably retired. She nibbled on her popcorn and all at once her throat went dry. The entire world seemed to be made up of people in love. She envied the other women in the audience their long-lasting relationships, their forever loves, which was what she and Gary should have had. She wanted another chance. She was attractive, well-off, a nice person—and alone. Falling in love again was first on her list of wishes. But she didn't want another relationship unless she could find a man like Gary and there didn't seem to be many of those.

Until the other widows had started talking about those stupid wishes, Barbie's life had seemed to be trudging along satisfactorily enough. Her mother's list was nearly complete. Not Barbie's. She'd written down a few things besides falling in love. She wanted to learn how to belly dance. She and Gary had seen a belly dance performance during a brief stopover in Cairo years before and she'd been intrigued by the sensuous, feminine movements. She'd listed something else, too. She wanted to go snorkeling in Hawaii and shopping in Paris and sightseeing in London—all of which she'd done with Gary and enjoyed. But she didn't want to do them alone.

At the moment, her desire to fall in love again seemed an illusion beyond her grasp. But she wasn't exactly *looking* for a relationship. If she truly wanted to love and be loved, she had to be receptive to love, open to it, willing to risk the pain of loss.

She shook her head, telling herself there was no point in believing that a man might one day love her the way Gary had. Love *her*. Not her money, not her beauty. Her.

All of a sudden tears welled in her eyes and she dashed them angrily away. She didn't have a thing to cry about. Not a single, solitary thing. Dozens of women, hundreds of them, would envy her life. She had no money problems, her children were responsible adults, and at forty she didn't look a day over thirty. The tears made no sense whatsoever, and yet there was no denying them.

Reaching for her purse, Barbie pulled out a pack of tissues, grabbed one and loudly blew her nose.

The previews for upcoming features were still flashing across the screen. They were apparently comedies because the audience found the clips amusing. Sporadic laughter broke out around her.

Sniffling and dabbing her eyes, she noticed a man in a wheelchair approaching the row. He was staring at her, which wasn't uncommon. Men liked to look at her. Only it wasn't appreciation or approval she saw in his gaze. Instead, he seemed to be regarding her with irritation.

Maneuvering his chair into the empty space beside Barbie, he turned to glare at her. "In case you weren't aware of it, you're sitting in the row reserved for people with wheelchairs and their companions."

"Oh." Barbie hadn't realized that, although now he'd mentioned it, she saw the row was clearly marked.

"You'll need to leave." His words lacked any hint of friendliness.

He must have someone with him and wanted the seat for that person. No wonder he frowned at her as if she'd trespassed on his personal property.

Retrieving her large purse, she draped it over her shoulder, grabbed her popcorn and soft drink and stood. Instead of walking all the way through the empty row, she tried to get past him.

In an effort to give her the necessary room, he started to roll back his wheelchair and somehow caught the hem of her pants. Barbie stumbled and in the process of righting herself, dumped the entire contents of her soft drink in his lap.

The man gasped at the shock as the soda drenched his pants and ice cubes slid to the floor.

"Oh, I am so sorry." Barbie plunged her hand in her purse for the tissue packet and managed to spill her popcorn on him as well.

"I…I couldn't be sorrier," she muttered, more embarrassed than she'd ever felt before.

"Would you kindly just *leave*."

"I—"

He pointed in the direction he wanted her to go, then shook his head in disgust.

Barbie couldn't get out of the row fast enough. Feeling like a clumsy fool, she rushed into the empty lobby. She yanked a handful of napkins from the dispenser and hurriedly returned to the theater.

The man was still brushing popcorn off his lap when she offered him the napkins.

"Can I get you anything else?" she asked in a loud whisper.

His intense blue eyes glared back at her. "I think you've already done enough. The best thing you could do is *leave me alone*."

"Oh."

He didn't need to be so rude. "I said I was sorry," she told him.

"Fine. Apology accepted. Now if it's possible, I'd like to enjoy the movie."

Barbie gritted her teeth. She felt like dumping another soft drink on his head. It wasn't as if she'd purposely spilled the soda. It'd been an accident and she'd apologized repeatedly. She felt her regret turn into annoyance at his ungracious reaction.

Because he'd made it abundantly clear that he wanted her far away, Barbie took an empty seat on the aisle five rows back from the wheelchair section. She made a determined effort to focus her attention on the movie, which had started about ten minutes earlier.

It was a comedy, just as she'd requested, only now she wasn't in any mood to laugh. Instead, she tapped her foot compulsively, scowling at the unfriendly man seated below her. When she saw that her tapping was irritating others, she crossed her legs and allowed her foot to swing. In all her life she'd never met anyone so incredibly rude. He *deserved* to have that soda dumped in his lap!

The rest of the audience laughed at the antics on the screen. Barbie might have, too, if she'd been able to concentrate. Almost against her will, her eyes kept traveling to the man in the wheelchair. The little girl in her wanted to stick her tongue out at him.

He'd asked her to move and yet no one sat next to him. In fact, the entire row was empty. He hadn't come with anyone; he just didn't want *her* sitting next to him.

What exactly was wrong with her? Lots of men would have welcomed her company. And they would've been more polite about that little accident, too. She was tempted to give that…that Neanderthal a piece of her mind. He had a lot of nerve asking her to leave. It was a free country and she could sit anywhere she darn well pleased.

Barbie left halfway through the movie, pacing the lobby in her exasperation. Where did he get off acting like such a jerk—and worse, making *her* feel like one? The teenager who'd sold her the ticket watched her for several minutes.

"Is everything okay?" she called out.

Barbie whirled around, her agitation mounting. "I was just insulted," she said, although there wasn't anything the girl could do about it. "Without realizing it, I sat in the wheelchair seating and this man told me to move."

The girl looked down, but not before Barbie caught her smiling.

"Do you think that's funny?" she asked.

"No, no, I'm sorry. You didn't have to move if you didn't want to."

"I didn't know that at the time. I assumed there was someone with him and I'd taken his or her spot."

"He was alone."

"So it seems. Furthermore, I didn't mean to spill my drink on him. It was an accident."

The girl's eyes widened. "You spilled your drink? On him?"

"In his lap."

The teenager giggled and covered her mouth with her hand. "Did he get mad?"

"Well, yes, but it was an accident. The popcorn, too."

Another giggle escaped. "Oh, my gosh."

Barbie raised her eyebrows at this girl's amusement. "I have never met a more unreasonable or ruder man in my entire life," she said pointedly.

"That's my uncle Mark," the girl explained, grinning openly now.

"He's your…uncle." Barbie seemed to leap from one fire into another. Every word she'd said was likely to be repeated to "Uncle Mark." Well, good. Someone should give that arrogant, supercilious hothead a real talking-to. Who did he think he was, anyway?

"Unfortunately, he can be a bit unreasonable," the girl said.

"Tell me about it."

"You shouldn't let him bother you."

Barbie opened her mouth to argue and then decided the girl was right. She'd paid for her ticket, the same as he had, and could sit wherever she pleased. If she chose to sit in the wheelchair area, that was her business, as long as no one legitimately needed the seat. And no one did.

"Why don't you go back in?" the girl suggested. "It's a very funny movie, you know."

"Thanks—I will." Barbie marched into the theater, determined to sit where *she* wanted.

And lost her nerve.

It just wasn't in her to create a scene. Instead she walked over to her previous seat. She slipped into it, balancing her purse on her lap, and stared at the screen. Whatever was happening in the movie bypassed her completely.

Giving up on the film, she studied the back of the man's head. He must've sensed her watching him because he shifted his position, as though he felt uncomfortable. Fine with her.

In another thirty minutes, the movie ended and the lights came on. The theater emptied, but Barbie remained in her seat. Mark whatever-his-name stayed where he was, too. When the last person had walked out, he wheeled his chair toward the exit.

"Are you always so rude?" she asked, striding after him.

He wheeled around and for an instant seemed surprised to see her.

"I'm rude when the situation calls for it," Mark informed her.

In the darkened theater Barbie hadn't gotten a good look at him. She did now and almost did a double take. The man was gorgeous. Mean as a snake, though. Gary would never have talked to a woman the way this man did. He'd always been respectful. Polite.

"I wish I hadn't apologized," she muttered. "You didn't deserve it."

"Listen, you do whatever you want. All I ask is that you stay out of my way."

"Gladly." She marched ahead with all the righteousness she could muster. But before she left the building, Barbie decided to stop at the ladies' room.

She'd just emerged when she saw Mark wheel himself into the theater lobby.

"He was pretty annoyed," his niece said in a low voice, joining Barbie.

"I told him exactly what I thought of him."

The girl smiled gleefully. "Did you really?"

Barbie nodded. "And then some." Although she was beginning to suspect she'd overreacted.

"People tiptoe around him."

"Not me." She and Gary had believed in treating people equally. Anything else was a form of discrimination, of seeing the disability and not the person.

"It's because everyone in the family feels sorry for him and he hates that."

"Oh." Well, she certainly hadn't shown him any pity—but maybe she'd been somewhat rude herself.

"I don't, though," the girl went on, "which is one reason he stops in here on the evenings I'm working."

"Does he come to the movies often?" Barbie wasn't sure what had prompted the question.

"Uncle Mark comes to the movies every Monday night." The girl held Barbie's look for an extra-long moment. "I'm Tessa, by the way, and Mark Bassett is my uncle's name." She thrust out her hand.

Barbie shook it. "And I'm Barbie."

"You'll come again, won't you?" Tessa asked.

"I live in the neighborhood." Well, sort of. It was a twenty-minute drive, but this theater was the closest multiplex in her vicinity.

"I wish you would," Tessa said, walking her to the glass doors that led to the parking lot. She held one open. "I'll see you soon, okay?"

"You will," Barbie said, removing the car keys from her purse. Sitting inside her vehicle, she let the conversation with Tessa run through her mind. Tessa was basically asking her to return the following Monday—and she'd more or less agreed. She'd need to give that some thought. She felt an undeniable attraction to this man, not to mention a sense of challenge and the exhilaration that came with it.

In fact, she hadn't reacted that strongly to anyone in…years. She didn't understand the intensity of her own response.

As she always did when she was upset or confused, Barbie phoned her mother. Lillie answered right away.

"Sweetheart, where were you?"

"I decided to go to the movies. I'm on my cell."

"I left you a message," her mother said. "I was hoping you'd come by the house and have dinner with me."

Suddenly ravenous, Barbie remembered that she hadn't eaten anything more than some toast and a few handfuls of popcorn all day.

"Thanks," she said. "Do you want me to pick anything up?"

"No, I got groceries earlier today."

"Do you have your car yet?" Barbie asked. The red-hot convertible had gone back to the dealership for the same problem as before. The shop had worked on the steering mechanism twice now.

"No, but I'm not worried."

"You're so calm about all this." Barbie marveled at her mother's patience. She hadn't complained even once.

"Is everything all right, dear?" her mother asked. "You sound agitated."

"I am, a little." Barbie went on to explain what had happened—without, for some reason, mentioning that the man was in a wheelchair. To her dismay, her mother laughed.

"Mother!" she protested. "This isn't funny."

"I know…. It's just that I can't imagine you being so clumsy."

"It was his fault," Barbie insisted. "He's just fortunate I didn't land in his lap."

Instantly a picture appeared in her mind, and to her shock, it wasn't an unpleasant one. Barbie saw herself sitting on Mark's lap, her arms around his neck, their eyes meeting, their lips... She shook her head. She didn't know where that vision had come from because the man was so...unpleasant.

"You can tell me all about it once you're here," Lillie said.

"See you in a few minutes, then." Barbie was about to snap her cell phone shut when her mother's voice stopped her. "Barbie, listen, I almost forgot. Jacqueline Donovan invited us to a small gathering next Monday. You'll be able to attend, won't you?"

"Monday?" she repeated. "What time?"

"Around six."

"Sorry, Mom," she said, making her decision. "I'm afraid I've already got plans."

Mark Bassett wasn't going to get rid of her as easily as he no doubt hoped.

Anne Marie had been in emotional free fall ever since her Friday-night dinner with Melissa. She'd tried to push the conversation from her mind but hadn't succeeded. Robert's unfaithfulness hung over her every minute of every day—the betrayal, the pain, the anger. It wouldn't hurt as much if she hadn't so desperately wanted her husband's child. For him to adamantly refuse her and then fall into bed with another woman, a woman who now had a child that might be his, bordered on cruelty.

Another complication was her stepdaughter. Anne Marie didn't want to believe that Melissa had purposely set out to hurt and humiliate her, and yet she was suspicious. Still, she felt that Robert's daughter was distressed by her father's actions and had told the truth when she said she wasn't sure where else to turn. Anne Marie didn't understand, though, why Melissa hadn't confided in her brother. Surely Brandon would've been a more natural choice. Had she come to Anne Marie because she wanted to talk to another woman? Because she knew that no one else had loved Robert as much? One thing was certain; the instant Melissa had seen how badly she'd hurt Anne Marie, she was genuinely regretful. In the end, Melissa had been the one comforting *her*.

On Sunday Anne Marie hid inside her small apartment with only Baxter for company. She didn't answer the phone, didn't check her messages. How she managed to work even half of Saturday was a mystery. At about noon, she pleaded a migraine and left the shop in Theresa's hands. Thankfully, the store was closed on Sundays.

Anne Marie didn't leave her apartment other than to take Baxter for brief walks. She wandered from room to room with a box of tissues while she vented her pain and her grief.

How could Robert have let this happen? How could he betray her in such a fundamental way? The phone rang a number of times but she didn't answer. Her display screen showed that most of the calls were from Melissa, the last person she wanted to hear from. The messages accumulated until her voice-mail box was full. Anne Marie didn't care. As far as she was concerned, the less contact with the outside world the better.

Monday she had to work again. Intuitively, her staff—Theresa Newman and a college student named Cathy O'Donnell—seemed to understand she needed space. As much as possible, she stayed in the office at the back and shuffled through mounds of paperwork. She didn't feel capable of dealing with the public.

At twelve-thirty, Theresa entered the office. "Someone out front would like to see you," she said.

"A customer?"

"Umm…" Theresa acted uncertain. "I think it might be your stepdaughter."

Anne Marie tensed. If Melissa had come to the store, it likely wasn't a social visit. After Friday, Anne Marie was wary; she felt too fragile to deal with anything else her stepdaughter might have to tell her.

"Anne Marie?" Melissa pushed her way past Theresa and stepped into the office.

Theresa cast Anne Marie an apologetic look and excused herself.

Melissa stood awkwardly in the doorway. "Why didn't you answer the phone?" she demanded. "I called and called. It was like you dropped off the face of the earth."

She would've thought the answer was fairly obvious. "I…I wasn't up to talking to anyone."

"I've been worried about you. Brandon has, too."

"You told him?"

Melissa nodded. "He was furious with me. He said… he said I should never have told you."

Harsh words trembled on the tip of her tongue. How she wished Melissa had gone to her brother first. But she supposed that eventually she would've uncovered the truth on her own. Now or later, did it really matter?

Melissa seemed close to tears. "Brandon's right." Her voice was shaky. "I'm sorry, Anne Marie. At the time I…I felt I should tell you. I knew it would shock you like it did me, but I didn't realize how hurt you'd be. I was stupid and thoughtless. I'm so sorry."

It would've been easy to dissolve into tears all over again. Anne Marie made an effort to maintain the tight control she held on her emotions. "In a way you did the right thing," she said, trying to speak calmly. "I would've needed to learn about this baby at some point."

Melissa advanced one step into the room. "I still feel terrible."

"Let's put it behind us," Anne Marie said. The girl would never know what it cost her to make that offer. Instinctively she wanted to blame her for this pain, but Anne Marie discovered she couldn't do it. After years of trying to find some kind of connection with her stepdaughter, she didn't want to destroy the tenuous one they now shared.

"Can I do something to make it up to you?" Melissa pleaded.

Anne Marie shook her head.

"Can I get you anything?" she implored next.

She drew in a deep breath. "Do you have a new heart for me?" she asked, in a tone she hoped was offhand and witty. From the sad look in her stepdaughter's eyes, Anne Marie knew it hadn't been.

"Maybe I should just go." Melissa's shoulders slumped as she half-turned to leave.

"Why don't we have tea one day soon," Anne Marie suggested.

"You'd do that?" Melissa asked in disbelief.

"You're Robert's daughter and no matter what you think of me, I loved your father," Anne Marie said, unwilling to be dishonest.

"Even now?" Melissa asked. "Knowing he betrayed you?"

Love was difficult to explain. Robert's actions had devastated her, and while she wanted to confront him, force him to own up to his betrayal, that possibility had been taken away from her. And yet…she loved him.

"Dad hurt you badly, didn't he?"

"Yes, he did. I could hate him for that but—"

"I would," Melissa cut in, eyes narrowed.

"And what good would that do?" Anne Marie asked

her. "Believe me, I've been over this time and again. I could let the news bury me—and for a while it did."

"I know.… I blame myself for that."

"Don't worry. I meant what I said about putting this behind us. Anyway, I'm dealing with everything as best I can. At first I wanted to lash out, but I couldn't see how that would help. My pain and anger aren't going to change a thing, are they?"

Melissa stared at her for a long moment. "You're a better person than I am."

"I doubt that—just a bit more experienced, a bit more broken and bruised." She'd never expected Melissa to compliment her on anything. "You're still young. Life kicks us all in the teeth sooner or later." She didn't mean to sound so negative, but at this point it was difficult not to. "I appreciate that you wanted to check up on me."

"I felt so awful about what my father did. And the news about Rebecca's baby hit me hard. So I turned to you and I shouldn't have."

Anne Marie waved one hand airily. "Like I said, I'm beyond all that." It wasn't completely true; she didn't think she'd ever recover from Robert's betrayal—and the way she'd found out.

Melissa stayed a few more minutes and then left for her afternoon class. The invitation for tea was intentionally open-ended. Anne Marie would call her when she felt more…prepared.

An hour later, she felt composed enough to meet the public again. Cathy had gone for the day, and while Theresa took her lunch break, Anne Marie handled the cash register. Mondays were generally slow and she had only two customers, neither of whom needed help. She was emotionally off-balance, although she had to admit she felt better after talking to her stepdaughter. Melissa's concern, and Brandon's, had comforted her, at least a little.

The shop door opened and Elise Beaumont came inside. Her expression was speculative, but if she noticed that Anne Marie looked pale and drawn, she didn't mention it.

"Hello, Elise," Anne Marie said, trying to act cheerful. "I've put aside a couple of new titles you might like."

"Thanks." Elise walked up to the counter. "I came to see how it went with the Lunch Buddy program last week."

She'd almost forgotten about her volunteer project. "Oh, yes. It was fine."

"Did the school pair you up with a child?"

Anne Marie nodded. "Her name's Ellen Falk and she's in second grade." It took her a moment to conjure up Ellen's face, recalling how shy and awkward the young girl had been.

Elise picked up one of the books Anne Marie placed in front of her and flipped it open. "You don't seem too enthusiastic."

"I'm not sure Ellen and I are the best match." She went on to explain how the eight-year-old had barely said a word the entire lunch period. Over the weekend she'd lost whatever optimism she'd felt at the end of their previous session.

"It's early yet. Give it time," Elise urged.

"I will." However, Anne Marie still had her doubts about the project. She'd finish out the school year but then she'd look for a different volunteer organization. "I need to call the school counselor," she said. "The only real enthusiasm Ellen showed was when I talked about Baxter."

"The child's interested in dogs?"

"I think so. I thought if I got permission to bring him

to the school, Ellen would enjoy meeting him." Baxter was a good-natured dog who seemed to do well with children, and Anne Marie had no worries about his behavior.

"That's an excellent idea."

Elise decided to buy one of the books Anne Marie had recommended, a debut novel by a former journalist, and then wandered the store for a few minutes. With her own background as a librarian, she was an avid reader and a good customer. In fact, she often knew more about books and authors than Anne Marie did. With a second purchase in hand, Elise returned to the counter.

"Something's bothering you," she announced, studying Anne Marie.

"I—I'm fine," Anne Marie insisted.

"Actually, you aren't and that brave front is crumbling fast. You need someone to talk to. I'm available."

Elise liked to get to the point. She wasn't one to ease into a subject or look for a circumspect approach. Anne Marie usually appreciated her friend's directness. Now, however, she didn't feel ready to unburden herself.

"Well?" Elise pressed.

"I…I received some shocking news last Friday," she began. "But I'm dealing with it."

Elise waited patiently for her to continue.

Anne Marie glanced over her shoulder at another customer who'd just walked in and was scanning the shelves. "I don't want to talk about it here."

Elise patted her hand. "That's understandable. We'll just wait until—"

The shop door opened, as if on cue, and Theresa came back from lunch. "The French Café has a fabulous squash soup today," she said breezily. "You should try it."

"I might do that." Anne Marie hadn't eaten much of

anything since Friday night. She was too thin as it was but she wasn't hungry, and this latest incident had contributed to her lack of appetite.

"I was thinking of having a bite to eat myself. Join me," Elise said.

It was more of a decree than a request; still, Anne Marie agreed. Elise was probably right—it would help to talk about this *and* to eat. With the glimmer of a smile she recalled Elise's advice at the Valentine's get-together. Theresa took over for her, and Anne Marie collected her purse and walked out with Elise.

"You should be wearing more than a sweater," the older woman told her.

Anne Marie shrugged half heartedly. "You're beginning to sound like my mother," she murmured.

"From the look in your eyes, I'd say you need one."

That comment brought immediate tears, which Anne Marie struggled to hide as she returned to the office for her jacket. She grabbed a tissue to wipe her nose, then tossed it in the waste basket. She certainly couldn't talk to her mother about what she'd learned. Laura Bostwick would use it as an opportunity to harangue Anne Marie about the huge mistake she'd made in marrying Robert. Laura had disapproved from the start. Trapped in her own unhappiness, she seemed to take a malevolent pleasure in destroying other people's joy.

Elise linked arms with her as they crossed the street. "You're so thin now I'm afraid a strong wind will blow you away."

"Oh, come on, Elise. Don't exaggerate."

"It's a problem I wish I had," Elise muttered. "When Maverick died, I'm afraid I buried my sorrows in food. Isn't that ridiculous, considering how closely I watched his diet?" Unexpectedly she smiled. "He said he ate like a bird—flax seed, blueberries, wheat germ... Maverick had such a delightful sense of humor. I sometimes wonder if I'll ever stop missing him." She shook her head and brought her attention back to Anne Marie.

The French Café was the most popular restaurant on Blossom Street; even now, at almost two, it was crowded with lunchtime customers.

Alix Turner, who baked all the pastries, belonged to one of Anne Marie's reader groups and often recommended the bookstore to others.

When it was their turn to order, both Anne Marie and Elise chose the squash soup. While they waited for the server to deliver their order, they sipped their coffee.

"Tell me what's wrong," Elise said.

"Why don't we wait until after we eat?" Anne Marie murmured, not eager to discuss Robert's infidelity.

Elise looked at her sternly. "Don't put it off. Whatever happened is tearing you up inside. You'll feel better if you share it—if not with me, then someone else. Frankly, I'm your best option."

Anne Marie had to laugh; some of the things Elise said verged on egotistical. Fortunately she knew the other woman well enough not to take offense.

"Let's talk about our Twenty Wishes instead," Anne Marie said. "Are you working on your list?"

"I am." Elise smiled. "I'm determined to go on a hot air balloon ride. That one's at the top of my list." She hesitated. "I have another wish...."

"Which is?"

"You promise not to laugh or try to talk me out of it?"

"Of course." There was the matter of those red cowboy boots, for one thing.

"I'm going to get a tattoo."

What? Elise? Anne Marie nearly swallowed her tongue. "Have you decided where?"

"There's a tattoo parlor near the waterfront and—"

"No, I meant where on you. Your shoulder or—"

"Oh, I'm not sure yet. Maverick had one. On his right arm." The older woman looked flustered. "But that's enough about me. Tell me what's troubling you."

Anne Marie would rather avoid the subject altogether; at the same time she was grateful for the chance to talk about it with someone she knew and trusted. She sighed. "I had dinner with my stepdaughter Friday night."

"I take it the evening wasn't pleasant."

"No… Melissa had recently gone to Robert's office and discovered that his personal assistant had a baby."

Elise straightened her shoulders. "A baby…" she repeated. "Is it Robert's?"

Anne Marie shrugged. "I'd say it's highly probable."

Elise's eyes narrowed. "But you're not sure?"

"No."

"You're going to find out, aren't you?"

"I…I don't feel it's my place to say or do anything."

"Yes, it is!" Elise said adamantly. "Who better than you? Robert was your husband."

"But…"

"And so far the identity of the father is pure conjecture."

"Well, yes, to a certain extent. Apparently while Robert and I were separated, Melissa discovered her father and Rebecca in a, uh, compromising position. Nine months later, Rebecca turns up with a baby. What else am I to think?"

Elise pursed her lips. "It does seem suspicious. The only way to know for sure is to ask her."

Anne Marie saw the wisdom of confronting Rebecca, but she couldn't do it. She wasn't convinced she'd ever have the courage to speak to her. "Brandon and Melissa are the ones who need to know."

"You, my dear, were Robert's wife. Yes, I'm aware that you'd separated. He behaved badly, and I'm positive that if he was here, he'd tell you how much he regrets everything that happened."

"He didn't want another family.... Perhaps he just didn't want children with *me*."

"Don't say that," Elise said sharply. "Don't think it, either. If Robert was alive he'd be aghast at this news."

"You never knew him."

"But I know *you*," she came right back. "From what you've told me, Robert loved you."

"I thought he did." All of a sudden Anne Marie couldn't help wondering. She'd lost Robert to a heart attack and now the vision of the man she'd so desperately loved had been destroyed. Along with it, all her dreams of the future, the hopes and promises she'd hung on to during their separation, had fizzled out to nothing.

"Don't leap to any conclusions until you talk to Rebecca yourself," Elise warned her. "No one has more of a right to the truth than you."

Elise made it sound so simple, so straightforward and uncomplicated.

The server brought the steaming soup, and Anne Marie inhaled the gingery scent. For the first time in days she felt like eating. Elise reached across the table and clasped her forearm.

"Promise me you'll contact Rebecca and ask her. Do it for yourself," Elise said.

"I can't...."

"You can and you will," the other woman insisted. "Don't you remember what Scripture says? 'The truth shall set you free,' and until you know the truth you'll be held captive by your fears and doubts."

Anne Marie merely nodded as she tasted her soup. Delicious. They'd used coconut milk, she guessed, allowing herself to be momentarily distracted.

"Find out," Elise urged again as she picked up her spoon. "Don't accept all this conjecture and half-baked information. Melissa might have misjudged the situation entirely."

"I...don't think so." Naturally Anne Marie *wanted* to believe that Robert would never cheat on her. And yet she had to be realistic, too.

"But you don't know and you won't until you speak to this woman."

Anne Marie was forced to agree.

"You'll do it, then?"

Reluctantly she nodded. Not now, though, not when the pain was still so fresh and her heart was aching.

"You won't disappoint me, will you?" Elise held her gaze for a long moment.

"No," Anne Marie promised. "I'll get in touch with Rebecca and I'll ask her."

Then and only then did Elise smile. "Remember—the truth shall set you free."

Chapter

7

The minute Anne Marie entered the school grounds at Woodrow Wilson with Baxter on his leash, she was surrounded by children, apparently out for a late recess or an early lunch. Baxter looked up at her expectantly and, fearing the small dog might be overwhelmed, she lifted him into her arms.

"I brought Baxter to visit Ellen Falk," she explained as the children gathered around.

By the time Anne Marie had walked inside, Ellen's second-grade class had been dismissed for lunch. She found the little girl waiting in the hallway by the lunchroom. She stood with her back against the wall, staring down at the floor.

"Hello, Ellen." Anne Marie spoke softly so as not to alarm her.

Ellen glanced up and when she saw Baxter, a tentative smile slid into place. "You brought him!"

"I called and Ms. Mayer said it would be all right to bring Baxter so the two of you can meet." Bending down to her constant friend and companion, Anne Marie said, "Baxter, this is Ellen."

Ellen stared at the dog. "Hi," she said and offered him her hand to sniff. "Would it be okay if I petted him?" she asked, her eyes filled with longing.

"I'm sure he'd like that."

Even with Anne Marie's permission, Ellen hesitated as she raised her hand and gently touched the top of Baxter's head. As if he understood how badly this little girl needed a friend, Baxter licked her hand.

"He likes you," Anne Marie told her. "Would you like to hold him?"

The girl's eyes grew large. "I'm allowed to do that?"

"Of course. Let me show you how to carry him." She gave Ellen a demonstration of the way she tucked Baxter between her arm and her side, then handed her the dog.

Baxter wagged his tail, and Ellen couldn't stop smiling.

"Shall we get some lunch?" Anne Marie asked. "Ms. Mayer said we'll need to take our lunches back to the classroom. Is that okay?"

"Yes." Ellen looked at her anxiously. "Baxter can come, too, can't he?"

"Of course," Anne Marie assured her.

The Yorkshire terrier attracted lots of curious attention as Ellen waited outside the busy cafeteria. "I'll get our lunch while you watch Baxter," Anne Marie said.

The menu for the day was chili with corn bread, which Anne Marie remembered was Ellen's favorite. She chose a fresh salad and canned peaches for herself. Her appetite was improving. Since her lunch with Elise the day before, she'd actually felt the faint stirrings of hunger. Talking to her friend had made her feel calmer and more rational, although Anne Marie wasn't ready to confront Rebecca yet. She would in time, as soon as she was emotionally prepared to deal with the other woman's answer.

When she'd assembled their lunch, several of the children had gathered around Ellen, asking questions about Baxter.

"I have to go now," the child told the others, and Anne Marie grinned at the importance in her voice. Ellen dutifully followed her down the hall toward the classroom, carrying Baxter as though he was the most precious burden imaginable.

The door was open and Ellen led the way to her desk. Anne Marie set the tray down and pulled up a chair next to Ellen's.

"What about Baxter?" Ellen asked, carefully putting him down. "We can't eat in front of him, can we? That would be impolite."

"Yes, it would," Anne Marie agreed. She'd brought along a small can of gourmet dog food, which was a rare treat for him. As she retrieved it from her purse, Baxter practically did flips of joy.

Ellen giggled, covered her mouth with her hand, then giggled again. "He's so funny."

"Yes, he is," Anne Marie said, smiling, too.

Ellen seemed far more interested in Baxter than in eating her own lunch. She watched Baxter wolf down his food before she turned to her own plate. "One time my mama said she'd buy me a dog."

Knowing the girl lived with her grandmother, Anne Marie wasn't sure how to comment.

"What kind of dog do you want?"

Ellen looked up from her chili. "Any kind. But Mama made lots of promises she never kept. I live with my Grandma Dolores now."

"Does she have a dog?"

Ellen shook her head. "She said she's too old to take care of a dog."

"How old is she?" Anne Marie asked.

Ellen contemplated the question. "Really, really old. I think she's over fifty."

Anne Marie managed to suppress a smile. *"That old,"* she said, exaggerating the two words. Her guess was that the woman was actually quite a bit older.

"She sleeps a lot."

"Oh."

"I don't mind. She lets me watch TV as long as I do my homework."

"She loves you very much, doesn't she?"

Ellen swallowed a bite of her corn bread. "And I love her," she mumbled.

"I'm sure you do."

"Only…only Grandma's too old to get me a dog."

"I'll be happy to share Baxter if you'd like," Anne Marie offered.

Ellen's eyes lit with pure joy. "You promise?"

"We're Lunch Buddies, aren't we?"

Ellen nodded enthusiastically. She took another spoonful of her chili and paused long enough to pet Baxter.

"We should probably take him for a walk once you finish your lunch." Anne Marie had brought a plastic bag in case nature called while Baxter was outside.

"I'm full," Ellen said decisively, planting both hands on her stomach.

"Positive?"

"Yup. Can we go now?"

They brought their trays back to the cafeteria and Anne Marie carried them inside, remembering how Ellen had instructed her the week before. When she'd put them away, the two of them walked outside.

Typical for early March in the Pacific Northwest, a fine mist was falling. Most of the children ignored the rain, as did Ellen and Anne Marie. They strolled through the yard, making for a small grassy area nearby, and Anne Marie let Ellen take the leash. A dozen kids trailed behind as if she were the Pied Piper. They would've followed Ellen and Baxter out of the yard if the recess monitor hadn't intervened.

As usual, Baxter sniffed every inch of territory as he trotted toward the gate. "How's school?" Anne Marie asked, walking side-by-side with the eight-year-old. She hoped to become better acquainted with Ellen, although it seemed she had to drag every morsel of information from the child. Thus far, all she knew was that Ellen liked dogs and lived with her grandmother, who was over fifty and therefore "really old."

"Good."

That wasn't a lot of help. "Do you have any problems with math?"

"Nope. Grandma Dolores says I'm smart."

"I'll bet you are."

Ellen seemed to have nothing else to say. She concentrated on the dog, praising him and periodically bending down to stroke his silky ears.

Anne Marie had hoped for more progress today. She'd seen some, thanks solely to Baxter, but now that she'd made a commitment to the program and to Ellen, she was eager for the next breakthrough.

"Does he know any tricks?" Ellen asked.

"Baxter? He can sit on command."

Ellen seemed pleased. She stopped walking. "Sit," she said sternly.

Her Yorkie immediately complied, and Ellen beamed. "He's smart, too."

"Yes, he is. He doesn't know how to roll over, though. I've tried, but I can't make him understand what I want him to do."

"I'll get a book from the library and teach him," Ellen instantly volunteered. "Can you bring him next Wednesday?"

Anne Marie had discussed the situation with Ms. Mayer. The point of the Lunch Buddy program was for Ellen and Anne Marie to become friends. Helen Mayer's concern was that Ellen would bond with Baxter and not Anne Marie. She'd suggested the dog only visit once a month.

"I'm afraid Baxter won't be able to come next week," Anne Marie explained. "But I'll bring him again soon."

The light seemed to go out of Ellen's eyes and she docilely accepted the news. "I like Baxter," she said a few minutes later.

"I can tell, and he certainly likes you."

Her returning smile was fragile, as if she'd long ago learned to accept disappointments.

What little conversation they exchanged after that was focused on Baxter. When the lunch bell rang, Ellen lingered on the playground.

"I'll see you next week," Anne Marie promised.

Ellen lowered her eyes and nodded.

Ellen was obviously accustomed to adults making promises they didn't or couldn't keep. Anne Marie wanted to reassure the youngster that if she said she'd be at the school, she would be, but actions spoke much louder than words. She hoped that over time Ellen would come to trust her.

"Can I hug Baxter goodbye?" she asked.

"Of course."

Crouching, Ellen petted the dog, then picked him up and gently gave him a hug. "Thank you for bringing him."

"You're welcome, Ellen."

With that, the girl raced toward the school building. She was the last one to enter and Anne Marie hoped she wouldn't be late for class.

"Well, Baxter," she murmured to her pet, "you were a real hit." As she hurried toward the parking lot, Helen Mayer stepped out of the school, walking purposefully in Anne Marie's direction.

"How'd it go with Ellen this week?" she asked, quickly catching up. She wore only a sweater and shivered as she wrapped her arms around her waist.

"Okay, I think," Anne Marie said. "Thanks mostly to Baxter."

"I thought that might be the case."

"She wanted me to bring him next week, but I told her he couldn't come and it would only be me."

"That little girl's been through a great deal," Ms. Mayer said, lowering her voice. "As I explained earlier, the privacy laws prohibit me from saying any more, but rest assured that Ellen badly needs a friend."

Anne Marie immediately felt guilty for wishing she'd found another volunteer effort. But she'd made a four-month commitment and she planned to see it through. She had no intention of being another adult in Ellen's life who broke her promises.

The school counselor walked her to the parking lot. "Next Tuesday afternoon, the school's putting on a play."

Anne Marie nodded.

"Ellen has a small role in it."

"Ellen?"

"She's in the chorus."

"How nice. She didn't mention it."

Ms. Mayer didn't seem surprised. "She wouldn't. She's such a shy child. She's gifted vocally, you know."

"It'll help her self-esteem if she excels in singing." Anne Marie didn't say that she used to sing, too. Or that she hadn't sung in months...

"Oh, I agree, this opportunity is wonderful for Ellen," Ms. Mayer said. "She's the only second-grader in the chorus."

"That's terrific." There was no personal reason for Anne Marie to feel proud of Ellen, but she did.

"Her grandmother won't be able to attend."

Despite herself, Anne Marie grinned. "Her Grandma Dolores is really old, you know. She's at least fifty."

The school counselor smiled, too. "It would mean the world to Ellen if you could come for the program."

"Me?" She thought about her Tuesday schedule. She had meetings with a publisher's rep, plus an appointment with her bookkeeper. As much as possible, Anne Marie tried to handle all her paperwork on Tuesdays.

"Ellen won't have anyone there if you don't come," the counselor continued. "If there's any chance at all..."

Anne Marie swallowed a sigh. "What time is the program?" she asked.

"Tuesday afternoon at two."

She'd been afraid of that. "I have an important meeting then. I'm sorry, but I can't make it."

Ms. Mayer didn't hide her disappointment. "Are you *sure* you can't? It would mean so much to Ellen."

Anne Marie hesitated as she tried to work out how to rearrange her schedule. Perhaps she could ask Theresa to meet with the sales rep, who was coming in at two.

"I know you're busy...."

"Everyone's busy," Anne Marie said. Her life was no different from anyone else's. She wondered if Ms. Mayer championed the other children as diligently as she did Ellen. "I'll see if I can get one of my employees to cover for me. Exactly how long will this take?"

Ms. Mayer smiled broadly. "An hour at the most. And after the program, the children will be serving cookies and juice."

"I'll see what I can do," Anne Marie promised.

"That would be wonderful." She clasped her hands together. "Thank you so much." Glancing over her shoulder, she said, "I have to get back to the school now, but thank you again."

Anne Marie opened her car door and placed Baxter inside. This volunteering was demanding more time and commitment than she'd originally assumed. The problem was, she couldn't say no.

When she returned to Blossom Street, she brought Baxter up to the apartment, then hurried into the bookstore to see Theresa. The part-time position suited Theresa well, since she had three children, one in high school and the youngest two attending junior high.

"How's it going?" Anne Marie asked.

"Fine." Theresa smiled. "I wasn't expecting to see you."

Anne Marie tried to stay as far away from the business as she could on Wednesdays. The moment she made herself available, she invariably ended up spending half her day in the store. "I'm only here for a few minutes," she said, vowing that would be the truth.

"Ms. Higgins was looking for you," Theresa told her. "She forgot this was your day off."

"Were you able to help her?"

"Oh, she wasn't interested in buying anything. She wanted to talk to you."

Anne Marie would give Lillie a call later on. She hoped the situation with her car had been settled.

"There was a phone call for you, too. A man at one of the distributors. He asked if you'd call him back at your convenience. I wrote down all the information."

"I'll take care of that in the morning," Anne Marie said. She lingered, procrastinating because she hated to ask her employees to fill in for her. "Theresa, I was wondering if you could work next Tuesday afternoon."

"Next Tuesday?" Looking pensive, Theresa bit her bottom lip. "I think so. Can I get back to you to confirm?"

"Of course. The thing is, I'm a Lunch Buddy for this little girl named Ellen and…well, she's going to be in a school production and the counselor seemed to think it would help if I could be there." She didn't know why she was rattling on about this when it really wasn't Theresa's problem.

"I'm sure it'll be fine but I'll need to check with Jeff first."

"Thanks, and if it doesn't work out, don't worry. I'll try Cathy or Steve."

Anne Marie went upstairs to her apartment. Baxter, who was asleep in his small bed, didn't so much as stir. Apparently the excursion to the school had tired him out. "Some watchdog you are, Mr. Baxter," she muttered.

Her plan for the afternoon was to work on her list of wishes. Since her dinner with Melissa, Anne Marie hadn't really given it much thought.

1. Buy red cowboy boots
2. Learn to knit

3. Volunteer—become a Lunch Buddy
4. Take French lessons

Then, because it seemed so unlikely and yet necessary, she added the first wish, the one she'd crossed out earlier.

5. Find one good thing about life

She took out the binder she'd purchased and assembled the scrap-booking supplies and the few pictures she'd already cut out. Red cowboy boots from a catalog. A hand-knit sweater from a magazine. A photo of the Eiffel Tower. She'd need to get a picture of Ellen and... Suddenly it seemed pointless to go on, in light of what she'd discovered about Robert and her own pitiful life.

Rather than allow herself to sink into further depression, she reached for her phone and called Lillie. They arranged to meet for dinner at a Thai place they both liked.

That evening Lillie arrived at the restaurant before Anne Marie did and had already secured a table. "I'm so glad you phoned," Lillie said, kissing her cheek. "I've got lots to tell you."

"I can't wait to hear."

"It's that list."

"The Twenty Wishes?" Earlier, just reading her list had depressed her. She'd been convinced she'd never feel like dreaming again, not when she'd obviously been so wrong about her entire life.

"That list's given me a whole new burst of energy," Lillie said. "I've told my friends about it and now they're all writing their own lists."

"Really?"

"Lists are big these days. Who would've believed it?" Lillie's eyes twinkled with merriment. "I've been adding to mine nearly every day, thinking about all the things I want to do. Things I haven't considered in years. It all started when I bought that red convertible."

"Speaking of which…"

Lillie waved the question aside even before Anne Marie could ask it. "Just a minor glitch and that nice man from the service department is taking care of everything."

"You mean to say you're still driving a loaner?"

"Yes, but it doesn't matter. Everything's under control and I haven't been inconvenienced in the least."

"You shouldn't be inconvenienced. You bought their car!"

Lillie studied her menu. "I'm starved. How about you?"

Anne Marie needed to think about it, then realized she actually was hungry. "I am, too."

"Great. The way I feel right now, I'm tempted to order everything on the menu. Let's begin with the assorted appetizers, and then a green mango salad.…"

"And pad thai. I love their pad thai," Anne Marie said, entering into the spirit of the evening.

Between the perfectly spiced food and Lillie's invigorating company, dinner was a welcome reprieve from the low-grade depression that had been hanging over Anne Marie. Back in her apartment a few hours later, she came across the binder and the scrapbooking supplies spread out on the kitchen table.

She sat down again and read over her list. Maybe her wishes weren't so impossible, after all.

Lillie Higgins paid extra-close attention to her makeup Friday morning, chastising herself as she did. Anyone who even suspected that she was preening and primping for the service department manager at a car dealership would be aghast.

Lillie had nothing to say in her defense. She just found Hector Silva appealing; he was kind and generous and unfailingly polite. He seemed so natural, while the men who usually set out to charm her came across as self-conscious, trying too hard to impress. Not Hector Silva. His work ethic, his dignity and decency... She couldn't praise him enough.

They'd exchanged two brief conversations, and after each one Lillie had walked away feeling good. More than good, elated. She *liked* him—it was that simple—and she enjoyed talking to him. Both times she'd wished the conversations could've been longer.

Now that her car was repaired to Hector's satisfaction, she didn't have an excuse to chat with him anymore. So she'd decided to make the most of today's encounter, which would likely be their last.

Lillie arrived at the dealership with the loaner at the precise time Hector had indicated. She wore a pink linen pantsuit with a silk floral scarf tied around her head. She'd struggled with that, not wanting to look like a babushka or some latter-day hippie, and she'd finally managed to arrange it in an attractive style. Desiree, the temperamental French hairdresser she and Jacqueline Donovan shared, had insisted that if Lillie was determined to drive a convertible, she take measures to protect her hair.

When Lillie pulled into the parking space outside the service area, Hector immediately stepped outside as if he'd been standing by the door, waiting for her.

"Good morning, Ms. Higgins," he said with the slightest bow.

"Good morning, Mr. Silva."

"Please call me Hector."

"Only if you'll call me Lillie. After everything we've been through with this car, I believe we've become friends, don't you? And friends call each other by their first names." Referring to him as a friend might be presumptuous, but she couldn't seem to stop herself.

He grinned, and his dark eyes glinted with pleasure. "I feel the same way." After the briefest of hesitations, he added, "Lillie." She loved how he said her name, placing equal emphasis on each syllable. She'd never heard anyone draw it out like that. He made it sound…sensuous. Completely unlike the blunt "Lil" her husband used to call her.

"Your vehicle is ready." He gestured toward the red convertible parked near the service area.

"Did you ever find out what the problem was?" she asked, although in truth she didn't really care.

"As far as I can tell, the hydraulic hose had an air bubble in it. I worked on it myself and I had my best mechanic check it, too. He assures me the problem has been fixed. You shouldn't have any steering troubles from now on."

"Then I'm sure I won't." Instinctively, she felt certain that Hector's pledge was the only guarantee she needed.

"I've taken it out for a test drive and in my estimation it runs beautifully. However, if you'd like, the two of us could go for a short ride."

Lillie knew this was above and beyond anything that was necessary. Nevertheless she nearly squeaked with joyful anticipation. Oh, she was behaving badly, wasn't she? And she intended to go on doing it.

"I'd appreciate that very much," she told him earnestly. "But only if it won't keep you from your duties."

"You are our customer, Lillie, and it is the goal of the dealership to exceed your expectations."

"Oh." His dedication to duty dispelled the notion that he was doing this for her and her alone. In fact, he seemed to be quoting from a policy manual. That gave her pause. Perhaps what she felt toward him was imaginary, something she'd dreamed up—but she knew it wasn't. The real question was whether Hector reciprocated her feelings.

Hector held open the driver's door for her.

Lillie slipped behind the wheel as he walked around the vehicle and joined her in the passenger seat. "You're sure you have time for this?" she asked again.

"Yes, Lillie, I'm very sure." He encouraged her with a smile.

She turned the key and the engine instantly surged to life. "Is there any specific place you'd like me to drive?" she asked, hoping he'd suggest a route.

"Green River in the Kent Valley should be a good test."

That was where the vehicle had broken down the first time. It was also twenty minutes away. This was more than a short test drive, she thought excitedly. More than business.

Still, they didn't exchange a single word as she drove down the freeway. It wasn't until they neared the river that Hector spoke.

"The car is in perfect running condition," he told her in a solemn tone.

"You can tell just by the sound?"

"Oh, yes. My wife, when she was alive, used to tease me. She said I could read cars better than I could people, and she was right."

"You're a widower?" Lillie had noticed that Hector wasn't wearing a wedding band, but she'd assumed it was because of his work.

"Yes, almost ten years now."

"I'm sorry." Lillie knew the pain of losing a life partner, even when the marriage wasn't ideal.

"Angelina was a good woman and a good wife," Hector said. "And a devoted mother. We have three beautiful children."

"My husband died in a plane crash three years ago."

"I'm sorry for your loss."

Lillie focused her attention on the road, although their conversation was of far greater interest. She might be seeing more here than was warranted but she sensed that Hector wanted her to know he was a widower. She wanted him to realize she was unattached, too.

"How old are your children?" she asked, not reacting to his sympathy, which made her a little uncomfortable. David had been an excellent provider and an adequate husband, but he'd had his weakness. Unfortunately that weakness involved other women.

For years Lillie had turned a blind eye to David's wanderings. It was easier to pretend than to confront the ugly truth of her husband's infidelities. During the last ten years of their marriage, there had been no real intimacy between them. Lillie had swallowed her pride and pretended not to know about her husband's affairs—as long as he remained discreet.

"My children are all grown now," Hector said. "They have graduated from college and taken advanced training in the fields of their choice. Manuel is an attorney. Luis is a physician and my daughter, Rita, is a teacher." His pride in his family was evident.

"My goodness, all three of your children are accomplished professionals."

"Their mother and I believed in higher education." Hector looked at her as she slowed the car's speed to take a sharp curve. "You have children?"

"A daughter. Her husband was with mine in the plane crash."

"He died, as well?"

"Unfortunately, yes."

Hector's eyes grew dark with compassion. "In one day your daughter lost both her husband and her father."

"Yes." It had been a horrific day and not one Lillie wanted to think about. David and Gary had been flying home from a business trip. The pilot and co-pilot had died, too.

The FAA had investigated, and after a thorough exploration of the facts had determined the cause of the accident—sudden, catastrophic engine failure. But that knowledge didn't take away the shock and the grief.

"Your daughter has a family?"

"Two sons."

"You are a grandmother, then?"

The boys were the very joy of her life. "I have twin grandsons, Eric and Kurt. They're in their first year of college."

"No." Hector wore an astounded expression. "It isn't possible you have grandchildren of that age."

"Both my daughter and I married young." Although she wasn't in the habit of divulging her age, she felt she could with Hector. "I'm sixty-three."

Again his eyes widened. "I would have said you were closer to your mid-forties."

With another man, Lillie might have expected such flattery. But Hector was what David would've called a straight shooter. He didn't give compliments for any reason other than that he meant them.

"I'm sixty-four," Hector admitted. "I will be retiring in a few months."

"How long have you been with the dealership?"

"Thirty-four years."

"With the same dealership?" That was practically unheard of these days.

"I started as a maintenance man and attended night school to become a mechanic. When I had my certificate, the service manager at the time offered me a job. I worked hard and within ten years I was the chief mechanic."

"When did you take over as manager?"

Hector didn't need to think about the answer. "Almost twelve years ago. I would have retired sooner, only the expenses of college made that impossible." He grimaced comically. "Private colleges."

"All of your children were in college at the same time?"

"Yes. Thankfully, each one received financial assistance through scholarships and grants. But I have to tell you the costs were staggering."

Lillie knew that from what Barbie had told her about Eric and Kurt's tuition costs and the other expenses associated with getting them started in school. Even now, Lillie could hardly believe it. She was impressed that Hector had managed to put three children through school on what he earned as a service manager. He no doubt made good money, but still...

"Your daughter attended college?" he asked.

Lillie shook her head. "Barbie married young, just as I did. Both David and I were disappointed but ultimately we approved of the marriage. She knew her own mind, and I will say she and Gary were very happy."

"That's how it is with love sometimes, isn't it?" he said, glancing in her direction. "Sometimes the heart really does know what's best."

Her own heart was speaking loud and clear at that very moment.

They'd been gone for more than an hour, and it was time to return to the dealership. Both grew quiet. Minutes earlier, their conversation had been animated; now, reality set in and there didn't seem to be anything else to say.

When Lillie pulled into the dealership's parking lot, she experienced a pang of regret. This was it; the ride—and her relationship with Hector—was over. There was no further reason to see him. It wasn't as though they'd ever encounter each other in the normal course of their lives.

He told her where to park, pointing at an empty slot.

"Thank you, Hector, for everything you've done," she said, forcing a smile.

"My pleasure."

They sat in the car, and he seemed as reluctant to move as she was.

"I should get back to work," Hector finally said.

"Yes, of course."

His hand was on the door handle. "It isn't every day I get to ride with such a beautiful woman," he said with quiet gallantry. He climbed out and gently closed the door. His eyes avoided hers. "Goodbye, Lillie."

"Goodbye, Hector."

He was a service manager for a car dealership and she was a wealthy widow. She accepted that their paths would likely never cross again. Despite that, she could do him one good turn. When she got home, Lillie phoned the dealership, leaving a message for the owner, Steve Sullivan. She praised Hector's efforts on her behalf and stressed to Steve that he had an outstanding employee.

That way, at least, she could play a small, if benevolent, role in Hector's life.

It wasn't enough but it would have to do.

Monday evening, Barbie showed up at the theater a little later than she had the previous week. Tessa Bassett was selling tickets again, and when she saw Barbie, her face lit up.

"Should I recommend another movie?" the girl asked cheerfully.

"Please do." The ill-tempered Mark Bassett was the sole reason Barbie had come back. In the last week she'd spent a lot of time thinking about him. She felt strangely invigorated by the challenge he offered, but it was more than that. She was attracted to him, not only because of his looks but because she saw in him the same loneliness she'd experienced since her husband's death. Once she made it past the barrier he'd erected against the world, perhaps they could be friends. Perhaps even more. The fact that he was physically disabled didn't bother her, nor did she find it especially daunting. She knew it didn't define or describe the person inside, any more than *her* appearance did.

Tessa mentioned a movie Barbie had never heard of and handed her the ticket, as well as her change.

"You're sure this is a good movie?" Barbie asked.

Tessa's eyes held hers. "It's the perfect movie."

Barbie was willing to take the girl's word for it. In the theater lobby, she once again purchased a small bag of popcorn and a cold soda, then walked into the dimly lit theater.

She saw that Mark was already in one of the wheelchair spaces. Tessa had been right; this was the perfect movie.

Without hesitation, Barbie moved around the back and entered the row from the opposite direction. She sat down, leaving one empty seat between her and Mark.

The instant she did, he turned to glare at her. "This space is reserved for wheelchair seating."

"Yes, I know," she said as she crossed her legs. She started to eat her popcorn as if she didn't have a care in the world. Feeling both silly and daring, she tossed a kernel in the air and caught it in her mouth. Proud of herself, she grinned triumphantly at Mark.

Clearly he wasn't impressed with her dexterity. "Would you kindly move?"

His voice was even less friendly than it had been the last time.

"I have every right to sit here should I choose to do so," she returned formally. She held out her bag of popcorn. "Here," she said.

He frowned. "I beg your pardon?"

"I'm offering you some of my popcorn."

"What makes you think I want your popcorn?"

"You're cranky. My boys get cranky when they're hungry, so I figured that might be your problem."

He looked pointedly away.

"If you're not interested, the proper response is *no, thank you.*"

He ignored that, and Barbie munched her popcorn, swaying her leg back and forth.

"Stop that."

"What?"

"Swinging your leg like a pendulum."

She crossed the opposite leg and swung it, instead.

Mark groaned.

The theater darkened, and the previews appeared on the screen. Barbie finished the small bag of popcorn. Her hands were greasy, but in her rush to get into the theater she'd forgotten to pick up a napkin. She'd also forgotten to replace the tissues she kept in her purse. She stood up to go back to the lobby. Rather than march all the way down the row, she leaned over to nudge Mark.

"Excuse me."

"You're leaving?" He actually seemed pleased.

"No, I need a napkin. Can I get you anything while I'm up?"

"No," he muttered.

She sighed audibly. "Are you always this rude or is it just me you don't like?"

"It's you."

She refused to feel insulted; instead she interpreted his response as an admission that he was aware of her. Aware and interested.

"You act as if that pleases you," he said, sounding surprised.

"Well, it doesn't hurt my feelings if that was your intent. Now, can I get by? Please?"

With exaggerated effort, he rolled back his wheelchair, allowing her to exit the row.

Barbie pushed the sleeve of her soft cashmere sweater up her arm and hung her purse over her shoulder. "Don't get too comfortable," she told him. "I'll be back in a few minutes."

"Don't hurry on my account."

"I won't."

When she entered the lobby again, she saw that Tessa was working behind the concession stand. The girl looked curiously in her direction and Barbie nodded. She grabbed some napkins to wipe her hands, then walked over to wait her turn. She made an impulsive purchase, smiling as she did.

"How's it going?" Tessa asked, handing her the change.

"He wants me to leave."

Tessa seemed worried. "You're not going to, are you?"

Barbie shook her head. "Not on your life."

Tessa nearly rubbed her hands together with glee. "This is so cool."

"What is?"

The teenager shrugged. "Well, you know. You and my uncle Mark. He needs someone in his life. He doesn't think so, but...well, it'd just be so cool if that someone was you."

"Don't get your hopes up, Tessa." Barbie felt obliged to warn her. "I'd better get back. The movie's about to start."

"Don't let him give you any crap," the girl advised. "Oops, I mean attitude."

Barbie grinned and gave her a thumbs-up.

Attitude was the right word, she mused as she made her way into the theater. It wasn't hard to figure out that his surliness was an attempt to protect himself from pain and rejection. If there was one thing she knew about, it was dealing with the insecurities of the adolescent male. And if she had her guess, he'd reverted to that kind of negative behavior after his accident. Beneath all the hostility, he was as lonely and lost as she was.

The film was just beginning as she reached their row. She stood in the aisle, waiting for him to roll his chair back.

"Excuse me," she said when he pretended not to notice. "I'd like to sit down."

"Must you?" he asked sarcastically.

"Yes, I must." Taking the initiative, she raised her leg and attempted to climb over his lap. He got the message fast enough when she presented him with an excellent view of her rear. He shot back with enough force to bolt into the empty space two rows back.

Barbie reclaimed her seat, then tossed him a chocolate bar. "Oh, here," she said. "I thought this might sweeten your disposition."

He tossed it back. "My disposition is as good as it gets. Chocolate isn't going to change it."

"Fine. I'll eat it then."

From that point on, she ignored him and he ignored her.

The movie, another romantic comedy, was delightful and Barbie quickly got involved in the plot. She and Mark didn't exchange a word until the credits were rolling and the lights came back on.

"That was really good," she said to no one in particular.

"It was sappy," Mark muttered.

"Naturally you'd say that," she protested. "Don't you believe in the power of love?"

"No."

So why had he chosen this movie? "Well, I happen to believe in it," she told him.

"Good for you." He wheeled back and started out of the theater, with Barbie keeping pace five steps behind him. Tessa, still at the concession stand, glanced at her eagerly. She gave the teenager another thumbs-up, and the girl returned a huge grin.

Just outside the complex, he unexpectedly wheeled around and confronted her. "Are you going to make a habit of this?" he demanded. The corners of his mouth curled scornfully.

"Of what?" she asked, playing dumb.

"Monday night at the movies. The only reason you're here is to irritate me."

"I didn't realize I had to pay money to do that. Couldn't I just sit out here and do it for free?"

He pinched his lips tightly closed.

"I enjoy the movies and Monday's a good night for me."

"Come another night," he said.

"I don't want to."

Frustration showed in his face. "Why are you doing this?"

"Doing what?" she asked, again feigning innocence. "You mean coming to the movies two weeks in a row on a Monday night?"

"Yes."

"Well, like I said, Monday evenings are good for me and movies are my favorite form of entertainment."

One look told her he didn't believe a word of it. "Then how come you picked the same movies I did?"

She tried to pretend she was bored with the subject. "If memory serves me, I was seated first last week. You're the one who invaded my space."

He frowned as if he'd forgotten that. "Maybe so, but this week was no accident."

"You seem to have an inflated opinion of your charms." His mouth opened and he seemed about to launch a comeback, but she didn't give him a chance. "Now, if you'll excuse me, I'm going home. Good night, Mark."

He frowned. "How do you know my name?"

"I asked. I'm Barbie, by the way. Barbie Foster."

"Barbie," he repeated and snickered. Then he laughed outright. "Barbie. It figures. You're about as plastic as they come."

"And you're about as rude as any man I've ever had the misfortune to meet."

"Then stay away from me and we'll both be happy."

"Maybe," she said flippantly as she reached for her car keys, buried deep inside her giant purse. "And maybe not. I haven't decided yet." She left him then, with a decided sway to her hips. It was an image she hoped would stay with him for a long time.

T uesday was a good sales day at the bookstore, which wasn't typical. Anne Marie had worked out a careful method of maintaining inventory, balancing the number of mainstay and classic books she kept on the shelf with the new ones. It was crucial to have a wide range of titles. Relatively new to the business, she was learning as she went. Past experience had come from a part-time job at the University of Washington campus bookstore. Her previous career, as a customer service rep at a national insurance company, had taught her some valuable skills, too—but she hadn't loved it and was glad enough to give it up, at Robert's suggestion, to work in the bookstore, with an eye to eventually buying it.

The store was independent and needed an edge to compete with the large chains. Each bookstore, whether a chain store or an independent, was important in its own way. Blossom Street Books served the community. Over the past four years, since the renovations to the entire neighborhood, the store had developed a following and earned the loyalty of local residents. Anne Marie hadn't wanted to specialize, like some independents did, in mystery fiction or cookbooks or children's books; she preferred to meet all her customers' book-buying needs. She ordered books for them, ran several reading groups, offered competitive discounts on bestsellers and provided a cozy, intimate atmosphere. She'd made the store as inviting as possible, with comfortable chairs, a gas fireplace and warm lighting.

Her clientele depended on Anne Marie for recommendations and updates on authors and publishing houses. She'd managed the store before she bought it, to make sure she really wanted to take on her own business, and in the process familiarized herself with the industry.

Even as a child, Anne Marie had been an inveterate reader. She'd found her adventures in the pages of a book. Never outgoing, or one to stand out in a crowd, she'd been her husband's opposite in personality. Robert had been gregarious and sociable, and they'd complemented each other well. He was fun to be around, and that had attracted her from the beginning. Their age difference had never concerned her because he didn't seem older. Except when it came to having another child...

Rather than sink into depression again, Anne Marie focused on creating a fresh display for the front table. Bookstores were a low-margin business, and the real profits came from notecards, stationery, games and other accessories. She was working on a St. Patrick's Day exhibit, featuring books like *How the Irish Saved Civilization* and fiction by Maeve Binchy, Marian Keyes, Edna O'Brien and other popular Irish novelists. Around the books she arranged packages of greeting cards with shamrocks on them, green candles and St. Patrick's themed paper napkins. She stepped back, pleased with the result.

The previous owner, Adele Morris, had a bookstore in the Fremont neighborhood, and when there was an opportunity for a second store on Blossom Street, Adele took it. Because of the renovation, she'd been offered a favorable rent and for the first couple of years she'd divided her time between the two stores. That proved to be too difficult, and Anne Marie had joined as manager soon afterward; later she purchased the business with Robert's encouragement. In her husband's eyes, the bookstore, like Baxter, was a solution to their dilemma. If Anne Marie was preoccupied with the store, she might forget about having a baby.

At one-thirty Theresa came in and for an instant Anne Marie couldn't remember why she'd shown up for work on a Tuesday.

"Ellen!" She said the child's name aloud as the memory rushed in. She was supposed to be at the school for Ellen's performance.

Theresa nodded. "You told me your Lunch Buddy was in some function at the school that you wanted to attend."

"Right." Rushing into the office, she grabbed her purse and threw on her jacket. She gave Theresa some last-minute instructions for her meeting with the children's book sales rep. Then she hurriedly left the shop via the back entrance, where she'd parked her car.

Thankfully the school was relatively close, and it only took her ten minutes to drive there. But when she arrived she discovered that the parking lot and nearby streets were jammed with vehicles and she wondered if every parent in a three-state area had decided to come for the performance. After another ten minutes she located a parking space three blocks from the school. She locked the car and ran toward Woodrow Wilson Elementary.

The music had already started by the time she entered the large gymnasium, sweaty and out of breath. The place was packed with parents and students, and if there was an available seat she couldn't find it.

Every adult in the room seemed to be in possession of a camera. Anne Marie hadn't even thought to bring one and wanted to kick herself. Ellen's grandmother would've appreciated a photograph of her granddaughter on stage.

Muttering her excuses, Anne Marie slipped past several people until she squeezed herself into a tight space where she had a good view. Sure enough, she could see Ellen standing on a riser with the other members of the chorus. She wore her Sunday best—a dress one size too small and white patent leather shoes. The stage set consisted of two large painted trees and a castle. The artwork had apparently been done by the students, as well. If she'd been told the name of the production, Anne Marie didn't remember. Clearly, though, it was the retelling of some classic fairy tale.

Anne Marie watched Ellen, who looked awkward and uncomfortable standing front row center, with two rows of children behind her.

As if she felt Anne Marie's eyes on her, Ellen glanced in her direction. When she saw Anne Marie, the girl's entire face was transformed by the beauty of her smile. Seeing how happy her presence had made Ellen, Anne Marie was glad she'd taken the trouble to show up. She sent the girl a small wave. Ellen waved back.

The music died down as the singing director stepped in front of the choir and raised both hands. The children on the risers instantly came to attention.

The performance, which turned out to be a rather inventive version of "Snow White," lasted forty minutes. No one was going to mistake it for professional theater. But the dwarves were hilarious and the singing was lively. Anne Marie nodded her head to the beat.

When the performance was finished, the principal came forward and announced that juice and cookies would be served in the children's rooms. Anne Marie checked her watch. She really should be getting back to the store. Then again, a few more minutes wouldn't hurt.

As she started toward Ellen's classroom, she nearly bumped into Helen Mayer, the school counselor.

"Anne Marie!" she exclaimed. "I'm so glad you could make it."

"Yes, the play was very well done," she said warmly. "Thank you for telling me about it."

"No, thank *you*, thank you so much."

With a quick smile, she hurried off in the opposite direction.

Anne Marie was standing by Ellen's desk when the child walked into the room, her eyes bright with happiness. "Did you hear me?" she asked. "Did you hear me sing?"

Anne Marie hadn't been able to discern Ellen's small voice among so many others. But in this case she figured a white lie was appropriate. "I did, and you were terrific."

Ellen blushed at the praise.

"You didn't tell me you like to sing."

Ellen nodded. "Mrs. Maxwell said I have a good voice. She's the music teacher."

"How many other second-grade students were part of

the choir?" Anne Marie asked, although she already knew the answer.

"I was the only one."

"Just you?" Anne Marie feigned surprise.

"Yup, just me. Mrs. Maxwell said maybe by the time I'm in fourth or fifth grade I might get to sing a solo."

"Ellen, that's wonderful. Congratulations." Anne Marie had never seen the girl this excited.

The classroom had begun to fill up with other children and parents.

"Would you like some juice?" Ellen asked politely. The juice and cookies were set up on a table in the front.

Anne Marie noticed that the other students were delivering refreshments to their parents.

"That would be very nice. Thank you, Ellen."

The child waited for her turn and poured Anne Marie a small paper cup full of juice, which appeared to be some watered-down fruit punch. She also brought her two small cookies, definitely a store-bought variety.

"You didn't get anything for yourself," Anne Marie said.

"That's because you're supposed to serve your guests first," Ellen informed her solemnly.

"Of course," Anne Marie murmured. "I must've forgotten my manners."

Silently Ellen stood next to her.

Anne Marie bent down and whispered, "What's going to happen next?"

"Nothing," Ellen said. "You're supposed to drink your juice and eat your cookies."

"Okay." Anne Marie sampled a cookie, which crumbled in her mouth at the first bite. She washed it down with a gulp of juice that was far too sweet. Ellen waited until

Anne Marie had finished before she returned to the re-
freshment table and poured a second cup of juice and
took two small cookies for herself.

"Baxter wanted me to tell you hello," Anne Marie said
when she came back.

Ellen swallowed the cookie she was chewing and nodded. "He's a good dog."

"A little spoiled, though."

"I'll teach him how to roll over the next time you bring
him to school," Ellen promised. "I got a book from the library and I read about teaching dogs tricks. Baxter's smart
and I know how to get him to roll over."

"I hope you can show me how to teach him, too."

"I will," Ellen said.

"I've tried to teach Baxter new tricks, but he doesn't
seem to understand the concept." Anne Marie felt it only
fair to warn Ellen; she didn't want to discourage the girl,
nor did she want her to think it would be an easy task.

One of the other mothers glanced speculatively at Anne
Marie and Ellen and moved toward them. "Are you Ellen's
mom?" she asked Anne Marie.

"Actually, no, I'm her friend."

"Anne Marie is my Lunch Buddy," Ellen explained
proudly. "She brought her dog for me to meet."

"Oh." The other woman drew a tiny long-haired girl
close to her side. "I'm Shelly Lombard and this is my
daughter, Cassie. She's friends with Ellen."

"Hi, Shelly, Cassie," Anne Marie said, smiling. "It's
nice to meet you."

"I wanted to ask if Ellen could come over for a play
date one afternoon. Would that be possible?"

This wasn't something Anne Marie could answer.
"You'll have to ask her grandmother."

"Ellen lives with her grandmother, then?"

Anne Marie nodded.

"Oh…well, I don't know if that would work. I was actually hoping we could exchange play dates once in a while."

"I see."

"It's just that occasionally I have an appointment after school and it's difficult to find someone to look after Cassie for just an hour or two."

"You could always ask her grandmother," Anne Marie said a second time.

"Yes, of course, but if she couldn't arrange to come for Ellen's performance, it's unlikely she'd be up to looking after an extra child."

Shelly had a point. Anne Marie remembered Ellen's saying that her grandmother slept a lot, which made her wonder if the woman was ill.

Shelly drifted away to chat with another parent. Anne Marie wanted to leave but she could tell that Ellen was desperate for her to stay. She searched for a topic of conversation.

"Would you like to show me your schoolwork?" Anne Marie asked. She remembered that during her brief orientation, this was an option suggested for Lunch Buddies.

"Okay." Ellen sat in her small chair and opened her desk to retrieve a notebook. Everything inside was impeccably organized.

Ellen set the notebook on top and flipped it open for Anne Marie to examine. On nearly every page the teacher had written a comment praising Ellen's work.

"You're an excellent student," Anne Marie said.

"Grandma Dolores makes me study every night." Ellen didn't seem happy about this.

"That's good, isn't it?"

"I guess." Ellen shrugged.

"Then you get to watch TV, right? That's what you told me before."

She bobbed her head. "We watch shows on the religion channel."

"What about cartoons?"

"Grandma Dolores doesn't think cartoons are good for kids. She saw *South Park* once and got upset. She hid my face in her apron and started praying to Jesus."

Anne Marie bit her lip, trying not to smile.

A buzzer rang, announcing the end of the school day. In short order, children and parents began to vacate the classroom. Ellen looked up at Anne Marie. "I need to catch my bus."

"Would you like me to walk outside with you?"

"Yes, please."

While Ellen put on her coat and gathered her things, Anne Marie went to introduce herself to the teacher. Ms. Peterski smiled at Anne Marie. "I'm so pleased you could come."

"I am, too," she said and she meant it.

She and Ellen walked out to the schoolyard, negotiating their way through the laughing, shouting throngs.

"I'll see you tomorrow at lunchtime," Anne Marie said as they neared the area where the children lined up for their buses.

"You're coming tomorrow, too?"

"It's our lunch date, remember?"

Ellen blinked hard, apparently overwhelmed that Anne Marie would come to see her two days in a row.

"I can't bring Baxter, though," Anne Marie reminded her.

"That's okay." They approached the bus stop, and suddenly Ellen slipped her hand into Anne Marie's.

It felt as if the warmth of that small hand reached all the way to her heart.

Wednesday evening as Anne Marie prepared for bed, her phone rang. At the time she was brushing her teeth. Frowning, she turned off the tap and spit into the sink, then wiped her mouth before she went into the kitchen.

She couldn't even guess who'd be phoning after eleven o'clock. Caller ID told her nothing. It said Private Caller, which meant it was probably one of the widows. If Elise, Lillie or Barbie was calling her this late, that meant trouble of some kind, although she couldn't imagine what.

"Hello," Anne Marie answered cautiously. Nighttime phone calls usually brought bad news, and she'd had enough of that.

"This is Anne Marie?" The voice, that of an older woman, was barely audible.

"Yes."

"Anne Marie Roche?"

"Yes."

"I need…help." The woman, this stranger on the other end of the line, was close to panicking.

"Who is this, please?"

"Dolores. Dolores Falk."

"Who?"

"Ellen's grandmother."

Anne Marie sucked in her breath as a dozen disturbing possibilities ran through her mind. "Is Ellen all right?" she asked, fighting down a sense of panic.

"Yes…no. It's me who needs help… I wouldn't call you if there was anyone else." Each word seemed labored.

Anne Marie didn't know what she could possibly do. "Do you want me to call someone?" she asked, wondering how she might assist the older woman. Surely she had a neighbor or a friend she could contact. Anne Marie was a stranger.

"No, the aid car is on its way." The woman's breathing became harsh and irregular. "Just come…please. Hurry."

Anne Marie didn't understand. "Are you saying you want me to come to your house?"

"Please. Just…hurry."

"But…" How did Ellen's grandmother get her phone number? And what did she want? She was clearly in distress, but how could Anne Marie help?

"I don't have anyone else to take Ellen," Dolores gasped.

"*Me?* You want me to take Ellen? But I can't—" It was out of the question. Anne Marie didn't have room for a child.

"They're going to bring me to the hospital. Please. I'll refuse to let them unless you come."

Talk about emotional blackmail! In just the few minutes Anne Marie had been on the phone with Ellen's grandmother, she'd realized the older woman was badly in need of medical attention. As much as she resented this, Anne Marie didn't have a choice. She'd have to go and then try to sort out the situation later.

"What's the address?"

Dolores gave it to her with the added pressure of, "Hurry, please hurry."

"I'll be there as soon as I can." Exasperated, she replaced the phone and exhaled sharply. How had she ended up in this predicament? She'd volunteered to be a Lunch Buddy, not a…she didn't know what.

Pulling on jeans and a shirt, Anne Marie complained to Baxter, then promised to return as quickly as possible. With the address scribbled on a grocery-store receipt, she headed for her car. All she needed right now was to get attacked in the alley.

The alley was actually well lit, not that it would help her any if someone decided to leap out of the dark and mug her. Unlocking her car with shaking fingers, she climbed inside and started the engine.

Anne Marie considered herself the least capable person to deal with someone else's problems. If she'd had the school counselor's home number, she would've called Ms. Mayer and handed the whole mess over to her. Rescuing her Lunch Buddy in the middle of the night was *not* what she'd signed up for.

Dolores Falk's house was only about four miles away, but the neighborhood, an older working-class area, was unfamiliar. By the time Anne Marie arrived, the aid car was parked out front. A fire truck was there, too, plus paramedics. Several neighbors stood on their porches watching all the activity.

Anne Marie parked across the street, well away from the emergency vehicles. Purposefully she trudged over to the house.

The instant Ellen saw her, she bolted down the porch steps, then raced across the yard and threw her arms around Anne Marie's waist.

"What's going on?" Anne Marie asked, placing her hands on the child's shoulders.

"These men are taking Grandma to the hospital," Ellen sobbed, clinging to Anne Marie.

"But they're going to help her. Isn't that what we want?" she asked softly.

"N-o-o! She-e-e mi-gh-t *d-i-e*," the girl wailed.

"Let me talk to them," she said and gently loosened the child's arms. She walked Ellen back to the porch and left her sitting on the bottom step, still sobbing.

"Are you Anne Marie Roche?" an emergency medical technician asked as he stepped out of the house.

"Yes."

"Good. The grandmother refused medical treatment until you got here."

"Why me?"

"You'll have to ask her that yourself."

"Then let me talk to her."

He shook his head. "I'd prefer if you did that at the hospital."

"I only need a minute," she insisted stubbornly.

"The grandmother told us you'd be taking the child," the paramedic said as he started into the house.

"I'm her Lunch Buddy." She wanted to explain that her entire role in this child's life was to have lunch with her once a week. She'd met her exactly four times, if you included the brief orientation the previous month.

Lunch. That was supposed to be the full extent of her commitment.

No one had said anything about taking Ellen home

with her. That was probably against the rules, anyway, and there seemed to be a lot of those.

"Isn't there someone else?" she asked, following the EMT into the house.

"Apparently not." He hurried to a bedroom in the back, Anne Marie directly behind him.

She discovered Dolores Falk on a stretcher. The woman's complexion was sickly and gray, and every rasping breath seemed to cause her pain. Her hand rested on her heart, her eyes tightly shut. Ellen had said her grandmother was over fifty; in Anne Marie's observation, she had to be in her mid-sixties but looked older.

"Wheel her out," the EMT instructed the other two.

The woman's eyes flew open. "Wait."

"I'm here," Anne Marie rushed to tell her.

Dolores reached out and grabbed Anne Marie's hand in a grip that was shockingly strong. "Don't let them put Ellen in a foster home. I'll lose her if they do."

"But, Mrs. Falk… Where do you want me to take her?"

The woman's eyes closed again. "Home. Take her home with you."

"With me? I can't—"

"You have to…"

The EMT came in then and they rolled the stretcher down the hallway and out of the house. Anne Marie trailed behind, watching helplessly as the emergency crew loaded Ellen's grandmother into the aid car and drove off, sirens screaming.

With her hands covering her face, Ellen sobbed as she huddled on the steps, her shoulders trembling. Her pitiful cries were drowned out by the screeching aid car.

Anne Marie crouched so they were at eye level. "Your grandmother's going to see the doctors and they're going to make her well again." She prayed with all her heart that this was true.

Ellen nodded tearfully. "When will Grandma be back?"

"I don't know, sweetheart." She was so far out of her element here that she was breaking into a cold sweat.

"Where will I go?" Ellen asked.

"For tonight," Anne Marie said, mustering as much enthusiasm as she could, "you get to come home with me."

Ellen dropped her hands long enough to look up at Anne Marie. "With you?"

"Yes, that's why your grandmother called me."

"Is Baxter there?"

Anne Marie nodded. She should've thought of that sooner. Ellen loved Baxter and he'd help take the child's mind off what was happening to her grandmother.

"Baxter's waiting for us to get back to my apartment so he can see you. Didn't you say you wanted to teach him to roll over?"

"Yes-s-s." For the first time since Anne Marie had arrived, the eight-year-old stopped weeping. She bit her lip and managed to control her sobs.

"We should pack a few things for you."

"I have my backpack," Ellen said, looking small and lost and terrified.

"Good idea. We'll put what you need in there." Taking the child by the hand, Anne Marie went into the house. It was an older single-story home, probably built soon after the Second World War. The floors were linoleum and the furniture shabby and dated. The hallway led to three bedrooms.

Ellen's room was the farthest down the hall on the right-hand side.

It was furnished with a single bed, a dresser and a child-size desk and chair. The closet was narrow but more than big enough for Ellen's few clothes.

"Just get what you'll need for school tomorrow," Anne Marie said. In the morning she'd drive Ellen to Woodrow Wilson Elementary, then she'd talk to Ms. Mayer and find out what could be done for the child.

"I brushed my teeth already," Ellen said. Kneeling down on the braided rug next to her bed, she stuffed a pair of neatly ironed jeans and a pink sweater into her back-pack.

"Don't forget your shoes and socks," Anne Marie told her. Ellen was wearing bedroom slippers and well-worn pajamas over which she'd pulled a sweatshirt. "Did you have any homework?"

Ellen nodded and hurried to the kitchen, returning with a small binder. "It's math," she explained as she added that to the pouch, along with her tennis shoes and a pair of socks.

"This is way past your bedtime," Anne Marie said.

"Grandma said she wasn't feeling well when I got home from school," Ellen told her. "She said I could have cornflakes for dinner."

"Did you?"

Ellen shook her head. "I wasn't hungry."

Most likely Ellen was too worried about her grand-mother to have an appetite. "Who called for the aid car?"

"I did."

"You?"

"Grandma didn't look good and she didn't answer me when I talked to her and I got scared."

"That was a smart thing to do." Anne Marie had to credit the child with fast thinking. "Do you know how your grandmother got my phone number?"

"No." Ellen lowered her eyes. "But I'm glad she did."

Now that she'd seen the situation for herself, Anne Marie was glad, too, and grateful she'd come when she had. She could only imagine how much greater the trauma of this evening would've been for Ellen if she'd been handed over to Child Protective Services and placed in temporary foster care.

"Are you ready?"

Ellen nodded solemnly and reached for Anne Marie's hand. The child turned off all the lights on the way out. She stopped on the porch and took out the house key, hidden beneath a ceramic flowerpot, then locked the front door. When she'd finished, she replaced the key.

"Will I be able to visit Grandma in the hospital?" she asked, staring up at Anne Marie with huge eyes.

"I'll find out for you in the morning, okay?"

"Please," she whispered, and the plaintive little voice broke Anne Marie's heart.

As they drove back to Blossom Street, Anne Marie suspected the girl would fall asleep on the silent ride there, but Ellen appeared wide-awake. When they got to the bookstore, Anne Marie pulled into the alley behind it.

"This is your house?" Ellen asked.

"I have a small apartment above the bookstore."

"You live over a bookstore?" she whispered, as if Anne Marie resided in some enchanted castle.

"I do. I'll bet Baxter's standing by the door, too." The Yorkshire terrier seemed to recognize the sound of her car and waited eagerly by the back entrance.

Sure enough, the minute Anne Marie unlocked the door Baxter rushed forward, leaping up and down with excitement.

"Baxter!" Despite the anguish of the evening, Ellen

couldn't hide her delight at seeing the dog again. She fell to her knees and the terrier welcomed her, licking her hands and face.

"Ellen's spending the night," Anne Marie told him. Turning to the girl, she said, "Let me show you your bedroom."

"Okay." Reluctantly leaving the dog, Ellen followed Anne Marie through the apartment.

The second bedroom, which served as Anne Marie's home office, wasn't set up as guest quarters. But thankfully she had a sofa that folded out into a bed. Taking a set of sheets from the hall closet, she quickly made it up and added a couple of blankets and a pillow.

"Would you like some warm milk?" she asked when the bed was ready. "It might help you sleep."

Ellen made a face and shook her head.

"Sounds dreadful, doesn't it?" The only reason she'd offered was that her own mother used to give it to her. She hadn't liked it, either.

"Would it be okay…" Ellen hesitated.

"What is it, Ellen?"

"Could Baxter sleep with me?"

Anne Marie smiled. She should've suggested it herself. "That would be just fine."

"Thank you."

Anne Marie yawned. She was exhausted and knew Ellen must be, too. "Let me tuck you into bed," she said, "and I'll put Baxter up there with you."

"Thank you," Ellen whispered. She slipped off her sweatshirt and slippers and climbed into the newly made bed.

Once she was under the covers, Anne Marie folded them around her shoulders. She set her Yorkie on the bed.

As if understanding that the child needed a friend, Baxter immediately curled up next to her.

"Good night, Ellen," Anne Marie said, about to leave the room.

"Would you say a prayer with me?" the child asked.

"A prayer?" Anne Marie couldn't remember the last time she'd prayed.

"Grandma always does."

"All right, but you say the words."

"Okay." Ellen dutifully closed her eyes and although her lips moved, she didn't speak out loud. After a moment, she said, "Amen."

"Amen," Anne Marie repeated.

"I prayed for my grandma," Ellen told her.

"I'm sure God listens to little girls' prayers," Anne Marie said, choosing to believe that He did. She turned off the light, then realized she didn't know when Ellen was supposed to be at school. "Ellen," she whispered. "What time does school start?"

"Eight-twenty."

"I'll set the alarm for seven. That'll give us plenty of time."

"Okay."

Anne Marie left the room and eased the door partially closed so she'd hear if Ellen needed her during the night. She found a night-light for the bathroom and plugged it in.

Sitting at her small kitchen table, Anne Marie inhaled a deep, calming breath. Elise Beaumont had a lot to answer for—and she planned to let her know it. This Lunch Buddy business had become a far more complicated proposition than Anne Marie had been led to expect.

She liked Ellen and she was happy to help—well,

happy might be an exaggeration. She felt *obliged* to help, especially since the child's grandmother claimed she didn't have anyone else to ask. But in the morning, Anne Marie was driving Ellen to school and getting the name of the contact person listed for emergencies.

This was standard practice. The school would have the name of a responsible adult who'd take Ellen while her grandmother was in the hospital. Someone far more qualified than Anne Marie. Someone better equipped to look after a frightened child.

Anne Marie had her own problems. And as much as she wanted to help, she wasn't prepared to be the child's guardian for more than one night.

Chapter

12

Anne Marie woke before the alarm buzzed at seven and discovered Ellen sitting up in bed petting Baxter and talking to him in a voice that quavered slightly.

"Good morning," Anne Marie said as cheerfully as she could. She stretched her arms high above her head.

Ellen didn't respond.

"Would you like some orange juice?"

The girl shook her head.

"Are you sure?"

Ellen nodded.

"I'm going to take Baxter for a short walk. Do you want to come?"

"Okay." Ellen climbed out of bed and sat on the floor, where she'd left her backpack. While the child got dressed, Anne Marie prepared a pot of coffee and put on a pair of sweat pants and a fleece top.

Her usual morning routine was to take Baxter out while the coffee brewed, getting a few minutes of exercise at the same time. Their route never varied: down Blossom Street for two blocks, crossing over to a small park, going around the park twice and then back. The entire walk took twenty minutes. Once she was home again, Anne Marie always showered, changed clothes and did her hair and makeup. On a good day, everything could be accomplished in under an hour.

Ellen was ready by the time Anne Marie finished her first cup of coffee and pulled on her jacket.

"Would you like to hold the leash?" she asked.

"Yes, please."

As they headed outside, she asked Ellen a few more questions but the girl remained glum and uncommunicative. She wanted to ask Ellen what was wrong but figured it was obvious. The poor kid was worried about her grandmother, of course, and her own future. Anne Marie couldn't blame her for that, so she decided to tread carefully. If Ellen didn't want to talk, she shouldn't have to.

"When I take you to school this morning, I'm going to see the school counselor," Anne Marie said as they returned to the apartment.

"Okay."

"Do you have any relatives close by?"

"My aunt Clarisse."

That was a big relief, although Anne Marie had to wonder why Ellen's grandmother hadn't called her instead. Of course, there could be any number of reasons. Clarisse might've been out of town or at work or not answering her phone or…she ran out of excuses.

Anne Marie was confident that as soon as Clarisse learned that Dolores had been hospitalized, she'd be eager to have Ellen. Some of the tension left her now that she had the name of a responsible adult who'd step in and take care of the child.

When they entered the apartment, Anne Marie checked

her watch. Twenty-four minutes so far. That was good, especially with an eight-year-old in tow.

"What would you like for breakfast?" Anne Marie asked as they stepped into the kitchen.

Ellen shrugged.

"I don't have any kid cereals, but I do have shredded wheat. Would you like that?" Ellen had to be hungry, since she'd gone without dinner the night before.

"Okay. Thank you."

While Anne Marie got two bowls, the cereal and milk, Ellen made her bed and brushed her hair. It was straight and dark, parted in the middle with bangs that needed to be trimmed. If she'd had any little-girl hair clips, Anne Marie would've used them.

Ellen ate only a small portion of her breakfast and then placed her bowl in the sink. It was a bit early to drop her off at school, but Anne Marie wanted to be sure she had plenty of time to talk to the counselor.

"Are you ready to go?" she asked.

"Yes," Ellen replied. "Will you find out about Grandma Dolores?"

"I'll phone the hospital this morning," Anne Marie promised. She'd do it before ten, when the bookstore opened.

When Anne Marie and Ellen arrived at the school, the playground was already crowded with youngsters. The yellow buses had started to pull up, and students in bright jackets leaped down the few steps, like water cascading over a ledge. They all wore gigantic backpacks that threatened to topple them.

"Would you show me where the office is?" Anne Marie asked Ellen. She wanted the little girl to feel needed.

"Okay." Ellen silently led the way down the school's wide corridor.

"Would you like to play with your friends now?"

Ellen hesitated as if uncertain.

"Everything's going to be fine," Anne Marie assured her. "I'll call about your grandmother and let you know later."

Ellen's eyes brightened and she nodded, then ran off.

Watching Ellen join her friends, Anne Marie walked into the office; she asked to speak to Helen Mayer and within five minutes was escorted into the other woman's office.

"Is everything okay?" Helen asked immediately, a small frown between her eyes.

"Not really…" Anne Marie described the events of the night before.

Incredulous, the counselor stared at her. "Oh, my goodness."

"As you can imagine, this has all been a shock." Anne Marie pinned her gaze on the other woman. "I wonder how she got my phone number."

"Actually I gave it to her," Helen admitted a bit sheepishly. "She phoned last week and asked for it and I couldn't see any reason not to tell her. She said she wanted to talk to you about Ellen."

To be fair, the school counselor couldn't have known that Dolores would call in the middle of the night and place Anne Marie in such an awkward position. "I'm going to need the emergency contact number in Ellen's file," Anne Marie told her.

"Yes, of course." Helen turned to her computer and began to type. After a couple of minutes, she said, "The name is Clarisse McDonald." She reached for a pen and quickly wrote down the number.

Anne Marie took the piece of paper. As soon as she learned about Dolores's condition, she'd be in touch with Ellen's aunt.

"Do you know what hospital the paramedics took Dolores to?" the counselor asked.

At the time Anne Marie hadn't been thinking clearly enough to inquire, but she'd heard one of the EMTs mention Virginia Mason Hospital, which wasn't far from Blossom Street.

She was telling Helen Mayer this when a bell rang in the distance, indicating the start of classes. The sound caught Anne Marie off guard and she jerked in surprise.

"You get used to the bell," Helen said. "After a while you don't even hear it." She smiled. "You were telling me Dolores is at Virginia Mason?"

"Yes, I think so." Anne Marie would visit the hospital first. If she hurried, she should be able to make it there and get back to the store before ten.

She stood. "I'll call you as soon as I know anything."

"Thanks," Helen said as she walked Anne Marie to the office door.

Before Anne Marie left the building, she decided to check on Ellen. She stood by the classroom door and peeked in to see Ellen chatting with her friends as if nothing was awry. Relieved, she went out to the parking lot.

When Anne Marie reached Virginia Mason Hospital it was already nine-fifteen. She explained her situation to the woman at the information counter, who gave her Dolores Falk's room number.

She took the elevator to the correct floor and found Dolores alone in her room, hooked up to IV tubes. Her color seemed improved, Anne Marie thought. When she walked in, Dolores opened her eyes.

"How are you feeling, Mrs. Falk?" she asked as she approached the side of the bed.

"I'm doing better. How's Ellen?"

"She's fine. Don't worry about her."

Tears welled in the older woman's eyes. "I can't thank you enough for taking my granddaughter. I don't know what I would've done if you hadn't come."

"I'm glad to help out." Never mind that it wasn't entirely true.

Dolores's chest rose with a sigh. "The doctor says I'm going to need heart surgery."

Anne Marie squeezed the woman's hand. "They have excellent doctors here and—"

"I'm not worried for me," Dolores said, cutting her off. "My only concern is Ellen."

"You just concentrate on getting well. I have the number for Ellen's aunt Clarisse and—"

"No!" Dolores cut her off again. Her fingers tightened on Anne Marie's.

"She's the emergency contact you gave the school. So I—"

"Clarisse is in prison."

"Prison?" Anne Marie swallowed her gasp of shock.

"Fraud." Dolores closed her eyes again, as if admitting this to Anne Marie embarrassed her. Anne Marie was sure it did.

"What about Ellen's mother?"

Tears rolled from the corners of the woman's eyes and fell onto the pillow that supported her head. "Her mother is a drug addict. The state of California took Ellen away from her when she was three years old. I'd lost contact with my daughter—I didn't even know about Ellen. By the time I learned I had a granddaughter, Ellen had gone through a series of foster homes. It took me a year to get that child to sleep through the night. I won't put her back in the system. I won't do that to her."

"Oh, dear…" Anne Marie said weakly. There didn't seem to be an adequate response.

"Whatever happens to me, don't let them put her in foster care."

Her agitation grew and Anne Marie began to worry. "Promise me," she pleaded. "Promise me."

"Of course." What else could she say?

Dolores relaxed a little after that.

"What about her father?"

Dolores shook her head grimly. "My daughter probably doesn't even know who fathered this child."

"Oh."

"There's no one else."

"Perhaps her mother's clean and sober now." Anne Marie hated to sound desperate, but the options were dwindling fast.

"She's not. Last year she rescinded all rights as Ellen's mother."

"Oh." Anne Marie could feel what was coming. Dolores would ask her to watch Ellen while she was in the hospital. A rush of excuses, a dozen valid reasons she couldn't do it, were on the tip of her tongue. She couldn't make herself say them.

"That child is the only good thing I have in my life," Dolores whispered brokenly. "My daughters have both chosen paths that led to spiritual and emotional ruin. I pray for them every day."

"I'm sure you do, but—"

"I don't understand where I went wrong. Their father left us twenty years ago, and I raised them alone. I tried to show them the right way…"

Anne Marie murmured a few comforting words, although she knew there was no comfort to be had.

"I've been a proud woman all my life," Dolores continued. "I've never asked the government for help, even when I was entitled to."

With her free hand Anne Marie gripped the steel bar along the side of the bed.

"I'm asking for your help now."

Anne Marie swallowed. "But…I'm a stranger."

"Ellen talks about you constantly. You and Baxter." The faint hint of a smile came to her then.

Anne Marie was surprised she got a mention. She'd assumed the real attraction had been the dog. "But…I'm just her Lunch Buddy," she murmured.

"You're much more than that," Dolores told her. "Please take my precious Ellen and look after her for me."

"I…" Anne Marie didn't know what to say. Her place wasn't set up to take care of a child. She didn't even have a real bed for Ellen. After living alone all these months— more than a year now—she wasn't sure how she'd adjust to living with someone else. With a child.

At her obvious reluctance, Dolores said, "The doctor said once I have the surgery I should be good as new."

"You'll need recuperation time." Mentally Anne Marie tried to calculate how long that might be. A week? Two? Maybe a month. She couldn't possibly deal with this awkward situation for a whole month.

"Yes, I'll need time to heal," Dolores agreed, "but it'll go much faster if I know Ellen is well taken care of." She gazed up at Anne Marie with wide, imploring eyes. When Anne Marie didn't immediately respond, Dolores added, "Please. I'm asking you from my heart. I'm begging you not to let them take my granddaughter away from me."

Anne Marie couldn't refuse. "All right," she said, hoping she didn't sound begrudging—or afraid.

Dolores released a huge sigh. "Thank you, Lord." She pointed to the side table next to her bed. "I've written out a statement that gives you permission to see to any medical needs Ellen might have. I also wrote a statement authorizing you to keep Ellen while I'm in the hospital."

An orderly stepped into the room. "Ready, Mrs. Falk?" he asked far too cheerfully.

"Where are you taking her?" Anne Marie asked.

The young man raised his eyebrows. "Surgery."

"So soon?"

"I'll be fine," Dolores said. "Absolutely fine."

Anne Marie felt dreadful; she should've been the one consoling the other woman.

"I'll take care of Ellen," she promised with a sense of desperation. "Just get well."

The young man directed Anne Marie to the nurses' station, where she was given a phone number to check on Dolores's progress after the surgery. Anne Marie held on to that piece of paper as if it were a winning lottery ticket. "She'll be okay, won't she?" she asked the male nurse.

The burly man sent her a stoic look. "We're going to do everything we can to make sure she's home again as soon as possible."

That was supposed to reassure her? "Thank you," she said lamely. "I'll phone later this afternoon."

"I'll have an update for you then. Ask for Dana."

"I will. Thank you." She put the phone number, plus the signed papers Dolores had mentioned, in her purse and left the hospital.

By the time she got to her car, Anne Marie's stomach was so tense she actually felt nauseous. Yesterday afternoon she'd been working out at Go Figure, the women's gym on Blossom Street, with Barbie Foster. Less than a day later, she was responsible for the care and well-being of an eight-year-old child.

At the bookstore, Anne Marie turned over the Open sign and counted out cash for the register. She had a constant flow of customers until about one o'clock, when she called the school and spoke to Helen Mayer.

"What did you find out?"

"Ellen's grandmother had heart surgery this morning."

"How's she doing?"

"I don't know. I haven't spoken to anyone at the hospital yet. I wanted to update you, though—Ellen will be staying with me while her grandmother recuperates."

"With you? What about her aunt Clarisse?"

"Apparently she…she's moved and can't be reached." That was reasonably close to the truth and should spare Dolores some humiliation.

"I'm sorry to hear that."

Not nearly as sorry as Anne Marie.

"It's good of you to look after the child. I'm surprised you agreed to it."

As much as she'd like to see Ellen with someone else, Anne Marie couldn't tell a sick woman that she preferred not to take care of her only granddaughter.

"I'll be picking Ellen up from school this afternoon and making arrangements for her to catch the bus on Blossom Street."

"I can do that for you," Helen Mayer told her. "If you need me to do anything else, just let me know. I think it's wonderful that you're willing to help out like this."

Anne Marie ended the conversation and then called the hospital. Dolores had made it through surgery without a problem, Dana informed her. She was currently in recovery, and if there were any changes, he'd call. Anne Marie gave him her phone numbers.

Fortunately Steve Handley, who worked on Thursday afternoons, was able to come in an hour early despite the short notice, which freed Anne Marie to drive to the elementary school and get Ellen. The child's face brightened when she saw her.

"How's my grandma?" she asked.

"She's in the hospital, and the doctors and nurses are taking good care of her."

"When will she be home?"

"Soon." Anne Marie bent down to look into the little girl's eyes. "Until your grandmother's home again, would you like to stay with me?"

Ellen didn't answer right away. "I guess that would be okay."

It wasn't exactly an overwhelming affirmation, but it was good enough. "We'll need to stop by your house this evening and pack a bigger suitcase."

"Can Baxter come with us?"

"I think he'd like that."

"I brought the book from the school library," Ellen announced.

It took Anne Marie a moment to realize the book she meant was the one about dog tricks.

They drove back to Blossom Street in silence, Ellen staring straight ahead. After dinner, they'd return to the house and collect her things.

Dinner.

Anne Marie hadn't given it a moment's thought. No more skipping meals. No more pity parties, either. She had to be strong for Ellen's sake. She had to hold her life together for a couple of weeks. Anne Marie figured she could manage that.

Two weeks. Maybe three.

Four at the most.

The time would pass quickly. She hoped.

Chapter

13

Thanks to her list of Twenty Wishes, Barbie Foster was thinking harder, doing more and experiencing life with greater excitement. Her list was nearly complete, and she loved the way it helped her analyze what she really wanted. For years, her focus had been on Gary and the twins. But with her sons away at school, she'd been at loose ends, never quite adjusting to the change in her routine. She missed her husband so much, even now. He'd always be a part of her—and yet she was only forty, with a lot of life yet to be lived.

Instead of working at the dress shop this Saturday, Barbie decided to take a day off and go to the St. Patrick's Day concert in Freeway Park. Anne Marie Roche had arranged for the afternoon off, as well, and the two of them planned to make an occasion of it. They'd met a couple of times at Go Figure, and she'd enjoyed getting to know her better.

Her friendship with Anne Marie had deepened since their Valentine's gathering. Until that night, Barbie had viewed Anne Marie as reserved, a bit standoffish. All of that had changed when they started talking about their Twenty Wishes.

She'd begun to see Anne Marie as a kindred spirit and discovered a wry sense of humor. Her liking had turned to respect when she learned that Anne Marie was looking after eight-year-old Ellen Falk while her grandmother recuperated from heart surgery.

Barbie had met Ellen at the bookstore the day before, when she'd come in to buy a couple of romances. The child was sweet and unpretentious; she obviously idolized Anne Marie and was completely in love with her dog, Baxter. Barbie had watched with some amusement as Ellen struggled to teach the Yorkie to roll over, with no success.

Ellen was joining them for the St. Patrick's Day concert that afternoon. When Barbie met her and Anne Marie at the bookstore shortly after twelve, they were ready and waiting.

"Where are we going?" the little girl asked, fastening the buttons on her light green coat, which looked brand-new. Thankfully it wasn't raining; that was good news, since March was notorious for drizzle in the Pacific Northwest.

"We're attending a concert with Irish music," Anne Marie explained to the youngster. "Then afterward we're visiting my mother in Ballard."

"Will we visit Grandma Dolores, too?"

"Sure thing." Anne Marie buttoned up her own jacket. "Right after we see my mother."

The child nodded thoughtfully. "What's Irish music sound like?"

Anne Marie hesitated. "Well, it's usually pretty fast

and…" She shrugged, and Barbie laughed as she gave up trying to describe it. "Just wait. You'll hear it soon enough."

"Will I like it?" Ellen asked, tilting her head curiously.

"I do," Anne Marie told her. "I like it a lot."

Ellen nodded firmly. "Then I will, too."

Because Freeway Park was relatively close to Blossom Street, they decided to walk. The air was crisp, the sky clear and bright. They moved at a slow pace to accommodate Ellen's shorter steps. Barbie noticed that the child took in everything around her with huge inquisitive eyes.

When they reached Freeway Park, above Interstate 5, it was already crowded. Finding a spot to sit was difficult, although they eventually did when a couple of teenagers were kind enough to share their space. Anne Marie had remembered to bring a blanket, which she smoothed out on the grass. A platform had been built for the performance, and they had a good view of the stage.

Ellen sat cross-legged on the blanket. Barbie and Anne Marie arranged themselves close to her. Barbie hadn't done anything like this since before she'd lost Gary. It reminded her of family expeditions when the kids were little, and she felt a quiet joy, an awareness that she could be happy again.

After the accident, her primary concern had been for her children. Now that they were away at school, she was no longer insulated from the pain and the loss. It was this same loss she sensed in Mark Bassett, and one reason she was so drawn to him.

For her mother, widowhood had been a different story. They'd never really discussed it, but Barbie knew about her father's indiscretions. Lillie had chosen to ignore them. And because her mother said nothing, Barbie didn't, either. She knew that Lillie grieved for David. She'd loved him, but in some ways Barbie thought his death might have been a release for her mother—although she'd never so much as hint at such a thing.

"When will it start?" Ellen asked after sitting quietly for several minutes.

"Soon."

"Are you hungry?" Barbie asked.

The girl shook her head and tucked her hands beneath her thighs.

There was festive chatter all around them; everyone seemed to be in a cheerful mood, exchanging greetings, laughing, talking.

"Ellen likes to sing," Anne Marie told her.

"Do you?" Barbie asked, turning to the child.

At the question, Ellen's face grew red. "Anne Marie says I'm a good singer. She heard me sing in the school play." The child obviously put great stock in the compliment.

"Maybe Anne Marie can teach you a few Irish songs," Barbie suggested.

A look of such profound sadness flashed into her friend's eyes that Barbie instantly placed her hand on Anne Marie's forearm.

"I used to sing, but I don't anymore. I…can't," Anne Marie mumbled, staring down at the blanket. "I lost my voice after Robert died…. I thought it would return, but it hasn't yet."

"I'm sorry." Barbie felt she had to apologize because it so clearly upset her friend.

"No, no. I mean, for heaven's sake, it's not your fault." Recovering quickly, Anne Marie dismissed her concern with a quick shake of her head.

Ellen gazed up at her, frowning. "I didn't know you can't sing."

"Don't worry, Ellen," Anne Marie murmured. "I will again."

Because Ellen was restless and maybe because Anne Marie wanted to change the subject, the two of them went for a short walk around the park before the music started. As they left, Barbie saw that the little girl stuck close to Anne Marie's side. Being with so many people was probably overwhelming for a child. Barbie had to credit her friend; it couldn't have been easy to bring this child into her home, even for a short while.

In fact, Barbie thought Anne Marie seemed softer now, less cynical and more open. Being with such a vulnerable child, having to take responsibility for her, meant that Anne Marie was less focused on her own sorrows. Wasn't that what Elise kept saying? Doing something for someone else made you feel better about yourself.

The group of Irish singers was introduced, and the crowd instantly broke into applause. Ellen and Anne Marie hurried back to the blanket just as the performance began.

The singers, the fiddlers and dancers were thrilling, and Barbie loved every minute. The music was infectious. And the dancing—it was so vigorous, yet disciplined, too. Ellen sat through the entire hour mesmerized. She seemed to absorb the music, every note of it. When the performance was over, her face glowed.

"That was so good," she said, looking at Anne Marie and Barbie. "I want to sing like that someday. Do you think I can?" she asked plaintively.

"Yes," Anne Marie told her in a confident voice. "I'm sure you can."

People had started to leave the park. The exodus was

slow moving, but Barbie wasn't in any hurry. Besides, her feet hurt. That was what she got for wearing designer shoes; she'd chosen them because they were the perfect complement to her black linen pants and green silk blouse. The sun warmed the day, and she'd left her raincoat open, the belt dangling at her sides.

As she and Anne Marie waited patiently for the crowd to thin, Barbie saw a flash of chrome from the corner of her eye. She turned to look and then caught her breath. She grabbed Anne Marie's elbow.

"It's him…." She could barely get the words out. Feeling self-conscious, she dropped her hand.

"Who?" Anne Marie asked.

Barbie couldn't tell her because she hadn't told anyone about her attraction to Mark Bassett, the man in the wheelchair. She looked again, just to be sure. He was alone, or appeared to be, apparently waiting for the crowd to disperse. Maneuvering his wheelchair would be difficult with so many people pressing in around him.

"You know someone here?"

"Not really." Barbie tried to calm the wild beating of her heart. This was an unexpected surprise, a bonus. She was pleased now that she'd taken care with her outfit and makeup.

If Mark had seen her—and she couldn't tell either way—he refused to acknowledge her. Barbie bit down hard on her lower lip to keep from raising her hand and calling out to him.

"Do you know that guy in the wheelchair?" Anne Marie asked.

"I…not exactly. I bumped into him recently." She didn't mention the part about emptying her soda in his lap.

"He's certainly a striking man."

He was. Barbie had trouble taking her eyes off him.
The crowd had mostly disappeared by then and only a few
stragglers remained.

"Can we go see Grandma Dolores soon?" Ellen asked.

Anne Marie smiled at the girl. "After we visit my
mom, okay?"

Her patience with Ellen impressed Barbie.

"I think I'd better head out," Anne Marie said, glancing down at her watch. "We're meeting my mother for a
late lunch, and after that we're going to the hospital."

"Of course, no problem," Barbie told her. "I've got
plans myself."

They left, which worked out well because now she
was free to confront Mark. Barbie didn't have a single idea
as to what she'd do or say once she reached him. She'd
figure that out when the time came.

He'd managed to leave Freeway Park and was moving
steadily down the sidewalk. Barbie raced after him, having some difficulty with her shoes. "Hello, again," she
called out cheerfully.

He ignored her.

"Remember me?"

At her second attempt, Mark spun his wheelchair
around. "What are *you* doing here?"

"I came to enjoy the music, just like everyone else."

"I didn't know there'd be a concert," he grumbled.

"In other words, you wouldn't have come if you had."

"Right."

"But you enjoyed it, didn't you?"

"No."

Barbie didn't understand him—and she didn't believe
he hadn't been affected by the music. "Why are you such
a grouch?" she asked.

"I like being a grouch."

"Yes, Oscar."

He frowned. "What?"

"Oscar the Grouch from *Sesame Street*." Her sons had often watched it when they were young. She planted herself directly in front of his wheelchair, blocking him off.

He wasn't amused.

She'd never been so rude in her life, but Barbie wasn't about to let him escape.

"What is it you want?" he demanded.

Now that he'd asked, she wasn't entirely sure. To get his attention, yes, but she couldn't admit that. "To talk, I guess."

He tried to wheel around her, but once again she hindered his progress. "I'm not interested in talking, nor am I the least bit interested in you."

Barbie sighed deeply. "That is *so* refreshing."

"I beg your pardon?"

She smiled down at him. "You wouldn't believe how many guys constantly hit on me. Not you, though, and yet we seem to like the same movies. You know, we might actually have something in common."

He wagged his index finger at her. "I'm on to your game. You and Tessa are in cahoots—you have to be. That's how you knew which movie I'd be watching last week. Well, that won't happen again."

Barbie felt her blood surge with excitement. "I wouldn't count on it. You can't tell me which movie to see or not to see."

He scowled back at her. "Don't count on me being there."

"That's no guarantee we won't bump into each other somewhere else," Barbie said, changing tactics. "We met here, didn't we? I think it must be fate."

"I think it's bad luck."

"Oh, Mark, honestly."

His scowl grew darker.

"Your niece seems fond of you," Barbie said conversationally.

His hands were on the wheels of his chair. "I'd like to get out of here if you don't mind."

"I wanted to talk, remember?"

"I don't."

"Fine." She raised both hands in a gesture of defeat. "Have it your way."

"Thank you," Mark said gruffly and as soon as she stepped aside, he wheeled past her.

Despite his dismissive tone, Barbie followed him. "Can I ask you something?" she began.

Mark disregarded her, apparently a habit of his. His speed was surprising and in an effort to catch up with him, Barbie was nearly trotting. Her heel caught on a crack in the sidewalk and she went flying forward, landing hard on her hands and knees.

"Damn!" she cried at the sudden sharp pain. Momentarily stunned, she sat back and brushed the grit from her hands. Blood seeped through her pants and tears smarted her eyes.

Mark stopped, then reluctantly spun around to face her. "What happened?" he asked, none too sympathetically.

"I tripped."

"Are you hurt?"

"Yes. Look, there's blood."

"Should I call 911?"

He was making fun of her, but Barbie didn't care. She peeled up her pant leg to examine her knee.

"That's what you get for wearing those ridiculous shoes."

She let the insult pass.

"Do you need help getting up?"

"No, I can manage." When she scrambled to her feet, she discovered that she'd broken the heel off her left shoe. "Would you *look* at this?" she cried. "If you knew what I paid for these shoes, you'd be as outraged as I am."

"Next time don't go chasing after me," he said. "I'm not interested, understand?"

"Okay, fine," she snapped.

"Fine with me, too." He started to roll away from her.

Barbie sniffled and limped off. She'd made an idiot of herself and now she was paying the price. So much for this supposed bond between them. He wanted nothing to do with her. Well, she got his message, loud and clear.

Her progress was slow with her knee aching and her broken shoe.

"Miss, Miss."

Barbie turned to find a woman with a first aid kit in her hand. "I heard that you fell."

"Who told you?"

"A man in a wheelchair stopped in my store and said you might need help."

"Really." So Mark wasn't as hard-edged as he'd like her to believe. He *was* concerned about her but he didn't want to show it. "I'm okay. My pride hurts a lot more than my knee. It was my own fault."

"Are you sure I can't help you?"

Barbie thanked the woman with a smile. "I think I'll just go home." She'd call her mother for sympathy and then have a cup of hot tea.

"The man told me you'd probably say that. If you'll sit down, I'll take a look at your knee."

"I don't suppose you have any glue, do you?" she asked, holding up her broken shoe.

"No, sorry."

Barbie thanked her again and left, hobbling back to Blossom Street, where she'd parked her car. The injury to her knee was nothing more than a scrape but the blow to her pride would take much longer to heal.

Her one consolation was the fact that, despite everything, Mark had sent someone to check on her. It wasn't a lot, but it was *something*. A tiny fracture in his resistance. It gave her hope.

By Monday evening, Barbie's knee was healing nicely. Although she didn't need to, she wore a huge bandage over it and a short skirt, short enough to reveal her bandaged knee.

Tessa was at the ticket window when Barbie approached.

"So, which movie should I see?" Barbie asked, the same as she had the week before.

Tessa's dark brown eyes searched hers. "He isn't here."

"You mean not yet, right?"

"Uncle Mark's not coming, period."

"Why not?" Barbie couldn't have disguised her disappointment if she'd wanted to.

"He figured out that I was the one feeding you information." Tessa sounded as disgruntled as Barbie felt.

Because she was holding up the line, Barbie stepped aside until there was a break.

"I'm sorry," Tessa murmured. "He told me he won't be coming to the movies again and that I should make sure you knew it."

"Oh," Barbie murmured. "Do you see him outside the movies very often?"

Tessa shrugged. "Sometimes."

"Next time you do, tell him I think he's a coward."

Tessa's jaw dropped. "You're not serious."

"Yes, I am," Barbie insisted. "Tell him that for me."

She purchased her ticket, plus popcorn and a soda. Although she sat through the entire movie, she couldn't remember a single scene.

Chapter

14

Monday evening Anne Marie put a meat-loaf-and-potato casserole in the oven. It was a favorite recipe of her mother's and one she hadn't made in years. The meat mixture baked with sliced raw potatoes, both covered in tomato sauce. Anne Marie had liked it when she was around Ellen's age and she hoped Ellen would, too.

As she closed the oven door, she noticed Ellen approaching the large oak desk where she kept the scrapbooking materials for her Twenty Wishes book.

"What's this?" Ellen asked, looking over her shoulder.

"My Twenty Wishes."

"Twenty Wishes," the girl repeated. "What are those?"

"Well, on Valentine's Day, my friends and I had a small party. We started talking about all the things we'd wished for in our lives and then we each decided to make a list."

"Just twenty?"

Anne Marie laughed. So far, coming up with twenty had been hard enough, and in fact, she was only halfway there. "This is fine for now. I'll think of more later on," she said. "In fact, I'm still working on my first twenty." She had a total of nine: the five she'd written earlier, plus her most recent additions.

6. Find a reason to laugh

7. Sing again

8. Purchase a home for me and Baxter

9. Attend a Broadway musical and learn all the songs by heart

She was considering a line dancing class, which was a wish she'd erased earlier. The St. Patrick's Day performance had inspired her interest in dancing again.

"The wishes don't need to be practical," Anne Marie went on to explain. "That's why they're called wishes instead of resolutions or goals."

"What do you mean?"

"Well, I don't necessarily expect them all to happen."

"If you don't expect them to happen," Ellen asked, regarding her quizzically, "then why are you writing them down?"

"Because they're *wishes*," Anne Marie said. Finding a pen, she added a wish she'd erased two or three times.

10. Travel to Paris with someone I love

That encompassed the essence of what she sought—love, adventure, new experiences.

Ellen stared down at the recently entered wish. "Can anyone make a list like this?"

"Of course." Anne Marie set the timer on the oven. They'd gotten into a routine, the two of them, during the past five days. It felt as if Ellen had been with her much longer. One obvious difference in her life was that Anne Marie now regularly cooked dinner.

Ever since she'd started living alone, she'd fallen into

the habit of grabbing something quick and easy or skipping dinner altogether, which she could ill afford to do. But Ellen needed regular nutritious meals and a daily structure. With everything else in the girl's life in upheaval, Anne Marie could at least offer her that.

The phone rang and Anne Marie picked it up immediately before Call Display could even register the number. She was expecting to hear from Elise, whom she'd been trying to reach all afternoon. "Hello." She figured Elise wanted to share her news, which Anne Marie had already heard via the neighborhood grapevine. Elise had taken a part-time job working for Lydia at A Good Yarn.

"Anne Marie, it's Brandon."

"Brandon! It's great to hear from you," she said with genuine pleasure.

"I've been meaning to call you for a couple of weeks," he went on. "Melissa told me what she did. I can't believe my sister sometimes. And as for my father…"

"Don't worry, I'm fine." That was mostly true.

"You're sure?" Brandon pressed. "To be fair to Melissa, I doubt she realized how hard you'd take that business about Dad. And she was pretty devastated herself."

"Really, it's okay," Anne Marie lied, brushing off his concern. The last thing she wanted was to talk about her husband's indiscretion—or even think about it. She felt a rush of pain whenever she remembered and constantly guarded herself against the image of Robert with Rebecca. In his office, on the couch…

"You're *sure?*" he asked again. He didn't seem convinced.

"Yes. Positive." As much as possible she made light of the incident.

Her stepson hesitated a moment, then blurted out, "Let

me take you to dinner tonight. I know it's short notice, but we could talk and—"

"I can't." She hoped he'd take her at her word, not force her to explain.

"Why not?" Brandon's voice fell with disappointment.

"I have a visitor."

Her announcement was met with a short silence. "Anne Marie, are you seeing someone?" he asked somberly.

"No, of course not!" The question amused her. "Melissa asked me the same thing."

"Of course not? Why say it like that? You're young and beautiful and—"

"I'm with a…friend."

"Ah, the mystery intensifies."

"It's not a mystery," she said, smiling at his teasing banter. "It's Ellen. She's eight and she's living with me for the next week or two."

"You have an eight-year-old living with you? Is she a relative of yours?"

"No, I met her through a nearby school—the Lunch Buddies program. Why don't you join us?" she said impulsively. "I just put a casserole in the oven and it won't be ready for another forty minutes."

"You made dinner?"

"Don't sound so shocked. I did a lot of cooking in my time."

"Okay, I'd like that. Thanks. Give me a few minutes to finish up here and I'll drive straight over. You're still living above the bookstore, right?"

"For now." She really did hope to purchase a house, and soon. Spring, especially May and June, were the best months to look. As soon as Ellen was back with her grandmother, Anne Marie had every intention of beginning her search.

"Brandon, one thing…Melissa's and my conversation…"

"Yes?"

"I don't want to discuss it, all right?"

He hesitated. "If that's what you want."

"It is," she told him, keeping her voice firm.

Anne Marie hung up the phone and turned around to discover Ellen perched on a chair at the kitchen table, staring blankly into space. She had the end of a pencil clamped between her teeth.

"Are you doing your homework?" Anne Marie asked.

Ellen shook her head. "I'm making a list of Twenty Wishes."

"Oh, really?"

Ellen nodded. "Do you want to hear what I have so far?"

"I would." Anne Marie pulled out the chair next to her and sat down.

"One," Ellen announced with great formality. "Plant a garden."

"What kind of garden?"

"Flowers," Ellen said. "I read the book you gave me about that garden, remember?"

Anne Marie smiled approvingly. Of course. On Sunday she'd given her a copy of the Edwardian children's classic, *The Secret Garden*. Ellen was an advanced reader and had no difficulty with comprehension. Occasionally she'd asked about the meaning of a word. She'd loved the idea of the walled garden, hidden from the world, and had instantly identified with the story's orphaned young heroine.

"Is there any other kind of garden than flowers?"

"Vegetables."

"You can grow tomatoes?" Ellen asked in an excited voice. "I like tomatoes a lot, especially when they're warm from the sun. I like them with salt."

Anne Marie looked at her curiously. "Did you ever have a garden before?"

Ellen lowered her gaze. "No… Grandma Dolores told me about warm tomatoes with salt. I've never had one but I know they'd be really good because my grandma said so."

"I like tomatoes, too." Anne Marie closed her eyes at the memory of working in her garden at the house she and Robert had owned. The smell of earth, the sun warm on her back… "Last summer I grew tomatoes right here, on my balcony."

The child seemed thoroughly confused by that.

"It was a container garden because I didn't have any-where to plant an actual garden."

"What about corn?"

"That might be a challenge, but I'll check into it. If you like, we can plant seeds in egg cartons and then once your grandmother's home again, I'll help you clear a small space in her yard for your very own garden."

"Really?" The girl's face shone with uncomplicated joy. "A garden," she breathed.

"Anything else on your list?"

Ellen nodded. "I want to bake cookies with Grandma Dolores."

"I bet she'd like that."

"She always said we could, but then she'd get tired or she wouldn't be feeling well and we never got to do it."

Anne Marie slipped her arm around Ellen's shoulders. "When your grandmother's back from the hospital, she'll be feeling much better and have a lot more energy, and I'm sure she'll want to bake cookies with you then."

"Oatmeal and raisin are my favorites." Ellen set the pencil down. "I couldn't think of anything else."

"What about something whimsical?"

Ellen turned to her, expression blank.

"Whimsical means fanciful—a wish that's not…serious, I suppose you could say. Something lighthearted, just for fun."

The end of the pencil returned to Ellen's mouth. "Do *you* have anything whi-whimsical on your list?" she asked.

Good question. "Not yet. Let's think about it." She stood to get three plates from the cupboard.

"Like what?"

"Well…" Anne Marie murmured. She looked at the child, then walked over to her own list.

11. Dance in the rain in my bare feet

"What did you write?" Ellen asked.

Anne Marie told her.

Ellen started to giggle. "That's silly. Aren't you afraid your clothes will get wet? Or you'll get mud between your toes?"

"I wouldn't care, especially if I was dancing with someone I loved." She opened the refrigerator and removed a bag of romaine lettuce and other salad ingredients. Anne Marie occasionally made salad for dinner; she wasn't afraid to add unusual ingredients, like walnut bits, cranberries, raw green beans, Chinese noodles, sunflower seeds, pickle slices, beets… Her inventions weren't always successful—the chopped anchovies came to mind—but they were usually interesting.

For Ellen's sake, she chose more conventional makings of cherry tomato, shredded carrots, cucumber and green pepper. Then, because she couldn't resist, she added crushed pretzels, guessing Ellen would enjoy that.

As she and Ellen set the table with a white cloth and some leftover St. Patrick's Day napkins, Anne Marie explained who Brandon was and said he'd be coming over for dinner. Ellen seemed a little nervous about meeting him and, perhaps, about sharing their private time together.

There was a knock at the back door just as Anne Marie put the salad in the middle of the table. "That's Brandon," she said, walking to the door as Baxter barked excitedly.

Her stepson entered the small apartment, both hands behind his back. With a sweeping gesture, he produced a bottle of her favorite wine and a bouquet of flowers. He kissed Anne Marie on the cheek, then presented the flowers to Ellen. "You must be the lovely Ellen. These are for you," he said.

Ellen gave Brandon a tentative smile. "I like flowers."

"Are you going to thank Brandon?" Anne Marie asked.

"Thank you."

"You're welcome, Ellen."

Anne Marie found a vase and helped Ellen arrange her bouquet. Then, without being asked, Ellen opened the silverware drawer and counted out what they needed, while Anne Marie got two wineglasses and one for juice. When they sat down to dinner, Ellen said grace.

Brandon's eyes met Anne Marie's as he bowed his head, and he murmured "Amen." Ellen insisted on saying a prayer before all their meals. Her grandmother had taught her that and it always made Anne Marie wonder how the woman's two daughters, presumably raised the same way, had turned out so badly.

Brandon raved about the casserole. "This is *really* good."

"Secret family recipe," Anne Marie told him with a smirk.

"Will you give it to my grandma?" Ellen asked, scraping up the last of the casserole from her plate.

"If you want me to."

"I'd like it, too," Brandon added. "Hey, I'll give it to one of my girlfriends to make."

"Hey, make it yourself."

"Fine," he laughed. "I will."

They finished their wine; then Brandon and Ellen cleared the table, while she made a pot of coffee.

"May I go to my room and read?" Ellen asked. She'd just started the Laura Ingalls Wilder series and Anne Marie knew she was eager to return to *Little House in the Big Woods*.

"Yes, Ellen, you may."

They watched as Ellen retreated to her bedroom, Baxter close behind. Brandon turned to Anne Marie, leaning casually back in his chair. "You'd make a good mother," he said thoughtfully.

"Thanks," she said, but it was a moot point. If she was going to have a child, there had to be a father, and she was nowhere near ready for another relationship. In a few months she'd be thirty-nine and soon after that it would simply be too late. She had no intention of doing what a few women she'd heard of had done—get pregnant via a willing "sperm donor," a man who would play no role in their babies' lives.

When the coffee had brewed she filled a mug for Brandon and one for herself before joining him at the table.

"Have you talked to Rebecca yet? My dad's assistant?" he asked.

He certainly hadn't delayed in getting to the point, even though she'd explicitly said she'd prefer not to discuss it. Anne Marie let the question slide for a moment as she busied herself with the cream and sugar.

"You don't have to tell me if you don't want to," Brandon said with teasing sarcasm.

She sighed, giving up. "The short answer is no. The long answer is I'm not sure I ever will. If she comes forward and acknowledges the child is Robert's…then I'll deal with it. Not before."

"I can understand that," Brandon said after a long moment. "I want you to know that Mel genuinely regrets what happened."

Anne Marie shrugged it off. "How is your sister?"

"We talk every now and then. I have to say she seems a lot more serious now. More mature, you know?" He frowned. "When I called her last week, she told me she's on the outs with Mom."

That surprised Anne Marie. As far as she knew, Melissa and Pamela were close. Robert's ex-wife lived in England, where she worked for an international hotel chain. According to Robert, her devotion to her career had led directly to their divorce. Pamela had accepted a position that involved frequent travel, even though Robert had asked her to wait until the children were out of school. She'd refused and left him and their family for months on end.

"What's wrong between Melissa and her mother?"

Brandon shrugged. "She wouldn't tell me. When I pressed the issue, she changed the subject. She obviously doesn't want to talk about it, but she made it sound like she's busy with school and she probably is."

"She's graduating this year, isn't she?"

Melissa was completing an MBA program; she then planned to follow in her mother's footsteps, moving into hotel management.

Robert had always been proud of his children, and he'd

often said they were the only good thing to come out of his marriage to Pamela.

"Yeah, she should be done in June."

"Is she still seeing Michael?"

"As far as I know. He's a good guy. I like him better than any of the other guys my sister's gone out with. Some of them were…well, put it this way." He reached for his coffee again. "Melissa's made some strange choices."

Before Anne Marie could respond, Ellen stepped into the room, a pad and pencil in her hand. "Is having a gold-fish a wish or a goal?"

"Well, it's a little of both, I'd say."

"Okay."

"I thought you were reading," she said.

Ellen looked down, a tendency she had when she was afraid she might be in trouble. "I was reading, but then I thought of another wish. I want twenty, the same as you."

"I only have eleven written down so far."

Ellen nodded. "Can I put dancing in the rain with bare feet on my list, too?"

"Sure." Anne Marie grinned. "Just remember, there's no need to rush. Think carefully about each wish."

"Okay." Ellen returned to the bedroom, muttering quietly to herself.

That interruption generated a series of questions about Anne Marie's Twenty Wishes. She didn't mind Brandon's interest; in fact she was grateful for the change of subject and explained in detail what she and the other widows were doing.

A half hour later, after Brandon had finished his coffee, he left. It was eight-thirty, time to get Ellen ready for bed.

"Grandma sounded tired when I talked to her this afternoon," Ellen said, sliding her nightgown over her head, thin arms raised.

"She'll be tired for a long time. Heart surgery takes a lot out of a person. She's going to need plenty of rest."

Ellen seemed distressed by that. "But—"

"You'll be able to go home to your grandmother soon," Anne Marie promised quickly. She received daily updates on Dolores's condition and everything was progressing exactly as it should. In two or at most three weeks, she'd be back in her own home, with a visiting nurse to look in on her. Ellen would be returning to the only stable life she'd ever known.

Pulling back the sheets, Anne Marie tucked the child into bed.

"Can we say our prayers?" Ellen asked sleepily.

"Of course."

"Should I say the words out loud or should I just say them in my heart?" Ellen murmured. Most nights she'd prayed in silence, mouthing the words as Anne Marie watched.

"What do you usually do with your grandmother?"

"She likes me to say them out loud."

"Then do it like that," Anne Marie said. The child's simple faith touched her, reminding her of a time when she, too, had prayed. Anne Marie couldn't remember when she'd stopped or why. She'd just…gotten out of the habit, she supposed.

Ellen studied her. "You're supposed to hold my hands and close your eyes. That's what Grandma Dolores does."

"All right." She clasped Ellen's hands in hers and shut her eyes.

Apparently she'd satisfied Ellen, because the youngster began to speak. "God, it's me, Ellen, again." She prayed for her grandmother and thanked God for her teacher and her friends and went through a long list of subjects, from hoping she'd do well on tomorrow's spelling test to thanking God for her new green raincoat.

Anne Marie didn't want to interrupt, but *she* was the one who'd supplied the coat, not God.

"And thank you most of all for Anne Marie, so I didn't have to go to a foster home and amen," Ellen whispered.

"Amen," Anne Marie echoed. Her knees had started to hurt and she rose awkwardly to her feet. On impulse she bent over and kissed Ellen's forehead. "Good night, sweetie."

"Good night."

About ten, she took Baxter for a five-minute walk, keeping the apartment in sight. When she got back, the phone rang; it was Elise Beaumont. "I wondered when we'd connect," Anne Marie said after her initial greeting.

"Sorry to call so late."

"That's okay."

"The last couple of times I stopped by the bookstore, you were busy."

"I know."

"I wanted to ask how the conversation with Rebecca Gilroy went."

"Oh." That question just didn't seem to go away. "I heard you're working for Lydia now," she said instead.

"Don't try to distract me. Have you spoken to Rebecca?"

Anne Marie didn't understand why everyone seemed to think it was her responsibility to confront the other woman.

"You *have* spoken to her, haven't you?"

"No." She had good reasons for not contacting the woman who'd been sleeping with her husband—reasons that were no one's business but her own.

Why would she *want* to talk to this woman, who'd likely given birth to Robert's child?

Anne Marie tossed and turned all that night, and when she got up at seven, she doubted she'd had even two hours' sleep. Whenever she started to drift off, she'd jerk awake, unable to escape the image of Robert and his assistant together, arms and legs entwined. Anne Marie had only met Rebecca Gilroy a few times but remembered her well. Tall and curvy, auburn-haired and in her twenties. As she struggled to sleep, all she could see was the other woman with her swollen belly. Pregnant.

With Robert's child.

Ever since the dinner with Melissa, Anne Marie had tried hard to keep busy, not to think, not to dwell on the pain that threatened to swallow her whole. But then it would come back, refusing to leave until she acknowledged it.

No, she wouldn't confront Rebecca Gilroy. She couldn't see the purpose of exposing herself to that reality if she could avoid it.

With Baxter on his leash, Anne Marie walked Ellen to

the bus stop, where a small group of youngsters waited, her eyes smarting from lack of sleep. She took her dog home and did a few household chores before going down to the bookstore at ten and officially opening it.

Lillie was there at five after. As soon as she saw Anne Marie, she frowned. "You look terrible."

"Thanks," Anne Marie said wryly. "Good morning to you, too."

"Is something wrong?" Lillie asked.

"I didn't sleep very well last night."

"Anything I can do to help?"

"No, but thanks for offering." She wasn't going to discuss this with one more person, even a friend as caring and sympathetic as Lillie.

Anne Marie turned on her computer to do an inventory check while Lillie roamed the shelves. A little while later, she brought an armload of books to the counter; she was a voracious reader and usually purchased hardcovers. Anne Marie could count on Lillie to buy as many as ten books a month. Her most recent selection included a couple of romances. This was a switch; her friend tended to read mysteries and thrillers. Anne Marie added up her purchases, which Lillie paid for with a debit card.

"Have you spoken to Elise lately?" Lillie asked as she slipped her card back into her wallet.

"She called last night."

"Did she mention her Twenty Wishes?"

Elise and Anne Marie had chatted about a number of things; however no topic had stayed in her mind beyond the first one Elise had brought up. "Not really."

Lillie shook her head. "We really need to meet again and update one another. I've taken action and I know you have, too. Sharing our lists would be an encouragement, don't you think?"

Anne Marie wasn't convinced of that, but arguing about it required more energy than she had. Lillie suggested a day and time, and Anne Marie agreed. "We'll meet at my house next Thursday, the twenty-seventh," Lillie said, consulting an elegant little calendar she pulled out of her purse.

Anne Marie agreed to that, too.

"Barbie told me you're looking after a young girl," Lillie said next. "That's wonderful!"

Anne Marie was beginning to feel guilty accepting all this praise. The fact was, had there been any other alternative for Ellen, she would've been grateful.

"My wishes are coming along nicely," Lillie said, continuing the conversation. "I'm taking this very seriously, you know. It was exactly what I needed." She sighed. "I find myself thinking more and more about the things I'd like to do, to experience." She placed one hand over her heart. "I have a sense of...of *expectation* that I haven't felt in years. It's like I've finally given myself permission to do what *I* want."

Anne Marie hadn't felt any of that. Most of her wishes had to do with recovering from Robert's death. To sing, to laugh, to dance. None of those had come to pass yet and in her current frame of mind, she wasn't sure they would.

Feeling obliged to say something, she said, "Did I tell you I bought scrapbooking supplies and a binder for my wish list?"

Lillie straightened. "You did, and I like the idea very much. I've been planning to do it myself."

"You should," Anne Marie urged. She didn't hold an exclusive on the idea.

"I think we'd *all* profit from making a Twenty Wishes binder, don't you?"

Anne Marie nodded with a tired smile.

Lillie left a few minutes later, carrying two large bags, and the day crawled from that point on. Anne Marie could hardly make the effort to smile. She could've phoned Theresa to fill in for her but didn't. Ever since Ellen had come to live with her, she'd called on her three part-time employees again and again. Since her other two were college students, they were in class on and off during the day. She didn't want to take advantage of Theresa's kindness, although she would gladly have gone upstairs and crept into bed, craving the oblivion of sleep.

When the school bus dropped Ellen on Blossom Street, the girl dashed into the bookstore, her eyes sparkling. "I got an A on my spelling test!"

Anne Marie tried to show her how pleased she was and wondered if she'd succeeded.

Ellen didn't seem to notice her exhaustion. "Can I show my grandma?" she asked eagerly.

"I…"

"You said we could visit her again on Tuesday, remember?"

Unfortunately Anne Marie did. "Sure," she said, taking a deep breath. Too many promises made to Ellen had been broken, and she refused to be guilty of that herself. Robert had promised to take her to Paris one day. And he hadn't. He'd promised to love her and be faithful. He hadn't done that, either.

She allowed Ellen to bring Baxter down to the store, and the two of them curled up in one of the big chairs. Ellen spelled each of the words from her test for the Yorkie, who appeared to listen intently.

At four Steve Handley arrived. He usually worked from four to six Monday to Wednesday and four to eight on Thursday and Friday. He often closed for her, and Anne Marie trusted him implicitly.

As soon as she'd handed everything over to Steve, she, Ellen and Baxter retreated to the apartment. Not up to making dinner, Anne Marie heated yesterday's leftover casserole for Ellen, adding an apple and a store-bought oatmeal cookie. Her own appetite was nonexistent.

The child ate silently, then placed her dishes in the sink.

"Are you ready to go?" Anne Marie asked.

Ellen turned to face her, eyes wide and hopeful. "I *can* visit Grandma Dolores?"

"You certainly can." God would bless her for this, Anne Marie told herself.

Ellen raced into her room and hurried back with her spelling test clutched in one hand. All the way to the hospital Ellen talked excitedly, about Baxter's progress with his new tricks and how she'd almost spelled *puzzle* with one *z* and a hundred other things she planned to tell Dolores. Anne Marie felt wretched. She'd been so consumed by her own troubles that she'd failed to realize how desperately the child missed her grandmother.

Ellen needed reassurance that Dolores was on the mend and that everything would soon return to normal. Anne Marie wasn't the only one whose life had been disrupted. The child must feel so lost and adrift without her grandmother's love and guidance.

Anne Marie had kept in touch with Dolores Falk by phone, and she'd called the hospital every day for information on the older woman's condition. Dolores was improving at a steady rate. The last time she'd spoken with the head nurse, Anne Marie had learned that Dolores would be transferred from the hospital to a nursing facility for at least a week before she went home.

Anne Marie was fortunate enough to find a parking space on the street and decided to view that as a reward for thinking of Ellen's needs rather than her own. Holding the child's hand, she walked briskly toward the hospital's main entrance.

"Will Grandma be able to talk more?" Ellen asked.

On their first visit the previous Saturday Dolores had a tube in her throat that prevented her from speaking in anything other than a hoarse whisper. "The tube's out, so she should be able to talk normally again," Anne Marie explained.

Dolores had slept through most of that visit, and afterward Ellen had seemed quieter than usual. The contrast between the child who'd listened to the Irish singers and the child who'd walked out of the hospital later that afternoon was striking. Anne Marie had tried to tell her that Dolores was doing well, but all Ellen saw was a very sick woman.

"Your grandmother's going to be so proud of you for getting an A," Anne Marie told her now.

"I know," Ellen said solemnly.

They passed the gift shop.

"Should we bring her flowers again?" Ellen asked, looking at the floral arrangements displayed in the window.

"I'm sure the ones we brought on Saturday are still fresh." After the concert on Saturday, they'd purchased white tulips and yellow daffodils from Susannah's Garden, the flower shop next to the bookstore. Dolores had hardly seemed aware of the bouquet, which, given the circumstances, was understandable.

They walked directly to the elevator and Ellen pushed the button for the fifth floor, which was reserved for surgical patients. The doors opened in front of the nurses' station.

When they entered the room, Dolores was sitting up in bed, watching the television mounted on the wall. The flowers in their vase rested on the stand beside her bed. Although the room was a semi-private, she was the only patient. The moment she saw Ellen, Dolores's expression changed to one of rapture. "Oh, my little Ellen, my little love."

"Grandma! Grandma!" Ellen rushed toward the hospital bed with such enthusiasm she bounded into the mattress.

"Oh, Ellen, it's so *good* to see you." Dolores turned off the TV, focusing on her visitors, and held out both arms.

Anne Marie lifted Ellen up for a moment so she could gently hug her grandmother. She was moved almost to tears by the deep affection between them. This was love in its purest form. A child and her grandmother.

"I got an A on my spelling test," Ellen said, thrusting the paper at Dolores.

"Oh, Ellen! I'm so pleased."

"She studied hard," Anne Marie said.

"This was all the spelling words since Christmas, too."

"*All* the words?" Dolores's eyes widened with appreciation.

"Yup, and Stevie Logue and me were the only kids who got an A."

"That's excellent, honey." Dolores reached for her pitcher of water. "Ellen," she said, "could you do me a favor? Would you please go to the nurses' station and ask if I can have some more ice?"

The little girl nodded and took the pitcher, obviously

gratified to be performing this important task for her grand-mother.

"How's she doing?" Dolores asked urgently.

Anne Marie smiled at her. "Really well."

"I knew I could trust you," Dolores said as tears filled her eyes. "I hadn't even met you, but I knew you were the one from everything Ellen had to say about you."

"I'm happy to help." Anne Marie discovered this was the truth, that it had *become* the truth.

"Ellen likes you."

"I like her, too."

"If anything happens to me…" Dolores continued, leaning forward to clasp Anne Marie's arm.

Shock bolted through her. "You haven't had bad news, have you?" Surely the medical staff would've told her if that was the case. Still, she wasn't family, and she didn't know how liberal the hospital's policies were in regard to non-relatives.

"No, no, I'm doing well, according to the doctor," Dolores said.

"Oh, thank goodness!" Anne Marie couldn't hide her relief.

"But I'm not a new dishwasher." Dolores smiled, releasing her grip on Anne Marie. "That's what the young woman who operated on me said. I don't come with a guarantee that all my parts are going to work perfectly for the next five years."

"Of course not. No one does."

"But…I feel better than I have in months."

That definitely boded well.

"Still," Dolores said thoughtfully, "one never knows."

Anne Marie swallowed. She wondered if Robert had any premonition when he woke up that it would be the last day of his life. She wondered if he'd experienced any warning signs. Had there been any pain? Nausea? Tingling in his fingertips? Had his left arm ached? Did he assume the pressure in his chest was just heartburn? If she'd been living with him at the time, would she have recognized what was happening and been able to help?

Anne Marie didn't have the answers to any of those questions and they would forever haunt her.

"One never knows," she echoed bleakly.

"I gave birth to two daughters," Dolores told her.

"I know."

"I tried to be a good mother after their father left me."

"I know," Anne Marie said again.

Once more there were tears in the older woman's eyes. "I have no idea where I failed and there's no going back. Candace and Clarisse," she whispered. "Such beautiful girls. And now…"

"I understand." Anne Marie spoke soothingly, seeing how distressed Dolores was.

Dolores seemed to reach some decision. She turned to Anne Marie and took her arm again. Her eyes were fierce. "You have to *promise* that if anything happens to me you won't let Ellen go back to her mother."

"But it's not up to—"

"She's on meth," Dolores broke in. "The last time I saw her was in court. Her hair was falling out and her teeth were rotting in her head and she's barely thirty years old. My daughter is killing herself."

"You have sole custody of Ellen?"

"Yes. Promise me you won't let Ellen go back to her."

"I'm sure the Child Protective Services wouldn't—"

"*Promise me,*" Dolores insisted, her hand tightening on Anne Marie's forearm.

"But I—"

"I won't rest until I know Ellen will be with someone who loves her. Promise me."

Anne Marie could see that it would do no good to argue. "I promise." She suspected the state would never allow it, but she had to calm the woman down and there was no other way to do it.

Dolores relaxed her hold on Anne Marie's arm. "Thank you," she breathed.

"*You're* the one who's going to raise Ellen," Anne Marie said. "You're going to get well and Ellen will go home...."

"Clarisse." Dolores's voice cracked.

Anne Marie already knew the second daughter was in prison.

"She's as bad as her sister."

"I'm so sorry."

Dolores looked away. "Maybe I should've had tighter control of them when they were teenagers."

"I..."

"I did my best but it wasn't enough. They got in with a bad crowd and before I knew it, they dropped out of school and started doing drugs...."

"I'm so sorry." Anne Marie wished she could think of something else to say. Something more useful.

"The state might try to give Ellen to Clarisse once she's out of prison. Ellen can't go with her, either. Understand?"

"I won't let that happen." Anne Marie had no idea how she was supposed to prevent it, should the state make that decision. She decided not to worry about any of this, since Dolores would probably live for years and would be taking care of Ellen herself.

As though suddenly exhausted, Dolores closed her eyes and fell back against the pillow.

Just then Ellen returned, escorted by one of the nurses, who left right afterward. Ellen held the plastic pitcher filled with ice and carefully set it on the stand next to the flowers. "Is Grandma sleeping?" she asked in a loud whisper.

When Dolores didn't open her eyes, Anne Marie figured she'd either drifted off or was close to it. Their conversation had drained her of strength; she was, after all, recovering from surgery. And—perhaps even more of a factor—she'd been recalling the bitterest regrets of her life.

"I got ice," Ellen said.

"She'll thank you later," Anne Marie told the girl. "But at least you had a chance to show her your test. Didn't you see how proud she was of you?"

Ellen nodded reluctantly.

"We should let her sleep."

"Okay." Still Ellen didn't seem ready to leave. "Would it be all right if I sat with her for a few minutes?"

There was only one chair by the bed, and Anne Marie was sitting there. Soon Ellen had climbed onto her lap. The even rise and fall of Dolores's chest, the regular cadence of her breathing, lulled Anne Marie into closing her eyes, too.

She didn't know how long she'd been dozing there when her head slumped forward and she realized Ellen had cuddled up in her arms with one cheek pressed against her shoulder. The child's weight was warm and oddly comfortable, and she would've been content to stay that way for a while.

"Did Grandma Dolores tell you who my daddy is?" Ellen asked.

Anne Marie wondered what had prompted that question. "No…"

"Oh." She sighed with disappointment.

"Do you remember him, Ellen?"

"No." Ellen sounded so sad that Anne Marie wrapped her arm more securely around the girl's thin shoulders. "He's on my wish list."

"Your daddy?"

"Yes, I want to see him."

Dolores had said that Candace, Ellen's mother, probably didn't even know who the father was. Anne Marie didn't want to encourage Ellen to pursue something that would bring her more unhappiness. But as Dolores had also said, you never knew. The man just might make an appearance in the child's life when she needed him most.

"We can look on your birth certificate, I guess." Perhaps the school had a copy, although Anne Marie wasn't sure they'd show it to her.

"I have six wishes now," Ellen stated proudly.

Six wishes.

Six reasons to hope.

"Are you ready to go home?" Anne Marie asked. It was nine o'clock now, and she was surprised they hadn't been told that visiting hours were over.

"Okay."

Ellen climbed down from her lap. "Thank you for bringing me to see Grandma Dolores."

"You're welcome."

"Thank you for telling me about the wishes, too."

Anne Marie nodded. For some reason the gloom of depression had lightened and the image of Robert and Rebecca had receded. Holding this child in her arms made everything else seem less important, less immediate.

Dolores snored softly on. Anne Marie held Ellen's hand and flicked the switch, darkening the room, and they returned to Blossom Street.

Chapter

16

Lillie Higgins was meeting Jacqueline Donovan, her dearest friend on earth, for lunch. She wore a beige linen skirt and a jacket that showed off the pearls David had bought her in Hong Kong. Lillie was well aware that some transgression had elicited her husband's generosity.

The three-strand necklace was a guilt offering. She didn't know what had happened while he was in the Far East—or with whom—and she preferred it that way. Her husband generally gave her expensive gifts when he felt remorseful about something. That *something* always involved another woman.

Lillie had rarely worn the pearls until after David's death. Now it didn't seem to matter. They really were lovely and it didn't make sense to hide them in a drawer. She had no reason to feel guilty, so she'd begun to wear them regularly.

As she fastened the matching pearl earrings the phone rang. Lillie hesitated, tempted to let it ring. But Jacqueline was usually ten or fifteen minutes late, so Lillie decided to take the call.

"Hello?"

"Ms. Higgins?"

Lillie instantly recognized the voice of Hector Silva, the service manager at the BMW dealership.

"Hello, Mr. Silva," she said, unable to disguise her pleasure.

"I hope you don't mind that I'm phoning you."

"On the contrary, I'm delighted." And that was the truth. She hadn't expected him to contact her, and this came as a marvelous surprise.

"I'm calling to thank you for speaking to Mr. Sullivan."

"I'm sorry, who?"

"Mr. Sullivan owns the dealership. You phoned and left a message about me and the good service you received."

"Oh, yes." Lillie remembered that now. "You went above and beyond my expectations, and I wanted Mr. Sullivan to realize what a valuable employee he has in you."

"Thank you again."

"Mr. Silva, please, I'm the one who's indebted to you."

"Hector," he said. "We agreed to use first names," he reminded her.

She smiled at the genuine warmth in his voice. "And I'm Lillie."

"I wanted to inform you, Lillie, that as a direct result of your comments I was named employee of the month for February."

"Which you deserved."

"I…ah…" He hesitated and seemed about to say something more. "I know it's not—" Again he paused, as if unsure how to proceed.

"Yes?" Lillie's heart was in her throat. It might be presumptuous of her, but she had the distinct feeling that he was about to suggest they meet again.

"I hope you have a pleasant day," he finished in a rush.

"You, too." She didn't bother to hide her disappointment. Then, hoping to encourage him and let him know she'd welcome an invitation, she added, "Was there anything else, Hector?"

Her question was followed by a long pause. "Not really."

"Oh." She swallowed.

"Calling Mr. Sullivan was very nice of you," he said, rushing his words again. "I hope you're enjoying your new car."

"Very much, thank you, Hector."

"Goodbye, Lillie."

"Goodbye."

He didn't hang up right away and neither did she. Lillie closed her eyes, willing him to speak, willing him to suggest they see each other again. He didn't, and after a short pause she heard him disconnect. Her heart sank about as far as it could go. Well, that was that, she supposed. It was probably for the best—although it didn't *feel* that way—but she had to be reasonable. His social status was too different from hers and financially they were worlds apart. Hector understood that even if she didn't.

If she had a relationship with him, her friends would think she'd lost her mind. Well, maybe she had. Maybe she was tired of all the pretense that surrounded her life. She'd loved her husband, but her marriage had been a sham. When David had his affairs, she'd politely turned her head and looked the other way. Lillie had carried the knowledge and the shame that the man she loved, and had been completely faithful to, treated his marriage vows as if they were merely suggestions.

She fingered the pearls at her throat. She remembered the night David had given them to her. He'd stood behind her as she sat at her dressing table and draped them around her neck. In that moment it was all she could do not to rip them off. Although David could well afford the pearls, their price had been too high.

She and Hector Silva were little more than acquaintances, but Lillie instinctively recognized that this man would never cheat on his wife. Unlike David, Hector was a man who took his emotional commitments seriously. Anyone might ask how she could possibly know this about a man she'd only met a few times. But Lillie knew. Call it intuition or whatever you wanted. She just *knew*.

Feeling melancholy, she sighed, removing the car keys from her purse. She wouldn't hear from him again.

Lillie left the house to meet Jacqueline Donovan at the exclusive Seattle Country Club. The two of them had been members for years. They'd worked on any number of charitable projects together and been co-chairs of the Christmas Ball more times than she could recall.

When the news came that David and Gary had been killed in the plane crash, the one person Lillie had turned to for solace and advice had been Jacqueline. Barbie had her own intense grief to cope with and her sons to comfort; those three had formed a closed circle in the weeks after the accident. Jacqueline had stayed by Lillie's side for days, helping her deal with the multitude of immediate decisions. Her love and concern didn't end there, either. Jacqueline remained her friend while others had drifted away. She was also the only person Lillie had confided in about David's affairs.

Lillie pulled up in front of the club building and was instantly greeted by a valet. He didn't give her a voucher. None was necessary. Every employee of the club recognized her and her new vehicle.

"Ms. Donovan arrived two minutes ago," the valet told her.

"Thank you, Jason," she murmured and headed inside.

Sure enough, Jacqueline sat at their usual table, glancing over the menu. She wore her hair in her customary French roll and had chosen a Venetian glass necklace in teal and gold that stood out against her black pants and jacket. She put down the menu and smiled.

"It's not like you to be late," she said when Lillie joined her. She'd already ordered a bottle of their favorite wine, a New Zealand sauvignon blanc that the club kept in stock primarily for them.

"I answered the phone on my way out the door," she said, reaching for the menu, although she practically had it memorized. Naturally, it didn't include prices. A bill would be mailed at the end of the month with the accumulated charges.

"Nothing important, I hope."

"Not really." Lillie considered mentioning Hector but quickly dismissed the thought. Of all her friends, she trusted Jacqueline most, and yet…

A waiter came to their table, and Lillie decided on the Oriental salad; Jacqueline ordered blackened scallops.

"So," Jacqueline said, swirling the wine in her glass. "What's new with your Twenty Wishes?"

The last time they'd met for lunch, Lillie had been full of enthusiasm about her list. She'd talked nonstop, extolling the idea and describing her wishes—starting with the red BMW.

"The widows are meeting next week. We're going to talk about our wishes and the progress we've made."

Jacqueline sipped her wine. "This is *such* a good idea. Ever since you told me about it, I've been planning to make my own list."

"You should," Lillie said, nodding vigorously.

"It's certainly given you a new lease on life."

"You think so?"

"I haven't seen you look this happy in years."

She wanted to say something, to explain that the Twenty Wishes weren't entirely responsible for this "new lease on life." But she couldn't; she had to accept that she wouldn't be hearing from Hector again. Ironically, he was the embodiment of her most longed-for wish—a decent, honorable man. A man she couldn't have. A wish she might never fulfill. The whole thing was just so hopeless.

"Actually, I've met someone." Lillie couldn't imagine what had made her say that. She wanted to snatch the words back the instant they left her mouth.

Jacqueline nearly tipped over her wineglass. "When? How? Who? You haven't said anything about this!"

"It's just that, well…"

"Yes?"

Lillie drew in a deep breath. "Before you get all excited, let me say this isn't anyone you know."

"He's not a member of the club?" Much of their social life revolved around the country club, although both had plenty of friends outside it.

"No."

"Why the big secret?"

"It isn't a big secret. I should never have mentioned it." Lillie could feel her face heating up. "Forget I said a word."

"No way! You're dying to tell me, I know you are. Spill it."

"I've never been on a date with him, so there's nothing to tell you."

Jacqueline frowned. "Why not?"

"For one thing, he hasn't asked and…and, well, I doubt he ever will, so there's no point in discussing this any further."

"He's younger, isn't he?"

"No! Wipe that silly grin off your face, Jacqueline Donovan. It's not like that at all. He's shy and I think he'd like to ask me out but he hasn't." She was saying more than she'd ever intended and desperately wished she'd never introduced the subject.

Jacqueline leaned back in her chair and a smile quivered at the edges of her mouth. "You should ask *him* out," she insisted.

"What?"

"I mean it. In our generation, the men always did the asking, but times have changed. If you're interested in this man—this stranger you refuse to tell your very best friend about—then all I can say is that you need to take the initiative."

Lillie stared across the table at her, and Jacqueline stared right back. "I can't!"

That went against every dictate of her upbringing. Ask Hector out on a date? It was a preposterous suggestion. An outrageous idea.

Jacqueline simply shook her head. "You *can* ask him, and you will. Or…or I'll do it for you."

At that Lillie giggled. "Don't be ridiculous."

"Why is that ridiculous? You're a lovely woman, Lillie, and a beautiful person besides. You deserve happiness. Isn't that what the Twenty Wishes are all about? Going after the things you want in life. Places to see, people to meet, experiences to live. Don't hold back now. Go for it!"

Could Jacqueline be right?

Without some kind of pledge from her, Jacqueline was never going to shut up about this. "I'll consider it." That was the best she could do for now.

"Good." Jacqueline nodded, obviously pleased.

Lillie relaxed, wondering how she could possibly approach Hector. Oh, for heaven's sake, what did she know about such things? If David was alive, she could've asked him. He was the one with all the dating experience. That thought produced a hysterical giggle that she tried, unsuccessfully, to swallow.

Jacqueline regarded her closely. "What's so funny?"

Embarrassed, Lillie shook her head. "You—saying I should contact my friend and ask him out." She waved her hand. "I was just thinking— Never mind."

"No, tell me," Jacqueline insisted.

So she did, and soon they were both laughing.

The waiter brought their lunch and automatically refilled their glasses.

"You're going to do it," Jacqueline said firmly, leaving no room for argument. She reached for her fork.

"I—"

"Yes, you are," Jacqueline returned. "You want to see this man, don't you?"

Lillie gave a barely perceptible nod.

"If you need me standing by to encourage you, then that's what I'll do."

Lillie felt a moment's hope. No, it was impossible. Even if she did find the courage, she didn't know what kind of outing to suggest. Perhaps a movie? Barbie seemed to be going to the movies a lot these days.

"You're looking serious now," Jacqueline said.

"He might refuse," she blurted out. "I might've completely misread him."

"So what?" Her friend shrugged as if this was an insignificant concern. "Nothing ventured, nothing gained."

"But…"

"Would you *stop*," Jacqueline said.

Lillie had yet to try a single bite of her salad. "You're right, you're right." She picked up her fork, then laid it down again. "The problem is, we don't have a single thing in common."

"Except for the fact that you're attracted to him and I assume he is to you."

"Jacqueline, I *am* attracted to him. I really am. I go to sleep thinking about him. I yearn for him…" Her face flushed with embarrassment.

"Have you analyzed what's so attractive about him?"

She knew the answer immediately. "Oh, yes—he's kind and gentle and honorable. He loves his children and I'm positive he was a faithful husband." Just talking about Hector was enough to bring tears of longing to her eyes.

"You want him?" Jacqueline whispered the question.

"Not the way you're thinking." This wasn't merely physical desire, although he was a good-looking man and late at night she'd fantasized about his mouth, his hands…. What she felt was, above all, *emotional*—that need for true kinship, that recognition of another's soul. She tried to explain her feelings to her friend.

"This must be one helluva man," Jacqueline commented.

"He is."

"Then don't wait, Lillie," Jacqueline said earnestly. "I wasted too many years of my life before Reese and I…"

She let the rest fade, but Lillie knew what she meant. Jacqueline's marriage had been like her own. She and Reese had lived as strangers for years. Jacqueline had reason to believe her husband had a mistress, and as a result she'd moved into a spare bedroom. They'd remained stiffly polite, ignoring each other as much as possible in the privacy of their home, acting like a loving couple outside it. No one had suspected. No one knew the truth about them.

Except Lillie.

She'd been able to identify the signs because she'd lived the same scenario.

Then, shortly before David's accident, something changed between Jacqueline and Reese. Almost overnight they set aside their differences and became lovers again. They'd even traveled to Greece on a second honeymoon. The love was back in their marriage and in their lives.

Lillie never learned exactly what had brought about the change, although she suspected that Tammie Lee, the Donovans' daughter-in-law, had something to do with it. When Paul had first brought home his young bride from Louisiana, Jacqueline had been horrified. Tammie Lee, with her southern drawl, wasn't the daughter-in-law she'd wanted.

Personally, Lillie had instantly liked the young woman. She was sweet and genuine and good-humored, even if she did talk about recipes for pickled pigs' feet and boiled peanuts.

It'd taken Jacqueline months to accept the idea of her only son married to Tammie Lee and then gradually, the relationship between the two women had undergone a shift. Not long after that, the relationship between Jacqueline and her husband had improved, too. She and Reese had clearly achieved some sort of reconciliation.

For a while Lillie had been jealous. She wanted the same happiness Jacqueline had rediscovered in her marriage. She'd hoped for that kind of turnaround in her own—but it never happened.

And yet, it wasn't too late for a change in her *life*. It wasn't too late to fulfill a wish...

She'd do it.

She'd defy her upbringing and find a way to ask Hector Silva on a date.

Monday morning Anne Marie was finishing up the sale of a hardcover novel for one of her favorite customers, Larry Barber, a retired accountant, when Lillie and Barbie entered Blossom Street Books. Mother and daughter had never looked better. In fact, Anne Marie caught herself staring. A transformation had taken place in both women and while it might not be apparent to anyone else, Anne Marie noticed. Trying to discern what was different about them, she decided it was a new sense of *life*. They seemed to shimmer with it.

They talked animatedly to each other while Anne Marie completed the sale.

As Larry signed the charge slip, Anne Marie smiled a warm welcome at her friends. He wasn't in any hurry to leave. Since his wife had died, he was lonely and came to the store for conversation with Anne Marie as much as he did for reading material. When business was slow, Anne Marie didn't mind. She knew what it was to be alone and to crave companionship. This morning, however, she was impatient to be with Lillie and Barbie.

Larry must have realized the other women wanted to talk to her and, reaching for his purchase, thanked Anne Marie and headed out the door.

The minute he left, Barbie shimmied up to the counter. "I found a belly dancing class," she announced and threw her arms in the air as if she was about to give a demonstration.

"Belly dancing?" Anne Marie repeated. "You talked about that during our Valentine's get-together."

"It's on my wish list," Barbie informed her. "I'm so excited I can hardly stand it."

Lillie rolled her eyes playfully. "My daughter sometimes shocks even me."

Barbie waved off her mother's comment. "Oh, honestly, Mother, I've wanted to learn how to belly dance for ages."

"You never said anything to me."

"I know—I thought about it a lot, but it seemed so…oh, I don't know, silly, I guess. Then, when I read about a class at the Fitness Center, I decided to learn how to do this. I'm not putting it off any longer."

"That's great," Anne Marie said. She had an announcement of her own. "I signed up for a knitting class." On Saturday, Lydia had told her that the long-delayed beginners' class would start the following Wednesday at twelve-thirty.

"Is Elise teaching the class?" Lillie asked.

Anne Marie nodded. Elise had told her that working at the yarn store had been one of her wishes.

These lists of Twenty Wishes were influencing all their lives—and those of others, too. For instance, Elise's wish had been a solution to Lydia's problem of teaching too many classes. It seemed that every time Anne Marie went into A Good Yarn, Margaret was complaining that Lydia shouldn't be taking on as much as she did. Now Elise would fill in as sales help when necessary and teach three classes. In addition to the beginners' class, she'd be teaching a session on knitting with beads and another on felted purses.

For the moment, Anne Marie was content with the beginners' class. Once she learned the basics, she'd venture out into more complicated techniques and projects.

Already she could see that knitting was something she'd enjoy. At noon she'd go and choose her yarn for the first class, two days from now. Timing would be tight, since Anne Marie was still joining Ellen for lunch on Wednesdays, even though the girl was living with her.

"I was thinking I'd have everyone over for dinner," Lillie murmured, breaking into Anne Marie's thoughts.

She must have responded with a blank look because Lillie immediately said, "For our meeting? To discuss our wishes."

"Oh, right."

"Is Thursday still okay?"

"Yes…" Anne Marie returned with some hesitation. "But remember Ellen will be with me."

"That's fine," Lillie said.

"You're sure?" Anne Marie could probably find someone to watch Ellen if she had to.

"Of course she can come," Lillie was quick to tell her. "We'd love to have Ellen."

Anne Marie grinned, remembering how intently Ellen had worked at compiling her own list. "Did I mention Ellen's got Twenty Wishes, too?"

Mother and daughter shared a smile.

"My friend Jacqueline Donovan is making one, as

well," Lillie added. "The minute I mention the idea to any-one, they decide to make their own. Jacqueline told me the first wish on her list is to ride a camel in Egypt and see the pyramids. She also wants to sleep under the des-ert sky."

"Jacqueline? Camping?" Barbie said incredulously. "That woman likes her luxuries."

"I know, that was my reaction, too." Lillie shrugged in amusement. "Who knew?"

"I guess it's like me and belly dancing," Barbie said. "It was in the back of my mind, just a vague…whimsy, I guess you could say, but it didn't enter my consciousness until I started working on my list of wishes. Sometimes I think we're afraid to admit we want certain things. Espe-cially things that contradict the image we have of our-selves."

"Or the way others think of us," Lillie said.

"Right."

A customer walked in the door.

Lillie glanced over her shoulder and then back at Anne Marie. "Thursday night, then. Shall we say six?"

"I'm looking forward to it." And she was. Her list of wishes was growing and it wasn't as difficult to come up with ideas as it had originally been. All at once a whole world of wishes, of desires and possibilities, had opened up to her, ideas that had seemed beyond the scope of her imagination only a few weeks ago. Perhaps her heart had finally, gradually, begun to mend. She had fifteen wishes now.

12. Take a cake-decorating class and bake Ellen a huge birthday cake

13. Practice not-so-random acts of kindness at least once a week

14. Ride the biggest roller coaster in the world at Six Flags in New Jersey

15. Visit the Civil War battlefield in Gettysburg and then go to Amish country

That evening as Anne Marie put away the dinner dishes, Ellen sat at the kitchen table, doing her homework. Her ankles were demurely crossed, her entire demeanor intent.

Ellen had spoken with her grandmother before dinner, and so had Anne Marie. Dolores was regaining her strength. It wouldn't be long before Ellen could return to her home and all that was familiar.

"Do you need me to go over your spelling with you?" Anne Marie offered as she wiped the countertop.

"No, thank you."

"Don't you have a test tomorrow?"

"Yes, but I already memorized all the words," Ellen said proudly.

"On the very first day?"

Ellen nodded. "I did that after I took Baxter for a walk when I got home."

Anne Marie wasn't sure how Baxter would do when Ellen went back to her grandmother's. The eight-year-old had completely spoiled him. Every day after the school bus dropped her off, Ellen ran up the stairs to their apartment and lavished Baxter with love and attention. Anne Marie walked her dog twice every day and once in the evening. Now, however, Ellen took over for her in the afternoon. Anne Marie used to take him to the alley behind the store. Not Ellen. She paraded him up and down Blossom Street with all the ceremony of visiting royalty. Needless to say, Baxter loved their excursions.

Cody, Lydia's stepson, had recently come by with his

dog, Chase, and the two dogs and children had quickly become friends. The dogs made quite a pair; Chase was a hefty golden retriever and Baxter was tiny by comparison. Like many small dogs, Baxter wasn't intimidated by the bigger dog's size, and the two of them marched side by side, looking for all the world like Laurel and Hardy.

In addition to Lydia, Ellen had made friends with several of the other business-owners. It wasn't uncommon for her to return with a carnation given to her by Susannah or a cookie from Alix at the French Café.

"Can I watch TV when I finish my homework?" Ellen asked, glancing up from her arithmetic.

"Okay." Anne Marie was looking forward to sitting down in front of the TV, too. She'd gone to A Good Yarn on her lunch break to buy the necessary knitting supplies. With the extra classes and increased business, the shop was now open six days a week. Elise had helped Anne Marie select her yarn and needles. The choices seemed endless, and after much debate, she'd decided on a soft washable wool in lavender to make a lap robe for Dolores Falk.

Because Anne Marie was so eager to learn, Elise had taught her how to cast on and showed her the basic knit stitch. To her delight, Anne Marie had picked it up without a problem.

An hour later, Anne Marie and Ellen sat together on the sofa, watching the Family Channel. Wanting to practice what she'd learned, Anne Marie took out the needles and a skein of yarn.

"What are you doing?" Ellen asked.

"Knitting," she said, adding "I hope," under her breath.

"Grandma Dolores used to knit."

Anne Marie nodded.

"She said she'd teach me."

Again Anne Marie acknowledged the comment with a slight inclination of her head as she concentrated on casting on stitches.

"Is it hard?"

"Not really."

"Can I watch you?"

"Sure."

Ellen scooted closer and stared fixedly as Anne Marie attempted what Elise called a knitted cast-on. The term didn't mean anything to her; all she wanted to do was get stitches onto the needle.

"That's knitting?" Ellen said.

Anne Marie paused. "I think so."

Ellen removed her shoes and stood on the sofa to get a better view.

Suddenly she bounded off the sofa and dashed into her bedroom. She was back an instant later with a pad and paper.

"What's that for?" Anne Marie asked as the youngster skidded to a stop, barely missing the coffee table.

"My Twenty Wishes. I want to learn how to knit, too."

Anne Marie grinned at her. "How about if I teach you what I know?"

"You mean *now?*" The girl's eyes grew round.

"Why not?"

"Okay." Ellen leaped back onto the sofa, sitting close beside Anne Marie.

"According to Elise…"

"Mrs. Beaumont?"

"Yes, Mrs. Beaumont. There are actually only two basic stitches. The first is called the knit stitch and the second is a purl stitch."

"Okay," Ellen said again, nodding sagely.

"So far, I just know how to do the knit stitch. I'll learn how to purl in the first class."

"You haven't taken a real class yet?"

"No. I'm signed up, though."

"Oh."

"In other words, Ellen, I don't know all that much, but I'm willing to show you what I can do. If you like it, I'll take you to the yarn shop and let you pick out your own needles and yarn."

"Really?" Ellen was beside herself with excitement. "*Really?* Really?"

"Yes, really," Anne Marie responded, smiling.

For the next forty minutes, the television show was forgotten as Anne Marie showed Ellen what Elise had taught her that day. Ellen didn't catch on as easily as Anne Marie, but she was, after all, only eight.

Anne Marie was pleased with the child's determination to learn. Before the evening was over, Ellen was every bit as proficient as Anne Marie.

"I want to knit something for Grandma Dolores," Ellen stated. "Something pretty."

"What about a scarf?" Anne Marie suggested. She'd seen several exquisite ones at the shop. Elise had explained that these elaborate scarves had been knit using only the basic stitch she'd taught Anne Marie that very day.

"For her to wear to church," Ellen continued excitedly. "I'll give it to her when she comes home from the hospital."

"That's a very good idea."

Ellen finished the row and was about to start another when Anne Marie noticed the time. The evening had simply slipped away. It used to be that the hours she spent in the apartment moved so slowly she seemed aware of every passing minute.

"It's nine o'clock," Anne Marie said. "You should've been in bed half an hour ago."

"Is it nine *already?*" Ellen protested, but she couldn't hold back a yawn.

"I'm afraid so. We'll knit again tomorrow night," Anne Marie promised.

Ellen set down the needles and yarn and stumbled toward her room, yawning every step of the way.

"Call me when you're ready for your prayers, sleepyhead."

"Oh…kay."

A few minutes later, Ellen called out that she was ready. They followed the same routine as when Ellen lived with her grandmother, which meant that Anne Marie listened to the girl's prayers. She'd been saying them aloud for the last while.

Kneeling by the sofa bed, Anne Marie propped her elbows on the mattress, closed her eyes and bowed her head. Ellen's prayers didn't vary much. First, she asked God to help her grandmother get better. Then she asked Him to bless a number of people, Anne Marie and Baxter included, with lengthy descriptions of each. Finally, she gave thanks for the day's small triumphs. At the end of the seemingly interminable list, she said "Amen."

"Amen," Anne Marie echoed. "Good night, Ellen." She drew the covers more firmly around her and was about to get up when Ellen threw both arms around Anne Marie's neck and hugged her tightly. "Thank you for teaching me to knit."

"You're welcome," she said, hugging her back. Getting

to her feet, she turned out the light, then tiptoed out of the room. As was her habit, she left the door slightly ajar.

It wasn't until Anne Marie had stepped into the hallway that she realized something—this was the first time Ellen had actually hugged her. The night Anne Marie had gone to Dolores Falk's home, Ellen had fallen weeping into her arms, but that wasn't a real hug. Not like the one she'd just received.

Anne Marie stood right where she was and savored the moment. She felt loved and needed in a way she never had before.

It was like the return of warmth after the coldest winter of her life.

"**Y**ou came!" Tessa Bassett said with unrestrained glee when Barbie stepped forward to purchase her movie ticket. The teenager's face was flushed with excitement, and she leaned forward, lowering her voice. "He's here."

"Mark?" Barbie could hardly believe it. She hadn't expected this, but it shouldn't surprise her. Mark was definitely intrigued, even if he resisted her. Despite his hostility he hadn't really wanted to scare her off. Or maybe he was testing *her* interest. At any rate, Barbie saw the first substantial crack in that impervious exterior of his.

Mark was back. For that matter, so was she.

"He said he wasn't going to ever come on a Monday again—but now he has. I wanted to call you but I didn't know your last name, so I couldn't. I just hoped you'd be back and you are." This was all said in one breath. While she was speaking, Tessa slipped her the movie ticket and held out her hand for the money.

"What movie am I seeing?"

"A horror flick."

"Oh…"

"You don't like horror movies?"

"Not particularly."

"Oh." Tessa's face fell. "Do you want to see something else? You don't, do you? Because I think my uncle Mark likes you. Only he's afraid 'cause, after the accident, his wife divorced him and he's never heard from her again. You're the first woman he's even noticed since then."

Barbie stared at her, appalled. This ex-wife of his sounded like a shallow, selfish woman. Whatever happened to "for better, for worse"? Marriage vows didn't become null and void if one of the partners got sick or hurt. She knew without a second's hesitation that she and Gary would have stuck by each other, regardless of circumstances.

She sighed. "I suppose I could watch a horror movie," she said. "How bloody is it?"

Tessa grimaced, wrinkling her nose. "Real bloody."

"Are there dismembered body parts?" That was the worst, in Barbie's opinion.

Tessa nodded reluctantly. "But he came back! That's big."

Undecided, Barbie chewed on her lower lip. Tessa was right; neither of them had expected Mark to return. Barbie wasn't sure why she'd come—force of habit? Hope?

"Just go," Tessa urged. "Don't look at the screen. Close your eyes and plug your ears. That's what I do."

Other than the thought of having to watch the dispersing of gore and guts, choreographed to loud, pounding music, Barbie couldn't have said what was stopping her. So the movie wasn't exactly her choice. So what? She'd be with Mark and wasn't that the whole point of being here?

"Okay," she said with a deep breath. "I'll do it."

"Terrific!"

She just prayed she wouldn't have nightmares for the rest of her life.

"Let me know what happens, okay? With Uncle Mark, I mean," Tessa said. "My parents and my grandmother want to know, too."

"Okay." That meant the whole family was in on this, which was encouraging.

Barbie took her time, waiting until the last possible minute before slipping into the darkened theater. She purchased her popcorn and soda and lingered in the lobby until the show was about to start.

When she walked into the theater, the previews had already begun. She made her way to the row where Mark had parked his wheelchair, the same as usual. As she had previously, she sat one seat away from him.

He turned and stared at her in feigned surprise. "What are you doing here?" he whispered.

She could act as well as he could. "Oh, hello," she said brightly. "Is that you again?"

For a moment she suspected the hint of a smile. She turned back to the screen just in time to see an ax-murderer heave his weapon of death into a wall next to a trembling woman's head. Unable to stop herself, she gasped aloud and nearly dropped her popcorn.

"Frightens you, does it?" Mark asked in a far too satisfied tone. "Might I remind you these are only the previews."

"Yikes." She gritted her teeth.

Mark laughed, causing a woman behind him to make a shushing sound. *"Yikes,"* he repeated, lowering his voice. "Is that the best you can do?"

"Might I remind you I have sons."

"And you're a *lady*, right?" He spoke as if he intended that to be an insult.

"As a matter of fact, yes," she said stiffly. "I know all the words you do. I merely choose not to say them."

"I doubt it," he muttered, then settled back in his wheelchair to watch the movie, which was just getting started.

He gave every appearance of enjoying it, but as far as Barbie was concerned, this was torture. She'd always avoided being around when her sons watched horror DVDs with the bloodthirsty gusto of teenage boys. Now she squirmed in her seat, covered her face frequently and dashed out of the theater twice. It was even worse than she'd expected. Special effects being what they were, little was left to the imagination.

Barbie knew very well that Mark had planned this. He'd guessed—and guessed right—that she'd hate a movie like *The Axman Cometh* and had intentionally subjected her to an hour and a half of disgusting violence. The more she thought about it, the more irritated she became. And yet, she was determined to prove she could take it. Even if she couldn't.

After her second escape, when she'd hurried into the foyer to avoid watching another horrific scene, Mark leaned toward her and asked, "Are you going to finish that popcorn?"

"How can you possibly eat?" she snapped.

His grin seemed boyish as he reached for her bag and helped himself to a huge handful. Oh, yes, she thought grimly, he was enjoying her discomfort.

The movie wasn't actually all that long but it seemed to drag on for hours and hours and hours. The music, the tension, the blood, the *stupidity* was simply too much. By the time the movie ended, Barbie felt drained. The lights came up and the twenty or so viewers filed out of the theater. Mark stayed put and so did Barbie.

Finally she turned to him. "You did that on purpose, didn't you?"

"Did what?" he asked innocently.

Barbie wasn't fooled. She also decided that if this was a test, she'd failed. He knew she wanted to be with him, and because of that she'd endure this…this torture. She began to wonder if Tessa and her family had it all wrong. Maybe Mark wasn't attracted to her. Maybe he was just trying to punish her. Barbie began to mistrust her own intuition, her certainty that he reciprocated her interest. If he meant to signal that he didn't want her to bother him again, perhaps she should listen.

Fine. She would.

Barbie stood and, without another word, walked out of the theater.

Tessa, who'd been busy both times she'd fled into the lobby, was waiting for her.

"Well?" the girl asked anxiously.

"I don't care if I ever see your uncle again," Barbie said flatly.

Tessa's mouth fell open.

"What?"

"You heard me. He's rude and arrogant and…and…" She tried to think of a word that adequately described him. "Mean," she concluded. Making her sit through that debacle of a movie was downright mean.

"What did he say?" Tessa demanded, trotting alongside her.

"He didn't have to say anything. I got the message."

"Tell me," Tessa pleaded. "My mom and grandma are gonna bug me if I don't tell them what happened."

"Let me put it succinctly. Mark isn't interested. Period. If you think he is, then you and your family are sadly mistaken." Hearing his wheelchair behind her, Barbie whirled around to face him, ignoring the curious bystanders arriving for the next movie. "Isn't that right, Mark?"

Mark was silent.

"You like her, don't you, Uncle Mark?"

"I came to see a movie," he responded, his voice impassive. "If I wanted to find my perfect match, I would've gone online. She *is* right. I'm not interested."

Barbie tossed the girl an I-told-you-so look and stalked out. She was all the way to the exit when Mark called her name.

"What?" she asked angrily. "Don't worry," she told him before he could say a word. "I won't make the mistake of sitting next to you again—at any movie."

He blinked, then shrugged as if it made no difference to him. "Whatever."

Over the years, Barbie had come to hate that word and its connotation of teenage apathy. With as much dignity as she could gather, she continued toward the parking lot.

She was surprised when Tessa ran out of the building after her. "He didn't mean anything," she said breathlessly. "How would he know you hated scary movies? He just wanted to find out if you'd be willing to see something besides a romantic comedy. The least you can do is give him another chance."

"Why are you trying so hard?" Barbie asked. She was willing to accept that she'd made a mistake and move on. As attractive as she found Mark, she wasn't going to invite his rejection over and over again.

"You *have* to give him another chance," Tessa said.

"Why?"

Tessa paused, then answered on a heavy sigh. "Because my uncle Mark deserves to be loved." Her eyes pleaded with Barbie's. "This is new to him. He married his high school girlfriend and never loved anyone else and then she dumped him after the accident…." She gulped in a breath. "I'm positive he likes you—only he doesn't know how to show it."

Barbie hesitated. If anything about this entire evening astonished her, it was that Mark hadn't come outside and insisted his niece mind her own business. Delving inside her purse, she searched for a business card. "Okay, fine. Give him this and tell him the next move is his."

Tessa's face shone with eagerness as she nodded. "Great! Thank you so much. Thank you, thank you. You won't be sorry, I promise."

That remained to be seen.

Feeling wretched, Barbie did what she always did when she needed solace—she drove to her mother's house.

Lillie opened the door and immediately asked, "What's wrong?" Without delay she led her into the kitchen. "It isn't the boys, is it?"

Barbie swallowed hard and shook her head.

Hands on her hips, Lillie stood in the middle of her beautiful, gleaming kitchen. "Should I put on coffee or bring out the shot glasses?"

Barbie managed to smile. "This time I think I need both."

Lillie took a whiskey bottle from the small liquor cabinet in the kitchen, then started a pot of coffee. That involved first grinding beans, a production Barbie lacked the patience to bother with.

"So, tell me what happened," Lillie said when she'd

made two Irish coffees. She sat on the stool at the counter next to Barbie and they silently toasted each other with the mugs.

"I saw Mark again."

Her mother nodded. "The man you met at the theater."

"Yes." She hadn't told Lillie much about him, and with good reason. As soon as her mother learned he was in a wheelchair, she'd find a dozen reasons to dissuade her from pursuing him.

Barbie already knew a relationship with Mark wouldn't be easy. She'd done her homework. All right, she'd looked up a few facts about paraplegics on the Internet. Even his anger with the world wasn't unusual. Until this evening, she'd assumed she was prepared to deal with it. Apparently not.

Lillie gestured for Barbie to continue. "And…"

"And he…he isn't interested."

Lillie cast her a look of disbelief. "That can't be true. You're gorgeous, young, accomplished—and a lovely person. Is something wrong with him?"

"Not really." A half truth.

"He's not…"

"No, Mother, he's not gay. Or married." Barbie wondered how much more she should explain.

Lillie studied her and raised one elegantly curved eyebrow. "What aren't you telling me?"

Barbie should've known her mother would see straight through her prevarication.

Lillie's voice grew gentler. "What is it, honey?"

Barbie sighed. "If I tell you, I'm afraid you'll discourage me, and I don't think I could bear that just now."

For a long moment her mother didn't respond. "It's odd you should say that, seeing I have something I wanted to discuss with you and…and haven't, for the very same reason."

"What?" Barbie's curiosity was instantly piqued. She couldn't imagine her mother keeping anything from her. They were each other's support system, especially since David and Gary had died. But then, she'd never supposed she'd ever hide secrets from Lillie, either. Obviously they were both guilty of deception.

Lillie cleared her throat. "I…I recently met someone myself."

Barbie was stunned. "You haven't said a word."

Her mother avoided eye contact. "I'm afraid if I mention…my friend, *you'll* discourage *me*." She picked up her coffee and took a deep swallow. "This man I met—I believe we're both afraid of what others will think," she added. "Jacqueline urged me to ask him out, since he seems reluctant to approach me. But women of my generation don't do things like that. Yet I find the idea so appealing, I'm willing to put aside everything I've had ingrained in me all these years just for the opportunity to spend time with him again."

Lillie's cheeks were flushed and her hands trembled slightly as she raised the mug to her lips. It might've been the whiskey, but Barbie doubted that. There was more to this. Her mother was the most competent, composed woman she'd ever known and her being so flustered and unnerved over a man was completely out of character.

"Mom, you don't need to worry what I think."

"But I do. You're my daughter and, well…okay, I'm just going to blurt it out." Lillie straightened her shoulders. "He's the service manager at the car dealership."

Barbie couldn't help it; her jaw dropped. Her mother was attracted to a mechanic—a man with grease under his fingernails? Lillie Higgins, society matron, and a *me-*

chanic? Instantly warnings rose in her mind. This man must know that her mother had money. Lillie was lonely and vulnerable, easy prey. Her usual common sense had evidently deserted her, and she needed protection from this gigolo or whatever he was.

Barbie saw that her mother was waiting for her reaction, so she said, "I…see."

Lillie downed the last of her Irish coffee."His name is Hector Silva."

This was as shocking as the fact that he was a mechanic. "He's Hispanic? Is he legal?"

"Yes! Of course! Hector's a citizen. He's decent and hardworking and kind." She hiccuped once, then covered her face. "This is even worse than I thought it would be," she moaned.

"No, Mom, really, I apologize. That was a stupid question. It's just…I don't know what to think." She'd assumed her mother couldn't surprise her; she'd assumed wrong. Of all the men who'd love to date her mother, Lillie had fallen for a mechanic?

Lillie dropped her hands. "I believe I know what you're trying to say," she said in a cold voice. "And I'm disappointed in you."

"I'm sorry," Barbie mumbled. But the image of her mother with this man refused to take shape in her mind.

Her mother motioned toward her. "It's your turn."

"But…"

"Tell me what the problem is with this Mark. Why you didn't want to say anything. Is he too old? Too young? Some kind of addict?"

"None of those." Like her mother, she squared her shoulders and expelled her secret in a single breath. "He's paralyzed from the waist down."

Lillie closed her eyes briefly. "Oh, Barbie."

It was just as she'd expected. Annoyed, she slid off the stool. "I knew it! I should've realized you'd react like this. I wish I hadn't said anything." Her annoyance turned to disillusionment and then just as quickly to pain. "You're the one person in the world I trust to understand me and all you can say is *Oh, Barbie?*"

"You weren't exactly a great encouragement to me, either," Lillie muttered.

"Oh, please. A mechanic? You want to ask a mechanic out for a date and you expect me to *cheer?*" All the frustration and anger of the evening burst from her. She stood with her hands knotted into fists at her sides. "You didn't tell Jacqueline who this man is, did you? What's the appeal? Do you think he'll be good in bed? Is that it?" Her own words shocked her, but not nearly as much as they did her mother.

Lillie stood frozen, her eyes wide with horror. Then she did something she'd never done in her whole life.

She slapped her.

Stunned into silence, Barbie pressed her hand to her cheek. Tears sprang to her eyes.

When her mother spoke, her voice shook with fury. "At least Hector could take me to bed."

Barbie gasped at the implication, grabbed her purse and shot out of the house. Over the years they'd quarreled—every mother and daughter did—but never anything like this.

A sick feeling engulfed her as she drove to her own house, less than two miles away. Pulling into the garage, she sat in her car and hid her face in both hands. The urge to break into heaving sobs of rage and pain and regret nearly overwhelmed her. But she refused to give in to the swell of grief, refused to allow the ugliness that had come between them to disintegrate her emotions any further.

Barbie didn't sleep that night or the next.

Nor did she speak to her mother. Ten times at least she reached for the phone. Normally they spoke every day, often more than once. Now the silence was like a vast emptiness.

As far as Barbie was concerned, her mother owed her an apology. Lillie had struck her—her own daughter.

By the end of the second day, Barbie could hardly stand it. She missed her mother. She *needed* her.

The dinner for the widows' group was scheduled for Thursday night. Barbie was determined to go, but as Tuesday passed and then Wednesday, that resolve weakened.

This was ridiculous, she told herself. They'd both been at fault.

They'd both said things they regretted. It was time to apologize and put this behind them.

Late Thursday afternoon, a floral delivery truck parked in front of her dress shop just as Barbie was about to close for the day. The man carried in a huge floral arrangement from Susannah's Garden. This had to be a hundred-dollar order. It took up nearly half the counter space.

The driver handed her a clipboard, and Barbie signed her name as a rush of relief came over her. She didn't need to look at the attached card to know her mother had sent the flowers. Like her, Lillie was sorry. She was apologizing, trying to restore what they'd lost. Smiling, Barbie removed the small envelope and opened it.

She was wrong; Lillie hadn't sent the flowers.

Only one word was written on the card.

Mark.

Chapter

19

$\mathbf{A}$nne Marie and Ellen were both looking forward to dinner at Lillie's that night. Earlier, Anne Marie had called to ask what she could contribute to the meal.

"Nothing," her friend had insisted. "Just bring yourselves." As she replaced the receiver, Anne Marie thought that Lillie didn't sound like herself. Ever since they'd made their wish lists, Lillie's spirits had been high. But following their conversation, she wondered if she'd misread Lillie's feelings. Her voice had been flat, emotionless, devoid of her usual enthusiasm.

Anne Marie was afraid this dinner might be too much work for her. Later in the day she phoned Lillie again, to make sure everything was all right.

"Everything's perfectly fine," Lillie said, although her tone belied her words. "Actually, I'm really enjoying myself. It's been a long time since I've cooked for a dinner party." Anne Marie heard a timer in the distance, and Lillie told her she had to get off the phone.

Still, Anne Marie wondered. She sensed that something was off, but Lillie obviously wasn't going to tell her. All she could do was accept her word and hope that if there *was* a problem, it would soon be resolved.

The school bus rolled past the shop window and Anne Marie knew Ellen would appear in a few minutes.

"It's tonight, isn't it?" Ellen said happily as she bolted into the store. She released one strap and allowed her backpack to slip carelessly over her shoulder.

"Tonight's the night," Anne Marie concurred. Being invited to someone else's home for dinner seemed to be a new experience for the eight-year-old. Although Ellen had always displayed good manners, Anne Marie reviewed them with her, just to be on the safe side.

"I won't talk with my mouth full or interrupt the conver…conver—" she stumbled over the word "—the conversation."

"Excellent." Anne Marie smiled at her. "You can bring your knitting if you want."

At that suggestion, Ellen raced up the stairs to the apartment as if they were heading out the door that very moment. Such exuberance made Anne Marie smile again.

They were both making progress with their knitting. Anne Marie's first official class the day before had gone well. In teaching Ellen, she'd learned more about the basic knit stitch than she'd realized. After school on Tuesday, Anne Marie had taken Ellen to A Good Yarn and allowed her to purchase yarn and needles of her own. Lydia had chatted with Ellen for quite a while; by now, as Lydia said, the two of them were old friends. That evening, after the dinner dishes and Ellen's homework, they'd sat side by side, helping each other. Anne Marie couldn't avoid reflecting that this was something she'd never had the chance to do with her stepdaughter. Even as a ten- or eleven-year-old, Melissa had rejected all her attempts to work on projects together, whether it was reading or baking or gathering autumn leaves for a scrapbook. Whatever Anne Marie suggested was deemed "stupid" or "boring." The memory had produced a sadness she found hard to forget.

In the knitting class, Anne Marie had learned how to purl and she had about three inches of the lap robe finished. Ellen was half done with the scarf for her grandmother; the girl had a good eye for color and had chosen a soft pink yarn and a peach. The combination was lovely. They were colors Anne Marie would never have thought to put together.

Lydia had praised her color choice, too, and Ellen glowed with pleasure at the compliment.

"Are you bringing your Twenty Wishes binder?" Ellen asked now.

"Yes, I think so."

Ellen slipped her knitting into her backpack. "Should I bring my list?"

Anne Marie hesitated, a little worried that Ellen might inadvertently dominate the conversation. "Maybe next time, okay? For tonight I want you to sit and listen."

"Okay." Running up the stairs with her backpack, Ellen collected an excited Baxter for his walk, the requisite plastic bag tucked into her jeans' pocket.

At four, Steve Handley came into the shop for his shift. Anne Marie didn't have time to shower, but went upstairs to refresh her makeup. The day was overcast, so she decided to put a forest-green knit vest over her cream-colored long-sleeved blouse.

Ellen was modeling the new denim skirt Anne Marie had bought her when the phone rang.

"Want me to answer?" Ellen asked.

Anne Marie hesitated. "Let me check who it is first." She glanced at the phone as Caller ID flashed Melissa's name and number.

Instinctively Anne Marie backed away. She still hadn't recovered from her last conversation with her stepdaughter. Another heart-to-heart might just finish her off.

The phone rang again and then again. After the fourth ring, voice mail came on. Anne Marie listened to the brief message. Melissa identified herself, then said, "Call me," without explaining why.

"Anne Marie?" Ellen spoke tentatively, staring up at her with worried eyes.

"Hey," she said, forcing some enthusiasm into her voice. "I thought we had a dinner date. Are you ready?"

Ellen nodded eagerly.

"Me, too. Let's go."

On the short drive to Lillie's, they sang camp songs. Or rather, Ellen sang. Anne Marie *tried* to sing and once again her voice sounded as if someone was strangling her. After the first few lines, she stopped and simply listened. Ellen truly was gifted and she loved to sing. After the first song, she immediately started a second one—"This Little Light of Mine," a song she told Anne Marie she'd learned in church.

Which reminded Anne Marie that one thing she hadn't done was take Ellen to church. It wasn't part of her normal practice, not that she had anything against religion. Although, at the moment, she didn't exactly feel God had dealt her a fair hand. Yet she realized that if she was going to maintain the routine Ellen had with Dolores, she should probably be taking her to Sunday-school class.

Just as Ellen's song came to an end, Anne Marie pulled

up outside Lillie's house. This was the first time she'd been invited here. She parked in the circular drive, gaping at the sprawling Tudor-style house, which must have seven or eight thousand square feet. The outdoor lighting revealed a sweeping, verdant lawn and, closer to the house itself, an arrangement of flower beds filled with tulips of all colors, daffodils and delicate narcissus.

"Wow," Anne Marie whispered.

"Does Mrs. Higgins live in a castle?" Ellen asked in a hushed voice.

"It seems so."

Barbie arrived then, pulling into the drive behind them, and they all walked into the house together, followed a moment later by Elise. As soon as Barbie greeted her mother, Anne Marie could tell that something was amiss, although both Barbie and Lillie struggled to hide it. Instead of the usual camaraderie, the teasing and joking, they were stiffly polite with each other.

They must've had an argument or a falling-out. No wonder Lillie had seemed upset.

Lillie had arranged a small buffet with everything on a sideboard in the formal dining room. The buffet started with a selection of cheeses, olives, brie-stuffed dates and three different salads—a seafood pasta, a Caesar with home-made croutons and a fruit salad. For the entrée, Lillie presented them with ricotta-filled chicken breasts and scalloped potatoes.

Elise shook her head. "My goodness, Lillie, you must've been cooking for days."

"Mother is a tremendous cook," Barbie said quickly.

Lillie turned to her daughter. "Thank you. I enjoy spending time in the kitchen—it takes my mind off other concerns."

The comment seemed to be directed at Barbie, whose cheeks flushed as she looked away.

Anne Marie helped Ellen prepare her plate and then served herself. The five of them assembled around the table, which seated twelve. Anne Marie noticed that Lillie didn't have much of an appetite; for that matter, neither did Barbie. They barely seemed to touch their meals. Anne Marie, Ellen and Elise, however, savored every bite.

Conversation was general at first, with everyone asking Ellen about school and which classes she liked best. Reading, spelling and math, she'd answered, providing examples of what she'd recently learned.

"Speaking of classes, did you sign up for belly dancing?" Anne Marie asked Barbie.

Barbie jerked her head up, apparently caught unawares. "Belly dancing?"

"You said it was one of your wishes."

"Oh, yes. No, I haven't. Not yet at least. I will, though…probably." She sat straighter in her chair, chasing the food around her plate before she set her fork aside.

"What else is on your list?" Elise asked her.

Barbie reached down for her purse and withdrew a sheet of paper. "I started a binder like Anne Marie but left it at home this morning. I have my list here, though."

"I left my binder at home, too. At Anne Marie's house," Ellen said in a comforting voice as if to reassure Lillie's daughter.

"You know, I thought skinny-dipping would be fun."

"I've always thought that would be fun, too," Anne Marie murmured. She'd forgotten all about it until now.

Ellen tugged at her sleeve and when Anne Marie bent close, the girl whispered, "Is skinny-dipping a new diet? Grandma Dolores talked a lot about diets."

DEBBIE MACOMBER   231

Anne Marie wasn't sure how to answer. "It's, uh, something like that."

For the first time that evening a smile tweaked the edges of Barbie's mouth. "I *am* going to do it."

"Do what?" Elise asked as she and Lillie entered the dining room with dessert—platters of brownies, cookies and tarts.

"Skinny-dipping."

"Barbie!" her mother gasped.

"At night, Mother. In what you'd call a controlled environment."

"In the moonlight," Elise added softly. "Maverick and I—" She stopped abruptly and her face turned bright pink.

"You and Maverick went skinny-dipping?" Barbie asked.

"It was years ago...." She paused. "Well, to be honest, it happened shortly after we reunited." Elise shook her head fondly. "That man was full of crazy ideas."

"No wonder you loved him so much," Barbie whispered.

"Oh, I did, I did. I regret all the years we wasted. Maverick wouldn't let me talk about my regrets, though. He said we had to make up for lost time and we did everything in our power to squeeze thirty years of life into three." The expression on her face showed both happiness and loss and was almost painful to watch.

Anne Marie's eyes filled with tears, and she stared down at her binder. "What about you, Lillie?" she asked, wanting to draw attention away from Elise so the older woman could compose herself.

"You first," Lillie insisted, offering Ellen some dessert. The girl studied the platter carefully and chose a blueberry tart.

Anne Marie smiled, then glanced down at her binder again. The sheet she'd turned to had a picture of the Eiffel Tower. "I want to go to Paris with someone I love."

"That's so nice," Barbie murmured.

Anne Marie didn't mention that this was one of Robert's promises. She'd felt the lure of France, of Paris in particular, from her high school days, when she'd taken two years of French. Robert had spun wonderful stories of the adventures they'd have…someday. It was always in the future, always around the corner. Next month. Next season. Next year. And whenever they made tentative plans, his job interfered.

She tried to dismiss the thought. Her life was her own now and if there was happiness to be found, it was up to her to seek it. She couldn't, wouldn't, rely on anyone else ever again.

Because she'd loved and supported her husband, Anne Marie had never complained. Now it became clear that she'd lived her entire marriage based on tomorrow—on well-intentioned promises, directed toward the future.

"I believe you talked about that one before," Elise reminded her. "It must be important to you."

"It is."

"What's stopping you?"

"I don't want to go alone."

Ever practical, Elise said, "Okay, it's not just about seeing Paris. It's also about falling in love."

"Yes, that's true. I want to be in love again."

"Good."

"I do, too," Lillie said quietly. Her gaze drifted down the table to her daughter.

Anne Marie was shocked to see tears glistening in Lillie's eyes. "I have Twenty Wishes but only one is important," Lillie said next. "None of the others means a thing without the first."

"What's that?" Elise asked. "As if I can't guess."

Lillie smiled briefly at Elise's remark. "I want to fall in love again," she said, "with a man who's honorable. A man respected by his peers. A man of principle who values me as a woman… A man who'll be my friend as well as my lover." A tear rolled down the side of her face. "I have lived most of my life trying to please others. I don't think I can do that anymore."

"Nor should you," Barbie said. "You deserve to find that man, Mom."

Lillie's voice shook. "So do you."

"I know."

Then to everyone's astonishment, Barbie burst into tears. "I'm so sorry, Mom, so sorry."

"I am, too."

Lillie pushed back her chair and a moment later, mother and daughter were hugging each other, weeping together.

Anne Marie looked at Elise, who shrugged. Once again Ellen tugged at the cuff of her blouse. "Why are they crying?" she asked in a loud whisper, leaning toward Anne Marie.

"I'm not sure."

"Will they be okay?"

Anne Marie placed her arm around Ellen's shoulders. "I think so," she said.

As quickly as Lillie and Barbie had burst into tears, they started to laugh, dabbing their eyes with the linen napkins, smearing their mascara and giggling like teenagers.

Ellen began to giggle, too. Soon Elise joined in. After

a while she got up and carried her dinner dishes into the kitchen and set them on the counter.

Anne Marie collected her plates and Ellen's and did the same thing. This evening had been cathartic for all of them in some way. Except for Ellen, but Anne Marie knew the experience had been valuable for her, too.

Before she left she picked up her binder and as she shut it, her gaze fell on the Paris postcard she'd glued next to the cut-out picture of the Eiffel Tower.

One day she *would* go to Paris—and she wouldn't go alone. Because the love of her life would be with her.

Chapter

20

When the official-looking woman in the no-nonsense suit walked into Blossom Street Books, Anne Marie knew she was the same one who'd called earlier in the day. She'd introduced herself as Evelyn Boyle, a social worker from Washington State Child Protective Services. She'd sounded calm, professional and reassuring; otherwise Anne Marie might have been alarmed. She had the paperwork Dolores Falk had given her before the surgery, and Ellen and Anne Marie spoke with Dolores frequently.

She didn't understand why a social worker was involved now. In a few days, Dolores would be released from the care facility and Ellen would return to her. If the state was concerned about Ellen, it was too little, too late.

"You must be Ms. Boyle," Anne Marie said as she stepped around the counter. Thankfully Theresa, who worked Friday afternoons, had arrived a few minutes earlier.

"And you must be Anne Marie." The social worker came forward and thrust out her hand. "Please call me Evelyn."

Despite the woman's tranquil demeanor, Anne Marie was nervous.

"Is there someplace private where we could visit?" Evelyn asked.

"Sure." Anne Marie momentarily left her and walked over to Theresa, who eyed her speculatively.

"Is everything all right?" Theresa whispered.

"It's fine," she whispered back. In slightly louder tones, she added, "I'll be upstairs if you need me."

Theresa nodded.

Anne Marie led Evelyn up the narrow stairway to the apartment. Now that Ellen was more comfortable living with Anne Marie, she'd left a pair of rubber boots on the steps. Anne Marie grabbed them on her way up the stairs.

Baxter stood there waiting for her, tail wagging wildly. He cocked his head to one side, as though curious about her unexpected appearance. After she'd paid Baxter the required amount of attention, he sniffed the social worker's shoes, then returned to his bed in a corner of the kitchen.

Without asking, Anne Marie walked to the stove and put on water for tea. Evelyn pulled out a chair at the table, then set her briefcase on it and withdrew a yellow legal pad.

"How did you know Ellen was staying with me?" Anne Marie asked. She assumed Dolores hadn't told Social Services, which meant it was either the hospital or someone from Woodrow Wilson Elementary, probably Helen Mayer.

"I received a call from Ellen's school," Evelyn said, confirming Anne Marie's guess as she dug around the bottom of her purse for a pen.

Anne Marie stood with her back to the kitchen counter,

hands behind her. "Ellen's grandmother wrote a statement that gives me full guardianship of Ellen while Dolores is recuperating." How legally binding that scribbled, almost illegible document was remained uncertain. Considering how desperate the poor woman had been for someone, anyone, to look after Ellen, she would've signed the girl's care over to practically anyone.

"I gather you were originally supposed to have Ellen for only a few days."

"Yes." Anne Marie wanted to say more but restrained herself. In instances such as this, the less said the better. "Dolores made me promise Ellen wouldn't go into a foster home."

Evelyn Boyle glanced up. "There are many excellent foster homes in this area."

"I'm sure there are…."

"But in essence, Anne Marie, Ellen is already in one."

"I'm someone Ellen knows and trusts," Anne Marie said quickly.

"That's true. It's exactly what I mean. You *are* her foster mother." Evelyn waited a moment. "I do understand the situation correctly, don't I? You and Ellen are not related in any way?"

"That's correct," she responded. But the question hovered in the air, swirling up doubts and fears.

The teakettle's whistle offered a welcome respite. Anne Marie concentrated on pouring water into the pot. She covered it with a cozy and set it in the middle of the table to steep while she got two matching cups and saucers.

Her good dishes were packed away in the storage unit, and the apartment cupboard was filled with mismatched place settings. It had never bothered her before, but it did now. Logically she knew that Social Services wouldn't pull Ellen from her care because her dishes didn't match. Still, Anne Marie discovered that she didn't want to take *any* chances.

She poured two mugs of tea, hating the way her hand trembled.

"I should tell you I stopped by the school before I drove over here."

Anne Marie couldn't decide if that was reassuring or not. "Did you speak to Ellen?"

"I did," Evelyn said as she reached for the sugar bowl and added a heaping teaspoon. "She had nothing but wonderful things to say about you. She told me about your visits to her grandmother and how you've bought her several pieces of clothing. Have you been to the house recently?"

"Twice," Anne Marie replied. "Ellen needed some of her stuff, and I told Dolores I'd check on the place for her."

"Excellent. I'm sure she'll appreciate that."

Some of the tension seeped away.

Evelyn raised her cup. "Is it true you taught Ellen to knit?"

"Actually, we sort of taught each other. Ellen's knitting a scarf for her grandmother and I've started a lap robe. The various colors don't match and Dolores won't be able to wear one near the other. Mine's a shade of lavender and Ellen went with a peach and pink combination. It's really lovely. I mean, who would've guessed…well, I suppose that isn't important." Anne Marie knew she was rambling and forced herself to stop. And yet, she couldn't resist bragging about Ellen's accomplishments.

Pushing back her chair, she hurried into the other room and got Ellen's scarf, still on the needles. "Look how even her stitches are," she said, displaying the child's efforts. "My own aren't half as neat. Ellen loves to knit and she's already taught three of her friends. Her teacher was so impressed she thought it might be a good idea for the whole class to learn."

Evelyn nodded approvingly. "Ellen's teacher mentioned that to me. She said knitting will help the children with math concepts and learning patience. It'll also give them a sense of achievement. I think it's a terrific idea."

"Really?" Anne Marie couldn't hold back a smile.

"When I spoke to Ellen, she also told me something about Twenty Wishes. What's that?"

"Ah…oh, it's nothing."

"Not according to Ellen. She has a book she drags to and from school in her backpack."

Anne Marie didn't realize Ellen brought it with her. "She does?"

"From what I understand, half the class is making lists as well."

"Oh…" Anne Marie took a sip of her tea. "A group of my friends and I decided it would be fun, that's all." She didn't want to explain anything beyond that; it was too complicated and too private.

"I love it," Evelyn said, her enthusiasm unmistakable.

Anne Marie's gaze shot toward the other woman. "You do?"

"Why, yes. In fact, I immediately started thinking about what I'd put on my own list."

Anne Marie relaxed a little.

"When I spoke to Ms. Peterski, she said there's been a marked improvement in Ellen in the last three weeks. Her grades have always been good but she had problems in other areas. Her social skills have vastly improved and she's making new friends and reaching out to others."

Anne Marie nodded. Although she had no personal

reason to feel such overwhelming pride, it was difficult not to.

"Ellen is happy, too. This arrangement has obviously worked out well," the social worker said.

"She's an easy child," Anne Marie told her. True, it had taken them a few days to find their footing, but they'd adjusted to living together with surprisingly few problems.

"Yes, she's done very well," Evelyn murmured.

"Did Ellen tell you she taught Baxter—my dog—to roll over?" Anne Marie asked. Ellen had worked with the dog for weeks and had only recently accomplished that goal.

"As a matter of fact, she did," Evelyn said with a glance at Baxter, who snored softly in the corner.

"I believe I mentioned that I spoke to the staff at the nursing facility where Dolores Falk is currently residing, didn't I?" Evelyn continued.

In her nervousness, Anne Marie didn't recall. "I'm not sure. Dolores tells me she's recovering nicely. She said she'd be released sometime next week. Wednesday, she thought."

Ms. Boyle hesitated before responding. "I understand Mrs. Falk is making excellent progress. She confirmed that you and Ellen visit frequently. And she waits every day for that brief telephone chat with her granddaughter."

"We see Dolores as often as we can."

"I'm aware of that, and I applaud your conscientiousness."

"Three to four times a week," she added. She made the effort to fit those visits into her schedule because she appreciated how important it was for Ellen—and, of course, Dolores.

"Very good."

"Thank you. I'm doing my best."

Evelyn sipped her tea. "I can see that, and the proof is in Ellen. Her teacher's delighted. Ms. Mayer, the school counselor, sang your praises, too."

This conversation wasn't nearly as difficult as Anne Marie had feared it would be. She was beginning to relax.

"Getting back to Mrs. Falk…" The words hung in the air like an unanswered question.

"Yes?" Anne Marie put down her cup.

"Did I hear you say she's going to be released next week?"

"Yes. Ellen and I were by on Wednesday after school and Dolores said she'd talked to one of the nurses about it."

"I'm afraid that's wishful thinking on Mrs. Falk's part," Evelyn Boyle said.

"What? How do you mean?"

"I spoke with the doctor's office as well as the head nurse."

A chill raced down Anne Marie's spine. "She's going to be all right, isn't she?"

"Oh, yes," Evelyn assured her. "The healing process is coming along well. But don't forget she had major heart surgery."

"Yes, of course." Fortunately there didn't seem to be any significant complications.

"However…"

"Are there problems with her recovery?"

"Not exactly problems."

"What is it, then?"

Mrs. Boyle's hand lingered on her cup and she ran her index finger along the rim. "Unfortunately it will be some weeks before she'll be able to return to her own home."

"Weeks?" Anne Marie repeated. This was a shock and she knew Ellen would be terribly disappointed.

"I'm sorry."

"How…many weeks?" Anne Marie asked, wondering how she'd explain this to Ellen. "Can you tell me how much longer it'll be before Dolores can go home?"

"I'm not a physician."

"What did the doctor say?"

"Two weeks."

"That's what I was told," she said. "But you mean an *additional* two weeks, right?" Anne Marie exhaled slowly.

"Yes. Are you okay with that?"

"Definitely. I'm just afraid this is going to be upsetting for Ellen. The child loves her grandmother very much."

"I know."

"Ellen's been marking off the days until she can move back in with Dolores."

"I understand this will be a setback for the child. I also understand that it's far and above what you agreed to when Ellen came to stay with you," Evelyn said. "If you feel it's too much, I could probably find a temporary home for Ellen."

"That would upset her even more," Anne Marie said, dismissing the offer out of hand. "It was difficult enough for Ellen to be separated from her grandmother. Placing her in another completely foreign environment would be doubly traumatic."

"I couldn't agree with you more."

At least they saw eye to eye on that, Anne Marie thought with relief.

"Then you won't mind keeping Ellen for another two weeks?"

"Of course I don't mind." Any other option wasn't worth considering.

"In that case, I'd like to leave some forms for you to complete."

"What kind of forms?" Anne Marie didn't like the sound of this.

Evelyn Boyle took a sheaf of papers out of her briefcase. "Since Ellen's been with you for more than two weeks already and is likely to remain for an additional two, I'd like you to apply for your license."

"My license for what?"

"To be Ellen's foster parent," she said as if this was perfectly logical.

An automatic objection rose in her throat, but Anne Marie bit down on her tongue rather than argue. The best thing to do was to appear compliant. However, she had no intention of becoming a foster parent. What was the point? By the time she finished applying, Ellen would be back with her grandmother and it would be irrelevant.

"Thank you," Anne Marie said, accepting the papers.

She stood and took the teapot and cups to the sink. "I appreciate your coming by," she said, since the interview was clearly over.

"My pleasure."

Baxter got up from his dog bed and walked them to the stairs, as though that was one of his prescribed duties. He stood silently at the top while the two women climbed down.

Anne Marie was saying goodbye to the social worker when she noticed a lone figure in the overstuffed chair, her head drooping, hair half-covering her face. The woman appeared to be asleep. Anne Marie glanced at her again, and suddenly realized who she was.

Her stepdaughter, Melissa Roche.

As if aware of Anne Marie's scrutiny, Melissa opened her eyes and sat up, looking self-consciously around. Anne Marie wished now that she'd returned her phone call. Even from a distance, she could see that Melissa was in distress.

For the moment she ignored her and accompanied Evelyn Boyle to the door. She thanked her for the visit and agreed to read over the paperwork—and read it was all Anne Marie intended to do. Evelyn obviously didn't want to remove Ellen from her temporary custody any more than Anne Marie wanted to let the child go.

She had to admit she felt ambivalent about this latest information concerning Dolores. On the one hand, she knew Ellen would be disappointed; on the other hand, she herself wasn't unhappy about the girl's extended stay.

By the time Evelyn Boyle had gone and Anne Marie turned back to the shop, Melissa was standing uncertainly beside the chair. She seemed to be waiting for Anne Marie.

Anne Marie spoke with Theresa for a few minutes about some special orders, then walked toward her stepdaughter. "Hello, Melissa."

"You didn't return my phone call. I left a message."

"I was out last night. I didn't get in until late."

Melissa seemed confused. "You're not dating anyone, are you?"

Why would she ask that question again? Anne Marie couldn't even *think* about another relationship so soon after losing Robert. "No. I was with a group of women friends, although that isn't really any of your concern," she said brusquely. "I intended to call back this evening." Actually, Melissa had phoned more than once. Caller ID had shown three calls, all from her stepdaughter, although she'd left only the one message.

"Could I buy you lunch?" Melissa asked in a surprisingly tentative voice.

"Thank you, but I've already eaten."

Melissa blinked as if she hadn't expected that despite the fact that it was nearly two in the afternoon.

"The truth is, I'm not eager to visit another restaurant with you."

Melissa blanched. "I said I was sorry about that."

Anne Marie nodded. "Yes, you did."

"And I am, I really am! Sometimes I do stupid stuff. I don't know why I thought I should tell you what I saw. Brandon about bit my head off. He said—well, never mind."

Anne Marie could see this wasn't going to be a quick visit, so she motioned for Melissa to sit down again, then took the chair next to hers. The shop wasn't busy and Theresa was handling what business there was without a problem.

The two women sat silently for several seconds. Anne Marie was determined not to speak first. After all, Melissa had sought *her* out. She was the one with the agenda and frankly, Anne Marie was curious as to what it might be.

"Brandon said he came by a little while ago."

"Yes." She didn't elaborate.

"He said there's a child living with you."

"It's a temporary situation."

Melissa acknowledged the comment with a slight nod. "He said that, too, and that—Allie, is it?"

"Ellen."

"He said that Ellen's a real sweetheart."

"She is." Anne Marie wished Melissa would get to the point. "But I'm sure you aren't here to discuss my child-care activities."

"No," her stepdaughter agreed, fidgeting nervously with her hands. "Did Brandon tell you Mom and I aren't getting along?"

"He mentioned it."

"Mom's really upset with me."

"Is there any particular reason?"

She responded with a shrug. "Several, actually. For one thing, it doesn't look like I'll be able to graduate on time."

"Melissa!" Anne Marie couldn't help the gasp of shock. Robert had bragged about Melissa's making the Dean's list; Anne Marie doubted that the problem, whatever it was, had anything to do with her grades.

"I…I dropped out of school."

Anne Marie's mouth fell open. "But…why?" she asked incredulously.

Melissa didn't answer.

Anne Marie saw that the younger woman's eyes had

filled with tears. Gazing down at her hands, Melissa murmured something Anne Marie couldn't quite hear.

"I beg your pardon?"

Melissa raised her chin. She inhaled and then said more loudly, "I'm pregnant."

Anne Marie sagged against the back of the chair. "Pregnant," she repeated, doing her best to hide her stunned surprise.

"Mom is furious with me."

Anne Marie could well imagine. Pamela had big plans for her daughter. Melissa's career path had been paved very nicely by her mother, who worked in upper management for an international chain of hotels based in London. According to Brandon, Pamela had secured a middle management position for Melissa as soon as she received her MBA. With the position came the opportunity to live in England.

"She wants me to get rid of it—that's how she put it. To have an abortion."

Anne Marie hardly knew what to say. She couldn't believe Pamela wanted to "get rid of " her first grandchild. Robert would've been horrified by that, she thought.

"The baby is your boyfriend's? Michael's?" Anne Marie asked as she tried to sort through her own emotions.

Melissa nodded.

"What does he want you to do?"

Melissa closed her eyes. "I...I haven't told him yet."

Leaning forward, Anne Marie clasped Melissa's hand.

Sobbing, her stepdaughter held on tightly. "I broke up with him."

In her fear and panic, Melissa suddenly seemed very young to Anne Marie. "Michael has a right to know," she whispered gently.

Melissa sniffed piteously. "I realize that, and I will tell him. It's just that... Everything's so messed up, and I'm not sure what to do. I didn't tell him because I was afraid he'd try to influence me one way or the other, and I didn't want that."

"So you broke off the relationship?"

Melissa bit her lip. "Stupid, wasn't it?"

Uncertain how to respond, Anne Marie squeezed her hand.

"I've never missed my dad more than I do right now. I'm so confused, and my mother's so angry with me."

"What does she say?" Anne Marie asked.

"She e-mails me two and three times a day with what she calls advice, except it reads more like a court order. I made a mistake, according to Mom, but that mistake doesn't need to screw up the rest of my life. She told me to make an appointment at one of those clinics and terminate the pregnancy before it's too late. She said if I lose this chance to work in England, I'll always regret it—that I'll never get a chance like this again."

Anne Marie had to struggle to keep from saying what she thought of that advice.

"She made it sound like I wouldn't regret making a hasty decision about my baby. I don't think I can do it, Anne Marie." The tears made wet tracks down her pale cheeks.

"What do *you* want?"

"I don't know," she whispered, still clutching Anne Marie's hand. "The thing is, I talked to a lady at the Pregnancy Crisis Clinic, and there are a lot more options than I thought."

"Don't you think you should explore all your options before you make such an important decision?"

"That's just it," Melissa sobbed. "I only have a few more days while it's still legal to have an abortion."

"How far along are you?"

"Over three months." She pulled her hand free and scrabbled in her purse for a tissue. "At first I didn't believe I could possibly be pregnant. I mean, I've never had regular periods, anyway, and there wasn't any reason to…to think I might be. Michael and I used protection and, well, apparently it wasn't a hundred percent effective, because here I am." She gestured weakly, then wiped her nose.

"You've been to a doctor?"

She looked away. "Not yet. But a technician at the pregnancy clinic did an ultrasound and I actually saw my baby move."

"Have you talked to your friends? Or Brandon?"

Melissa shook her head. "No one knows, other than you and my mother. I just couldn't face anyone else."

"How can I help you?" Anne Marie asked, wondering why Melissa had turned to her. But the reasons for her stepdaughter's change of heart didn't matter, Anne Marie told herself. She would do whatever she could.

"I need…I need someone who can help me decide."

Melissa had difficulty making decisions; that was clear, since she'd made a number of spectacularly bad ones. But perhaps some of them could be reversed.

"Okay," Anne Marie began, taking a deep breath. "First, I don't see that there's any reason to drop out of school, especially this close to graduation."

"I know. That was just as stupid as breaking up with Michael, wasn't it?"

"Do what you can to get back on course for graduation. Your father would've wanted you to complete your education."

Melissa nodded; she seemed to appreciate the advice. "Several of my professors have asked to talk to me, so I don't think that'll be a problem."

"Good."

"What about my mother?" Melissa asked, looking anxiously at Anne Marie.

"This is your decision, not hers."

She nodded again, as if she needed to be reminded of that. "If I don't go to England..."

"Why can't you go?"

"Mom said I couldn't have the job unless I aborted the baby."

"I'm sure she doesn't mean that," Anne Marie said. "The news shocked her, that's all." She remembered Robert confessing that he'd gotten Pamela pregnant before they were married. Apparently she was afraid her daughter would repeat her own mistakes by marrying too young—and in Pamela's view, marrying the wrong man.

"I should tell Michael right away, shouldn't I?"

"That would be a good idea." Anne Marie could see this was something Melissa wanted to do. "The two of you can talk it over together. Do you love him?"

"Yes, but... A friend told me she saw him with someone else." She paused, tears running unchecked down her face. "If he loved me, he wouldn't be dating again so soon, would he?"

"Who knows why men do anything?" Anne Marie asked, hoping to inject a bit of humor into the conversation.

Melissa responded with a wobbly smile. A moment later, she whispered, "Thank you, Anne Marie. I never thought I'd turn to you for anything and now I feel you're the only person I can talk to."

There'd been a time, a long time, when Anne Marie would've done anything to win her stepdaughter's approval. Little did she realize it would come after Robert's death.

They hugged and arranged to meet for lunch the following week. As they broke apart, Anne Marie recognized that Melissa wanted to say something else. She looked away and then back at Anne Marie, her eyes intent.

"I *am* sorry about the last time we met—you know that, right?"

Anne Marie nodded.

"Have you…?" She didn't complete the thought, almost as though she was weighing the advisability of even asking.

"Have I what?"

Melissa shrugged. "Contacted Rebecca? Have you asked her about the…baby?"

"No." Anne Marie kept her voice as flat as possible.

Her stepdaughter accepted that without further comment. With a wave and a "See you next week," she headed for the door.

Anne Marie waited until Melissa had left the bookstore before she collapsed onto the overstuffed chair and pressed one hand over her eyes. This nightmare that had become her life just wasn't going away. *She* was the one who wanted a child.

Not Rebecca.

Not Melissa.

Anne Marie.

Her longing for a baby had led to her separation from Robert—a desperate attempt to impress on him how serious she was. Not that it had done her any good. Instead,

Robert's personal assistant now had a baby, most likely his, and his daughter had turned to Anne Marie for advice about an unwanted pregnancy.

But there was no baby for her.

No love, either.

She sensed someone at her side and opening her eyes, found Theresa standing there. Her employee rested one hand on Anne Marie's shoulder.

"Bad news?" she asked.

Forcing a smile, Anne Marie shook her head. "That was Robert's daughter."

Knowing the history between them, Theresa stared at her. "Melissa? Is she okay? Are *you?*"

"She…she misses her father."

So did Anne Marie, even more than she'd thought possible.

Chapter

22

On Monday evening Barbie purposely stayed away from the movies. It wasn't easy, but she felt she had no option. Last week she'd left her business card with Tessa; now Barbie felt the next move had to come from him.

In a way Mark *had* made the next move by having flowers delivered, although she considered that an indirect, even cowardly approach. The flowers were a lovely gesture, but she'd been looking for more—like an apology or an invitation to meet again. By ordering the floral arrangement he'd managed to communicate his interest, yet keep his pride intact.

Maybe…the gesture *was* enough. For the moment.

She recognized what he was trying to tell her. He'd made a move in this elaborate game of theirs; the next one was hers.

She knew a little more about him after a Google search. He was an architect with an independent practice and lived in a downtown condo he'd designed himself.

Barbie felt encouraged by his interest. No, she was ecstatic. Still, she had to restrain herself, not let him have the upper hand. She decided she'd return to the movies again, but not right away.

Tuesday afternoon, she thought she'd register for the belly dancing class being held at the Seattle Fitness Center. This was her first trip here, and she was surprised to find an Olympic-size pool, along with a huge gymnasium and several activity rooms. As she walked down the hallway to the office, she passed a shop that sold workout clothes, swimsuits and other exercise paraphernalia.

After filling out the paperwork and paying her fee, Barbie began to leave the building, feeling positive and determined. She was making her wishes come true. Smiling to herself, she rounded the corner and stopped abruptly as a man in a wheelchair moved toward the pool.

Mark Bassett.

Coincidence? Fate? Barbie wasn't about to question it. Her heart felt as if it had shot all the way up into her throat. Without conscious thought she did an about-face and headed back, toward the shop. Within five minutes, she'd purchased a swimsuit and towel. Gaining entrance to the pool was a bit more difficult; before she was allowed to swim, she had to buy a six-month fitness membership. She slapped her credit card down on the counter, impatient to get into the water before Mark.

He *had* to believe this meeting wasn't staged—which, in truth, it wasn't. Okay, so her showing up at the pool might be a bit manipulative, but when life presented you with an advantage, you had to grab it with both hands.

Barbie changed into the swimsuit, a sleek blue one-piece, in the women's dressing room and walked out as though she was strolling along a Caribbean shore. The suit, thankfully, was a perfect fit. She squared her shoulders and silently thanked her mother for every lesson she'd taken at that expensive charm school.

Using the railing, she lowered herself into the water and cringed at the temperature. Her own swimming pool was kept at a comfortable eighty-five degrees. This was eighty, eighty-one maximum, and in her opinion downright cold.

When she'd entered the pool area, the attendant had explained that this was the adult lap swim. As soon as she got into the water, Barbie realized these noontime swimmers were serious about their workout. They wore goggles and bathing caps, and to her they resembled nothing so much as a bunch of insects with their smooth shiny heads and large round eyes.

Barbie refused to allow her hair to get wet. She needed to go back to work right afterward, and she couldn't arrive with dripping hair.

The minute she broke away from the side of the pool, another swimmer streaked past her, quickly followed by a second. It was quite apparent that no one appreciated her rather lazy form of breaststroke.

A third swimmer went by, kicking wildly, splashing her face and hair. Barbie swallowed a mouthful of chlorinated water and choked violently. She felt like she was about to cough up her tonsils. So much for making a sophisticated appearance.

She muttered a curse, treading water for a moment while she caught her breath. When she could breathe normally again, she wiped the water from her eyes. She'd given up even trying to keep her hair dry.

As she finally reached the other end of the pool, she saw Mark hanging on to the side, watching her. Clearly her antics were a source of amusement to him. His gaze found hers and he actually smiled.

Mark had *smiled* at her!

Since her hair was already ruined, Barbie stopped worrying about it and started swimming for all she was worth, face fully in the water. Her mascara was probably running, too, but she no longer cared.

Mark was waiting for her.

"Fancy meeting you here," she said, hoisting herself up on the side of the pool. She gave him what she hoped was a dazzling smile.

"Yeah, some coincidence."

"Come here often?" she asked.

"Every day."

"Me, too," she lied. "I can't believe we've never run into each other before."

"Every day?" He arched his eyebrows in disbelief. "Since when?"

She wasn't fooling him, so she might as well own up to the truth. "Since today."

He not only smiled at that, he laughed. The sound was deep, pleasant to the ear. She had the impression that he hadn't done a lot of laughing in the last few years.

"I come here to swim," Mark told her. "Keep up if you can."

"Hey, you're going to have to catch *me*," she shouted after him.

That was a joke if there ever was one.

Mark took off and with impressive upper body strength sliced through the water. His ease and grace were mesmerizing. Barbie didn't make the slightest effort to catch up with him. When he'd passed her twice, he stopped, waiting for her in the deep end. By then, Barbie had swum two laps and was too exhausted to swim anymore. Her breath came in shallow gasps. She grabbed the edge of the pool and felt her heart pounding hard against her ribs.

"Are you going to tell me why you're really here?" he asked.

"Belly dancing."

"I beg your pardon?" He sounded incredulous, and she wondered if he thought she was making fun of him.

"I signed up for a belly dancing class."

"At the Fitness Center?"

Propping her elbow on the ledge, Barbie pushed the hair away from her face. "Just as I was leaving, I saw you and had the overwhelming urge to take a dip."

"You're a member?"

"I am now."

Almost everyone had left the water. Barbie looked around, astonished to discover that only the two of them were still in the pool. When she glanced back at Mark, she saw him frowning.

"Why?" he asked.

Barbie didn't know how to answer him. "Don't *you* feel it?" she asked him instead. Judging by his puzzled expression, he either didn't understand or didn't want to, so she continued. "That first night in the theater… I've never experienced anything like it."

"You're imagining things," he snapped.

"No, I'm not." She wasn't going to let him lie to her, let alone himself. The attraction between them was too intense to ignore.

"I realize you're not happy about this," she whispered. "You've made that pretty obvious."

"Then leave me alone."

"I wish I could," she said, "but I just…can't." She

hadn't meant to reveal so much, but the words slipped out before she could stop herself. Their eyes met, and she could see the warring emotions inside him.

Using his free arm, he reached for her and slid his hand behind her neck and then slowly, as if fighting her every inch of the way, he brought his mouth to hers. He gave her ample opportunity to pull back.

She didn't.

Barbie wanted his kiss, hungered for it. She opened her mouth to welcome him, and then she was crushed in his embrace, arms and legs entwined, mouths joined.

Their kiss was better than she ever would have dreamed. They abandoned the effort to stay afloat and started to sink. Clinging to each other they sank far below the surface, their mouths straining, searching, devouring.

By the time they broke the surface again, Barbie was gasping for air. Because his legs were paralyzed, Mark needed to move his arms to remain afloat. All Barbie needed to do was kick her feet.

The sheer exhilaration of his kiss overwhelmed her. But he didn't seem to share her enthusiasm. His look was fierce, angry…afraid. He glared at her in much the same way he had the evening they'd first met.

"That was wonderful," she said reverently. It was very different from any sensual experience she'd shared with Gary, and she felt no guilt, no regret. Only gratitude.

He didn't respond.

"Mark?" His name was a soft plea on her lips. She couldn't bear it if he said or did something to destroy what she'd found to be an intensely moving experience.

Without speaking he kissed her a second time, and they sank into the clear blue water, wrapped in each other's arms. After a moment, they bobbed to the surface again.

Mark released her and Barbie sagged breathlessly against him, her head on his shoulder. It'd been so long since a man had held her or kissed her....

"I don't think this is a good idea," he whispered, but even as he spoke the words, he caressed her wet hair. He was braced against the side of the pool and she held on to him.

"It's a brilliant idea! Stop arguing with me." The sheer joy of being in his arms rang in her voice.

"Barbie—"

She shushed him with a kiss. "I mean it. Stop arguing."

He laughed again, the sound echoing in the cavernous room.

"I'm turning into a prune," she said, "and I love every second of it and all because I'm with you."

"You're very beautiful," he murmured.

"Even with wrinkled skin and mascara running down my face?"

"If you only knew…" Then, seeming to reach some kind of decision, he slowly exhaled. "Listen, Barbie, this is all very flattering, but—"

She interrupted him again, kissing him full on the mouth, using her lips and tongue to steal his very words. After coming this far, she didn't plan to let him get cold feet and a cold heart now.

His eyes were still closed when she broke off the kiss.

"You don't know what you're letting yourself in for. You—"

"I won't tolerate a man making decisions for me. If you think I'm going to allow you to decide what I do and don't know, then you're sadly mistaken."

The edges of his mouth quivered with the effort of suppressing a smile. "So you know everything there is to know about my disability."

"Of course I don't."

He ignored her response. "You read a few things on the Internet and you think you know it all."

"Well…okay, I read a few things."

His eyes narrowed. "Like what?"

A flush rose in her cheeks. "Mainly, I was interested in how we'll make love."

Mark gasped—or perhaps it was a groan, she couldn't tell which. "You're getting way ahead of yourself."

"Probably,"she admitted."But that's what I was most curious about."

His face somber and apprehensive, he smoothed a wet tendril from her cheek. "I should tell you…I haven't… since the accident."

"Then it's about time." Barbie couldn't believe they were having this conversation. Even more unbelievable was the fact that she could speak so openly and boldly about lovemaking with a man she barely knew.

Mark held her gaze a long moment. "Where do we go from here?"

"Where do you want to go?"

A smile twitched his lips. "Now, that's a leading question if I ever heard one."

She slapped his shoulder. "What I mean is we should probably get to know each other a little better."

"Must we?" he asked with pretended chagrin.

"Yes!"

"We can't go to bed first and ask questions later?"

"I'm not that kind of woman." Although considering the way he kissed, she might think about converting.

"I was afraid of that."

"You swim every day?"

"Every day," he assured her. "You too, right?"

"Right." This schedule change was going to take some adjustment. "Except Monday and Wednesday, when I'll be in my belly dancing class. Okay, I'll swim two or three times a week."

"Right."

"I could meet you afterward." Her staff was going to be putting in a lot of extra hours. That wasn't a problem; Barbie had been planning to give them more hours, anyway.

"You're sure about this?" Mark didn't seem convinced.

"I'm positive and if you ask me once more, I'll—"

"If you want to punish me, all you have to do is press that perfect body of yours against mine."

"That's nice to know." She moved closer and slid her right leg between his thighs. Her breasts brushed his chest as she spread eager kisses along his jaw.

"I suggest you stop now," he muttered. "There's a seniors' class coming in soon."

"Can't. I'm thanking you."

"For what?"

"The flowers you sent." She wouldn't have found the courage to confront him this afternoon if he hadn't made that move.

Mark went very still. "I didn't send you flowers."

"But…the card had your name on it."

He muttered something she couldn't completely hear; she caught the gist of it, though. Mark's sister or perhaps his mother was responsible for that bouquet.

"So, you *didn't* send the flowers," she confirmed.

Mark wound his fingers into her hair and dragged her mouth to his. "Let's just pretend I did."

Barbie was more than willing to do exactly that.

Chapter

23

Lillie Higgins stared at the phone, then groaned in frustration and turned away. This should be easy. Everyone seemed to think there was nothing to it. But try as she might, Lillie couldn't make herself call Hector.

In desperation, not knowing how else to manage this, she'd contacted the dealership instead, with a list of imaginary complaints about her car. The receptionist she spoke with made her an appointment for Thursday morning at ten. By the time she arrived at the service department, her stomach was tied up in knots a sailor couldn't untangle.

A man she didn't recognize came out to discuss the trouble her car had supposedly been giving her.

"Could you explain again what the problem is?" he asked, studying his clipboard.

Lillie had a panicky moment before she remembered what she'd told the receptionist yesterday when she'd made the appointment. "There seems to be a hesitation…."

"Coming from a full stop?" he asked, glancing up from his notepad.

"Yes, that's it. From a full stop."

"How often has this happened?"

She didn't want to overplay the situation. "A couple of times."

He jotted that down. "Just twice."

"No, more. Four or five times." Her hands were clammy and her mouth had gone dry and she had the most compelling urge to turn tail and run. If she hadn't handed her car keys over to the mechanic, she would've made an excuse and left before she looked like an even bigger fool.

He wrote something else on the chart.

"This won't take long, will it?" she asked.

"Not at all," he assured her.

Inside the waiting area, Lillie got a cup of coffee from the machine and picked up that day's paper. Although she'd come for the express purpose of seeing Hector again, now she prayed she wouldn't. How could she possibly explain what she'd done?

Lillie liked to think of herself as mature and sensible. Never in all her life, not even as a teenager, had she indulged in such a ridiculous deception over a man. Her face burned with mortification. She'd lied about her car—told an outright lie in a futile effort to see Hector Silva again.

Fifteen minutes later, the receptionist came to tell her that her vehicle was ready. She immediately went to pay the bill but found there was no charge. Eager to be on her way, Lillie hurried out of the building to the lot, where her car was waiting for her.

She nearly stumbled when she saw Hector standing next to it.

"Lillie," he said, his smile warm. "I've personally checked out the car and I can't find anything wrong with it. I thought if we took a drive, the problem might reappear and I could analyze it."

The offer to spend time with him was tempting, but she'd frittered away enough of his morning. "If you say it's in fine working order, then I'm sure it is. I trust you."

"I wouldn't mind, Lillie."

"Hector." Her face shone as brightly as a lighthouse beacon. "There's nothing wrong with my vehicle," she said, making a spontaneous decision. "I apologize. I shouldn't have wasted your time." The most important thing at the moment was getting away with her dignity— or what remained of it—intact.

Hector nodded. "We don't need to test your car, then?"

"We don't."

He opened the driver's door for her, and she climbed in. Her hand trembled as she inserted the key in the ignition. The door was still open.

"Do you…have you ever gone bowling?" The words came at her in a rush.

"Bowling?" she repeated, frowning. "Oh, sure, of course." This must be a day meant for lies. In her entire life, she'd never even stepped inside a bowling alley.

"I know it's short notice… I hope you don't mind my asking…"

"I don't mind." How eager she sounded. Her heart did a silly dance while she tried to disguise her excitement at his invitation.

"Tonight?" he asked.

"Yes."

"Six?"

"Okay." She mentally reviewed her closet filled with Misook and St. John suits. What did one wear bowling? Barbie would know.

Barbie would help her. Then she realized she couldn't tell Barbie.

Not yet. Later maybe, after she'd gone out with Hector.

He grinned. "Perhaps we should meet there?"

"That's fine." *Anywhere* was fine.

With a verve that was almost boyish, he shut her door, but not before giving her the address of the bowling alley. "I'll look forward to seeing you this evening," he said.

"Yes." Lillie didn't know if she was going to dissolve into tears or giggles. Either way, her actions today had been embarrassing—but she didn't care.

She was going to see Hector tonight, and they wouldn't be talking about cars, either.

That evening, thanks to her navigation system, Lillie located the bowling alley and got there at ten minutes to six. She wore beige linen slacks and a soft teal cashmere sweater with a floral silk scarf around her neck. Earlier in the afternoon, she'd purchased tennis shoes and white cotton socks. With her makeup she'd gone for a light, natural look, and she'd worn her hair neatly tied back. Every detail of her appearance had been closely scrutinized.

Hector, dressed in a suit and tie, was waiting outside the entrance, and when he saw her, his eyes lit up. Lillie knew exactly how he felt, because she felt the same happiness at seeing him.

As she approached, he held out his hands to her, and for a moment neither of them spoke. "Thank you for agreeing to meet me."

"Thank you for asking me." A little worried, she glanced at his suit. He looked as if he was about to attend a wedding. Lillie hadn't realized bowling was such a formal sport.

"Should I change my clothes?"

Hector shook his head. "No, no, you look perfect."

"But you're wearing a suit…."

His cheeks reddened slightly. "My daughter said I should never have suggested bowling. She said you must think me a buffoon. Would you care to dine with me, Lillie? I apologize if I offended you by offering to take you bowling. It's been many years since I invited a woman out. I don't know how such things are done now."

All Lillie really wanted was to be with Hector. It didn't matter to her if they were in a five-star restaurant or knocking down pins in a bowling alley. "Hmm. That's quite a decision."

"Shall we have coffee and then decide?"

She nodded. "That's an excellent idea."

Next they needed to figure out where to have coffee. They chose the café in the bowling alley, since that was the simplest alternative. Hector led her to a booth in the corner. The menu was shaped like a bowling pin and the salt and pepper shakers were empty beer bottles. Lillie was enchanted.

He slid into the booth across from her as she glanced happily around. The atmosphere reminded her of a fifties diner, the kind of place where the day's special was a double bacon cheeseburger with greasy fries. When she read the chalkboard, she saw that the special here was actually two cheese enchiladas with rice and beans.

Hector raised his hand and the waitress brought coffee.

Lillie leaned forward. "There was nothing wrong with my car." She'd told him as much earlier, but she wanted him to understand the reason for her pretense.

"Oh, I knew that all along."

"You did?" That made it even more embarrassing. "I wanted to see you again," she said bluntly.

Hector spooned sugar into his coffee. "I wanted to see you, too."

"But you didn't phone, you didn't ask.... My daughter and my friends urged me to contact you. They say that's how it's done nowadays."

"You *didn't* call me."

She avoided eye contact. "I wasn't sure how. I've never called a man—well, other than a professional or a friend."

"My daughter said if you agreed to see me after tonight, it would be a miracle. She crossed herself when I told her I invited you to go bowling."

Lillie laughed. The cheerful clatter from the bowling alley made her curious and she noticed that everyone seemed to be having fun. "I have another confession to make."

"Two confessions in a single night?"

"Two," she said with a smile. "I've never been bowling."

This didn't appear to surprise him, either. "Would you like to learn?" he asked.

"Only if you're going to be my teacher."

From across the table he grinned at her and she was mesmerized. He'd captured her imagination and her senses with his unfailing courtesy, genuine charm and with his kindness.

When they finished their coffee, Hector procured them a lane, fitted her with rented shoes, and then proceeded to show Lillie how to bowl.

By the end of the evening, Lillie had to admit she hadn't laughed this much in twenty years. It was gratifying—and completely unexpected—to discover that she had a certain knack for the sport. What they both found nothing short of hilarious was the fact that her ball rolled at the speed of an earthworm. She'd release it just the way Hector instructed, return to her seat and wait while the bowling ball slowly but surely trundled down the narrow lane. After what seemed like minutes, the ball would connect with the pins. They'd fall lazily over, one at a time, almost in slow motion, knocking into one another.

People stopped to watch when the ball finally made contact and the pins started to tumble. Once she managed to knock down nine pins, and the people in the next alley actually broke into applause.

Hector—who was obviously an accomplished bowler, as his succession of strikes made clear—claimed he'd never seen anything like it. Apparently, no one else had, either. The place was growing crowded, and Lillie was unaccustomed to all the attention, which embarrassed her. All she could do was laugh.

And when she laughed, Hector did, too.

In the last frame of their final game, Lillie achieved her first strike. It took nearly a minute for all the pins to fall and when the last one spun around and around and eventually toppled, she jumped up and down like a schoolgirl. Hector hugged her and then self-consciously stepped back.

After that, they turned in their bowling shoes and balls. She could hardly remember a time she'd enjoyed more. When they left the lanes, it seemed the most natural thing in the world for him to take her hand.

"Are you hungry?" he asked, dangling his suit jacket over his shoulder.

"Famished."

"I am, too. Do you have a favorite restaurant?"

"Yes, I do," she said and smiled over at him. "It's right here."

"Lillie, please allow me to take you to a real restaurant."

"This one looks real."

He hesitated. "My daughter suggested Gaucho's. She said they have an extensive wine list and a pianist who plays classical music."

"Are you planning to wine and dine me?"

"Yes, it's what you deserve. I want only the best for you."

It was such a sweet gesture, but she couldn't let Hector spend that kind of money. Besides, she wasn't dressed for anyplace formal. "I'd love some cheese enchiladas," she told him.

Hector squeezed her fingers. "If you insist, but if my daughter asks, please tell her the choice was yours, not mine."

"Rita?"

"Yes. Before I left this evening, she gave me a long list of things I should and shouldn't do. When I dated my wife, it was nothing like this." He stopped abruptly. "I don't mean to imply…"

"I know," she assured him. "I feel just as nervous as you."

"Really?"

She laughed. "You mean you can't tell?"

"No." He seemed genuinely surprised. "I have a confession to make."

"You?" Well, she'd already made two of her own.

"When I look at you," he said in a low voice, "I forget to breathe."

She wondered if he realized the effect his words had on her—or that she felt the very same way.

"Me, too," she whispered. She might have said more but a booth became available and they slid inside.

They each ordered the enchiladas and lingered over coffee, chatting until after one in the morning. Only when Lillie couldn't hold back a yawn did Hector suggest they call it a night.

He walked her to her car, which was one of the four or five still in the parking lot. The entire time Lillie prayed Hector would want to see her again. When he didn't mention it, she was sure this would be their one and only date.

"I had a lovely evening," she said, fumbling for her keys.

"So did I."

"Thank you for everything." She opened the car door and got in.

He nodded, stepping away as she started the engine.

Lillie's heart was in her throat.

"Saturday," he blurted out just as she was ready to drive away.

"Pardon?"

"Would you like to attend a lecture at the museum with me this Saturday?"

The relief was so overwhelming, she nearly broke into tears. "That would be wonderful, Hector."

Wonderful didn't begin to describe it.

Chapter

24

Wednesday afternoon, the sun was shining and the wind off Puget Sound was warm. This was a perfect spring day, and Anne Marie suddenly realized she felt…good. She'd almost forgotten what that was like. The comfort of the sun, the freshness of the breeze, the company of others—they all contributed to her sense of well-being. Most of all, though, she felt a contentment she hadn't experienced since before her separation from Robert.

She'd just finished a knitting class with Elise Beaumont. Three other women had signed up, and the session had been fun, with plenty of banter as Elise reviewed their work.

While she was at the yarn store, Colette Dempsey came by with her infant daughter. At first Anne Marie had been afraid that seeing her friend with the baby would be painful; it wasn't. Even though the world seemed to be full of surprise pregnancies and secret ones, too, she managed to distance herself from destructive emotions like envy.

She found she could genuinely delight in Colette's joy. They talked for an hour, and the visit passed with barely a ripple of pain.

Anne Marie was saddened by the news that Colette and her husband, Christian, would be moving to California at the end of June for business reasons. Christian owned a successful importing firm and would be opening a second office in San Diego.

She recalled that only a year ago, Colette had been a widow like her, and that was something they'd had in common. But Colette had been hiding a pregnancy and she'd struggled with a painful dilemma that had been dramatically resolved.

Nothing dramatic had happened to Anne Marie in the last two months. Nothing had really changed, either; certainly not her circumstances, other than the fact that Ellen was living with her but that was only a temporary situation. The only difference was in Anne Marie's attitude.

She still had to make an effort to maintain that attitude. Her Twenty Wishes had helped, because she now felt she had some control over her emotions. Doing something for someone else—Ellen—had, without a doubt, made the biggest difference.

To be honest, avoiding the question of Rebecca Gilroy had helped, too. One day soon, she'd ask who had fathered her son. But not until she felt ready to accept the answer.

After her knitting class, Anne Marie did a few errands and then collected Baxter so the two of them could meet Ellen's school bus. The girl's spirits had been low since she'd learned that her Grandma Dolores wouldn't be home as quickly as they'd hoped.

Anne Marie decided a leisurely walk, maybe stopping somewhere for something to eat, might improve Ellen's

mood. As she waited on the street corner, the big yellow
bus rumbled down Blossom Street. When it stopped and
Ellen hopped out, Baxter strained against his leash, whin-
ing excitedly.

"How was school?" Anne Marie asked.

"Good."

The noncommittal reply was typical. Generally it wasn't
until later in the evening, usually over dinner, that Ellen
began to talk more freely. She appeared to need time to as-
similate the day's events and perhaps figure out how much
to share.

"I thought we'd take Baxter for a walk together."

"Okay."

Ellen rarely showed much enthusiasm when Anne
Marie suggested an outing. She revealed her pleasure
in other ways. Anne Marie suspected she was afraid to
let anyone know she was happy about something, for fear
that the object of her happiness would be taken away.

"If you want, you can leave your backpack in the shop
with Theresa."

"Okay." Ellen raced ahead and Anne Marie watched
her employee place the heavy bag behind the counter. A
minute later, Ellen was back.

"You ready?" Anne Marie asked.

"Ready."

Baxter certainly was. Her Yorkie pulled at the leash;
apparently Anne Marie wasn't walking briskly enough to
satisfy him. The dog had places to go, territory to mark
and friends to greet, especially the friends who kept special treats just for him.

"Let's walk down to Pike Place Market and have dinner at one of the sidewalk cafés," Anne Marie said. "Does
that sound good?"

Ellen shrugged. "I guess so."

"Or would you rather go down to the waterfront and get fish and chips?"

Ellen didn't seem to have an opinion one way or the other. "What do you want?"

Anne Marie had to think about it. "Pizza," she finally said. "I haven't had it in ages."

"What kind?"

"Thin crust with lots of cheese."

"And pepperoni."

"Let's see if we can find a restaurant where we can order pizza and eat outside."

"Okay."

On a mission now, they trudged down the steep hill toward the Seattle waterfront. Pike Place Market was a twenty-minute hike, but neither complained. Baxter didn't, either, although this was new territory for the dog. Once they reached the market, Anne Marie picked him up and cradled him in her arms.

Ellen and Anne Marie strolled through the market, where they watched two young men toss freshly caught salmon back and forth for the tourists' benefit. Ellen's eyes grew huge as she gazed at the impromptu performance. When they left the market building, the little girl slipped her hand into Anne Marie's and they walked down the Hill Climb stairs to the waterfront. Baxter was back on his feet by then, taking in all the fascinating smells around him.

"Have you ever been to the Seattle Aquarium?" Anne Marie asked when the building came into view.

"My class went."

"Did you like it?"

Ellen nodded eagerly. "I got to touch a sea cucumber and

it felt really weird and I saw a baby sea otter and a real shark."

She'd liked it, all right, if she was willing to say this much. After investigating several stores that catered to Seattle tourists, Anne Marie located a pizza place. While they waited for their order, they sipped sodas at a picnic table near the busy waterfront. They both ate until they were stuffed and still had half the pizza left over.

"Shall we save it?" Anne Marie asked. Ellen agreed. It seemed a shame to throw it out, but she suspected that if they brought it home, it would sit in the refrigerator for a few days and end up in the garbage, anyway.

Anne Marie carried the cardboard box in one hand and held Baxter's leash in the other. They'd just started back when Ellen noticed a homeless man sitting on a bench, his grocery cart parked close by. Tugging at Anne Marie's arm, she whispered something.

"What is it?" Anne Marie asked. "I couldn't hear you."

"He looks hungry," Ellen said a bit more loudly. "Can we give him our pizza?"

"What a lovely idea. I'll ask if he'd like some dinner." Impressed with Ellen's sensitivity, Anne Marie gave her the leash and approached the man on the bench.

He stared up at her, disheveled and badly in need of a bath. Despite the afternoon sunshine, he wore a thick winter jacket.

"We have some pizza," Anne Marie explained, "and we were wondering if you'd like it."

The man frowned suspiciously at the box. "What kind you got?"

"Well, cheese and—"

"I don't like them anchovies," he broke in. "If you got anchovies on it, I'll pass. Thanks, anyway."

Anne Marie assured him the pizza contained no anchovies and handed him the box. He lifted the lid and frowned. "That's all?" They'd gone about a block when the absurdity of the question struck her. She started to giggle.

Clearly puzzled, Ellen looked up at her.

That was when Anne Marie began to laugh, really laugh. Her shoulders shook and tears gathered in her eyes. "That's all?" she repeated, laughing so hard her stomach ached. "And he didn't want anchovies." Why she found the man's comments so hilarious she couldn't even say.

Ellen continued to study her. "You're laughing."

"It's funny."

"That's on your list, remember?"

Anne Marie's laughter stopped. Ellen was right. She wanted to be able to laugh again and here she was, giggling hysterically like a teenage girl with her friends. This needed to be documented so she pulled out her cell phone and had Ellen take her picture.

Then she dashed back and piled all the change and small bills she had—four or five dollars' worth—on the pizza box.

Another wish—an act of kindness. The man grinned up at her through stained teeth and rheumy eyes.

It came to her then that she was *happy*.

Truly happy.

Deep-down happy.

Anne Marie had felt good earlier in the day, but that was the contentment that came from a sunny day, seeing old friends, spending a relaxing hour with her knitting class.

Granted, her newly formed optimism had a lot to do with these feelings. But her unrestrained amusement was something else—the ability to respond to life's absurdities with a healthy burst of laughter.

It meant the healing had begun, and she was well on her way back to life, back to being herself, reaching toward acceptance.

When they returned to the apartment, it was still light out. Ellen had a number of small tasks to perform. She watered the small tomato, cucumber and zucchini seedlings they'd planted in egg cartons last Sunday. Once Ellen was home at her grandmother's, Anne Marie would help the girl plant her own small garden. They'd already planted a container garden on Anne Marie's balcony, with easy-care flowers like impatiens and geraniums.

As soon as she'd finished the watering, Ellen phoned her grandmother. Anne Marie spoke to the older woman, too.

"I don't know why these doctors insist on keeping me here," Dolores grumbled. "I'm fit as a fiddle. Ready to go home."

"It won't be long now," Anne Marie told her.

"I certainly hope so." She sobered a bit. "How's Ellen doing? Don't whitewash the truth for me. I need to know."

"She misses you."

"Well, of course she does. I miss her, too."

Anne Marie smiled. "Actually, she's doing really well."

Dolores Falk sighed expressively. "God love you, Anne Marie. I don't know what Ellen and I would've done without you."

The praise embarrassed her. She was the one who'd truly benefited from having the child.

When Ellen had done her homework, the two of them knit in front of the television. Ellen had completed the scarf for her grandmother and started a much more ambitious project, a pair of mitts. After an hour's knitting, she had a bath and put on her brand-new pajamas. She crawled into her bed. Prayers were shorter than usual that night, since Ellen was especially tired, and then Anne Marie read to her. They were now on the third "Little House" book and rereading these childhood favorites gave Anne Marie great pleasure. Ellen fell asleep listening.

When Anne Marie got down from the bed, Baxter hopped up to take her place.

The small apartment was quiet now, and a feeling of peace surrounded her. As always, she kept the bedroom door partially ajar for Ellen, who was afraid of the dark.

Tiptoeing down the hallway to her own bedroom, Anne Marie opened her binder of Twenty Wishes. She wanted to document the fact that she'd laughed. As Ellen had said, that was, indeed, one of her wishes.

She turned the pages in the binder and reviewed her list. She added a few more.

16. Go to Central Park in New York and ride a horse-drawn carriage

17. Catch snowflakes on my tongue and then make snow angels

18. Read all of Jane Austen

Lillie and Barbie had both said they wanted to fall in love again. Anne Marie wasn't sure she did. Love had brought her more grief than joy. She'd loved Robert to the very depths of her soul, and his sudden death had devastated her.

Then to learn he'd had an affair with his personal assistant... The betrayal of it still felt like a crushing weight.

Anne Marie closed her eyes at the pain.

"Stop it," she said aloud. "Stop it right now."

She felt suddenly angry with herself. It was as if she'd

set out to dismantle the positive attitude she'd so carefully created and destroy all the happiness she'd managed to find by thinking of everything that had gone wrong. No, she wouldn't do it; she refused to let herself reexamine the pain of the last months.

Turning a page, she looked at the picture of the Eiffel Tower.

Someday she'd go.

Someday that wish, too, would be fulfilled.

Anne Marie entered the small neighborhood park at the end of Blossom Street, where she walked Baxter every morning. Her stepdaughter sat on a bench waiting for her.

The weather had taken a turn for the worse since Wednesday, when she'd gone to the waterfront with Ellen and Baxter. This afternoon the sky was overcast and the scent of rain hung in the air.

Melissa had suggested they have lunch in the park. Anne Marie appreciated not having any reminder of their last restaurant meeting.

"Hi," Melissa said as Anne Marie sat down beside her.

"Hi." Strange—after years of avoidance, they were now seeking each other's company. Anne Marie was curious about what Melissa had decided since they'd last talked.

"I'm having yogurt for lunch," Melissa announced. "I don't even like it, but Michael says it's good for me and the baby."

Anne Marie took out a tuna sandwich she'd slapped together that morning. "You told Michael, then."

Melissa peeled off the foil top on the yogurt container and discarded it in her sack. "I went to see him right after you and I met, just like you said, and I'm glad I did." She paused. "He was definitely shocked."

"No more than you were."

"True." She grew quiet. "He didn't believe me at first."

This angered Anne Marie on her stepdaughter's behalf. "Why not?"

"Remember I was the one who broke up with him, and I hurt him pretty badly. I should never have done that. I should've told Michael right away."

"We all make mistakes," Anne Marie said. She'd certainly made hers. Late at night she sometimes wondered if she'd been wrong to give Robert an ultimatum, if she should've tried to work things out in a different way.

"I could see he wanted to talk to me, but he was scared I'd hurt him again, so he kept looking away." Melissa's voice became more animated. "I kept asking him to look at me and he wouldn't."

It must have been terribly frustrating. "What did you do?"

Melissa appeared to be studying her yogurt, but Anne Marie could see she was smiling. "I kissed his neck."

"His neck?"

"Michael's a lot taller than I am and that's as far as I could reach. I said I loved him and that I was sorry. I put my arms around him and told him the reason I broke up with him was because I was afraid." She unwrapped the plastic spoon. "To tell you the truth I was even more scared then. My friend, the one who told me she'd seen Michael with someone else, had some other news, too, and I was sure I'd lost him for good."

"What news?"

"She said she'd seen him in a jewelry store, looking at diamond rings."

"For this other girl?"

"No, but I'm getting ahead of myself."

"Okay, go back. Tell me what happened when you told Michael about the baby."

Some of the happiness left Melissa's eyes. "Like I said, he didn't believe me. He thought I was making it up, which made me so mad I almost walked away. I'm glad I didn't, though."

"But why would he even think that?"

Melissa shrugged. "He didn't trust me, and I can't blame him." She met Anne Marie's eyes. "I said I'd never make anything like this up and to prove it I showed him the ultrasound I got from the pregnancy center."

"I'll bet that got his attention."

Melissa laughed. "It sure did. He nearly fainted. All he could do was look at that little picture of our baby and walk around in circles. He was so pale I thought he was going to pass out. It took him a few minutes to get used to the idea, and then he asked me what I intended to do. I said that was why I'd come to see him. I felt this was a decision we should make together."

Anne Marie nodded. She didn't speak, not wanting to interrupt the flow of Melissa's story.

"One thing I did, which I regret now, was to tell him that Mom wanted me to get rid of the baby. Michael got really, really upset. Basically I'd already decided that there are so many other options, an abortion would be my last resort."

Although she didn't remember Michael from the funeral, Anne Marie felt warmly disposed toward him, sure she'd approve of this obviously responsible young man.

"He said a baby needs a mother and a father," Melissa went on, "and naturally I agreed because I believe that, too." She dipped her spoon into the yogurt and put it in her mouth. "Do you want to know how bad my sense of timing is?" she said a moment later.

Anne Marie grinned at her wry tone. "I can't wait to hear."

"The night I broke up with Michael was the very same night he was going to propose to me. He had an engagement ring in his pocket and everything. The diamond he'd been buying when my friend saw him was for me. *And* the other girl was just a study partner."

"Oh, Melissa."

"We've been seeing each other practically every night. Oh, and I'm back in school." She smiled confidently. "You can bet he's doing all his studying with *me* now."

"That's great! But there's something I need to know. Are you going to marry Michael?" She knew marriage was what Robert would've wanted for his daughter.

"Yes. Yes, I am. But I'm getting ahead of myself again."

Anne Marie took the first bite of her sandwich. Robert would've been thrilled at the prospect of grandchildren. That was always his excuse when he refused to consider having a baby with Anne Marie. He said he didn't want children and grandchildren the same age. Nor, he claimed, did he want his children to be mistaken for his grandchildren.

But it was useless to review their arguments now. What she recalled most strongly was Robert's anticipation of his children's children. Unlike Pamela, he would've been thrilled with the news of this baby.

Apparently Pamela had yet to be won over. But Anne Marie had to assume it wouldn't be long before the baby stole his or her grandmother's heart.

"As soon as I talked to Michael, we decided to get engaged," Melissa said. "Then I called my mother." Melissa frowned. "Unfortunately, it wasn't a pleasant conversation."

"How did you tell her?"

"I tried to be positive. I told her I'd discussed the baby with Michael and right off, that upset her. Mom said the decision was mine and mine alone, and by dragging him into it I was only complicating what should be a simple decision."

"Oh, dear…"

"You know," she said slowly. "I've never seen this side of my mother before. She's totally convinced I should take that job and move to England."

"You still could, I guess."

She nodded. "But I'm not sure I ever really wanted a job in London. Mom keeps telling me what a fabulous opportunity this is, the chance of a lifetime and all that. But it would mean leaving Michael and I don't want that, baby or no baby."

"Couldn't he come with you?" With Pamela's resources, it might be possible to get Michael employment, as well.

"No. He's going to work in his father's carpet business. He'll eventually take over the company. It isn't like he could just pack up and follow me to another country. He'd like to go to Europe one day, but Seattle is home."

That made sense to Anne Marie.

"I don't think I ever realized how much Mom had her heart set on me joining her in the corporate world."

Anne Marie could understand Pamela's disappointment. Her marriage to Robert had disintegrated when Pamela accepted a position that often took her out of the country. The arrangement had suited Pamela and, although Robert was involved in his own career, he loved his children and willingly looked after their needs. Pamela cared about her kids, too, of course, but she had her own vision of what was best for them. Whether they actually agreed with that vision seemed irrelevant to her.

"I'm sure your mother's afraid you might be repeating her mistake—or what she sees as one," Anne Marie said as gently as she could. She didn't want to suggest the marriage had been a mistake, although in retrospect it probably was.

"She started yelling at me and said I'd regret this for the rest of my life."

Pamela's temper was legendary.

"I asked her if she regretted having Brandon and me and she said…" Melissa swallowed hard. "She said if she had to do it over again, she wouldn't have had either one of us because we've done nothing but let her down."

"She didn't mean that! She *couldn't* mean it." Anne Marie was horrified by such a cruel remark, even if Pamela had lashed out, unthinking.

"I know," Melissa said in a small voice. "Afterward she e-mailed me and apologized, but she still said she wanted no part of the wedding."

The wedding. Melissa and Michael were going to have a wedding. Of course they were.

"Will you come to our wedding?" Melissa asked tentatively.

"Absolutely! I wouldn't miss it for anything."

Melissa smiled, and Anne Marie saw tears in her eyes. "I still can't believe this, you know," she muttered.

"What your mother said?"

"No, me coming to you for advice. And support. A year

ago, even three months ago, I would never have done that. I…I thought I hated you."

"Let's try to forget all that, okay?"

"I blamed you for the divorce, although I know it wasn't your fault at all. Brandon and I had a long talk about you, and he's helped me figure things out. Emotions can become habits," she said haltingly as she wiped her eyes. "But habits can be changed."

There was a silence, and Anne Marie found herself blinking back tears of her own. "Can I help with the wedding?" she finally asked, diverting the subject from herself and their painful past.

"You'd do that?"

"I offered, didn't I?"

"Well, yes, but I never expected… I didn't think you'd have time."

"I'll make time." Anne Marie *wanted* to help Melissa. The possibility filled her with hope and a kind of exultation. For nearly thirteen years her relationship with her stepdaughter had been nothing less than turbulent. Then, for reasons she didn't completely understand, it had begun to change.

"I wasn't going to tell you, but…"

"Tell me what?" Melissa glanced at her suspiciously.

"I'm taking knitting classes and I bought yarn for a baby blanket."

Melissa smiled tremulously. "You did that for me?"

"I'm going to be a stepgrandma, aren't I?"

Melissa nodded, and tears coursed down her cheeks unrestrained. "I can't believe how wonderful you've been to me." Melissa reached for her hand and squeezed it. "Thank you, Anne Marie," she whispered.

Anne Marie put aside her half-finished sandwich.

"Ellen picked out the yarn," she said, clearing her throat. "It's a variegated one with yellow, pink, pale blue and lavender." Anne Marie was eager to start. As soon as she'd finished the lap robe for Dolores, she'd knit the blanket for Melissa and Michael's baby. Her stepgrandchild.

"I'll treasure it. And please thank Ellen. I hope I can meet her soon." Melissa used a napkin to mop her face as she spoke. Except for her reddened eyes, she looked as beautiful as ever.

"We'll arrange something," Anne Marie said. "Now, have you and Michael set a date?"

"July twelfth."

It was almost the middle of April. That didn't leave them much time, especially with Melissa and Michael both graduating from college in the next month.

"You're reinstated in your classes, but what about the work you missed? Have you caught up?"

"Yes, Mom," Melissa said with a laugh.

Anne Marie was beginning to feel like a parent, or rather Melissa was *letting* her feel like one. Melissa's stepmom. Darn it, she loved how that felt.

Robert would be so proud of them. This was what he'd always wanted for her and Melissa. How sad that it hadn't happened until he was gone. Somehow, though, she had the feeling he knew and approved.

"Your father would've wanted you to get your degree," she murmured.

"I know," Melissa said.

"Okay, let's discuss the wedding."

Melissa sighed. "It's a bit overwhelming. We have no idea where to start. Michael's mother said we should set a date and get the minister first, but I don't know any."

"I do."

"Really?"

"Sort of. Alix Turner, who works at the French Café, is married to a minister. Would you like me to get his phone number for you?"

Melissa nodded. "That would be great."

"What about your dress?" Anne Marie asked.

Another deep sigh. "I haven't even thought about that yet."

"I'll do some research—a friend of mine owns a dress shop, and if she doesn't have what you need, she'll know someone who does. Then I'll make an appointment with Alix over at the French Café so we can check out a catering menu and look at wedding cake designs."

"You're sure you have time for all this?"

"For you, Melissa, yes," Anne Marie told her. "Ellen will help, too. It'll be good to get her mind off her grandmother."

"Thank you," Melissa whispered with a watery smile.

She might not ever be a mother, Anne Marie realized, but she was a stepmom and she'd make a wonderful grandmother for Melissa's baby.

She'd learned two things from Melissa. The habits of a lifetime *could* be changed. And family could come about in the most unexpected ways.

Chapter

26

When Barbie approached the ticket window at the movie complex, she was pleased to note that Tessa was handling sales again that night. As soon as she'd advanced to the front of the line, Tessa broke into a huge smile.

"Uncle Mark left a ticket for you."

Barbie hesitated. "He bought my ticket?" So far, they'd met at the pool four times for the adult lap swim session—and that was it. Although they'd kissed that first day in the water, they hadn't since. Not for any lack of desire, at least on Barbie's part. But the circumstances weren't ideal; their privacy the first time hadn't been repeated, and she wasn't interested in giving the seniors' swim class an eyeful.

"It's a *date*," Tessa said, as if she needed to clarify.

"Please tell me we're not seeing another horror movie."

"No," Tessa assured her. "It's a courtroom drama. Lots of talking. You don't have to worry about being scared out of your wits."

But Barbie *was* scared. She'd fallen for Mark and fallen hard. The wheelchair didn't frighten her, but the man who sat in it did. Their relationship wouldn't be easy and the realities of a future with him were intimidating. Yet the strength of her attraction overcame her doubts.

He was slowly letting her into his life, and that thrilled her. As was her custom, she purchased popcorn and a soda and entered the theater.

"Howdy," she said as she slipped into the seat directly beside Mark's.

"Hi." He didn't look in her direction.

"Thanks for the ticket."

"My pleasure."

She tilted her bag of popcorn in his direction and he took a handful. "As Tessa pointed out, this is like a real date."

"Aren't we both a bit old for that nonsense?"

"I certainly hope not," Barbie said. "My mother has a male friend and *they're* dating."

"You make it sound like high school."

"Does it feel like that to you? In some ways it does to me." In good ways. She woke each morning with a sense of happy expectation. Mark was in her thoughts when she drove to the dress shop and then at noon when she dashed to the fitness center. He'd never asked for her phone number, which she would willingly, gladly, have given him. It would've been sheer heaven to lie in bed and talk to him on the phone, like she had with her high school boyfriend. And then with Gary…

"Yeah, it feels like high school." Mark snorted. "In all the stupid ways."

"Mark!"

"I'm not a romantic."

"No!" She feigned shock. "I never would've guessed."

"I'll be honest with you," he said, his voice clipped. "I don't expect this to last."

"You're obviously an optimist, as well," she teased.

Mark still wouldn't look at her. "I don't know what you expect to get out of this relationship, because I haven't got much to give."

"Would you stop?" This little speech of his sounded rehearsed.

"Let me finish."

"All right, have your say and then I'll have mine." She tilted the popcorn in his direction again.

He stared at it. "I can't eat very much of that."

"Why not?" He *had* eaten it earlier, so it wasn't a food allergy.

"I have a lot of limitations, Barbie. For instance, I can't eat whatever I want."

"Few of us can eat whatever we want. You know what? Everyone has limitations. Okay, so yours are more obvious than some people's. But I have several of my own, which I'm doing my best to keep under control."

"Let me guess." He pressed his index finger to his lips. "First, you have one hell of a temper."

She laughed outright at that. "How kind of you to remind me."

"What I'm trying to say," he continued, "is that this relationship is doomed. You apparently get some kind of emotional kick out of flirting with me, and that's fine. It's good for the ego, and mine's been in the gutter so long that this is a refreshing change. But I'm not a fool. A woman like you can have any man she wants."

"Mark, I—"

"I don't mean to be rude here. However—"

"Why not? It hasn't stopped you in the past."

He grinned. "True. Just let me finish, okay?"

She motioned with her hand. "Be my guest."

"This is the way I figure it. For reasons beyond my comprehension, you're attracted to me."

"Is it a one-way attraction?" For the sake of her own ego, she needed to find out. "Answer this one question and I promise I won't interrupt you again."

He raised his eyebrows. "You already know. You pretty much turn me inside out every time I see you."

Barbie clasped her hands, still clutching the bag of popcorn. "Do I really?"

"Barbie, please, you're making this difficult."

"Okay, sorry." But her heart was leaping with joy.

"Listen, for whatever reason, you feel safe with me. I'm not a threat to you. I don't know what went on in your marriage, and frankly, I don't want to know. Whatever happened then or since has really rattled you. So a guy in a wheelchair's a safe bet. Fine. The truth is, I can't seem to forget you and I'm tired of fighting my attraction to you."

"For your information, I had a very good marriage." Barbie wasn't sure where all this talk was leading. "I think you're looking for an excuse to avoid a relationship with me."

"Listen," he said again, exhaling slowly.

"You're ignoring what I just said."

For the first time he glanced in her direction. As soon as he did, his eyes softened. "Okay. For now I'm willing to do things your way."

"For now?" She wasn't sure what that meant either. "Could you explain that?"

He sighed loudly. "We play this by ear. When you want

out, you get out. Don't prolong it. Do me a favor and just leave, okay?"

Barbie had to think for a moment. "What about you?"

"If I want out, then I need you to respect my wishes in the same way that I intend to respect yours."

That sounded fair. Still, she hated the idea of planning their breakup. "Can't we just take this one day at a time?"

He didn't respond.

"One *date* at a time?" she said and then leaned over and kissed his ear, running the tip of her tongue over the smooth contours and nibbling on his lobe. "One kiss at a time?"

"Yes," he said, his voice husky with desire. "I told you earlier—you turn me inside out."

The theater darkened, and the previews started. Barbie put her popcorn and soda on the floor at her feet. To her surprise and delight, Mark reached for her hand and entwined their fingers. She leaned closer and a few minutes into the movie, rested her head on his shoulder.

Mark snickered softly. "Like I said, it's high school all over again."

"I don't care if it is or not."

In response, he raised their clasped hands to his lips and kissed the back of her hand. A chill of pleasure slid down her spine.

At the end of the movie, she walked next to his wheelchair as they left the theater. She half expected him to ask her out for coffee. The movie might be over, but that didn't mean their evening had to be. When he didn't suggest it, she did.

"I can't tonight." He didn't offer an excuse.

"Perhaps another night, then." She did her best to hide her disappointment.

"Perhaps."

"Will you call me?" she asked.

"When?" He didn't seem pleased by the prospect.

"Tonight before you go to sleep."

He frowned. "Barbie."

"It's a simple request. If you can do it, fine. If not…if not, I'll lie awake all night wondering why you didn't phone."

"Just like high school."

"Yes." She wasn't about to deny it.

He muttered something she couldn't quite decipher and from the gruffness of his voice, she figured she was better off not knowing. "Give me your phone number," he growled.

"Thank you." Reaching inside her purse for a pen and pad, she wrote out her home number and tore off the sheet.

Mark crumpled it and stuffed it in his shirt pocket. "If you're loitering here because you expect me to kiss you, don't. I never kiss in public."

She didn't remind him that he'd kissed her at the pool and at the theater. "Want to go someplace private, then?"

He shook his head. "I'll take a rain check on that."

"Okay." What she'd learned from their brief conversation before the movie was that Mark was as terrified as she was. He didn't *want* to be attracted to her. He'd rather chase her away, only he hadn't succeeded in that. Yes, he desired her, but he wouldn't risk any kind of dependence on her. Insofar as he was willing to get involved, he wasn't letting her any too close. She had a mental image of a heavily armed guard standing watch over his heart.

Several hours later, Barbie lay on her bed, waiting for Mark's call. It was almost eleven before the phone finally rang.

"Hello, Mark."

"Hi." His voice was impatient. "I phoned like you asked me to. Now what is it you want to know?"

She hadn't actually thought the call required a purpose. "What are you wearing?"

"I'm on to your game! You're asking me that because you want me to ask *you* the same question."

"I'm wearing an ivory silk gown."

"Short?"

"No, full-length. What about you?"

"I'm not saying," he muttered. "I don't understand the point of this, and I'm not interested in silly games."

"I want you to want me," she said. "That's all."

"That's *all?*" She heard him snort disbelievingly. "You don't need to work nearly this hard."

"That's very sweet."

"It wasn't meant to be."

"Mark," she whispered, nearly purring his name. "Loosen up. We hardly know each other. I thought we could use this time to talk, to get acquainted."

He didn't say anything for maybe ten seconds, although it felt more like ten minutes. "What do you want to know?" he asked again.

"What do you do?" She already knew, but wanted him to tell her, anyway.

"I'm an architect." He didn't elaborate or describe any of the buildings he'd designed. Nor did he say where he worked or where he'd gone to school or anything else regarding his professional life. Barbie was beginning to understand him. The less he revealed about himself, the less likely she was to hurt him.

"I have twin sons," she said, moving the conversation into more personal realms.

"Identical?"

"Yes. My husband was killed in a plane crash three years ago."

"I know. You aren't the only one who uses the Internet. Both your husband and your father worked for a huge perfume conglomerate. You never said anything about that."

"Why should I? Mom and I don't really have anything to do with the company."

"You do smell good most of the time."

"*Most* of the time?" she flared.

"Chlorine isn't one of your better scents."

"I'll have you know you're the only man in the world who could get me into a public swimming pool. I live in mortal fear that my hair's going to turn green in that over-chlorinated water and it'll be entirely your fault."

"Then don't come."

"Uh-huh. And miss getting splashed by you? It's the highlight of my week!"

He laughed, and in her mind she saw the mercenary who stood guard over his feelings lay down one weapon in his arsenal.

They spoke for two hours. Before they said good-night, Mark admitted he'd been in bed a full thirty minutes before he phoned. He wouldn't have called at all, he said, if not for the fact that he couldn't sleep. Every time he closed his eyes, the thought of her waiting in bed taunted him until he couldn't tolerate it anymore.

On Thursday afternoon, after her belly dancing class, Barbie met her mother for lunch. Lillie had already arrived at the upscale hotel restaurant and was reviewing the menu when Barbie joined her. Lillie did an immediate double take.

"My goodness, you look wonderful! I know it's a cliché, but you're positively glowing."

"It's just sweat. This belly dancing is hard work."

"No, it's more." Lillie set the menu aside. "Is it that... man?"

"His name is Mark and yes, now that you mention it, he and I have been talking."

"You really like him, don't you?"

Barbie was crazy about him, but she wasn't ready to let her mother know that. She didn't want to ruin their lunch; so far, Lillie had been accepting of the situation and Barbie wanted it to stay that way.

"I didn't come to talk about Mark. I want to know how things are developing between you and Mr. Silva."

Lillie smiled, her eyes warm. "If I tell you something, you have to promise not to laugh."

"Mom, I won't laugh."

"We've been bowling twice in the last week."

"Bowling?"

"I'm good at it."

"My mother's a bowler. I'm calling Jerry Springer," Barbie teased.

"Oh, stop it," Lillie said and blushed.

"All you do is bowl?"

"Oh, heavens, no. We've gone for long walks and we attended a lecture at the Seattle Art Museum and signed up for a Chinese cooking class."

"What I mean is, has he kissed you?"

Lillie lowered her eyes. "Yes. We might be over sixty, but we aren't dead."

"That's for sure." In fact, her mother looked more alive than she had in years. "I think this is great!"

"What about you and Mark?"

There wasn't much to tell. "We've talked for the past three nights." Mark had confessed he generally didn't enjoy chatting on the phone. Still, they'd talked nearly two hours every time. Gradually, he was opening up to her and he became as engrossed in their conversations as she did.

When Barbie returned from lunch that afternoon, a large floral arrangement had been delivered. "Who sent the flowers?" she asked.

"Don't know. The card's addressed to you," one of her employees announced.

Eagerly she removed the small envelope and tugged out the card. Mark had written his name, together with a short note. *This time the flowers really are from me.*

Chapter
27

The big day had finally come. Dolores Falk was going home after nearly a month away, first in the hospital and then a nursing facility. According to her physicians, the heart surgery had been a complete success and Dolores had many good years left.

Certainly Anne Marie had noticed a definite improvement in the older woman. Every day Dolores seemed to regain more of her strength and her spirit. She was as eager to get home as Ellen was to join her there.

Thursday morning at breakfast, Ellen talked incessantly about moving back with her grandmother. The instant she got home from school, she ran upstairs to pack her bag. Anne Marie could hear her telling Baxter that she'd visit him soon. Ellen had him repeat the tricks they'd practiced—rolling over and playing hide-and-seek with his tennis ball—a few times for good measure. "So you won't forget," she told him sternly.

Anne Marie drove the child to her old neighborhood. "Remember, your grandmother's been very sick," she cautioned her.

"I know. I won't do anything to upset her," Ellen promised.

She glanced at the girl sitting in the passenger seat, the dog on her lap. "You can come see Baxter whenever you want," she said.

"Can I see you, too?"

"Of course."

"Will you still be my Lunch Buddy?"

Ellen must've asked the same questions ten times since they'd been told that Dolores was being released. "Of course," she said again.

"Goody." And then as if she'd almost forgotten something important, Ellen added breathlessly, "What about Lillie and Barbie and Mrs. Beaumont? What about Lydia and Margaret and Susannah and Theresa?" she asked. "Will I be able to visit them, too?"

"I'm sure that can be arranged." Her friends and neighbors didn't know yet that Ellen was moving back with her grandmother. As soon as they heard, they'd send their love to Ellen, and to Dolores.

"I'll still knit every day," Ellen assured her. She had a knitting bag now, the same as Anne Marie's. Young as she was, the child had proven to be an adept knitter.

"Me, too," Anne Marie said. She'd finished the lap robe for Dolores earlier and had given it to her during their most recent visit; she'd completed Melissa's baby blanket, as well. For her third project she planned to knit Ellen a sweater and had chosen a simple cardigan pattern. The girl had picked out a soft rose-colored yarn. Ellen was working on a pair of mittens. She wanted to knit Anne Marie a sweater but Lydia had wisely suggested she knit one for Baxter first and then try a larger project. Ellen had agreed.

As she neared the street where Dolores lived, Anne Marie examined the neighborhood more closely than she had before. It consisted of mostly older homes, many of them in ill repair. Now Anne Marie couldn't help wondering if this was a safe place for Ellen—or Dolores for that matter.

It'd been weeks since she'd seen the Falk home, which seemed even shabbier and more run-down now that she really looked at it. The front porch tilted, indicating the foundation had eroded on one side. The roof had a plastic tarp over part of it. Funny, Anne Marie hadn't noticed that before. The yard needed some serious attention; the flower beds sprouted weeds and a lone rosebush struggled for survival, choked off by the encroaching lawn. A pang went through Anne Marie at leaving Ellen here. Yet, this was her home....

"After we say hello to your grandmother, I'll need your help carrying in the groceries." Before heading over to Dolores's house, Anne Marie and Ellen had picked up some necessities. She didn't think Dolores would be up to a trip to the grocery store anytime soon.

"Okay," Ellen agreed. She'd already put Baxter in the back and unfastened her seat belt.

With a smile, Anne Marie watched Ellen dash out of the car and fly across the yard. She threw open the front door, then barreled inside. By the time Anne Marie entered the house, she found Ellen in her grandmother's arms, both of them a little teary. For an instant Anne Marie felt like an intruder.

Dolores Falk looked up at Anne Marie. "I can't thank you enough for taking care of my girl."

"I was glad to do it," Anne Marie said simply.

Holding on to her grandmother, Ellen said, "Anne

Marie's still going to be my Lunch Buddy and she said I can see Baxter anytime I want. We're growing seeds and she taught me to knit and we knit every night after dinner when I'm finished my homework."

Dolores had heard about Ellen's knitting at least a dozen times. The child was more animated today than Anne Marie had ever seen her.

Breaking away from her grandmother, Ellen raced toward the hallway. "I want to see my room!"

"I didn't have an actual bed for her at my place," Anne Marie explained. "She slept on a pull-out sofa." She wished now that she'd purchased a bed for Ellen, but it hadn't seemed logical at the time. She couldn't possibly have known the girl would be with her a full month.

Obviously fatigued, Dolores sank into her recliner. "I'm just so grateful for everything you did."

"I'm going to miss her." The apartment, tiny though it was, would seem empty without her.

Ellen tore back into the living room. "Should we bring my clothes in now?"

"Sounds like a plan," Anne Marie said briskly. For a moment she'd forgotten about Ellen's bags and the groceries. "We got a few things we thought you'd need for the first couple of days," she told Dolores. "Enough to last until you can get to the grocery store."

Dolores seemed about to weep. "God bless you."

Anne Marie shrugged off her appreciation and, with Ellen at her side, returned to her vehicle. Baxter, lying in the backseat, didn't seem pleased to be left out of the action.

"Can I take Baxter for a walk?" Ellen asked as she pressed her nose to the car window.

"Help me first and then you can take him. Just be sure his leash is secure."

"Okay."

They collected Ellen's various bags, unloaded the groceries and brought everything inside. Anne Marie sorted through the cartons of milk and juice, the vegetables, cereal, cheese and bread, and organized them as logically as she could so Dolores wouldn't have any problem locating what she needed.

"Is there anything else I can do for you?" she asked Dolores once Ellen had come back with the dog.

"No, no—you've done far more than I would've thought to ask."

Anne Marie moved toward the front door, reluctant to leave. "Ellen, finish your homework, okay?"

"I will."

"See you soon," Anne Marie said, trying to swallow the lump in her throat.

"Okay." Ellen hugged Baxter goodbye, then ran across the living room to throw her thin arms around Anne Marie, holding tight. Her shoulders trembled with her sobbing.

"Hey," Anne Marie said, bending down. "This is your home, remember? You're back with your Grandma Dolores. Isn't that great?"

"Yeah, but…" Ellen sniffled. "I'm going to miss you."

"I'll miss you, too, but we'll see each other often."

"You promise?"

"I promise, and I always keep my promises," Anne Marie said. "You know that, right?" She rubbed Ellen's back gently as the child nodded. "In fact, why don't I stop by tomorrow evening to see how everything's going?" Glancing over at Dolores, she asked, "If that's okay with your grandmother?"

"That would be just fine," Dolores said.

Anne Marie left a few minutes later. As she drove away

from the bedraggled little house, she experienced an over-whelming sense of loss. For one wild moment, she felt a compelling urge to turn back. She couldn't imagine what she'd say if she did. Ellen belonged with her grandmother; Dolores deeply loved this child. *So did she*. Anne Marie realized it with a shock that galvanized her.

She understood now that what she'd seen as affection, caring, a feeling of responsibility—all emotions she'd readily acknowledged—added up to one thing. *Love*.

She loved this little girl and wanted to be part of her life for as long as she could.

"Well, Baxter," she murmured, sighing loudly. "It's just you and me again."

Her Yorkie, who'd been sitting up in the backseat, turned in a circle several times, then dropped down. He curled up, nose to tail, and Anne Marie thought he seemed as despondent as she was.

When she reached her quiet apartment, she roamed from room to room, feeling restless. Dissatisfied. Living here was only supposed to be a temporary situation. The apartment was empty when she'd separated from Robert and it had seemed the logical place to live while they sorted out their differences. It really was time to look for a house, a home for her and Baxter. She might see if she could find one in the same area as Ellen, a fixer-upper she could keep for a while and then sell for a nice profit.

As she moved into the kitchen to prepare a sandwich, Anne Marie stopped abruptly, recognizing something about herself. She was different than she'd been a month ago. She'd gradually changed into a woman who could make her own wishes come true. A woman who was ready to move on with her life. This was the gift Ellen had given her. She'd opened Anne Marie's eyes to the many ways she'd been blessed, despite her losses, and the many possibilities that still existed.

Preparing for bed, she paused in the doorway to Ellen's room. The bed was a sofa again, and Baxter had nestled on the cushion and gone to sleep, as if he expected the little girl to return.

The room was neat and orderly. Nothing of Ellen remained, and yet Anne Marie felt her presence. Many a night she'd stood right here, watching Ellen sleep. That ritual would come to an end now. But she couldn't be sad about it because Ellen was where she wanted and needed to be, with the grandmother who adored her.

"Sleep tight, sweetie," she whispered, then went to her own room to read before turning out the light.

L illie was as nervous as a bride the night before her wedding. Hector was coming to pick her up and bring her to his place for dinner. They'd seen each other a number of times, but this was different.

Hector had invited her to his *home*.

Lillie felt as if she'd passed some test, and that the invitation to visit his home was Hector's way of saying he trusted her and was willing to reveal more of his life.

When the doorbell rang, she pressed her hand over her heart and took a deep breath before walking to the front door and opening it. As always Hector was punctual.

"Good evening, Lillie," he said, bowing his head slightly. "I hope you had a pleasant afternoon."

"No. I mean, yes, I had a lovely afternoon." Rather than explain her initial response, she gathered up her sweater, made of silver-blue cashmere, and her purse.

She'd agonized over whether to ask him in, self-conscious about her wealth and her luxurious house. But it wasn't an issue, since he immediately asked if she was ready to leave.

After she'd locked her door and set the alarm, he led her to his car, parked in her driveway. His manners were impeccable as he escorted her and made sure she was comfortably seated. His courtesies came from a soul-deep regard for others, a true considerateness; she knew that with absolute certainty. This was nothing as superficial as charm. It was a mark of respect.

"I hope you're hungry," he was saying once he'd joined her in the vehicle.

Lillie was far too nervous to be hungry. "I've been looking forward to this all day," she told him.

He glanced over at her, his dark eyes intense. "I have, too."

Her stomach pitched. From the first moment they'd met, he'd had an unprecedented effect on her. She felt things with him that she hadn't felt before. David had never shared much with her; he'd been what her women's magazines now referred to as "emotionally inaccessible." His affairs were part of that, of course. It wasn't until after his death that she'd recognized how withdrawn she'd become through the years. There had been a price to pay for ignoring his betrayals, for turning a blind eye to his short-comings as a husband and lover. The price had been much higher than she'd realized. Only now was she beginning to understand how repressed her feelings had become. She'd learned to subdue her own emotions as well as her expectations.

Hector was talking about dinner, and she shook off her pensiveness.

"You made everything yourself?" she asked.

"My daughter offered advice."

Hector and his daughter seemed to be especially close. Like everything else about this man, she found that endearing—and she couldn't help comparing it to David's relationship with Barbie. At first he'd been disappointed not to have a son, but Barbie had quickly wrapped him around her little finger. He'd accepted Lillie's inability to have other children and lavished his attention on his daughter. David could be generous and loving; he'd certainly shown Barbie that side of himself. But Lillie considered him both uncommitted and morally weak in his emotional life. Yet he'd been a scrupulously honest businessman.... She supposed that was a result of his skill at "compartmentalizing," which men were said to be good at, again according to her magazines.

"I need to mention something about my home," Hector said, looking straight ahead as he concentrated on traffic. "I don't live in a fancy neighborhood."

"I understand that."

"Your home is beautiful, Lillie."

"Hector, are you telling me you're ashamed of your home?" she asked bluntly.

"No, I'm not."

"Then please don't apologize for it."

"You apologized for yours, remember?"

She had. She'd feared that once Hector saw her opulent home, the differences between their financial situations would discourage him. She'd been wrong. He wasn't easily intimidated. At least, she didn't think so until he'd brought up the subject of his own neighborhood.

"People might talk about us, Lillie," Hector added. "However, Rita's aware that I'm seeing you and has been most supportive."

"My daughter has been, too." Lillie didn't mention their initial conflict and the painful few days that had followed their disagreement.

"I haven't told my sons about you yet."

"Oh?"

"They might not be as understanding as Rita."

Lillie glanced at him. "Will their opinion make a difference?" she asked.

He didn't answer right away. "I would like to tell you it wouldn't. The truth is, I don't know. My family is important and I trust that my children love me enough to want to see me happy. And you, Lillie, make me happy."

"Oh, Hector." His sincerity touched her heart. "You make me happy, too," she whispered in return.

Hector reached for her hand.

As soon as they turned onto Walnut Street, Lillie knew instantly which home was his. The yard was beautifully maintained, the flower beds splashed with brilliant color. When he pulled into the driveway of the house she'd guessed was his, it was all Lillie could do not to congratulate herself.

Hector helped her out of the car and led her to the front door of the white-painted two-story house. The first thing she saw inside was a multitude of family pictures. They covered the walls and the top of the piano. The wall next to the stairs was another gallery of photographs. Lillie's gaze went to a portrait of Hector and his deceased wife. Angelina, maybe fifty in the picture, had been a slender, elegant woman.

"These are my children," he said, pointing to college graduation photos of his daughter and his two sons. "This is Manuel," he said, tapping the picture of his oldest son.

"The attorney," Lillie murmured. The young man in the cap and gown, proudly displaying his diploma, had serious eyes and a fierce look. Lillie could picture him in a courtroom vanquishing his opponent.

"Luis," he continued, tapping one finger on the next photograph.

"The doctor." Unlike his brother, Lillie observed, Luis had gentle eyes that reminded her of Hector's. "He looks the most like you."

"Yes," Hector said. "Angelina and I always knew he'd work in the health field. From the time he was a little boy, he wanted to help anyone in pain."

Yes, this son was most like his father.

"And Rita," Hector said, going down the line of photographs.

His daughter was a true beauty who resembled her mother. There was an engaging warmth in her smile.

"She's lovely," Lillie whispered. "I'm sure she's a popular teacher."

The smells coming from the kitchen were enticing, and suddenly Lillie felt ravenous. Once she'd torn her gaze from the photographs, she noticed that Hector had set the dining table with his best dishes; a small floral centerpiece sat in the middle.

"What's for dinner?" she asked.

"You'll see." He escorted her into the dining room.

"What can I do?"

"Nothing. You're my guest."

"Hector, I *want* to help."

He hesitated but finally agreed. "If you insist. You can cut the bread."

"You never did tell me what you're serving."

"It's a classic Mexican dish," Hector teased as he opened a drawer and pulled out a bread knife. "It's spaghetti. My daughter gave me the recipe. She even went to the store with me and chose the ingredients."

"You didn't need to go to all that trouble," Lillie said,

although she was flattered that he had. "We could eat potato chips and it would taste like ambrosia to me because I'm here with you."

Hector grinned, then took a step closer. "I have been lonely for a long time," he said in a low voice.

Lillie had spent most of her marriage being lonely. "I have, too," she told him.

For just a moment it seemed that he was about to kiss her. Their first kiss had been the evening they'd attended the Frida Kahlo lecture at the Seattle Art Museum. At the end of the evening, he'd dropped her at home; he'd declined coffee but walked her to the door. It'd been an awkward moment and by unspoken agreement they'd each leaned forward and kissed. Lillie was eager to repeat the experience. Their kiss had been polite, almost chaste, but very satisfying....

"I'm just reheating the sauce," he said, wielding a large wooden spoon.

"Hector?" Lillie drew in a deep breath before plunging ahead.

The way she said his name seemed to alert him to the fact that she had something important to say.

"Yes?"

"I want you to know…"

"Yes, Lillie?"

"When it feels right to you, I hope you'll kiss me again." She didn't want him wondering—or worrying—about what her response might be.

"Thank you." His eyes sparkled with delight. "I shall keep that in mind."

She picked up the bread knife and carefully sliced the loaf of French bread, arranging the pieces neatly on a serving plate.

After stirring the sauce, Hector boiled the spaghetti noodles; when they were ready, he placed them inside a beautiful hand-painted ceramic dish. Next he poured the meat-and-tomato sauce over the noodles. The salad, waiting in the refrigerator, was already mixed. Lillie put it on the table, along with the bread.

Hector opened a bottle of red wine that he told her Rita had recommended. Then he seated her at the table and sat across from her.

They toasted each other, touching glasses, and began the meal. She discovered that Hector preferred his food spicier than she did but he'd made the sauce fairly mild, adding chili peppers to his own. Another example of his thoughtfulness.

In the beginning their conversation was tentative. But it wasn't long before the hesitation dissolved and they found any number of topics to discuss. They agreed on political issues and surprisingly had enjoyed some of the same films and novels. Hector bragged about his grandchildren and she told stories about her grandsons. The conversation flowed naturally from one subject to the next as they lingered over their wine. Afterward, despite Hector's protests, Lillie helped with the dishes. Her shoes hurt her feet, so she took them off and tucked a dishtowel into her waistband as she moved effortlessly around his kitchen.

Hector put on some easy-listening music from the '70s, and soon they were dancing about the room, twirling and laughing. He kissed her once, twice, and it was as natural as breathing. His touch left her with the most inexplicable urge to weep. Rather than allow him to see the effect his kisses had on her, she buried her face in his shoulder.

Hector released her and they both went back to cleaning the kitchen, dancing around each other as they did.

He was about to kiss her again when the back door opened and Manuel walked inside. Lillie recognized him from his photograph. "Dad, I need to borrow your—" He stopped abruptly. "Dad!" he barked, shouting to be heard above the music.

Instinctively Lillie stepped closer to Hector. He leaned over to turn off the CD player on the counter, and the resulting silence was almost shocking.

Hector straightened, putting his arm around Lillie's waist. "Son, this is my friend, Lillie Higgins. Lillie, this is Manuel."

Manuel nodded politely in her direction but addressed his father. "I didn't realize you had a woman friend."

"Your father's told me quite a bit about you," Lillie said, feeling guilty although she wasn't sure why.

"Funny, he hasn't said a word about you." Manuel gave her a cold look.

Hector placed one hand on her shoulder and spoke gently. "If you'll excuse me, I will talk to my son privately." He ushered Manuel out of the room.

She nodded and finished wiping the kitchen counter. She rinsed and wrung out the cloth, then draped it over the faucet and removed her makeshift apron. Slipping on her shoes, she stood in the kitchen and waited for Hector.

Manuel left without saying anything else to Lillie and she could see from the look in Hector's eyes that the conversation hadn't gone well. "I'm sorry," she whispered and walked into his arms.

"I'm the one who owes you an apology," he murmured, holding her close. "My son was inexcusably rude."

"What did he say?" she asked.

Hector shook his head, obviously unwilling to repeat what his son had said. Lillie closed her eyes and remembered Barbie's immediate response when she'd told her she was interested in a man who worked for a car dealership.

"Give him time to adjust to the idea," she urged.

"Perhaps that's the best thing to do."

He took Lillie home soon after that; they were both quiet during the drive. When they reached her house, Hector walked her to the door. She thanked him for dinner, they kissed goodnight and then he left. Not until she was inside did she realize that he hadn't asked to see her again.

Lillie felt sick.

This was the end; she was sure of it. His family and their opinion mattered more to Hector than his own happiness. Even if he was torn, and she knew he was, Hector would appease his children rather than fight for a relationship with her.

When she didn't hear from him the next day or the day after that, Lillie decided to make this as painless as possible for them both. She wrote him a letter.

She'd only intended to write a brief note but by the time she finished she'd written three full pages. She described her list of Twenty Wishes and said that one of her wishes had been to meet an honorable man. She'd found that man in him.

In the last paragraph she explained that she had no desire to damage his relationship with his children and felt it was best that they not see each other again.

With tears in her eyes, she dropped the letter off at the post office. After a quick phone call to Barbie, she booked a trip to the coast.

Chapter

29

Anne Marie was fortunate enough to nab a spot in the Woodrow Wilson parking lot. She was back to the routine of Lunch Buddy dates and had brought Baxter with her for this visit.

Yesterday, when Ellen and Anne Marie had spoken, the eight-year-old had told her how much she missed her canine friend.

Bringing Baxter today was a surprise, and Anne Marie could hardly wait to see the child's face light up.

The transition from Anne Marie's home to her grandmother's had gone smoothly. Anne Marie wished she could say Ellen's departure had been as straightforward for her. The apartment just wasn't the same without Ellen; it was still too quiet, too empty. Her life felt that way, too. The child had found a vulnerable space in her heart, and Anne Marie had discovered how much she craved love. She wanted to give it as well as receive it.

Of course, she was in regular touch with Ellen and would continue to be so. She'd already arranged a trip to

Woodland Park Zoo for that Saturday. Anne Marie was looking forward to it and she knew Ellen was, too. Dolores needed the break and seemed to appreciate Anne Marie's interest in the girl.

Ellen's grandmother was a pitiable woman. She blamed herself for what had become of her two daughters, and all the love in her heart was reserved for her only granddaughter. Dolores lived for Ellen. Since she'd been released from the hospital, Dolores had told Anne Marie that she'd searched for more than a year before she'd found Ellen. Once she did locate the child in the California foster care program, it had taken nearly another year to convince Child Protective Services to grant her custody of the little girl. Ellen had been with her grandmother for three years now, and the older woman had given her the security she so desperately needed.

As Anne Marie headed into the school, with Baxter on his leash, she saw Helen Mayer, the counselor, hurrying toward her.

"I can't tell you how grateful I am that you're here," Helen said. "The woman who answered the phone at the bookstore told me you were on your way."

"Is something wrong? Is Ellen okay?"

"Please—come into my office."

With a growing sense of panic, Anne Marie picked up Baxter and followed Helen into the building. She couldn't imagine what had happened, but all her instincts said it was bad.

The counselor waited until Anne Marie was inside her office, then closed the door and walked slowly to her desk. She sat down and turned to look at Anne Marie.

Tension twisted Anne Marie's stomach as she lowered

herself into a chair. "What is it?" she asked, placing Baxter on the floor near her feet.

"We received word this morning that Dolores Falk died sometime last night."

Anne Marie gasped and covered her mouth with her hand. "Did…Ellen find her?"

"Apparently she overslept because her grandmother didn't wake her for school. According to what I learned, she made her own breakfast and decided to let her grandmother sleep. On her way to school she met a neighbor who inquired about Dolores. Ellen explained that her grandmother wasn't feeling well and that she was still in bed. A short while later, the neighbor went to check on her and when she couldn't rouse Dolores, she called 911."

At least the child was spared the trauma of discovering the body. Anne Marie thanked God for that. But this probably meant Ellen didn't know yet.

"Is she here?" Anne Marie knew how hard Ellen would take the news. Although she dreaded telling her, Anne Marie thought she was the best person to do so. Poor Ellen.

"I'm afraid not," Helen Mayer said.

Anne Marie barely heard her. "I'll make arrangements to get her things and bring her home with me." She wondered if Dolores had made funeral arrangements; she'd have to find out.

After burying Robert, Anne Marie had some experience in such matters. The staff at the funeral home had been both kind and respectful. Anne Marie would like them to handle the arrangements for Dolores, too, if that was possible.

"I don't think you heard me," the other woman said. "Ellen isn't at school."

Anne Marie stared at her, uncomprehending. "Where is she then?"

Helen Mayer placed her hands on the desk and leaned forward. "About an hour ago, Child Protective Services took her away."

The words hardly made sense. "What? What do you mean they took her away?"

"I mean they came to the school, told us Ellen's grandmother had died and that they had to find a home for her."

"But…"

Helen Mayer gestured helplessly. "The only other relatives Ellen has are her mother, who has relinquished all parental rights, and her aunt, who is apparently incarcerated."

Anne Marie was well aware that Ellen had no one else. That was the very reason the child had come to live with her while Dolores was hospitalized.

"She's been placed in a foster home."

Anne Marie couldn't believe it. "Already?"

"Yes. I realize it's a shock. I tried to contact you but you didn't answer your cell. The woman at the bookstore said you'd be here soon."

Anne Marie felt disoriented but she had to focus on Ellen. The girl must be terrified. She had to get to her, reassure her that everything would be all right. "I have the name of the social worker assigned to her. She gave me her business card."

"What are you going to do?" Helen asked.

"I'll bring Ellen back to live with me." There was no question about that.

The counselor sighed with relief. "I'm so glad to hear it."

Anne Marie was on her feet, ready to take action. She

committed half a dozen traffic violations in her rush to get back to the bookstore. She prayed she hadn't thrown out the social worker's card. In her agitation she couldn't even recall the woman's name.

With Baxter at her heels, Anne Marie ran up the stairs. Heart pounding, she stood in the middle of her kitchen while she tried to remember where she'd put the woman's card.

Suddenly she remembered. She hurried into her bedroom and jerked open the top drawer of her nightstand. Yes—it was there, and the woman's name was Evelyn Boyle. She collapsed onto her bed and grabbed the phone.

Her hand trembled as she punched out the number. She listened to an automated system that requested the extension, which Anne Marie dutifully supplied. The phone rang five times before Evelyn's voice mail came on.

"This is Anne Marie Roche," she said. "I'm calling about Ellen Falk. Please contact me at your earliest convenience." She gave three phone numbers—home, work and cell—afraid the woman would give up too easily if she couldn't reach her on the first try.

The waiting was intolerable.

Anne Marie paced, she cleaned out drawers, then paced some more. When she couldn't stand it any longer, she drove to Dolores's house. The place was locked up. The neighbor who'd found her said the coroner's office had already removed the body. No one knew anything about Ellen or where she might be. Anne Marie gave the woman her numbers, desperate to learn whatever she could.

When her cell phone finally did ring, it was after four and Anne Marie nearly ripped it out of her purse in her haste.

"This is Anne Marie Roche," she said, the words tumbling over each other.

"Anne Marie, this is Evelyn Boyle returning your call."

"Where's Ellen?" she cried. The child must be frantic. Anne Marie was close to panic herself. Ellen needed her and Anne Marie needed to be with Ellen.

"It's unfortunate, but the only thing I could do was place her in a temporary foster home. It's a short-term solution until I can find a permanent home for her."

"I'll take her," Anne Marie blurted out. "Bring her to me."

"I wish I could. If you recall, when I visited the bookstore I suggested you apply for a license to become Ellen's foster parent. I didn't hear from you after that."

Anne Marie wanted to kick herself for not following through. Had she been able to look into the future, of course, she would've started the paperwork that very day. How was she to know? Dolores had been doing so well.

"I promised Dolores Falk that Ellen would never go back into the foster care system. What can I do now? How long will it take to be approved?" Her fear was that the paperwork would still take months. By then, Ellen might have been moved any number of times. Ms. Boyle had said the home where she was currently placed was temporary, which implied that Ellen would be transferred soon.

She remembered Dolores Falk telling her it had taken a year to find Ellen once she'd learned she had a granddaughter, although Anne Marie didn't know how much of that time had been spent searching in other states.

"We can have a background check done on you in twenty-four hours."

"Then Ellen can come and live with me?"

"Yes. We want what's best for Ellen and I feel that's you."

The relief was enough to flood her eyes with tears. "Thank you. Thank you."

The social worker explained the process. Anne Marie tried to pay attention but her mind kept darting off in different directions. One thing that did register was that there'd be a home study, which hadn't been scheduled yet. The apartment, small as it was, hardly seemed suitable. That would mean an immediate move. Anne Marie didn't care. She'd do whatever was necessary.

"If everything checks out, I should be able to deliver Ellen to you sometime tomorrow afternoon."

Anne Marie tried to recall any possible blemish on her record. She had a speeding ticket, but thankfully, nothing of any real importance.

All the next day, Anne Marie waited. The tension was almost more than she could bear. She left three messages for Evelyn Boyle, wanting to make sure there weren't any problems with her background clearance. The social worker didn't return any of the calls.

Had Anne Marie known where Ellen was staying, she would've driven there and parked outside the house.

When she hadn't heard anything by five o'clock on Thursday afternoon, Anne Marie was positive something had gone wrong. She'd been useless the entire day, too nervous and jittery to concentrate.

Just as she was about to give up in despair, the door to the bookstore opened and Evelyn Boyle came in with Ellen at her side.

Ellen looked at Anne Marie and burst into tears as she bolted toward her.

Anne Marie fell to her knees, her arms open for Ellen.

They clung tearfully to each other. "You promised, you promised," Ellen sobbed against her shoulder. "You said—you said…"

"It's all right," Anne Marie whispered, brushing Ellen's

hair. "You're here now, and no one's going to take you away from me."

Ellen sniffled. "Grandma Dolores went to live with Jesus."

"I know."

"I don't have anyone who loves me."

"I love you, Ellen," Anne Marie whispered, tears streaking her face. "You're going to be my little girl from now on."

"I can live with you?"

Anne Marie couldn't speak, so she just nodded.

"I don't have to go back to the foster house?"

"No, not ever again."

Still sobbing, Ellen tightened her arms around Anne Marie's neck. "Everyone I love goes away."

"Not anymore, Ellen," she promised. "Not if I can help it."

"I loved my mommy and she…she did bad things and she left me and then Grandma Dolores d-died and then you left me."

"I didn't leave you," Anne Marie insisted. "I would never leave you."

They continued to hold each other until Baxter started to bark at the foot of the stairs. Anne Marie released Ellen who ran to open the door. The dog immediately did a dance of joy at the sight of his friend.

Wiping the tears from her face, Anne Marie stood to find Evelyn Boyle watching her.

"I believe we have a good placement for Ellen," she said, her own eyes moist.

Anne Marie wasn't going to make another mistake. "I've decided I don't want to be Ellen's foster parent."

A look of shock broke out across the other woman's face. "I beg your pardon?"

"I want to adopt her," she said. "I want to make Ellen my legal daughter." The child was already her daughter in every way that mattered. It was time to make that official.

"**M**om," Anne Marie said, speaking softly into the receiver. It was late Monday evening, and Ellen had just gone to sleep. The poor kid still wasn't sleeping well, so Anne Marie didn't want to risk waking her. Every night since Dolores's death, Ellen had ended up crawling into bed with Anne Marie and crying herself to sleep. The girl had suffered yet another loss. Being taken out of school, informed that her grandmother was dead and then shuffled off to a foster home hadn't helped.

"Anne Marie?" her mother murmured. "My goodness, I haven't heard from you in weeks. Is something wrong? There must be if you're phoning me this late."

"I should've called earlier." Handling the funeral arrangements and looking after Ellen had kept her busy. But the truth was, it hadn't occurred to her to contact her mother until that night.

Even now she hesitated, fearing her mother's reaction once she learned that Anne Marie was going to adopt

Ellen. Her mother had made her disapproval known when she decided to marry Robert. She'd been equally negative when Anne Marie purchased the bookstore. Laura wasn't a risk-taker and she'd been convinced that Anne Marie would be throwing away her investment. She generally believed in living a cautious, conventional life, although she wouldn't have put it in those terms.

Despite her mother's reactions in the past, Anne Marie felt compelled to seek her out. Perhaps it had to do with becoming a mother herself….

Might as well just blurt it out. "I thought I should tell you that you're about to become a grandmother."

A strained silence followed her announcement.

"You're…pregnant?" Once again, Laura Bostwick's reproach was evident. "I know you want a baby, Anne Marie, but I don't think you have any idea what life's really like for a single mother. Oh, dear…"

"It isn't…I'm not—" Anne Marie didn't get the opportunity to explain before her mother interrupted her.

"If you don't mind me asking, who's the father? No, don't tell me. Obviously there's a problem, otherwise you would've married him. You *aren't* secretly married, are you?"

"No, I—"

"I don't need to know any more about him. He's married, I suppose?"

"Mom!"

"Sorry, sorry. I said not to tell me and then like a fool I ask. It's none of my business. Well, you're going to have a child. When are you due?"

"It's a bit more complicated than that," she began.

"For heaven's sake, you haven't done anything stupid, have you?"

"What do you mean?" Anne Marie asked, a little taken aback.

"Artificial insemination, that's what. I heard about it at the hairdresser's. Apparently a lot of women are using artificial methods to get pregnant. Please don't tell me you went to one of those fertility clinics and—"

"Mother, I'm *adopting*."

She'd finally shocked her mother into total silence.

"Remember Ellen Falk?"

"Who?"

"I was her Lunch Buddy. You met her the Saturday before St. Patrick's Day. We had lunch with you." Surely her mother hadn't forgotten.

There was another silence. Then Laura said, "Let me see if I have this straight. This second-grade girl you agreed to have lunch with once a week is the one you're going to adopt?" Her mother sounded incredulous.

"Yes, Mom. She came to stay with me, remember?"

"Well, yes, and I told you I thought it was rather nervy of that girl's grandmother to call you in the middle of the night."

"Dolores Falk died."

This information appeared to unsettle Laura. "Oh... dear. That is a shame."

"Ellen doesn't have anyone else," Anne Marie said.

"You're fond of the child?"

"I love her as though I'd given birth to her myself," Anne Marie confessed. "I've already talked to the social worker and asked to be considered as Ellen's adoptive mother." She closed her eyes, certain her mother would discourage her, as she had with every important decision Anne Marie had ever made, from the school she'd chosen to the man she'd married.

"Oh, Anne Marie…"

She waited for it.

"I think that's a wonderful thing to do."

Her jaw fell so fast and hard, Anne Marie was surprised she hadn't dislocated it. "You…think I'm doing the right thing?"

"My dear girl, you're old enough to decide what you want to do with your own life. If this child means so much to you, then by all means bring her into the family."

As far as Anne Marie could remember, this was the first time in her adult life that her mother had supported her choices. She didn't understand it, other than to assume the child had won over her mother's heart in the hour or two they'd spent together.

"There won't be any legal problems, will there?" Laura went on to ask.

"I don't know." Evelyn Boyle had to do a search for Ellen's birth certificate and find out who was listed as the father. He would need to be contacted and given the opportunity to state his wishes.

Anne Marie was pretty sure Ellen's biological father didn't even know she existed. But if Evelyn managed to track him down… He could decide to declare his parental rights and Anne Marie would have no option but to relinquish Ellen. The thought made her feel ill.

"What about her biological mother?"

"She gave up all rights to her daughter three years ago when Ellen went to live with her grandmother."

"Does that mean the mother can't change her mind?"

"It's too late for that. Anyway, if it wasn't for Dolores, Ellen might've been put up for adoption years ago."

"Oh."

"The social worker was encouraging." The fact that

Ellen was living with Anne Marie and that they'd so obviously bonded was a hopeful sign. However, the issue of Ellen's biological father still had to be resolved.

Anne Marie suddenly remembered something. "The wishes."

"I beg your pardon?" her mother said. "Stop mumbling, Anne Marie. How many times do I have to tell you? Speak up."

"Sorry, Mom. I was just thinking out loud."

"What was that about wishes? That's what you said, isn't it? It certainly sounded like *wishes*."

"Ellen has a list of wishes. Twenty wishes." Anne Marie had no intention of referring to her own list or those of the other widows. Her mother would no doubt throw scorn on the idea or dismiss it as childish.

"Children do that sort of thing," her mother said, confirming her suspicion. "I wouldn't give it any mind. I suppose she wished for a mother and father?"

"No, no…nothing like that." Then, because she felt she had to explain after bringing it up, she said, "Ellen wants to meet her father."

"Every child wants that. My guess is she's well rid of him."

The rest of the conversation made no impact on Anne Marie. A few minutes after she ended the call, she wandered into Ellen's tiny bedroom and watched the child as she slept, one hand flung out and resting on the dog, who was cuddled up close beside her. The poor kid was exhausted and seemed to be lost in her dreams.

Earlier, in between working at the store and looking after Ellen, Anne Marie had called the school. She'd updated Helen Mayer, who'd cheered when Anne Marie told her about adopting Ellen. She'd even offered a character reference should any be needed in the adoption process.

Anne Marie was just afraid the proceedings might not get that far.

On Saturday morning, three days after Dolores's death, they'd visited the funeral home and arranged for a small private service. A short obituary written by Anne Marie appeared in the paper. Several neighbors stopped by on Sunday to pay their respects.

The house was a rental property and Anne Marie had until the end of the month to get it cleaned out and ready for the next tenants.

That afternoon, with a few friends gathered around, Anne Marie and Ellen had laid Dolores Falk to rest. Throughout the service, Ellen stayed by Anne Marie's side. She didn't weep, although her eyes filled with tears more than once. Afterward, they'd returned to the apartment alone.

"I think Grandma Dolores was ready to live with Jesus," Ellen had said calmly as she reached for her knitting bag. She seemed to find solace in knitting.

"What makes you say that?"

She'd glanced up. "I saw it in her eyes. She told me she was tired."

Anne Marie had thought her heart would break.

Late Tuesday afternoon, Anne Marie and Ellen were in the apartment, planning a visit to Dolores's house to sort out what to keep and what to give away, when the phone rang. It was Cathy in the bookstore. "The social worker's here to talk to you. Should I send her up?"

"Yes, please." Evelyn Boyle had said she'd hoped to attend the memorial service the previous day; she'd also said she had a court date and wasn't sure how long that would last.

Anne Marie waited anxiously for her at the top of the stairs.

"How did everything go yesterday?" Evelyn asked, taking the steps one by one.

"It was very nice." Several of Dolores's neighbors had attended, and Helen Mayer from the school had been there, too, along with Lydia, Elise and Lillie. Dolores had requested that her remains be cremated; Anne Marie and Ellen would receive the ashes at a later date.

"I'm so sorry I couldn't be there."

Anne Marie bit her lip until it hurt. "Do you have news?"

"I do." The middle-aged woman paused on the landing and placed her hand over her heart. "Stairs are God's way of telling me I'm not getting any younger."

Anne Marie resisted the urge to shake her by the shoulders and demand to know what she'd learned. "Come in, please," she invited, doing her best to disguise her nervousness.

The social worker stepped into the kitchen. Ellen sat at the table knitting, with Anne Marie's notes for the disbursement of Dolores's belongings scattered about. "My goodness," Evelyn murmured, "who taught you to knit so well?"

"Anne Marie," Ellen said without looking up. "I'm sorry, Ms. Boyle, but I can't talk now. I'm counting stitches."

"Perhaps you could move into the living room so Ms. Boyle and I can chat. Okay?" Anne Marie said.

"Okay." With the ball of yarn under her arm, Ellen carried her wool and needles into the other room and, Anne Marie hoped, out of earshot.

Evelyn Boyle pulled out a kitchen chair and sat down as Anne Marie gathered up her notes and put them in a loose pile. Evelyn placed her briefcase on the table and opened it, then ceremoniously removed Ellen's file.

Anne Marie sat across from her. Waiting…

"I located a copy of Ellen's birth certificate and the father is listed—"

Anne Marie's heart slammed hard against her ribs. She hadn't expected this. "You have a name?" Okay, she'd deal with it. No matter what, Anne Marie would find a way to be part of Ellen's life and she didn't care what it cost.

Evelyn frowned. "If I'd been allowed to finish, you would've heard me say that Ellen's father is listed as unknown."

"That means…" Anne Marie was too excited to complete the question.

"It means that as far as the State of Washington is concerned, you're free to adopt Ellen Falk."

"Thank you," Anne Marie whispered, her throat thickening with emotion. "Thank you so much."

"Have you said anything to Ellen?"

Anne Marie hadn't felt she could until she had all the facts. "Not yet."

"Then let's tell her now." The social worker called out to the eight-year-old. "Ellen, would you please join us in the kitchen?"

Ellen immediately came inside and sat down in the chair next to Anne Marie.

"Hello, Ellen."

The child regarded the social worker suspiciously. Anne Marie didn't blame her; it was Evelyn Boyle who'd taken her out of class and uprooted her entire life with the news of her grandmother's death.

Hoping to reassure Ellen, Anne Marie leaned over and gently touched her arm.

"What would you think if Anne Marie became your mother?" Evelyn asked. "Would you like that?"

Ellen didn't answer right away. Then she turned and looked at Anne Marie. "Would I call you Mom?"

"If you wanted," Anne Marie said. "Or you could call me Anne Marie. Whatever you prefer."

"Could I have play dates with my friend Cassie if you were my mom?"

"Yes, of course." Anne Marie remembered the day of the school concert, when she'd been approached by the mother of Ellen's friend about a possible exchange of play dates.

Ellen looked from Anne Marie to the social worker. "Would it mean no one could ever take me away again?"

"No one, not ever," Anne Marie promised.

Ellen shrugged. "I guess it would be all right."

"You *guess?*" Anne Marie teased. "You guess?"

Ellen's face lit up with a huge smile. "I'd like it a whole lot."

"I would, too," Anne Marie told her.

Ellen bounded out of her chair and threw her arms around Anne Marie's neck.

"Wonderful," Evelyn Boyle whispered. "This is just perfect. It's cases like this that make everything else worthwhile." She opened the file again. "I have all the paperwork with me. Be warned, though, the process will take about six months."

Anne Marie didn't care how long it took. The paperwork was a mere formality.

She already had her daughter, and Ellen had her mother. Nothing would ever come between them again.

Chapter

31

"Tell me where we're going," Mark said, wheeling his chair alongside Barbie on 4th Avenue. They'd left Seattle Fitness and, after some pestering on her part, Mark had agreed to join her. She refused to allow his mood to taint this lovely May afternoon. The sun was shining, and she was in love. Mark loved her, too, although he wasn't ready to admit it yet.

"It's a secret. But we're going to meet a couple of my friends first," she explained. He knew that and had already agreed. "Stop acting so cranky."

He was quiet for a moment. "You might not have noticed, but I don't do well with most people."

"I promise you'll like Anne Marie and Ellen."

"What makes you so sure?"

"Mark, please, we've been through this." She found it difficult to hold back a smile.

"You cheated," Mark grumbled. "You lured me here under false pretenses, telling me you had a surprise for me."

"I do have a surprise for you," she said, ignoring his protests. "Besides, a deal is a deal."

Mark slowed his pace. "I might be in a wheelchair, but…"

"A wheelchair doesn't have anything to do with this." They'd struck a bargain, and she was going to ensure he kept his part of it. She'd promised him dinner and an evening for just the two of them—after he'd met her friends. She hadn't told him yet that dinner would be at her house.

"You don't play fair," he muttered.

"Doesn't matter. You agreed."

"Might I remind you that you had your legs wrapped around my waist at the time?"

"Oh, did I?" She loved being in the pool with Mark, especially when they had the entire area to themselves. It was never more than ten or fifteen minutes at the end of a session, and it didn't always happen. But when it did… The water seemed to free him, allowing him to show his need for her in ways he never would while sitting in his chair. They played in the water, teased and kissed and chased each other. Gradually, the barriers Mark had erected against her, against the world, were coming down.

"These are two of your gal pals who also have a list of Twenty Wishes, right?"

"Right. Anne Marie has a list and I believe Ellen's got one, as well."

Mark still wasn't satisfied. "But why do I have to meet them?"

She sighed. "Do you need a reason for everything?"

"Well, yes, I do," he said with a chuckle. "That's just how I am."

"I don't understand why you're making such a fuss."

"Okay, okay, but at least tell me where we're going now."

"If you *must* know," she said, and smiled down at him, "we're meeting them at a Burger King." She'd been looking forward to introducing him to Anne Marie for quite a while.

He frowned. "I don't eat fast food."

Barbie knew Mark was a real stickler about his diet. For one thing, he had to be careful about his weight.

"We aren't eating there. I'm making dinner at my place."

Mark's frown deepened. "I can't get into your house," he muttered.

"Mark," she said, coming to a halt. "Would I invite you if you couldn't get your wheelchair into my home?"

He studied her closely. "You have a ramp?"

She nodded.

His eyes revealed his shock. "You're serious about us. You must be, if you're going to all this trouble."

"Are you finished arguing with me now?" She started walking again and had gone several feet before she realized he hadn't budged. "Are you coming or not?"

Slowly, he wheeled toward her. "You really know how to get to a guy."

"I'm happy you think so." The joy that coursed through her was enough to send her dancing through the streets.

When they reached the Burger King restaurant, Barbie held open the door. As soon as they were inside, Ellen skipped toward her. "Barbie! Barbie—" She stopped abruptly when she saw Mark.

"Ellen, this is my friend Mark."

"Hello," Ellen said and solemnly held out her hand, which Mark shook. "I saw you before."

"Did you? Where?"

"At the St. Patrick's Day party in Freeway Park."

"Did I see you?"

Ellen shrugged. "You were watching Barbie."

That wasn't the way Barbie remembered it. "You were?"

"You didn't even watch the singers," Ellen elaborated, studying Mark. "The whole show, all you did was look at Barbie."

Mark shifted uncomfortably and was saved from having to respond by Anne Marie who'd just joined them.

"You must be Mark," she said. "Barbie's told me about you."

"Has she really?" He twisted around to stare up at her.

"She's only said the most flattering things," Anne Marie told him with a grin.

True, Barbie thought; she hadn't made a secret of how she felt about Mark.

"We're just finishing our meal." Anne Marie led them to the table littered with the remains of their dinner. They'd evidently ordered hamburgers and fries.

"Anne Marie and I signed up for karate lessons," Ellen explained, her excitement unmistakable.

"Karate?" Barbie repeated. "How come?"

"It's on my list."

"And I decided I might as well join her," Anne Marie said.

"We already had one lesson. We're going to the karate place right after we have our drinks." She pointed at a carton of chocolate milk.

"Karate, huh? I wouldn't want to meet either of you alone in a dark alley," Mark teased. "I can picture it now. You'll warn me off by telling me you've had two—count 'em, two—karate lessons. I'll be shaking in my boots."

Ellen giggled.

Barbie noticed that Mark was grinning, too. He so rarely showed any emotion, and it pleased her to know he liked her friends. But then, she'd predicted that he would.

Mark turned to Anne Marie. "You're the one who started this Twenty Wishes business."

"Four of us—all widows—came up with the idea together," Anne Marie said.

"Do *you* have any wishes?" Ellen asked him.

"Yes, indeed," Mark said. "Several."

"Have you ever made a list?"

"I can't say I have, Ellen. Do you recommend it?"

"Oh, yes," Ellen returned seriously. "It's helpful if you have a real list. Otherwise you might forget."

"That's true," Mark concurred.

"Your heart has to let your head know what it wants," the child added.

"You sound very wise for one so young," Mark said, raising his eyebrows. "Where did you learn this?"

"Anne Marie told me. It's true, too. I didn't even know how much I wanted a mom until I put it on my list of Twenty Wishes."

"You wrote that down?" Anne Marie asked, apparently surprised by this revelation.

Ellen nodded, her eyes downcast.

"You never showed me that."

"I know," the girl said. "I wrote your name in pencil beside my wish 'cause if I could choose my own mom, I wanted you."

Anne Marie slid her arm around Ellen. "If I could have any little girl in the world, it would be you."

"Anne Marie's adopting Ellen," Barbie explained for Mark's benefit.

"We'd better scoot." Anne Marie smiled. "Like Ellen said, we're on our way to karate."

"Karate Kid and Mom, the sequel," Barbie joked.

"After that, we're going to see Melissa and help her work on wedding plans," Ellen said excitedly.

"That's my stepdaughter," Anne Marie told Mark.

"I might get to be in the wedding! Melissa said she needs a little girl to help serve the cake and Anne Marie said what about Ellen and Melissa said she thought that was a good idea."

"I think it's a grand idea myself." Barbie knew the difficult relationship Anne Marie had with her stepdaughter and was delighted by the way things had changed.

They left, and Barbie sat down in one of the chairs vacated by her friends. "So," she murmured, "you only had eyes for me last March, huh?" She reached for a leftover French fry and dipped it in ketchup.

Mark avoided her gaze. "I didn't think you'd let that pass."

"That was just the third time we met." If it took all night, she'd force him to admit how he felt about her.

"And?"

"And you're crazy about me," she insisted.

"I already told you I'm willing to go along for the ride, however long it lasts." His voice didn't betray a hint of sentiment.

"Monday-night movies."

He shrugged casually. "Sure."

"Lap swims on Tuesday and Thursdays."

He sloughed that off, as well. "We could both use the exercise."

"Dinner at my house tonight."

He hesitated. "Sure. Why not?"

Barbie took a crumpled hamburger wrapper and smoothed it out. Then with the ketchup-dipped fry, she drew a heart. "What am I getting out of this relationship?" she asked in conversational tones. "So far, I seem to be the one doing all the giving."

Mark tensed. "I've asked myself that from the start. I told you anytime you want out, all you need to do is say the word."

"Just like that?" she asked and snapped her fingers.

"Just like that," Mark echoed, snapping his own.

"No regrets?"

"None," he assured her.

"No explanations?"

He shook his head.

"No looking back, either."

"Not on my end."

"What if that isn't enough for me?" she asked.

His face tightened and his eyes went hard. "Let's clear the air right now."

"Fine by me."

"Exactly what do you want from me?" he demanded, none too gently.

Taking the same French fry, she scribbled out the heart. This discussion wasn't one she'd intended to have and yet she couldn't stop herself. Her pulse raced. She was afraid that by pressuring him for a response she'd put everything on the line. She'd chosen the one sure way to lose Mark.

"I'm not sure what I want," she replied, unable to look at him.

"Yes, you are," he countered, "otherwise we wouldn't be having this conversation."

"My list of wishes…" she said, and her voice faltered.

"Oh, yes, those Twenty Wishes you and your friends have." His tone had a mocking quality, which made her furious.

"You might think they're silly, but they're not!" she insisted.

"I didn't say they were," he said calmly. He could be so difficult to talk to sometimes. Squaring her shoulders, she met his eyes. "Okay, I'll tell you what I want."

"Good. I was hoping you would."

He wouldn't like this. The truth would probably scare him off. Still, it was a risk Barbie had to take. "I want to be loved," she said. There, it was out.

"By me?" he asked.

"You're the one I love." She might as well go for broke, and he could either reject her right now or accept her.

For a long time Mark didn't say anything, and when he did, regret weighted each word. "I don't want to love you," he said slowly.

So that was how it was going to be.

Barbie swallowed painfully. Hard as it was, she'd rather he was honest. "Thank you for not leading me on," she managed to say through quivering lips. She stood up to leave.

Mark caught her hand. "I don't *want* to love you," he repeated, "but I do."

"You love me?" She could hardly believe it, yet she knew it was true. He let his love shine from his eyes and his fingers tightened around hers.

"I have practically from the first moment I saw you at that theater."

"You tried to kick me out, remember?"

"That's because you scared me to death," he said wryly. "But regardless of what I said or did, you wouldn't go away."

She offered him a shaky smile and sat back down, dragging her chair close to him, their knees touching.

"Then before I knew it," Mark muttered, his eyes closed, "I was dreaming about you."

Barbie savored every word.

"For the first time since the accident, I'd wake up each morning with a sense of…hope. I'd go to the movies and hope you'd stay away and at the same time, I'd hope you'd show up—and then I'd curse myself for being so stupid. Acting like that, I was just looking for more heartache."

Breathless, Barbie didn't trust herself to speak. This was everything she'd craved, everything she wanted to hear.

"I'm grateful you came into my life," Mark said and all his intensity was focused on her. "I can't say it any plainer than that."

"You mean *forced* my way into your life, don't you?"

He laughed and then grew serious again. "You want my heart? You've got it, Barbie. You've had it all along." Then he did something completely out of character. Reaching for the paper crown left behind by a birthday group, he placed it on his head and leaned over to kiss her.

Barbie leaned back and stared at him as a chill raced down her arms. With tears blinding her eyes, she held both hands to her lips. Despite all her efforts, she doubled over and started to weep.

"Barbie?" Mark touched her back. "What's wrong?"

She straightened and noticed that the paper crown sat crookedly on his head. *Her wish.* She'd wanted to be kissed by a prince. She'd known it was a ridiculous request—yet it had been fulfilled.

Mark was her prince. He loved her.

And she loved him.

Slipping her arms around his shoulders, she hugged him with such exuberance she nearly toppled his wheelchair. "I'm going to love you for the rest of my life."

"I certainly hope so," he muttered. "Now do you think we can get out of here?"

"What's the matter? Is the aroma of those burgers getting to you and weakening your resolve?"

"The only thing getting to me is you. I think it's time you showed me what you learned on the Internet."

Her eyes widened. Mark didn't need to remind her what she'd looked up weeks ago.

All she needed to know was that he loved her.

As much as she loved him.

Everything else they'd figure out with a little inventiveness and a lot of time.

Chapter

32

Thursday afternoon, Anne Marie waited for her pulse to slow before she called Robert's office. Even after nearly two years, the number was ingrained in her memory.

Anne Marie knew she finally had to see Rebecca Gilroy. She didn't want to show up without warning, so she'd decided to phone Robert's assistant and make a formal appointment first.

She had to know the truth before she could put this behind her—or at least in perspective.

Was the child Robert's? If so, she wondered why Rebecca hadn't come forward. Robert's son deserved part of his estate, was entitled to an inheritance. Despite the circumstances, that was only right.

Her heart in her throat, she made the call. A moment later, she heard Rebecca's voice.

"This is Rebecca Gilroy. How may I help you?" The young woman, now presumably an assistant to one of the other partners, sounded businesslike and professional.

Anne Marie took a deep breath. "Hello, Rebecca," she said, speaking quickly. "It's Anne Marie Roche, Robert's wife."

Rebecca's tone softened instantly. "Anne Marie, of course. How are you?"

"Better." Which was true. "What about you?"

"Busy."

Anne Marie couldn't tell if this was a brush-off or an indication that Rebecca couldn't speak now.

"I won't keep you then," she said, following the other woman's lead. "I was hoping we could get together soon. Would that be possible?"

"You and me?" Rebecca didn't bother to conceal her surprise, or her reluctance.

"Could we meet for lunch? When it's convenient for you…"

"Well, I suppose lunch would work. How about tomorrow?"

A strange calm settled over Anne Marie. A day from now she'd know the truth, whatever it might be. She'd make this as painless as she could for all involved. Two months ago, when Melissa had told her about this, she'd wanted to hate Robert's assistant, to view her as the manipulative other woman. She still tended to see Rebecca as a gold digger who saw her big chance when Robert and Anne Marie separated. And yet…she'd never approached the family for child support.

Rebecca suggested a small, upscale restaurant close to Pike Place Market. Anne Marie knew it well; Robert had taken her there on a number of occasions. It catered to businessmen who wanted privacy to conduct negotiations over lunch or dinner—and the deals they negotiated obviously weren't all business.

Rebecca said it would have to be an early lunch and asked if eleven-thirty was okay. Anne Marie agreed.

Rebecca must know why Anne Marie had called her. The choice of restaurant told her so. Anne Marie tried not to imagine the younger woman and Robert at the dark corner table, the one he used to reserve for their intimate lunches.

On Friday Anne Marie arrived at eleven-fifteen, fifteen minutes early. Theresa had promised to substitute for her at the bookstore for the rest of the day. In an effort to pack as much into one free afternoon as she possibly could, Anne Marie was going shopping with Melissa after lunch.

The wedding plans consumed every free moment Melissa had and much of Anne Marie's time, as well. Unfortunately Melissa's mother continued to shun her, but Anne Marie believed that once the baby was born, Pamela would have a change of heart. How could she *not* love her very own grandchild?

The hostess led Anne Marie to a quiet table near the window. The restaurant typically wasn't busy until noon and she appreciated the privacy. So far, only one other table was filled, with three men and a woman engaged in some intense discussion. Anne Marie ordered iced tea while she waited. She nervously squeezed lemon into the tea as she rehearsed her remarks.

Rebecca got there right at eleven-thirty and was escorted to the table. "Hello again," the other woman greeted her. She pulled out the chair across from Anne Marie.

What struck her all over again was how very young Robert's assistant was. Young and lovely. Her hair was a rich auburn, shoulder-length and naturally thick. She wore an olive-green skirt and matching jacket with a white silk blouse. An antique cameo—a family heirloom? a gift from Robert?—was pinned at her throat.

"Thank you for taking the time to join me," Anne Marie said, keeping her voice neutral.

Rebecca didn't respond; she opened the menu and scanned it, saying, "Perhaps we should order first."

"Good idea," Anne Marie said, eager to do anything to delay this uncomfortable conversation. "By the way, this is on me."

"That's not necessary," Rebecca said with cool politeness, "but thanks."

Anne Marie amended her assessment of Rebecca Gilroy. She might be young and vulnerable-looking, but she had a self-confidence that wouldn't have been out of place in someone much older.

When the waitress came to take their order, they both chose a soup and salad combination.

"I expect you're here to discuss what happened between Robert and me," Rebecca said, leaping headfirst into the conversation Anne Marie had been avoiding—until today.

"Yes."

"I thought so." Rebecca kept her eyes lowered and toyed with the spoon, belying the confidence she'd shown just moments before.

"Did Robert lead you to believe we were divorced?" Anne Marie asked bluntly.

"No."

"Had you been…physically involved before the two of us separated?"

Rebecca shook her head. "No. We…we weren't actually involved at all."

"What do you mean?"

"Well, physically—as you put it—we were." Rebecca shrugged. "I knew the two of you were going through some difficulties and that you were working toward a reconciliation. Mr. Roche didn't share much of his personal life with me, or anyone else for that matter. I learned you were living apart quite by accident."

"I see." Her own fingers moved to the silverware. She caressed the tines of the fork as she listened.

"We were both working lots of extra hours."

The muscles in Anne Marie's throat tightened, in nervous anticipation of what Rebecca was about to tell her.

"It was a bad time emotionally for us both. I'd recently broken up with my boyfriend, and I knew you and Robert weren't living together anymore."

That was no excuse for what they'd done! Anger and pain raged within her, but Anne Marie dared not let either emotion show.

The waitress chose that moment to bring their meals. The soup, tomato basil, smelled delicious and was accompanied by a Caesar salad with homemade croutons. Anne Marie waited until Rebecca reached for her spoon before she did.

"As I was saying," Rebecca said, picking up the conversation. "Both Robert and I were at a low point in our lives."

"And spending a lot of time together," Anne Marie added.

"Yes."

"So it was…natural for you to be attracted to each other."

She shrugged again. "I suppose."

Any appetite Anne Marie might have had vanished.

"I'm not proud of what happened," Rebecca said, "and I believe Robert was…ashamed of it."

"How long did this affair last?" Anne Marie didn't know what had prompted the question other than the fact that she was obviously looking for more pain. "How... many times did you—"

"Does it matter?" She stared down at the table.

Well, yes, it does, she wanted to say but didn't. That night she and Robert had slept together, shortly before his death—was he still involved with Rebecca then?

"Afterward everything changed between us," Rebecca was saying. "We'd had a great working relationship and that was completely ruined by the affair. We tried to keep it quiet and except for that one time when Melissa walked in on us, I don't think anyone knew."

She lowered her head and Anne Marie could see that this was as embarrassing for Rebecca as it was for her.

Rebecca raised her head. "I'm surprised Melissa told you. That's how you found out, isn't it?"

"She...she was very upset."

"Robert was, too. He was mortified. His biggest fear was that you'd learn the truth."

That news was of little comfort. "Had...did he..."

"Did he what?" Rebecca pressed.

It was increasingly difficult even to speak. "Did he see other women? Were there others?" As his personal assistant, Rebecca was in a position to know.

Her hesitation said it all.

"How many?" She would never have believed it. She felt shocked, *grieved,* that she'd misjudged him so completely.

"One, I think," Rebecca admitted reluctantly. She seemed unwilling to divulge any more.

"Please," Anne Marie said urgently. "I need to know."

"He had me make a reservation at a hotel by the ocean under a different name."

"Redford?" she asked.

Rebecca's gaze widened. "You know about her?"

Her throat muscles relaxed. "That was me. Us. We... played this game." A smile came and went, tinged with humor and relief. Memories of their getaway weekend immediately came to mind. Happy, playful memories that were in stark contrast to what she'd just experienced.

"Okay, well, like I said, that's the only other time. And it turns out he *wasn't* cheating on you."

"Thank you," Anne Marie whispered, and she meant it.

"I should tell you that the night Melissa caught us was the last time." She paused. "Deep down, I know that if we could do everything over again, neither of us would've done it." Her eyes held Anne Marie's. "I'm not just saying that, either. It's the truth. If Robert were here, he'd agree."

"Was there..." The moment had come, and still Anne Marie couldn't make herself ask the question. "Did he ever tell you why we'd separated?" she asked, taking another route to the question that burned in her heart.

Rebecca looped a strand of thick auburn hair around her ear. "Actually, we didn't talk about you very often."

That made sense. "Robert was a private person," Anne Marie murmured.

"Yes, he was."

"I wanted a baby," Anne Marie said.

Rebecca looked away. "I didn't know that. I guess Robert didn't want another child."

"No. He...he was opposed to starting a second family and I felt that if I could show him how important this was to me, he'd change his mind."

"But you were getting back together," Rebecca said.

Anne Marie suddenly realized something. She knew why Robert had left after that night they'd spent together. He'd been gone in the morning, and the callous way he'd simply disappeared without a word or even a note had devastated her. For the first time, Anne Marie understood why he'd done it. Robert had been overcome with guilt. He was sorry about the affair with Rebecca. He'd probably wanted to tell her and ask her forgiveness, and at the last second he'd backed down. She assumed the affair was over by then; if not, she felt certain he would've ended it.

"You had a baby," Anne Marie said without flinching.

"A son. I named him Reed."

"Is the baby's father—is this Robert's child?" The question was out at last. Much as she feared the answer, she needed to know.

"Robert's?" Rebecca repeated, looking stunned. "No!"

"No?"

"Of course Reed isn't Robert's! Oh, my goodness, that's what this lunch is all about? You thought I'd had Robert's child. No, no, no. Reed's father is my ex-boyfriend. Denny cheated on me and I found out the same week I discovered I was pregnant. I should've explained. The only reason I slept with Robert was because I was trying to hurt Denny. It was just so twisted and stupid."

"Denny knew about Robert?"

"Yes."

"And he knows about Reed?"

"Of course, and so far he's been a good father."

"You're getting married?"

"No way! I'm not an idiot. If Denny couldn't keep his pants zipped before the wedding, he won't afterward. I'm seeing someone else now."

"Oh." Anne Marie had to resist hugging the other woman and thanking her for not giving birth to Robert's son.

"You must've heard about Robert and me and then learned I was pregnant and thought—"

Anne Marie nodded. "That's exactly what happened."

"But if that was the case, don't you think I would've contacted his attorney? I mean, Reed would've been a legal heir once paternity was established."

"I wondered why you hadn't."

"Well, it was for a very good reason. Reed isn't Robert's son."

Anne Marie's heart soared with relief and, even more than that, with joy.

"I didn't know Robert all that well," Rebecca told her. "But I know one thing about him—he loved you."

"He loved *me*," Anne Marie said.

"He did," Rebecca concurred with a smile.

That was all that mattered. Robert wasn't going to win any Husband of the Year award. The pain of his betrayal would always be with her but *he'd loved her*. It was what she'd known all along, despite the doubts and the mistakes they'd both made.

Yes, that was all that mattered.

The past wouldn't change.

But the way Anne Marie saw the future would.

Lillie estimated that it'd been three weeks since Hector had received her letter. She hadn't heard from him, and after all this time, she didn't expect to. The last thing she wanted was to cause problems between Hector and his children.

After mailing the letter, Lillie had spent a week by the ocean and found solace. The ocean had always been her escape. Whenever she learned about another of David's affairs, she'd booked a visit to her favorite ocean resort. She'd gone there three or four times every year, often enough that she had her own room, and the staff knew her by her first name. Although it'd been well over three years since her last visit, she'd been greeted warmly. Her regular room was ready and waiting for her.

She'd regained her emotional equilibrium walking along the beach. Every morning, she'd strolled in the sand, letting the waves lap against her bare feet, thinking, meditating, praying. After a while, the ache would gradually diminish as she was reminded that her worth as a woman, as a human being, didn't depend on David. His actions couldn't demean her. Her husband, sad though it was to admit, was a man without honor.

When she returned from the ocean, Lillie carefully sorted through the mail, searching for a response from Hector. There was none. She'd hoped he'd answer her note, although she hadn't really expected it.

On Monday, the twelfth of May, Lillie spent the morning working in her garden. She loved her Martha Washington geraniums, and with the rhododendrons in full bloom and the azaleas as well, her garden had never looked better. Her neighbors hired landscape specialists and Lillie had a company that performed the more demanding physical tasks, such as mowing. The flower beds, however, were her domain. Her personal joy.

At noon, she took a break and went inside for a glass of iced tea. The mail had been delivered and, as she drank, she leafed through the few advertisements and set the bills aside. A hand-addressed envelope caught her attention.

She didn't immediately recognize the writing. Curious, she opened it to discover an invitation to a retirement party for Hector Silva.

Lillie read it twice.

The party was planned for that very evening and when she studied the handwriting a second time, she realized the envelope had been personally addressed to her by Hector. She recognized his penmanship from the work order on her car.

All the necessary details were there. Date. Time. Place.

Lillie inhaled sharply. The party would be held at the dealership at seven that night, and she had every intention of attending.

By six forty-five, Lillie was dressed in a semiformal knee-length linen dress with a cropped jacket. Barbie phoned just as she was about to walk out the door and Lillie explained where she was going.

"I was *sure* Hector would be in touch," her daughter said in that gleeful way of hers when she knew she was right. "Have a wonderful evening, Mom."

"I will," Lillie promised.

They spoke for a few more minutes and then it was time for Lillie to go. Although she was nervous, she had a strong intuition that this was going to be one of the most magical evenings of her life. Happiness spread through her and she felt so light it was as though she could float.

When Lillie arrived at the dealership, the retirement party was in full swing. The showroom floor was decorated with banners and balloons, the counters spread with bottles of champagne and trays of lovely hors d'oeuvres. Surrounded by his children, customers and coworkers, Hector didn't see her right away. As soon as he did, his eyes flew wide open and he said something to his daughter, whom Lillie recognized from the photographs she'd seen in his home. Hector broke away from the group and hurried toward her.

"Lillie." He held out both hands.

"Hello, Hector."

"I'm so pleased you came." His gaze seemed to devour her, and she couldn't doubt the sincerity of his words.

Her own eyes were equally hungry for him. "Thank you for the invitation, and congratulations on your retirement." Words hardly seemed necessary. All she wanted to do was stare at him.

"Thank you." His hands firmly clasped hers. Then, as

if he'd forgotten himself, he asked, "Can I get you some champagne?"

"I'd like that."

But Hector didn't need to leave her. Rita, his daughter, brought over a champagne flute and offered it to Lillie.

"I'm Rita," she said unnecessarily.

"You're as lovely as your pictures," Lillie said as she accepted the flute and impulsively hugged his daughter. Rita hugged her back, her expression welcoming.

"This is Andy, my fiancé," she said, introducing the man at her side. "Dad wasn't sure you'd come. I told him you would."

"I don't think I could've stayed away if I'd tried," Lillie confessed.

Hector stood close by as his two sons, Manuel and Luis, walked toward them, their progress hindered by the crowd. Manuel studied Lillie, his eyes devoid of emotion; that, to her way of thinking, was an improvement over the hostility he'd shown at their previous meeting. A lovely red-haired young woman—obviously pregnant—was with him.

"You came," Manuel said, not bothering with any form of greeting.

"Manuel," Hector warned in low tones. "I won't have you disrespecting Lillie."

His oldest son conceded with a nod. "Welcome, Ms. Higgins," he said. He introduced his wife, Colleen.

"Thank you, Manuel. Nice to meet you, Colleen. Both of you, please call me Lillie."

"This is my son Luis," Hector said, gesturing toward the second young man.

Luis and Lillie exchanged a friendly greeting under Hector's watchful eye—and Manuel's.

"You made my father very happy by accepting his invitation," Manuel told her when Luis had drifted off to talk to someone else.

"He made me happy by sending it to me."

Manuel gave her a tentative smile.

Lillie smiled back. She hoped that in time the two of them could be friends.

"We're all going to dinner after the party," Hector said, leaning closer. "Can you join us?"

Lillie readily agreed.

"Allow me to introduce you to my friends," Hector said and led her away. As they moved from one group to another, she became even more aware of how greatly he was respected and loved. His coworkers told story after story about Hector, embarrassing him since he was a modest man. Lillie enjoyed every word. If she needed confirmation that this man was everything she'd imagined, then she received it tonight, many times over.

The party started to break up at about eight-thirty. She'd remained at Hector's side, either clasping his hand or with his arm about her waist. There could be no doubt that they were together.

At the Mexican restaurant a little later, Lillie met more members of his family. A brother and sister, nieces and nephews, various in-laws. The table seated at least thirty, and the names flew past her, although she made a determined effort to remember each one. She sat between Hector and Manuel.

Music and laughter filled the room. Children ducked under the table and raced around the chairs while their parents—Hector's nieces and nephews—did their best to contain them. Although Hector introduced Lillie to everyone in his extended family, it seemed they already knew her. They accepted her without question and seemed genuinely pleased to make her acquaintance.

When the food arrived, it was served family style. Manuel passed Lillie the first dish, holding the heavy platter of rice while she helped herself. Next came *chilaquites*, which seemed to be some kind of tortilla casserole, followed by corn tamales, chili rellenos and another dish Lillie didn't hear the name of.

"My father's in love with you," Manuel said quietly.

"I beg your pardon?"

"He showed me your letter. You love him, too, don't you?"

Lillie could see no reason to deny it. "Very much."

"He's a man of strong feelings," Manuel said. "His family is important to him."

"I know." That was the reason she'd decided to break off the relationship; she refused to place Hector in the impossible position of choosing between his family and her.

Manuel acknowledged her statement. "Yes. He was willing to give you up for our—*my* sake."

"He already had." She couldn't resist asking, "Can you tell me what changed?"

Rita slapped her brother's arm. "Hey, Manuel, what's the holdup here? You're supposed to be passing the food."

"Sorry." Manuel handed the dish to his sister.

"I've never seen my father this miserable," Manuel informed Lillie. "Even when Mom was ill, the entire family counted on our father to keep up our spirits and he did. He nursed Mom, cared for her, held her when she breathed her last and loved her to the very end."

"He still loves her." Lillie blinked hard as tears welled up in her eyes. This was how she wanted to be loved.

"After Mom died it was Dad who held our family together. Don't misunderstand me—he grieved for our mother. But her death was also a release from terrible pain. Dad understood that better than anyone. He was lonely and lost but he found ways to cope. Through work, family—and now you."

Mesmerized by Manuel's words, Lillie passed plate after plate without serving herself.

"After receiving your letter, my father wept." Luis, who'd been listening avidly, spoke from across the table while Hector was busy talking to a nephew on his other side.

"He...did?"

Manuel frowned at him, but he acknowledged Luis's words. "It's true." He paused to take a gulp of his Corona. "I knew then that you were no ordinary woman," he resumed, "and that I'd made a mistake. If my father loves you, then I need to be willing to look past my own prejudices and give you a chance, as well."

That did it. Tears spilled down her cheeks. "Thank you, Manuel."

He nodded and passed her another dish.

Lillie wondered why Hector had waited so long to reach out to her. As if reading her thoughts, Manuel added, "Rather than repeat what happened with me, Dad decided he needed to let the family know. So he went to everyone and explained that he'd met someone very special."

No wonder his family behaved as though they already knew her.

"My fear was that you'd break his heart," Manuel murmured. "Unfortunately my attitude toward you was what did that. I hope you can look past our rather...difficult beginning and start again."

"Of course," she said and when the next dish was

handed to her, she scooped up a huge helping of *chili conqueso* and placed it on Manuel's plate before serving herself.

Manuel grinned and then winked at her.

After dinner, the music began, and Hector took Lillie's hand and led her onto the dance floor. "You and Manuel seemed to be deeply involved in conversation," he said as he turned her into his arms.

"Hmm." She leaned her forehead against his and closed her eyes, grateful that this was a slow number.

"He apologized?"

"He said you loved me."

Hector exhaled noisily. "I never expected to fall in love a second time and certainly not like this."

"I didn't, either. Blame the wishes if you want."

"Your Twenty Wishes?"

"I wrote that I wanted to be loved by an honorable man."

"You are loved, my Lillie. By me."

"I love you back."

His hold on her tightened briefly. "Manuel still has doubts that it'll work out between you and me."

"At least he's agreed not to interfere."

Hector nodded. "He said he'd be willing to wait and see—after he told me there's no fool like an old fool."

"Shall we be foolish together?"

Hector laughed. "I was hoping you'd say that."

Lillie lifted her head. "Are we a pair of fools, Hector?"

"I can't think of anyone I'd rather be foolish with than you, my Lillie."

"Me, neither."

With their eyes closed, they continued dancing until the music ended. When the last note faded, they reluctantly

broke apart in order to applaud politely. To Lillie's astonishment, the entire Silva family had formed a circle around
them and started to clap.

Lillie blushed profusely and Hector laughed.

It was a relief when the musicians began again. This
time, his family joined them on the dance floor. The ma-
riachi music was lively, punctuated by slow, plaintive
songs. Hector and Lillie danced every dance and stayed
until the restaurant was ready to close.

Hector drove them back to the dealership, where Lillie had left her car. There, in the shadows, he kissed her.
Lillie slipped her arms around him and leaned into his embrace, letting her actions tell him of all the love in her
heart.

"How did you know I'd come to the party?" she asked.

"I prayed you would."

"And if I hadn't?"

"Then I would have come to you. Most men don't find
a love this good, this pure, once in a lifetime—let alone
twice. I wasn't letting you go, Lillie, not without a fight."

"But you already had," she reminded him.

"No," he said swiftly. "I needed time to regroup and to
reason with my son. You were always with me, always in
my heart." He took her hand and pressed her palm against
his chest. "You inspired me, my Lillie."

"I did?"

"Yes. I have my own list of Twenty Wishes now."

"Really?"

"Oh, yes." He paused to kiss her again. "And every one
of those wishes is about you."

Chapter 34

The small chapel adjacent to the Free Methodist Church off Blossom Street reverberated with the traditional wedding march as Brandon Roche escorted his sister down the center aisle.

Standing in a pew at the front of the chapel, with Ellen at her side, Anne Marie felt her heart swell with joy. When Melissa walked past her, she turned to look at Anne Marie and mouthed the words "Thank you."

Robert would've been so proud of them, she thought, not for the first time. So proud and so delighted by the change in their relationship. For all the pain his betrayal with Rebecca had caused, it had a positive—if inadvertent—effect. It had brought Anne Marie and Melissa together.

Anne Marie gazed after her stepdaughter. The wedding was small, with only a maid of honor and best man. As promised, Ellen would serve wedding cake at the reception, a role to which she attached great importance.

Melissa looked lovely in her pale pink floor-length

dress. A halo of flowers adorned her head, with flowing white ribbons cascading down her back, and she carried a small bouquet of white roses. The pregnancy was just starting to show.

Anne Marie's one disappointment was Pamela. It would've meant so much to Melissa if her mother had relented enough to attend her wedding. Unfortunately, she remained upset and angry, and Anne Marie couldn't help thinking she should have put her daughter's needs ahead of her own feelings. But then, Pamela hadn't come to Robert's funeral, either, although her children could have used her support. Anne Marie hoped they'd eventually be able to resolve their differences.

With the maid of honor and the best man, both close friends, standing beside them, Melissa and Michael approached the young minister. Jordan Turner, Alix's husband, would be performing the ceremony. He'd agreed as long as Melissa and Michael were willing to participate in marriage counseling classes. Even with all the busyness of college graduation, the couple had gone to every session, which boded well for their marriage, Anne Marie thought.

When the ceremony began, Ellen leaned forward, absorbing every word. This was her first wedding and she didn't want to miss a single detail. Anne Marie had enjoyed watching Ellen line up her dolls and stuffed animals the night before and then carefully choose two—a Barbie and a panda—to march down the makeshift aisle. Later, Anne Marie had found her at the kitchen table writing furiously in a tablet.

"What are you doing?" she'd asked.

"I'm putting a new wish on my list," Ellen explained. "I'm going to have a big wedding with lots of people and a dress with lace and pearls and a long veil."

"What about your husband?"

Ellen chewed on the end of her pencil. "He'll be handsome."

"Is that important?"

The eight-year-old considered her response carefully. "I want to marry a man who's handsome on the *inside,* too," she'd said.

"And if he's good-looking on the outside, that would be a bonus, right?"

"Right," Ellen had said.

Now as Melissa and Michael exchanged their vows, Ellen studied them attentively, dreaming of her own wedding one day.

Anne Marie gazed protectively at this child who would legally become her daughter. The greatest desire of her life was to be a mother and that wish had been fulfilled—but not in the way she'd expected.

As the ceremony continued, Anne Marie felt Dolores Falk's presence. The older woman had nurtured the child to the best of her ability. She'd given her love and security. In the end, Ellen's grandmother had handed her over to Anne Marie.

Anne Marie believed Dolores had recognized that the two of them belonged together, that they needed each other. Once the bond between them was established, and Anne Marie had promised to keep Ellen if anything happened to her grandmother, Dolores had been able to die in peace, knowing Ellen would be loved and cared for by Anne Marie.

Anne Marie placed her hand lightly on Ellen's shoulder. The child joined fervently in the applause when Jordan Turner pronounced Michael and Melissa husband and wife.

"This is so nice," Ellen whispered as the music crescendoed. Then Melissa and Michael walked back down the aisle together, their arms linked and their faces bright with joy.

"It is lovely," Anne Marie agreed, struggling to hold back tears. It wasn't only Dolores Falk's presence she felt, but Robert's, too. She knew Melissa felt him there, as well.

"What happens next?" Ellen stared up at Anne Marie with wide, curious eyes.

"Now we go to the wedding reception."

"Oh, goody! When do I serve the cake?"

"Not till later on."

The reception was at a restaurant on Lake Washington. Lillie had secured the banquet room, and Melissa's friends had decorated it and prepared everything for the small reception. Anne Marie had volunteered to help and Melissa was grateful but said she'd already done so much. Besides, her friends had everything under control.

When they arrived, Anne Marie could see that was true. The room, which was separate from the main part of the restaurant, had a sweeping view of Lake Washington. Sailboats with their multicolored spinnakers glided across the choppy waters. A lush green lawn sloped from the restaurant down to the waterfront, bordered by rows of blooming perennials. Double-wide French doors opened onto a stone patio.

Because the day was overcast, Melissa and Michael had decided to hold the reception indoors. Lillie had chosen the perfect location, Anne Marie thought gratefully.

The room itself was strung with white streamers, twisted from the center of the ceiling, where a large paper wedding bell hung. The streamers fanned out in every direction. White and silver balloons were tied behind each chair.

The cake, topped with the traditional bride and groom, sat on a table with an array of gifts surrounding it. Alix Turner had baked and decorated it herself. The restaurant had supplied an elegant buffet, for which Brandon, Anne Marie and Michael's parents had split the expense.

Entering the room with her husband, Melissa looked radiantly happy.

"It's hard to think of my little sister as married," Brandon said, claiming the chair next to Anne Marie and Ellen. "Mom's going to regret not flying over for the wedding."

Anne Marie nodded. Pamela would have to accept that Melissa had her own path to follow. "In time I believe she will."

"Who?" Ellen asked, then added, "Is this for adult ears or kids' ears?"

Brandon laughed outright. "It's for adult ears."

"Okay."

He grinned at Anne Marie.

The buffet line formed, and after they'd filled their plates and sat down again, Michael's parents, Jim and Paula Marshall, joined them.

"This turned out to be such a lovely wedding," Paula said, watching her son with pride. "I wasn't sure what to expect. What is it with children these days? So much happening at once. Michael graduated from college, married and a father-to-be. It's enough to make my head spin."

Anne Marie agreed. "I didn't know what to knit first, a baby blanket or a garter for her wedding."

"You knit?" Paula asked with real interest.

"I'm only just learning. I've been taking classes."

"It was one of our wishes," Ellen told her gravely.

Anne Marie explained and marveled anew at the changes in Ellen since the child had come to live with her. When they'd first met, only three months ago, the youngster had barely spoken a word. These days it was difficult to get her to stop.

"Have I introduced you to Ellen?" Anne Marie asked Jim and Paula and tucked her arm around the child's waist. "This is my daughter, Ellen Falk."

"I'm getting a new name soon," Ellen said, looking at Anne Marie.

"A new name?" Paula repeated. "What do you mean?"

"I'm adopting Ellen," Anne Marie said, "and when I do her last name will be Roche, the same as mine."

"Congratulations to you both," Jim said, sampling the lobster salad.

"Anne Marie's going to be my new mother," Ellen said amicably. "I have an old one, but my grandmother told me my real mom couldn't take care of a little girl, which is why I get a new one. I'm glad my new mother is Anne Marie."

The conversation moved on to pets when Ellen lovingly described Baxter, now "our" dog, and lauded his intelligence. The Marshalls contributed stories about their own badly behaved but much-loved dog, Willow. Everyone laughed a great deal, and Anne Marie was thrilled that Ellen responded so naturally and well to adult company.

Halfway through the reception, Lillie Higgins and Hector Silva came in. Anne Marie had met Lillie's friend a couple of times previously. He was everything Lillie had promised and obviously adored her.

"I'm sorry we're late," Lillie began as she approached Anne Marie.

"It's my fault," Hector said. "My oldest grandson had a soccer game and wanted me there to see him play."

"Can I play soccer, too?" Ellen asked.

Anne Marie nodded. "Once we've moved into the new house, we'll see about signing you up for soccer."

Ellen clasped her hands, her expression rapturous. "And I want to join Girl Scouts."

"One thing at a time, Ellen," Anne Marie said gently. She didn't want to squelch the child's enthusiasm, but didn't want her overwhelmed by too many activities, either. She already took karate lessons and if she added soccer *and* Girl Scouts to that, there wouldn't be enough time just to sit and read or knit or play imaginative games with her dolls.

"I got your message," Lillie said. "You found a house?"

"Yes! We move August first." The house was in a good neighborhood close to Woodrow Wilson Elementary, which meant Ellen wouldn't need to change schools. With all the upheaval in the child's life, Anne Marie had wanted to keep her there.

"I'll have a real bedroom, too," Ellen inserted.

"And a real bed," Anne Marie said. Ellen hadn't complained once about sleeping on the fold-down sofa in the tiny apartment. One of her first purchases would be a bedroom set for Ellen, with a matching dresser, bookcase and computer desk.

She looked forward to getting her own things out of storage. She'd delayed for a long time, preferring to live in the small apartment rather than move. Her fear was that the household goods that had belonged to her and Robert would trigger too many memories.

A few months ago, Anne Marie hadn't felt strong enough to deal with the past. Her grief had been too raw, too close to the surface. She'd purposely kept the furniture in storage, convinced she'd never find the courage to sit at the table where she'd shared so many meals with her husband. Every item, everything she'd so carefully packed away, was linked to Robert.

But the memories of her life with him no longer tormented her. Even knowing of his betrayal, she continued to love him and always would.

Michael and Melissa ceremonially cut the first slice of cake, and Melissa beckoned to Ellen.

"Can I serve cake now?" Ellen asked, eyeing the slices Melissa's friends were placing on colorful plates. "Is it time?"

"Looks like it," Brandon said. They stood up and headed for the table, where the plates had been set out.

"Aren't you going to throw the bouquet?" Alicia, the maid of honor, asked Melissa.

"Oh, my goodness, I almost forgot." Melissa turned her back to her group of friends and hurled the bouquet over her shoulder.

Anne Marie hadn't been part of that group. She didn't mean to participate, but when the bouquet shot directly at her, she instinctively grabbed it.

"Anne Marie!" Melissa cried, laughing delightedly. "You caught the bouquet!"

Ellen squealed with excitement, a plate of cake in each hand.

"This means," Melissa told her, "that Anne Marie will be the next one to get married."

"I don't think so." Anne Marie tried to pass the bouquet to one of Melissa's college friends, who refused to take it.

"Don't be so sure," Melissa chided good-naturedly. "You never know when love's going to tap on the door."

Frankly, Anne Marie wasn't interested in falling in

love again. She had everything she needed for happiness. Ellen was part of her life now, and she'd made peace with the past. She had Blossom Street Books. Her eyes fell on Lillie and Hector, and she immediately added dear friends to her list of blessings.

"Look," Ellen said, tugging at her sleeve. She pointed at the French doors.

Music swirled in from the piano player in the nearby bar.

Anne Marie bent down. "What am I supposed to be looking at?"

"It's raining."

It was more of a mist than rain but Anne Marie didn't point that out. "Yes?"

"Your wish," Ellen reminded her.

What wish?

Then Anne Marie remembered.

Taking Ellen's hand, she walked out to the small patio. With the music playing softly in the background, they removed their shoes and stepped onto the wet grass.

Ellen slipped one arm around Anne Marie's waist and together they spun 'round and 'round.

"We're dancing barefoot in the rain." Ellen giggled.

Throwing back her head, Anne Marie giggled, too.

The music grew louder and their movements became more sweeping as the rain fell and people gathered at the open doors to watch them.

She saw Brandon giving them a thumbs-up and Michael and Melissa waving. Lillie and Hector smiled.

Anne Marie Roche had made Twenty Wishes and they'd brought her love.

Epilogue

November

The courtroom was crowded as Anne Marie and Ellen waited patiently for their turn to come and stand before the judge. When their names were called, Anne Marie stepped forward with Ellen beside her. Evelyn Boyle, Ellen's social worker, moved to the front of the court.

Judge Harold Roper read over the paperwork, which included a home study and background check. This was actually a formality; Child Protective Services had already approved the adoption. The six months had passed quickly. They'd moved into their new home, and Ellen was a third-grader now, getting top marks in her classes.

"So, Ellen, you're going to have a new mother," Judge Roper said.

"Yes, Judge," Ellen answered politely.

"Your Honor," Anne Marie whispered.

"Your Honor," Ellen repeated.

She placed her hand in Anne Marie's and edged closer to her side.

"Congratulations," the judge said and signed his name at the bottom of the document.

"That's all there is?" Ellen asked in a whisper.

Anne Marie was surprised herself. "Apparently so."

Anne Marie's mother wept noisily at the back of the courtroom. The next name was called, and Anne Marie and Ellen hugged and left the room. Laura Bostwick continued to sob, dabbing at her eyes with a tissue as they walked out into the hallway. The heavy door closed behind them.

"Are you sure you can't come to the party, Mom?" Anne Marie asked.

"I'll come by later if that's okay."

"Of course. I want you to meet my friends."

Catching her off guard, Laura awkwardly hugged Anne Marie. "You're going to be a wonderful mother."

"Thanks, Mom."

"My name is Ellen Roche," Ellen announced to a guard who strolled past.

"That's a nice name," the uniformed man told her.

"Ellen Dolores Roche," she said. "Dolores was my grandmother's name. She's with Jesus now."

The man smiled at Anne Marie and kept on walking.

"This is my new mother," Ellen called after him. "She loves me a lot."

"Ellen," Anne Marie murmured. "He's busy."

"I just wanted to tell someone I have a new name," she whispered, lowering her head.

"Would you like to tell Barbie and Mark?"

The girl nodded eagerly. "Lillie and Hector, too?"

"They'll all be at the party."

"What about Mrs. Beaumont and Lydia and Cody and all my friends from Blossom Street?"

"They wouldn't miss it."

"Melissa and Michael, too?"

"Yes." This shouldn't be news to Ellen, who knew all about the party at Blossom Street Books after court.

But Anne Marie understood. Ellen was happy and excited, and she had to express that happiness. She mattered to all these people, belonged to their community as Anne Marie did. Her daughter... Anne Marie's mind came to a sudden halt.

Ellen was her *daughter*.

Her *daughter*.

Unexpected tears gathered in her eyes.

"Anne Marie?" Instantly Ellen was concerned. "Are you okay?"

"Yes—I'm just happy."

"Like Grandma Laura?" she asked.

Anne Marie squeezed her hand. "Just like Grandma Laura."

By the time they got to the bookstore, it seemed the entire street was there to celebrate. Susannah from the flower shop had come, but could only stay briefly. She brought a number of small floral bouquets to commemorate the adoption and a pretty pink corsage for each of them to wear.

Soon after their arrival, Alix Turner carried in a tray of freshly baked cookies, compliments of the French Café. Lydia and her sister, Margaret, came over in turns, so as not to leave the store unattended. They'd brought several bottles of champagne for the adults—Veuve Clicquot, of course—and sparkling lemonade for the kids. Lydia's husband, Brad, dropped in later, bringing their son, Cody.

Michael and a heavily pregnant Melissa showed up, too, and Ellen was doubly excited.

"It's going to be a girl, right?"

"Right."

"Can I be her big sister?"

"I'm counting on it," Melissa said. She was due anytime and Michael remained close to her side. They didn't stay long as they had a birthing class to attend, their final one before her due date.

Theresa, Cathy and Steve, her part-time employees, helped serve, and Ellen mingled with the crowd, reminding everyone that she had a new name and a new mother.

Late in the afternoon, both Evelyn Boyle and Anne Marie's mother stopped by, but could only stay for a few minutes.

"We were thinking of holding a mommy shower for you," Barbie said around five o'clock. Almost everyone had come and gone by this point. The ones who remained were the original members of the widows' group—with the addition of Mark and Hector. It was hard to believe nearly a year had passed since that bleak Valentine's evening, when they'd started their lists of Twenty Wishes.

She'd completed her list last spring.

19. Karate classes with Ellen

20. To live happily ever after

Twenty wishes, nearly all of them a reality now.

Anne Marie had found a pair of red cowboy boots in a secondhand store for a fraction of the cost. They fit perfectly and she wore them often.

Then one Sunday in July, shortly after Anne Marie had begun attending church with Ellen, she'd spontaneously sung a hymn. She was well into the second verse before she remembered that she couldn't sing anymore and yet here she was…. Now not a day went by without her belting out one song after another.

Anne Marie's gaze fell on Barbie, who sat next to Mark, holding his hand.

Anne Marie had met him only half a dozen times, but Barbie had spoken of him often enough to make her feel as if she knew him.

"When we made our lists of wishes, did you ever dream it would come to this?" Elise asked, joining the circle of friends.

"We haven't talked about our lists recently," Lillie said, sitting in the overstuffed chair with Hector standing behind her, his hands on her shoulders. "Has anyone completed any wishes lately?"

"I have," Elise said, looking down at her plastic glass of champagne. "I've set up a charitable foundation in memory of Maverick."

"Elise, that's wonderful!"

The older woman struggled to hide her emotion. "That's not all. I took my two grandsons on a hot-air balloon ride. That was something Maverick and I always intended to do. We put it off—and then it was too late."

"Was it as exciting as you thought it would be?" Anne Marie asked.

Elise smiled warmly. "Even better than I imagined. When I closed my eyes, I could almost feel Maverick's arms around me again," she said in a low voice. "It was the most thrilling sensation to be that high above the ground. He would've loved it."

"I completed one of my wishes, too," Barbie volunteered.

"Which one?" Lillie asked.

Eyes dancing, she glanced at Mark. "I went skinny-dipping."

Lillie frowned. "I have a feeling you weren't alone."

Barbie giggled like a schoolgirl. "As it happens, I wasn't."

Mark shifted uncomfortably in his wheelchair. "I believe that falls under the heading of too much information."

"You went with *Mark*." Lillie feigned shock.

Barbie laughed and leaned over to kiss his cheek. "I'm not telling."

Mark couldn't quite restrain a smile.

"What about you, Anne Marie?" Barbie asked, diverting attention away from her and Mark.

"I'm about to accomplish one of my most heartfelt wishes."

"About to?" Hector asked. "I thought the adoption was finalized this afternoon."

"It was, and Ellen's now my daughter in the eyes of the law. But this is another wish." She opened her purse and removed a thick envelope and showed it to the group.

Ellen dashed over to her side. "Can I tell everyone?" she pleaded.

"Go ahead," Anne Marie told her.

"Mom," she said, and looked at Anne Marie. "Is it okay to call you Mom?"

"Absolutely."

"Mom bought tickets for us to fly to Paris for our first Christmas together."

"Paris," Elise repeated slowly. "What a perfect idea."

Anne Marie slipped her arm around Ellen. "I'm going to Paris with someone I love."

Barbie's eyes were soft. "That's just beautiful." She glanced at Mark, who grumbled something about not getting any ideas. She ignored him and reached for the brochure Anne Marie handed her.

"Barbie, I'm warning you right now, I'm *not* going to

Paris." Mark hesitated. "Go if you like. I'll even encourage it. But I'm staying right here."

"Yes, Mark."

"I mean it, Barbie."

"I know you do." Apparently she had no intention of arguing with him. "I'm perfectly capable of traveling to Europe for two weeks on my own."

"Two weeks?" Mark said, frowning. "That long?"

"It would hardly be worth my while to travel all that way for less than that."

Mark groaned. "Why do I have the feeling that I'm going to be staring up at the Eiffel Tower and wondering how I got there?"

Everyone smiled.

Ellen walked over to where Anne Marie was sitting and climbed onto her lap. "One of my wishes came true, too," she told the group.

"Which one was that?" Hector asked kindly.

"I found a mom," Ellen announced. "I thought Anne Marie would just be my Lunch Buddy but now she's my mom. Forever and ever."

"Forever and ever," Anne Marie repeated.

It was a solemn moment, broken only by Ellen's happy shout. "Hey, Mom! You have to start a *new* list of Twenty Wishes now, don't you?"

Anne Marie smiled. This truly wasn't the end but a new beginning for them all.

* * * * *

Spend the summer on Blossom Street!
Turn the page to read the first chapter of the next
Blossom Street story by #1 New York Times *bestselling*
author Debbie Macomber.
In SUMMER ON BLOSSOM STREET, you'll revisit
A Good Yarn and sit in on Lydia's new knitting class—
"Knit to Quit." You'll also follow the ongoing story of
Ellen and Anne Marie, as well as that of Alix. And
you'll meet brand-new characters Phoebe Rylander and
Bryan Hutchinson, who have their own reasons for
joining Lydia's class…
SUMMER ON BLOSSOM STREET—it's a vacation
in a book!
SUMMER ON BLOSSOM STREET is available
wherever books are sold.

I

CHAPTER

I

Wednesday morning, a not-so-perfect June day, I turned over the Open sign at my yarn store on Blossom Street. Standing in the doorway I breathed in the sweet scent of day lilies, gladiolas, roses and lavender from Susannah's Garden, the flower shop next door.

It was the beginning of summer, and although the sky was overcast and rain threatened to fall at any moment, the sun shone brightly in my heart. (My husband, Brad, always laughs when I say things like that. But I don't care. As a woman who's survived cancer not once but twice, I feel entitled to the occasional sentimental remark. Especially today…)

I took a deep breath and exhaled slowly, enjoying the early-morning peace. I just don't think there's anyplace more beautiful than Seattle in the summer. All the flowers spilling out of Susannah's Garden are one of the benefits. The array of colors, as well as the heady perfume drifting in my direction, makes me so glad A Good Yarn is located where it is.

Whiskers, my shop cat, as Brad calls him, ambled across the hardwood floor and leaped into the window display, nestling among the skeins of pastel yarns. He takes up residence there most days and has long been a neighborhood favorite. The apartment upstairs is an extra storeroom for yarn at the moment; perhaps one day I'll rent it out again but that isn't in the plans yet.

The French Café across the street was already busy, as it is every morning. The windows were filled with pastries, breads and croissants warm from the oven, and their delectable aroma added to the scents I associate with summer on Blossom Street. Alix Turner is usually there by five to bake many of these wonderful temptations. She's one of my dearest friends—and was among my first customers. I'm so proud of everything she's accomplished in the past few years. It's fair to say she reinvented her life—with a little help from her friends. She has an education and a career now, and she's married to a man who seems completely right for her.

Blossom Street Books down the street was ready for business, too. Anne Marie Roche and her staff often leave the front door open as a welcoming gesture, inviting those who wander past to come inside and browse. Anne Marie and her daughter, Ellen, would be coming home from Paris later today.

Nearly every afternoon Ellen walks their Yorkie past the window so Whiskers and Baxter can stare fiercely at each other. Ellen insists it's all for show, that the cat and dog are actually good friends but don't want any of us to know that.

I grinned at Whiskers because I couldn't resist sharing my joy and excitement—even with the cat. In fact, I wanted to tell the whole world my news. Yesterday, we found out that we'd been approved for adoption. I hadn't yet shared this information with anyone, including my sister, Margaret. We've been through the interviews, the home test and fingerprinting. And last night we heard.

We're going to adopt a baby.

Because of my cancer, pregnancy is out of the question. While the ability to conceive has been taken from me, the desire for a baby hasn't. It's like an ache that never quite goes away. As much as possible I've tried to hide this from Brad. Whenever thoughts of what cancer has stolen from me enter my head, I try hard to counter them by remembering all the blessings I've received in my life. I want to celebrate every day, savor every minute, without resentment or regret.

I have so much for which to be grateful. I'm alive and cancer-free. I'm married to a man I adore. His son, Cody, now nine years old, has become my son, too. And I have a successful business, one that brings me great pleasure and satisfaction. When I first opened A Good Yarn, it was my way of shouting to the world that I refused to let cancer rob me of anything else. I was going to *live* and I was going to do it without the constant threat of illness and death. I was determined to bask in the sunshine. I still am.

So A Good Yarn was the start of my new life. Within a year of opening the store, I met Brad Goetz and we were married the following spring. Because of what I'd been through in my teens and again in my twenties, I didn't have a lot of experience with men or relationships. At first, Brad's love terrified me. Then I learned not to reject something good just because I was afraid of its loss. I learned that I could trust this man—and myself.

How blessed I am to be loved by him and Cody. Each and every day I thank God for the two men in my life.

Even with all I have, my arms ached to hold a baby. *Our* baby. Brad, who knows me so well, understood my need. After discussing the subject for weeks on end, after vacillating, weighing the pros and cons, we'd reached our decision.

Yes, we were going to adopt.

The catalyst for all this happened when Anne Marie Roche adopted eight-year-old Ellen.

I realized the wait for a newborn might be lengthy but we were both prepared for that. Although we'd be thrilled with an infant of either sex, I secretly longed for a little girl.

I heard the back door close and turned to see my sister, Margaret. She's worked with me almost from the first day I opened the shop. Although we're as different as any two sisters could be, we've become close. Margaret is a good balance for me, ever practical and pragmatic, and I think I balance her, too, since I'm much more optimistic and given to occasional whimsy.

"Good morning!" I greeted her cheerfully, unable to disguise my happiness.

"It's going to pour," she muttered, taking off her raincoat and hanging it in the back storeroom.

My sister tends to see the negative. The glass would always be half-empty to Margaret. Or completely empty—if not shattered on the floor. Over the years I've grown accustomed to her attitude and simply ignore it.

When she'd finished removing her coat, Margaret stared at me, then frowned. "Why are *you* so happy?" she demanded. "Anybody can see we're about to have a downpour."

"Me? Happy?" There wasn't much point in trying to hold back my news, even though I knew Margaret was the one person who wouldn't understand my pleasure. She'd disapprove and would have no qualms about imparting her opinion. It's her pessimistic nature, I suppose, and the fact that she worries about me, although she'd never admit that.

Margaret continued to glare. "You're grinning from ear to ear."

I made busy work at the cash register in order to avoid eye contact. I might as well tell her, although I dreaded her response. "Brad and I have applied for adoption," I blurted out, unable to stop myself. "And our application's been accepted."

A startled silence followed.

"I know you think we're making a mistake," I rushed to add.

"I didn't say that." Margaret walked slowly toward me.

"You didn't need to *say* anything," I told her. Just once I wanted Margaret to be happy for me, without the doubts and objections and concerns. "Your silence said it all."

Margaret joined me at the counter next to the cash register. She seemed to sense that her reaction had hurt me. "I'm only wondering if adoption's a wise choice for you."

"Margaret," I began, sighing as I spoke. "Brad and I know what we're doing." Although Margaret hadn't said it openly, I could guess what concerned her most. She was afraid the cancer would return. I'm well aware of the possibility and have been ever since its recurrence ten years ago. It was a serious consideration and one that neither Brad nor I took lightly.

"Brad agrees?" My sister sounded skeptical.

"Of course he agrees! I'd never go against his wishes."

Margaret still didn't look convinced. "You're *sure* this is what you want?"

"Yes." I was adamant. Sometimes that's the only way to reach her. "Brad knows the risks as well as I do. You don't need to spell it out, Margaret. I understand why you're afraid for me, but I'm through with living in fear."

Margaret's eyes revealed her apprehensions. She studied me and after a moment asked, "What if the adoption agency doesn't find you a child?"

This was something Brad and I had discussed and it could certainly happen. I shrugged. "Nothing ventured, nothing gained. We'll take the chance."

"You want an infant?"

"Yes." I pictured a newborn, wrapped in a soft pink blanket, gently placed in my waiting arms. I held on to the image, allowing it to bring me comfort, to fill me with hope.

To my surprise Margaret didn't immediately voice another objection. After a thoughtful minute or two, she muttered in low tones, "You'd be a good mother...you already are."

I'm sure my jaw fell open. The shock of Margaret's endorsement was almost more than I could take in. This was as close as Margaret had ever come to bestowing her approval on anything regarding my personal life. No, that wasn't fair. She'd been partially responsible for Brad and me getting back together when I'd pushed him away—a reconciliation that led directly to our marriage.

"Thank you," I whispered and touched her arm.

Margaret made some gruff, unintelligible reply and moved to the table at the back of the store. She pulled out a chair, sat down and took out her crocheting.

"I put up the poster you made for our new class," I told her, doing my best to conceal the emotion that crept into my voice. The last thing I'd expected from Margaret had been her blessing, and I was deeply touched by her words.

She acknowledged my comment with a nod.

The idea for our new knitting class had been Margaret's. "Knit to Quit," she called it, and I loved her suggestion. Since opening the yarn store five years earlier, I'd noticed how many different reasons my customers—mostly women but also a few men—had for learning to knit. Some came looking for a distraction or an escape, a focus to take their minds off some habit or preoccupation. Others were there because of a passion for the craft and still others hoped to express their love or creativity—or both—with something handmade.

Four years ago, Courtney Pulanski, a high school girl, had signed up for my sock-knitting class, which contributed to her successful attempt to lose weight. Hard to believe Courtney was a college senior now and still a knitter. More importantly, she kept off the weight she lost that summer.

"I hope Alix takes the hint," Margaret said, cutting into my thoughts.

I missed the connection. "I beg your pardon?"

"Alix is smoking again."

It wasn't as if I'd missed that. She smelled of cigarettes every time she walked into the store. There was no disguising the way smoke clung to her clothes and her hair. And yet Alix seemed to think no one noticed, although of course everyone did.

"My guess is she'd like to quit."

"Then she should sign up for the class," Margaret said emphatically. "She could use it."

How typical of Margaret to feel she knew what was best for everyone. Currently, though, I was more amused than annoyed by her take-charge attitude.

My first customer of the morning—a woman I'd never

met before—stepped into the shop and fifteen minutes later, I rang up a hundred-dollar yarn sale. A promising start to the day.

As soon as the door closed, Margaret set aside her project, an afghan for our mother who resides at a nearby assisted living complex. "You know what's going to happen, don't you?"

"Happen with what?" I asked.

"This adoption thing."

I froze. I should've known Margaret wouldn't leave the subject alone. At least not until she'd cast a net of dire predictions. I understood that this impulse was one she couldn't resist, just as I understood that it was motivated by her protectiveness toward me. I didn't need to hear it right now.

"What's that?" I asked, hoping my irritation didn't show.

"Have you talked to a social worker yet?"

"Well, of course." I'd spoken to Anne Marie, and she'd recommended Evelyn Boyle, the social worker who'd been assigned to Ellen and had handled her adoption. Anne Marie and Ellen fit so perfectly together that their story had inspired me to look beyond my fears. So Brad and I had approached Evelyn.

Margaret shook her head, which annoyed me even more.

"Anne Marie gave me the phone number of the woman who helped her adopt Ellen," I said.

Margaret's brows came together in consternation and she tightened her lips.

"What now?" I asked, trying to remain calm.

"I wouldn't recommend that."

"Why not? It's too late anyway."

"This social worker deals with foster kids, right?"

"I guess so." I knew so, but didn't see how that was relevant. "Why should that matter?"

My sister rolled her eyes, as though it should be obvious. "Because she's got children in her case files," Margaret said with exaggerated patience. "She probably has lots of kids and nowhere to place them. Mark my words, she'll find a reason to leave some needy child with you. And not a baby, either."

"Margaret," I said pointedly, "Brad and I are going to adopt an infant. This social worker, Evelyn, is helping us through the process, nothing more."

Margaret didn't respond for several minutes. Just when it seemed she was prepared to drop the subject, she added, "Finding an infant might not be that easy."

"Perhaps not," I agreed, unwilling to argue. "We'll have to wait and see what the adoption agency has to say."

"It might be expensive, what with lawyers and everything."

"Brad and I will cross that bridge when we come to it."

Margaret looked away, frowning slightly, as if she needed to consider every negative aspect of this process. "There are private adoption agencies, too, you know?"

I did know about them, but it made better financial sense to approach the state agency first.

"What about adopting from outside the country?"

Margaret was apparently trying to be helpful, but I wasn't convinced I should let down my guard.

"We're holding that in reserve," I said.

"I hear it's even more expensive than private adoptions."

"Yes, well, it's another option to investigate...."

Margaret's shoulders rose in a deep sigh. "Are you going to tell Mom?"

With our mother's fragile health and declining mental condition it wasn't something I'd considered doing. "Probably not…"

Margaret nodded, her mouth a tight line.

"Mom has a hard enough time remembering that Cody's my stepson," I reminded her. On our last visit she'd asked copious questions about the "young man" I'd brought with me.

My sister swallowed visibly. "Mom didn't recognize Julia when we went to see her a few days ago."

I felt a jolt of pain—for Margaret, for her daughter Julia, for Mom. This was the first time Margaret had mentioned it. Our mother's mental state had declined rapidly over the past two years and I suspected that in a little while she wouldn't recognize me anymore, either. Margaret and I shared responsibility for checking in on her and making sure she was well and contented. These days my sister and I had taken over the parental role, looking after our mother.

I could pinpoint exactly when that role reversal had taken place. It'd been the day Mom's neighbor found her unconscious in the garden. She'd collapsed while watering her flowers. Everything had changed from that moment on.

Our mother had ceased to be the woman we'd always known. Living in a care facility now, she was increasingly confused and uncertain. It broke my heart to see Mom struggling so hard to hide her bewilderment at what was happening to her.

"Mom will be happy for you," Margaret mumbled. "At some point her mind will clear and she'll realize you have an infant."

I smiled and hoped this was true, although I had my doubts…and I knew Margaret did, too.

The bell above the door chimed before we could discuss it further, and I glanced up at an attractive young woman who'd entered the shop. I hadn't seen her before.

"Hello," I said, welcoming her with an encouraging smile. "Can I help you?"

The woman nodded and toyed nervously with the cell phone in her hand. "Yes…I saw the notice in the window for the Knit to Quit class."

"Do you know how to knit?"

She shook her head. "No…well, some. I learned years ago but I've forgotten. Would this class be too advanced for someone like me?"

"Not at all. I'm sure you'll pick it up in no time. I'll be happy to help you refresh your skills." I went on to explain that there'd be seven sessions and told her the price of the class.

She nodded again. "You can sign up for the class no matter what you want to quit?" She stared down at the floor as she spoke.

"Of course," I assured her.

"Good." She set her bag and cell phone on the counter. "I'd like to pay now." She handed me a credit card and I read her name—Phoebe Rylander.

"You're our very first class member," I told her.

"So the class starts next week?"

"Yes."

"The sign said Thursdays from six to eight?"

"Yes. I'm keeping the store open late. I've never had a night class before."

I processed her payment and wrote her name on the

sign-up sheet. "What are you trying to quit?" I asked in a friendly voice.

"Not what, who," she whispered.

"Oh…" Her answer took me by surprise.

"There's a man I need to get over," she said with tears in her eyes. "A man I…once loved."

#1 *NEW YORK TIMES* BESTSELLING AUTHOR

DEBBIE MACOMBER

Summer on Blossom Street

$24.95 U.S./$29.95 CAN.

$2.⁰⁰ OFF

MIRA®

A brand-new Blossom Street novel from #1 *New York Times* bestselling author

DEBBIE MACOMBER

Summer on Blossom Street

Available April 28, 2009 wherever hardcover books are sold!

$2.⁰⁰ OFF the purchase price of SUMMER ON BLOSSOM STREET by Debbie Macomber.

Offer valid from April 28, 2009, to May 31, 2009.
Redeemable at participating retail outlets. Limit one coupon per purchase.
Valid in the U.S.A. and Canada only.

52608650

5 65373 00082 3 (8100)0 11602

Join #1 *New York Times* bestselling author

DEBBIE MACOMBER

at

CEDAR COVE DAYS

Taking place in
Port Orchard, Washington
From
August 26–30, 2009

Debbie Macomber fans from across the country are invited
to the real Cedar Cove (and Debbie's hometown!) to attend
this exciting multiday literary festival, which will feature a
wide variety of fantastic Debbie-inspired events!

For more information, please visit
the Cedar Cove Association at

www.CedarCoveDays.com

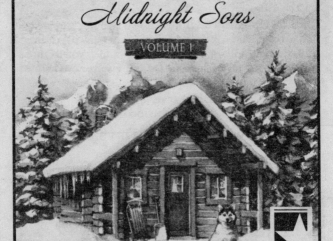

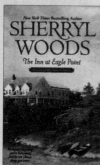

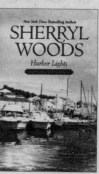

REQUEST YOUR
FREE BOOKS!

2 FREE NOVELS
FROM THE ROMANCE/SUSPENSE
COLLECTION PLUS 2 FREE GIFTS!

YES! Please send me 2 FREE novels from the Romance/Suspense Collection and my 2 FREE gifts (gifts are worth about $10). After receiving them, if I don't wish to receive any more books, I can return the shipping statement marked "cancel." If I don't cancel, I will receive 4 brand-new novels every month and be billed just $5.49 per book in the U.S. or $5.99 per book in Canada, plus 25¢ shipping and handling per book plus applicable taxes, if any*. That's a savings of at least 20% off the cover price! I understand that accepting the 2 free books and gifts places me under no obligation to buy anything. I can always return a shipment and cancel at any time. Even if I never buy another book from the Reader Service, the two free books and gifts are mine to keep forever.

185 MDN EF5Y 385 MDN EF6C

Name	(PLEASE PRINT)	
Address		Apt. #
City	State/Prov.	Zip/Postal Code

Signature (if under 18, a parent or guardian must sign)

Mail to **The Reader Service:**
IN U.S.A.: P.O. Box 1867, Buffalo, NY 14240-1867
IN CANADA: P.O. Box 609, Fort Erie, Ontario L2A 5X3

Not valid to current subscribers to the Romance Collection,
the Suspense Collection or the Romance/Suspense Collection.

Want to try two free books from another line?
Call 1-800-873-8635 or visit www.morefreebooks.com.

* Terms and prices subject to change without notice. N.Y. residents add applicable sales tax. Canadian residents will be charged applicable provincial taxes and GST. Offer not valid in Quebec. This offer is limited to one order per household. All orders subject to approval. Credit or debit balances in a customer's account(s) may be offset by any other outstanding balance owed by or to the customer. Please allow 4 to 6 weeks for delivery. Offer available while quantities last.

Your Privacy: Harlequin is committed to protecting your privacy. Our Privacy Policy is available online at www.eHarlequin.com or upon request from the Reader Service. From time to time we make our lists of customers available to reputable third parties who may have a product or service of interest to you. If you would prefer we not share your name and address, please check here. ☐

BOB08R